Deltas of the World

Series Editor

Abdelazim Negm, Zagazig, Egypt

Worldwide, over 600 million people live in deltas. Extensive human activities in deltas, their supplying rivers, and related maritime currents increase the pressure on deltas' natural resources and habitat equilibrium, leading to degradation of the available resources and jeopardizing inhabitants' livelihoods – and in some cases even their lives.

This book series is intended to promote a sustainable future for deltas around the globe and the communities that live in them. Accordingly, we publish studies focusing on various aspects of deltas, including hydrogeology, soil pollution, water pollution, environmental hydrology, environmental hydraulics, groundwater engineering and management, morphology, anthropology, socioeconomics, wastewater management, water resources engineering and management, agriculture, greenhouse agriculture, irrigation and drainage engineering, hydraulic structures, geophysics and geology, sedimentology, mineralogy, remote sensing and GIS studies, climate change variabilities and impacts, marine-delta interactions and sustainability, and solid waste management. The findings presented here support the SDGs and Agenda 2030 with regard to helping countries with deltas and related environments to preserve their natural resources for future generations.

The editors invite scientists, researchers and scientific communities to contribute to the series by submitting delta atlases, edited books, monographs, and high-quality conference proceedings focused on deltas around the world.

Editors
Mohamed Abdelghany Khalifa
Department of Geology
Faculty of Science
Menoufia University
Shebeen El-Kom, Egypt

Khaled S. Gemail
Department of Geology
Faculty of Science, Environmental Geophysics
Research Center of Excellence (ZEGL)
Zagazig University
Zagazig, Egypt

ISSN 2731-832X ISSN 2731-8338 (electronic)
Deltas of the World
ISBN 978-3-031-96593-7 ISBN 978-3-031-96594-4 (eBook)
https://doi.org/10.1007/978-3-031-96594-4

This Springer imprint is published by the registered company Springer Nature Switzerland AG
The registered company address is: Gewerbestrasse 11, 6330 Cham, Switzerland

Mohamed Abdelghany Khalifa ·
Khaled S. Gemail
Editors

Geology of the Nile Deltas and Natural Resources

Challenges and Sustainable Management

Springer

Preface

The Nile River has forged several delta formations throughout its geological history, most notably the modern Mediterranean Nile Delta and the ancient El Fayoum Delta. The contemporary Mediterranean delta, distinguished by its nutrient-rich silt and clay, supports extensive agricultural activities but is increasingly threatened by rising sea levels and pollution. As a cradle of civilization and a hub of ecological complexity, the current Nile Delta stands at a critical juncture, grappling with the multifaceted challenges of sustainable development. With global population growth and climate change accelerating, the sustainable management of natural resources has become one of humanity's most pressing concerns. This urgency is particularly acute in deltaic systems like the Nile Delta, where geological history, resource abundance, and human vulnerability intersect in intricate ways.

This book, *Geology of the Nile Deltas and Natural Resources: Challenges and Sustainable Management*, seeks to unravel these complexities, offering a geoscientific lens to navigate the delicate balance between resource exploitation and environmental stewardship in one of the world's most dynamic and vulnerable regions.

Geology is not only the study of Earth's past, but also a roadmap for its future. From deciphering sedimentary processes that have shaped the Nile Delta over millennia to mapping modern groundwater aquifers and hydrocarbon reserves, geoscientific inquiry provides the foundational knowledge necessary to mitigate risks and optimize resource use. The Nile Delta, formed by the interplay of fluvial and marine forces from the Miocene to the Holocene, serves as a living archive of geological evolution. Its thick alluvial deposits, ancient distributaries, and subsurface stratigraphy govern the distribution of critical resources: fertile agricultural soils, freshwater aquifers, offshore natural gas fields, and coastal ecosystems. Yet, these resources are under unprecedented strain from population growth, sea-level rise (SLR), saltwater intrusion (SWI), and anthropogenic activities.

Home to 44% of Egypt's population and the source of 65% of its agricultural output, the Nile Delta is both a lifeline and a cautionary tale. Ranked as the second-most vulnerable delta globally to SLR-driven displacement and the fifth-most impacted by coastal urbanization, it epitomizes the cascading consequences of climate change. Rising seas compound land subsidence in the north, while upstream damming and over-extraction degrade freshwater quality and quantity. Saltwater encroachment threatens soil productivity, transforming fertile lands into saline wastelands. Meanwhile, the delta's offshore domains-rich in natural gas and condensate reserves, highlight the tension between energy security and environmental preservation.

This book synthesizes decades of multidisciplinary research to confront these challenges. We unravel the spatial and temporal dynamics of the Nile Delta's resources by integrating geological mapping, hydrogeological modeling, geophysical surveys, and machine learning. Beyond academic analysis, this work emphasizes actionable solutions. It advocates for geoscience-driven policies: managed aquifer recharge to combat SWI, predictive models to optimize agricultural practices, and adaptive frameworks to safeguard coastal communities. The lessons from the Nile Delta resonate globally, offering blueprints for deltaic systems from the Ganges-Brahmaputra to the Mississippi, where similar food-energy-water nexuses strain under climatic and demographic pressures.

The book is merely a geological treatise and a manifesto for sustainable development. It argues that understanding Earth's processes is not a luxury but a necessity for survival. As we confront the realities of a warming planet, the Nile Delta reminds us that resilience lies at the intersection of science, policy, and community engagement. By harnessing geoscience to decode the past and navigate the future, we can transform vulnerability into vitality, ensuring that global deltas remain engines of life for the next generations. Consequently, we welcome constructive comments and future contributions for the next edition.

The structure of this book is designed to guide readers through its comprehensive exploration of the Nile Delta. Part I delves into the geology of the Nile Delta, addressing the geological processes and origins, including sedimentary environments, delta landscapes, geological history, stratigraphic settings, and paleo-environments. Part II shifts focus to implementing these geological processes, concentrating on water resource assessment challenges and sustainable management strategies. This part covers critical issues such as groundwater degradation, the impacts of sea-level rise and saltwater intrusion on water resources, and the role of geoinformatics in monitoring and assessing these challenges. Finally, the concluding part explores the hydrocarbon potential of the region, highlighting the geological factors that influence the accumulation of oil and natural gas. It also examines the application of machine learning modeling and geomechanics in assessing these valuable hydrocarbon resources. The book provides a holistic understanding of the Nile Delta's complex interplay of geology, resources, and sustainability through the explained structured approach.

We invite researchers, geoscientists, policymakers, and national stakeholders to explore these pages, not just as observers of the Nile Delta's story, but as architects of its sustainable future.

We want to express our heartfelt gratitude to **Prof. Abdelazim Negm**, the editor of the *Deltas of the World* series, for his invaluable review of all manuscripts, ensuring the quality and rigor of this work. We extend our appreciation to all the authors and co-authors for their contributions, as well as the peer reviewers of the initial submission for their insightful feedback. Finally, we acknowledge the entire team of Springer for their support and dedication throughout this project. Their collective efforts have been instrumental in bringing this book to fruition.

Menoufia, Egypt — Mohamed Abdelghany Khalifa
Zagazig, Egypt — Khaled S. Gemail
April 2025

Contents

Part I
Geology of the Nile Deltas

1 Overview of the River Nile and the Evolution of the Nile Delta

Mohamed Abdelghany Khalifa

Abstract

This chapter provides the readers with the original definition of the Nile River that considered the father of the African rivers and the longest river in the world. The route of the Nile River in the Nile Basin countries in central Africa and its ancient branches on the Mediterranean coast have been described. The Nile River has resulted in the formation of two deltas along its history (Cretaceous-Holocene), the Fayoum Nile Delta and the Mediterranean Nile Delta. The Fayoum Nile Delta was formed at first (during the Middle Eocene, more than 40 M.y). It covers a wide area in the northwestern desert, including the Fayoum province, Sadat City, the Wadi Natrun, the El Bahr Depression northeast of the Baharyia Oases, and the south Qattara Depression. The Mediterranean Nile Delta was formed later (Miocene and Quaternary) and occurs on the Mediterranean coast and was initiated during the Lower Miocene up to the Holocene. Different types of deltas have been mentioned, such as wave-dominated delta, fluvial-dominated delta, tidal marine delta, tidal freshwater delta, abandoned delta, inland delta, Gilbert delta, truncated delta, blocked delta, and mega delta. The importance of deltas for humanity has also been mentioned.

Keywords

Nile River • Fayium • Mediterranean • Wadi El Natron • Qattara • El Bahr • Qattara • Middle Eocene • Miocene • Quaternary

M. A. Khalifa (✉)
Geology Department, Faculty of Science, Menoufia University, Shibin El Kom, Egypt
e-mail: khcycle1949@gmail.com

M. A. Khalifa and K. S. Gemail (eds.), *Geology of the Nile Deltas and Natural Resources*, Deltas of the World, https://doi.org/10.1007/978-3-031-96594-4_1

1.1 Introduction

Recently, the current Nile Delta characteristics and challenges have been covered by numerous publications (i.e., Frihy and Gamai 1991; Stanley and Warn 1998; Stanley and Crown 2013; Rashid et al. 2018; Abu Hashish and Ali 2021). On the other hand, the geological history of the Nile River and Nile Delta was reviewed by many eminent geologists of Hull (1896), Lyons (1906), Gregory (1920), Blanckenhorn (1910, 1921), and Awad (1928). Further recent studies were carried out on the origin of the Nile River which must be mentioned, e.g., Sandford and Arkel (1929, 1933, 1939), Huzayyin (1941), Sandford (1934), Butzer and Hansen (1968), Wendorf and Schild (1976), Salem (1976), Said (1975, 1981), and Zaghloul et al. (1977). This chapter provides an overview of (1) the Nile River, (2) old branches of the Nile River, (3) the Nile Deltas, (4) the geomorphological features of deltas, (5) the types of deltas, and (6) the importance of deltas for humanity.

1.2 The Nile River

The term Nile is derived from the Greek Neilos (Latin: Nilus), which probably originated from the Semitic root naḥal, meaning a valley or a river valley (Encyclopedia Britannica). The ancient Egyptians called the river Ar or Aur (Coptic: Iaro), "Black," in allusion to the color of the sediments carried by the river when it is in flood. Nile mud is black enough to have given the land itself its oldest name, Kem or Kemi, which also means "black" and signifies darkness. In the Odyssey, the epic poem written by the Greek poet Homer (seventh century BCE), Aigyptos is the name of the Nile (masculine) as well as the country of Egypt (feminine) through which it flows.. In recent times or nowadays, the Nile River in Egypt and Sudan is now locally called Al-Nīl, Al-Baḥr, and Baḥr Al-Nīl or Nahr Al-Nīl. In addition, the Nile River is known as the father of African rivers. The Nile River has three major tributaries: the White Nile, the Blue Nile, and the Atbara (Fig. 1.1). The White Nile (the longer tributary) is traditionally considered to be the headwaters stream that rises in the Great Lakes region. It has two branches in the Great Lake regions, the first starts from Kivu Lake in Rwanda, Edward, and Albert Lakes in Uganda (Fig. 1.1), and the second branch begins from Victoria Lake (Southern Uganda) and Kyoga Lake (northern Uganda). Both branches are connected at the Albert Lake (northeastern Uganda), forming what is called the Albert Nile (northeast of Uganda). The Albert Nile enters Southern Sudan and extends north of Southern Sudan, forming the White Nile. At the latter locality, the White Nile is connected with the Bahr El Ghazal (northwest of South Sudan) and with the Sobat, which comes from Ethiopia and Southern Sudan. The White Nile passes northwards, meeting the Blue Nile at Khartoum, the capital of Northern Sudan (Fig. 1.1). The Blue Nile is the second important source of most of the water of the Nile downstream, containing 80% of the water and silt. The Blue Nile begins at the Tana Lake in northeast

Ethiopia and flows into southeast Sudan until it meets the White Nile at Khartoum. The third tributary of the Nile River is the Atbara, stemming from northwest Ethiopia and passes through southeast Sudan until it connects with River Nile at Atbara, southeast of Sudan (Fig. 1.1). The Nile River is widely recognized as the longest river in the world, originating south of the Equator and flowing northward through northeastern Africa before emptying into the Mediterranean Sea (Said 1981, 1990) (Fig. 1.1). With an approximate length of 4,132 miles (6,650 km), it drains a vast area estimated at 1,293,000 square miles (3,349,000 square kilometers) (Abdel Aziz et al. 2024). The river's basin encompasses regions in Tanzania, Burundi, Rwanda, the Democratic Republic of the Congo, Kenya, Uganda, South Sudan, Ethiopia, Sudan, and the agricultural lands of Egypt (Fig. 1.1). The Nile traverses five distinct geographic regions, each characterized by unique geological histories, landforms, and environmental conditions (Said 1981; Risha 2024) (Fig. 1.2):

1. The Lake Plateau and Highlands of Ethiopia, which includes the southern headwaters of the White Nile at Lake Victoria in Uganda, along with the origins of the Blue Nile and Atbara River.
2. The Sudd Swamp, extending from Juba in South Sudan to the confluence of the White and Blue Niles at Khartoum in central Sudan.
3. Central Sudan stretches from Malakal to Northern Sudan (Khartoum).
4. The Cataracts Region features larger channel belts and low-gradient floodplains that run from Northern Sudan (Khartoum) to southern Egypt (Wadi Halfa).
5. The Egyptian Nile Region, beginning at Aswan and extending to the Mediterranean delta, is characterized by a low-gradient complex where the river meets the Mediterranean Sea (see Fig. 1.2).

1.3 Ancient Branches of the Nile River

In ancient times (Ptolemy time), Cooper (2014) reported that the Nile River has seven distributaries or branches, from east to west are: the Pelusium or Pelusiac, the Tanitic, the Mendesian, the Phatnitic (or Phatmetic) (Hayes 1964), the Sebennytic, the Bolbitine, and the Canopic (also called the Herakleotic and the Agathodaemon (Cooper 2014; Ali et al. 2024) (see Fig. 1.3). Modern Egyptologists suggest that in the Pharaonic period, there were only five main branches (Ian and Paul 1995; Bunson 2009; Abdel-Raheem et al. 2024), the Pelusiac, the Sebennytic, the Canopic, the Damietta, and the Rosetta. The first three have dried up over the centuries due to flood control, silting, and changing relief, while the last two still exist today. There are now only two main branches, due to flood control, silting, and changing relief: the Damietta (corresponding to the Phatnitic) to the east, and the Rosetta (corresponding to the Bolbitine) in the western part of the delta (Fig. 1.3).

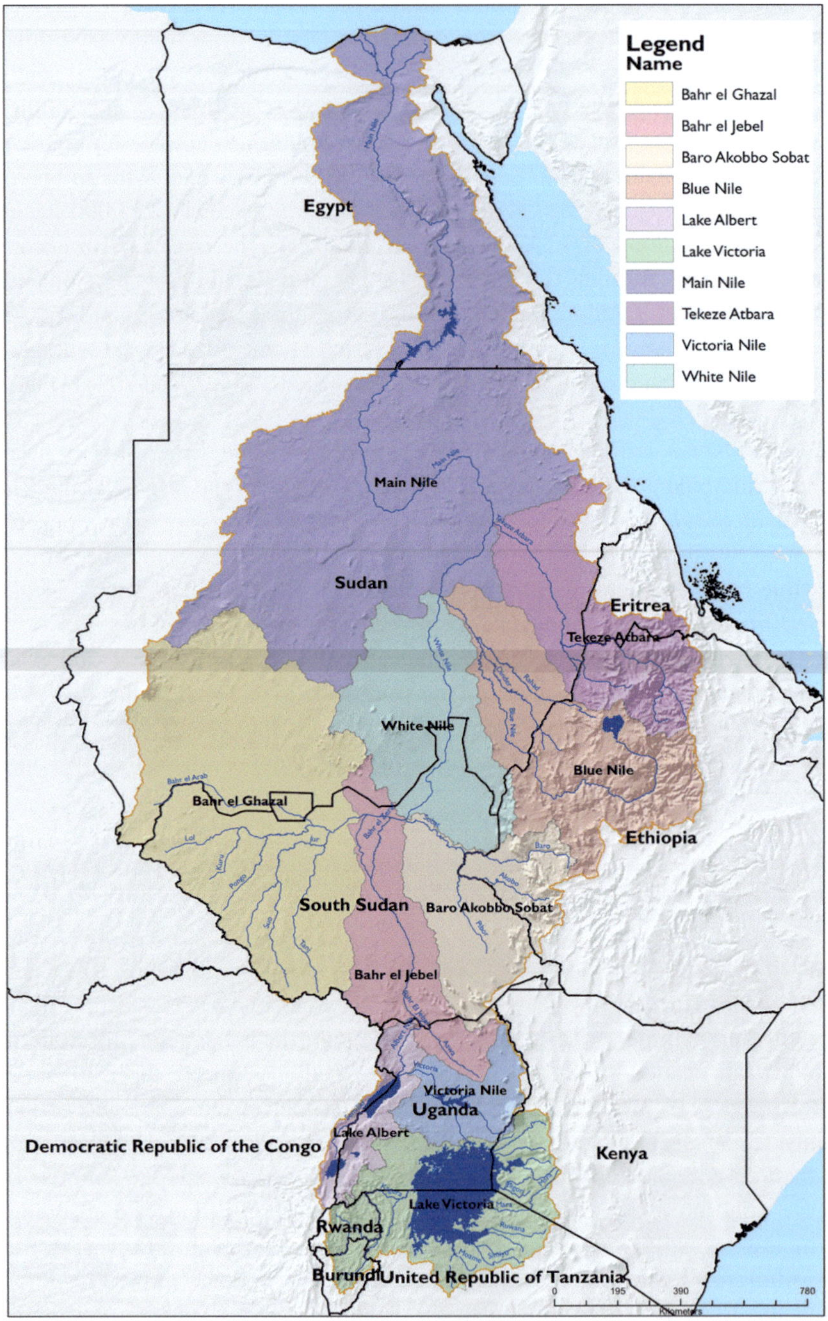

Fig. 1.1 Geographic map of the Nile River Basin, illustrating the route of the Nile River, its main distributaries, and the states of the Nile Basin (NBI 2016)

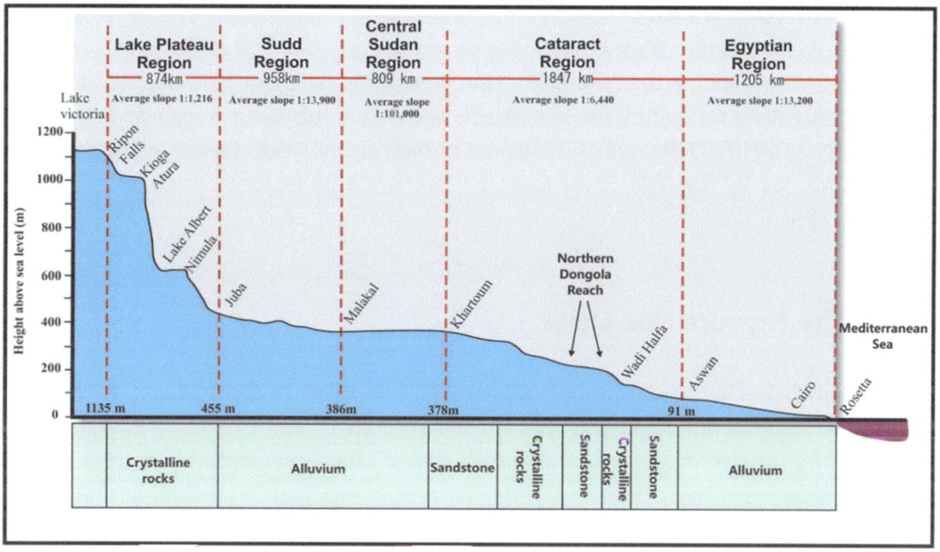

Fig. 1.2 Schematic diagram showing five geographic regions of the Nile River from its upstream (Ethiopia) to downstream (northern Egypt) (After Said 1981)

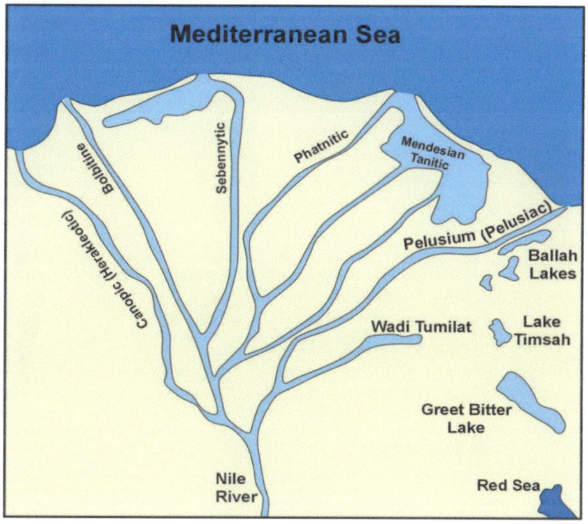

Fig. 1.3 Schematic diagram showing the general view of the old tributaries of the River Nile that occur in the Mediterranean Nile Delta. Nile Delta (Cooper 2014)

1.4 The Nile Deltas

Most of the researchers in the Nile Delta believe that the Nile Delta on the Mediterranean Sea is the only delta formed by the Nile River, but in fact, this delta is regarded as younger than the Fayoum Nile Delta, which is named in the current work. Hence, we can describe the deltas formed by the Nile River into: the Fayoum Nile Delta, and the Mediterranean Nile Delta (Fig. 1.4).

1.4.1 The Fayoum Nile Delta

The Fayoum Nile Delta starts from Latitude 27 30 N (west of Beni Mazar, Menia Governorate) to Latitude 29 30 N just west of the Nile River and extends westwards up to longitude 31 E. It covers an area of about 40 square kilometers and includes the Fayoum

Fig. 1.4 Landsat image demonstrating the Fayoum Nile Delta and the Mediterranean Nile Delta. The Fayoum Nile Delta exhibits a fan-shaped orientation in the northwestern desert of Egypt. It includes Fayoum province, Wadi El Natron, Sadat City, Moghra Oasis, Bahr depression northeast of the Bahariya Oases, and the south Qattara Depression

province, Wadi El Natron, Sadat City, Moghra Oasis, and the south Qattara Depression (Fig. 1.4).

1.4.2 The Mediterranean Nile Delta

The Mediterranean Nile Delta occupies about 2.4% of the total area of Egypt. Without the Nile Delta, Egypt is mainly a desert country (Fig. 1.4). It accounts for 63% of arable lands in Egypt, 50% of the country's industrial production, 65% of agricultural production (Hamza 2006), and 60% of fish catch in the country (El-Fishawi 1993). It was formed in Lower Egypt, where the Nile River spreads out and drains into the Mediterranean Sea (Dumont 2009). It is one of the world's largest river deltas, extending from Alexandria in the west to Port Said in the east; it covers 240 km of the Mediterranean coastline and is a rich agricultural region (Negm 2017). From south to north, the delta is approximately 160 km in length. The delta begins slightly downriver from Cairo (Zeidan 2006).

The Mediterranean Nile Delta is situated in the middle of the Egyptian Mediterranean coastline of approximately 1,000 km (16% of the total Mediterranean coast) and supports many natural and man-made types of habitats. The delta has an onshore area of about 22,000 km^2 and about an equal area offshore down to the 200 m isobaths (Nashaat 1992; Nabawy and Shehata 2015; Khalaf 2016; Fisher 2018). The Mediterranean Nile Delta has coastal lakes such as Manzala Lake in the northeast, and Burullus, Idku, and Mariout Lakes in the northwest Nile Delta. The Nile Delta is considered to be an "arcuate" delta (arc-shaped), as it resembles a triangle or flower when seen from above. Some scholars, such as Aristotle, have written that the delta was constructed for agricultural purposes due to the drying of the region of Egypt (Holz 1969).

1.5 The Geomorphological Features of Deltas

The geomorphological features of deltas have several distinguished shapes, e.g., arcuate delta, bird-foot delta, and Cuspate Delta.

1.5.1 Arcuate Delta

The term delta was advocated by the Greek philosopher Herodotus about 490 B.C. to describe the arcuate-shaped alluvial plain formed at the mouth of the Nile River (Fig. 1.5) (Boggs 2006). The arcuate or fan-shaped delta has a bowed or curved delta with a convex margin facing the body of water like the Ganges Delta in India (Fig. 1.5). Nevertheless, the term (alluvial) delta is still applied to any deposit, subaerial, or subaqueous formed by fluvial sediments that build into a standing body of water (Boggs 2006). In another

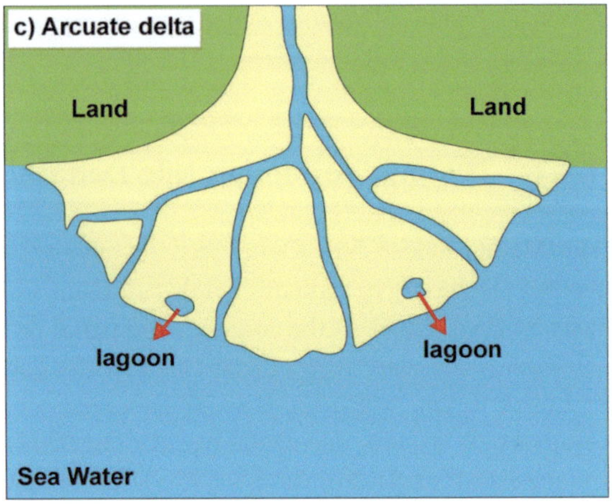

Fig. 1.5 General view of the arcuate deltas that are like an arc of a circle or a bow and are of lobate form in appearance, wherein the middle portion has maximum extent toward the sea, whereas they narrow down toward their margins

explanation, a river delta is defined as a landform shaped like an arcuate, formed by deposition of alluvial sediment that is carried by a river and enters slower-moving or stagnant water (Miall 1979; Elliot 1986; Boggs 2006) (Fig. 1.5).

1.5.2 Bird-Foot Deltas

Bird-foot deltas, characterized by their resemblance to a bird's foot, form through the deposition of finer materials that remain suspended in river water, which is lighter than seawater (Fig. 1.6). As these materials settle, they divide the river into smaller distributaries, each of which features a lobe at its terminus (Fig. 1.6). A prominent example of a bird-foot delta is the Mississippi River Delta in the USA.

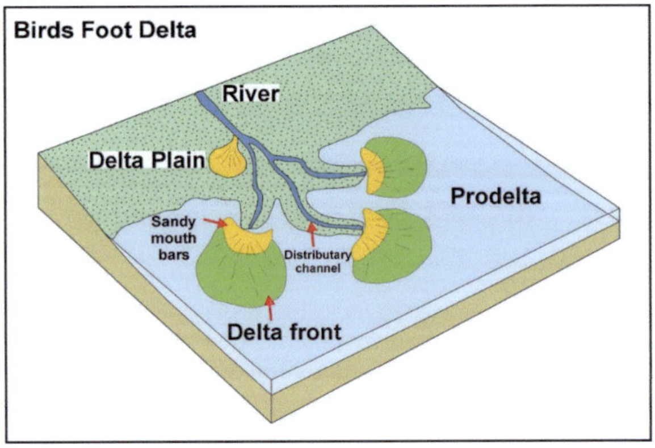

Fig. 1.6 Schematic diagram showing the bird-foot delta and or finger-like delta. In such a type of delta, each tributary is individual and its end is like a foot, like the Mississippi Delta

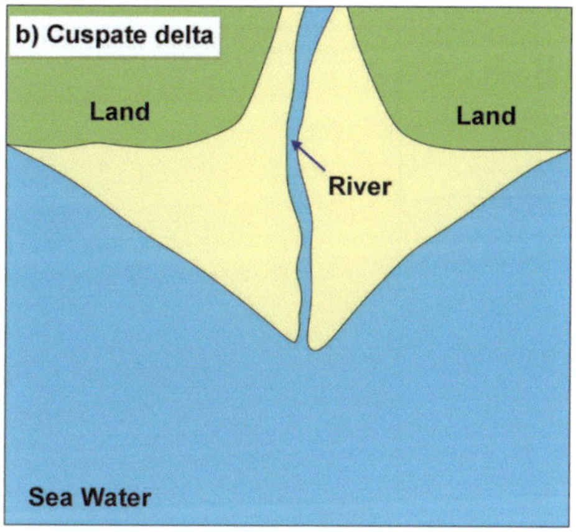

Fig. 1.7 Schematic diagram showing the general view of the cuspate. It is a tooth-shaped delta formed when a single distributary flows through and deposits its load on its either sides, such as the Tiber River in Italy

1.5.3 Cuspate Delta

This type of delta has a tooth-shaped delta formed when a single distributary flows through and deposits its load on either side (Fig. 1.7). The size and shape of a delta are controlled by the balance between watershed processes that supply sediment and the receiving basin processes (Blum and Törnqvist 2000; Pasternack et al. 2001). An example of the Cuspate Delta occurs in the Tiber River of Italy.

1.6 Types of Deltas

Deltas are typically classified based on river supply, wave supply, and tidal processes (Galloway 1975). The other two factors that play a major role in the formation of deltas are the landscape position and the grain size distribution of the source sediment entering the delta from the river (Orton and Reading 1993). These types are: Wave-dominated delta, fluvial-dominated delta and tidal marine-dominated delta, tidal freshwater delta, Gilbert delta, abundant delta, truncated delta, inland delta, estuarine delta, lacustrine delta, blocked delta, and mega delta.

1.6.1 Wave-Dominated Deltas

The wave-driven sediment transport controls the shape of the wave-dominating deltas, and much of the sediment emanating from the river mouth is deflected along the coastline (Fig. 1.8) (Galloway 1975). The relationship between waves and river deltas is quite variable and largely influenced by the deep-water wave regimes of the receiving basin. With high wave energy near shore and a steeper slope offshore, waves will make river

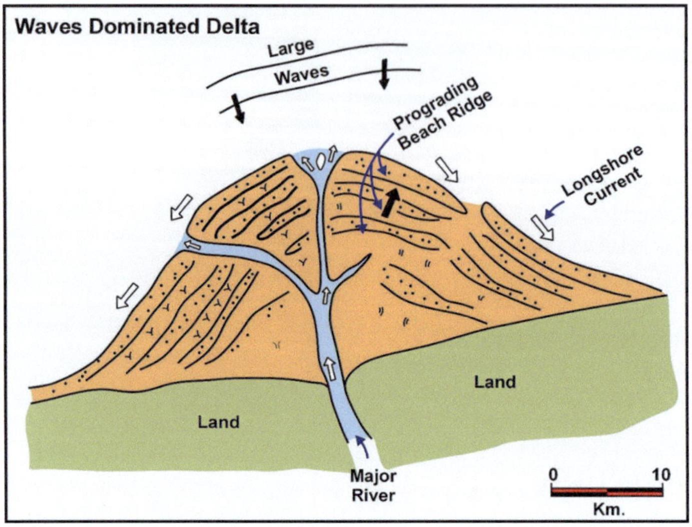

Fig. 1.8 Schematic diagram showing the general view of the wave-dominated delta, e.g., the Okavango Delta in Botswana

deltas smoother. Waves can also be responsible for carrying sediments away from the river delta, causing the delta to retreat. The Okavango Delta in Botswana is one example.

1.6.2 Fluvial-Dominated Deltas

Fluvial-dominated deltas are recorded in areas that have low tidal range and low wave energy (Boggs 2006), where the river water is nearly equal in density to the basin water. Fluvial-dominated delta is distinguished by homosocial flow, in which the river water rapidly mixes with basin water and abruptly dumps most of its sediment load (Boggs 2006) (Fig. 1.9). The fluvial-dominated deltas have usually, delta, delta front, and delta plain (Fig. 1.9) In case of, the river water has higher density than basin water, typically from a heavy load of sediment, the delta is characterized by hyperpycnal flow in which the river water hugs the basin bottom as a density. This allows fine sediments to be carried a considerable distance before settling out of suspension (Boggs 2006). The Nile Delta and the modern Mississippi River Delta represent an ideal type of a fluvial-dominated delta whose outflow is buoyancy-dominated.

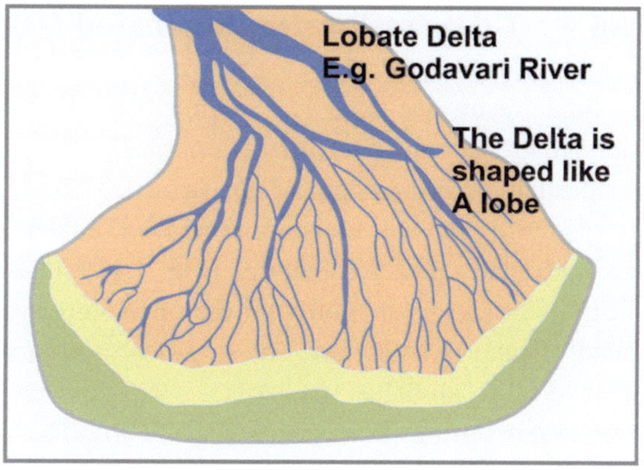

Fig. 1.9 Schematic diagram showing the general view of the fluvial-dominated delta, e.g., the Nile and the Mississippi Deltas. It usually includes prodelta, delta front, and delta plain

1.6.3 Tidal Marine-Dominated Deltas

Tidal-dominated deltas differ from the fluvial-dominated and the wave-dominated deltas, as they have a few main distributaries (Fagherazzi 2008). Erosion is also an important control in tide-dominated deltas (Fagherazzi 2008) (Fig. 1.10). Such deltas have two deposits separated by sand or barrier bars and ridges on both sides. The first occurs landward embedded in the lagoon, called a flow delta, and the second occurs seawards, which is named an ebb delta (Fig. 1.10). In a tidal delta, new distributaries are formed during times when there are a lot of floods or storm surges. The Ganges River Delta and Saint Lawrence River in Ontario, Quebec, and New York are examples of tide-dominated deltas.

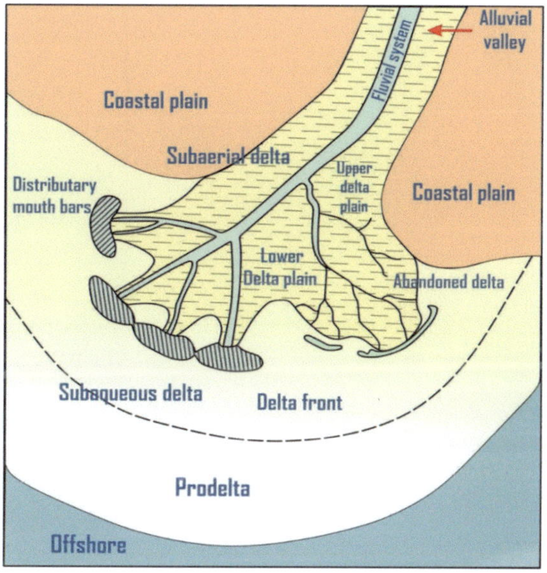

Fig. 1.10 Schematic diagram showing the general view of the tidal marine-dominated delta, e.g., Ganges River Delta and the Saint Lawrence River in Ontario, Quebec, and New York are examples of tide-dominated delta

1.6.4 Tidal Freshwater-Dominated Deltas

A tidal freshwater delta is a sedimentary deposit formed at the boundary between an upland stream and an estuary (Pasternack 2017) (Fig. 1.11), in the region known as the "subestuary" (Pasternack 1998). Drowned coastal river valleys inundated by rising sea levels during the late Pleistocene and subsequent Holocene tend to have dendritic estuaries with many feeder tributaries (Fig. 1.11). Each tributary mimics this salinity gradient from its brackish junction with the main stem estuary up to the fresh stream feeding the head of tidal propagation. The origin of a tidal freshwater delta involves processes that are typical of all deltas (Pasternack 1998) as well as processes that are unique to the tidal freshwater setting (Pasternack et al. 2000). The combinations of processes that create a tidal freshwater delta result in a distinct morphology and unique environmental characteristics. Many tidal freshwater deltas that exist today are directly caused by the onset of or changes in historical land use, especially deforestation, intensive agriculture, and urbanization (Pasternack and Grace 1998). These types of deltas are well manifested by the many tidal freshwater deltas prograding into Chesapeake Bay along the east coastline of the USA.

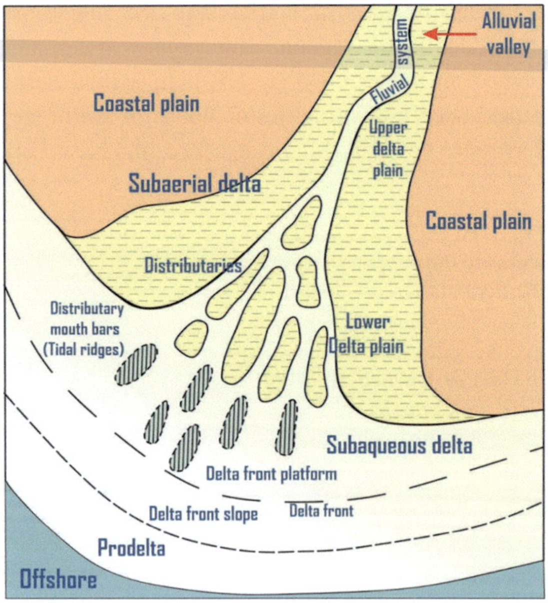

Fig. 1.11 Schematic diagram showing the general view of the tidal freshwater-dominated delta in the Gulf of Papua, New Guinea

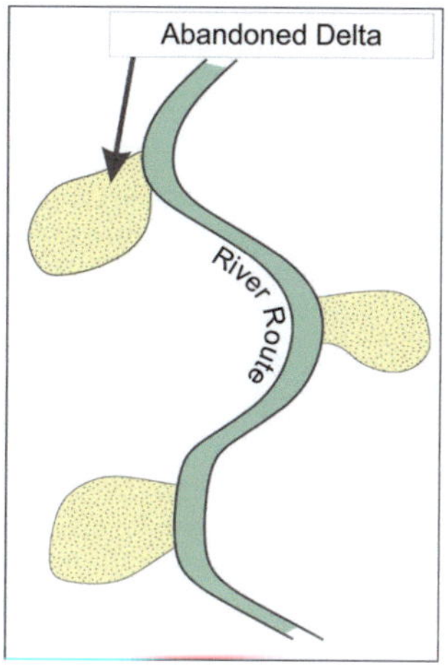

Fig. 1.12 Schematic diagram showing the general view of the abandoned deltas. Notice the occurrence of abandoned deltas on both sides of the meander river routes

1.6.5 Abandoned Delta

Several abandoned deltas can occur on both sides of the Main route of rivers (Fig. 1.12). When the rivers shift their mouths into the seas and oceans, new deltas can be formed, while the previous deltas are left unnourished. Such deltas are called abandoned deltas (Fig. 1.12). The Yellow River of China (formerly Hwang Ho) has changed its mouth several times and thus has formed several deltas. For example, the present delta of the Yellow River occurs to the north of the Shantung Peninsula, while the previous delta was deposited to the south of this peninsula. Moreover, the Mississippi River was switched, resulting in the formation of seven or eight distinct deltaic lobes over at least the past 5,000 years.

1.6.6 Gilbert Deltas

This type of delta was first identified by the geologist Grove Karl Gilbert, who described mountain streams feeding ancient Lake Bonneville (Fig. 1.13). Gilbert described delta river deposits that are characterized by the occurrence of large, coarse sediments and are usually confined to rivers emptying into freshwater lakes. Gilbert delta consists of three parts: bottom set (suspension fine sediments), fore set (coarse sediments (gravity gliding), and top set (traction sediments) (Fig. 1.13). They are usually steeper than the normal flat plain of a wave-dominated or tide-dominated delta (Fig. 1.13). Gilbert deltas are typically found in mountainous regions, where mountain rivers deposit sediment into lakes (Allgar el al. 2024).

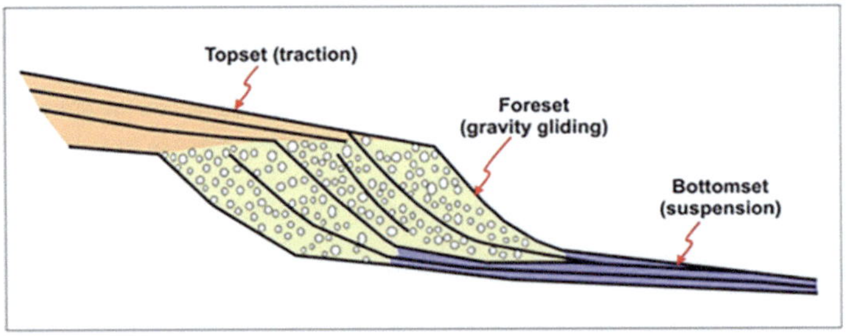

Fig. 1.13 Schematic diagram illustrating the general view of Gilbert deltas that includes three parts, bottomset (suspension fine sediments), foreset (coarse sediments), and topset (traction sediments)

1.6.7 Inland Deltas

The term inland delta was first coined by Alexander von Humboldt for the middle reaches of the Orinoco River, which he visited in 1800 (Meade 1994; Beracho et al. 2025). Their formations are more commonly seen in mountainous areas where mountain rivers pour sediment into lakes. In the case if the river delta occurring inside a large valley, it is called an inverted river delta (Fig. 1.14). Sometimes, it is called an inland delta and can be divided into multiple branches and often occurs on former lake beds. Other prominent examples of inland delta include the Inner Niger Delta, (Dadson et al. 2010; Sabokruhie et al. 2024), Peace–Athabasca Delta (Leconte et al. 2001), and the Sistan Delta of Iran. (Van Beek et al. 2008). The Okavango Delta in Botswana is probably the most well-known inland delta and is considered one of the "Seven Natural Wonders of Africa."

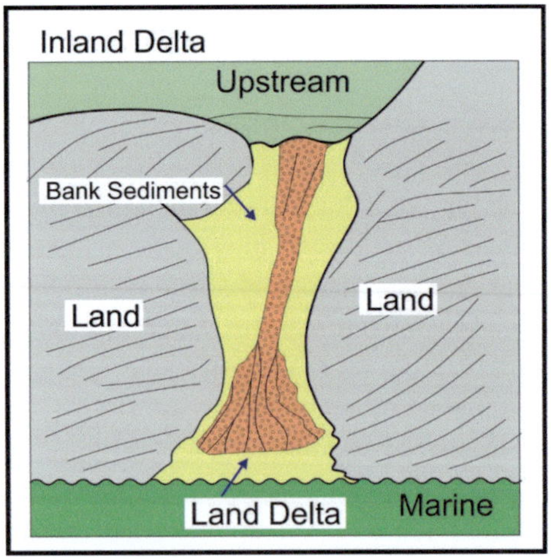

Fig. 1.14 Schematic diagram illustrating the general view of the inland delta that usually occurs in a plain, before reaching the Sea or Ocean

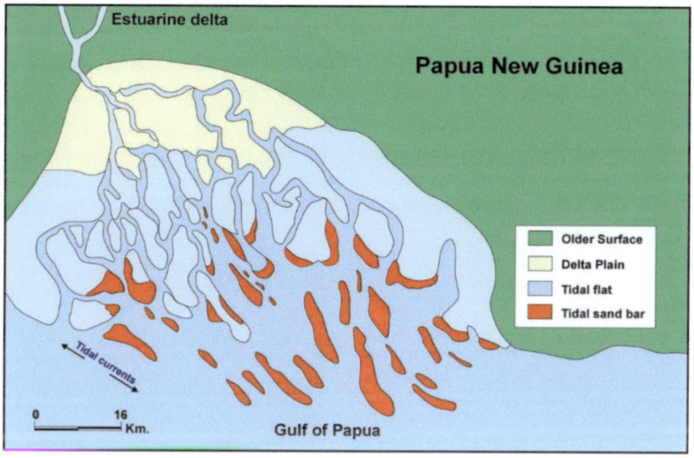

Fig. 1.15 Schematic diagram manifesting the general view of the estuarine delta that exhibits the contact between the river sediments with the seawater

1.6.8 Estuarine Deltas

The term estuarine delta refers to subaqueous and subaerial deltaic deposits that have accumulated within the semiconfined and protected environment of an estuary (Fig. 1.15) (Wright 1978; Wright et al. 1975). Two genetically distinct types of the estuarine delta can be recognized: (1) deltas which wholly or partially fill open-ended, funnel-shaped estuaries, e.g., Ord, Fly, Colorado, and Amazon deltas (Wright 1978; Wright et al. 1975) and (2) deltas which wholly or partially filling barrier lagoons or estuaries impounded by coastal barrier formations, e.g., the Mitchell delta in the Gippsland Lakes, Australia (Bird 1962), The Seine River of France and the deltas of Narmada and Tapi rivers of India.

1.6.9 Lacustrine Delta

A lacustrine delta is an accumulation of alluvium (i.e., gravel, stones, or boulders). Laid down where rivers flow into lakes, such as the Lough Leanne River Delta in Ireland. In moving from a river to a lake, water movement slows down, and with it, the water's lowered capacity to carry sediments creates a river delta (Postama 1990; Adams et al. 2001) (Fig. 1.16). The low, nearly flat, alluvial tract of land at or near the mouth of a river, commonly forming a triangular or fan-shaped plain of considerable area. Most deltas are partly subaerial and partly below water, such as the Lough Leanne River Delta in Ireland.

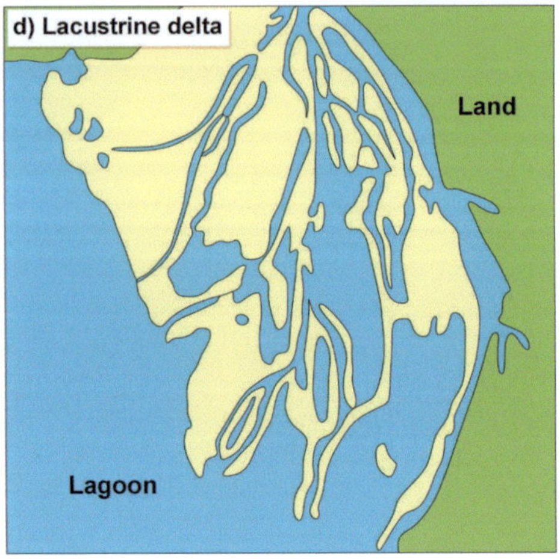

Fig. 1.16 Schematic diagram manifesting the general view of the lacustrine delta that empties its sediments into lakes

1.6.10 Truncated Delta

A truncated delta can be formed when sea waves and ocean currents modify and even destroy deltas deposited by the river through their erosional work. Thus, eroded and dissected deltas are called truncated deltas (Fig. 1.17). In another meaning, the truncated delta has been formed when there is a discontinuous supply of sediments by the river, and concurrent activation of the marine water currents can lead to the removal and transport of river sediments. This type of delta can be formed along three stages (Fig. 1.17). The first stage shows the effect of currents and waves on opposite sides of the delta, stage two gives rise to separating a part of the delta that is called the truncated delta, the third stage can displace the truncated part of the delta away from the main delta (Fig. 1.17). Example of the truncated delta is the Lough Leanne River Delta in Ireland.

1.6.11 Blocked Delta

Blocked deltas are those whose seaward growth is blocked by sea waves and ocean currents through their erosional activities. The progradation of deltas may also be hampered by a sudden decrease in the supply of sediments consequent upon climatic change or management of catchment areas of concerned rivers. The modern Danube Delta began to form after 4000 BCE (in Romania) in a bay of the Black Sea when the sea rose to its present level. A sandy barrier blocked the Danube Bay where the river initially built its delta (Giosan et al. 2006; Lenlihan et al. 2024).

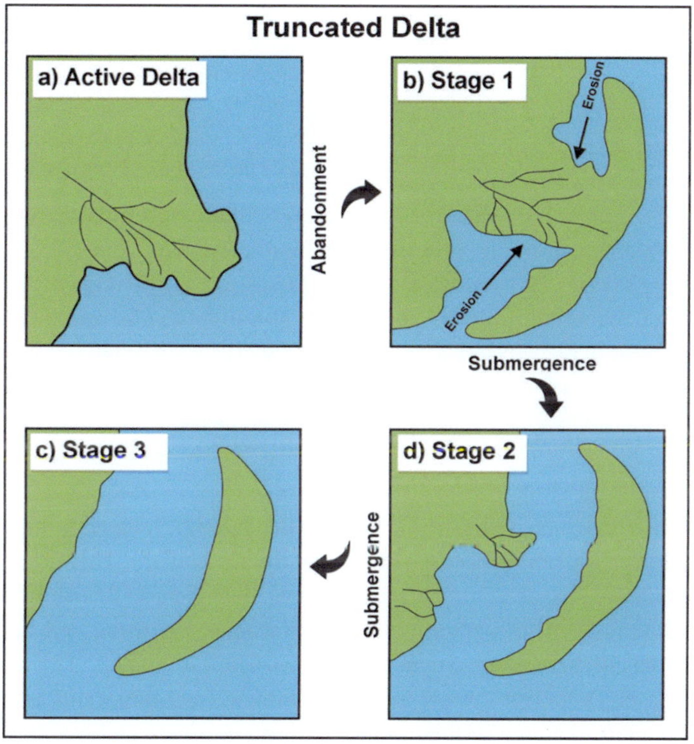

Fig. 1.17 Schematic diagram manifesting the general view of the truncated delta that can be formed in three stages. The Lough Leanne River Delta in Ireland is an example

1.6.12 Mega Deltas

The term mega delta can be used to describe the very large Asian river deltas, such as the Yangtze, Pearl, Red, Mekong, Irrawaddy, Ganges-Brahmaputra, and Indus Deltas (Seto 2011; Darby et al. 2016; Chan et al. 2024).

1.7 Importance of River Deltas to Humanity

1. The delta is home to a significant portion of the world's coal reserves and numerous major petroleum reservoirs, including the Neogene and Quaternary deltas of Egypt (Mediterranean Nile Delta).
2. Deltas are vital for absorbing runoff from river floods and storm surges from lakes and oceans. They effectively filter water as it flows through their complex networks of channels, helping to reduce the effects of upstream pollution.

3. Deltas are essential wetland habitats, supporting a variety of plant species, including lilies and hibiscus, as well as numerous herbs used in traditional medicine.
4. Deltas significantly influence the human geography of their areas, acting as key hubs for trade and commerce. For instance, Vancouver, British Columbia, is located on the Fraser River delta, which serves as a major port for exporting goods from Canada and importing products from around the world.
5. Additionally, deltas offer valuable land for agriculture and industry while managing access to water for drinking, industrial use, and irrigation.
6. As low-lying regions at the mouths of rivers, deltas are fertile and attractive for human habitation and economic activity, yet they are also highly vulnerable. Today, deltas are among the most densely populated areas in the world, with approximately 500 million people living in or near them. With their rich soil, abundant fishing opportunities, and ideal locations for large ports and harbors, deltas have become hubs of industry and trade. However, their low elevation—often just above or below sea level—exposes them to significant risks from natural disasters.
7. The Pearl River Delta, also known as the Guangdong Delta, is a highly urbanized region and one of China's fastest-growing economic centers. This delta encompasses both of China's special administrative regions: the former British colony of Hong Kong and the former Portuguese colony of Macau. Hong Kong and Macau are open to western business, serving as gateways to the Chinese market. The rapid growth of the Pearl River Delta often leads to labor shortages, as immigrants from the interior of China flock to the area in search of better opportunities and higher wages.
8. Due to significant silt deposition, deltas are typically fertile agricultural regions. The Ganges–Brahmaputra Delta, which flows into the Bay of Bengal and spans Bangladesh and India, is the largest in the world, with nearly all of Bangladesh. Key agricultural products from this delta include fish, seafood, rice, and tea.
9. Deltas remain vital to humanity today, providing essential resources such as sand and gravel, crucial for constructing infrastructure, buildings, and roadways.
10. Although they comprise only 1% of the world's land, deltas are home to over half a billion people and support rich ecosystems, including mangroves and marshes.
11. Globally, delta regions are critical sources of agricultural and fishery resources for local, regional, and global communities.
12. For centuries, river deltas have been essential to human civilization due to their nutrient-rich soils. Great ancient civilizations, such as those along the Nile and Tigris-Euphrates rivers, thrived in these areas, with communities adapting to the natural flooding cycles.
13. Delta lands are also valuable for agriculture. For example, the Sacramento-San Joaquin Delta in California is one of the state's most productive and diverse agricultural regions, yielding a wide variety of crops, including tangerines, alfalfa, and kiwis.

14. River deltas host some of the planet's most biodiverse ecosystems. These unique environments are home to numerous plant, animal, insect, and fish species, many of which are rare, threatened, or endangered, underscoring the need for their continued preservation.
15. In addition to supporting biodiversity, deltas and wetlands act as natural buffers against hurricanes. As storms approach larger, more populated areas, these open lands can help mitigate their impact. For instance, the Mississippi River Delta plays a crucial role in reducing the effects of powerful hurricanes in the Gulf of Mexico.

References

Abdel Aziz MAH, Tawfik FS, Attia AH (2024) Geological and environmental considerations for radioactive waste disposal in Egypt. J Egypt Geophys Soc 2(1):197–206

Abdel-Raheem KH, Khalil MM, Abdelhady AA, Tan L (2024) Anthropogenic–induced environmental and ecological changes in the Nile Delta over the past half-century. Sci Total Environ 926:171941

Adams EW, Schlager W, Anselmetti FS (2001) Morphology and curvature of delta slopes in Swiss lakes: lessons for the interpretation of chloroforms in seismic data. Sedimentology 48:661–67

Ali ES, Mohamed W, Mohsen A (2024) Delineating Plio-Pleistocene channels (gas reservoirs) via seismic attributes in the Baltim Gas Field, Nile Delta, Egypt

Awad M (1928) Some stages of the evolution of the River Nile. In: Proceedings of the international geography congress. Cambridge, pp 267–288

Beracko P, Kubalová S, Matečný I (2025) Aquatic macrophyte dynamics in the Danube Inland Delta over the past two decades: homogenisation or differentiation of taxonomic and functional community composition? Environ Monit Assess 197(3):332

Bird ECF (1962) The river deltas of the Gippsland lakes. Proc R Soc Victoria 75:65–74

Blanckenhorn MLP (1910) Neues zur Geologie Palastinas und des aegyptischen Nildelta. Zeitschr D Geol Ges (Berlin) 62:405–461

Blanckenhorn MLP (1921) Handbuch der Regionalen Geologie Aegyptens. Heidelberg, p 244

Blum MD, Törnqvist TE (2000) Fluvial responses to climate and sea-level change: a review and look forward. Sedimentology 47:2–48

Boggs F (2006) Principles of sedimentology and stratigraphy, 4th edn. Printed in the United States of America, p 673

Bunson M (2009) Encyclopedia of ancient Egypt. Infobase Publishing, p 98. ISBN 1438109970

Butzer KW, Hansen CL (1968) Desert and river in Nubia. University of Wisconsin Press, Madison, p 562

Chan FKS, Paszkowski A, Wang Z, Lu X, Mitchell G, Tran DD, Zhu YG et al (2024) Building resilience in Asian mega-deltas. Nat Rev Earth Environ 5(7):522–537

Cooper J (2014) The Medieval Nile: route, navigation, and landscape in Islamic Egypt. The American University in Cairo Press, p 76

Dadson SJ, Ashpole I, Harris P, Davies HN, Clark DB, Blyth E, Taylor CM (2010) Wetland inundation dynamics in a model of land surface climate: evaluation in the Niger inland delta region. J Geophys Res 115:D23

Darby SE, Hackney CR, Julian L, Matti K, Lauri H, Daniel R, Best JL, Nicholas AP, Rolf A (2016) Fluvial sediment supply to a mega-delta reduced by shifting tropical cyclone activity. Nature 539:276–279

Dumont HJ (2009) The Nile: origin, environments, limnology and human use. Springer Science & Business Media, p 88

El-Fishawi NM (1993) Sea level changes and their consequences for hydrology and water management. In: Proceedings of the international workshop, Noordwijkerhout; session 4, p 3

Elliot T (1986) Deltas. In Reading HG (ed) Sedimentary environments and facies. Backwell Scientific Publications, Oxford

Fagherazzi S (2008) Self-organization of tidal deltas. Proc Natl Acad Sci 105:18692–18695

Galloway WE (1975) Process framework for describing the morphologic and stratigraphic evolution of deltaic depositional systems. In: Brousard ML (ed) Deltas, models for exploration. Houston Geological Society, Houston, Texas, pp 87–98

Giosan L, Donnelly JP, Constantinescu S, Filip F, Ovejanu I, Vespremeanu-Stroe A, Vespremeanu E, Duller GAT (2006) Young Danube delta documents stable Black Sea level since the middle Holocene: morphodynamic, paleogeographic, and archaeological implications. Geology 34:757–760

Gregory JW (1920) The African rift valleys. Geogr J 56:13–47, 327–328

Hamza W (2006) Estuary of the Nile. In: Wangersky P (ed) Estuaries. Handbook of Environmental Chemistry. Springer Verlag, Heidelberg 5H, pp 149–173

Hayes W (1964) Most ancient Egypt. J near East Stud 23:73–114

Holz RK (1969) Man-made landforms in the Nile delta. American Geographical Society, OCLC, p 38826202

Hull FG (1896) Observations on the geology of the Nile Valley and the evidence of the greater volume of the river at a former period. Quat Geol Soc (Lond) 32:308–319

Huzzayin SA (1941) The place of Egypt in Prehistory. Inst Egypte Mem 43:440

Ian S, Paul N (1995) The dictionary of ancient Egypt. British Museum Press, London, p 83

Khalaf A (2016) Mud logging quality control and pore pressure prediction using geological and geophysical data in selected fields, Nile Delta, Egypt. Ph.D thesis in Geology, Faculty of Science, Mansoura Univ., p112

Leconte R, Pietroniro A, Alain DL, Prowse TD (2001) Effects of flow regulation on hydrologic patterns of a large, inland delta. Regul Rivers Res Manage 17:51–65

Lenihan HS, Reed DC, Vigo M, Leiphardt C, Hofmiester JK, Gallagher JP, Miller RJ et al (2024) Regional differences in fishing behavior determine whether a marine reserve network enhances fishery yield. Sci Rep 14(1):1242

Lyons HG (1906) The physiography of the River Nile and Its Basin. Egypt Surv Dept, p 411

Meade RH (1994) Suspended sediments of the modern Amazon and Orinoco rivers. Quatern Int 21:29–39

Miall AD (1979) Deltas. In Walker RG (ed. Facies models. Geological Association of Canada, Hamilton, Ontario

Nabawy BS, Shehata AM (2015) Integrated petrophysical and geological characterization for the Sidi Salem-Wakar sandstones, offshore Nile Delta. J Afr Earth Sci 110:160–175

Nashaat M (1992) Geopressure and geothermal studies Nile Delta, Egypt. M.Sc, Faculty Science, Al Azhar University, Cairo, p 133

NBI (2016) Nile Basin water resources Altes. https://nilebasin.org/sites/default/files/2024-08/NBWR%20Atlas%202016.pdf

Negm AM (2017) The Nile Delta. Springer, p 36. ISBN 978-3-319-56124-0

Orton GJ, Reading HG (1993) Variability of deltaic processes in terms of sediment supply, with particular emphasis on grain size. Sedimentology 40:475–512

Pasternack GB, William BH, Brush GSB (2000) Biogeomorphology of an upper Chesapeake Bay river-mouth tidal freshwater marsh. Wetlands 20(3):520–537

Pasternack GB (1998) Physical dynamics of tidal freshwater delta evolution (PhD dissertation). The Johns Hopkins University. OCLC 49850378

Pasternack GB (2017) Watershed hydrology, geomorphology, and ecohydraulics: tidal freshwater deltas. pasternack. ucdavis.edu. Accessed 12 June 2017

Postma G (1990) Depositional architecture and facies of river fan deltas: a synthesis. International Association of Sedimentologists

Risha M (2024) Coastal evolution of the Nile, Indus, and Yellow River Deltas: historical analysis, and machine learning prediction for future shoreline. North Carolina State University

Sabokruhie P, Duffy A, Mohammadiazar A, Carriere G, McKay K, Lindenschmidt KE, Bradford L et al (2024) Listening to each other: a collaborative 2D hydraulic model to examine water and sediment availability in an inland delta

Said R (1990) The geology of Egypt. A. Balkema Publishers, New York

Said R (1981) The geological evolution of the River Nile. Springer, New York, p 151

Said R (1975) The geological evolution of the River Nile. In: Wendorf F, Marks AE (eds) Problem in prehistory Northern Africa and the Levant. Southern Methodist University Press, Dallas, Texas, pp 1–44

Salem M (1976) Evolution of the Eocene-Miocene sedimentation pattern in northern Egypt. AAPG 60(34):64

Sandford KS (1934) Paleolithic man and the Nile Valley in Upper and Middle Egypt. Chicago Univ Orient Inst Publ 3:1–131

Sandford KS, Arkell WI (1939) Paleolithic man and the Nile Valley in Nubian and Upper Egypt. Chicago Univ Orient Inst Publ 36:1–105

Sandford KS, Arkell WI (1929) Paleolithic man and the Nile-Faiyum divide. Chicago Univ Orient Inst Publ 1:1–77

Sandford KS, Arkell WI (1933) Paleolithic man in Nubia and Upper Egypt was divided. Chicago Univ Orient Inst Publ 17:1–92

Seto KC (2011) Exploring the dynamics of migration to mega-delta cities in Asia and Africa: contemporary drivers and future scenarios. Glob Environ Chang 21:S94–S107

van Beek E, Bozorgy B, Vekerdy Z, Meijer K (2008) Limits to agricultural growth in the sistan closed inland delta, Iran. Irrig Drain Syst 22:131–143

Wendorf F, Schild R (1976) Prehistory of the Nile valley. Academic Press, New York and London, p 404

Wright LD (1978) River deltas. In: Davis RA (ed) Coastal sedimentary environments. Springer-Verlag, New York, pp 5–68

Wright LD, Coleman JM, Thom BG (1975) Sediment transport and deposition in a macrotidal river channel, Ord River, Western Australia. In: Cronin LE (ed) Estuarine research, vol 2. Academic Press, New York, pp 309–321

Zaghloul ZM, Andrawis SF, Ayyad SN (1977) New contribution to the stratigraphy of the Tertiary sediments of Kafr El-Dawar well no. 1, North West Nile Delta, Egypt. Ann Geol Surv Egypt 4:292–307

Zeidan B (2006) The Nile Delta in a global vision. Sharm el-Sheikh, archived from the original on 10 July 2020

2 Stratigraphy of the Nile Deltas

Mohamed Abdelghany Khalifa

Abstract

The sediments that have accumulated over 40 million years by the Nile River are identified in the present chapter, specifically in the Fayoum Nile Delta and the Mediterranean Nile Delta. The Fayoum Nile Delta has a fan-shaped northwest and occurs in the northern Western Desert, west of the Fayoum province. It covers wide regions including Wadi El Natron, Sadat City, the Bahr Depression (northeast of the Bahariya Oases), and the southern part of the Qattara Depression. It comprises the Gehannam, Qaroun, Qasr El Sagha, Qatrani, El Khashab, and the Moghra formations. These rock units are mostly made up of clastics (sandstone, siltstone, mudstone, and paleosol) with thin limestone beds enriched with vertebrate remains, e.g., whale skeletons, sea cows, fish bones, snakes, turtles, crocodiles, shark teeth, and ancestors of elephants. The ancestral Nile River deposited these rock units over 40 million years. The second delta is called the Mediterranean Nile Delta, occurring along the Mediterranean Sea coast in the central Egyptian coast. It is separated into two superimposed deltas, the Neogene Delta at the base and the Quaternary Delta at the top, due to the effect of the Messinian unconformity. The Neogene Delta comprises six rock units in ascending order: the Upper Oligocene Tineh Formation, the Lower Miocene Qantara Formation, the Middle Miocene Sedi Salem Formation, the Qawasim Formation, the Upper Miocene Rosetta, and the Abu Madi Formation. The Quaternary Nile Delta includes the Lower Pliocene Kafr El Sheikh, the Upper Pliocene Al Wastani, Pleistocene-Holocene Mit Ghamr, and the Holocene Bilqas formations.

M. A. Khalifa (✉)
Geology Department, Faculty of Science, Menoufia University, Shiben El Kom, Egypt
e-mail: khcycle1949@gmail.com

M. A. Khalifa and K. S. Gemail (eds.), *Geology of the Nile Deltas and Natural Resources*, Deltas of the World, https://doi.org/10.1007/978-3-031-96594-4_2

Keywords

Nile Deltas • Neogene • Quaternary • Gehannam • Qaroun • Qasr El Sagha • Qatrani • El Khashab • Moghra • Tineh • Qantara • Sedi Salem • Qawasim • Abu Madi • Kafr Shiekh • Mit Ghamr • Bilqas

2.1 Introduction

The Nile Delta's stratigraphy was traversed by several episodes. The first episode takes place in the 1940s and 1950s when seismic, magnetic, and gravitational geophysical technologies were initially used to explore the Nile Delta. Since no structural elements could be found using the geophysical methods used above, no articles have been published at this time regarding the aforementioned insufficient data. In addition to the above, the Nile Delta was covered with agricultural soils, and its stratigraphic succession was not indicated by any exposures. The second episode occurred in the early 1960s and involved the first comprehensive seismic survey carried out in the Abu Madi region in the heart of the Nile Delta. Certain structural features were revealed by this seismic survey. The Mit Ghamr-1 well and the Kafr El Sheikh-1 well both have two structural elements drilled on top of them, but they are both dry wells. Subsequently, the first well was dug in 1967 in the central Nile Delta region of Abu Madi, which was enriched with a gas reservoir. The first deep drilling well in the northwest offshore zone of the Mediterranean Sea, at Abu Qir Bay, the offshore portion of the Nile Delta, was commenced by the Philips Company in 1969. In the third episode of the 1970s, the first published articles by Salem (1976), Zaghloul et al. (1977, 1979), Ryan (1978), Rizzini et al. (1976, 1978), Barber (1981), Said (1981), El Heiny (1982), and Barakat (1982) were based on the above-drilled wells. These published papers are considered the main platform of the stratigraphic classification of the Oligocene, Miocene, and Quaternary rock units of the Nile Delta. The fourth episode started during the eighties and nineties when the 2-D and 3-D seismic stratigraphy methods appeared. These methods helped in the interpretation of the occurrence and declaration of structures, unconformities, distributary channels, and the nature of the vertical and lateral relationships among beds. Since then, tenth published papers by several geologists, e.g., Abu El-Ella (1990); El Heiny et al. (1990); El Heiny and Morsy (1992); El Heiny and Anani (1996); Kamel et al. 1998; Abdel Aal et al. (2000); Sarhan and Hemdan (1994); Farouk et al. (2014); Leila et al. (2016); Leila and Moscariello (2018, 2019); Sarhan and Safa (2017, 2019); Sarhan (2021a, b, c) among others. In the present study, we will discuss the stratigraphy of the two deltas of the Nile River. The old and the first deposited delta is the Fayium Nile Delta, while the second is the Mediterranean Nile Delta. The stratigraphic sequence of the Fayium Nile Delta is exposed and includes the Gehannam Formation (Middle Eocene), Qaroun Formation (Middle Eocene), Qasr El Sagha Formation (Late Eocene), Qatrani Formation (Oligocene), El Khashab Formation, and the Moghra Formation (Early Miocene). The Mediterranean Nile Delta all rock units

Table 2.1 Correlation of the rock units of the Fayoum Nile Delta and the Mediterranean Nile Delta

Age	Fayoum Nile Delta	Mediterranean Nile Delta
Holocene		Bilqas Formation
Pleistocene Holocene		Mit Ghamr Formation
Pleistocene		Al-Wastani Formation
Pliocene		Kafr El Shiekh Formation
Late Miocene		Rosetta–Abu Madi Formations
Middle Miocene		Sedi Salem–Qawasim Formations
Early Miocene	El-Khashab–Moghra Formations	Qantara Formation
Oligocene	Qatrani Formation	Tineh Formation
Late Eocene	Qasr El Sagha formation	
Late Middle Eocene	Qaroun Formation	
Middle Eocene	Gehannam Formation	
	El-Gharaq Formation	

are not exposed and encountered in the subsurface, beginning with the Oligocene Tineh Formation, Early Miocene Qantara Formation, Middle Miocene Sedi Salem Formation, Early Miocene Qawasim Formation, Late Miocene Abu Madi and Rosetta formations, Pliocene Kafr El Shiekh Formation, Plio-Pleistocene Al Wastani Formation, Pleistocene Mit Ghamr Formation, and the Holocene Bilqas Formation (see Table 2.1).

2.2 Stratigraphy of the Fayoum Nile Delta

The stratigraphic sequence of the Fayium Nile Delta includes the Middle Eocene Gehannam Formation, the Upper-Middle Eocene Qaroun Formation, and the Upper Eocene Qasr El Sagha Formation. The Oligocene rock units are the Qatrani Formation, while the Lower Miocene rock units include the El Khashab and the Moghra formations (Fig. 2.1). These rock units are exclusively exposed in and around the Fayium Province and its environs (Table 2.1).

2.2.1 Eocene Rock Units

Among the geologists who examined the Eocene rock units in the Fayum region are Blanchenhorn (1900, 1903), Beadnell (1901, 1905), Andrews (1904, 1906), Sandford and Arkell (1924), Ansary (1955), Said (1962), Bown and Vondra (1974), Kortland (1980), Strougo and Haggag (1984), Strougo (1985a, b), Fleagle (1986a, b), Strougo (1986, 1988), Bown and Kraus (1988), Kappelman (1991), and Sweden (1992) (Table 2.1).

2.2.1.1 The Gehannam Formation

The term Gehannam Formation was established by Said (1962) to describe the exposed Middle Eocene at Qaret Gehannam (Fig. 2.2). This formation was previously named by Beadnell (1905) as Ravine Beds (claystone beds) that were exposed in ravines (small valleys) in the southern Fayium depression. However, because the beds are exposed very well at Qaret Gehannam with a defined base and top, the name Gehannam Formation is more commonly used here instead of the Ravine Beds. In nearly all locations, the bottom limit of this formation is unexposed (Fig. 2.2). On the other hand, it displays angular or diagonal nonconformity with the underlying Middle Eocene El Gharaq Formation (nummulitic limestone), as indicated by the vertical stratigraphic connections (Fig. 2.4; Iskander 1943; Khalifa 1981). Fish scales, shark and ray teeth, and marine animal bones are abundant in the thick layers of glauconitic sandstone and marls that define the base of this formation, devoid of large types of benthic Forams (Zalmout 2008). The upper limit of this formation manifests unconformable contact with the overlying Qaroun Formation, which may be seen near its upper limit (Fig. 2.3). The upper boundary occurs at the bioturbated Camp White Layer with shrimp burrowing tubes of the uppermost Gehannam Formation (Gingerich 1992) and the gray claystone of the Basal Qarun Formation. The Mangrove layer contains whale skeletons, sea cows, fish bones, and shark teeth. There are also many traces of *teredolites ichno* sp. and bioturbations filled by celestite and red iron oxide. The

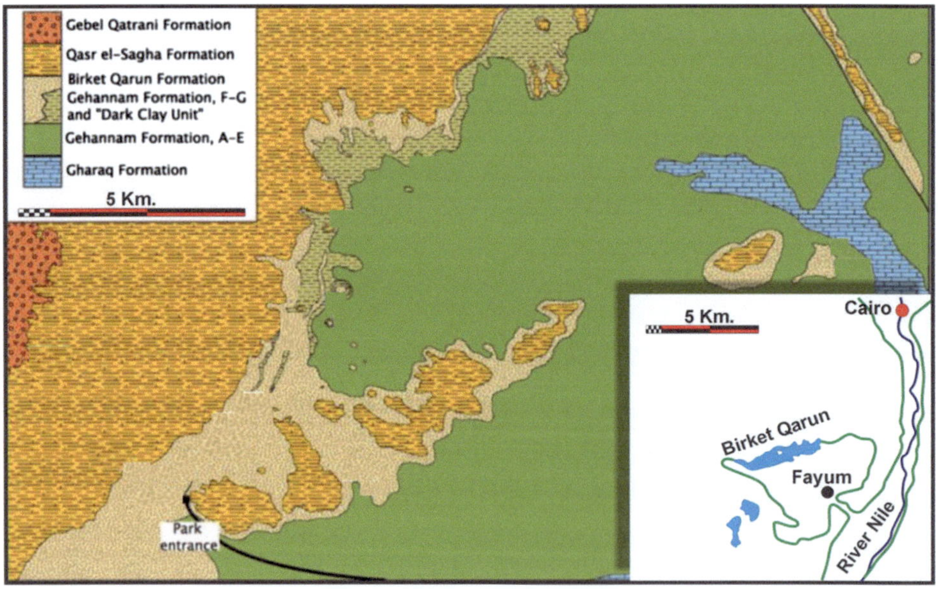

Fig. 2.1 Geologic map of the Fayium province showing the rock units that constitute the Fayium Nile Delta (After Underwood et al. 2011)

Fig. 2.2 Field photograph of the unexposed base of the Gehannam Formation at its type locality, southwest Fayium depression (After Abu El Kheir 2010)

Fig. 2.3 Field photograph showing the Lower unexposed part of the Gehannam Formation, including the glauconitic (gl), overlain by intercalation of sandstones and mudstones of the Birket Qarun Formation (After, Underwood et al. 2011)

boundary probably should be extended deeper in the valley to the top level of the calcareous sandstone that weathers into spherical masses demarking the lowermost of the Birket Qarun Formation (Gingerich 1992). At Qaret Gehannam, the formation comprises intercalation of marl and sandy claystone in the basal part, marl and limestone in the middle part and marl and claystone in the upper part (Fig. 2.6). Northwestward of the Qaret Gehannam and just to the west side of the Birket Qarun near the village of Guta, Minqar El hut and Sandouk El Bornita the Gehannam Formation comprises intercalation of burrowed sandstone, marl and claystone (Fig. 2.6) (Gameil et al. 2016). At Sandouk El Bornita, the formation consists of two lithofacies, sandstone and claystone, with the absence of marl and limestone. In the above localities (Fig. 2.6), many skeletons of marine mammals can be observed (Fig. 2.7). At Gebel Naaloun and Gebel Lahoun, southeastern part of the Fayium depression, the main lithological components are intercalation of marl, marly limestone and sandy limestone (Wanas 1991) (Fig. 2.8). The marl is gray, moderately hard, with several gypsum veinlets. The marly limestone beds are thicker than marl, usually grayish white, and each bed ranges in thickness from 1 to 1.5 m. The Gehannam Formation has a variable thickness; its maximum thickness is about 50 m near the Qaret Gehannam plateau (Wanas 1991). It decreases in thickness eastwards to 37 m and 31 m at Gebel Lahoun and Gebel Naalon, respectively (Wanas 1991).

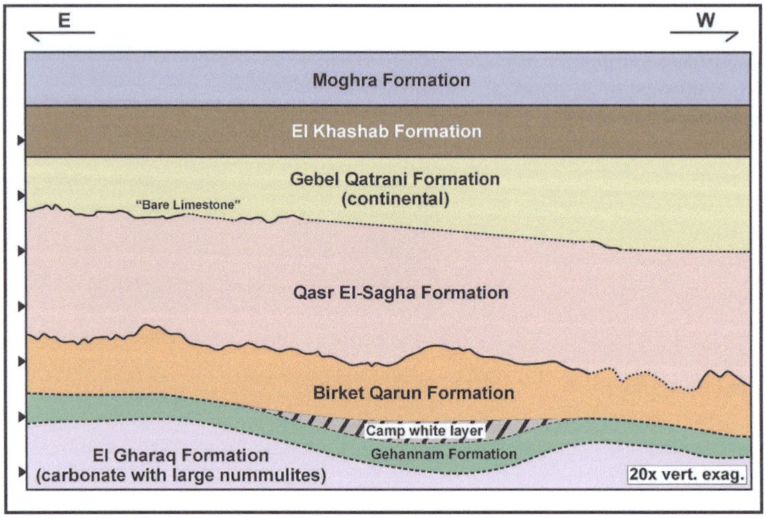

Fig. 2.4 East–West schematic cross-section illustrating the distribution of the El Gharaq, Gehannam, Qarun, Qasr El Sagha, and Qatrani Formations in the Fayium region (After, Underwood et al. 2011, with modification)

Fig. 2.5 Field photograph showing the bioturbated White Camp Layer with shrimp burrowing tubes that discriminate the topmost limit of the Gehannam Formation and its contact with the overlying Qaroun Formation

Based on its foraminifera composition (Planktonic foraminifera), this formation dates back to the late middle Eocene (Abdu and Abdel-Kareem 1975; Strougo and Haggag 1984; Haggag 1988, 1990). A conspicuous mangrove root from the Late Eocene Camp White Layers at the top of the Gehannam Formation at Wadi Hitan, and west of the Fayoum depression, was identified by El Sadawi (2005). According to him, these kinds may survive in brackish water.

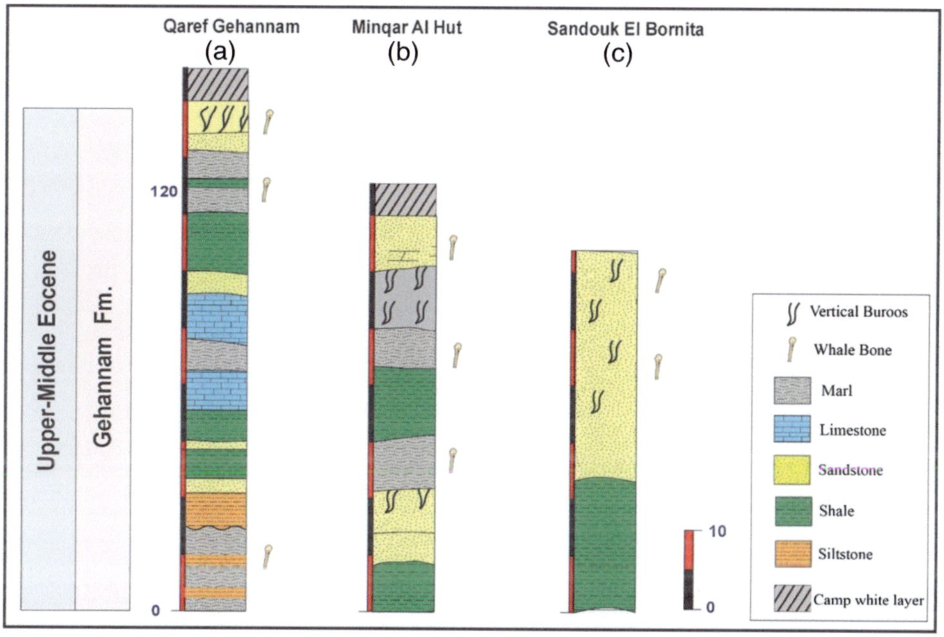

Fig. 2.6 Measured lithostratigraphic section of the Gehannam Formation at its type locality (Qaret Gehannam), Minqar El Hut, and Sandouk El Bornita (After, Gameil et al. 2016)

Fig. 2.7 Field photograph illustration of the whale vertebrate fauna occurring in the Gehannam Formation, at Wadi Hitan, southwest of the Fayium depression

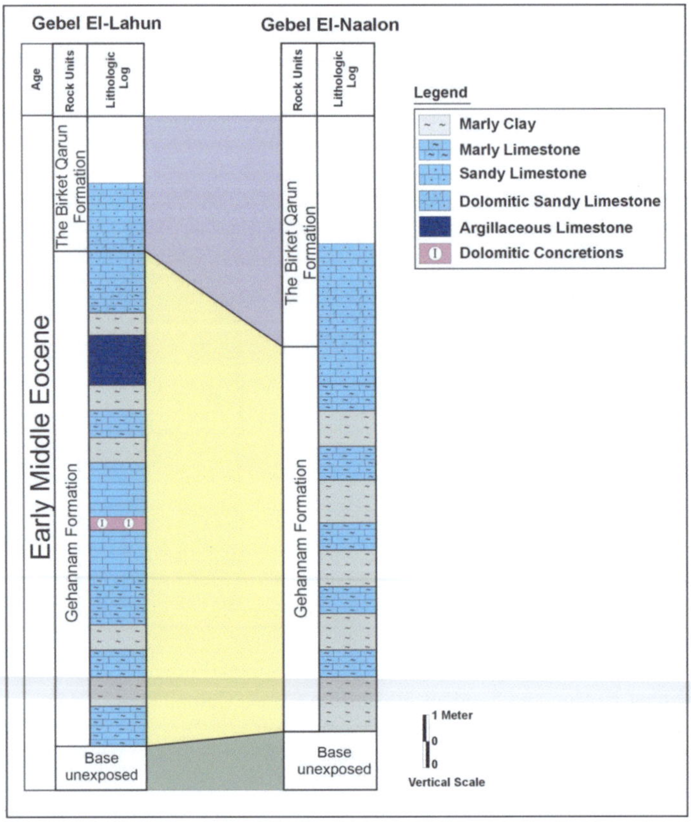

Fig. 2.8 Measured sections of the Gehannam Formation at Gebel Lahoun and Gebel El Naaloun, illustrating the main lithology of this formation (After, Wanas 1991)

2.2.1.2 The Qaroun Formation

Beadnell (1905) named this formation to denote the intercalation of sandstone and claystone succession. According to Beadnell (1905), its type section is exposed at the western extremity of Qaroun Lake (Fig. 2.9). According to Abu El Kheir (2010) and Gameil et al. (2016), this formation unconformably overlies on top of Gehannam and beneath the Qasr El Sagha Formations (Fig. 2.10).

The lithology of the Qaroun Formation exhibits facies changes from southwest to northeast (Fig. 2.11). At the southwest of Wadi Hitan, the Qaroun Formation comprises vertical cliffs of calcareous sandstone (Fig. 2.11), forming a single sand body that splits to the northeast into four separate sand units separated by claystone units (Fig. 2.11) (Gameil et al. 2016). At the southwest of Wadi Hitan, the formation entirely consists of burrowed fine to coarse-grained sandstone (Fig. 2.12a). At Qaret Gehannam, the formation comprises mainly burrowed sandstone with thin interbeds of marl and claystone. It

Fig. 2.9 Field photograph manifesting the lithological component of the Qaroun Formation at its type locality (northwest of Birket Qaroun). Notice, the intercalation of gray claystone and page sandstone throughout the formation (After Gameil et al. 2016)

Fig. 2.10 Field photograph manifesting the stratigraphic position of the Qaroun Formation that is restricted between the Gehannam Formation below and the Qasr El Sagha Formation above, north of Qaroun Lake (After Abu El Kheir 2010)

begins with reddish yellow, ferruginous, bioturbated sandstone at its base, overlain by one meter of dark gray laminated marl and bioturbated shale. Such a sequence is repeated vertically to the end of this formation (Fig. 2.12b). The sandstones are coarse-grained, burrowed with minor siltstone bands, long spired gastropod *Turritella* sp. At Minqar Al Hut and Sandouk El Bornita, the Qaroun lithology is closely similar in facies association and is composed of an intercalation of sandstone and mudstone and rarely marl (Fig. 2.12c, d). The sandstones are usually yellow, fine to medium-grained, calcareous, burrowed, and rich in the bivalve *Lucina* sp. The mudstones are gray, massive to bioturbated with severely bioturbated bases by *Ophiomorpha* and *Thalassinoides* (Gameil et al. 2016). At the latter locality, it is capped by yellowish-white, bioturbated, fossiliferous limestone containing *Pecten* sp., echinoids, and *Nummulite* sp. (Gameil et al. 2016). North of Minqar Al Hut, the Qaroun Formation begins with yellowish-white, fine-grained

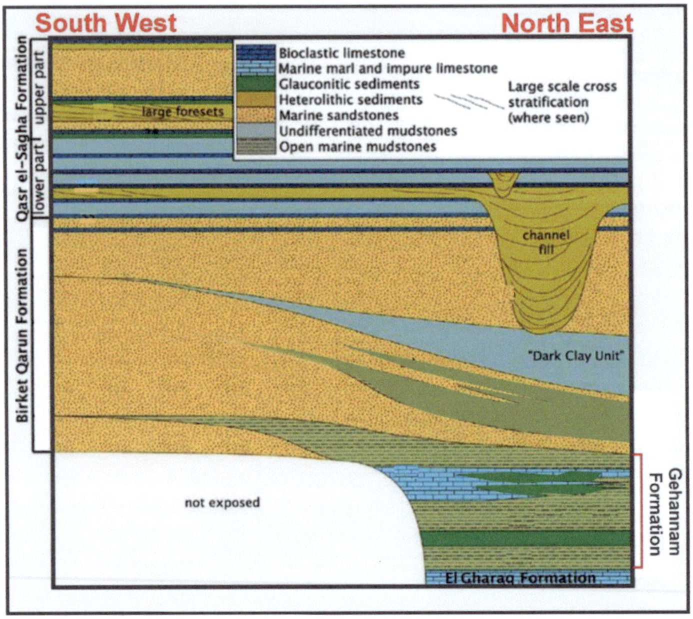

Fig. 2.11 Northeast-southwest cross-section showing the lateral and vertical facies changes of the Gharaq, Gehannam, Qaroun and Qasr El Sagha formations in the western part of the Fayium Province (After Underwood et al. 2011). Notice the thick sandstone sequence of the Qaroun Formation at the southwest and split into sandstone and claystone units northeastwards

sandstone containing some plant remains and vertebrates, fossils of turtles, crocodiles, fish bone, shark teeth, sea cow, ancestors of elephants, snakes (Gingerich 1992) (Fig. 2.12d). Added to the above, oysters and gastropods as well as some vertebrate fossils belonging to *Saghacetus Osiris* and sea cow skeletons are recorded in all measured sections of the Qaroun Formation, Gameil et al. (2016). Gingerich (1992) stated that the Birket Qarun Formation represents a series of barrier bar deposits trapping clay and silts in small lagoons.

Detailed measured section of the Qaroun Formation by Seiffert et al. (2008) declares that this formation is made up of two primary lithological units (Fig. 2.13). The lower unit that constitutes one-third of this formation consists of claystone and pebbly conglomerate, while the upper unit forms about two-thirds of the formation and consists of interbeds of pebbly conglomerate, sandstone, siltstone, and mudstone, Seiffert et al. (2008) (Fig. 2.13). The lower unit exhibits fining-upward cycles, each begins thin (few centimeters) bioclastic granule and pebbly limonitic conglomerate, capped with thick (3–4 m) claystone or mudstone (Fig. 2.13). The upper unit also comprises fining-upward cycles, each of which begins with a thin conglomerate bed, followed upward by medium to fine-grained yellow sandstone, sandy mudstone to siltstone, and capped by gray mudstone, Seiffert et al.

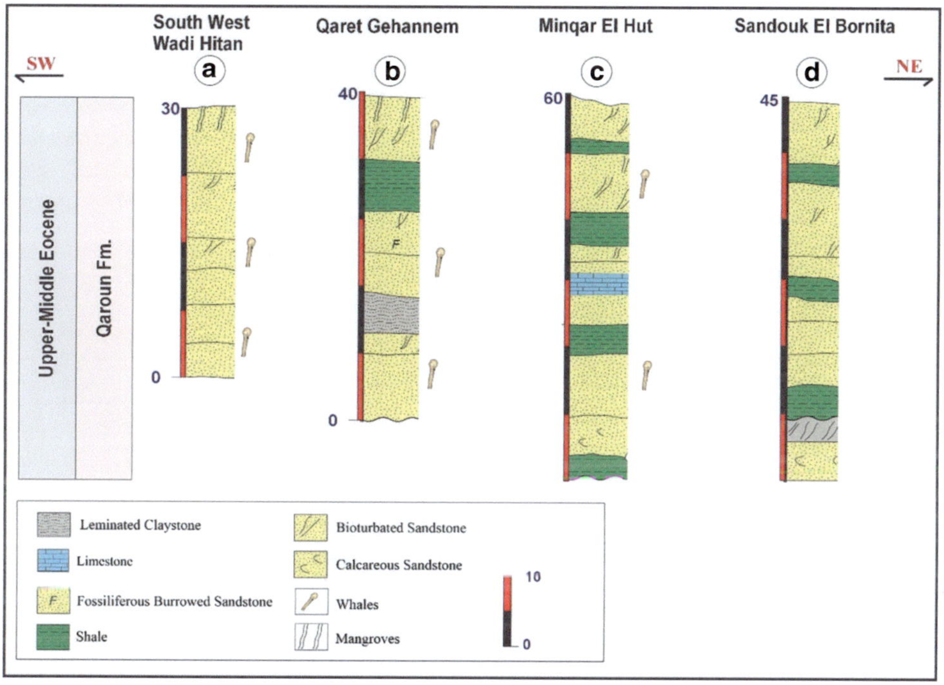

Fig. 2.12 Measured sections of the Qaroun Formation at, southwest of Wadi Hitan (**a**), Qaret Gehannem (**b**), Minqar El Hut (**c**) and Sandouk El Bornita (**d**), Fayium depression (After Abu El Kheir 2010); Gameil et al. 2016)

(2008) (see Fig. 2.13). In some cycles either in the lower unit or upper unit, the cycle is capped with oblate spheroidal nodules sandstone named biscuits, Seiffert et al. (2008) (Fig. 2.14). When such spheroidal sandstone is exposed on the surface by weathering, it becomes look-like melon-sized concretions (Fig. 2.14). The thickness of the Qaroun Formation assumes 30 m at Wadi Hitan, 40 m at Qaret Gehannam, and 45 m at Sandouk El Bornita. The maximum thickness occurs at Minqar El Hut (60 m).

The fossil marine mammals and other vertebrates of the Qarun Formation were collected from many levels, from the lagoonal shale and the sandstone barrier zone (Fig. 2.15). Some whale skeletons of *Basilosaurus isis* (Andrews 1904) and *Dorudon atrox* (Andrews 1906) are common and well preserved at its base (Fig. 2.16). The top of the formation at the northern escarpment of Wadi Al Hitan contains subaquatic and terrestrial fossils like snake bones, *Moeritherium* sp., and other species of the ancestors of Elephants mixed with marine fossils like shark teeth, crocodiles, turtles, whales, foraminifera, and ostracodes (Seiffert et al. 2008). Vertebrate remains of shark teeth, fish remains, and many skeletons of *Basiliosaurus isis* and *Dorudon atrox* are found at the topmost calcareous sandstone layer (Seiffert et al. 2008). Uhen (2004) and Zalmout (2008) described

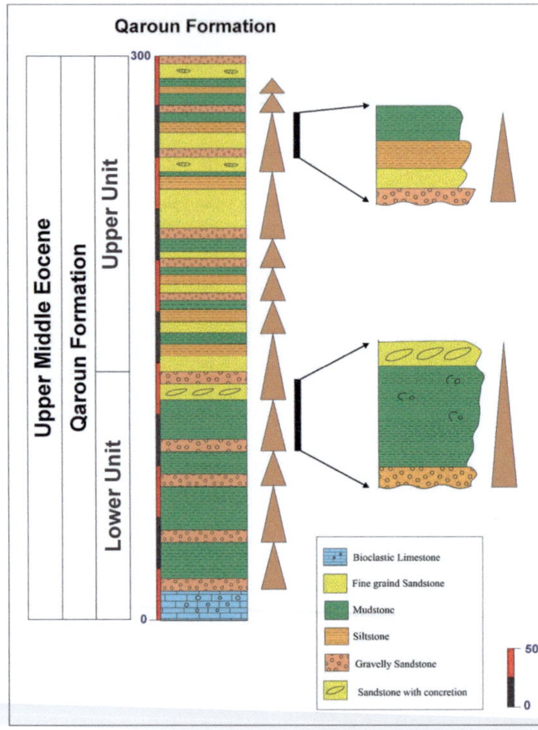

Fig. 2.13 Measured lithostratigraphic section of the Qaroun Formation which comprises two units, the lower unit that constitutes the lower one third of this formation consists of claystone and pebbly conglomerate, while the upper unit that forms about two upper thirds of the formation consists of interbeds of sandstone, siltstone, mudstone, and pebbly conglomerate (After Seiffert et al. 2008). Notice the presence of fining-upward cycles in each unit

Fig. 2.14 Field photograph showing the sandstone concretions occurring usually on the weathered top cycles of the Qaroun Formation (After, Abu El Kheir 2022)

Fig. 2.15 Field photograph of the whale vertebrate remains (whale skeletons of Basilosaurus isis, Andrews 1904) in the Qaroun Formation at Minqar El Hut, southwest of the Fayium depression (After Underwood et al. 2011)

Fig. 2.16 Field photograph illustrating some whale skeletons of Dorudon atrox (Andrews 1906) are common and well preserved at the base of Qaroun Formation at Minqar Al Hut, Fayium Province (After Underwood et al. 2011)

cetaceans and sirenian skeletons from this layer. The irregular echinoid *Schizaster libycus* is common in the calcareous sandstone of the lower part of the formation.

Abundant benthonic Forams and rare planktonic Forams exist in this calcareous sandstone. The marine mammal-bearing beds from the top of the Gehannam Formation and the lower third of the Birket Qarun Formation at Wadi Al Hitan are early Priabonian in age based on the overlapping of calcareous nannoplankton zone and planktonic foraminifer's zone (*Globigerinatheka semiinvoluta*) (Zalmout 2008). A collection of fossil fish teeth and other elements from the Qaroun Formation north of Birket Qarun is reported (Murray et al. 2010). Such small teeth were not previously reported from the Fayoum depression. Most of these teeth remain, represent fully freshwater fish (Murray et al. 2010).

Fig. 2.17 Field photograph showing general view of the Qasr El Sagha formation that includes the lower Temple and the upper Abu Lifa Members, north of Fayium depression (After, Abu El Kheir 2022)

2.2.2 The Qasr El Sagha Formation

The sediments of the Qasr El Sagha were first described as a series by Beadnell (1905), who described them as the *Carolia* beds in Fayoum province. Later, Vondra (1974) replaced Beadnell's term "series" with Qasr El Sagha Formation. Its type section is located at Qasr El Sagha temple, north of Qaroun Lake (Fig. 2.17). The Qasr El Sagha Formation is regarded as the last and thickest Eocene rocks in the Fayium depression, which may exceed 180 meters in total thickness. This formation conformably overlies the Qaroun Formation and unconformably underlies the Oligocene Qatrani Formation (Gameil et al. 2016) (Fig. 2.18). The base of the Qasr El Sagha Formation was a matter of debate among different geologists. It occurs at the base of the lowest clay unit, several meters above the lowest *Carolia* bed (Gingerich 1992). On the other hand, the base occurs at the first recognition of *Carolia placunoides* (Peters et al. 2009). Its top is marked by a disconformity contact with the varicolored fluvio-marine and terrestrial conglomeratic sandstones of the Oligocene Qatrani Formation (Fig. 2.19). The Qasr El Sagha Formation is divided into two members in its type locality (Bown and Kraus 1988). The lower is called the Temple Member (limestone, shale, sandstone, and *oyster* banks intercalations), while the upper is called Dir Abu Lifa Member (cross-bedded sandstone, shale, limestone) (Bown and Kraus 1988) (Fig. 2.19). The lower Temple Member refers to Qasr El Sagha Temple, while the Dir Abu Lifa Member is proposed from Dir Abu Lifa and represents an ancient Coptic Christian church about 3 km above and northeast of the Qasr El Sagha Temple (Bown and Kraus 1988). Gingerich (1992) divided the lower part of the formation of the Temple Member into the Qaret Umm Rigl Member interbedded clays, sands, and thin shelly calcareous sandstone beds, and the overlying Harab Member (monotonous clays).

The lower Temple Member consists of an intercalation of glauconitic arenaceous and bioclastic limestone with thicker laminated sandy and carbonaceous claystone (Fig. 2.20)

Fig. 2.18 Field photograph showing the unconformable contact of the Qasr El Sagha formation with the overlying Oligocene Qatrani formation, north of the Fayium depression (After Gameil et al. 2016)

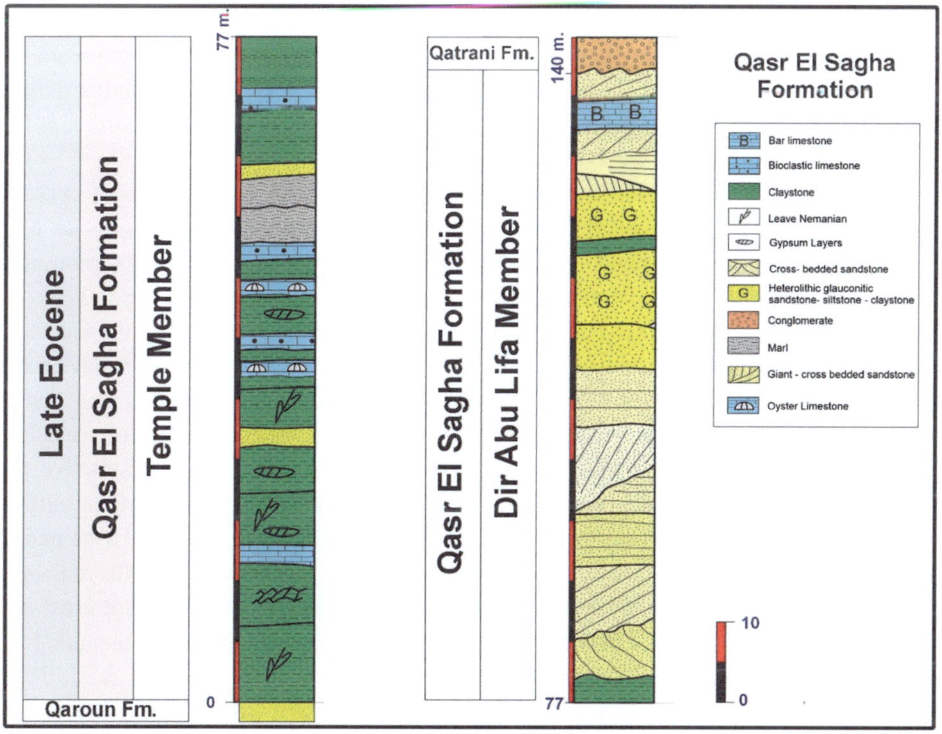

Fig. 2.19 Measured lithostratigraphic section of the Qasr El Sagha Formation at its type locality (Qasr El Sagha temple). It is subdivided into two members, the lower is the Temple Member, while the upper is the Abu Lifa member (Modified from Bown and Kraus 1988), north of Birket Qaroun Lake, Fayium Province

Fig. 2.20 Field photograph showing the lithology of the lower Temple Member of the Qasr El Sagha Formation in which the giant cross sets of the Abu Lifa Member occur above erosion surfaces

(Bown and Kraus 1988). The bioclastic and sandy limestone is usually gray, green, and brown, and each bed has an average thickness of a few decimeters and contains molluscan fauna (Beadnell 1905). The sandy limestone and sandstones form numerous ledges all along the lower part of the Qasr El Sagha escarpment, where more gentle slopes of claystone units separate them. The claystone is green-gray, thick-bedded, and enriched with gypsum streaks. The gypsum occurs as finely disseminated fibrous crystals, as platy masses developed along fractures and bedding planes. Leave remains are also common in the claystone (Vondra 1974; Bown and Kraus 1988). The upper Abu Lifa Member comprises several lithofacies, e.g., giant cross-bedded sandstone, green and gold mudstone sandstone, green sandstone, and gypseous sandstone (Fig. 2.21). This giant-bedded sandstone sequence is the thickest facies type and was named by Bown and Kraus (1988), because the large cross sets greatly exceed 3 m in thickness. This type of sandstone occurs in the basal, middle, and upper parts and is regarded as the most distinctive feature in the Qasr El Sagha Formation. It is characteristically developed as vertical cliffs in and surrounding the type area above Qasr El Sagha Temple (Fig. 2.22). It consists mainly of unusually thick sets of cross-stratified white sandstone that contain· smaller intra-sets of alternating cross-bedded sandstone and carbonaceous mudstone. The giant cross sets occur above erosion surfaces that appear to be channel scours varying in breadth from 40 to 50 m and from 3 to 25 m in depth (Bown and Kraus 1988; Legler et al. 2013) (Fig. 2.22). Therefore, the entire giant cross-bedded sandstone sequence comprises several superposed channel deposits. The giant cross-bedded sandstone includes *Carolia* and oyster fragments, crab claws, shark, skate, and ray teeth, sea, snake vertebrae, and rare turtle, crocodile, and mammal bones (Bown and Kraus 1988). Bones and teeth of the mammals *Brachyodus*, *Moeritherium*, *Barytherium*, *Apterodon Eotheroides*, and, possibly, *Arsinoitherium* are associated with the latter remains in a similar coquina developed

Fig. 2.21 Field photograph showing the general view of the lithological characteristics of the upper part of the Qasr El Sagha Formation (Abu Lifa Member) of the Qasr El Sagha Formation, north of Fayium depression (After, Underwood et al. 2011)

Fig. 2.22 Field photograph showing the giant cross-bedded sandstone of the Abu Lifa Member, the upper part of the Qasr El Sagha Formation, north of the Fayium depression. It appears as channel scours on top of the Temple Member (After Underwood et al. 2011, Legler et al. 2013)

locally near the top of the sequence (Bown and Kraus 1988). The green and gold mudstone sandstone is thin around Qasr El Sagha Temple but thickens to its maximum (about 6 m) near Dir Abu Lifa. The base of the sequence is gradational with underlying rocks or has a mildly erosional relationship to them. It is thin or absent in other areas. It comprises interbedded green and gold medium sandstone that becomes muddier upwards. Internal stratification is poor owing to disruption by abundant anastomosed burrows, including Ophiomorpha and burrows of crabs.

The green sandstone succession is intercalated with the green and gold mudstone sandstone unit and consists of 8 m of fine-medium green muddy sandstone with prominent gold mottles. This unit is burrowed with disposed-branched crab burrows and Ophiomorpha (Bown and Kraus 1988). Near the top of the Abu Lifa Member, north of the Qasr El Sagha Temple sequence, there is a limestone sequence named "bare limestone" by Beadnell (1905). This sequence derives its name from its upper surface, which is exposed over a wide area at the top of the Qasr El Sagha Escarpment. It is restricted between cross-bedded sandstone with gradational contact. The bare limestone consists of arenaceous biomicrosparudite (Bown and Vondra 1974). This sequence contains occasional shark and ray teeth, fragments of bone and silicified wood, and Late Eocene marine mollusks. Its maximum thickness reaches up to 10 m near Madwar El Bighal at the far western extremity of the Fayium Depression. However, it measures 6 m, 6 km east of the basalt mine. The "bare limestone" sequence is not only considered a marker bed between Qasr El Sagha and the Qatrani formations (Vondra 1974; Bowen and Vondra 1974), but also the Eocene/Oligocene boundary in northern Egypt (Said 1962).

The Qasr El Sagha Formation contains oyster banks characterized by the abundance of bivalves like *Ostrea* sp., *Carolia placunoides,* and gastropods like *Turritella* sp (Gameil et al. 2016). Calcareous nannofossils, planktonic Forams, and *nummulites* are also abundant. The formation contains some fragments of whale bones belonging to *Saghacetus osoris*. Strougo assigned the formation to the Priabonian (Upper Eocene). The vertebrate fossils (Sirenian skeletons) are distributed in the successive vertebrate horizon in the uppermost layer of the Dir Abu Lifa Member. This horizon contains some other associated fossils such as turtles, crocodiles, shark teeth, and abraded invertebrate shells and gastropods (Abu El Kheir 2022) that represents the unconformable contact with the overlying Gebel Qatrani Formation.

2.2.2.1 The Qatrani Formation

Beadnell (1905) named the sediments that build up Gebel Qatrani to the fluvio-marine series and dated them to the Late Eocene-Early Oligocene. Later, Said (1962) named the fluvio-marine rocks at Gebel Qatrani the Qatrani Formation (Fig. 2.23). This formation unconformably overlies the Priabonian Qasr El Sagha Formation with apparent erosional unconformity (Bown and Kraus 1988). This contact is placed between the bare limestone bed forming the uppermost Qasr El Sagha Formation and the gravelly sandstone of the basal Qatrani Formation. The bare limestone bed contains marine mollusks and different types of vertebrates (Beadnell 1905; Bown and Kraus 1988). Its upper boundary unconformably underlies the Widan El Faras basalt (Fig. 2.24). In some localities, its upper boundary exhibits undulatory and uneven surfaces with the overlying Oligocene Widan El Faras Basalt (Fig. 2.25).

Lithologiaclly, the Qatrani Formation comprises intercalation of cross-bedded reddish sandstone, pebbly sandstone, brown conglomerate, variegated claystone, and reddish paleosol (Fig. 2.26). This formation includes six fining-upward cycles (Fig. 2.25). The lower cycle (1) measures about 60 m thick and bounded from the base with the unconformable contact with the Abu Lifa Member of the Qasr El Sagha Formation and consists of

Fig. 2.23 Field photograph showing the general view of the Qatrani Formation, at its type locality (Gebel Qatrani), north of the Fayium depression

Fig. 2.24 Field photograph showing the Qatrani Formation that unconformably underlies the Widan El Faras basalt, north of Fayium depression

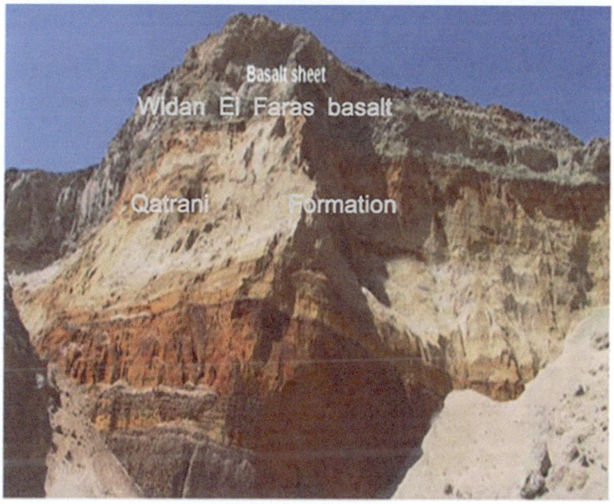

Fig. 2.25 Field photograph showing the undulatory and uneven contact of the upper part of the Qatrani Formation with the overlying Widan El Faras basalt, north of the Fayium depression

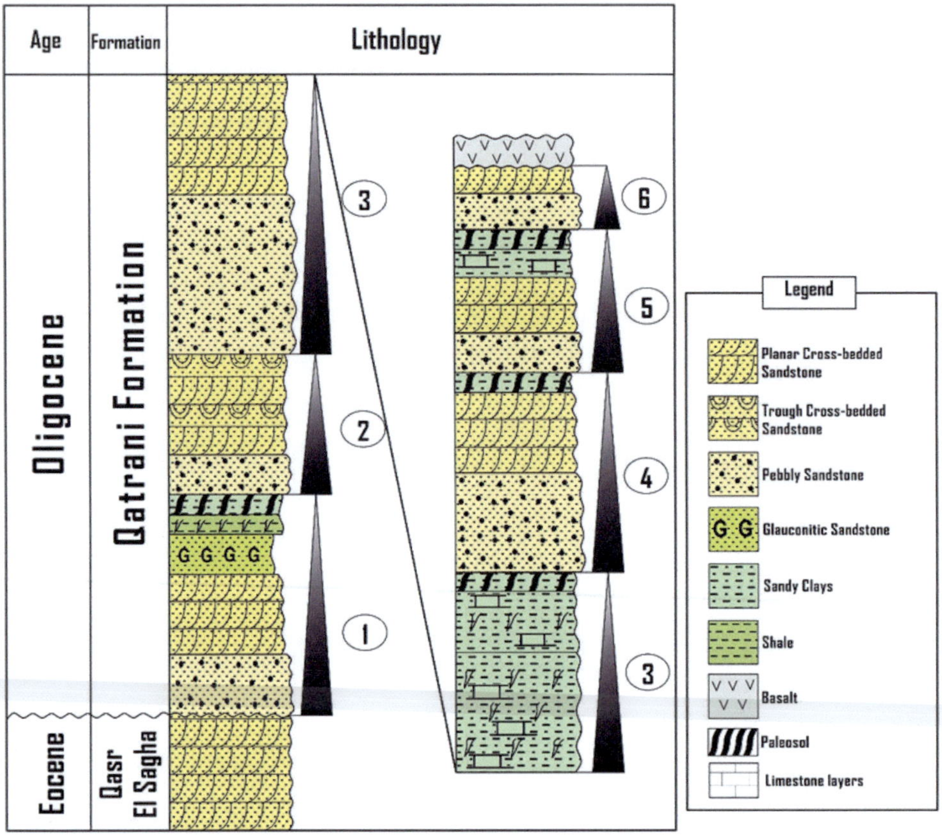

Fig. 2.26 Measured lithostratigraphic section of the Qatrani formation at its type locality that comprises fining-upward cycles. Each consists of an intercalation of pebbly conglomerate, pebbly sandstone, siltstone, mudstone, and thin beds of paleosol (After Bown and Kraus 1988)

gravelly sandstone at the base, followed-upward with reddish cross-bedded sandstone, green variegated sandstone–siltstone interbeds and capped by reddish paleosol (Fig. 2.26). However, some cycles (cycle 2) are incomplete and consist of two lithofacies, gravelly sandstone at the base, capped by cross-bedded sandstone at top, others include gravelly sandstone, sandstone and capped by reddish paleosol (cycle, 5). The gravelly sandstone unit at the basal cycles is composed of white and yellow medium to coarse sand with occasional stringers, lenses, and interbeds of gravel. Yellow mottling is present in some white sands, and red mottling is encountered more rarely. Both the coarsest and the finest sand in the gravelly sandstone sequences are composed dominantly of quartz, chert, and feldspar. Grain roundness varies from angular to moderately well-rounded. Most of the pebbles and granules of chert are moderately well-rounded. The chert pebbles are commonly associated with mudstone clasts and occur at the bases of the large-scale trough and tabular cross-bed sets (Bown and Kraus 1988). Forests are common in the gravelly sandstone sequences. Taxa of Oligocene woods of Egypt were studied by Krausel (1939) and discussed further by Wing and Tiffney (1982) and Bown et al. (1982). The silicified woods occur both as logs and smaller fragments. Fossil vertebrate remains, represented by white coprolites, occur locally in the middle and upper gravelly sandstone sequences. Rhizoliths and animal Bioturbation are common.

The sandstones that typically lie above the gravelly sandstone or conglomerate are characterized by their distinctive brick-red coloration resulting from contamination with oxidized iron minerals. However, these sandstones often exhibit laterally abrupt color transitions, shifting from red to green or red to brown (Bown and Kraus 1988). They commonly display primary structures, including cross-bedding, trough cross-bedding, and massive arrangements. Additionally, these sandstones are locally abundant in silicified wood and chert pebbles (Bown and Kraus 1988).

The green sandstone that occurs at the top of cycles is coarse-grained, massive glauconitic with claystone strings It is a very persistent unit near the base of the Qatrani Formation about 8 km northeast of Widan El Faras and continues south and west, marking the top of the Qasr El Sagha Formation above Qaret Umm El Zalat (Bown and Kraus 1988). The green sandstone lies with an erosional contact a top of the red sandstone sequence over much of this area. However, it is poorly exposed and/or) absent over large regions east of Widan El Faras and west of Qaret El Esh. The green sandstone in many cycles ranges in thickness from 10 to 13 m. but it thins to about 5 m some 6 km northeast of Widan El Faras (Bown 1982). Such type of sandstone exhibits a set of troughs or tabular cross-stratification. Azimuths on these cross beds indicate flow directions to the southwest and west-southwest (Bown 1982). The silicified wood or vertebrate remains are not observed in the green sandstone (Bown 1982).

The barite sandstone (white to pistachio green sandstone) was named by Bown and Kraus (1988) as it is well cemented by barite mineral that separated the section of the Qatrani Formation into lower and upper parts (Fig. 2.26). It has a very persistent 10–15 m thick and is developed about 152–162 m above the formation base. The barite sandstone is cliff-forming and massive in appearance and is composed of well-indurated green, white, and pinkish-white medium- to coarse-grained and locally granular sand. Along the line of section, the barite sandstone grades upward into a green, fine to medium-grained sandstone. It yields a large collection of fossil compound fruits of the *liane Epipremnum* (Wing and Tiffney 1982; Bown and Kraus 1988).

The variegated claystone beds occurring above the green sandstone and below the reddish paleosol in most cycles consist of green, gray, yellow, fine to coarse variegated sandstones, muddy sandstone and sandy mudstone, and sandy thin limestone. Most sandstones are friable, but pervasive cementation by calcium carbonate or dolomite is commonly encountered (Bown and Kraus 1988). The variegated sequences show fining-upward cycles as defined by foggy zones of brown iron oxides, cemented sand and mud intraclasts at the top of each cycle. Poorly preserved fossils such as wood, large mammal bones, well-preserved leaf fossils and at least 25 different kinds of fossil fruits were recorded (Wing and Tiffney 1982).

The paleosol sediments usually occur at the uppermost cycle (Fig. 2.26) and have different colors ranging from red to red mottled with orange. Such color variation may reflect differential horizon development related to differing groundwater regimes (Bown and Kraus 1988). They contain rhizoliths and trace fossils produced by sediment surface or near-surface organisms. The rhizoliths regarded as excellent paleosol indicators contain root and stem fossils of many small bushy plants, reedy plants, and trees as well as traces of animals (Bown 1982). Most such animals include termites, ants, crabs, crayfish, oligochaete worms, bees, and vertebrates (Bown and Kraus 1988).

The Qatrani Formation has a wide geographic distribution, extending from northwest Fayoum to the northeast as far as El Bahr Depression, 50 km northeast of the Bahariya Oases. In the Fayium depression, the exposure and topographic expression of the formation is controlled largely by the presence or absence of the Widan El Faras Basalt and by the development of large east-west-trending normal faults (Bown and Kraus 1988). The Qatrani Formation has a maximum thickness of about 350 m at its type locality (Gebel Qatrani). In places, the maximum thickness can occur in case it is capped by basalt and vice versa because the basalt well preserves the Qatrani Formation from weathering. Also, this formation can be found between Widan El Faras and Cairo in the Eastern Desert. Such an area is equivalent to the Gebel El Ahmer Formation (Barron 1907; Shukri 1953; Said 1962).

Long trunks of silicified wood are abundant throughout the formation (Fig. 2.27). Mangrove rhizoliths are abundant, too. Fossil vertebrate remains represented by white coprolites occur locally in this bed. Rhizoliths and animal Bioturbation are common and are most obvious \and best preserved in the cemented portions of these units.

2.2.2.2 The El Khashab Formation

The El Khashab Formation was coined by Bown and Kraus (1988). It derives its original name from the innumerable silicified wood trunks at Gebel El Khashab, north of the Fayoum Depression (Dixon 1873; Cuvillier 1940). This formation contains strewn and scattered wood trunks across broad areas of the desert north of the Fayium Depression (Dixon 1873; Cuvillier 1940). The above nomenclature contradicts Issaw et al. (2009), who mentioned that this formation was named Gebel Khashab Formation by Said (1962). Neither Said (1962) nor Said (1990) named these rocks the Gebel Khashab Formation, but he named them Gebel Khashab red beds. Its type locality occurs at Gebel Khashab north of Gebel Qatrani. It nonconformably overlies the Oligocene Widan El Faras basalt. The lower contact is markedly erosional, and the presence of incised channels in the top of the basalt, which are filled with coarse sand, sometimes containing basaltic debris, and chert pebble and silicified wood trunks (Bown and Kraus 1988). In most exposures, its upper

Fig. 2.27 Field photograph manifesting the distribution of long trunks of silicified wood in the sandstone of the Qatrani Formation, north of the Fayium depression

boundary is not covered by any sediments; however, near the Moghra Oasis, it unconformably overlies the Lower Miocene Moghra Formation. This formation comprises an intercalation of pebbly conglomerate, reddish siltstone, siliceous sandstone, pebbly reddish sandstone, silicified wood, chert, and basal debris (Figs. 2.28 and 2.29). The chert pebble and cobble conglomerates constitute about the lower ten meters in the basal El Khashab Formation. Such a conglomerate has produced a planket desert pavement over much of the desert north of Gebel Qatrani and even south of Qasr El Sagha exposures (Bown and Kraus 1988). In the absence of Widan El Faras basalt, the formation directly unconformably overlies the Qatrani Formation. The discrimination between them is difficult, but in some localities, the contact can be distinguished due to the concentration of the chert pebbles at the basal part of the El Khashab Formation. This formation is characterized by the presence of numerous wood trunks, which usually show elongation north to northwest direction, forming a look-like forest (Fig. 2.30). The wood trunks cover an area of about 30 km^2 in the El Khashab Formation. The forest was broken by several meandering streams that debouched westward and northwestward into the Fayium depression (Bown and Kraus 1988). Such wood trunks usually appear on the surface along the October City and the Bahariya Oases Road (El Sadawi et al. 2014; El Mahdy et al. 2020) (Fig. 2.31).

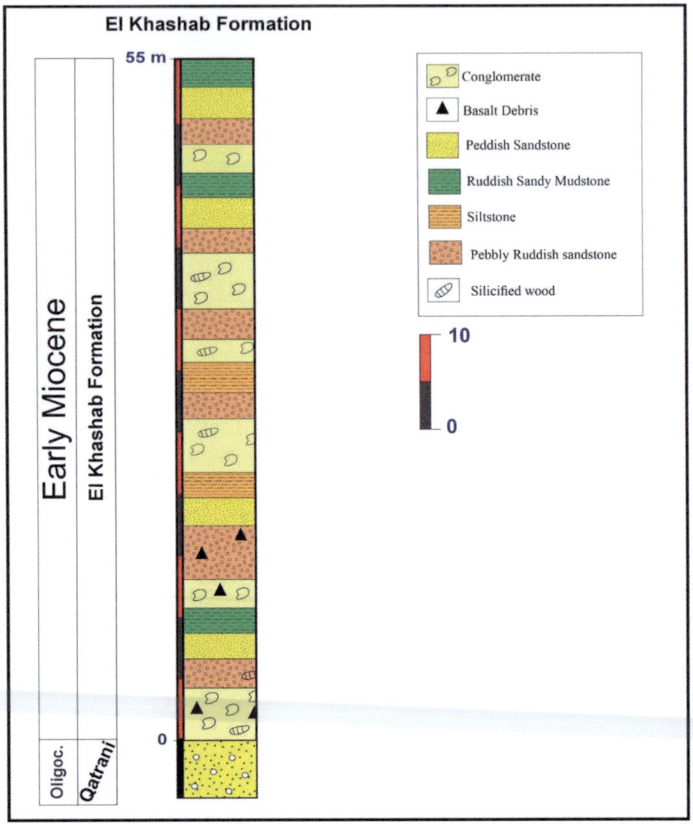

Fig. 2.28 Measured section of the El Khashab formation at its type locality north of Widan El Faras Hill, north of Fayium Province

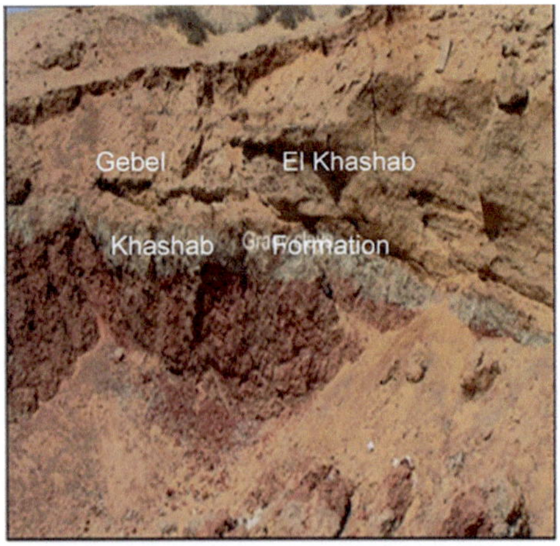

Fig. 2.29 Field photograph showing the lithological characteristic of the El Khashab Formation that comprises interbeds pebbles, sandstone, siltstone, and mudstone at its type locality (Gebel El Khashab)

Fig. 2.30 Field photograph illustrating the occurrence of silicified wood trunks embedded in the coarse clastic facies of the El Khashab Formation. Notice the elongation of wood trunks toward the north–south direction and represent wood forests in the El Khashab Formation, north of the Fayium depression

Fig. 2.31 Field photograph illustrating brocken silicified wood trunks scattered on the weathered rocks of the El Khashab Formation. The broken woods were affected by several meandering streams which debouched westward and Northwestwards the Fayium depression northwest Fayium depression

The El Khashab Formation has a wide distribution. It is poorly exposed north of Jebel Qatrani, where much of the area between Jebel Qatrani and Wadi Farigh is predominantly covered by dune deposits. The best exposures of the base of the Khashab Formation are preserved above the north, northwest, and west of Widan El Faras. Said (1990) mentioned that these rocks grade northwards at the latitude of Wadi El Farigh, southwest of Wadi Natrun, to form delta Palin sediments of the Moghra Formation. In Wadi Farigh, continental rocks probably belonging to the El Khashab Formation are developed in low ridges paralleling the extension of the Wadi El Farigh, and on a few outliers at Jebel Roussa, Garet El-Raml, and Garet El Hamir (Bown and Kraus 1988). Moreover, the El Khashab

Formation occurs at latitude east of Gebel Abu Roash and is exposed 15 km west of the El Giza Pyramids (Issaw et al. 2009). It extends from Gebel Khashab, northeast of Fayium Depression, up to the foot slopes of the Gebel El Hagif and Gebel Khashm El Qaud at Wadi El Natrun (Issaw et al. 2009). Also, it stretches from the Moghra Oasis at the southeastern of the Qattara Depression up to the Nile Delta in the east. In the latter locality, it occurs as dissected buttes and low hills, before entering below the Nile Delta (Issaw et al. 2009). This formation is 50 to 60 m thick and consists of yellow-brown coarse to gritty-sized sandstone. The sandstone is intercalated with conglomerate and ferruginous gypseous claystone. Miocene fossil vertebrates were found at Wadi Farigh. In some localities, the formation comprises gravelly sandstones, red and orange medium to coarse-grained sandstones, and red and drab mudstones. Ironstone- and silicified sandstones form resistant caps on some isolated buttes that rise above the serir, notably Garet Hamid, Garet El-Raml, and Garet El Hamir.

Freshwater shells and numerous marine *Scutella* remains are recorded from these beds. The latter fossils are especially abundant in Gebel Ruzza, a conspicuous hill west of the Giza Pyramids plateau (Said 1962). On top of the Upper Eocene rocks at Tell-el-Zalat is a conspicuous basalt flow, followed by gravel and sand deposits that make up the enormous plain stretching around the entire area.. The sands and gravels include abundant fossil wood. The beds are specially developed to the west of the Giza Pyramids, twenty kilometers west of these pyramids. Also, a spectacular locality lies in the Western Desert of Egypt, where large fossil logs are preserved in sediments bordering the Cairo-Bahariya Oasis Desert Road. Such petrified wood occurs abundantly in the Lower Miocene Gebel El Khashab Formation, which is composed mainly of loose, moderately to poorly sorted sands and gravel deposited by an ancient river. The fossil wood occurs as fragmented tree trunks that are found mingled or scattered on the sand surface, indicating the fluviatile character of the deposit. The age of these sands and gravels is difficult to ascertain. The beds look like the Oligocene Gebel Ahmer sands and gravels, but are assigned to the lower Miocene only because of their position above the basalt flow. The Khashab Formation is equivalent of the El-Raml Member of the Moghra Formation as suggested by Omara and Sanad (1975).

2.2.2.3 The Moghra Formation

The Moghra Formation was established by Said (1962) to describe 230 m of sandstone, siltstone, and calcareous shale at the Moghra Oasis. It is a type of section that occurs at the extreme eastern tip of the Qattara Depression (Fig. 2.32). It overlies and underlies the El Khashab and Marmarica formations, respectively. It is composed of fluvio-marine sediments that mainly contain gravelly sandstone, coarse sands, sandstone, and clay interbedded with vertebrate remains and silicified wood (Fig. 2.32). Three members were recognized within the Moghra Formation: El-Raml, Bait Oweina, and Monqar El-Dowi (Omara and Sanad 1975) (Figs. 2.33 and 2.34). Toward the North at the Minqar El Mankhafad area, the Moghra Formation shows lateral variation in lithology grading

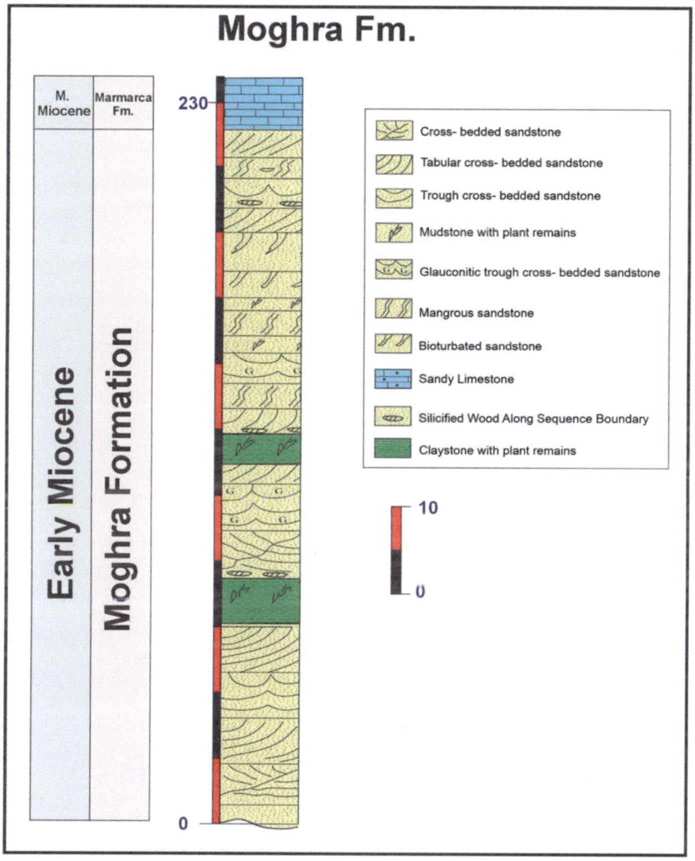

Fig. 2.32 Measured lithostratigraphic section of the Moghra Formation at Qattara depression that comprises cross-bedded sandstone (trough cross-bedded, planar cross-bedded), bioturbated sandstone (El-Barkooky et al. 2009)

into predominantly shaly, as transitional facies to the Mamoura Formation (Farouk and Khalifa 2010). El-Raml Member covers a wide region of the northern part of the Western Desert, between the scarp of Gebel Qatrani and the Moghra depression. It consists of alternating multi-colored sands, sandstone, and gravel of different compositions with scarce intercalating clay bands. The sandstone is calcareous, gypsiferous, and ferruginous. The maximum thickness of the El-Raml Member reaches 500 m (Marzouk 1970). It lies unconformably on basaltic flow at different depths. The El-Raml Member is continental to the lagoonal environment of sedimentation due to the abundance of well-sorted, rounded, and false-bedded sands included within most of this member, together with gypseous and salty sandstone (Omara and Sanad 1975).

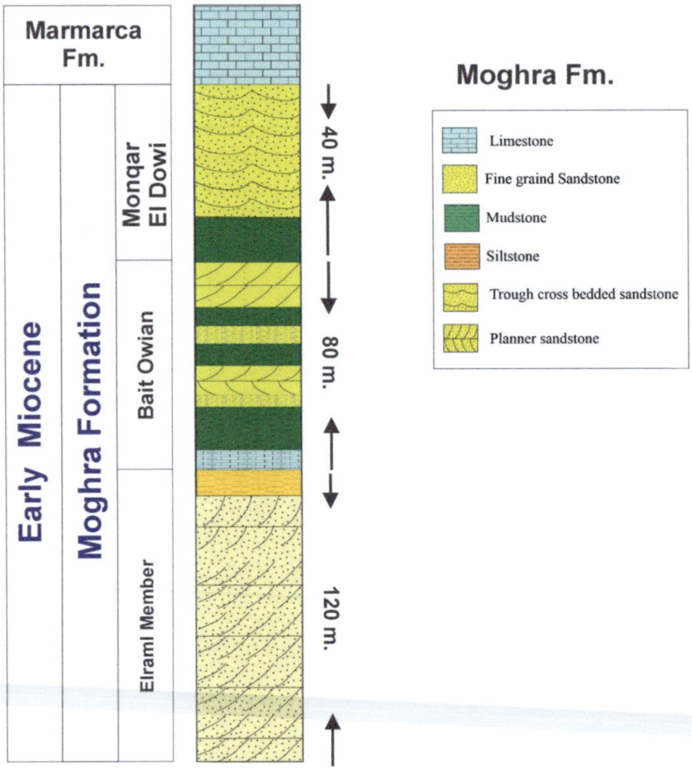

Fig. 2.33 Measured section of the Moghra Formation at Moghra Oasis which is divided into three members, from to top: El-Raml, Bait Oweina, and Monqar El-Dowi (After Omara and Sanad 1975)

Fig. 2.34 Field photograph showing the Moghra Formation that consists of three formal members, El-Raml, Bait Oweina, and Monqar El-Dowi members (After Omara and Sanad 1975)

The Bait Oweina Member conformably overlies and underlies the El-Raml Member and Monqar El-Dowi Member, respectively. It comprises pebbly, cross-bedded, and thin claystone layers (Figs. 2.33 and 2.34). The cross-bedded sandstone is about 40 m thick and is coarse to medium-grained, cross-bedded, and ferruginous. The claystone is varicolored with bands of argillaceous limestone containing spots and patches of iron oxides. In addition, it contains streaks, spots, and bands of gypsum and halite. The isopach contour map of this member manifests an increase in thickness eastwards and westwards of Moghra Lake with minimum thickness at the center. This reflects that Bait Oweina was deposited on the horst fault (Omara and Sanad 1975). A thin bed of limestone occurs at the base, which consists of lime-mudstone.

The Monqar El-Dowi Member occurs above the Bait Oweina Member and contains sandy limestone and calcareous sandstone. The sandstone has cross lamination and is calcareous, varying in color and changing laterally to sandy limestone at Minqar Laban. It contains patches, spots, and streaks of iron oxides in some parts. Four microfacies have been recognized, e.g., calcareous quartzarenite, glauconitic calcareous quartzarenite, ferruginous calcareous quartzarenite, and sandy biomicrite from west to east. The thickness of this member is structurally controlled and ranges from 30 to 40 m. However, it can reach up to 70 m (Omara and Sanad 1975). Albritton et al. (1990) mentioned that the Moghra Formation comprises fine- to medium-grained, reddish-brown, usually friable sandstone and siltstone. Many of the sandstones are cross-bedded, and some contain numerous small- to medium-sized, well-rounded quartz and chert pebbles. Large concentrations of petrified logs, 20–75 cm in diameter and as much as 10 m in length, are observed in several stratigraphic horizons (Albritton et al. 1990). The trunks lie nearly parallel, with a general east-northeast orientation. In the upper part of the Moghra Formation, several vertebrate fossils (*hippopotamus*) are present with micro-vertebrate remains, including rodents (Albritton et al. 1990).

Hassan (2013) and El Barkooky et al. (2009) studied the Moghra Formation and subdivided it into repeated cycles. The cycles reached thicknesses of up to 45 m each, and are bounded by regional erosion surfaces that are partially armored by thin ferruginous pebble conglomerate containing petrified wood and bone fragments. The lower part of each cycle consists of unconsolidated cross-stratified sand and sandstone with locally abundant vertebrate fossils, petrified logs, Thalassinoides, and Ophiomorpha burrows.

Lizard and snake remains have been recorded from the Early Miocene (Burdigalian) of the Moghra Formation (Georgalis et al. 2020). These fossils indicate the fluvial origin of the Moghra sediments. Strontium isotope analysis of macrofossil fragments recorded in the Moghra Formation has indicated an age for the Moghra Formation ranging from 20.5 Ma at the base to 17 Ma at the top, placing most of the study area within the Burdigalian (Hassan 2013).

2.3 Lithostratigraphy of the Mediterranean Nile Delta

The stratigraphy of the Mediterranean Nile Delta was studied by many authors such as Salem (1976); Rizzini et al. (1978); Ross and Uchupi (1977); Ryan (1978); Barber (1981); Said (1962); Zaghloul et al. (1979, 1989, 1990); El Heiny (1982), Barakat (1982); EGPC (1994); Sarhan and Hemdan (1994); Nashaat (1998, 1999), Harms and Wray (1990); El Heiny and Morsy (1992), Sestini (1989, 1995); El Heiny and Anani (1996), Kamel et al. (1998); Leila and Moscariello (2018, 2019) among others. The Mediterranean Nile Delta is divided into Neogene Delta (Tineh, Qantara, Sedi Salem, Qawasim, Rosetta, and Abu Madi formations) and the Quaternary Delta (Kafr El Shiekh, El Wastani, Mit Ghamr, and Bilqas formations) (see Fig. 2.35).

2.3.1 Stratigraphy of the Neogene Delta

2.3.1.1 The Tineh Formation

The term "Tineh Formation" was introduced by the International Egyptian Oil Company (IEOC) in 1967. Its type section is found in the Qantara well-1, at depths ranging from 3110 to 4196 m (El Heiny and Morsy 1992). The base of the formation is encountered in the southern Delta block (south hinge zone), particularly in the Tineh-1 well (depths of

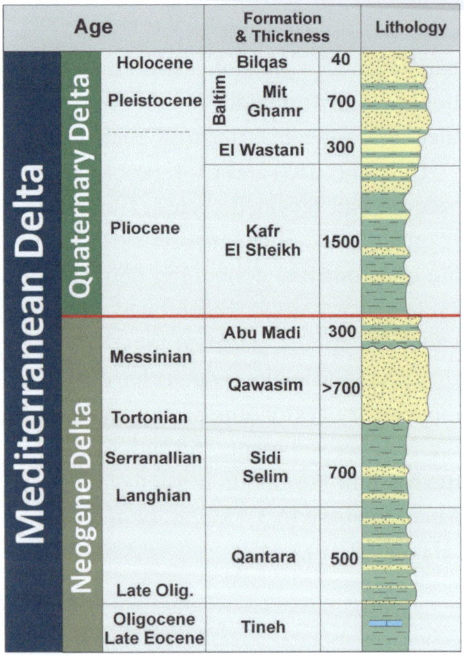

Fig. 2.35 Geological column showing the Neogene and Quaternary rock units of the Mediterranean Delta (After, Zaghloul et al. 1977, El Heiny 1982 with modification

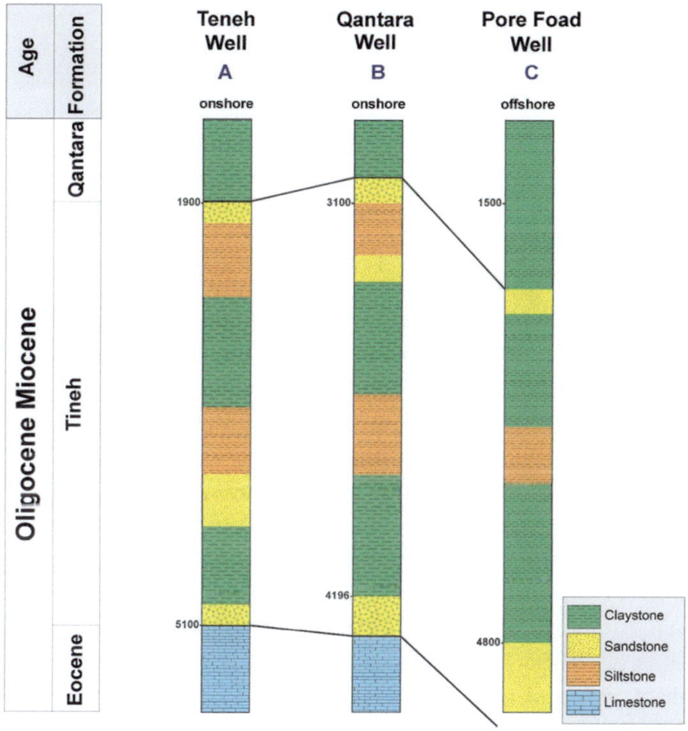

Fig. 2.36 Onshore and offshore lithologic logs and stratigraphic setting of the Tineh Formation at Tineh well and Qantara wells (**a**, **b**) and Pore Foad well (**c**)

1900 to 5100 m) and the Pore Foad well-1 (depths of 1500 to 4800 m). In these wells, the Tineh Formation unconformably overlies the Eocene rocks (El Heiny and Morsi 1992). However, in the El-Fayrous well, Rommana well-IX, and Tanta well, the base is not penetrated (Farouk et al. 2014) (Fig. 2.36). The Tineh Formation unconformably underlies the Qantara Formation, with a 20-m-thick layer of conglomeratic sand containing carbonate intraclastic rocks separating the two (El Heiny and Morsi 1992). Its lower part consists of an intercalation of greenish gray, fissile, shale, silty sandstone, and siltstone with intercalation of thin beds of arenitic sandstone that are cemented by calcite (El Heiny and Morsi 1992) (Fig. 2.36). Its upper part of includes gray to dark gray shale with intercalation of thin siltstone (El Heiny and Morsi 1992) (Fig. 2.36).

This formation has a wide distribution in the onshore northeast Delta and north Sinai (onshore), its top occurs at 4200 m depth (Sedi Salem well-1), and occurs between 3070 and 2725 m depths in the El-Fayrouz borehole. In the Rommana-1X borehole, it is encountered between 2805 and 2555 m. In the northwest delta, it is completely missing south of the Hinge zone at the Tanta-1 Well, where the Burdigalian Moghra Formation directly overlies the Ruperlian Dabaa Formation above a major unconformity (Farouk et al. 2014). In offshore areas, it is not detected in several boreholes, e.g., Abu Qir fields, Temsah, Baltim, Wakar, and Por Foad.

The calcareous planktonic microfossil content of the formation indicates that its age is Middle Oligocene to the Early Miocene (El Heiny and Morsi 1992). It is nearly barren of planktonic foraminifera below the top of Zone NN2 at the top of the Tineh Formation, equivalent to Event I of El Heiny and Morsi (1992) in the Qantara-1 borehole. This formation changes laterally eastwards to the Moghra and Dabaa formations.

2.3.1.2 The Qantara Formation

In 1976, the Qantara Formation was named by the International Egyptian Oil Company (IEOC) in the Nile Delta succession. The type section occurs in QantaraWell-1 (from depth 2577–3110 m), on the northeastern side of the Nile Delta area (Fig. 2.37). In the El-Fayrouz and Rommana Wells, its lower boundary is unconformable with the underlying Tineh Formation, while its upper boundary also unconformably underlies the Sidi Salem Formation at the above-mentioned well (El Heiny and Morsi 1992; Farouk et al. 2014). In the south-to-north profile, the formation is detected in Sindy well-1 from 1600–1900 m, 3200–3700 depth in E. Delta well-1, the same interval in Tarif well-1, and 4500–4700 in El Matariya well-1. The top of the Qantara Formation is encountered only by the Naf well-101 in the northwestern delta, which did not reach the base of this formation (Ismail et al. 2010). Further south, in its type locality, this formation consists of shale interbedded with argillaceous sand and limestone, as indicated by drilling data from other wells (Fig. 2.37). Northwestwards (offshore wells), the limestone of this formation decreased, and was replaced by clastics (shale and sand) at Naf well-101 (Ismail et al. 2010) (Fig. 2.37). The formation extends from 2725 to 2200 m with a total thickness of 525 m in the El-Fayrouz borehole, while in the Rommana-1X borehole, it is detected from 2000 to 2555 depths with an average thickness of 555 m. Based on calcareous nanofossils recorded in this formation, Faris et al. (2007), mentioned that the Aquitanian–Burdigalian was attributed to the Qantara Formation in the Rommana-1X borehole.

Added to the above, this formation is dated as Early Miocene in age, according to the presence of *Globigerinoides primordius* (Ismail et al. 2010). Moreover, this age assignment is confirmed by the presence of the Early Miocene *Dentoglobigerina altispira* and *Globorotalia obesa*. Also, there are several other planktonic species associated with these fossils (e.g., *Globigerinoides bisphericus, Neogloboquadrina continuosa, Globorotalia archeomerandii, Globigerina brazieri, Globigerina praebulloides, Orbulina* cf. *suturalis, Orbulina universa,* and *Globigerinoides trilobus immaturus*). Moreover, there are several benthic species associated with the previous planktonic species, e.g., *Bolivina dilatata, Gyroidinoides soldanii, Nonionina scapha, Bulimina elongata, Lagena sulcata, Globocassidulina elongata, Uvigerina semiornata, Nonionella auris, Lenticulina cultrata, Cibicides refulgens* and *Neoeponides* sp. (Ismail et al. 2010).

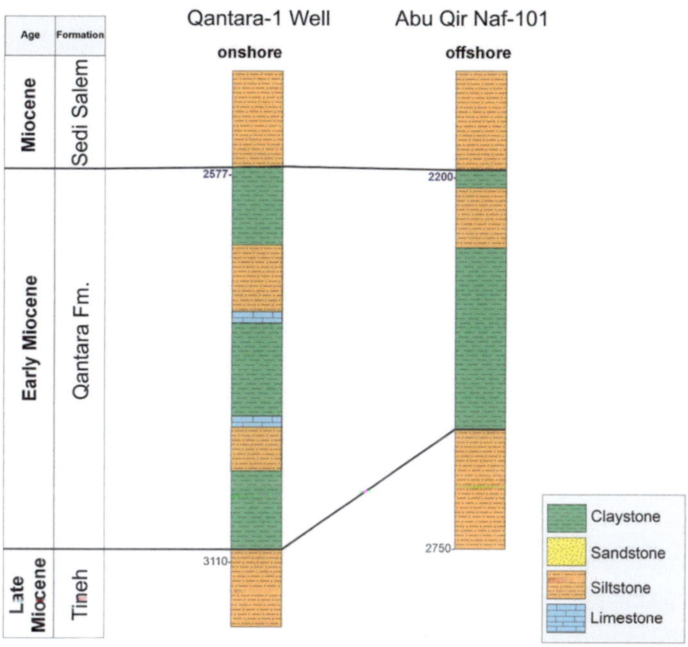

Fig. 2.37 Lithologic logs for the Qantara formation onshore at its type locality (Qantara well-1) (**a**) and in offshore at Naf well-101 (Abu Qir field) northwest of the Mediterranean Delta (**b**)

2.3.1.3 The Sedi Salem Formation

This formation was named by the Egyptian Petroleum Corporation (IOEC 1969). The type section of this formation occurs at the Sidi Salem well-1 (located south of Lake Burullus) (Fig. 2.38). It is encountered from 3092 to 4038 m depth (Rizzini et al. 1978). In onshore wells (Tanta, Sidi Salem, Fyrouz and Rommana wells), it unconformably overlies the Lower Miocene Qantara Formation and underlies the Upper Miocene Qawasim Formation. In El-Fayrouz well, the lower boundary is between the limestone of the upper part of the Qantara Formation and the green claystone of the basal Sedi Salem Formation (Farouk et al. 2014). The upper boundary with the overlying Qawasim Formation is generally similar in all drilled wells and is defined by the presence of the thick conglomeratic bed. This contact is demarked by the absence of fauna (in the Qawasim Formation), and just the appearance of sands, of the overlying Qawasim Formation (Rizzini et al. 1978) (Fig. 2.38). In Tanta and Rommana wells, the lower boundary lies between the sandstone of uppermost Qantara Formation and claystone of the basal Sidi Salem Formation (Fig. 2.38a). The upper boundary in the later wells is similar to that found in the Fayrouz well. In offshore area, in the northwestern delta (Naf well-101) this formation overlies the Qantara or Moghra Formation and underlies the Qawasim Formation (Ismail et al. 2010), while in the eastern side of delta (Akhen field), this formation overlies the Qantara formation and underlies the Pliocene Kafr El Sheikh Formation (Khaled et al. 2012). In the latter case, the transition is marked mainly by faunal changes, rather than by changes in sediment type (Rizzini et al. 1978). The bottom of this formation was reached in Naf well-101, but it was not reached in two wells (Naf-2 and Naf-3) (west of Abu Qir fields (Fig. 2.38b). Its maximum thickness (400 m) occurs in well Naf-101 (Ismail et al. 2010). In onshore areas, this formation is composed of intercalations of shale, siltstone, and thin limestone beds with rare occurrences of white, bioclastic limestone and dolomitic marl (Rizzini et al. 1978). The claystone is green and consists mainly of kaolinite with montmorillonite and a little illite. In offshore regions, the formation comprises an intercalation of sandstone, siltstone, claystone, and thick limestone beds. The sandstones are commonly bioturbated, consisting mainly of quartzose, cemented by argillaceous and calcite (Rizzini et al. 1978). This limestone occurs less frequently and gradually changes to claystone toward the northeast of the study area (Ismail et al. 2010) (Fig. 2.38b). In offshore, at Akhen fields, Khalid et al. mentioned that this formation comprises four sandstone channels intercalated with claystone and siltstone from depth 4350–4554 m. The vertical stacking pattern of the above-mentioned lithofacies suggests superimposed or vertical repetition of fining-upward cycles. Each cycle begins with channelized sandstone, followed by siltstone and capped by massive claystone (Fig. 2.38a, b).

The thickness of the Sidi Salem Formation varies from one place to another according to its stratigraphic position in the tectonic setting and to the amount of truncation of the Qawasim Formation. It assumes 740 m (2200 to 1460 m) in the El-Fayrouz well-1 and 125 m (2000–1875 m) in the Rommana well-1X. It shows sudden decreases in thickness in the Tanta well-1, reaching up to 65 m only (1935–1870 m). In the latter well, it consists mainly of sandstone interbedded with claystone. The sandstone is usually white to yellowish-gray. Occasionally, it is pinkish white, with fine to coarse grains, poorly sorted, a clay matrix, calcite–silica cemented with traces of pyrite, glauconite, and kaolinite (Farouk et al. 2014). The claystone is olive-gray to brownish-gray, hard, massive, slightly calcareous, and carbonaceous. In Sedi Salem, well it has a thickness of 1200 m from intervals 3330–4200.

The formation is of Middle Miocene age, based on the presence of *Globigerinoides bollii* at the top of this formation (Ismail et al. 2010), which could represent a useful datum in the Mediterranean region, where the *Globigerinoides fohsi* lineage is not recorded (Ismail et al. 2010). However, the absence of the three upper biozones of the Middle Miocene age (*Globigerinoides ruber*, *Neogloboquadrina mayeri,* and *Globorotalia menardii*) biozones indicates a hiatus between the Middle Miocene and the Late Miocene as shown on the distribution chart of the Naf-101 well. (Ismail et al. 2010), However, the presence of *Praeorbulina glomerosa curva* and *Praeorbulina glomerosa glomerosa* species supports the age assignment of this formation, which directly underlies the Upper Miocene sediments (Ismail et al. 2010). This also indicates a similar hiatus in the formation as shown in the Naf well-3. Accordingly, this formation is dated from the early Middle Miocene to Late Miocene.

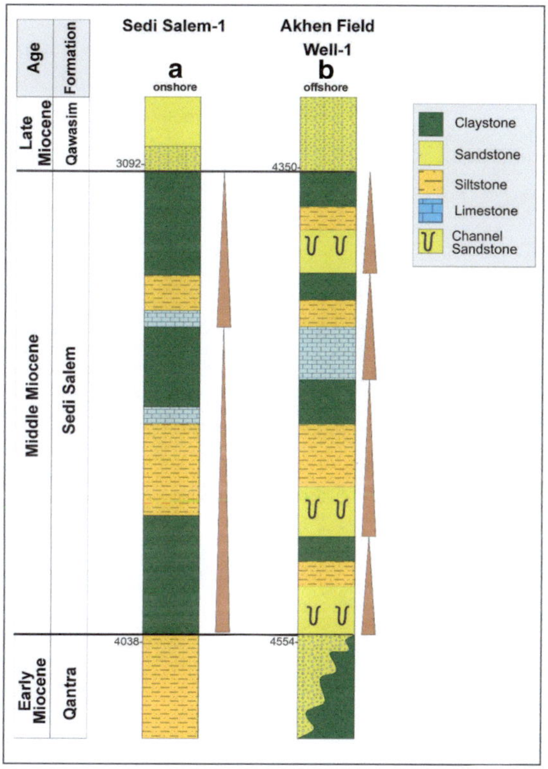

Fig. 2.38 Onshore lithologic log and stratigraphic position of the Sedi Salem Formation at Sedi Salem well (**a**), the central part of Nile Delta and its lithologic log in offshore (Naf well-101) in the northwest Delta (Abu Qir fields) (**b**)

2.3.1.4 The Qawasim Formation

The type section of the Qawasim Formation lies at the Qawasim well-1 from 2800 to 3733 (Fig. 2.39a) (Rizzini et al. 1978). It unconformably overlies the Sidi Salem Formation. The lower stratigraphic boundary between the Qawasim Formation and the underlying Sidi Salem Formation is marked by radical vertical facies changes from the green-gray claystone topping the Sidi Salem Formation below and the coarse-grained sandstone of the basal Qawasim Formation above (Fig. 2.39a). It underlies the Abu Madi or Rosetta Formations (Fig. 2.39a). The upper boundary of the Qawasim Formation is rather difficult to define based on the subsurface data alone. In most cases, it is defined by using several lines of evidence such as faunal association and sedimentological criteria (Rizzini et al. 1978). The paleontological criteria are the most easily applicable, where the benthonic faunas of the Qawasim Formation are replaced by the Pliocene marine faunas of the Abu Madi Formation (Rizzini et al. 1978). In general, the unconformity surface is the most applicable phenomenon. This formation features a somewhat irregular succession of thick layers of sands, sandstones, and conglomerates interbedded with clay layers (Rizzini et al. 1978) (Fig. 2.39a). This formation features a somewhat irregular succession of thick layers of sands, sandstones, and conglomerates interbedded with clay layers (Rizzini et al.

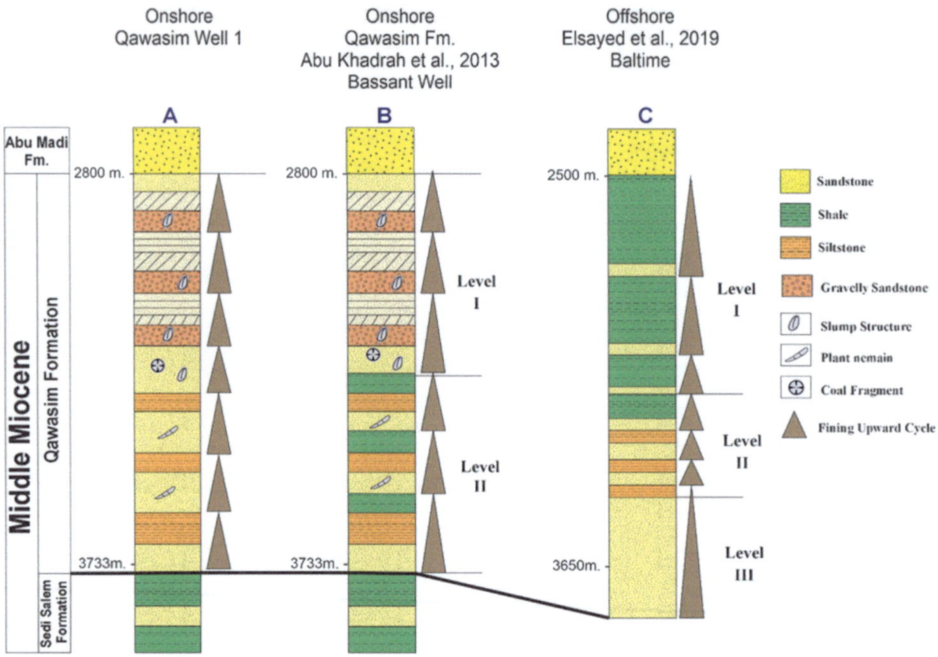

Fig. 2.39 Onshore lithologic log and stratigraphic position of the Qawasim Formation at its type locality, Qawasin well (**a**) and in Basent well-1 (**b**), central part of Nile Delta, and offshore Baltime well-1 (**c**)

1978) (Fig. 2.39a). Predominantly lithic, the sandstones exhibit significant grain size variations, primarily being loosely cemented by a clayey-calcareous matrix. In some sections, loose sands are also present (Rizzini et al. 1978). The sandstones are either massive or characterized by planar cross-beddings and parallel laminations, and they contain a rich assemblage of plant fragments (Rizzini et al. 1978). The conglomeratic layers exhibit a lenticular shape, with massive structures and planar cross-bedding, often displaying slump features. Additionally, layers of peat and coal fragments are frequently present. In the El Basant well-1 in the onshore eastern delta, the Qawasim Formation can be divided into two major facies: level-1 and level-11 (Abu Khadra et al. 2018) (Fig. 2.39b). The lithology of level-1 measures about 95 m thick and comprises fining-upward cycles, each of which begins with conglomerate, followed by cross-bedded sandstone, capped by horizontal bedded fine-grained sandstone (Abu Khadra et al. 2018) (Fig. 2.39b). While the lithology of level-11 measures 65 m thick and consists of fining-upward cycles, each one begins with conglomerate, pebbly sandstone, and siltstone and is capped by mudstone (Fig. 2.39b). The sandstones are ripple laminated, cross-laminated, cross-bedded, massive and consist of quartz wacke and quartzarenite At El Tamad Field, onshore Nile Delta, two main facies were distinguished in the studied formation: clastic and carbonate facies. The clastic rocks include conglomerate, sandstone, and shale, while the carbonate rocks include limestone and dolostones. In the offshore Baltim area, the Qawasim Formation shows different lithological as compared with onshore. It consists of three levels (El Sayed et al. 2019), from top to base, level-1 consists of intercalation of claystone and siltstone, level-11 comprises mainly of sandstone with thin intercalation of siltstone, level-111 consists entirely of medium to coarse-grained sandstone (Fig. 2.39c).

The Qawasim Formation assumes about 170 m, at Qawasim well-1 (between 1870 and 1700 m depths), while it is 445 m thick in Rommana-1X in west Dkirnis (between 1875 and 1430 m depths) (Nabawy et al. 2023). It is missing in the El-Fayrouz borehole. In Rommana well-1, this formation consists mainly of sandstone intercalated with sandy limestone and claystone. The sandstone is white, fine to coarse-grained, associated with cherty, glauconitic, and pyritic. The sandy limestone is white to yellowish-gray and hard.

The fauna in this sandy formation is generally scarce and is frequently absent. Sporadically, clusters of foraminifera can be found, rich in individuals but poor in species. The foraminifera species are benthonic, mostly *Cibicides, Bolivina, Eponides, Ammonia,* and *Elphidium* are recorded (Rizzini et al. 1978). Some levels rich in ostracodes (genus *Cyprideis*) are noticed. This assemblage is typical of the Messinian in the Mediterranean Basin (Rizzini et al. 1978). Also, on the western side of the delta (Naf-101 well), there are several planktonic species associated with the above species (e.g., *Globigerinoides pseudobesa, Globigerina ciperoensis ciperoensis, Dentoglobigerina venezuelana, Globigerina woodi, Globigerina nilotica, Orbulina universa, Orbulina bilobata, Globigerinoides trilobus immaturus, Globigerinoides obliquus extremus* and *Globigerinella obesa*). Moreover, there are several benthic species associated with the planktonic taxa listed above (e.g., *Quinqueloculina bosciana, Bolivina hebes, Ammonia beccarii, Bulimina elongata, Asterigerina planorbis, Nonionellina cf. labradorica, Gyroidinoides soldanii, Pullenia quinqueloba, Elphidium macellum,* and *Neoeponides* sp.) (Ismail et al. 2010). Most of the above-mentioned fauna suggest Late Miocene (Tortonian–Messinian).

2.3.1.5 The Rosetta Formation

This formation was named by Rizzini et al. (1978). Its type section occurs in Rosetta well-2 between intervals 2678 and 2720 with an average thickness of 42 m (Fig. 2.40a). The well is located offshore, northeast of the mouth of the Rosetta Nile branch. At Rosetta well-2 and Timsah well-1, the lower boundary of the Rosetta Formation unconformably overlies the sandstone facies of the Qawasim Formation, while its upper boundary unconformably underlies the pebbly sandstone of the basal Abu Madi Formation (Fig. 2.40a). In case of the absence of the Abu Madi Formation, it becomes directly below the Pliocene shaly facies of the Kafr El Shiekh Formation. In the latter case, the contact lies between the evaporitic facies of the uppermost Rosetta Formation and the shaly facies of the lowermost Kafr El Sheikh Formation (El-Gharabawy 2018). In onshore, at Fayrouz well-1 (Farouk et al. 2014), the Rosetta Formation overlies the Qawasim Formation and underlies the Abu Madi Formation (Fig. 2.40b). Abdel Aal et al (1994), stated that the Rosetta

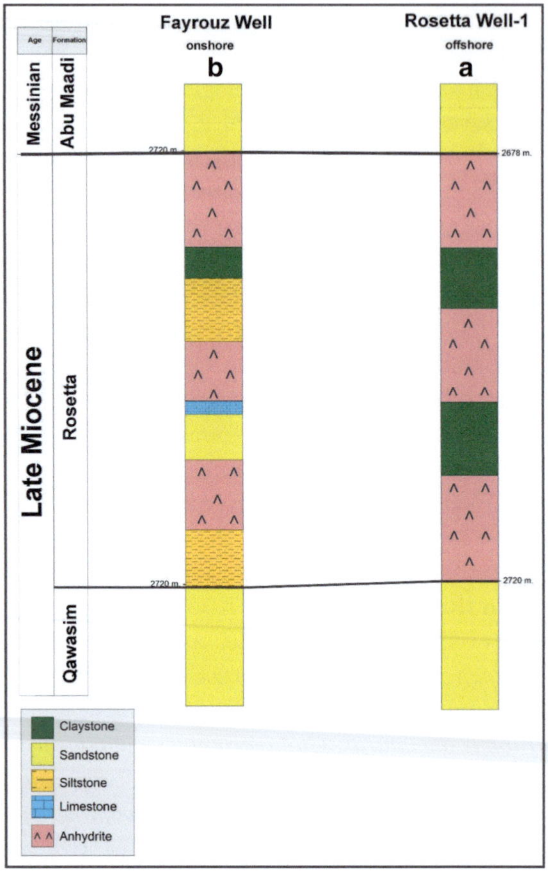

Fig. 2.40 Offshore and onshore lithologic log and stratigraphic position of the Rosetta Formation at its type locality, Rosetta well-2 (**a**), and Fayrouz well-1 (**b**)

anhydrite is fairly widely distributed across the Nile Delta. It locally overlies the Upper Qawasim Formation and has been recorded as far south as the Neogene hinge line.

At its type locality and Rosetta wells (offshore), the formation consists of large layers of anhydrite interbedded with thin claystone. However, in some wells, it consists mainly of shale and siltstone with sandstone and argillaceous limestone streaks, intersected with evaporate beds (Fig. 2.40a) (Farouk et al. 2014). Anhydrite is creamy white, moderately hard to hard, and cryptocrystalline. An anhydrite body caps the Messinian classics of sandstone and shale. This reflects a restricted shallow water lagoon during periods of a semi-arid climate, high clastic input, and relative sea-level rise. The shale is dark gray, sub-blocky to sub-flaky, moderately soft, from non-calcareous to slightly calcareous in parts. The sandstone is grayish-white, fine to very fine-grained, occasionally grading to siltstone. It is barren of diagnostic planktonic microfossils (Farouk et al. 2014; Abdel Aal et al.1994).

The Rosetta Formation is widely extended all over the study area, and its thickness is variable due to Tortonian tectonics and the existing topographic surface of the Wakar Formation. It is encountered in most wells drilled in the embayment at the same level recorded beneath the Mediterranean Sea. The Rosetta Formation is well developed in the northern part of the Nile Delta, but it is generally of restricted and patchy occurrence elsewhere. The absence of Rosetta anhydrite in certain areas of the Delta may be due to local shoaling or perhaps erosion by a later Messinian event. The thickness of the Rosetta Formation ranges from 11 m. to 245 m. in the study area. The thickness of this formation decreases northward (99 m in Naf well-101 and 28 m in Naf well-2) (Ismail et al. 2010). Also, its thickness is 50 m in Naf well-3 It is composed of sand with occurrences of anhydrite and clay. Kamel et al. (1998) mentioned that the top of the Rosetta Formation is equivalent to the Qawasim and Abu Madi formations in the central basin of the Nile Delta, respectively.

In the western side of the delta (Naf well-101, Naf, -well-1, 3), some planktonic foraminiferal species have been recognized (e.g., *Globigerinoides tenellus, Globigerinoides obliquus extremus, Globigerina nilotica, Orbulina bilobata,* and *Orbulina univers).* Ismail et al. 2010). Also, there are a few benthic species associated with the planktonics (e.g., *Gyroidinoides soldanii, Alabamina sp., Uvigerina semiornata and Loxostomum pseudodigitale).* The age of the Rosetta Formation is Late Miocene (Messinian).

2.3.1.6 The Abu Madi Formation

The Abu Madi Formation was introduced in the subsurface Nile Delta by the National Committee of Geology Science (NCGS) in 1974. The type section occurs in Abu Madi well-1 from 3007 to 3329 m (Fig. 2.41a). This formation exhibits different stratigraphic positions. In its type section (Abu Madi Well-1), the Abu Madi Formation unconformably overlies the Sidi Salem Formation (El Heiny et al. 1990). However, in normal cases, it unconformably overlies the Rosetta and the Qawasim Formation (El Heiny and Morsi 1992; Ouda and Obaidalla 1995). Its lower contact is situated between the pebbly sandstone of the lowermost Abu Madi Formation and the evaporates of the uppermost Rosetta Formation (Fig. 2.41a) (El Heiny and Morsi 1992; Ouda and Obaidalla 1995). In some instances, the lower boundary consists of conglomerate mixed with sandy overlying Sedi Salem Formation. Its upper boundary is placed between the sandstone of the uppermost Abu Madi Formation and the green claystone of the basal Kafr El Shiekh Formation (Fig. 2.41a, b).

The sediments of the Abu Madi Formation are composed of large thick layers of rarely conglomeratic sandstone and interbedded with clay layers, which became thicker and more frequent in the upper part of the formation (Rizzini et al. 1978). In more detail, the vertical arrangement of this formation indicates fining-upward cycles, each of which starts with pebbly and conglomeratic sandstone, followed by siltstone and capped by claystone (Abdel Fattah and Slatt 2013) (Fig. 2.41a, b). The conglomerate is an oligomict conglomerate that shows a variable in grain size and is almost loose. The sandstone in the

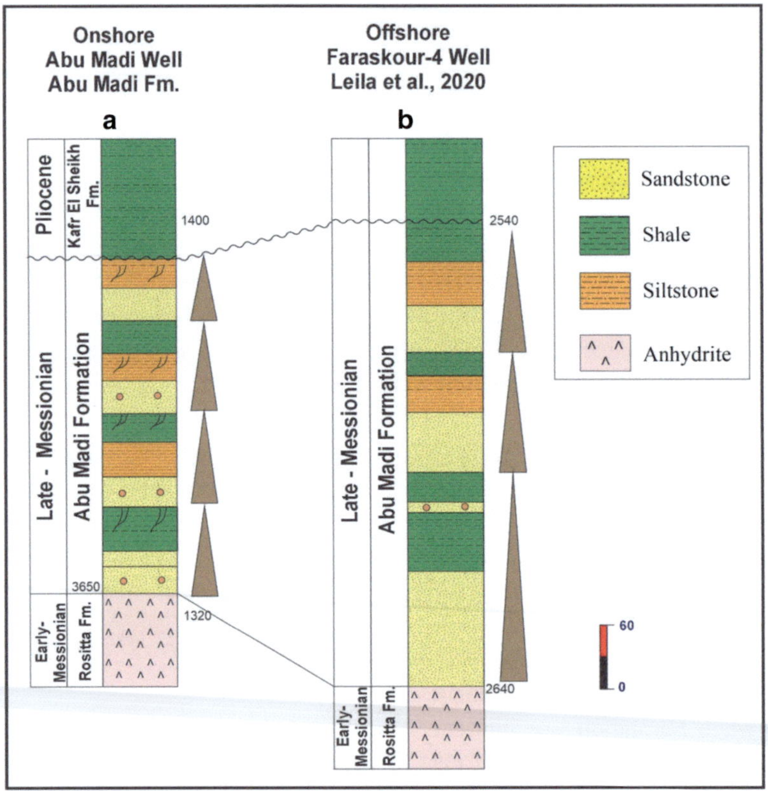

Fig. 2.41 Onshore lithologic log and stratigraphic position of the Abu Madi Formation at its type locality (**a**) central part of Nile Delta and in Faraskour well-4 (offshore, **b**)

lower parts of cycles is characterized by large cross-bedding, massive and poorly sorted, and shows angular clay fragments (Rizzini et al. 1978). In offshore (Faraskour well-4, (Fig. 2.41b), the sand is commonly quartrtzarenite, lithic quartz arenite. The claystone in the cycle top is gray-green, massive and exhibits frequent bioturbation structures (Rizzini et al. 1978). The Abu Madi reservoir has been consistently proven to be the best reservoir in the Nile Delta as it has a high porosity with an average of 21%. The majority of fields had been produced from the Abu Madi Formation (Schlumberger 1984, 1998). These sandstones were considered the main gas-producing horizons in the Nile Delta area.

In the offshore (Baltimore north). The Abu Madi Formation was described by Abdel Fattah (2014), El Sayed et al. (2019), and Leila et al. (2020) as it includes several superimposed intervals (level-1, level-11, level-111). Each interval is mainly made up of coarse to fine-grained sandstone at the base with thin beds of siltstone and shale above. Abdel Fattah (2014) mentioned that the sandstones in the different levels of the Abu Madi Formation (level-1, level-11, level-111) are composed of fluvial braided-channel deposits, which are characterized by the sandstones exhibiting fining-upward patterns. The above-mentioned levels are closely similar to the fining-upward cycles in Abu Madi well. However, the cored interval of level III is composed of sand bar deposits characterized by massive sandstones at the base, gradually fining up to siltstones (Abdel Fattah 2014).

The maximum thickness of this formation reached about 592 m in Kafr El Sheikh Well-1 (depth intervals between 2738 and 3330 m, in Tanta-1 borehole is 120 (between 1700 and 1680), in the El-Fayrouz borehole, it assumes140 m (between 1460 and 1320 m) at while at Rosseta-7 borehole is 90 m thick (between 3420 to and 3330 m), with a minimum thickness of approximately 225 m between 1580 and 1345) at the Rommana-1X borehole.

The age of the Abu Madi Formation is a controversial debate in the stratigraphy of the Nile Delta. This controversy is due to the absence of index fossils, especially in the lower part, in addition to its obvious lateral facies, and because no sharp vertical facies change from the overlying lower Pliocene Kafr El Sheikh Formation can be distinguished. However, the age of the Abu Madi Formation belongs to the early Pliocene (Zaghloul et al. 1977; Said 1990). Recent investigations were attributed to the late Miocene (Messinian) (El Heiny and Enani 1996; El Sisi et al. 1996).

2.3.2 Stratigraphy of the Quaternary Delta

The stratigraphic studies of the Quaternary Delta (Pliocene-Holocene rock units) were studied by numerous geologists (e.g., Rizzini et al. 1978; Zaghloul et al. 1977, 1979, 1989; Gheith 1985; El-Shahat and Kora 1986; Ayyad et al. 1987; Sestini, 1989; Abu Ella 1990; Mosconi et al. 1996; Shaaban et al. 2006; Sarhan and Hemdan 1994; Samuel et al. 2006; Leila and Moscariello 2018, 2019; Leila et al. 2015, 2019; Butzer 1997; El-Asmar et al. 2015; El Asmer and Wood 2000; Chen and Stanley 1993; Chen et al. 1992; El-Asmar and Gheith 1995). This was followed by the investigation of Nile Delta subsidence (Aly et al. 2012; El-Bastawesy et al. 2017; Stanley 1988, 1990, 2005; El Fawal et al. 2016, El Gharabawy 2018; Stanley and Corwin 2013; Stanley and Toscano 2009; Stanley and Warne 1997). Then, the focus has been conducted on the Late Quaternary stratigraphy and sedimentation in response to paleoenvironment and paleoclimatic conditions with geoarchaeological implications (El-Asmar et al. 2015).

The lower marine Pliocene lies, sometimes with relevant angular unconformity, on the Late Miocene Qawasim and Abu Madi formations (Fig. 2.42). These rock units include, from base to top as follows: the Kafr El Sheikh, El Wastani, the Mit Ghamr and the Bilqas formations (Fig. 2.35).

2.3.2.1 The Kafr El Sheikh Formation

The Kafr El Sheikh Formation was introduced in the subsurface of the Pliocene rocks in the Nile Delta by the National Committee of Geology Science (NCGS) in 1974. The

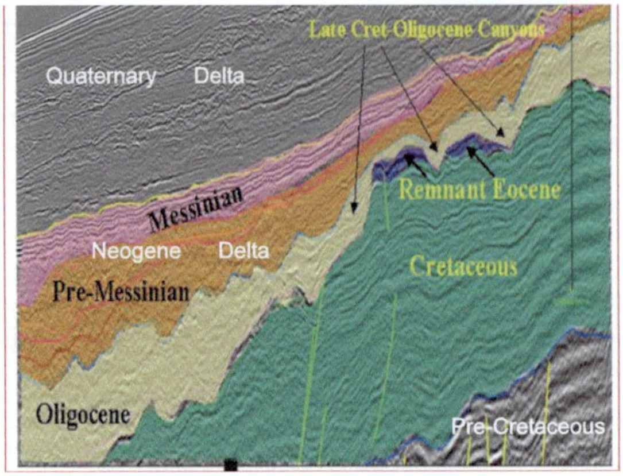

Fig. 2.42 Schematic cross-section deduced from the seismic image showing the contact between the pre-Cretaceous, Cretaceous, Eocene, Oligocene, pre-Missinian, Missinian (Neogene Delta), and the Quaternary Delta (After El-Barkooky et al. 2009)

type section of this formation occurs at Kafr El Sheikh well-1 in intervals from 1255 to 2735 m (Rizzini et al. 1978) which is located at the south-central part of the onshore Nile Delta area (Fig. 2.43a). It unconformably overjacent the Abu Madi Formation and unconformably subjacent the El Wastani Formation (Zaghloul et al. 1977). The time gap between the Kafr El Sheikh and El Wastani Formation, which extended from 2.4 to 3.2 million years, is missing (Ryan 1978). The lower boundary exhibits variation from one area to another. In the southern parts (onshore), it shows an irregular and uneven surface with the underlying Qawasim Formation, in case of missing the Abu Madi Formation. Its upper boundary can be marked by the appearance of sandstone of the basal El Wastani Formation that is associated with the littoral fauna of *Ammonia* and *Elphidium* sp. (Rizzini et al. 1978). While on the northern parts of the Delta (offshore), it also exhibits irregular contact with the underlying Abu Madi Formation (Fig. 2.43b), while its upper boundary can be easily identified by the appearance of the overlying sand sheets belonging to the El Wastani Formation (Rizzini et al. 1978).

In onshore regions, this rock unit comprises intercalation of thick shales and clays and a few thin sandstone bands (Fig. 2.43a). The clays are composed of kaolinite and montmorillonite with traces of illite (Rizzini et al. 1978). The sandstone is mostly quartzose, cemented by clay minerals (Rizzini et al. 1978). It is intercalated with fine sand beds indicating a periodical lowering of the sea level (Zaghloul et al. 1977). In offshore areas (e.g., Abu Qir wells), the succession comprises an intercalation of claystone, dolomitic limestone, and dolostones (Ismail et al. 2010) (Fig. 2.43b). The claystone is green-gray, compact, while dolomites are pale yellow, hard, massive (Ismail et al. 2010).

This formation was widely distributed in the whole area of the Nile Delta and was considered as the thickest rock unit in the Neogene succession. This formation covers the whole delta and migrated southwards, covering the Nile River till Aswan (Issaw et al. 2009). In onshore wells, the Kafr El Sheikh Formation attains a maximum thickness of about 1820 m in Abu Madi Well-4 (Fig. 2.43a), 1400 m in Ras-El Barr Well-11, 992 m in Qantara Well-1, and Faraskor Well-1. In the above location, the formation consists

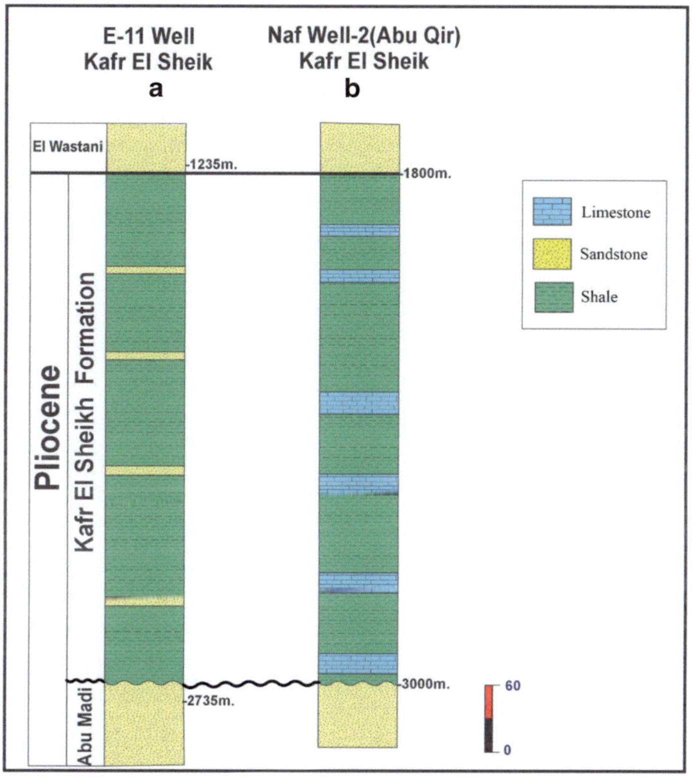

Fig. 2.43 Onshore lithostratigraphic section and stratigraphic setting of the Kafr El Sheikh Formation at Kafr El Sheikh well-1. It mainly comprises thick claystone beds with thin sandstone layers (**a**). Offshore lithostratigraphic section of the Kafr El Sheikh Formation at Naf well-2 (Abu Qir field). It mainly comprises thick claystone beds with thin limestone layers (After Ismail et al. 2010) (**b**)

of thick claystone beds with thin intercalations of fine sands. Eastwards in the delta, the facies are exclusively of intercalation of thick claystone with thin siltstone, In onshore wells it shows minimum thickness, e.g., Rosetta 300 m (3200–3500), 100 m (1400–1500 m depth) at Tanta well-1, 300 m (1200–1500 depths) at El-Fayrouz well-1 and 150 m (1200–1350 depths) at Rommana well-1. In offshore at Abu Qir fields, this formation exhibits a maximum thickness ranging from 1121 m in the Naf well-101, 1149 m in the Naf well-2, 1400 m (1600–3000m depths) in the Naf well-3 (Ismail et al. 2010). In such a location, the formation shows lithological changes as compared with the onshore sites; it consists of thick beds of claystone with thin beds of marly limestone to dolomitic limestone (Ismail et al. 2010) (Fig. 2.43b).

Several planktonic species were recognized in this formation (e.g., *Globigerinoides obliqua, Globigerinoides trilobus trilobus, Globigerina falcon, ensis, Globigerina nilotica, Globigerina apertura, Turborotalia humilis, Globorotalia exilis, Globorotalia acostaensis*

acostaensis, Bella praedigitata, and *Catapsydrax parvulu* (Ismail et al. 2010). Moreover, there are several benthic species associated with the planktonics (e.g., *Nonion affinis, Anomalina grosserugosa, Bulimina elongata, Gyroidinoides soldanii, Bolivina hebes, Ammonia beccarii, Cibicidoides sp., Cassidulina brocha Pullenia osloensis* and *Uvigerina semiornata)* (Ismail et al. 2010). Added to the above, it comprises the *Globorotalia margaritae* and the *Globorotalia puncticulata* zones of the Lower Pliocene and the *Globorotalia aemiliana* and *Globorotalia crassaformis* zones of the Middle Pliocene (Rizzini et al. 1978; Barakat, 1982; Badran 1996; Debis et al. 1986; Effat and El Gezeiry 1986). These fossils indicate Early- Middle Pliocene (Ismail et al. 2010).

2.3.2.2 The El Wastani Formation

El Wastani Formation was introduced in the Nile Delta by EGPC (1994). Its type section occurs in El Wastani well-1 from 1009 to 1132 m (Rizzini et al. 1978) (Fig. 2.44a). This formation unconformably overlies and underlies the Kafr El Sheikh and Mit Ghamr formations, respectively (Ismail et al. 2010) (Fig. 2.44a). Its upper boundary is placed where the formation exhibits to be completely sandy facies for several tens of meters (Rizzini et al. 1978). In addition, the upper boundary is defined by the almost complete disappearance of littoral marine *Ammonia* and *Elphidium* faunas, since they are related to the clays. In onshore regions, this formation consists of thick sandstone beds interbedded with siltstone and thin clay beds. The vertical staking patterns exhibit fining-upward cycles (Fig. 2.44a). The sandstone shows thinning toward the top of the formation (Fig. 2.44a). In offshore areas (Abu Qir and, Baltim wells), the formation consists of intercalation of thin sandstone, thick claystone and thin some carbonate beds (dolomite and limestone) occur toward the top and they increase northwards (Ismail et al. 2010) (Fig. 2.44b). The vertical arrangement of the above-mentioned lithofacies manifests fining-upward cycles (Fig. 2.44b). The sands are quartzes coarse to medium-grained with little feldspars and little rock fragments. The clays are soft and very sandy. This formation is a transitional state between the shelf faces of the Kafr El Sheikh Formation and the coastal and continental sands of the Mit Ghamr Formation above (Rizzini et al. 1978). In onshore areas, at El Wastani well-1, the thickness of this formation assumes about 121 m (1009–1132 depths), while in offshore areas, at Naf well-3 and Naf well-101, it has an average of 156–200 m. It increases in thickness northwards, reaching 301 m at Naf well-2 (Ismail et al. 2010). This formation contains several species of planktonic foraminifera. such as *Globigerinoides trilobus sacculifer, Globigerina rubescens, Globigerina bulloides, Orbulina universa,* and *Globigerinella obesa.* There are also several benthic species associated with the planktonic forms (e.g., *Nonion asterizans, Ammonia tepida, Natlandia cf. secasensis, Eponidella* cf. *libertadensis, Lenticulina reedi, Bulimina elongata, Cassidulina brocha,* and *Ammonia umbonata* (Ismail et al. 2010). According to the above-mentioned fossils, it is attributed a Late Pliocene time (Rizzini et al. 1978; Badran 1996; Ismail et al. 2010).

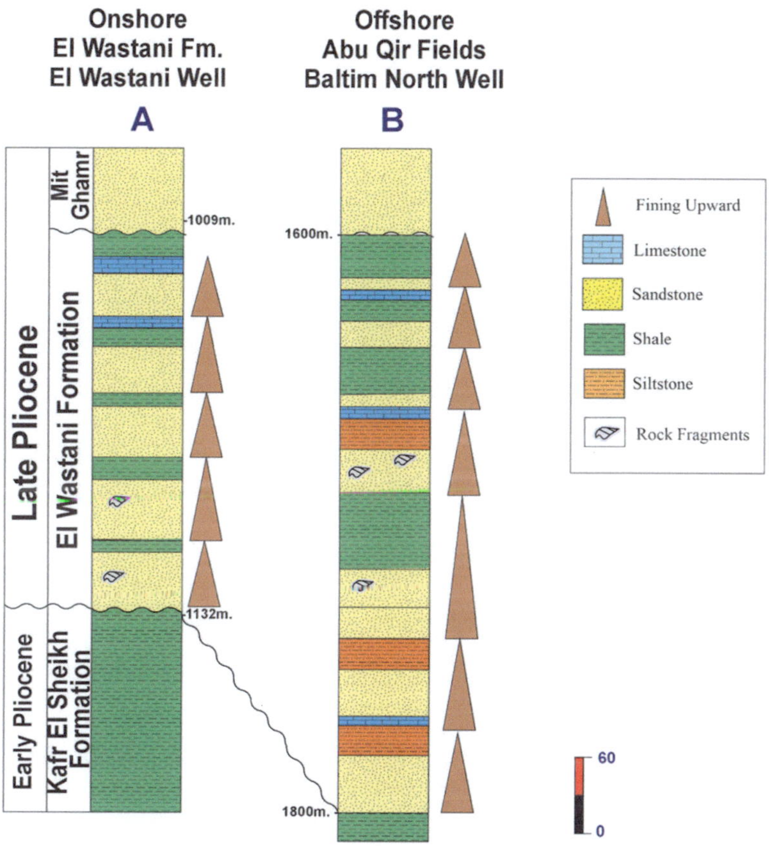

Fig. 2.44 Onshore lithostratigraphic column and stratigraphic setting of El Wastani formation at El Wastani well-1 (**a**). It mainly comprises thick sandstone, thin siltstone, and claystone that is arranged vertically into fining-upward cycles. Offshore lithostratigraphic column of the Al Wastani formation at Abu Qir and Baltim wells. It consists of fining-upward cycles, each is made up of thin sandstone at the base, capped by thick claystone and limestone at the top (**b**)

2.3.2.3 The Mit Ghamr Formation

The term Mit Ghamr Formation was coined by EGPC (1994) to describe the claystone intercalated with siltstone and thin beds of limestone in the Nile Delta. Its type section occurs at Mit Ghamr well-1, located in the southern part of the Delta on the eastern side of the Damietta branch and encountered between 20 and 483 m (Rizzini et al. 1978) (Fig. 2.45a). This formation shows a constant stratigraphic position either onshore or in offshore, in both sites it overlies the El Wastani Formation and underlies the Bilqas Formation (Fig. 2.45a–c). Its lower contact is sharp to unconformable with the underlying Al Qawasim Formation. The boundary is placed at the last appearance of *Globigerinoides obliquus* (Bolli) at the top of the El Wastani Formation and the first appearance of the

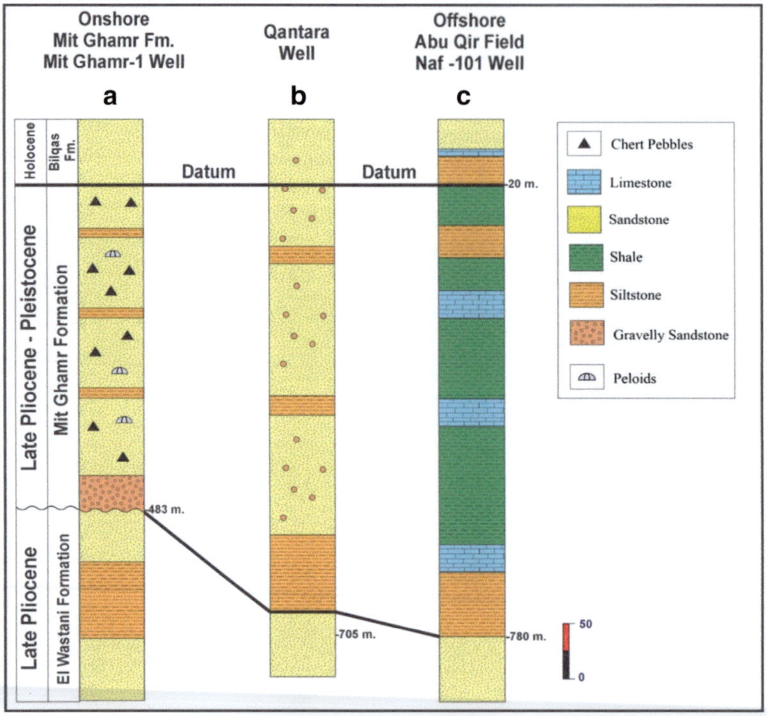

Fig. 2.45 Onshore lithologic log and stratigraphic setting of the Mit Ghamr Formation at Mit Ghamr and Qantara wells (**a**, **b**). It mainly comprises an intercalation of intercalation of pebbly sandstone and siltstone. Offshore (Naf well-101, Abu Qir field), it comprises intercalation of thick claystone, thin siltstone, and limestone (**c**)

planktonic foraminifera *Globigerina calida* (calida) at the lower part of Mit Ghamr Formation (El-Bassiouni and Ayyad 1986). Its upper boundary with the overlying Bilqas Formation is distinguished by thin beds of clay, silt, and peat containing coastal or lagoonal fauna (Shaaban et al. 2016). In onshore areas (at Mit Ghamr and Qantara wells), the formation contains thick sands and pebbles with a limited thickness of clay intercalation (Rizzini et al. 1978). The sands are medium to coarse-grained sands, while the pebbles consist of quartzite, chert, and dolomites (Rizzini et al. 1978) (Fig. 2.45a, b). Some occasionally silicified limestone and chert pebbles are found. The sands can also contain shells of bivalve mollusks. Farther eastwards, at the Qantara well, the formation comprises gravelly sandstone intercalated with coarse-grained sandstone (El Heiny and Morsi 1992) (Fig. 2.45b). The Mit Ghamr Formation exhibits radical facies changes offshore at the Naf well-101(Abu Qir field), where it essentially consists of green claystone with thin interbeds of siltstone and limestone (Ismail et al. 2010, Fig. 2.45c). In general, the lithofacies distribution of this formation all over the delta shows distinct lateral lithologic variations ranging from sandy facies covering most of the southern delta to sandy shale in the northern marginal zones of the delta. Between these two types of facies, a narrow belt of shaly sand facies occurs (Serag El-Din 1998). Penningtona et al. (2017) classified the Mit Ghamr Formation into two members, the Zagazig Member at the base and the Minuf Member at the top. They mentioned that the Zagazig Member consists of coarser sediments and is mineralogically different from the upper Minuf Member. As they did not define the facies characteristic of each member, the upper and lower boundaries of these members, and the thickness of each unit, it is difficult to generalize this classification in describing the Mit Ghamr Formation in the delta region. The Mit Ghamr Formation has an average thickness of about 463 in the northern delta (onshore), increasing northwards to 700 m offshore (Abu Qir fields) (Zaghloul et al. 1979; Ismail et al. 2010). In general, the thickness of the Mit Ghamr Formation gradually increases toward the north and northeast. It ranges from 100 to 400 m in the south, from 500 to 700 m in the middle, and reaches a maximum thickness exceeding 900 m in the north (El-Fayoumi 1987; Serag El-Din 1989). The riverine sediments of this formation appear in outcrops as Turtle backs in the agricultural fields, representing the higher parts of the eroded surface of this complex (Said 1981).

No foraminifera have been recorded in this formation, which may be partly due to the small number of studied samples. However, the age is controversial among researchers, it is dated to the Late Pliocene to Pleistocene (Rizzini et al. 1978; Badran 1996). However, Ismail et al. (2010) dated this formation from the Pleistocene to Holocene throughout the western Nile Delta, based on its correlation with the neighboring areas.

2.3.2.4 The Bilqas Formation

The term Bilqas Formation was established by EGPC (1994). Its type section occurs in the Bilqas well-1 between 0 and 25 m (Rizzini et al. 1978) (Fig. 2.46a). It is considered the youngest rock unit in the Nile Delta. It unconformably overlies the Mit Ghamr Formation. In the south delta (onshore), the Bilqas Formation is relatively thin in thickness ranging between 0 and 40 m (Fig. 2.46a). In onshore regions, this formation comprises medium- to fine-grained sand intercalated with thin clay beds, enriched with pelecypod, gastropod and ostracode fragments (Rizzini et al. 1978) (Fig. 2.46a). In offshore areas at Naf well-1(northwest delta, Abu Qir fields) the formation comprises intercalation of claystone, siltstone with thin beds of sandy limestone (Ismail et al. 2010) (Fig. 2.46b). The clays contain vegetal matter and peat remains and carbonaceous matter (Ismail et al. 2010). The formation occurs on the whole delta area mixed with the present continental formations (Rizzini et al. 1978; Ismail et al. 2010). Toward the north (offshore) it becomes thicker, reaching more than 450 m at Abu Qir field, northwest Nile Delta (Ismail et al. 2010) (Fig. 2.45b). In general, the lithofacies of Bilqas Formation at the Nile Delta show lithological variations ranging from clay and silt at the eastern and north central parts of the Nile Delta, while in the south (coastal zone), it comprises entirely of dunes to sandy

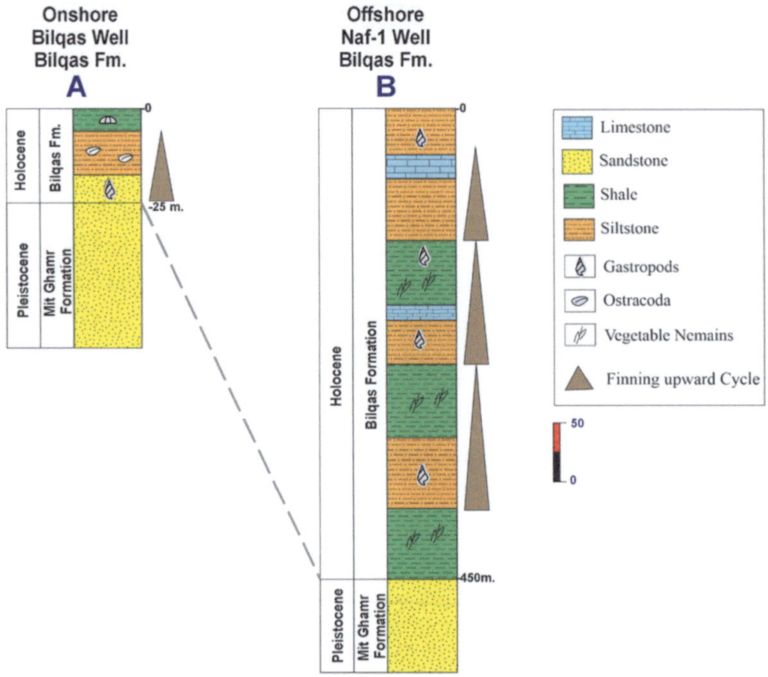

Fig. 2.46 Onshore lithologic log and the stratigraphic setting of the Bilqas Formation at Bilqas well-1. It mainly comprises an intercalation of pebbly sandstone, sandstone, and siltstone (**a**). Offshore at Abu Qir fields, (Naf well-1) the Bilqas Formation comprises an intercalation of pebbly sandstone, sandstone, siltstone, and limestone (**b**)

facies (El-Fayoumi 1987). These facies changes and their distribution may be attributed to the irregularity of the basin bottom during deposition and the density of the old Nile distributaries and hence their energy (El-Fayoumi 1987).

Recently, Stanley et al. (1996) investigated a large number of boreholes taken from the Nile Delta region, > 3 m deep, many of which have been carried out since 1991 (Butzer 1998, 2002). This project drilled 87 radiocarbon-dated cores within the coastal fringe of the delta. Based on the above core samples taken from boreholes, Penningtona et al. (2017) have divided the Bilqas Formation into two units, Bilqas-1 and Bilqas-2. They comprise fining-upward cycles, most of which begin with fluvial sands, followed by back swamp mud and peat, capped by floodplain mud or aeolian sands. This classification cannot generalize on the Holocene sediments in the Quaternary Delta for the following reasons. (1) The core samples were taken from very shallow portholes that are not enough to represent the whole thickness of the Bilqas Formation (0–50 m), (2) The boundaries of each unit (Bilqas-1, Bilqas-2) are not defined, (3) No index faunal association is demarked in each unit. This formation has been attributed Holocene age (Rizzini et al. 1978; Barakat 1982; Badran 1996). However, Ismail et al. (2010) mentioned that this formation ranges from Pleistocene to Recent. According to Said (1981), the Bilqas Formation may be included within the Neonile sediments assigned Late Pleistocene to the Holocene.

References

Abdel Aal A, Price RJ, Vaitl JD, Shrallow JA (1994) Tectonic evaluation of the Nile Delta, its impact on sedimentation and hydrocarbon potential. In: 12th exploration and production conference, vol 1. Egyptian Geological Petroleum Company, pp 19–34

Abdel Aal A, El Barkooy A, Meyer GH, Schwander M, Zaki H (2000) Tectonic evolution of the eastern mediterranean basin and its significance for the hydrocarbon prospectively in the ultra-deep water of the Nile Delta. Lead Edge 19:1086–1102

Abdel Fattah ML (2014) Petrophysical characteristics of the Messinian Abu Madi Formation in the Baltim east and north fields. Offshore Nile Delta, Egypt. J Pet Geol 37:183–195

Abdel Fattah ML, Slatt RM (2013) Sequence stratigraphic controls on reservoir characterization and architecture: case study of the Messinian Abu Madi incised-valley fill, Egypt. Cent Eur J Geosci 5:497–507

Abdou HF, Abdel-Kireem MR (1975) Planktonic foraminiferal zonation of the middle and upper Eocene rocks of Fayium Province, Egypt. Rev Esp Micropaleontol 7:15–64

Abu El Ella R (1990) The Neogene-quaternary section in the Nile delta, Egypt; geology and hydrocarbon potential. J Pet Geol 13:329–340

Abu El Kheir GA (2022) A contribution of the Late Eocene Sirenian fossils at the north of Lake Qaroun, Fayium, Egypt. Ann Geol Surv V.XXXV111:29–43

Abu El-kheir GA (2010) Paleontological studies and sedimentological evolution in Wadi El-Hitan District, Fayium Governorate, Egypt. Msc. Thesis, Cairo University

Abu Khadrah AM, Selim SS1, Abdel Baset AA (2018) Sedimentology and seismic geomorphology of the tidal dominated estuarine incised valley fill, Late Miocene Qawasim reservoir, onshore east, east Nile Delta. Egypt J Geol 62:103–118

Albritton CC, Brooks JE, Issawi B, Swedan A (1990) Origin of the Qattara depression, Egypt. Geol Soc Am Bull 102:952–960

Aly MH, Klein AG, Zebker HA, Giardino JR (2012) Land subsidence in the Nile Delta of Egypt observed by persistent scattered interferometry. Remote Sens Lett 3(7):621–630

Andrews CW (1906) Descriptive catalogue of tertiary vertebrata of the Fayoum. Egypt British Museum Nat Hist, London, p 324

Andrews CW (1904) Further notes on the mammals of the Eocene of Egypt. Geol Mag Lond Ser 5(1):211–215

Ansary SE (1955) Report on the foraminiferal fauna from the Upper Eocene of Egypt. Bull Inst Desert 6:1e160 (Egypt)

Ayyad S, El-Bassiouni AE, Ziko A (1987) Benthonic foraminifera, Bryozoan and the paleoenvironments of the quaternary succession in north west Rosetta of shore well No.1, Nile Delta, Egypt. In: 6th Symposium on quaternary and development. Mansoura University, Egypt

Badran AM (1996) The contribution of logging analysis and seismic stratigraphy in the reservoir valuation and lithology determination of the north western Delta Abu Qir Offshore Egypt. MSC. Thesis, Geophysics Department, Faculty of Science, Ain Shams University, Cairo, Egypt, p 112

Barakat MG (1982) General review of the petroliferous provinces of Egypt with special emphasis on their geological setting and oil Potentialities. Geology Department, Faculty of Science, Cairo University, Cairo, Egypt, p 87

Barber PM (1981) Messinian subaerial erosion of the Proto-Nile delta. Mar Geol 44:253–272
Barron T (1907) The topography and geology of the district between Cairo and Suez: Cairo. National Printing Department, p 133
Beadnell HJL (1901) The Fayium depression: a preliminary notice of the geology of a district in Egypt containing a new Palaeogene vertebrate fauna. Geol Mag Decade IV 8:540–546
Beadnell HJL (1905) The topography and geology of the Fayium Province of Egypt. Survey Department Egypt, Cairo, p 101
Blanckenhorn M (1903) Neue geologisch-stratigraphische Beobachtungen in Aegypten. Sitzungsberichte der Mathematisch-physikalischen Classe der Königlichen bayerischen Akademie der Wissenschaften., München 32:353–433
Blanckenhorn M (1900) Neues zur Geologie und Paläontologie Aegyptens II Das Palaeogen. Zeitschrift der deutschen geologischen Gesellshaft, Stuttgart 52:403–479
Bown BW, Vondra CF (1974) Paleoenvironmental interpretations of the Oligocene Gabal el Qatrani Formation, Fayum Depression, Egypt (essays on African paleontology). Ann Geol Surv Egypt 4:115–138
Bown TM (1982) Ichnofossils and rhizoliths of the nearshore fluvial Jebel Qatrani Formation (Oligocene), Fayum Province, Egypt. Palaeogeogr Palaeoclimatol Palaeoecol 40:255–309
Bown TM, Kraus MJ (1988) Geology and paleoenvironment of the Oligocene Jebel Qatrani Formation and adjacent rocks. Fayium Depression, Egypt, United States Geological Survey Professional Paper 1452, pp 1–60
Bown TM, Kraus MJ, Wing SL, Fleagle JG, Tiffney BH, Simons EL, Vondra FC (1982) The Fayum primate forest revisited. J Hum Evol 11:603–632
Butzer KW (1997) Late Quaternary problems of the Egyptian Nile:stratigraphy, environments, prehistory. Paléorient 23(2):151–173
Butzer KW (1998) Late Quaternary problems of the Egyptian Nile: stratigraphy, environments, prehistory. Paleorient 23:151–173
Butzer KW (2002) Geoarchaeological implications of recent research in the Nile Delta. In: van den Brink ECM, Levy TE (eds) Egypt and the levant: interrelations from the 4th through the early 3rd millennium BCE. Leicester University Press, London, pp 83–97
Chen Z, Stanley DJ (1993) Alluvial stiff muds (Late Pleistocene) underlying the lower Nile Delta plain: Petrology, stratigraphy and origin. J Coast Res 9(2):539–576
Chen Z, Warne AG, Stanley DJ (1992) Late Quaternary evolution of the northwestern Nile Delta between the Rosetta promontory and Alexandria, Egypt. J Coast Res 18(3):527–561
Cuvillier J (1940) L'Oligocene du Gebel Kashab (Desert Lybique): Comptes Rend us Academie des Sciences Paris, vol 210, pp 340–342
Debis S, Futan ART, Ince DM, Morley RJ, Seymour WP, Thmpson S (1986) The stratigraphy framework of The Nile Delta and its implications with respect to the region's hydrocarbon Potential. In: Eighth exploration conference centenary of first oil well, EGPC, p 15
Dixon W (1873) A petrified forest in the Libyan desert. Nature:363–364
Effat A, El Gezeiry M (1986) New aspects for Abu Qir stratigraphic and sedimentologic model. Wepco, P.O. Box 412, Alex, Egypt
EGPC (Egyptian General Petroleum Corporation) (1994) Nile Delta and North Sinai: field discoveries and hydrocarbon potentials (a comprehensive overview) EGPC Cairo, p 387
EI-Heiny I, Morsi S (1992) Stratigraphic correlation of Neogene sediments in the eastern Nile delta and Gulf of Suez. In: 11th Exploration and production conference, Cairo, pp 166–193
EI-Heiny E, Enani N (1996) Regional stratigraphic interpretation of Neogene sediments, northern Nile Delta. In: 13th EGPC petroleum conference exploration, Cairo, pp 270–290

El-Barkooky A, Hamdan M, Hassan SM, Christie-Blick N, Goldstein S, Cai Y (2009) Sequence stratigraphy of the lower miocene moghra formation in the qattara depression, northwestern desert, Egypt. In: AAPG annual convention, Denver, Colorado
El Gharabawy S (2018) Geological and structural evaluation on temsah gas field, NE-offshore Nile Delta, Egypt. Int J Sci Eng Appl Sci (IJSEAS) 4:18–29
El Heiny I (1982) Neogene stratigraphy of Egypt. Nwesl Stratigr 2:41–52
El Heiny I, Rizk R, Hassan M (1990) Sedimentological model for Abu Madi reservoir sands, Abu Madi field, Nile Delta, Egypt. In: Proceedings of the tenth exploration Conference. EGPC, Cairo
El Mahdy M, Abioui M, Mustoe GE (2020) Miocene petrifed trees at Bahariya Oasis, Egypt. Int J Earth Sci 109:2869–2870
El Sadawi WE (2005) A fossil rhizome at the mangrove site of Wadi Hitan, Egypt. Taeckholimia 25:129–136
El Sayed AA, El Sayed NA, Lala A, Metwally M (2019) Reservoir rock genetic types using log curve shapes: AbuMadi formation, offshore Nile Delta—Baltim gas field, Egypt. IOP Conf Ser Earth Environ Sci 221:012045
El Sharief AWM (2020) Reservoir characterization of the Messinian Abu Madi Formation at West El Manzala development lease, Onshore Nile Delta, Egypt. M. Sc. Thesis, Menoufia University, Egypt, p 128
El Sisi ZA, Sharaf LM, Dawood HY, Hassouba AB (1996) Source of clastic supply to the Sidi Salem and Abu Madi Reservoirs of the Nile Delta, Egypt. In: E.G.P.C. 13th petroleum exploration production conference, vol 1, pp 291–296
El-Asmar HM, Gheith AM (1995) Stable isotope stratigraphy and paleoclimatology of the offshore subsurface Quaternary sediments, Nile Delta, Egypt. Sedimentol Egypt 3:39–49
El-Asmar HM, Wood P (2000) Quaternary shoreline development: the Northwestern coast of Egypt. Quat Sci Rev 19:1137–1149
El-Asmar HM, Taha MMN, El-Kafrawy SB, El-Sorogy AS (2015) Control of Late Holocene Geo-processes on the sustainable development plans of the Tineh Plain, NW Sinai coast, Egypt. J Coast Conserv 19:141–156
El-Bassiouni AE, Ayyad SN (1986) Quaternary Foraminifera of El-Temsah well-2, offshore Nile Delta, Egypt. El Mansoura Sci Bull 13(1):133–188
El-Bastawesy M, Cherif OH, Sultan M (2017) The geomorphological evidences of subsidence in the Nile Delta: analysis of high resolution topographic DEM and multi-temporal satelliteimages. J Afr Earth Sc 136:252–261
El-Fawal FM, Sarhan MA, Collier REL, Basal A, Abdel Aal MHA (2016) Sequence stratigraphic evolution of the post-rift megasequence in the northern part of the Nile Delta basin, Egypt. Arab J Geosci 9:585
El-Fayoumi IF (1987) Geology of the quaternary succession and its impact on the groundwater reservoir in the Nile Delta Region. Bull Faculty of Science, Mansoura University, Egypt
El-Sadawi W, Kamal El-Din MM, Wheeler E, Osman R, El-Faramawi MW, El-Noamani Z (2014) Early Miocene woods of Egypt. IAWA J 35:35–50
El-Shahat A, Kora MA (1986) Size analysis and paleoenvironments of the subsurface sediments, west of Nile Delta. El Mansoura Sci Bull 13(2):185–207
Faris M, Samir AM, Shabaan M (2007) Calcareous nannofossil biostratigraphy of the subsurface Miocene sequence, northeast Nile Delta, Egypt. In: The fifth international conference of the geology of Africa, vol 1, pp VII-1–VII-31
Farouk S, Khalifa MA (2010) Facies tracts and sequence development of the Middle Eocene—middle Miocene successions of the southwestern Qattara depression, northern Western Desert, Egypt. Freib Forsch 536:195–215

Farouk S, Ziko A, Eweda SA, Said AE (2014) Subsurface Miocene sequence stratigraphic framework in the Nile Delta, Egypt. J Afr Earth Sci 91:89–109

Fleagle JG, Bown TM, Oobradovich JD, Simons EL (1986a) How old are the Fayium primates? In: Else JG, Lee PC (eds) Primate evolution. Cambridge University Press, Cambridge, pp 3–17

Fleagle JG, Bown TM, Oobradovich JD, Simons EL (1986b) Age of the earliest African anthropoids. Science 234:1247–1249

Gameil M, Al Anbaawy M, Abdel Fattah M, Abu El-Kheir G (2016) Lithofacies and biofacies characteristics and whales skeletons distribution in the Eocene rock units of Fayoum Area, Egypt. J Afr Earth Sci 116:42–55

Georgalis GL, Abdel Gawad MK, Hassan SM, El-Barkooky AN, Hamdan MA. Oldest co-occurrence of Varanus and Python from Africa—first record of squamates from the early Miocene of Moghra Formation, Western Desert, Egypt. Peer J 8:e9092

Gheith AM (1985) Sedimentological and mineralogical investigations of the subsurface Quaternary succession of Temsah offshore well-2, Nile Delta Coast. Egypt J Geol 29:20–50

Gingerich PD (1992) Marine mammals (Cetacea and Sirenia) from the Eocene of Gebel Mokattam and Fayum, Egypt: stratigraphy, age, and paleoenvironments, vol 30.University of Michigan Papers on Paleontology, pp 1–84

Haggag MAY (1988) Middle Eocene planktonic foraminifera from Fayum area, Egypt. Revista Espanola De Micropaleontologia 17:27–40

Haggag MAY (1990) Globigerina pseudoamphiapertura Zone, a new late Eocene planktonic foraminiferal zone (Fayoum area, Egypt). Neues Jahrbuch für Geologie und Palaeontologie. Monatshefte 5:295–307

Harms JC, Wary L (1990) Nile delta. In: Said R (ed) The geology of Egypt. Balkema, Rotterdam, pp 329–344

Hassan SM (2013) Sequence stratigraphy of the lower Miocene Moghra Formation in the Qattara Depression, North Western Desert, Egypt. Springer Briefs in Earth Sciences, Springer

International Egyptian Oil Companies (IEOC) (1969) Stratigraphical reports. I.E.O.C. Wells drilled in the Nile Delta. Unpublished report, Cairo

Iskander F (1943) Geological survey of the Gharag el Sultani sheet no. 68/54. Standard Oil Company, Egypt S.A. Reports 51, pp 1–29

Ismail AA, Boukhary M, Abdel Naby AI (2010) Subsurface stratigraphy and micropaleontology of the Neogene rocks, Nile Delta, Egypt. Geol Croatica 63:1–26

Issaw B, Francis MF, Youssef AA, Osman RA (2009) The Phaneorozoic geology of Egypt: a geodynamic approach. Special publication No.81. Minsitry of Petroleum. The Egypt Minsitry Research Authority, p 589

Kamel H, Eita T, Sarhan M (1998) Nile delta hydrocarbon potentiality, Egypt. In: Proceedings on 14th EGPC exploration and production conference, vol 1. Cairo, Egypt, pp 485–503

Kappelman J (1991) The age of the Fayoum primates as determined by paleomagnetic reversal stratigraphy. J Hum Evol 22:495–503

Khaled KA, Sharaf E, Abushady AI (2012) Resevoir characterizeion of Sidi Salem Formation, Akhen fields, offshore Nile Delta Egypt. Egypt J Geol 55:178–190

Khalifa MA (1981) Geological and sedimentological studies of west Beni—Mazar area, south El Fayoum province, Western Desert, Egypt. Ph.D Thesis, Cairo University

Kortland A (1980) The Fayum forest: did it exist? J Hum Evol 9:277–297

Krausel R (1939) Ergbnisse der Forschungreisen Prof. E. Stromer's in den Wusten Agyptens, IV. Die fossilen Floren Agyptens: Abhandlungen der Bayerischen Akademie der Wissenschaften Mathematische-Naturwissenschaften Abteilungen. Neue Folge 47:1–140

Legler B, Johnson HJ, Hampson GJ, Massart BYG, Jackson CL, Jackson MD (2013) Facies model of a fine-grained, tide-dominated delta: lower Dir Abu Lifa Member (Eocene), Western Desert, Egypt. Sedimentology. https://doi.org/10.1111/sed.12037

Leila M, Kora M, Ahmed M, Ghanem A (2015) Sedimentology and reservoir characterization of the Upper Miocene Qawasim Formation, El-Tamad oil Field onshore Nile Delta, Egypt. Arab J Geosci 9:1–13

Leila M, Moscariello A (2018) Depositional and petrophysical controlson the volumes of hydrocarbons trapped in the Messinian reservoirs, onshore Nile Delta, Egypt. Petroleum 4:250–267

Leila M, Moscariello A (2019) Seismic stratigraphy and sedimentary facies analysis of the pre- and syn-Messinain salinity crisis sequences, onshore Nile Delta, Egypt: implications for reservoir quality prediction. Mar Petrol Geol 101:303–321

Leila M, Moscariello A, Šegvić B (2019) Depositional facies controlson the diagenesis and reservoir quality of the Messinian Qawasim and Abu Madi formations, onshore Nile Delta, Egypt. Geol J 54:1797–1813

Leila M, Kora MA, Ahmed MA, Ghanem A (2016) Sedimentology and reservoir characterization of the Upper Miocene Qawasim Formation, El-Tamad oil field onshore Nile Delta, Egypt. Arab J Geosci 9:1–13

Leila M, Eslam A, Abu El-Magd A, Alwaan L, Elgendy A (2020) Formation evaluation and reservoir characteristics of the Messinian Abu Madi sandstones in Faraskour gas field, onshore Nile Delta, Egypt. J Pet Explor Prod Technol

Marzouk I (1970) Rock stratigraphy and oil potentialities of the Oligocene and Miocene in the western desert of Egypt. In: 7th Arab petroleum congress, vol 54, pp 1–37

Mosconi A, Rebora A, Venturino G, Bocc P, Khalil MH (1996) Egypt-Nile Delta and North Sinai Cenozoic tectonic evolutionary model: a proposal. In: Proceedings on 13th EGPC exploration and production conference. Egypt, Cairo, pp 203–123

Murray AM, Cook TD, Attia YS, Chatrath P, Simons EL (2010) A freshwater Ichthyofauna from the Late Eocene Birket Qarun Formation, Fayum, Egypt. J Vertebr Paleontol 30:665–680

Nabawy BS, Abd El Aziz EA, Ramadan M, Amer Shehata A (2023) Implication of the micro- and lithofacies types on the quality of a gas-bearing deltaic reservoir in the Nile Delta, Egypt. Sci Rep 13 (Article number: 8873)

Nashaat M (1998) Seal assessment in the Nile Delta, Western Desert, and the Gulfof Suez, Egypt, In: Proceedings of the 14th petroleum conference EGPC, vol 1, pp 492–510

Nashaat M (1999) Petroleum system in the eastern part of the Nile Delta. In: Abstract, 2nd conference, Mansoura University, p 1

Omara S, Sanad S (1975) Rock stratigraphy and structural feature of the area between Wadi El Natrun and the Moghra depression (western desert), Egypt. Geol Jahrb 16:45–73

Ouda KH, Obaidalla N (1995) The geologic evolution of the Nile Delta area during the Oligocene-Miocene of Egypt. Egypt J Geol 93:77–111

Penningtona BT, Sturtb F, Wilsonc P, Rowlandd J, Browna AG (2017) The uvial evolution of the Holocene Nile Delta. Quat Sci Rev:1–22

Peters SE, Antar MS, Zalmout IS, Gingerich PD (2009) Sequence stratigraphic control on preservation of Late Eocene whales and other vertebrates at Wadi Al Hitan Egypt. Palaios 24:290–302

Rizzini A, Vezzani F, Cococcetta V, Khalil M (1976) Stratigraphy and sedimentation of the Neogene-Quaternary Section in the Nile Delta Area. In: A.R.E. 5th exploration seminar. Cairo, pp 327–348

Rizzini AF, Coccetta VV, Milad G (1978) Stratigraphy and sedimentation of a Neogene-quaternary section in the Nile Delta area. Mar Geol 27:328–348

Ross D, Uchupi E (1977) Structure and sedimentary history of northeastern Mediterranean Sea-Nile cone area. Am Assoc Pet Geol Bull 61:872–902

Ryan WFB (1978) Messinian badlands on the SE margin of the Mediterranean Sea. Mar Geol 27:389–363

Said R (1962) The geology of Egypt. Elsevier, Amsterdam, p 377

Said R (1990) Cenozoic. In Said R (ed) The geology of Egypt. AA Balkema, Netherlands (NLD), pp 451–486

Said R (1981) The geologic evolution of the river nile. Springer- Verlag, New York, p 151

Salem RM (1976) Evolution of Eocene-Miocene sedimentation patterns in parts of northern Egypt. Am Asso Pet Geol Bull 60:34–64

Samuel A, Kneller B, Raslan S, Sharp A, Parsons C (2006) Prolific deep-marine slope channels of the Nile Delta, Egypt. Am Assoc Pet Geol Bull 87:541–560

Sandford KS, Arkell WJ (1924) Paleolithic man and the Nile-Fayoum divide, vol 3. Chicago University, Oriental Institute Publications, pp 1–77

Sarhan M, Hemdan K (1994) North Nile Delta structural setting and trapping mechanism, Egypt. In: 12th EGPC petroleum conference exploration, Cairo, pp 1–18

Sarhan MA, Safa MG (2019) 2D seismic interpretation and hydrocarbon prospects for the Neogene-Quaternary succession in the Temsah Field, offshore Nile Delta Basin, Egypt. J Afr Earth Sci 155:1–12

Sarhan MA, Safa MG (2017) Application of seismic attributes for detecting different geologic features within Kafr El Sheikh Formation, Temsah concession, Nile Delta basin. Sci J Damietta Fac Sci (SJDFS) 7:26–34

Sarhan MA (2021a) New prospective gas plays in Pliocene sands, offshore nile Delta basin: a case study from Kamose-1 well at north Sinai concession, Egypt. Petrol Res 10:1016

Sarhan MA (2021b) Assessing hydrocarbon prospects in Abu Madi formation using well logging data in El-Qara field, Nile Delta Basin, Egypt. J Petrol Explor Prod Technol 11:2539–2559

Sarhan MA (2021c) Gas-generative potential for the post-Messinian megasequence of Nile Delta basin: a case study of Tao field, north Sinai concession, Egypt. J Pet Explor Prod. https://doi.org/10.1007/s13202-021-01354-4

Schlumberger (1984) The essentials of log interpretation practice services techniques Schlumberger, France, 58 p.

Schlumberger (1998) Log interpretation charts. Schlumberger Well Services, New York, U.S.A, 152p

Seiffert E, Bown TM, Clyde WC, Simons E (2008) Geology, paleoenvironment, and age of birket qarun locality2 (BQ-2), Fayum depression, Egypt. In John G, Fleagle JG, Gilbert CC (eds) Elwyn Simons: a search for origins. Springer Science + Business Media, LLC

Serag El-Din H (1989) Geological and hydrogeological studies on the Qauternary aquifer. Ph. D. dissertation, Faculty of Science, Mansoura Univ.ersity

Sestini G (1995) Egypt. In: Kulke H (ed) Regional petroleum geologyof the world, part II, Africa, America, Australia and Antarctica (Beiträge zur regionalen Geologie der Erde 22. Gebrüder Borntraeger Verlagsbuchhandlung, Stuttgart, pp 66–87

Sestini G (1989) Nile Delta: a review of depositional environments and geological history Geolog. Soc London Spec Publ 41:99–127

Shaaban H, El-Qady Q, Al-Sayed E, Ghazala H, Taha AI (2016) Shallow groundwater investigation using time-domain electromagnetic (TEM) method at Itay El-Baroud, Nile Delta, Egypt. NRIAG J Astron Geophys 5:323–333

Shukri NM (1953) The geology of the desert east of Cairo: institute desert Egypte. Bulletin 3:89–105

Stanley DJ (1990) Recent subsidence and northeast tilting of the Nile delta, Egypt. Mar Geol 94:147–154

Stanley DJ (2005) Submergence and burial of ancient coastal sites on the subsiding Nile delta margin, Egypt. Méditerranée 1(2):65–73

Stanley DJ, Corwin KA (2013) Measuring strata thicknesses in cores to assess recent sediment compaction and subsidence of Egypt's Nile Delta coastal margin. J Coast Res 29(3):657–670
Stanley DJ, Toscano MA (2009) Ancient archaeological sites buried and submerged along Egypt's Nile Delta coast: gauges of Holocene delta margin subsidence. J Coast Res 25(1):158–170
Stanley DJ, Warne AG (1997) Holocene sea-level change and early human utilization of deltas GSA. Geol Soc Am Gas Today 7(12):1–7
Stanley DJ, Mcrea JE, Waldron JC (1996) Nile Delta drill core and sample database for 1985–1994: mediterranean basin (MEDIBA) program. Smithsonian Institution Press, Washington, D.C.
Stanley DJ (1988) Subsidence in the northeastern Nile Delta: rapid rates, possible causes, and consequences. Science 240:497–500
Strougo A (1985a) Eocene stratigraphy of the eastern greater Cairo (Gebel Mokattam-Helwan) area. Middle East Res Center Ain Shams Univ Sci Res Ser 5:1–39
Strougo A (1985b) Eocene stratigraphy of the Giza Pyramids plateau. Middle East Res Center Ain Shams Univ Sci Res Ser 5:79–99
Strougo A (1986) Mokattam stratigraphy of eastern Maghagha-El Fashn district. Middle East Res Center Ain Shams Univ Sci Res Ser 6:33–58
Strougo A (1988) Mokattamian Bivalvia of greater Cairo: an annotated list. Earth Sci Ser Middle East Res Center Ain Shams Univ Cairo 2:29–48
Strougo A, Haggag MAY (1984) Contribution to the age determination of the Gehannam Formation in the Fayum Province, Egypt Neues Jahrbuch für Geologie und Palaeontologie. Monatshefte 1984:46–52
Swedan AH (1992) Contrinbution to the geology of Fayoum area, vol 34. Ph.D. Thesis. University of Michigan Press, Ann Arbor, p 84
Uhen MD (2004) Form, function and anatomy of Dorudon atrox, Mammalia, Cetacea): an archaeocete from the middle to late Eocene of Egypt, vol 34. University of Michigan Papers on Paleontology, p 222
Underwood CJ, Ward DJ, King C, Antar MS (2011) Shark and Ray fauna in the Middle and Late Eocene of the Fayum area, Egypt. Proc Geol Assoc 122:47–56
Vondra CF (1974) Upper Eocene transitional and near-shore marine Qasr el Sagha Formation, Fayum Depression, Egypt (Essays on African paleontology). Ann Geol Surv Egypt 4:79–94
Wanas H (1991) Sedimentological studies of the Middle Eocene rocks, Fayium depression, Western Desert, Egypt. M.sc.Thesis, Menoufia University
Wing SL, Tiffney BH (1982) A paleotropical flora from the Oligocene Jebel Qatrani formation of northern Egypt: a preliminary report. Misc Ser Bot Soc Am 162:67 (Bown TM, Kraus MJ, Wing SL, Fleagle JG, Tiffney BH)
Zaghloul ZM, Gheith AM, Abdel-Daiem AA, El Asmar HM (1989) The pattern of sedimentation of the Quaternary Nile Delta, Egypt. El Mansoura Sci Bull 16(2):393–413
Zaghloul ZM, Taha AA, Hegab O, E1-Fawal F (1979) The Plio-pleistocene nile subenvironments stratigraphic section and genetic class. In: The fifth conference on African geology, geological survey, Egypt, Cairo
Zaghloul ZM, Taha AA, Hegab O, El Fawal F (1977) The Neogene-quaternary sedimentary basins of the Nile Delta Egypt. Egypt J Geol 21:1–19
Zalmout IS (2008) Late Eocene Sea cows (Mammalia, Sirenia) from Wadi Al Hitan, the Fayium Basin, Egypt. Ph.D. Thesis, The University of Michigan, USA

Evolution of the Nile River and Depositional History of the Fayoum Nile Delta

3

Mohamed Abdelghany Khalifa

Abstract

This chapter provides further details on the geological information on the evolution of the Nile River and the geological history of the rock units that make up the Fayoum Nile Delta. Since the Cretaceous Period or before, the rifted Ethiopian Mountain was believed to be the site of the Nile River's origins. Originating from the Ethiopian Mountains, the Nile River carved its way across Sudan and Egypt, spanning from the Cretaceous to the lower Eocene, until it reached the west of Beni Mazar, nearly at latitude 27 50 N (now Minia Governorate). At this place and during the Middle Eocene, the Nile River moved westward and invaded the Fayoum Province and was carved through the Middle Eocene El Gharaq Formation and proceeded northwestwards, depositing the Middle Eocene Gehannam and Qaroun formations, Miocene El Khashab and Moghra formations, creating a fan-shaped delta oriented northwest direction (Fayoum Nile Delta) in the northwestern desert. A prodelta is where the Gehannam and Qaroun formations were deposited. During the Late Eocene time, the sea level declined, and subsequent northward regression resulted in the deposition of the Qasr El Sagha Formation in the delta front setting. In the late phase of the Fayoum Nile Delta, the Qatrani, El Khashab, and Moghra formations were deposited in the delta plain and/or at a riverine subaerial delta setting. The concentration of different vertebrate species at Wadi El Hitan and their absence in the equivalent rock units in the Eastern Desert (Qarara, El Fashn, and Beni Suef formations) and the Western Desert (Hamra Formation, Bahariya Oases) indicate that this region was the site of debouching the water of the Nile River. The concentration of vertebrates at the Fayoum Province in

M. A. Khalifa (✉)
Geology Department, Faculty of Science, Menoufia University, Shiben El Kom, Egypt
e-mail: khcycle1949@gmail.com

M. A. Khalifa and K. S. Gemail (eds.), *Geology of the Nile Deltas and Natural Resources*, Deltas of the World, https://doi.org/10.1007/978-3-031-96594-4_3

general and Wadi Hitan in particular, west of the Fayoum Province, may be attributed to the enrichment of the River Nile sediments with nutrients, enhancing the growth and recovery of these animals.

Keywords

Evolution • Nile River • Ethiopia • Wadi El Hitan • Fayoum • El Gharaq • Gehannam • Qaroun • Qasr El Sagha • Qatrani • El Khashab • Moghra

3.1 Introduction

The initial research on the evolution of the Nile River was carried out by some eminent geologists, e.g., Issawi (1976), Said (1981, 1993), and McCauley et al. (1982, 1986). Additional research on the Nile River's evolution was done by Williams and Adamson (1982), Issawi and Osman (2008), Talbot and Williams (2009), Woodward et al. (2015), Abdelsalam (2018), and Fielding et al. (2017). Despite the Fayoum (or Fayoum in some sources) region being challenged by numerous geologists in different specialized topics (stratigraphy, sedimentology, vertebrate and invertebrate fauna) over the past decades, no one has mentioned the presence of river sediments in the Fayoum region. In the current work, we introduced the term Fayoum Nile Delta to describe the fluvial sediments that constitute the Middle Eocene Gehannam, Upper-Middle Eocene Qarun and Upper Eocene Qasr El Sagha formations, Oligocene Qatrani Formation, the Lower Miocene El Khashab, and Moghra formations. In this work, these rock units were described and interpreted in detail, based on their lithological types, cyclic pattern, and faunal association. Our main concern in this investigation was the geological setting and the evolution of the River Nile, the depositional histories of the Fayoum Nile Delta (Oligo-Miocene).

3.2 Evolution of the Nile River

Numerous theories have been put forth to explain when the Nile River started to bring water into Egypt from the highlands of Ethiopia.

(1) For instance, the existence of Nubian Sandstone gravels (close to the second cataract) in a straight depression parallel to the main axis of the modern Nile Valley (Maley 1991). This suggests that the original Nile River was younger than the Upper Cretaceous Nubian Sandstone, which dates back up to 100 million years.
(2) Pik et al. (2003, 2008) mentioned that the high Ethiopian plateau had existed since the Oligocene and has a major impact on the hydrology of the Tekezze/Atbara and Abbai/Blue Nile River systems.

(3) Some authors (Said 1981; Gani et al. 2007; MacGregor 2012; Fielding et al. 2018) suggested that the Nile River was initiated during the Oligocene Epoch. (4) Shukri (1950) and Issawi et al. (2009) asserted that the Nile River's origins date back to the Pleistocene.

(4) Also, in this context, Issawi and McCauley (1992) asserted that Egypt was not drained by a single master stream during the Cenozoic Era, but rather by at least three distinct main drainages (Fig. 3.1). The first distinct drainage is the Gilf system, which originated in the Southern Western Desert and flowed northwards., The second drainage system is the Qena drainage system which was created during the Oligo-Miocene epochs by strong tectonic activity. In this context, they claimed that the Qena System predates the River Nile because the Qena River originated in the southern Galala and Red Sea highlands after the Red Sea rift (see Fig. 3.1).

The third Nile drainage system, which stretches from south to north, developed during the Late Miocene epoch. Issawi and Osman (2008) suggested that there was no Nile till the end of the Miocene, but many major rivers were flowing over Egypt older than the Nile River. The latter opinions of Issawi and McCauley (1992), Issawi and Osman (2008), and Issawi et al. (2009) are strongly disputed by several geologists who have studied the evolution of the Nile River, Said (1981), Salem (1976), Fielding et al. (2018). For

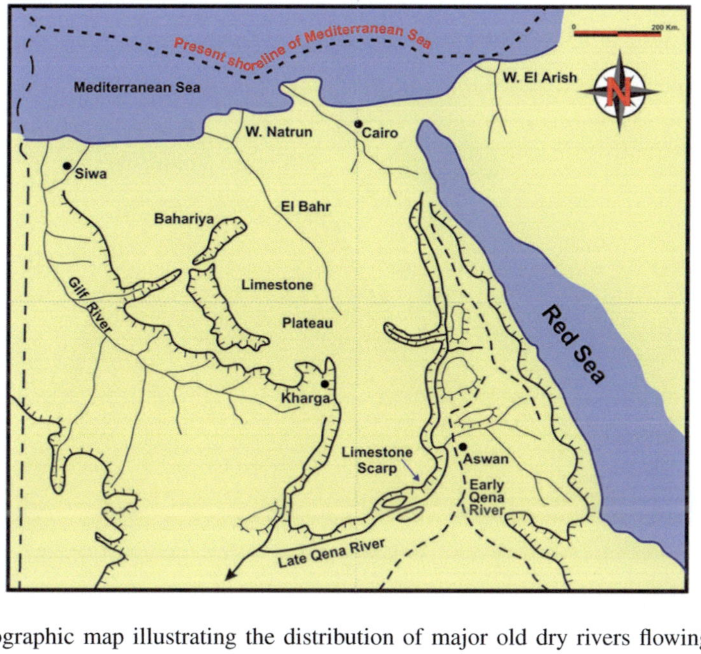

Fig. 3.1 Geographic map illustrating the distribution of major old dry rivers flowing over Egypt during the Cenozoic Era as suggested by Issawi and McCauley (1992)

instance, Fielding et al. (2018) analyzed certain samples from the Nile Delta sediments from the Oligocene to the Pleistocene epochs, using Sr–Nd, zircon U–Pb, and Hf. They concluded that the Nile River originated approximately 30 million years ago, during the Oligocene epoch.

Many factors are responsible for the evolution of the Nile River, the most important of them are the regional tectonics that affected the Ethiopian Mountains, the topographic relief of the Ethiopian Mountains to the southern Sudan, the structures that occurred on the Nile River route, and the formation of the Red Sea and the East African Rift System, all had an impact on the evolution of the Nile River, Talbot and Williams (2009), Woodward et al. (2015). In the present work, we will explain the above-mentioned factors.

(1) The regional tectonics that affected the Ethiopian Mountains is the rifting and dislocation of the Ethiopian Mountains that were most likely rifted during the Jurassic-Cretaceous periods (199–70 M.yrs) (Fig. 3.2). These rifting processes were responsible for the separation of the Pangea into certain continents and even formation of Atlantic Ocean. This viewpoint is consistent with that of Maley (1991), who proposed that the ancient Nile River was formed 100 million years ago.
(2) The Ethiopian Mountains were rifted toward the west and northwest directions, creating a lowlands side (see Fig. 3.2). If the Ethiopian Mountains rifted easterly or southeasterly, the collected rainfall might have flowed into the Red Sea. In such a case, there was neither the Nile River nor the Fayoum Nile Delta, nor even the Mediterranean Nile Delta. The Ethiopian Mountains were rifted toward the west and northwest directions, creating a lowlands side (see Fig. 3.2).

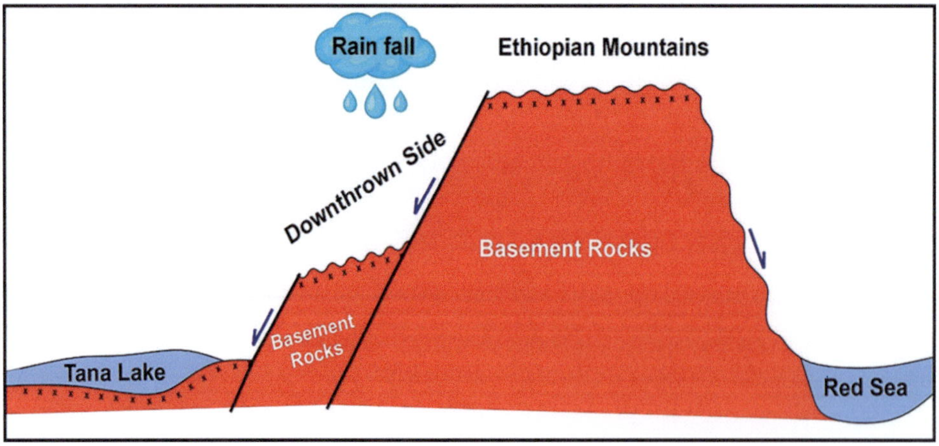

Fig. 3.2 Schematic diagram showing the initial refitting of the Ethiopian Mountains toward the southwest and northwest directions, creating lowlands on which the rainfall collected in the Tana and Victoria Lakes

(3) The topographic relief is important for the enhancement of water flow of the collected rainwater at the foot slope of the Ethiopian Mountains and along the route of the Nile River from Sudan to the Mediterranean Sea. Also, this factor is responsible for the erosion and transportation of the crushed rocks after the carving of the Nile water in the underlying rocks. In addition, the high relief can enhance the accumulation of rainwater in the southwestern regions, leading to the creation of Tana and Victoria Lakes (see Fig. 3.2).

The collected water in the previously mentioned lakes subsequently creates the white and Blue Nile branches. The name of the white-colored water in the White Nile branch may be attributed to the clear color of this water, the wide horizontal spread of the water, the shallow water depth, and the decontamination with fine detrital sediments. Conversely, the blue coloration of the Blue Nile Branch may result from several factors, including the deep water trapped in the channel, the settling of fine sediments, and the absorption of blue light by the water's surface. It is believed that during the Early Cretaceous, the White Nile, originating from Lake Victoria, merged with the Blue Nile, which comes from Lake Tana, to form the main Nile River near Khartoum, the capital of Sudan. Geological structures, such as faults, joints, and shear zones, play a crucial role in determining the orientation of the Nile River. From north of Khartoum to south of Qena, the Nile has cut through a shear zone running in a north-south direction, which is covered by Nubian Sandstone. During the Late Cretaceous, the Nile gradually moved northward, cutting through the shear zone on the basement rocks and subsequently eroding the Nubian Sandstone, as well as the Quseir, Duwi, Dakhla formations, and Tarawan Chalk in southern Egypt. In and around the Qena region, the Nile River changed to nearly east-west or north-northeast directions along the east-west strike-slip faults. From Qena up to the latitude of Beni Mazar (Minia Governorate, Latitude, 27–50 N), the River Nile directed north-south and path westwards down through normal fault and was carved through the Paleocene Esna, Lower Eocene Thebes, Middle Eocene rocks (Minia, Samalut, Midawara, and Sath El Hadid formations). The passing of the Nile River along a westward downthrown may raise an important question: Why did the Nile River enter the Fayoum Province at first and form the Fayoum Nile Delta? At the final stage at latitude 27–50 N, it was switched westward, cutting through the Middle Eocene El Gharaq Formation (nummulitic limestone) and entering the Fayoum depression. From the above-mentioned evolution stages, it is concluded that the initiation of the Nile River ancestors had been formed after the rifting of the Ethiopian Mountains in the Cretaceous Period and entered the Fayoum region after the deposition of the Middle-Middle Eocene El Gharaq Formation that was carved by the Nile River.

3.3 Historical Sedimentation of the Fayoum Nile Delta

Most geologists, if not all, believed that the well-known Nile Delta is the only delta formed by the Nile River. Nevertheless, little is known about the old Nile Delta near the Fayoum Province without clear and defined boundaries, such as Salem (1976), Said (1990), Sestini (1995), and Underwood et al. (2013). They stated that this delta originated during the Oligocene time, 30 million years ago. The current study confirms that the first and oldest delta (Fayoum Nile Delta) was formed in the northwestern desert, covering the Fayoum Province and its environs during the Late-Middle Eocene (more than 40 M.yrs.). When the Nile River turned northwest at latitude 27–30 N (in the latitude of Ben Mazar, Minia Governorate), sedimentation of this delta started. It was preceded in the same manner until Wadi Rayan, where it enters the southern portions of the Fayoum, and then it continues northwestwards until it reaches the southern Qattara Depression (Fig. 3.3). The Middle Eocene Gehannam Formation, Upper-Middle Eocene Qaroun Formation, Oligocene Qatrani Formation, and Lower Miocene El Khashab and Moghra formations were deposited after the carved the Lower Middle Eocene El Gharaq Formation (nummulitic limestone) (Fig. 3.3).

3.3.1 Depositional History of the Gehannam Formation

A few geologists talked about the Gehannam Formation's depositional settings. For instance, the Gehannam Formation was deposited in offshore open-sea environments (Gingerich 1992). The Gehannam Formation's depositional environment was interpreted by Abdel-Fattah et al. (2010) as being distal to an open-sea setting. The Gehannam Formation was deposited in a delta front setting (El Fawal et al. 2011). The Gehannam Formation

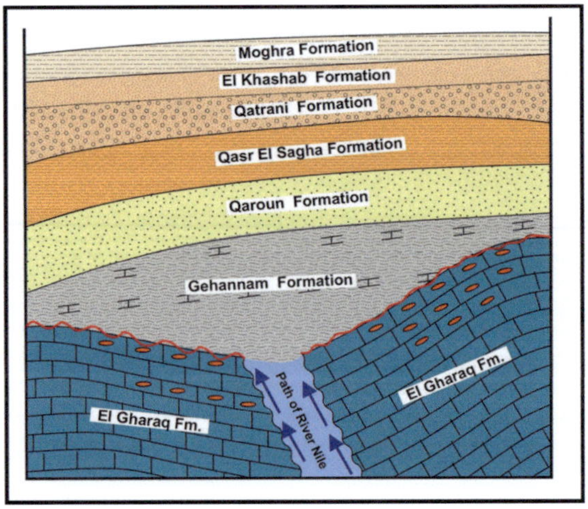

Fig. 3.3 Schematic diagram manifesting the initiation of the entry of the ancestor of the Nile River to the south of the Fayoum depression, cutting through the Middle Eocene El Gharaq Formation, resulting in deposition of the fluviomarine facies of the Upper-Middle Eocene Gehannam, Upper-Middle Eocene Qaroun, Upper Eocene Qasr El Sagha, Oligocene Qatrani, and the Lower Miocene El Khashab and Moghra formations

manifests lateral facies changes from the southwest to the northeast (see Fig. 3.4). Burrowed sandstone makes up the formation at the southwest of Wadi Hitan, while at Qaret Gehannam, Minqar El Hut, and Sandouk El Bornita, it comprises burrowed sandstone, siltstone, and claystone beds, making up the majority of the formation. Further eastwards at Jebel Lahoun and Jabel Naaloun, it consists of an intercalation of marl, mudstone, and marly limestone (Fig. 3.4). The following possibilities are the cause of these facies changes.

(1) Wadi Al-Hitan was considered the first place entered by the Nile River from the southeast and received sandstone sediments from the Nile River,
(2) Wadi Hitan was regarded as a structural high where strong currents and waves caused the finer sediments to winnow away, leaving the pure sandstone at Wadi Hitan, and
(3) The concentration of the marl and limestone at Jabel Lahoun and Jabel Naaloun indicates that such areas were structurally low and far from the mouth of the Nile River.

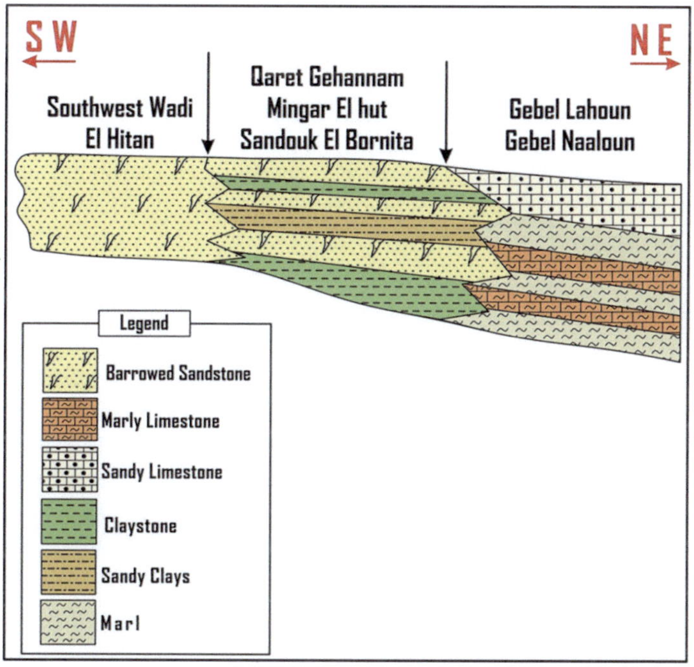

Fig. 3.4 Schematic cross-section showing the lateral and vertical facies changes of the Gehannam Formation from pure burrowed sandstone at Wadi Hitan, intercalation of burrowed sandstone, marl, claystone, and thin limestone beds at Qaret Gehannam, Minqar El Hut, and Sandouk El Bornita, to an intercalation of marl, mudstone, and marly limestone at Jabel Lahoun and Jabel Naaloun

A clear picture of the Gehannam Formation's depositional setting can be obtained from the measured sections at the Jabel Naalon and Jabel Lahoun areas (Fig. 3.5). The vertical stacking pattern of lithofacies in these sections closely resembles multiple superimposed shallowing-upward cycles (Fig. 3.5). Each starts with marl followed by laminated mudstone, which is capped by marly limestone. The marly lithofacies are enriched with whale fragments and represent a combination of siliciclastic sediment (clay and silt) and carbonate ooze, Scholle et al. (1983); James and Jones (2015). James and Jones (2015) mentioned that fine-grained sediment like marl can settle slowly in lakes and open waters. In addition, marl enriched with vertebrate remains might be found in river sediments transported to marine or lacustrine basins (Flugel 2004). The laminated mudstone occurring in the middle part of the cycle is thought to have been deposited in a tidal flat and is an indicator of variable tidal flow and sediment fallout during slack water (Reineck and Singh 1975). The lack of trace and body fossils preserved in the laminated mudstone may be due to freshwater influence on the tidal channels (Buatois et al. 2002). These mudstone lithofacies were mainly deposited out of suspension in a calm body of water at a relatively slow rate of deposition (Buatois et al. 2002).

The marly limestone capping cycles are normally composed of claystone and calcium carbonate. Most claystone is composed of kaolinite varieties (Wanas 1991). Calcium carbonate accumulated as an inorganic precipitate on the bioturbated, well-oxygenated lake bed (Kontopoulos 1991). The fragments of whales discovered in the marly lithofacies were probably carried there by southerly flowing Nile River water. The steadily declining marl and mudstone content and rising detrital sandstone content point to a progressive drop in water depth (Tucker and Wright 1990; Rehman et al. 2016). This phenomenon is present in the eastern regions of the Fayoum District, particularly in Wadi Hitan and Wadi El Hut.

The shallowing-upward cycles in the Gehannam Formation can be amplified by tectonics, sediment supply, and sea level fluctuations. These cycles represent mixed fluvial and marine facies. The river sediments (marl and claystone) that usually occur at the bottom of the cycle are enriched with vertebrate remains. This proves that these sediments were deposited by rivers that usually carried vertebrate fossils. The kaolinitic minerals composing the claystone (Wanas 1991) indicate that this mineral was derived from the continent and transported by river to the site of accumulation. The marine sediments (limestone) occurring at the top cycles indicate the inundation of marine water. The presence of planktonic foraminifera in limestone is considered the greatest evidence for this suggestion. The mixing of fluvial and marine environments in the Gehannam Formation explains that the marine environment was the prevailing environment and mixed with the Nile River water.

On top of the Gehannam Formation, the presence of mangroves at Wadi Hitan, west of the Fayoum (El Sadawi 2005), is suggestive of intermittent exposure and subaerial substrate during deposition (Dolan 1989; El Sadawi 2005). Mangroves can be formed as a result of the decrease in sea level (Dolan 1989). The root traces observed in the upper

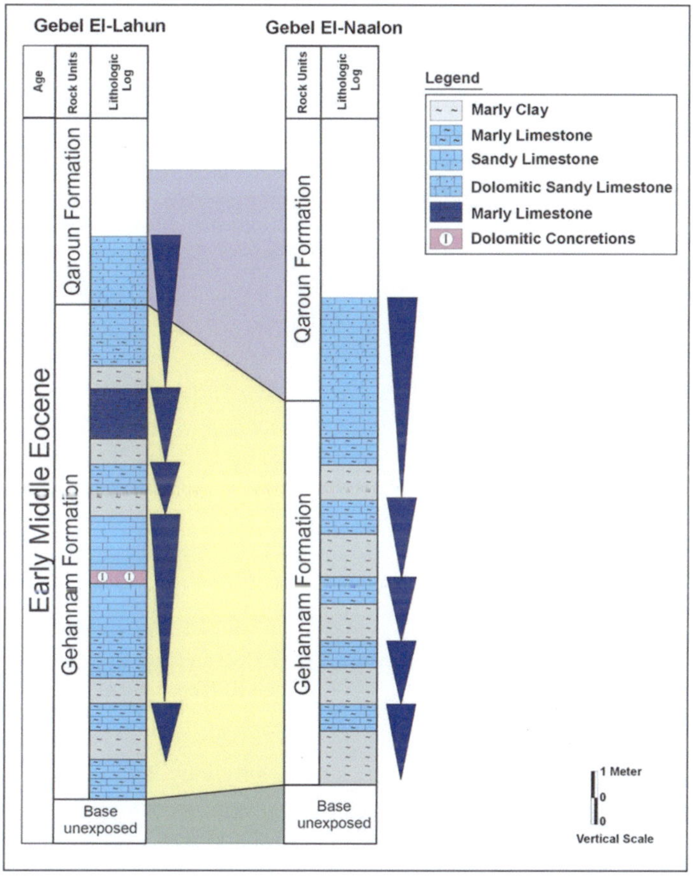

Fig. 3.5 Vertical staking pattern of the Gehannam Formation, exhibiting shallowing-upward cycles at Jabel Lahoun and Jabel Naaloun

parts of the sandstone facies at Wadi Hitan provide additional evidence for the subaerial exposure. El Sadawi (2005) stated that these types of mangrove roots grew in a combination of fresh and saltwater. The common existence of vertebrate remains, including whale skeletons (Basilosaurus isis, Andrews 1904), fish bones, shark and sawfish (Underwood et al. 2013), and sea cow remains (Gingerich 1992; Zalmout 2008) in the claystone and marly lithofacies indicates that the region was either near a river mouth, with very shallow marine water, or had a mixture of river and marine water. These animals most likely lived near the mouth of the ancient or ancestral Nile River, especially at Wadi Hitan, in the western Fayoum region. These vertebrate species may have increased in number and concentration due to the abundance of food nutrients from the Nile River. Based on the facies characteristics, cyclic types, faunal association, and presence of vertebrate remains, it is suggested that the possible depositional setting of the Gehannam Formation seems to

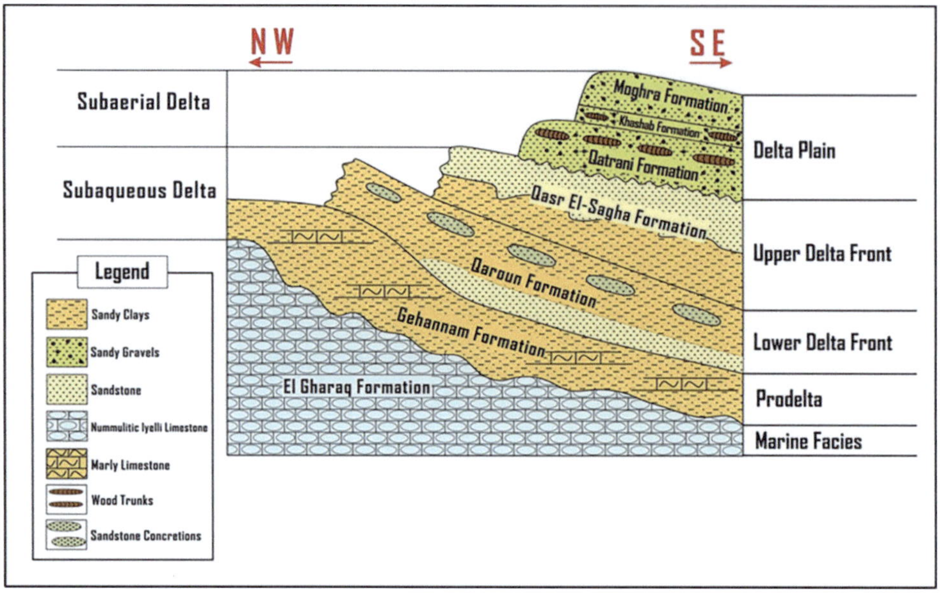

Fig. 3.6 Schematic diagram illustrating the depositional setting of the Middle Eocene Gehannam (prodelta), Qaroun formations (lower delta front), Upper Eocene Qasr El Sagha Formation (upper delta front), Oligocene Qatrani Formation, Lower Miocene El Khashab and Moghra

be prodelta (subaqueous delta) (Fig. 3.6). The silt, silty clay, and mud of the Gehannam Formation are closely similar to the sediments that were probably deposited beneath the wave base, traveling to the deep lacustrine mudstone (Ward and Ashley 1989). It was believed that silt-to-sand-sized particles make up the majority of the prodelta sediments in the majority of the delta, whereas clay-to-silt-sized particles predominate in the delta front and river or tidal channel sediments (Hori and Saito 2007). Benthonic and uncommon planktonic foraminifera occurring in the Gehannam Formation are characteristic of the prodelta setting (Fig. 3.6). This is comparable to benthonic foraminifera that were collected in the Gulf of Patti (Tyrrhenian Sea, Italy, NE Sicily), where a prodelta developed in front of the Mazzarrà River due to high rates of sedimentation.

3.3.1.1 Depositional History of the Qaroun Formation

The Qaroun Formation exhibits radical facies changes from southwest to the northeast (Fig. 3.7). Only burrowed sandstone associated with iron oxide granules makes up the Qaroun Formation at southwest Wad Hitan, while at Qaret Gehannam and Sandouk El Bornita, the thick sandstone sequences were divided into sandstone and claystone units (Gingerich 1992; Underwood et al. 2013) (Fig. 3.7). The sandstone concentration in southwest of Wadi Hitan may indicate that the area was the site of the overflow or pouring of the ancestors of the Nile River (2); the second possibility is that place was structurally

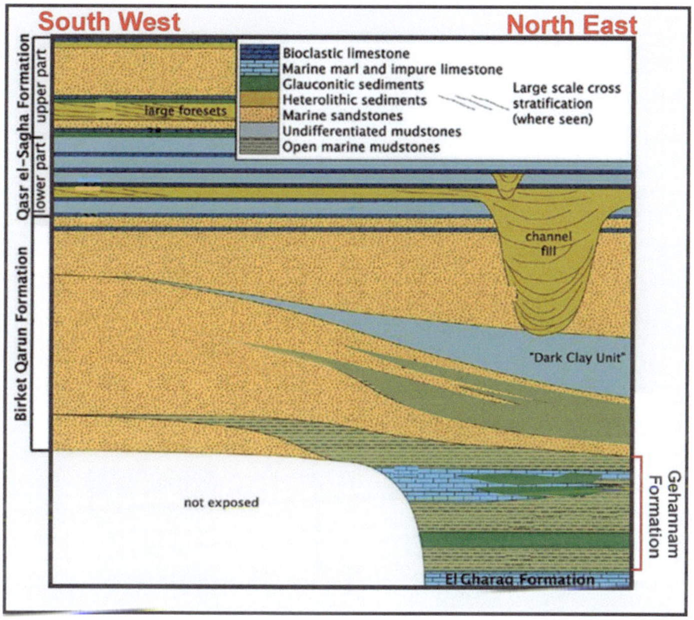

Fig. 3.7 Schematic cross-section showing the lateral facies changes of the Qaroun Formation from southwest to northeast direction (After Underwood et al. 2013), north of the Fayoum Province

high in which the water was very shallow and subjected to strong winnowing by currents and waves, resulting in moving the claystone eastwards intercalated with sandstone. This phenomenon of facies changes was noticed in the underlying Gehannam Formation.

The detailed measured section of the Qaroun Formation by Seiffert et al. (2008) northeast of Birket Qaroun Lake serves as the basis for important scientific addition for interpreting the depositional settings of this formation (Fig. 3.8). Seiffert et al. (2008) divided the Qaroun Formation into two units, the lower one comprises four cycles, each cycle ranges in thickness from 4 to 6 m., each of which starts with a conglomerate bed at base, followed upward by greenish-gray micaceous shale (Fig. 3.8). The upper unit includes eight fining-upward cycles; each begins with conglomerate bed (limonite pebbles and bioclastic granules), followed by medium- to fine-grained sandstone, sandy claystone, and gray laminated claystone at the top (Fig. 3.8).

The conglomerate beds that constitute the basal parts of most cycles in the lower unit of the Qaroun Formation consist of limonite pebbles and bioclastic granules (Fig. 3.9). Such constituent of the conglomerate was probably transported by agents such as rivers, streams, alluvial fans, or glaciers (Friedman 2003). These agents have enough energy to move and circle the clasts over long distances (Friedman 2003). The conglomerate usually occurs in high-energy environments like rivers, alluvial fans, or coastal areas

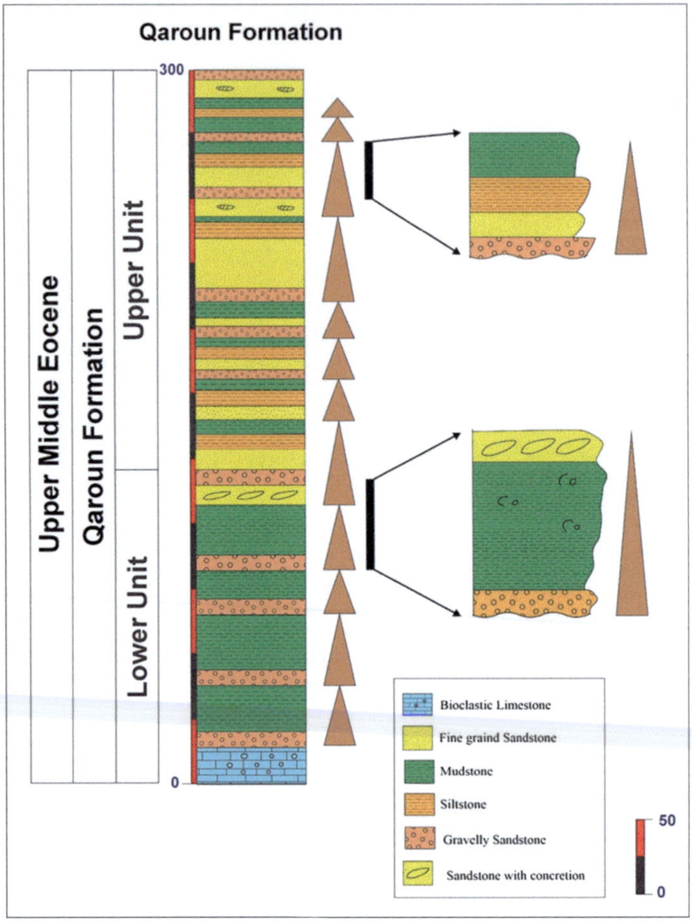

Fig. 3.8 Measured lithostratigraphic section of the Qaroun Formation that includes two informal units. The lower one comprises fining-upward cycles, each of which begins with a conglomerate bed (limonite pebbles and bioclastic granules), capped by greenish-gray micaceous shale. The upper unit also comprises fining-upward cycles; each begins with (1) conglomerate bed (limonite pebbles and bioclastic granules), followed by (2) medium to fine-grained sandstone, (3) sandy claystone, (4) gray laminated claystone (After Seiffert et al. 2008). Notice the usual occurrence of sandstone concretions near the cyclic tops

where the clasts are transported and deposited by water or gravity (Tucker 2003; Friedman 2003). Conglomerates are typically limited to the lowest portion of a channel filled in the sediments left behind by established rivers, where they are referred to as pebble lags (Neuendorf et al. 2005). Because the authigenic granule-pebble-sized limonitic ironstone clasts in the conglomerate are transported from other authigenic deposits, they are considered strong evidence for a fluvial origin (Friedman 2003; Tucker 2003). The thick

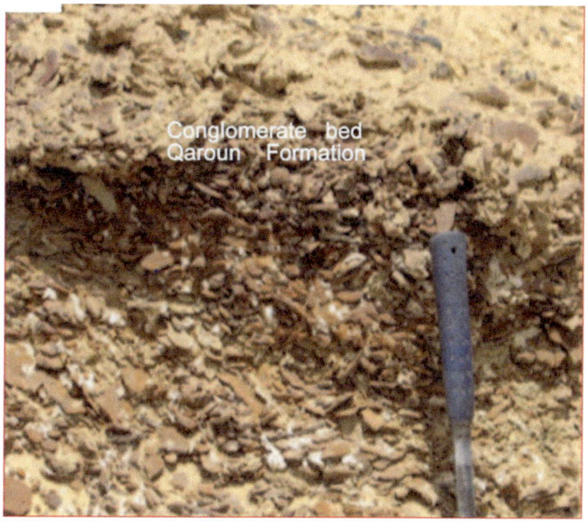

Fig. 3.9 Field photograph manifesting the conglomerate that constitutes the basal cycles and consists of worn fragments of different types of vertebrate fauna, Qaroun Formation, north of Birket Qaroun Lake, Fayoum Province (After Seiffert et al.2008)

mudstone occurring above the conglomerate indicates slag and quiet condition events at the end of the cycle.

The cyclic pattern of the upper unit differs in composition as compared with the lower unit; it comprises conglomerate at the base, followed by fine-grained sandstone in the middle, and sandy clay and laminated claystone at the top (Fig. 3.8). The conglomerates are closely similar to that of the lower unit both in its component and their depositional environments. The fine-grained sandstones in the middle part of cycles usually contain authigenic limonitic ironstone and rhizoliths and are cross-stratification, e.g., medium-scale planar-tabular sets and medium-large-scale trough cross-stratification, with set boundaries commonly containing accumulations of authigenic limonite, a material that also occurs throughout the upper unit (Seiffert et al. 2008). These data indicate a fluvial origin for the fine-grained sandstone. This is supplemented by the abundance of rhizoliths in sandstones and ironstones directly associated with the fossil-bearing beds. Other evidence for a fluvial origin of the fine-grained sandstone is provided by the presence of authigenic limonitic ironstone, which was transported as granule-pebble-sized rip-up clasts from other authigenic deposits. This indicates there was a terrestrial subaqueous source providing granule-pebble-sized particles during floods (Seiffert et al. 2008). The mudstone that makes up the cycle caps in the upper unit is rich in carbonaceous detritus. Such a type of mudstone is of marine environments; however, its occurrence can also occur in fluvial sediments (Seiffert et al. 2008). This evidence is supplemented by abundant carbonaceous debris, which is very rare in purely marine environments.

Elongated to rounded sandstone concretions typically occur at the cycle's top, either in the lower or in the upper units of the Qaroun Formation (Fig. 3.10, Seiffert et al. (2008). They are composed of medium-to-coarse grains of calcareous sandstone that have been bioturbated by closely spaced molluscan burrows. This kind of concretion ranges in size from 50 cm to 1.2. m (Fig. 3.10).

Some concretions are larger, looking like ball structures, ranging from 1.5 m to 2.5 m, can be found in this formation (Fig. 3.11) and consist of calcareous quartzarenite that was deposited in a fluvial channel environment (Seiffert et al. 2008).

Seiffert et al. (2008) found that the axis of these concretions in the Qaroun Formation is directed toward the NW and WNW, which is consistent with the directions that paleocurrent azimuths suggested for stream flow. Wanas (2008) distinguished two types of concretions in the Qaroun Formation: Type-1 occurs within the host sandstones at the

Fig. 3.10 Field photograph showing the oblate to disk-shaped sandstone concretions that occur on the weathered surface of the sandstone bed of the Qaroun Formation, north of Qaroun Lake, Fayoum Province (After Khalaf and Abu El Kheir 2022)

Fig. 3.11 Field photograph showing the large-sized sandstone concretions that derived from the Qaroun Formation, north of the Qaroun Lake, Fayoum Province (Khalaf and Abu El Kheir 2022)

northern scarp of Birket Qarun, while Type-2 exists at Jabel El-Lahoun. Both types of concretions consist of quartz arenite that is cemented by micritic and microspar calcites and deposited in shallow marine environments (Wanas 2008). Contrary to what was stated by Wanas (2008), we believe that there is only one type of sandstone concretion, which typically occurs at the end of cycles. The calcite cement between the quartz grains in the quartzarenite concretions was analyzed, using carbon and oxygen isotopes (Wanas 2008) indicating freshwater percolation on these concretions close to the earth's surface. This demonstrates that the concretions found at the apex of cycles are consistent with intermittent subaerial exposure. Al Dhwadi and Sallam (2019) studied these concretions in the Qarun Formation and gave these concretions the name "cannonballs", comparing them with concretions found in the Qazzun Formation in the Bahariya Oases. They mentioned that such concretions occur at the contact between the Lower Eocene Naqb and the Middle Eocene Qazzun formations, north of the Bahariya Oases. They stated that these concretions are composed of calcareous sandstone and were deposited in shallow marine environments. This comparison is not correct, difficult, and unacceptable due to the following facts: (1) the concretions are typically found within the Qazzun Formation rather than between the Naqb and Qazzun Formations; (2) the Qazzun Formation's concretions are entirely composed of various microfossil associations, such as algae and nummulites, cemented by microcrystalline silica, which are described as patch reefs by Khalifa (1981) in the Minia Formation (equivalent to the Qazzun Formation). (3) Additionally, the Qazzun concretions are dated back to the Lower Middle Eocene, while the Qaroun concretions date back to the Upper-Middle Eocene.

Generally speaking, the hard and compacted sandstone concretions encountered in sedimentary rock or soil are commonly formed by the precipitation of mineral cement between the particles' pores (Oard 2019). These concretions usually form early in the burial history of the sediment, before the rest of the sediment is hardened into rock. Such concretionary cement often makes the concretion harder and more resistant to weathering than the host stratum. The concretions were imitated when dissolved calcium carbonate in pore water of unconsolidated sand precipitated around a nucleus, forming a mass of calcium carbonate that cements the sand grains together, producing a concretion (Davis and Mozley 2005). Detailed studies on the concretions have declared that they can be formed after sediments are buried and before the sediment is fully lithified during diagenesis (Boles and Thyne 1989; Davis and Mozley 2005).

Many braided terrestrial mammals remain, including skate and ray mouthparts, shark teeth, and the teeth of a few marine fish, and are frequently found among the remains of terrestrial vertebrates in the Qaroun Formation. This indicates that such vertebrate fossil remain were derived from terrestrial sources (Bown and Kraus 1988; Peters et al. 2009). According to Bown and Kraus (1988) and Seiffert et al. (2008), these animals most likely originated in brackish water and dispersed throughout continental streams until the freshwater levels became unbearably high. Murray et al. (2010, 2014), who found freshwater teeth in the Qaroun Formation, is proven with the above conclusion. The Qaroun Formation's mammalian fossils were most likely deposited in broad rivulets of larger perennial rivers or close to shallow coastal streams (Bown and Kraus 1988). In this context, Seiffert et al. (2008) stated that the Qaroun strata were terrestrial, alluvial, and considered the oldest fully terrestrial deposits in Egypt since the discovery of the Cretaceous. According to Bowen and Kraus (1988), the Qaroun Formation's high concentration of arboreal monkeys indicates the existence of forested conditions at or very near riverine conditions. From the previously mentioned information, regarding the Qaroun Formation, it was deposited in a subaqueous delta at the lower delta front (Fig. 3.6), because the delta fronts are composed mostly of riverine sands (Davis and Mozley 2005).

3.3.1.2 Depositional History of the Qasr El Sagha Formation

The Qasr El Sagha Formation includes two formal members, the lower is the Temple Member and the upper is the Abu Lifa Member (Vondra 1974) (Fig. 3.12).

The vertical lithofacies stacking pattern of the lower member (Temple Member) consists of fining-upward cycles (Fig. 3.13a). Most of these cycles comprise brown bioturbated glauconitic and fossiliferous calcareous sandstone, followed by gypsiferous siltstone and sandy clays, capped by and carbonaceous gray laminated mudstone facies that usually contain root horizon (Fig. 3.13a). The individual bed is a few tens of centimeters thick

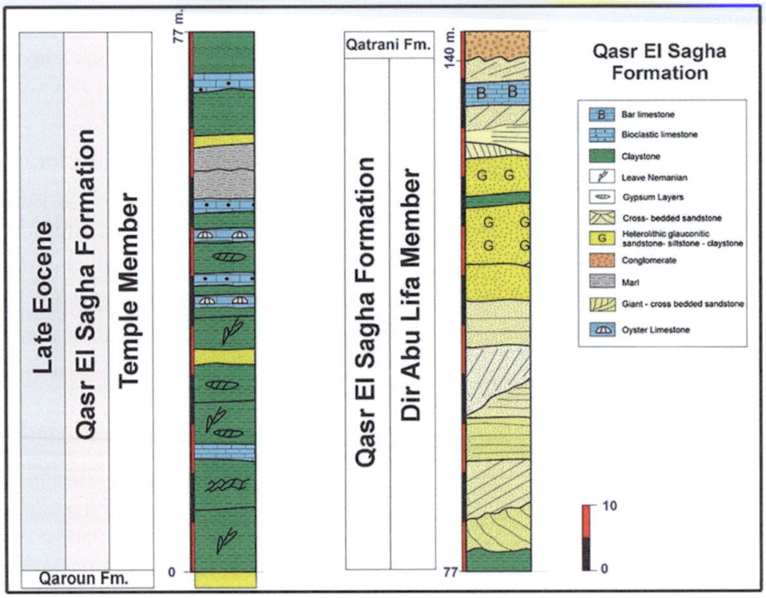

Fig. 3.12 Measured section of the Qasr El Sagha Formation manifesting the cyclic pattern of the lower Temple Member and the upper Dir Abu Lifa Member of Qasr El Sagha Formation, north of Fayoum Province (After Bown and Kraus 1988)

and is sometimes traceable for several kilometers (Bown and Kraus 1988). The glauconitic calcareous sandstone layers are highly fossiliferous with numerous molluscan shells (e.g., *Carolia placunoides* and *Exogyra frassi*) that are represented by pelecypodal-nummulitic wackestone microfacies (Bown and Kraus 1988; El Younsy and Soliman 2021). According to Vondra (1974) analysis, the lowest portion of the Temple Member, which consists of the bioclastic and arenaceous carbonate facies, was deposited in settings ranging from marine bar front to estuarine mouth, where the sea depth was typically less than 20 m. On the other hand, it appears that the upper portion of the Temple Member, which consists of siltstone facies and carbonaceous laminated mudstone, was deposited in shallow lagoons, presumably on the lee side of barrier islands (Vondra 1974). Ghandour (2020) studied the lower Temple Member, which consists of four vertically stacked coarsening-upward facies deposited in lagoonal deltas. He interpreted the strata in the upper part of the member as an incised valley infilled with fluvial, upper estuary, tidal flat, passing upwards into shallow and open marine facies. El Younsy and Soliman (2021) mentioned that the Temple facies were deposited in the intertidal flat.

The glauconitic calcareous sandstone occurring in the basal cycles of the Temple Member is enriched with molluscan shell debris and was deposited in fluvial front to estuarine mouth environments (Walker and Cant 1984a, b). The sandy clay and siltstone occur in the middle cycle overlie the green sandstone and underlie the carbonaceous mudstone range in thickness from 0.3 to 1.5 m and have both sharp and gradational contacts with the surrounding rocks (Fig. 3.13a). Such siltstone is dirty white, gray to black, and yellowish brown with argillaceous or carbonaceous matrix. They are massive, parallel, and cross-laminated burrows (ophiomorpha type), rootlets, loads, and other soft-sediment deformation structures are present. The parallel and cross-laminated siltstones suggest deposition on top of the bars in the abandoned channel and floodplain conditions during the periods of reduced discharge by vertical accretion or shallow water flow conditions (Walker and Cant 1984a, b; Collinson 1986; Collinson and Thompson 1989).

The calcareous mudstone lithofacies, capping the cycles, is enriched with carbonaceous matter, plant roots, plant remains, and coal streaks. Individual lithofacies unit ranges in thickness from 1.5 to 2.5 m. It is mainly massive to parallel laminated, dark gray, reddish brown, or carbonaceous materials. Rootlets and impressions of leaves are commonly present. The parallel lamination suggests that the mudstones were deposited in floodplain environments with frequent variations of energy conditions (Walker and Cant 1984a, b). The thick, dark mudstone associated with carbonaceous matter suggests deposition by vertical accretion in back-swamp or flood basin environments (Flores and Erpenbeck 1981; Walker and Cant 1984a, b; Guion et al. 1995).

Numerous whale remains have been found in the calcareous quartz sandstones and sandy mudstones. These remains have been linked to the bones of fish, turtles, sea snakes, crocodiles, and other vertebrate coprolites, as well as reworked molluscan shells (Bown and Kraus 1988; Gingrich et al. 2011). The concentration of such fossils suggests that they lived either as intertidal lags that occurred slightly offshore or as swash lags on beaches.

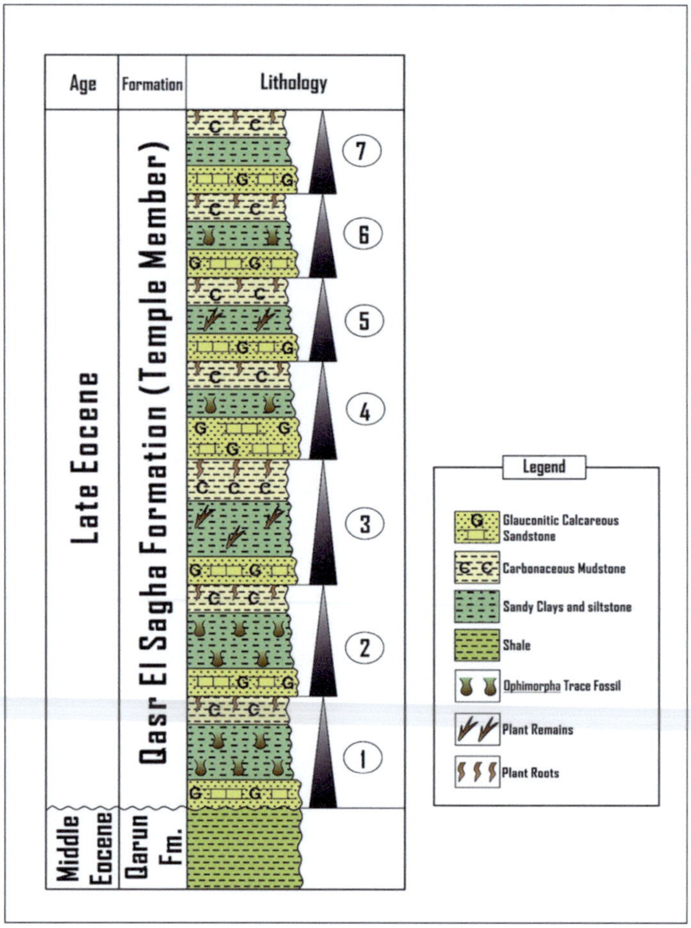

Fig. 3.13 **a** A cyclic pattern of the lower Temple Member of the Qasr El Sagha Formation, Fayoum Province. These cycles exhibit fining-upward, each of which comprises glauconitic calcareous sandstone at base, followed by sandy clay to siltstone with plant remains in the middle and carbonaceous mudstone with plant roots in the upper part. **b** Two types of fining-upward cycles are recognized in the upper Abu Lifa Member of the Qasr El Sagha Formation, Fayoum Province. The first type (cycles, 1, 3, 4, 5) exhibits fining-upward, each of which comprises conglomerate at base, followed by giant cross-bedded sandstone at top, while the second cyclic type comprises The second type of cycles (2, 6) comprises a conglomerate layer at base, followed by green-gold muddy sandstone, and gypseous sandstone at top

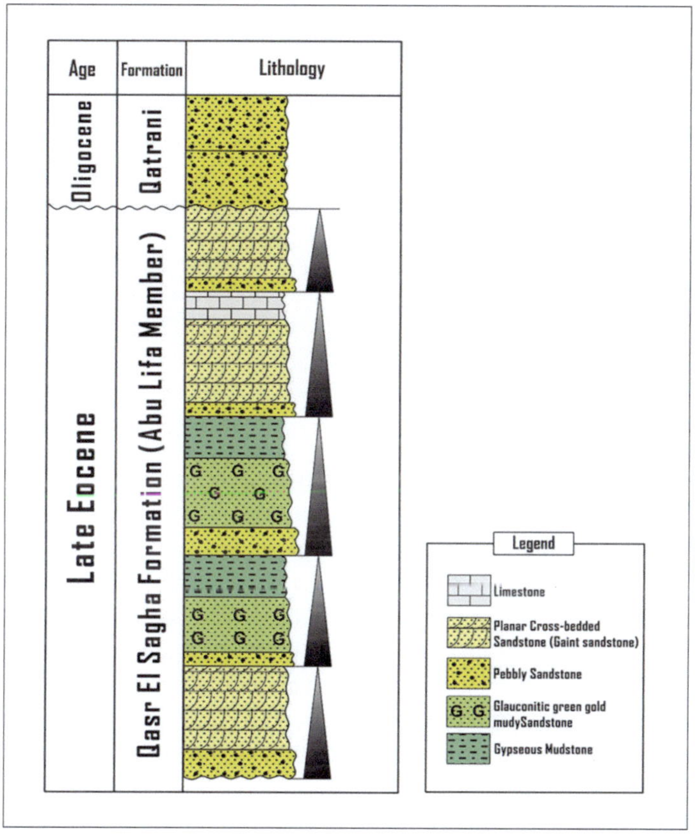

Fig. 3.13 (continued)

The presence of such faunal vertebrate remains is suggestive of near-shore accumulation, most likely caused by virtually unidirectional currents (Bown and Kraus 1988). They were probably accumulated in water depths, most likely less than 20 meters in several locations (Bown and Kraus 1988). In conclusion, even though the Temple Member of the Qasr El Sagha Formation was deposited mostly from near-shore marine environments, it is evident that the coast was never distant from the Fayoum region at any point throughout the Late Eocene time (Bown and Kraus 1981). Recent studies carried out on the palynomorph assemblages in the lower member declared that the lower part of the member was deposited in a lowland riparian environment with continued freshwater influx (Selim et al. 2018). In the current work, we suggest that the lithological types and vertical cyclic pattern of the lower Temple Member were probably deposited in flood plains in the upper delta front in the subaqueous delta (Fig. 3.6).

The upper Abu Lifa Member exhibits an angular or truncated boundary with the lower Temple Member and comprises two types of cycles (Fig. 3.13b). The first type consists

Fig. 3.14 Field photograph of the giant cross-bedded sandstone that shows angular or truncation of the Upper Abu Lifa Member with the underlying Temple Member, Qasr El Sagha Formation, Fayoum Province

of conglomerate at the base, followed by giant cross-bedded sandstone at the top (cycles, 1, 4, 5) (Fig. 3.13b). The second type cycle comprises a conglomerate layer at the base, followed by green-gold muddy sandstone, and gypsous mudstone at the top (cycles 2,3, Fig. 3.13b). The first types of cycles (cycles 1, 4, 5) are closely similar, and each one comprises a thin conglomerate bed at the base, capped by giant cross-bedded sandstone at top (Fig. 3.13b). The conglomerate ranges in thickness from 20 to 35 cm thick. It has a wide lateral extension (several kilometers) and comprises broken water-worn iron oxide nodules (sometimes containing coprolites), vertebrate coprolites, shark teeth, skate and ray mouthparts, fish bones of all kinds, reptile and mammal bones, teeth, and silicified wood (Bown and Kraus 1988).

The above specification of conglomerate suggests that they probably deposited in channel bases, transported from fluvial inundation. The cross-bedded sandstone forms the upper part of the cycle (cycles, 1, 4, 5) greatly varying in breadth from 40 to 50 m and in depth from 3 to 25 m (Bown and Kraus 1988) (Fig. 3.14). Such type of sandstone exceeds 3 m in thickness; therefore, they are named giant sandstone (McCabe 1977).

The giant cross sets are incised on top of the lower Temple Member above erosion surfaces that appear to be channel scours (Bown and Kraus 1988) (Fig. 3.15). The giant cross-bedded sandstones were described in the Upper Carboniferous rocks of northern England. They usually fill channels with depths up to 40 m and were developed in unusually deep distributary channels or deltaic slope sedimentation (McCabe 1977). In addition, the giant cross sets represent lateral accretion deposits that form within stream channels (Bown and Kraus 1981).

The second type of cycle (2, 3) in the upper Abu Lifa Member consists of a conglomerate layer at the base, followed by green-gold muddy sandstone, and gypseous mudstone at the top (Fig. 3.13b). According to Bown and Kraus (1981), the conglomerate is composed of a variety of materials, including fossil wood, frequently worn bones, tightly packed collections of shark and ray teeth, and abraded pieces of marine mollusks.

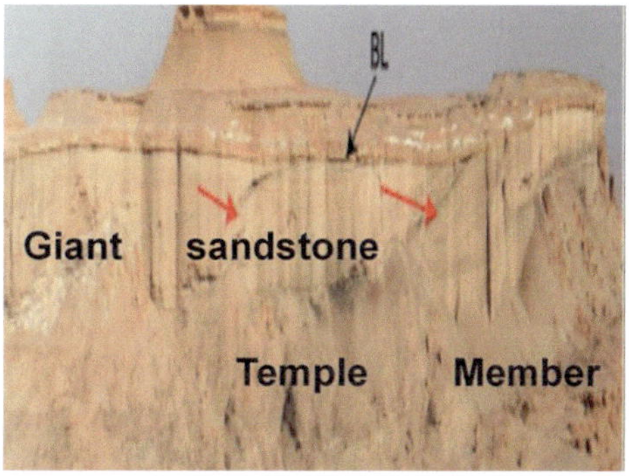

Fig. 3.15 Field photograph of the giant cross-bedded sandstone varying in breadth from 40 to 750 m and in depth from 3 to 25 m, and which appears to be channel scours on top of the Temple Member, the Upper Abu Lifa Member, Qasr El Sagha Formation, Fayoum Province

Furthermore, this conglomerate includes coprolites, abraded and/or rotten fossil wood, remnants of whales and sawfish, and lags from turtle carapaces (Bown and Kraus 1981). These data are indeed consistent with a delta debouching into a bay developed more or less to the north, as suggested by Vondra (1974). The green-gold muddy gypseous sandstone occurring in the middle of cycles is also massively enriched with shrinkage cracks filled with evaporites. Both types of sandstone were deposited in slag water, representing the levee delta (Vondra 1974; Bown and Kraus 1981). The presence of evaporites indicates dryness and subsequent evaporation at the end of cycles. The mudstones capping the second type of cycle are usually green-gray laminated and enriched with worn fossil debris. They were probably deposited in quite lagoonal water at the summit of the cycle.

The limestone lithofacies that occur at the top of cycle 4 are found above the giant cross-bedded sandstone. Beadnell (1905) was the first to describe this limestone as "bare limestone". In case of erosion of the uppermost cycle (cycle 5), such limestone bed became exposed at the summit of the Qasr el Sagha, covering a large region in the northern Fayoum depression. Its unique gray color makes the limestone appear polished and bare due to wind-driven sand polishing. This lithofacies is composed primarily of fossilized shell fragments from animals, which have been cemented with calcium carbonate to resemble calcareous duricrust. It most likely originated toward the end of the Qasr El Sagha Formation in the subaerial zone. This kind of facies is regarded as a diastem or hiatus between depositional cycles. In the absence of the uppermost cycle, this type of limestone can help recognize the upper limit of the Qasr El Sagha Formation. According to Vondra (1974) and Bown and Kraus (1988), the Qasr El Sagha Formation was deposited in a retreating shallow sea that was receiving detritus from the exposed hinterland. The influx of erogenous material became exceptionally high during the late Eocene when large rivers seemed to have discharged into the sea, forming submarine deltas.

Selim et al. (2018) stated that the depositional regime of the upper part in the Abu Lifa Member was swampy and dominated by freshwater. They added that the mixed angiosperms and pteridophytes dominated the woody forest vegetation and were the source for the formation of the Fayoum carbonaceous shale facies. The occurrence of palynomorph assemblage points to the prevalence of a tropical to a subtropical, warm, humid climate with heavy precipitation in the lowland, swampy areas that were near to the freshwater influx (Selim et al. 2018).

The vertically stacked pattern of the Abu Lifa Member (fining-up cycles) declares that they were probably deposited in the fluvial channel depositional setting (Miall 1977, 1978). The existence of fossil lag deposits containing higher proportions of land vertebrates remains in the Abu Lifa Member, emphasizing estuarine, lagoonal, as well as alluvial environments (Bown and Kraus 1988). The above-mentioned faunal associations (terrestrial vertebrates and leaf remains) that were reported in several cycles indicate that they most probably came from the south, where the freshwater of the Nile River came.

According to Bown and Kraus (1981), the source of clastic facies for Qasr el Sagha rocks must have derived from the Paleozoic and Mesozoic clastics as well as basement rocks from the area of El Galala El Bahariya, and Jebel Ataqa, some 125 km east of the Fayoum Depression in the Eastern Desert and adjoining the Red Sea. This assumption is not correct due to the following scientific facts:

The initiation of the uplifting and rifting of the Red Sea took place in the Oligocene time after the deposition of the Upper Eocene Qasr El Sagha Formation.

The uplifting of the Red Sea region may be responsible for providing the clastic rocks of the Qatrani and Dabaa formations and the younger formation (from Miocene up to the Holocene times).

The El Galala El Bahariya is mostly covered by carbonate rocks that can prevent the transportation of clastic derived from the Red Sea Mountains.

Jabel Ataqa consists exclusively of pure carbonate rocks, and there is no exposure of clastic rocks in and around it.

Moreover, the exposure of basement rocks in Southern and Northern Galalas is very limited. We suggest that the clastic sediments (sands, silt, and clays) composing the Qasr El Sagha Formation most probably came from the Nubian Sandstone, south of Aswan, and even from Sudan and the Ethiopian Mountains. They were most probably transported by the Nile River.

This is consistent with the opinion of Kortlandt (1980), who stated that the detrital clastics in the Fayoum region came from a north-flowing early Tertiary Nile River (Protonile). Concerning the depositional setting of the Qasr El Sagha Formation in the delta sequence, it is suggested that this formation was deposited in the upper delta front in the subaqueous delta (Fig. 3.6).

3.3.2 Depositional History of the Qatrani Formation

To decipher the depositional environments of the Qatrani Formation, the detailed measured section by Bown and Kraus (1988) is used herein to through more light on the depositional conditions of the Qatrani Formation. This formation is subdivided into vertical cyclic patterns that include fining-upward depositional cycles (Fig. 3.16). In such cycles, there are complete and incomplete cycles. The complete cycle begins with gravelly sandstone to sandy gravel at the base, followed by reddish coarse-grained sandstone, green sandstone and variegated claystone and paleosol at the top (cycles, 1, 3, 4, 5); the second incomplete cycle comprises sandy conglomerate at the base, followed by trough cross-bedded sandstone at the top (cycle 6).

The gravely sandstone forming the cyclic bases (cycles, 1, 3, 4, 5) ranges in thickness from 15 to 50 m thick. They are composed of white and yellow medium-to-coarse sand with occasional stringers, lenses, and interbeds of gravel. The sandstones are generally friable, but some beds have local calcium-carbonate cement. The sandstone lithofacies are large-scale troughs, tabular cross-stratification, and horizontal stratification, suggesting that they were deposited principally on point bars (Bown and Kraus 1988). However, the sandstone was probably deposited by streams as they have a wide distribution (Blanckenhorn 1900; Beadnell 1905; Said 1962; Kortlandt 1980; Rust and Koster 1984). Such lithofacies are enriched with silicified wood, occurring as logs and smaller fragments.

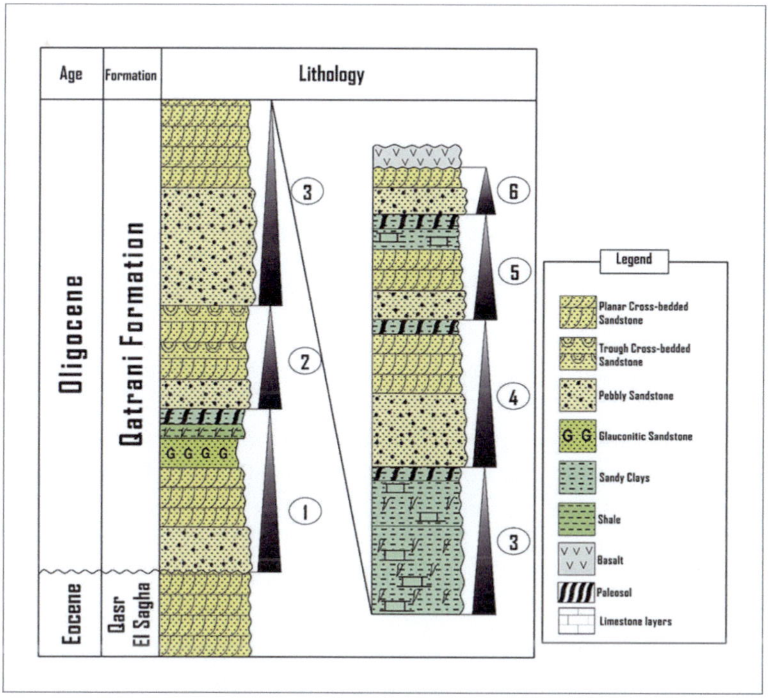

Fig. 3.16 Measured section of the lower Oligocene Qatrani Formation that contains seven fining-upward sedimentary cycles, Jabel Qatrani, north Fayoum Province (After Bown and Kraus 1988 with modification)

Moreover, fossil vertebrate remains, represented by white coprolites, were probably deposited in stream-channel deposits. This suggests that these fossils were derived from continental sources via streams. Paleocurrent measurements on the sandstone show that streams flowed principally to the southwest and west and, more rarely, to the northwest (Bown and Kraus 1988). The above-mentioned flow directions have coincided with the direction flow of the Nile River during its entrance to the Fayoum region.

The silicified woods that usually occur in the gravelly sandstone in the basal first cycles were deposited in a salient environment (Bown and Kraus 1988). The wood trunks were derived from a wetter, more tropical climate in Sudan and were transported to the Fayoum area by the ancestral Nile River ("Proto-Nile") (Kortlandt 1980). Central to the above-mentioned opinion, the woods were derived from in situ forests along Jabel Qatrani streams (Blanckenhorn 1900; Vondra 1967; Bowen 1970, 1971; Bowen and Vondra 1974). Bown et al (1982) stated that the wood does not occur at one specific horizon; rather, it occurs at different levels in the lower part of the formation, suggesting that their transportation was attributed to different types and sizes of streams (Bown et al. 1982).

The red sandstone sequences above gravelly sandstone and below green sandstone or variegated sandy clay (cycles 1, 3, 4, 5) are composed of fine to coarse sand with high amounts of coarse silt and gravel. They are brick red in coloration; however, these units exhibit laterally abrupt color changes from red to green or red to brown (Bown and Kraus 1981). The lithofacies contain large-scale troughs, cross-stratification, and horizontal stratification that grade upward into grouped sets of small-scale trough cross-stratification. The fining and thinning upward in this lithofacies attest to deposition during a regime of decreasing flow energy, typical of channel fills (Chakraborty 1996; Rust and Koster 1984). Poor sorting coupled with very coarse sand- to gravel-sized grains and the absence of bioturbation are the features that suggest tidal influence in the fluvial channel deposition (Bose and Chakraborty 1994; Chakraborty 1996). The usual reddish coloration in such lithofacies declares the oxidation of associated iron minerals in subaerial conditions with the percolation of freshwater.

Paleosols usually occur in the uppermost cycles of the Qatrani Formation (1, 3, 4, 5 cycles) (Fig. 3.16).

The most striking evidence of paleosol formation in the Qatrani sequence is the brilliant vegetation of the sediments and the presence of rhizoliths. Such paleosols persist for several kilometers until they are no longer exposed or have been cut out by erosion. Their color ranges from red and dark red to orange, which reflects differential horizon development related to differing groundwater regimes on local topography (Bown and Kraus 1988). Paleosols are of importance in the paleo-climatic study of the Oligocene deposits of the Fayoum area (Bown et al. 1982). Paleosols in the Qatrani Formation were very moist and formed on floodplains that frequently were inundated by waters from overbank depositional events and were periodically covered by broad areas of standing water. Similar depositional regimes are well known for other alluvial basins in tropical and subtropical areas and in which water is, at least for some part of the year (Richards

1966). We believe that the development of this type of sequence of superposed soils is to be expected in most ancient alluvial sequences (Bown and Kraus 1988). In places, the concentrations of iron and manganese are so great that nodular zones or flaggy duricrust horizons occur.

Rhizoliths present in the paleosols occur as straight to sinuous cylindrical structures formed of more or less well-cemented sediment, with lengths ranging from a few centimeters to several meters. They are also characterized by a circular cross-section, with diameters that range from 0.1 mm to about 20 cm (Fig. 3.17). Rhizoliths are organo-sedimentary structures that have been described from many parts of the world and are produced by the activity and decay of plant roots (Cohen 1984; Jones and Squair 1989; Klappa 1980; Mount and Cohen 1984). Rhizoliths are generally defined as ichnofossils (root traces) or fossils (petrified roots) in geology, taphonomy, and paleontology (Kraus and Hasiotis 2006; Barta 2011). Rhizoliths can be used as environmental markers of humidity, drainage, and local topography (Kraus and Hasiotis 2006). Many factors are responsible formation of rhizoliths. The high aridity and evaporation favor the condensation and crystallization of dissolved carbonates around roots and the formation of rhizoliths (Owen et al. 2008). (2) Rhizoliths can be formed around the vegetation region where water is available as they replace the modern dead roots (Barta 2011).

The second cyclic type in the Qatrani Formation is named incomplete cycles in the current work, as they include only two lithofacies as compared with the complete cycles (Fig. 3.16, cycles 2, 6). They show the fining-upward pattern, each of which begins with massive sandstone, capped by variegated claystone. Over much of this area, the sandstone lithofacies show an erosional contact atop the paleosol (top of the underlying cycle). It ranges from 10 to 13 m thick, but it thins out to about 5 m some 6 km northeast of Widan

Fig. 3.17 Field photograph showing the rhizoliths occurring in the paleosols in the upper cycles of the Qatrani Formation, Fayoum Province (After Khalaf and Abu El Kheir 2022)

El Faras (Bown et al. 1982). Such lithofacies are enriched with ichnofauna, particularly the hundreds of fossil nests of subterranean termites (Bown et al. 1982). These nests are preserved as innumerable chalcedony, quartz, and calcite geodes (Beadnell 1905). The upper part of this cycle is capped by variegated claystone lithofacies that consist of interbedded greenish-gray sandstone, sandy mudstone, marly sandstones, thin limestone beds, carbonaceous shales, and very thin low-grade coals (Bown et al. 1982). Such types of cycles were probably deposited as overbank (floodplain) settings (Bown 1982). Such sediments declare that the possible environment was characterized by abundant rainfall and both locally and periodically high water tables (Bown 1982). Added to the above, the presence of thin freshwater limestone, dolomite, and calcrete suggests that the overall soil environments of the Jabel Qatrani Formation were wetter than those of the subtropical (Bown and Kraus 1988). This condition is closely similar to the Willwood Formation Paleosols from the early Eocene of Wyoming (Bown and Kraus 1981).

The thin carbonate beds occurring within the overbank sediments in the upper part of the Qatrani Formation were probably formed under varied climatic conditions (Dean and Gorham 1975; Sellwood 1978; James 1979). For example, some carbonates in the Qatrani Formation yield abundant freshwater charophytes, indicating deposition in non-saline standing water (Bown et al. 1982). Moreover, another limestone and marly dolomitic limestone, encountered from five localities in the Qatrani Formation, contains brackish water mollusks, freshwater ostracode, and charophytes, suggesting the development of at least periodic large, shallow, low salinity floodplain ponds, and near-shore fluvial setting (Bown et al. 1982). The common occurrence of fossil vertebrates in the Jebel Qatrani fossil supports our contention that the Qatrani Formation paleoenvironment was damp, coastal, and densely vegetated. Poorly preserved gastropods and bivalves recovered from three localities in the lower part of the Jabel Qatrani Formation declared seasonally brackish influence in a dominantly freshwater fluvial system (Bown et al. 1982).

The Qatrani Formation contains diverse fossil vertebrates, e.g., shark teeth, crocodilians, turtles, snakes, lizards, sharks, lungfish, bony fishes, a variety of plants and trace fossils, skates, anthropoid primates, carnivorous mammals, and philomycid rodents (Andrews 1906; Simons 1968; Simons and Bown 1984; Simons and Rasmussen 1990; Kappelman et al. 1992). The above-mentioned vertebrate fossils are considered one of the earnest and most complete records of the continental Tertiary flora and fauna in North Africa (Simons and Rasmussen, 1990; Kappelman et al. 1992). Added to the above, the Qatrani Formation has yielded 13 other orders of mammals, including the only marsupial known from Africa (Bown et al. 1982; Simons and Bown 1984; Gingerich et al. 2011).

The recorded fishes in the Qatrani Formation, e.g., elasmobranch and osteichthyan faunas, indicate that this formation was not deposited entirely in fresh waters but had some marine influence, particularly in the lower part of the formation (Murray et al. 2014). As was shown above, the Qatrani Formation is exclusively fluvial in origin and was deposited by principally meandering streams on a lowland coastal plain (Bown et al. 1982). The aforementioned data (lithofacies types and vertebrate fauna) of the Qatrani

Formation strongly suggest the most suitable environments for this formation are meander streams in the delta plain (subaerial delta) (Fig. 3.6).

3.3.2.1 Depositional History of the El Khashab Formation

This formation unconformably overlies the Widan El Faras basalt and unconformably underlies the Lower Miocene Moghra Formation. It represents the transitional facies between the Oligocene and the Lower Miocene Moghra Formation. This formation consists of gravel and gravelly sandstone, chert pebbles, and numerous wood trunks (Fig. 3.18). They are well-cemented by reddish iron oxides. Petrified wood occurs abundantly in the El Khashab Formation along the October–Bahariya desert road. El Mahdy et al. (2020) mentioned that the El Khashab Formation, occurring along the October-Bahariya road, is composed mainly of loose, moderately to poorly sorted sands and gravel deposited by an ancient river. Fossil logs range in length from 0.5 to 30 m and include a variety of dicot taxa, and two species of palms can exist in the El Khashab Formation (El Sadawi 2005). The presence of wood trunks indicates that there were tropical or subtropical climates dominating during the Oligocene time (El-Saadawi et al. 2014). The sources of wood trunks were most probably derived from the underlying Qatrani Formation during weathering processes. The mode of occurrence of the fossil trunks in the form of fragmented trunks and logs and the absence of other plant remains such as twigs, roots, branches, barks, and soft parts suggest that these fossil woods were not preserved in their situ but transported from where they grew before silicification (El-Saadawi et al. 2014). Moreover, their missing wood trunks indicate that the wood fossilization occurred after long transportation from far areas in the south by the ancestral Nile River (Kortlandt 1980; Salama 2008). The predominance of woods with few, wide vessels is consistent with the warm, humid climate suggested for the early Miocene of the region (El Mahdy et al. 2020). The orientation of the petrified wood shows the water flow direction at the time of deposition (Fig. 3.18)

The reddish nature of El Khashab sediments is due to the presence of iron oxides, which are regarded as the common cementing material filling pores between gravel, sands, and basalt debris. The source of iron oxides in the pebbly sandstone is the siderite mineral that was subjected to oxidation on the surface (Collinson 1996). Displacive siderite nodules are common in peaty, methanic soils that develop within poorly drained, floodplain sediments (Ho and Coleman 1969; Moore et al. 1992). Iron oxide is considered the continental derivation and is also a good indicator of the dry conditions that occur on unconformable surfaces. The deposition of the El Khashab Formation was probably in the floodplain or delta plain setting in the subaerial delta (Fig. 3.6).

3.3.2.2 Depositional Environments of the Moghra Formation

Fourtau (1918) proposed two probable depositional environments for the Moghra clastics: a fluviatile terrestrial origin for the lower horizons, as evidenced by the presence of vertebrate fauna; and a marine origin for its upper horizons, which contain abundant

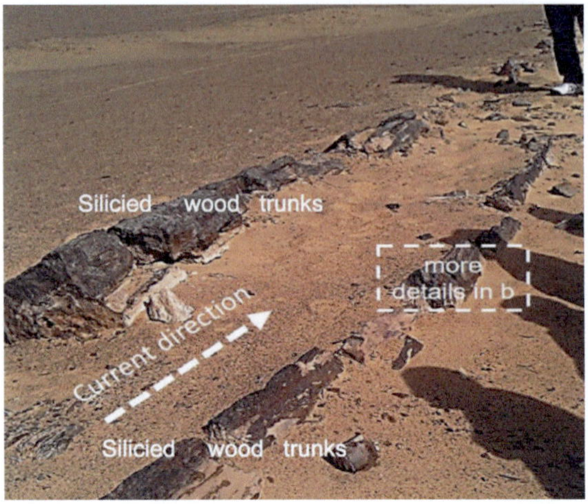

Fig. 3.18 Field photograph showing the silicified wood that is arranged in parallel to each other, along Cairo-Bahariya desert road, El Khashab Formation (After El Mahdy et al. 2020)

marine invertebrates. Added to the above, different depositional environments were advocated for the deposition of the Moghra Formation, e.g., fluvoxamine, semi-continental, and estuarine sediments (Abdallah 1966), fluviomarine sediments (Marzouk 1970), shallow marine to neritic, and as restricted mixed fluviomarine (Khalifa and Abu Zeid 1985). Hassan et al. (2012) stated that this formation comprises several transgressive-regressive units. In addition, Hassan (2013) mentioned that the Moghra Formation was deposited in three environments: tide-dominated estuary, open shelf, and tide-dominated delta estuarine complex. Such environments reflect repeated transgressive to regressive shoreline movements across the Burdigalian (early Miocene) coastal landscape. The measured section of this formation exhibits fining-upward cycles, each of which begins with glauconitic trough cross-bedded and planar cross-bedded sandstone, followed by mudstone enriched with plant remains. Some cycles comprise burrowed sandstone at the base, capped by mudstone at the top. The sandstones are enriched with wood trunks. The cross-bedded sandstone (trough, planar, and tabular) occurring in the basal cycles indicates fluvial channel deposits (Sigdel and Sakai 2016), while the mudstone facies enriched with plant remains at the end of the cycle can represent swampy and levee environments (Khan et al. 1997). The lithological and cyclic types of the Moghra Formation indicate fluvial sedimentation associated with tidal sedimentation processes. The vertical cyclic pattern in the Moghra Formation suggests that the Moghra Formation was deposited under fluvial and delta plain environments (Fig. 3.6).

3.3.3 Depositional Model of the Fayoum Nile Delta

The Fayoum depression attracted the attention of numerous geologists as it holds a rich heritage of paleontological, archeological, and geological exposures such as the Qasr El

Sagha (Temple of the Crocodile), Wadi El Ryan, and Wadi El Hitan (Whale Valley). Of these geologists are: Andrews (1901, 1904, 1906); Beadnell (1901, 1905); Bowen (1970, 1971); Bown and Vondra (1974); Fleagle et al. (1985); Bown and Kraus (1981, 1988); Simons (1968); Simons and Vondra (1982); Simons and Gingerich (1974); Simons and Rasmussen (1990), Zalmout (2008), Dolson et al. (2002); Zalmout et al. 2012; El Anbaawy et al. 2018, and El-Araby and El-Barkooky 2000. The Fayoum depression was formed on top of paleo-folding that probably began in the Middle Cretaceous and continued in subsidence up to the late Eocene time (Shata 1953; Bayoumi and El Gamili 1970). The Fayoum Basin was previously named before as the Fayoum-Gindi Basin, which has a 120 km wide graben filled with 2 km of Eocene sediments (Salem 1976; Zalmout 2008). It lies unconformably on top of the Upper Cretaceous strata. This basin is bounded from the north by El Kattaniya-Abu Rosh highlands, from the west by the Bahariya high or platform, and from the south by the Nashfa anticline south of Wadi El Rayan (Khalifa 1981) (Fig. 3.19). The latter forms the southern edge of the Syrian Arc System. Within the Fayoum basin, the middle and upper Eocene siliciclastic deltaic, lagoonal, and estuarine facies were deposited (Salem, 1976).

The Gehannam Formation (Middle-Middle Eocene), the Qaroun Formation (Upper-Middle Eocene), the Qasr El Sagha Formation (Upper Eocene), the Qatrani Formation (Oligocene), El Khashab, and the Moghra formations (Early Miocene) build up the Fayoum Nile Delta, that named in the current work. The Fayoum Nile Delta has a shape like a fan or an arcuate delta, with a convex curve that faces the Qattara Depression in the northwest (Fig. 3.20).

It is considered an abandoned or deserted delta after the Nile River switched northeastwards, leaving the Fayoum Delta after the Late Miocene time. The first deposited

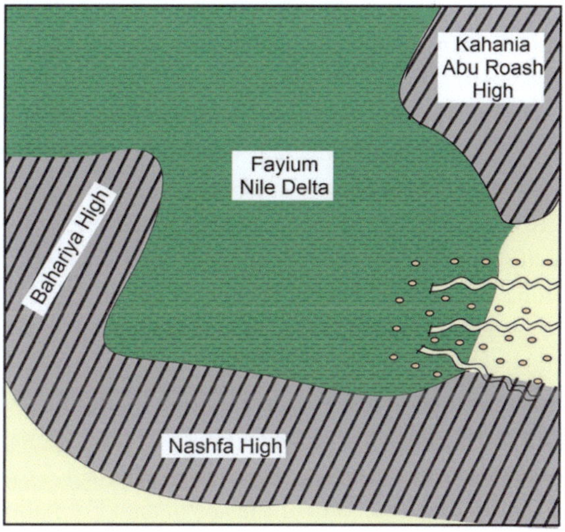

Fig. 3.19 Schematic diagram manifesting the structural setting of the Fayoum depression with the neighboring structural elements, e.g., Nashfa High from the south, Bahariya high from the west, Kattanyia and Abu Roash from the northeast

Fig. 3.20 Landsat image showing the fan-shaped Fayoum Nile Delta, starting west of Beni Mazar (Minia Governorate) and stretching northwestward, covering the Fayoum Province, Wadi El Natroun, Sadat City, El Bahr depression, and the Qattara Depression

formation in the Fayoum Nile Delta was the Gehannam Formation, which was carved on the top of the El Gharaq Formation, forming a canyon that cut several meters in depth and several kilometers wide (Fig. 3.21).

The Gehannam Formation comprises shallowing-upward cycles. It was deposited in mixed marine and fluvial conditions. The marine environment was responsible for sedimentation of marly limestone enriched with benthonic foraminifera, while the fluvial environment was responsible for the deposition of sandstone and claystone associated with vertebrate remains that were derived from the fluvial conditions, coming from south by the ancestor of the River Nile (Fig. 3.21). The top of this formation is capped by mangroves that were deposited in mixed marine-freshwater (or brackish water). This indicates that there was a time in-missing of sediments before the deposition of the Qaroun Formation. The remarkable concentration of sandstone facies of this formation in the southwest Wadi Hitan, associated with vertebrate remains, suggests that this site is the location of the old Nile River mouth. On the other hand, the concentration of the mudstone, marl, and marly limestone on the eastern side of the Fayoum at Jabel Lahoun and Jabel Naaloun can testify that this site was far from the pouring of river. In terms of delta setting, the Gehannam facies can represent pro delta deposited in a subaqueous type delta (Fig. 3.21).

After the deposition of the Gehannam Formation, there was a slight emergence concurrent with the retreat of seawater northwards that gave rise to the formation of the Qaroun

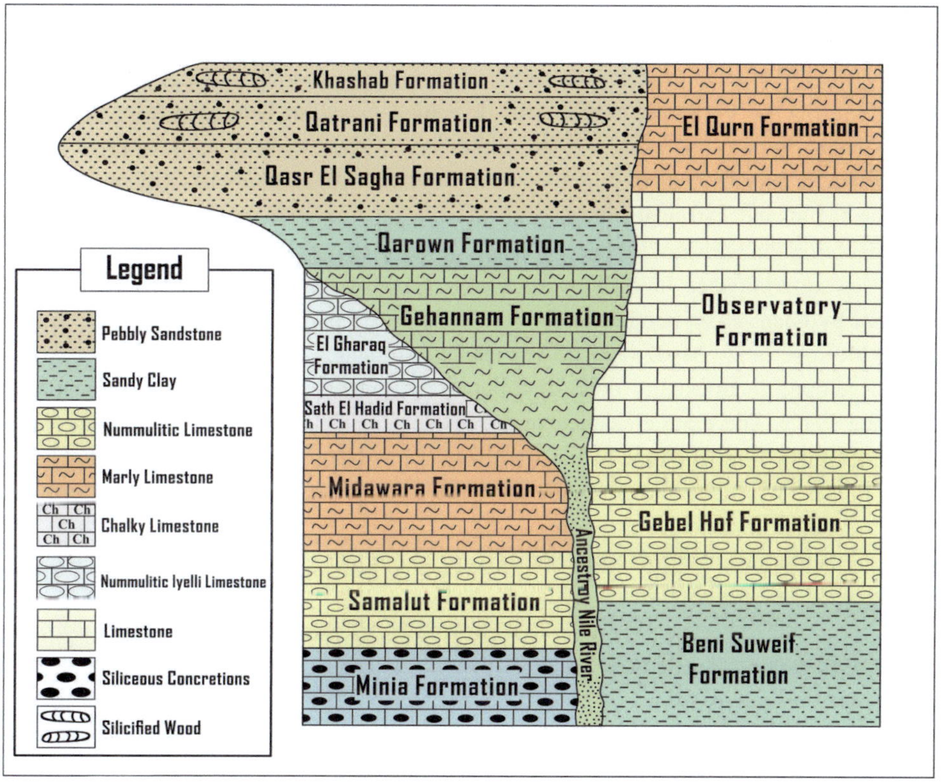

Fig. 3.21 Schematic illustration manifesting the possible depositional model of the rock units constituting the Fayoum Nile Delta

Formation. It includes two lithologic units; the lower one has four cycles with a thickness ranging from 4 to 5 m. The cycles start with a conglomerate bed and are capped by greenish-gray micaceous shale (Fig. 3.8). The upper unit includes eight fining-upward cycles; each begins with conglomerate, followed by a medium- to fine-grained sandstone, sandy claystone, and gray laminated claystone (Fig. 3.8). These cyclic patterns indicate that this formation was deposited under several phases of emergence-submergence or transgression-regression events that distinguished the fluvial regime. The presence of vertebrate remains in the Qaroun Formation may testify that they were swept from river tributaries (Bown and Kraus 1988). The presence of freshwater fish in this formation can support this opinion (Murray et al. 2014). The lithological constituents and cyclic patterns of the Qaroun Formation most probably resemble the lower delta front, which is dominantly fine- to locally medium-grained sandstone, and form units vary in thickness from 1 to about 10 m thick (Plint and Nummedal 2000). The presence of calcareous sandstone concretions at the top of the cycle suggests frequent subaerial exposure.

Further upward shallowing resulted in the deposition of the Qasr El Sagha Formation, which comprises two members (the Temple and the Abu Lifa members). According to Vondra's (1974) analysis, the lower Temple Member, which consists of the bioclastic and arenaceous carbonate facies, was deposited in settings ranging from marine bar front to estuarine mouth, where the sea depth was typically less than 20 meters. On the other hand, it appears that the upper portion of the Temple Member, which consists of siltstone facies and carbonaceous laminated claystone, was formed in shallow lagoons, presumably on the lee side of barrier islands (Vondra 1974). The upper Abu Lifa Member comprises superimposed fining-upward cycles. Each of these begins with conglomerate, followed by Giant cross-bedded sandstone. Such cyclic patterns suggest fluvial sedimentation (McCabe 1977; Miall 1977). They usually fill channels with depths up to 40 m and were developed in unusually deep distributary channels, or deltaic slope sedimentation (McCabe 1977). The giant cross sets in the lower part of the Abu Lifa Member represent lateral accretion deposits that formed within stream channels (Bown and Kraus 1981). In the present work, we suggest that the Qasr El Sagha Formation was most probably deposited in the upper delta front (Fig. 3.21).

At the beginning of the Oligocene, tectonic activities were widespread, changing the geography of the Tethys basin with the formation of the Paratethys to the north (Schultz et al. 2010) and the separation of Africa from the Arabian Peninsula as a result of the opening of the Gulf of Aden (Otero and Gayet 2001). The Oligocene was a time of uplift, regression, volcanicity, and continental sedimentation (Said 1990). During that period, the Qatrani Formation was deposited in a principally alluvial deposit composed of both single and multistory sand bodies, representing channel and near-channel deposits, and lesser amounts of mudstone, shale, and limestone, which formed on floodplains (Bown and Kraus 1981, 1988). Lower Qatrani rocks reflect the first major episode of terrestrial sedimentation in northern Egypt since the Late Cretaceous. This terrestrial episode was enhanced by the Late Eocene Tethyan regression that was recorded over all of the central and western Tethyan regions as well (Cavelier et al. 1981). These were deposited by high-sinuosity streams on a coastal alluvial plain, and most fine-grained floodplain deposits were removed by lateral reworking of the meandering channel systems. Paleocurrent analysis for the lower part of the Qatrani Formation indicates that these sediments were deposited by streams flowing essentially to the west or southwest. The Qatrani Formation was seemingly deposited in the subaerial part of a delta above the water table (Fig. 3.21). El Khashab and Moghra facies represent the delta plain and are considered the end of the Fayoum Nile Delta. They were deposited in the subaerial delta plain (see Fig. 3.21).

3.3.4 Evidence for the Recognition of the Old Fayoum Nile Delta

There is comprehensive evidence to document the formation of the Fayoum Nile Delta before the deposition of the Mediterranean Nile Delta (since more than 40 M.yrs).

(1) The Fayoum Nile Delta has fan-shaped stretches northwestward from latitude 27, 50 just west of the Mina Governorate. This delta contains the provinces of Fayoum, Wadi El Natrun, Sadat City, and the Qattara Depression (Fig. 3.20).
(2) The Gehannam Formation shows angular unconformity with the underlying nummulitic limestone of the El Gharaq Formation. This suggests that the Gehannam Formation was carved on top of the pre-existing El Gharaq Formation during the entry to the Nile River, creating its old passway (Fig. 3.21).
(3) The synchronous structural low, or rifted basin, that culminated during the Cretaceous and persisted until the Eocene is most likely what caused the River Nile to enter the Fayoum Province. Sediments brought over the River Nile were partially absorbed by this basin. This basin was encircled by the high relief of the Nashfa to the south, Bahariya to the west, Kattaniya and Abu Roash to the northeast, and so on (Fig. 3.19).
(4) Why is Wadi Hitan regarded extraordinary or unusual site in having such a large concentration of fossil whales in a relatively small area? (1) the concentration of whale vertebrates in this area may be attributed to the proto-Nile or other large river system debouched into the Hitan embayment (Dolson et al. 2002), and (2) there is no another site for the accumulation of vertebrate fossils in the Western Desert in the same age (Dolson et al. 2002). In our opinion, the concentration of vertebrate fossils in this area is attributed to the debauchery of the Nile River in this area during the Middle Eocene time.
(5) The presence of mangroves atop the Gehannam Formation and in the Qatari Formation indicates a mixture of fresh and salty water predominated during the Fayoum Delta (El Saadawi 1995).
(6) It is observed that most of the facies in the rock units that comprise the Fayoum Delta are clastic facies, which characterize the fluvial affinity (Allen 1965; Cant and Walker 1978). In contrast, the rock units occurring west of the Fayoum Province (e.g., Minia, Samalut, Midawara, Sath El Hadid, and El Gharaq) and those occurring east of Fayoum Delta (e.g., upper Lutetian Qarara, Bartonian El Fashn, Beni Sueif, and Priabonian Maadi formations) are entirely made of carbonate facies and do not include any sandy facies or vertebrate remains (Fig. 3.22).
(7) Peters et al. (2009) stated that whales are scarce in most offshore marine areas and more common in shallower seas. This can be explained by the possibility that the whale concentration requires ideal circumstances close to the river mouth or on a protected shelf.

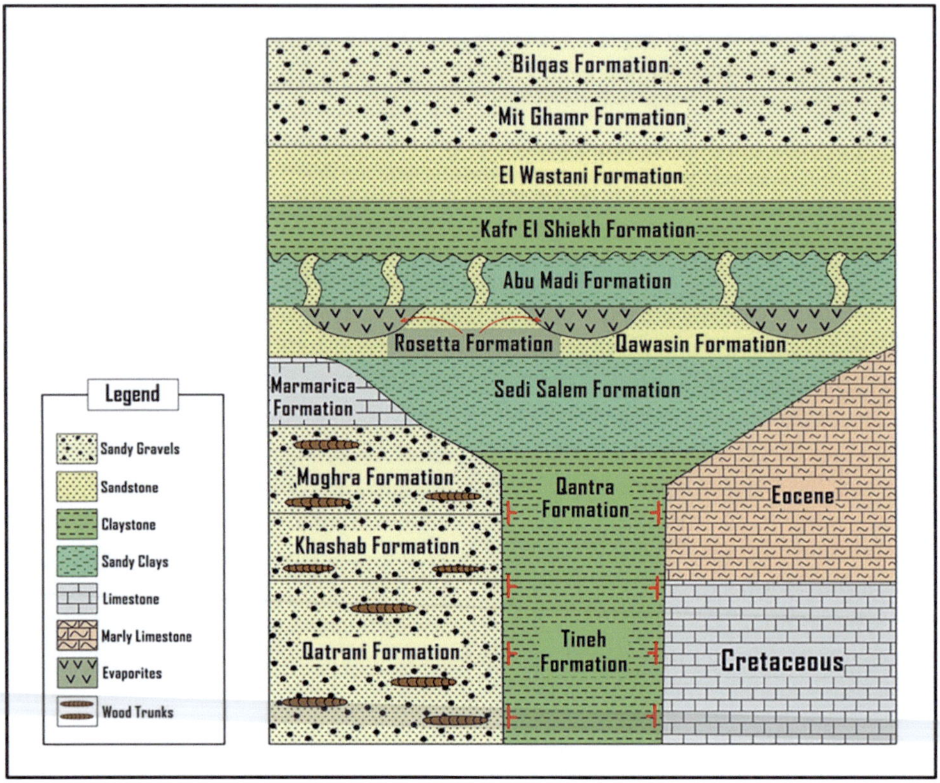

Fig. 3.22 Schematic diagram showing the correlation of the Nile River sediments in the Fayoum Nile Delta and the corresponding and equivalent rock units east and west of the Fayoum Delta

(8) Zalmout et al. (2012) investigated the vertebrate fauna of the Moghra Formation in the Minqar Tabaghbagh, which occurs on the western side of the Qattara Depression. They hypothesized that these fauna are indicative of a near-shore marine environment, perhaps a bay or estuary near land that was connected to rivers and deltas. They added that the diversity and richness of Lamniformes in the fauna of Tabaghbagh may be due to nitrogen input from rivers that flow northward from the African continent. The Nile River was most likely the source of nutrients that support the richness and prosperity of the vertebrate fauna in this area.

(9) The concentration and accumulation of diverse vertebrate species at Wadi Hitan suggest that Wadi Hitan is the site receiving the Nile River mouth, which supplies the region with sufficient nutrients to enhance the growth of those species.

(10) Murray et al. (2010, 2014) stated that the presence of freshwater vertebrate fauna (fish tooth fossils) in the Qaroun Formation provides strong evidence of the existence of a freshwater source during the deposition of the Upper Eocene Qarun Formation.

(11) The lower portion of the Qasr El Sagha Formation was deposited in a lowland riparian environment with continuous freshwater intrusion, as shown by the presence of palynomorph assemblages in the black shale of this formation (Selim et al. 2018). In addition, the top portion of this formation, which is crowded by woody forest vegetation and mixed angiosperms and pteridophytes, suggests that this facies sequence was marshy and dominated by freshwater.

(12) Holroyd et al. (1996) found new terrestrial animals from the Qasr El Sagha Formation. These, together with many specimens of Moeritherium, Barytherium, and anthracotheres, provide the earliest known record of a hyracoid in the Fayoum Province. They claimed that these terrestrial Eocene animals lived in continental habitats and were transported by the Nile River.

(13) The only concentration of vertebrate fauna in the Gehannam, Qaroun, Qasr El Sagha, Qatrani, El Khashab, and Moghra formations (mainly clastic facies), composing the Fayoum Nile Delta and their absence in the Middle Eocene El Gharaq Formation, which is found directly under the Gehannam Formation (nummulitic limestone) in the same area. It may suggest that the vertebrate fauna originated in and around the ancestor mouth of the Nile River.

(14) The presence of apes like Aegyptopithecus in the Oligocene Qatrani Formation, and their absence in the comparable Oligocene layers outside the Fayoum depression. This indicates the common presence of freshwater habitats near river mouths like the Nile River.

References

Abdallah AM (1966) Stratigraphy and structure of a portion in the north western desert of Egypt (El-Alamein- Dabaa- Qattara- Moghra area) regarding its economic potentialities. Geological survey of Egypt papers, vol 45, pp 1–19

Abdel-Fattah ZA, El Asmer H, El Fawal F (2010) Sedimentary environments and depositional characteristics of the Middle to Upper Eocene whale-bearing succession in the Fayum Depression, Egypt. Sedimentology 57:446–476

Abdelsalam MG (2018) The Nile's journey through space and time: a geological perspective. Earth Sci Rev 177:742–773

Al-Dhwadi Z, Sallam ES (2019) Spheroidal "Cannonballs" calcite-cemented concretions from the Fayoum and Bahariya depressions, Egypt: evidence of differential erosion by sand storms. Int J Earth Sci 108:2291–2293

Andrews CW (1901) Preliminary note on some recently discovered extinct vertebrates from Egypt (pt. 1). Geol Mag 8: 400–409

Andrews CW (1904) Further notes on the mammals of the Eocene of Egypt. Geol Mag Lond Ser 5(1):211–215

Andrews CW (1906) Descriptive catalogue of tertiary vertebrata of the Fayoum, Egypt. British Museum, National History, London p 324; Ball J (1939) Contributions to the geography of Egypt: Cairo, survey department, p 300

Barta G (2011) Secondary carbonates in loess-paleosol sequences: a general review. Open Geosci 3:129–146
Bayoumi Al, El-Gamili MM (1970) A geophysical study on the Fayum-Rayan area, regarding its subsurface structures. In: Proceedings of the 7th Arab petroleum congress, vol 2, Kuwait, pp 1–15
Beadnell HJL (1901) The Fayum depression: a preliminary notice of the geology of a district in Egypt containing a new Paleocene vertebrate fauna. Geol Mag:540–546
Beadnell HJL (1905) The topography and geology of the Fayum Province of Egypt. Survey Department of Egypt, Cairo, p 101
Blanckenhorn M (1900) Neues zur Geologie und Palaontologie Aegyptens. Zeitschrift der deutschen geologischen Gesellschaft 52:403–479
Bose PK, Chakraborty PP (1994) Marine to fluvial transition: proterozoic upper Rewa Sandstone, Maihar, India. Sediment Geol 89:285–302
Bowen BE (1970) The stratigraphy of the Jebel el Qatrani formation in the Fayum Depression, Egypt, U.A.R. Ph.D. Dissertation, Iowa State University
Bowen BE (1971) Paleoenvironmental interpretations of the Oligocene Jebel el Qatrani Formation, Fayum Depression, Egypt. U.A.R. Geological Society of America Abstracts with Programs, p 254
Bowen BE, Vondra CF (1974) Paleoenvironmental interpreations of the Oligocene Jabel El Qatrani formation, Fayoum Depression, Egypt, Cairo. Ann Geol Surv Egypt 4:115–138
Bown TM (1982) Ichnofossils and rhizoliths of the nearshore fluvial Jebel Qatrani Formation (Oligocene), Fayum Province, Egypt. Palaeogeogr Palaeoclimatol Palaeoecol 40:255–309
Bown TM, Kraus M (1981) Lower Eocene alluvial paleosols (‘Willwood Formation, northwest Wyoming, U.S.A.), and their significance for paleoecology, paleoclimatology, and basin analysis. Palaeogeogr Palaeoclimatol Palaeoecol 34:1–30
Bown TM, Kraus MJ (1988) Geology and paleoenvironment of the Oligocene Jebel QatraniFormation and adjacent rocks, Rayum Depression. U. S. Geological Survey Professional Paper 1452, pp 1–60
Bown TM, Kraus M, Wing S, Fleagle J, Tiffney BH, Simons E, Vondra CF (1982) The Fayum primate forest revisited. University of Nebraska—Lincoln, USGS Staff—Published Research, US Geol Surv 1:46–52, p 347
Buatoisa LA, Mángano GM, Alissaa A, Carr TR (2002) Sequence stratigraphic and sedimentologic significance of biogenic structures from a late Paleozoic marginal-to open-marine reservoir Morrow Sandstone, subsurface of Southwest Kansas, USA. Sediment Geol 152:99–132
Cant DJ, Walker RG (1978) Fluvial processes and facies sequences in the sandy braided Saskatchwan River, Canada. Sedimentology 25:625–648
Cavelier C, Chateauneuf J, Pomerol C, Maurice RD, Renard M, Grazzini CV (1981) The geological events at the eocene/oligocene boundary. Palaeogeogr Palaeoclimatol Palaeoecol 36:223–248
Chakraborty PP (1996) Sedimentary records of erg developments over braid plain, Proterozoic Dhandraul Sandstone, Vindhyan Supergroup, Son Valley. Memoir Geol Soc India 36:77–99
Cohen AS (1984) The significance of zoobenthic standing crop for laminae formation in tropical lake sediments. J Paleontol 58:499–510
Collinson JD (1986) Alluvial sediments. In: Reading HG (ed) Sedimentary environment and facies. Blackwell Scientific, Oxford, pp 20–62
Collinson JD, Thompson DB (1989) Sedimentary structures. Allen and Unwin, London, p 194
Davis JM, Mozley PS (2005) Internal structure and mode of growth of elongate calcite concretions: evidence for small-scale, microbially induced, chemical heterogeneity in groundwater. GSA Bull 117:1400–1412
Dean VE, Gorham E (1975) Major chemical and mineral components of profundal surface sediments in Minnesota lakes. Limnol Oceanogr 21:259–284

Dolan JF (1989) Eustatic and tectonic controls on the deposition of hybrid siliciclastic/carbonate basinal cycles: discussion with examples1. AAPG Bull 73:1233–1246

Dolson J, Barkooky AE, Wehr F, Gingerich P, Prochazka N, Shann M (2002) The Eocene and Oligocene paleo-ecology and paleo-geography of Whale Valley and the Fayoum basins: implications for hydrocarbon exploration in the Nile Delta and eco-tourism in the greater Fayoum basin: field trip guidebook for field trip No. 7, Cairo 2002. In: International conference and exhibition, sponsored by AAPG, EPEX, SEG, EGS, and EAGE,Cairo, Egypt, 27–30 Oct 2002

El Anbaawy MI, Helal SA, Abu El Kheir GA (2018) Contribution to the taphonomic significances of the Eocene marine mammals in Wadi El Hitan and northern Lake Qaroun. Egypt J Geol 62:401–422

El Mahdy M, Abioui M, Mustoe GE (2020) Miocene petrifed trees at Bahariya Oasis, Egypt. Int J Earth Sci 109:2869–2870

El-Araby A, El-Barkooky AN (2000) Stratigraphic architecture and facies evolution of the Oligocene Jabel Qatrani, Jabel Qatrani, North Fayum, Egypt. Geol Arab World:1175–1208

El-Fawal FM, El Asmer HM, Sarhan MA (2011) Depositional evolution of the Middle-Upper Eocene rocks Fayum area, Egypt. Arab J Geosci 6:749–760

El Sadawi WE (2005) A fossil rhizome at the mangrove site of Wadi Hitan, Egypt. Taeckholimia 25:129–136

El-Saadawi W, Kamal El-Din MM, Wheeler E, Osman R, El-Faramawi MW, El-Noamani Z (2014) Early Miocene woods of Egypt. IAWA J 35:35–50

El-Younsy ARM, Soliman AM (2021) Sequence stratigraphic framework of the Upper Eocene-Oligocene succession, northwest Birket Qarun, Fayum-Egypt. Environ Earth Sci 80:350

Fielding L, Najman Y, Millar I, Butterworth P, Ando S, Padoan M, Barfod D, Kneller B (2017) A detrital record of the Nile River and its catchment. J Geol Soc 174:301–317

Fielding L, Najmana Y, Millarb I, Butterworthc P, Garzantid E, Vezzolid G, Barfode D, Knellerf B (2018) The initiation and evolution of the River Nile. Earth Planet Sci Lett 489:166–178

Fleagle JG, Bown TM, Obradovich JD, Simon EL (1985) How old are the Fayum primates?. Congress of the International Primatological Society

Flores RM, Erpenbec MF (1981) Differentiation of delta front and barrier lithofacies of the upper Cretaceous pictured cliffs sandstone, southwest San Juan Basin, New Mexico. Mt Geol 18:23–34

Flügel E (2004) Microfacies of carbonate rocks: analysis, interpretation and application. Springer, Berlin

Fourtau (1918) Fourtau R. Contribution à l'étude des vertébrés miocènes de l'Egypte. Survey Department, Ministry of Finance Governement Press, Cairo

Friedman GM (2003) Classification of sediments and sedimentary rocks. In: Middleton GV (ed) Encyclopedia of sediments & sedimentary rocks, pp 127–135. Encyclopedia of Earth Science Series. Kluwer Academic Publishers, Boston, Massachusetts, p 821

Gani NDS, Gani MR, Abdelsalam MG (2007) Blue Nile incision on the Ethiopian Plateau: pulsed plateau growth, Pliocene uplift, and hominin evolution. GSA to-Day 17:4–11

Ghandour I (2020) The effect of early Priabonian sea-level change on the depositional architecture of the lower Qasr El-Sagha Formation, Kom Aushim, NE Fayum, Egypt. Int J Earth Sci (Environ Sci Geogr Geol)

Gingerich PD (1992) Marine mammals (Cetacea and Sirenia) from the Eocene of Jabel Mokattam and Fayum, Egypt: stratigraphy, age, and paleoenvironments. Univ Michigan Pap Paleontol 30:1–84

Gingerich PD, Antar MSM, Zamout LS (2011) Projection stratigraphy of the upper Eocene Gehannam, Birket Qarun, and Qasr el-Sagha formations and their fossil whales at the Wadi Al Hitan World Heritage Site, western Fayum Province (Egypt). Berichte Geol. B.A. 85 (ISSN 1017-8880) CBEP 2011, Salzburg

Guion PD, Fulton IM, Jones NS (1995) Sedimentary facies of the coal-bearing Westphalian A and B north of the Wales–Brabant High. In: Whateley MKG, Spears DA (eds) European coal geology. Geological Society, Special Publication, pp 45–78

Hassan SM (2013) Sequence stratigraphy of the lower Miocene Moghra formation in the Qattara depression, north western desert, Egypt. Springer Science & Business Media

Hassan SM, Steel RJ, El-Barkooky AN, Hamdan MA, Olariu C, Helper MA (2012) Stacked, lower Miocene tide-dominated estuary deposits in a transgressive succession, western desert, Egypt. Sediment Geol 2012(282):241–255

Ho C, Coleman JM (1969) Consolodation and cementation of recent sediments in the Atchafalaya basin. GSA Bull 80:183–192

Boles JR, Thyne GD (1989) Isotopic evidence for origin of the Moeraki septarian concretions, New Zealand. J Sediment Res 59:272–279

Holroyd PA, Simons EL, Bown TM, Polly PD, Kraus MJ (1996) New records of terrestrial mammals from the upper Eocene Qasr El Sagha formation, Fayûm depression, Egypt. Palaeovertebrata 25:175–192

Hori K, Saito Y (2007) An early Holocene sea-level jump and delta initiation. Geophys Res Lett 34. https://doi.org/10.1029/2007GL031029.issn:0094-8276

Issawi B (1976) An introduction to the physiography of the Nile Valley. In: Wendorf F, Schild R (eds) Prehistory of the Nile Valley. Academic, pp 3–22

Issawi B, Osman R (2008) Egypt during the Cenozoic: geological history of the Nile River. In: Yousef EAA (ed), Bulletin of the Tethys geological society, vol 3. Cairo University, pp 43–62

Issawi B, McCauley JF (1992) The Cenozoic rivers of Egypt: the Nile problem. In: Friedman RF, Adams B (eds) The followers of Horus, studies dedicated to Michael Allen Hoffman1944–1990, Oxbow monograph 20, pp 121–138.

Issawi B, Francis MF, Youssef AA, Osman RA (2009) The phaneorozoic geology of Egypt: a geodynamic approach. Special publication No. 81.Minsitry of Petroleum, The Egypt Minsitry Research Authority, p 589

James NP (1979) Facies models 10: shallowing upward sequences in carbonates. In: Walker RG (ed) Facies models. Geoscience Canada Reprint Series, no 1, pp 109–119

Jones B, Squair CHA (1989) Formation of peloids in plant rootlets, Grand Cayman, British West Indies. J Sediment Res 59:1002–1007

Jones NP, James B (2015) Origin of carbonate sedimentary rocks. American Geophysical Union, p 464

Kappelman JW, Simons EL, Swisher CC (1992) New age determinations for the Eocene-Oligocene boundary sediments in the Fayum Depression, northern Egypt. J Geol 100:647–668

Khalaf EA, Abu El-Kheir G (2022) The geological heritage and sustainable development proposed for the project geopark: an example from Gabal Qatrani, Fayoum Depression, Western Desert, Egypt. Geoheritage 14:22–30

Khalifa MA, Abu Zeid K (1985) Miocene environmental models in the area northwest of El Natrun, Western Desert. In: Abstracts of the 23D annual meeting of the geological society of Egypt, vol 1985, p 11

Khalifa MA (1981) Geologicla and sedimentological studies of the west Beni Mazar area, south of Fayoum province, Western Desert, Egypt. Ph. D. Thesis, Cairo University, p 320

Khan IA, Bridge JS, Kappelman J, Wilson R (1997) Evolution of Miocene fluvial environments, eastern Potwar plateau, Northern Pakistan. Sedimentology 44:221–251

Klappa CF (1980) Rhizoliths in terrestrial carbonates: classification, recognition, genesis and significance. Sedimentology:tb01651.x

Kontopoulos N (1991) The mineralogy and sedimentation environment of the Neogene marly limestones, Aitoloakarnania (Greece). Geol Balcanica 21:75–84

Kortlandt A (1980) The Fayum primate forest: did it exist? J Hum Evol 9:277–297
Kraus MJ, Hasiotis ST (2006) Significance of different modes of rhizolith preservation to interpreting paleoenvironmental and paleohydrologic settings: examples from Paleogene paleosols, Bighorn Basin, Wyoming, USA. J Sediment Res 76:633–646
Macgregor DS (2012) The development of the Nile drainage system: integration of onshore and offshore evidence. Pet Geosci 18:417–431
Maley J (1991) The African rain forest vegetation and paleoenvironments during late Quaternary. Clim Change 19:79–98
Marzouk I (1970) Rock stratigraphy and oil potentialities of the Oligocene and Miocene in the Western Desert of Egypt. In: 7th Arab petroleum congress, vol 54, pp 1–37
McCabe PJ (1977) Deep distributary channels and giant bedforms in the Upper Carboniferous of the Central Pennines, northern England. Sedimentology 24:271–290
McCauley JF, Schaber GG, Breed CS, Grolier MJ, Haynes CV, Issawi B, Blom R (1982) Subsurface valleys and geoarcheology of the eastern Sahara revealed by shuttle radar. Science 218:1004–1020
McCauley JF, Breed CS, Schaber GG, McHugh WP, Issawi B, Haynes CV, El Kilani A (1986) Paleodrainages of the eastern Sahara—the radar rivers revisited (Sir-A/B implications for a Mid-Tertiary Trans-African drainage system). IEEE Trans Geosci Remote Sens 4:624–648
Miall AD (1977) Lithofacies types and vertical profile models in braided river deposits: a summary. In: Miall AD (ed) Fluvial sedimentology. Geological Survey of Canada, Calgary, pp 597–604
Miall AD (1978) Lithofacies types and vertical profile models in braided river deposits: a summary. In: Miall AD (ed) Fluvial sedimentology. Canadian Society of Petroleum Geologists, Memoir, vol 5, pp 801–829
Moore SE, Ferrell RE Jr, Aharon P (1992) Diagenetic siderite and other ferroan carbonates in a modern subsiding marsh sequence. J Sediment Res 62:357–366
Mount JF, Cohen AS (1984) Petrology and geochemistry of rhizoliths Plio-Pleistocene fluvial and marginal lacustrine deposits, east Lake Turkana, Kenya. J Sediment Res 54:263–275
MurrayAM, Cook TD, Attia YS, Chatrath P, Simon EL (2010) A freshwater ichthyofauna from the late Eocene Birket Qarun Formation, Fayum, Egypt. J Verteb Paleontol 30
Murray AM, Argyriou T, Cook TD (2014) Paleobiogeographic relationships and paleoenvironmental implications of an earliest Oligocene Tethyan ichthyofauna from Egypt. Can J Earth Sci 51(10)
Neuendorf KKE, Mehl JP, Jackson A (2005) Glossary of geology, 5th edn. American Geological Institute, Alexandria, Virginia, p 779
Oard MJ (2019) A case for rapid formation of calcareous concretions. J Creat 33:7–9
Otero O, Gayet M (2001) Palaeoichthyofaunas from the Lower Oligocene and Miocene of the Arabian plate: palaeoecological and palaeobiogeographical implications. Paleogeogr Palaeoclimatol Paleoecol 165:141–169
Owen RA, Owen RB, Renaut RW, Scott JJ, Jones B, Ashley GM (2008) Mineralogy and origin of rhizoliths on the margins of saline, alkaline Lake Bogoria, Kenya Rift valley. Sediment Geol 203:143–163
Peters SE, Antar MS, Zalmout I, Gingerich P (2009) Sequence stratigraphic control on preservation of late Eocene whales and other vertebrates at Wadi Al-Hitan, Egypt. Palaios 24:290–302
Pik R, Marty B, Carignan J, Lavé J (2003) Stability of the Upper Nile drainage network Ethiopia deduced from U-Th./He thermo chronometry: implications for uplift and erosion of the Afar plume dome. Earth Planet Sci Lett 215:73–88
Pik R, Marty B, Carignan J, Yirgu G, Ayalew T (2008) Timing of East African rift development in southern Ethiopia: implication for mantle plume activity and evolution of topography. Geology 36:167–170

Plint G, Nummedal D (2000) The falling stage systems tract: recognition and importance in sequence stratigraphic analysis. Geol Soc Lond Spec Publ 172:1–17

Rehman SU, Mehmood K, Ahsan N, Shah MM (2016) Microfacies and depositional environments of upper cretaceous Kawagarh formation from Chinali and Thoba section, Northeastern Hazara Basin lesser himalayas, Pakistan. J Himal Earth Sci 49:1–16

Reineck HE, Singh IB (1975) Depositional sedimentary environments with reference to terrigenous clastics. Berlin, Heidelberg, New York, p 439

Richards PW (1966) The tropical rain forest: an ecological study. Cambridge University Press, Cambridge

Rust BR, Koster EH (1984) Coarse alluvial deposits. In: Walker RG (ed) Facies models. Geoscience Canada, reprint series 1, 2nd edn, p 53

Said R (1962) The geology of Egypt. Elsevier, Amsterdam, p 377

Said R (1981) The geological evolution of the River Nile. Springer, p 151

Said R (1993) The Nile River: geology, hydrology and utilization. New York

Said R (1990) The geology of Egypt. A. Balkema Publishers, New York

Salama AM (2008) Genesis, composition and geological significance of wood fossils in Maadi-qattamiya Area. Unpubl. M.Sc. Thesis, Cairo University

Salem R (1976) Evolution of Eocene-Miocene sedimentation patterns in parts of northern Egypt. AAPG Bull 60:34–64

Scholle PA, Bebout DG, Moore CH (1983) Carbonate depositional environments: AAPG Memoir (No 33)

Schultz RA, Okubo CH, Fossen H (2010) Porosity and grain size controls on compaction band formation in Jurassic Navajo Sandstone. Geo Phys Res Lett 37:L22306

Seiffert ER, Bown TM, Clyde WC, Simons E (2008) Geology, paleoenvironment, and age of Birket Qarun Locality 2 (BQ-2), Fayum Depression, Egypt. In: Fleagle JG, Christopher I, Gilbert C (eds) Simons E: a search for origins. Springer, pp 71–86

Selim SS, Gentizl T, Tahoun SS, El-Shafeiy M et al (2018) Depositional environment and characteristics of Late Eocene carbonaceous swampy tidal flat facies in the Fayoum Basin, Egypt. ResearchGate

Sellwood BW (1978) Shallow-water carbonate environments. In: Reading HG (ed) Sedimentary environments and facies, pp 259–313

Sestini G (1995) Egypt. In: Kulke H (ed) Regional petroleum geologyof the world, part II, Africa, America, Australia and Antarctica (Beiträge zur regionalen Geologie der Erde 22. Gebrüder Borntraeger Verlagsbuchhandlung, Stuttgart, pp 66–87

Shata A (1953) New light on structural developments of the Western Desert of Egypt. Inst Desert Egypt Bull 3:101–106

Shukri NM (1950) The mineralogy of some Nile sediments. Quart J Geol Soc Lond 105:511–534

Sigdel A, Sakai T (2016) Sedimentary facies analysis of the fluvial systems in the Siwalik Group, Karnali River section, Nepal Himalaya, and their significance for understanding the palaeoclimate and Himalayan tectonics. J Nepal Geol Soc 51:11–26

Simons EL (1968) Early cenozoic mammalian faunas, Fayûm Province, Egypt, part I: African oligocene mammals: introduction, history of study, and faunal succession. Bull Peabody Mus Nat Hist 28:1–22

Simons EL, Bown TM (1984) A new species of *Peratherium* Didelphidae: Polyprotodonta: the first African marsupial. J Mammal 65:539–548

Simons EL, Rasmussen DT (1990) Vertebrate paleontology of Fayum: history of research, faunal review and future prospects. In: Said R (ed) The geology of Egypt. A.A. Balkema, Rotterdam, Netherlands

Talbot MR, Williams MAJ (2009) Cenozoic evolution of the Nile basin. In: Dumont HJ (ed) The Nile. Monographiae Biologicae 89. Springer, pp 37–60

Tucker ME (2003) Sedimentary rocks in the field, 3rd edn. John Wiley & Sons Ltd., West Sussex, England

Tucker ME, Wright VP (1990) Carbonate mineralogy and chemistry. Carbonate Sediment:284–313

Underwood CJ, King C, Steurbaut E (2013) Eocene initiation of Nile drainage due to East African uplift. Palaeogeogr Palaeoclimatol Palaeoecol 392:138–145

Vondra CF (1967) Stratigraphy of the Upper Eocene and Oligocene deposits in the Fayum Depression, Egypt. Geological Society of America Abstracts with Programs, p 229

Vondra CF (1974) Upper Eocene transitional and near-shore martine Qasr el Sagha Formation: Fayum Depression, Egypt. Ann Geol Surv Egypt 4:79–94

Wanas HA (1991) Sedimentological studies of the Upper Eocene rocks, in the Fayoum area, Western Desert, Egypt. M.Sc. Thesis, Menoufia University

Wanas HA (2008) Calcite-cemented concretions in shallow marine and fluvial sandstones of the Birket Qarun Formation (Late Eocene), El-Faiyum depression, Egypt: Field, petrographic and geochemical studies: implications for formation conditions. Sediment Geol 212:40–48

Walker RG, Cant DJ (1984a) Sandy fluvial systems. In: Walker RG (ed) Facies models, 2nd edn. Reprint series, vol 1.Geoscience, Canada, pp 71−89

Walker RG, Cant DJ (1984b) Sandy fluvial systems. In: Walker RG (ed) Facies models, 2nd edn. Geoscience Canada reprint series 1. Geological Association of Canada, Newfoundland, pp 71–89

Ward LG, Ashley GM (1989) Introduction: coastal lagoonal systems. Mar Geol 88:181–185

Williams MAJ, Adamson DA (1982) A Land between two Niles: quaternary geology and biology of the central Sudan. Balkema, Rotterdam

Woodward JC, Macklin MG, Fielding L, Millar I, Spencer N, Welsby D, Williams M (2015) Shifting sediment sources in the world's longest river: a strontium isotope record for the Holocene Nile. Quatern Sci Rev 130:124–140

Zalmout IS (2008) Late eocene sea cows (Mammalia Sierenia) from Wadi Al Hitan in the Fayum Basin, Egypt. Ph.D. in the University of Michigan, USA

Zalmout IA, Antar MSM, Abdel-Shafy E, Metwally MH, Hatab EE, Gingerichi PD (2012) Priabonian sharkas and rays (late Eocene: Neoselachii) from Minqar Tabaghbagh in the western Qattara Depression, Egypt. Mus Paleontol, Univ Michigan 6:71–90

Depositional History of the Mediterranean Nile Delta

4

Mohamed Abdelghany Khalifa

Abstract

Because the Nile River created two deltas throughout its history, the first and the oldest is named in the current work, the Fayoum Nile Delta, which occurs in the northwestern Desert. In contrast, the second and the younger delta is called the Mediterranean Nile Delta, which arises on the Mediterranean coast. The later delta is divided into two superimposed deltas, the Neogene Delta at the base and the Quaternary Delta at the top. The Late Miocene Messinian unconformity separated them. The Neogene Delta shows angular unconformity with the underlying Cretaceous-Eocene rocks. It comprises, in ascending order, the Upper Oligocene Tineh, Lower Miocene Qantara, the Middle Miocene Sedi Salem, the Upper Miocene Qawasim, Rosetta, and Abu Madi formations. They were deposited in a narrow deep canyon on the shelf during the Oligo-Miocene in different depositional settings such as pro-delta, delta front, and delta plain settings. Following the sedimentation of the Neogene delta plain, the Atlantic Ocean and the Mediterranean Sea were sealed off at the area now known as the Jabel Tarek (Gibraltar) Strait. As a result, the channelized sandstone (Abu Madi Formation) and evaporite (Rosetta Formation) were deposited on dissected channels and lagoons. Following the opening of the Atlantic Ocean with the Mediterranean Sea, the Quaternary Delta that includes the Pliocene Kafr El Shiekh, Plio-Pleistocene El Wastani, Pleistocene Mit Ghamr, and the Holocene Bilqas formations was deposited. The Kafr El Sheikh Formation was deposited in open water and represents a prodelta. Meanwhile, the El Wastani Formation was deposited along the delta front. As water levels fell, the Mit Ghamr and Bilqas formations were deposited in the delta plain.

M. A. Khalifa (✉)
Geology Department, Faculty of Science, Menoufia University, Shiben El Kom, Egypt
e-mail: khcycle1949@gmail.com

M. A. Khalifa and K. S. Gemail (eds.), *Geology of the Nile Deltas and Natural Resources*, Deltas of the World, https://doi.org/10.1007/978-3-031-96594-4_4

Keywords

Mediterranean Basin • Nile Delta • Quaternary deposits • Prodelta • Delta plain • Neogene Delta • Qantara depression • Upper Oligocene Tineh Mit Ghamr formation • Holocene Bilqas Clay

4.1 Introduction

The interpretation of the Nile Delta sediments and even their depositional settings began with a focus on the Miocene rocks after the Abu Madi well was drilled in 1967 in the central part of the Delta. The primary data sources for this interpretation were ditches and a small number of core samples. The accurate interpretation of the Mediterranean Nile Delta with such data is insufficient. Three decades later, more onshore drilled wells produced accurate sedimentological data regarding the Nile Delta's depositional setting. Growing well drilling techniques, especially offshore at the start of the twenty-first century, have increased the accuracy of interpreting rocks from the Oligo-Miocene, Pliocene, Pleistocene, and Holocene periods. The largest clastic accumulation in the Mediterranean Sea by the Nile River led to the formation of the Mediterranean Nile Delta in an area with considerable Cenozoic tectonic activity (Abdel Aal et al. 1994). As shown in Fig. 4.1, the major Messinian unconformity gave rise to dividing the Mediterranean Nile Delta into two superimposed, independent, prograding clastic delta systems (Sestini 1989, 1995). The closing of the Jabel Tarek (Gibraltar) strait resulted in the Messinian (late Miocene) unconformity. This led to the formation of multiple incised valleys on the shelf and shelf slope and the deposition of thick evaporite sequences (mainly near the basin center) (Ryan 1978; Barber 1981; Abdel Aal et al. 2001).

Two superimposed deltas, the Neogene Delta (Oligo-Miocene) at the base and the Quaternary Delta at the top (Pliocene, Pleistocene, and Holocene), were formed as a result of the Messinian unconformity surface (Fig. 4.1). The Mediterranean Nile Delta was studied by many eminent geologists, among them. Salem (1976), Zaghloul et al. (1977, 1979, 1989), Cita et al. (1977), Ryan (1978), Rizzini et al. (1978), Summerhayes et al. (1978), El Heiny (1982), Barakat (1982), Gheith (1985), Coutellier and Stanley (1987), El Heiny and Morsi (1992), El Asmer (1986), El Bassiouni and Ayyad (1986), El Shahat and Kora (1986), Stanley (1988, 1990, 2005), Sestini (1984, 1989, 1995), Abu El-Ella (1990), Stanley and Warne (1993, 1998), Sarhan and Hemdan (1994), El-Asmar and Gheith (1995), El-Heiny and Enani (1996), Chen et al. (1992), Dolson et al. (2002, 2005), Stanley and Corwin (2013), Abdel-Fattah and Slatt (2013), Farouk et al. (2014), Abdel Fattah (2014), Abdel Aal et al. (1994, 2001), Sarhan (2015), Leila et al. (2016), Stanley and Clemente (2014, 2017), Leila and Moscariello (2018, 2019), El-Barkooky and Helal (2002a, b), among others.

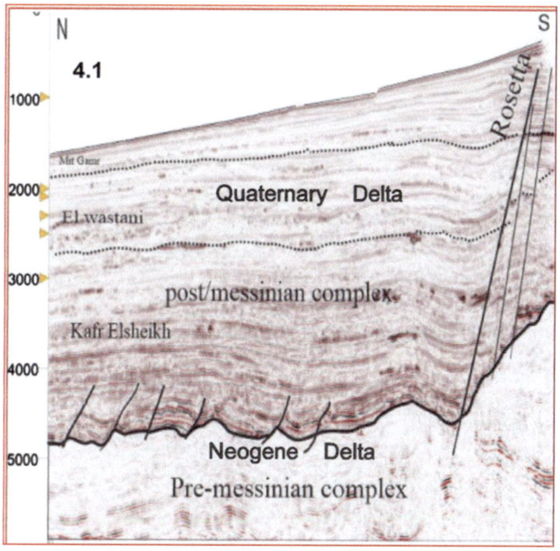

Fig. 4.1 Seismic line manifesting the Messinian unconformity that separated the Mediterranean Nile Deltas into the Neogene Delta at the base and the Quaternary Delta at the top (After El Shiekh 2021)

4.2 The Neogene Delta (Oligocene–Miocene)

The Neogene Delta includes the following rock units in ascending order: Tineh, Qantara, Sedi Salem, Qawasim, Rosetta, and Abu Madi formations (see Fig. 4.2).

4.2.1 Depositional History of the Tineh Formation

The Tineh Formation was encountered in the subsurface in a limited area, such as the eastern Nile Delta and northern Sinai. It was detected between 3070- and 2725-m depths in onshore at the El-Fayrouz borehole, and between 2805- and 2555-m depths in the Rommana-1X borehole (Farouk et al. 2014). The limited occurrences of the Tineh Formation in the eastern Delta and northern Sinai signify the initial incision of the Nile River and the earlier deposition of the Mediterranean Delta. This formation consists of pure claystone in its lower section, topped by a thin layer of conglomerate and some silty sandstone and conglomerate beds, which are approximately 20 m thick in the upper part. Notably, it is completely absent south of the Hinge Zone at the Tanta-1 well, where the Rupelian Dabaa Formation lies above the Burdigalian Moghra Formation (Farouk et al. 2014).

Furthermore, the absence of this geological formation on both the eastern and western sides of the Mediterranean Nile Delta may suggest the influence of faults that bound the delta in these areas (see Fig. 4.3). The presence of the Rupelian Dabaa Formation overlying the Burdigalian Moghra Formation further supports this observation. This explains why the graben on which the claystone of the Tineh Formation was created by the

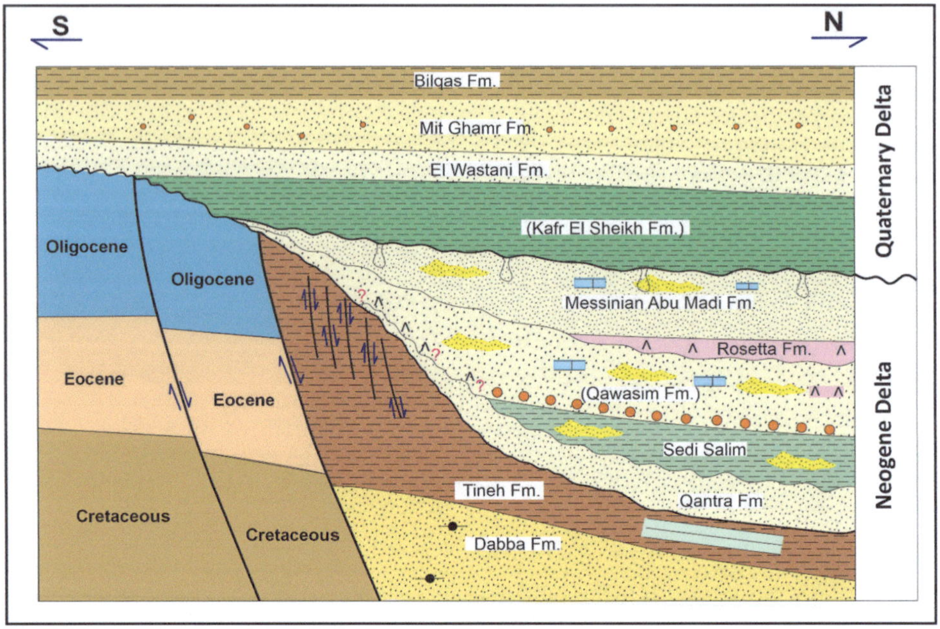

Fig. 4.2 North–South schematic cross-section of the Mediterranean Delta, showing the vertical and lateral distribution of the Neogene Delta (Tineh, Qantara, Sedi Salem, Qawasim, Rosetta and Abu Madi formations) and the Quaternary Delta (Kafr El Sheikh, Al Wastani, Mit Ghamr and Bilqas formations)

northeast and northwest delta faults, downthrowing from the east and west. However, because both the Oligocene Qatrani and the lower Miocene Moghra formations are made entirely of coarse clastics, they were deposited on the upthrown sides westward of the delta (Fig. 4.3). The presence of conglomerate beds intercalated with claystone and in the upper part of the formation strongly suggests that the initiation of rifting concurrent with deposition of collapse shoulders of the Mediterranean Nile Delta. This gave rise to the creation of a deep canyon concurrent with the initiation of the delta formation, resembling a deltaic fan. In this context, El Heiny and Morsi (1992) speculate that this fan conglomerate was formed tectonically and can be generated as a result of the constant dislocation on both sides of the Nile Delta. Rare to common agglutinated benthic foraminifera with limited species diversity can be found there. Also, Planktonic foraminifera hardly exist there (Farouk et al. 2014). According to El Heiny and Morsi (1992), the Tineh Formation's lack of arenaceous benthonic forams and planktonic foraminifera suggests an inner neritic environment. The nearly total absence of planktonic foraminifera in the major lower part of this Oligocene succession is a remarkable feature that made it difficult to divide into smaller units. The total absence of planktonic foraminifera in the Tineh Formation may be attributed to the high concentration of carbon dioxide that was emitted during

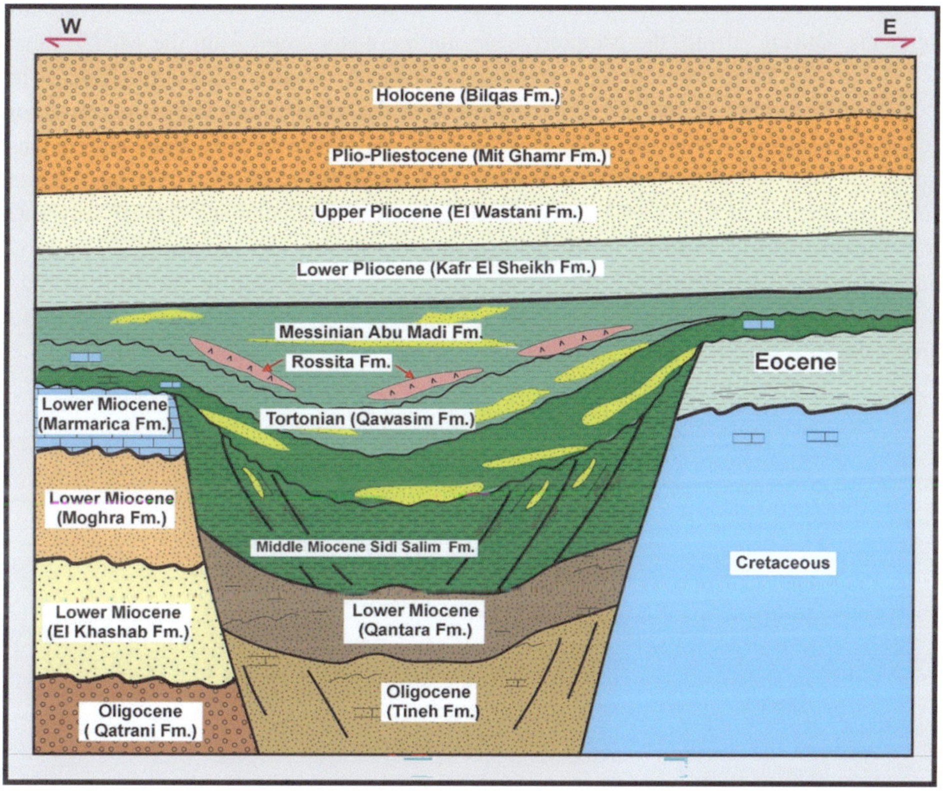

Fig. 4.3 East–West schematic cross-section showing the vertical and lateral distribution of the Neogene Deltas (Tineh, Qantara, Sedi Salem, Qawasim, Rosetta and Abu Madi formations) and the Quaternary Delta (Kafr El Sheikh, Al Wastani, Mit Ghamr and Bilqas formations); and their correlation with the older rocks on both sides of the Mediterranean Nile Delta

the Oligo-Miocene periods. This was most likely linked to volcanic activity, which has a poisonous effect on the growth of these types of fossils.

For the past fifty years, there has been disagreement over the provenance reconstruction of the Oligocene sedimentary rocks found in the Mediterranean Nile Delta. Two viewpoints have been discussed in this context. First, it was suggested that the Mediterranean Nile Delta originated from the Ethiopian plateau during the Late Messinian (Late Miocene) period (Chumakov 1967, 1973; Issawi and McCauley 1992; Said 1993). The opposing view indicates that the Nile Delta's Ethiopian source rocks date back to the Oligocene (McDougall et al. 1975; Burke and Wells 1989; Williams et al. 2000). According to our current research, the deposition of the Tineh Formation occurred in the Late Oligocene–Early Miocene, when the Mediterranean Nile Delta began to form.

The Oligocene Tineh Formation was deposited under mixed marine and fluvial environments. During rifting, the Mediterranean marine water came from the Mediterranean Sea as revealed by the presence of claystone enriched with benthonic foraminifera, while the fluvial environment that supplied the sands and conglomerate facies in this formation most probably came from the south via the Nile River water. Both mixed facies were deposited, filling the elongated canyon formed by rifting during Oligocene–Miocene times. It is clear that the Tineh Formation was deposited in a submerged deep canyon (on downthrown sides), while its equivalent Qatrani Formation was deposited on upthrown side during the Late Oligocene (Fig. 4.4). Concerning its position in the delta setting, it is proposed that it occurs in the delta as evidenced by presence benthonic foraminifera (Fig. 4.5).

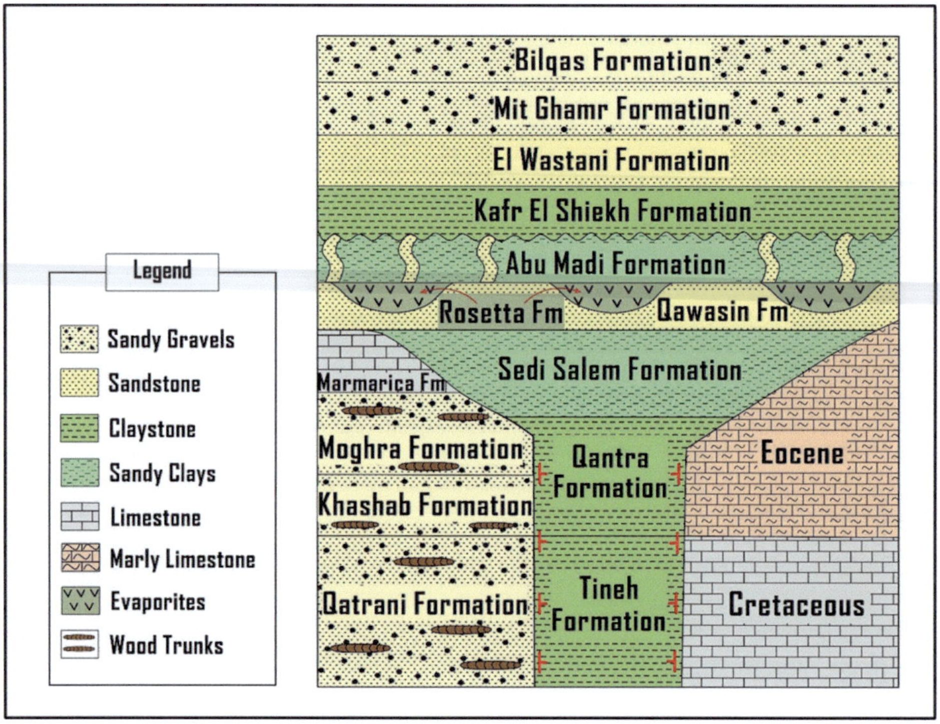

Fig. 4.4 Depositional setting and correlation between the Neogene rock units with their corresponding rock units occurring on both shoulders of the Mediterranean Nile Delta

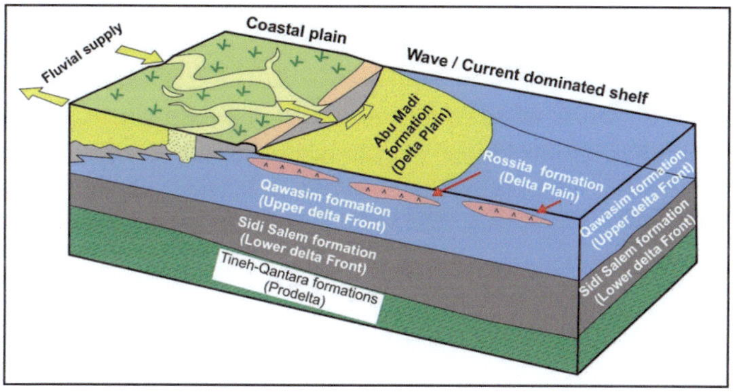

Fig. 4.5 Depositional model of the Neogene Delta manifesting the depositional setting of the Tineh, Qantara, Sedi Salem, Qawasim, Rosetta, and Abu Madi formations

4.2.2 Depositional History of the Qantara Formation

The north side of the Nile Delta's offshore region is home to a large geographic distribution of the Qantara Formation. The formation unconformably overlies and underlies the Tineh and Sidi Salem formations onshore at the El-Fayrouz borehole, respectively. In the Rommana-1X borehole, it is recorded from 2555 to 2000 m with a total thickness of 555 m, whereas at the El-Fayrouz borehole, it is encountered from 2725 to 2200 m depths with a total thickness of 525 m (Faris et al. 2007). Shale intercalated with argillaceous sandstone and limestone rich in assemblages of planktonic fauna make up the Qantara Formation. Because of the presence of limestone in this formation, the rocks were likely deposited on the outer shelf. This environmental setting is confirmed by the presence of planktonic foraminifera (e.g., *Globigerinoides bisphericus, Neogloboquadrina continuosa, Globorotalia archaeomerandii, Globigerina*brazieri, *Globigerina praebulloides, Orbulina* cf. *suturalis, Orbulina universa*, and *Globigerinoides illobusimmaturus*) and with some association of benthonic foraminifera (e.g., *Bolivinadilatata, Gyroidinoidessoldanii, Nonioninascapha, Buliminaelongata, Lagenasulcata, Globocassidulinaelongata*, Uvigerinasemiornata, Nonionellaauris, Lenticulinacultrata, Cibicidesrefulgens, and *Neoeponides*sp) (Ismail et al. 2010). The Lower Miocene El Khashab and Moghra formations, which are exposed on the western side of the delta, can be correlated with this formation (Figs. 4.4 and 4.5). The El Khashab and Moghra formations were deposited in the delta plain on the upthrown side of the Fayium Nile Delta in the Western Desert, whereas the Qantara Formation was formed in the prodelta on the downthrown side in the Neogene Delta (Fig. 4.4). This is based on the difference in lithology between El Khashab and the Moghra formations (coarse clastic) and the Qantara Formation (mainly claystone) (Fig. 4.4). This is demonstrated by the fact that the Qantara Formation, which is located in the Neogene Delta, has a greater total thickness (more than 500 m thick) more than the

Khashab and Moghra formations (85 m thick), which are equivalent and are deposited on the upthrown side (Fig. 4.4). The presence of a fault that separates between the Moghra and Qantara formations and trends northwest is most likely the cause of the Qantara Formation greater thickness when compared to the El Khashab and Moghra formations (Fig. 4.4). Because of the abundance of planktonic and benthonic foraminifera, the depositional setting of this formation in the Neogene Delta most likely represents prodelta similar to the Tineh Formation (Fig. 4.5).

4.2.3 Depositional History of the Sedi Salem Formation

The Sidi Salem Formation has a wide geographic distribution in the Neogene Delta as compared with the distribution of the Tineh and Qantara formations (Figs. 4.2, 4.3 and 4.4). It exhibits two facies sequences. In onshore (Sedi Salem well-1), the sequence is made up of claystone, siltstone, and limestone, which demonstrates deepening-upward (or fining-upward) cycles. Each of which begins with sandstone, followed by siltstone, and capped by claystone (Fig. 4.6a). In offshore areas (Akhen fields), the sequence comprises sandstone, siltstone, and claystone. Such a sequence demonstrates deepening-upward cycles. Each cycle comprises sandstone at the base, followed by siltstone, capped by claystone (Fig. 4.6b). At Temsah in the northeastern offshore area, the sandstone occurring at the base of cycles declares a sudden increase in thickness. This is thought to be channelized turbidite and is considered a reservoir target (Khalid et al. 2011). It is remarked that the mixed clastic and carbonate lithofacies occur onshore, and pure clastic lithofacies occur offshore. The spread of these sediments is unusual because the usual occurrence of pure clastic (sandstone–siltstone) is commonly deposited onshore, while the mixed clastic-carbonate usually occurs offshore. This can be explained according to the following reasons: (1) the presence of mixed clastic-carbonate sequence onshore may represent embayment and restricted area led to sedimentation of limestone and fine clastic, and (2) the presence of pure clastic offshore was most probably due to supply with sandstone–siltstone from onshore and were deposited in trenches on the shelf. In general, the Sedi Salem Formation's depositional environments are based on studied core samples from offshore at Akhen fields, west of the Nile Delta (Khalid et al. 2011). They stated that this formation, which ranges from 4350 to 4554 m, consists of four sandstone channels intercalated with claystone and siltstone. The lithofacies mentioned above have a vertical stacking pattern that suggests superimposed or vertical repetition of fining-upward cycles. Sandstone channels form the bases of each cycle and are composed of Pebbly quartz arenite, which is composed of 80% quartz, 15% clay minerals, and 5% feldspars (Khalid et al. 2011). Sand in certain beds is medium- to coarse-grained, somewhat sorted, and subrounded to subangular. The sandstones were deposited as deep marine turbidite channels on the continental slope (El Sisi et al. 1996; Khalid et al. 2011). The siltstone and claystone that make up the cycle tops comprise kaolinite and illite. These minerals

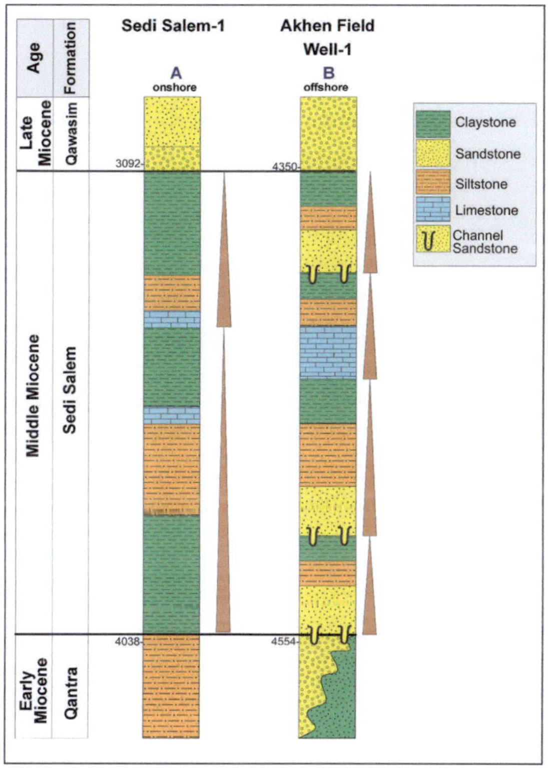

Fig. 4.6 Onshore subsurface log of the Sedi Salem Formation at Sedi Salem well.1 (**a**) manifesting fining-upward cycles, each one starts with siltstone, claystone, and is capped by limestone (**a**). In offshore at Akhen well-1, it consists of fining-upward cycles, each begins with bioturbated sandstone, followed by siltstone and capped by massive claystone (**b**) (after Khalid et al. 2011, with modification)

were most likely carried by a fluvial medium (the Nile River, most likely). These sediments were transported by turbidity currents and were deposited along the continental slope depositional environment (Middleton and Hampton 1973).

El Sisi et al. (1996) encountered several heavy non-opaque minerals of the Sedi Salem Formation, including garnet, tourmaline, zircon, rutile, monazite, and pyroxene. These minerals most likely originated in the Middle Miocene from metamorphic rocks found in the Red Sea Hills. The Sedi Salem Formation is equivalent to the Middle Miocene Marmarica Formation; but, the difference in their facies is due to the existence of a normal fault that separates them and trends northwest-southeast (Figs. 4.3 and 4.4). The faulting process was synchronous with sedimentation that gave rise to sedimentation of the Sedi Salem Formation on the downthrown side of the western side of the delta by the Nile River; while the carbonate facies of the Marmarica Formation were accumulated on the western upthrown side, deposited in shallow marine water (Fig. 4.4). In addition, the Sedi Salem Formation is thicker than (250 m) the Marmarica Formation (65 m), indicating a syn-depositional fault and ongoing sedimentation between them. Due to the occurrence of fining-upward cycles in the Sedi Salem Formation, this represents a lower delta front, deposited by the Nile River (Fig. 4.5).

4.2.4 Depositional History of the Qawasim Formation

After the deposition of the Sedi Salem Formation in the shelf slope in slightly outer-shelf or Abyssal environments, the sea-level fall may be concurrent with the Late Miocene and the beginning of the Messinian period. This gave rise to the deposition, of course, of clastics of the Qawasim Formation. This formation exhibits facies changes from onshore to offshore. In onshore setting at Qawasim well-1 and Basent well-1 (Abu Khadra et al. 2018), it mainly consists of sandstone, siltstone, and minor claystone (Fig. 4.7a, b), while in the offshore setting, it mainly comprises claystone with thin sandstone interbeds (Fig. 4.7c). In onshore at El Tamad Field, east of the Nile Delta, declared that the lithology of this formation comprises three main lithofacies: bioclastic limestone, dolomitic bioclastic sandstones, and massive and laminated sandstone (Leila et al. 2016; Ali et al. 2022).

In onshore at El Basant well-1 in eastern delta, the Qawasim Formation can be divided into two major facies, level-1 (in the upper part) and level-11 (in the lower part) (Abu

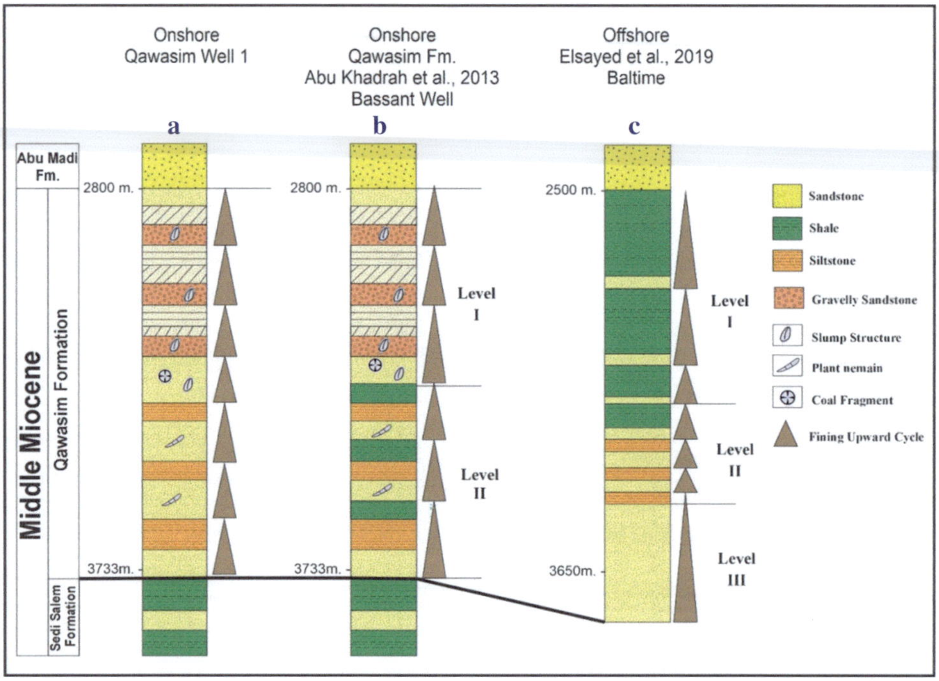

Fig. 4.7 Onshore vertical lithofacies of the Qawasim Formation that comprises fining-upward cycles (Qawasim well-1, **a**). It is divided into two levels (level-1 and level-11) at Basnet well-1 (**b**). In offshore at Baltime well-1, it is further divided into three level-1,11,11 (**c**) (after Abu Khadra et al. 2018)

Khadra et al. 2018) (Fig. 4.7b). Concerning level-1, the cycles comprise finning-upward cycles, each of which begins with conglomerate, followed by cross-bedded sandstone, capped by horizontal bedded fine-grained sandstone (Fig. 4.7b). Conglomerate/pebbly sandstone, present in basal cycle, occurs as small lens-like as well as moderately thick (2–3.5 m) channel-like bodies. Embedded clasts comprise rounded and well-rounded quartzite of pebble and cobble grades. The facies are generally massive, but with faint stratification and sand beds. It has been interpreted as channel lag deposits or longitudinal bar facies (Casshyap and Tewai 1984; Hota and Maejima 2004). It represents the product of highly concentrated currents flowing toward the toe of the delta front (Howell et al. 2005; Hampsons et al. 2011). Coarse-to-medium-grained sandstone occurring in the middle cycles represents channel to multistory and sheet-like bodies containing large-scale co-sets of planar and trough cross-stratified beds. The sediments have been interpreted as deposits of transverse/linguoid bars of low sinuous streams, whereas sheet-like bodies signify lateral accretion deposits in meandering streams (Casshyap and Tewai 1984; Maelmni et al. 2008). These channel-like bodies have been interpreted as vertical aggradation deposits as well as due to lateral shifting of coarse sandy channel bars (Tewari 1997; Maelmni et al. 2008).

The fine-grained sandstone facies that occur on top of cycles are commonly rippled, laminated, and horizontal lamination and correspond to the upper parts of sandy channel bars of low sinuous streams (Tewari 1997), whereas thick and persistent beds of vertical accretion are on stable and extensive overbank (levee) areas of meandering stream channels (Casshyap and Tewai 1984; Maelmni et al. 2008). Such types of cycles are the characteristic features of continental fluvial deposits (Miall 1978; Cant and Walker 1978).

In level-11, that occurs in the lower part of the Qawasim Formation, comprising fining-upward cycles. Each cycle consists of sandstone at the base, followed by sandy clays and capped by massive claystone (Fig. 4.7b). These fining-upward cycles represent the repetition of channel, overbank, and floodplain (lake or swamp) deposits in vertical order (Maelmni et al. 2008). It would imply a gradual decline in current competency from the lower to upper part in each cycle, which is considered a common phenomenon of river systems (Tewari 1997). It is generally believed that these fining-upward cycles are generated by the lateral migration of braid bars/meandering channels across the alluvial plain (Casshyap and Tewai 1984; Casshyap and Tewai 1984; Hota and Maejima 2004). The deposition of channel, overbank, and floodplain sediments should represent discrete and independent events of a fluvial system (Maelmni et al. 2008). Channel aggradation and lateral migration of stream channels are widely accepted causes for the development of fining-upward fluvial cycles (Miall 1978, 1988).

The basal part of the Qawasim Formation is more mature and was deposited under the more stable period of deltaic progradation, while its upper part is immature and was deposited during rapid cutting, following the rapid drop in sea level (Khalid et al. 2011; Abu Khadra et al. 2018). This formation represents both fluvial environments toward the landward side and shallow marine influences toward the basinward side, with scarce calcareous planktonic fauna. It represents the first major channel fill in the Miocene sequences of the Nile Delta.

This facies association suggests a significant supply of coarse-grained sediments by unidirectional currents. The development of steep depositional clinoforms is rather common in river-dominated transitional systems, typically characterizing fluvial delta-front settings (Hampsons et al. 2011). Additionally, the absence of trace fossils might reflect troubling conditions, which could be the result of mixing of marine and freshwater, episodic sedimentation, and/or a high proportion of suspended material in the water column (MacEachern et al. 2005; MacEachern and Bann 2008).

The above-mentioned lithofacies setting of the Qawasim Formation suggests that this formation resembles the upper delta front (Proximal delta front) of the Oligo-Miocene of the lower Nile Delta (Fig. 4.5). This is in harmony with the interpretation of Leila and Moscariello (2018). This formation has no equivalent rock units west of the delta, where the Middle Miocene Marmarica Formation is not covered by upper Miocene sediments. The absence of the upper Miocene sediments above the Middle Miocene Marmarica Formation may indicate the continuous subsidence and continuous sedimentation along the downthrown side of the western delta, during which the Qawasim Formation was deposited.

4.2.5 Depositional History of the Rosetta Formation

The main constituents of the Rosetta Formation (40–45 m thick) are gypsum and anhydrite intercalated with sandstone, siltstone, and sporadic thin limestone beds (Fig. 4.8). The evaporites occur mainly in the northern part of the Delta and offshore areas (Rashid et al. 2018), restricted between the Qawasim Formation below and the Abu Madi Formation above (Fig. 4.8). As seen in Rosetta well-1, the lithofacies mentioned above show shallowing-upward cycles, the majority of which start with claystone and end with evaporites (Fig. 4.8a). In the Fayrouz well, the cycles begins with sandstone, followed by siltstone and evaporites (Fig. 4.8b). The sandstone and siltstone that are found at the basal cycles are typically cross-bedded, reddish and laminated. They were deposited in the system's higher energy fluvial parts, which are similar to the Middle Miocene Fatha Formation in the Kurdistan Region of Northeastern Iraq (Aqrawi 1993). The evaporites that form on top of cycles are frequently laminated, usually with a thickness of a few millimeters. These evaporites are very similar to those that have been identified as Holocene evaporite sequences in the Red Sea's coastal hypersaline pools (Kushnir 1981), the Gulf of Suez, Egypt (Aref et al. 1997), and the Mediterranean basin's Messinian deep water evaporites (Hardie and Eugster 1971; Garrison et al. 1978). The shallowing-upward cycles manifest a high frequency of transgression/regression phases. The transgressive phase was responsible for the deposition of sandstone, siltstone, and limestone that make up the cycle's base were most likely deposited during a brief transgression phase, while the

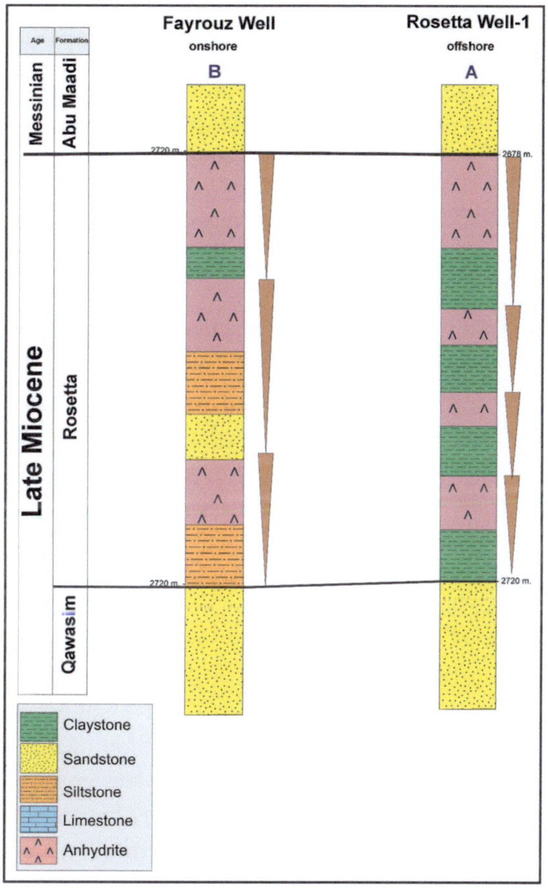

Fig. 4.8 Lithologic log of the Rosetta Formation that includes vertical stacking patterns of shallowing-upward cycles at its type locality (Abu Madi well-1, **a**) and in onshore (Fayrouz well-1, **b**), central Newgene delta (after Rizzini et al. 1978; El Heiny 1982 with modification)

regressive phase led to the formation of evaporites at the cycle's top. The current study interprets the subaqueous deposition of evaporites at the apex of a regressive cycle in delta plain tidal flats as the result of marine flooding (Fig. 4.5).

After the advent of the Qawasim Formation, there was a continuous fall in sea level, reaching the beginning of the Messinian time (Cita et al. 1977). During this time, the Mediterranean Sea was isolated from the Atlantic Ocean, causing the deposition of evaporites in the deeper basins. This phenomenon is known as the Messinian salinity crisis event (Rizzini et al. 1978). This isolation resulted in a "salinity crisis" in the Mediterranean, whereby the world's ocean lost 6‰ of its salinity because approximately one million cubic kilometers of evaporites were deposited on the floor of the deep Mediterranean basins (Hsü et al. 1978). The major drop in the sea level coupled with widespread erosion of the exposed formations, with subaerial incisions of deep canyons, has occurred (Barber 1981). Subsequent evaporation of the Mediterranean Sea led to the deposition of thick evaporite sequences (primarily near the basin center) and the formation of numerous incised valleys on the shelf and shelf slope (Barber 1981; Abdel Aal et al. 2001; Loncker et al. 2006). During this event (early Messinian), evaporites were developed in the depression where seawater remained isolated from the fluvial drainage (Fig. 4.5). The Rosetta Formation represents the time of closing of the Mediterranean Sea. While in the onshore area, the same time is represented by channel fill sediments of Abu Madi/Qawasim formations, which are composed of channelized sandstone and shale. In terms of delta setting, the Rosetta Formation was deposited in the lower delta plain (Fig. 4.5).

4.2.6 Depositional History of the Abu Madi Formation

The distribution and facies of Abu Madi Formation were controlled by paleotopography initiated by faulting, lowstand erosion, and rapid deposition of the Messinian age (El Heiny et al. 1990). According to the lithological types of the Abu Madi Formation, it contains three levels (1, 11,111) (Abdel Fattah 2014; Leila et al. 2020). Kassab et al (2015) recognized four lithofacies in the Abu Madi Formation in El Wastani field onshore Nile Delta; these are: F1 (Horizontally laminated, burrowed sandy-mudstone), F2 (very fine to fine, calcareous, argillaceous, cross-stratified to massive sandstone), F3 (Fine to medium, calcareous, argillaceous, tabular cross-stratified to massive, grey sandstone), and F4 (coarse, pebbly in part, calcareous, trough cross-stratified to massive sandstone).

The drilled sequence of the Abu Madi Formation in the subsurface, either onshore or offshore, gave rise to the subdivision of this formation into fining-upward cycles. In onshore, at Abu Madi well.1, the cycles consist of pebbly sandstone and conglomerate, followed by sandstone and capped by burrowed mudstone (Fig. 4.9a), while in offshore at Faraskour well.4, the cyclic pattern differs from Abu Madi well.1, where each cycle comprises sandstone at the base, followed by siltstone and capped by mudstone at the top (Fig. 4.9b). The subsurface well logging confirms these fining-upward cycles (Fig. 4.10). At the base of the cycle, the conglomerate assumes 10–30 m thick and is made up of 90% conglomerate and 10% sandstone (Abu Hashish et al. 2022). This facies is dominated by a clast-supported, structureless conglomerate that gives the impression of being massive. A poorly developed stratification is seen up-section in this area, and the size of the clasts changes as the sandstone matrix increases. The majority of the clasts are made of quartz and black and white chert, with minor sedimentary breccia (Abu Hashish et al. 2022). These particles are extra-basinal derivation plutonic varieties. Both proximal braided streams and stream-flow-dominated alluvial fans are consistent with a traction origin, which is supported by the conglomerate lithofacies (Eriksson 1978; Miall 1978; Rust and Koster 1984). The conglomerate beds, which are massive to coarsely stratified, are thought to be the result of bed load transport during flash flooding. According to Sohn et al. (1999), the abundant angular to subangular clasts in the area could be explained as debris flows.

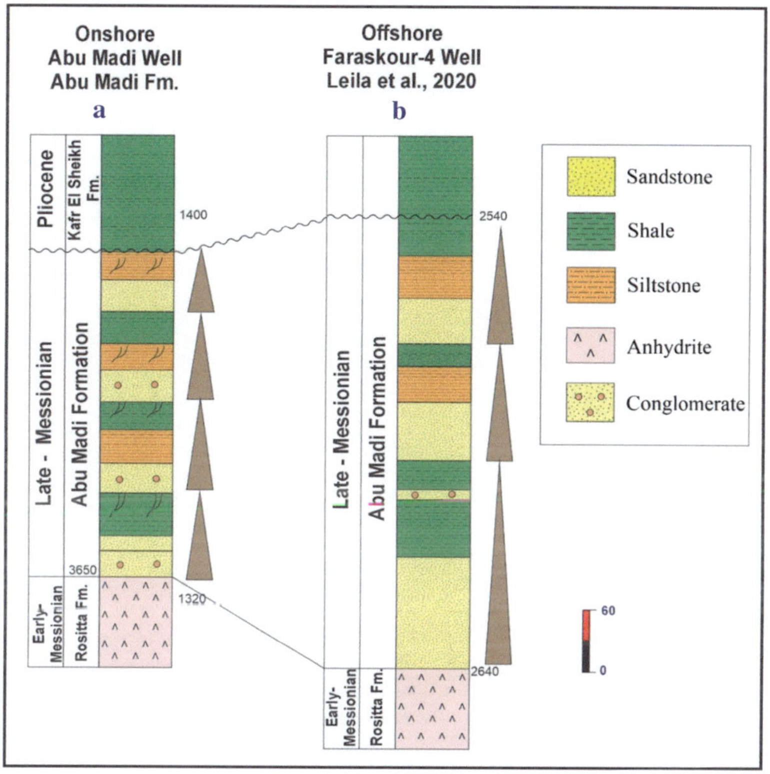

Fig. 4.9 Lithologic log of the Abu Madi Formation that includes vertical stacking patterns of fining-upward cycles at its type locality (Abu Madi well-1, **a**), and in the offshore (Faraskour well.4, **b**) (after Rizzini et al. 1978; El Heiny 1982 with modification)

The sandstone that covers the conglomerate and makes up the middle part of the cycles is typically cross-bedded, with a range of grain sizes from very coarse to fine sand. Its thickness varies from 15 to 25 m. Pebble stringers are observed to define set boundaries locally. Below the cosset boundaries, there are some locations where thin beds of laminated sandstone, siltstone, and mudstone, or wave and combined-flow ripple bed forms are still present. According to Rizzini et al. (1978) and Leila et al. (2020), quartzarenite, which is typically found in sandstones, is composed of quartz grains that are rounded to subrounded and are associated with subangular plutonic lithoclasts. Unidirectional flows altered by different tidal currents produced cross-bedded structures. Sets of trough and tabular cross-beds in the sandstone were formed by the migration of three-dimensional and two-dimensional bed forms (Simpson et al. 2002). The top of the cycles is defined by the laminated siltstone and mudstone lithofacies, which have a maximum thickness of 25 m. At the top of upward-fining cycles, these facies are often bioturbated and can occasionally be absent due to erosion at the overlying subsequent cycle. The thickness of this

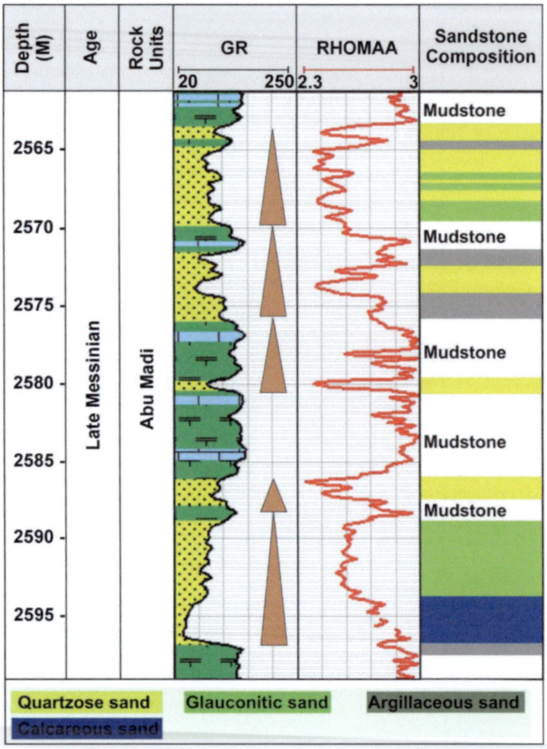

Fig. 4.10 Well log of the Abu Madi formation illustrating the fining-upward cycles. Notice the variation in thickness of the burrowed claystone forming the cycle tops as compared with the sandstone in the basal cycle (after, Leila et al. 2016)

facies varies from cycle to cycle, sometimes becoming thinner and other times thicker than the basal cycle sandstone (Fig. 4.10). Given that it has some vertical burrows, this lithofacies suggests an upper intertidal flat setting (Simpson et al. 2002). Channel river sediments have also been reported to contain similar facies (Sambrook Smith 1996; Sambrook Smith and Ferguson 1995).

The increasing influence of tidal effects over time at the expense of fluvial processes is manifested by the vertical succession of lithofacies within fining-upward cycles (Simpson et al. 2002). At the top of upward-fining cycles, there is an increase in the abundance of interlaminated sandstone–siltstone and mudstone, and the proportion of conglomerate decreases upward, reflecting this transition (Simpson et al. 2002). Relative sea-level rises associated with either "jerky" subsidence or, more likely, absolute sea-level fluctuations are responsible for the upward shift within cycles from fluvial to tidal settings (Simpson et al. 2002). Eriksson (1978) explained them as the result of lateral accretion in tidal channels with high sinuosity connected to a fluvial supply system. Paleogeographic mapping from the Late Miocene shows isolated basins with transient, continental, inner neritic, and deltaic subenvironments. The Abu Madi Formation was deposited during the Messinian period via a northward-entrenched channel that was confined by structural and topographic highs. According to El Heiny et al. (1990), the Abu Madi Formation was deposited under a variety of distinct depositional environments, beginning with fluvial conditions at its lowest point and moving up to lagoonal conditions and more marine conditions at the top. According to Salem et al. (2005), the Abu Madi Formation was deposited under estuarine depositional conditions, corresponding to the fluvial low-stand systems tract (LST) up to the transgressive systems tract (TST). It is remarked that the Abu Madi Formation manifests lateral facies from south to north. Hence, in the south (onshore, East Delta wells, Abu Madi, and El Qara), it has the thickest sandstone channels and the thinnest mudstone, while in the north (offshore, south and North Baltim wells, Fig. 4.11), it has the thickest mudstone and the thinnest sandstone (El-Heiny and Morsy 1992). These facies changes resulted from the Abu Madi Formation's deeper depositional setting, which goes farther north and allows for more mudstone to be added at the expense of sandstone.

Based on the previously stated information, the Abu Madi Formation was likely formed during a period of sea-level fall and was deposited as an incised valley fill toward the end of the Late Miocene (Messinian period) in a very short amount of time. The Abu Madi Formation is located in a delta plain setting according to the deltaic facies sequence mentioned above (Fig. 4.5).

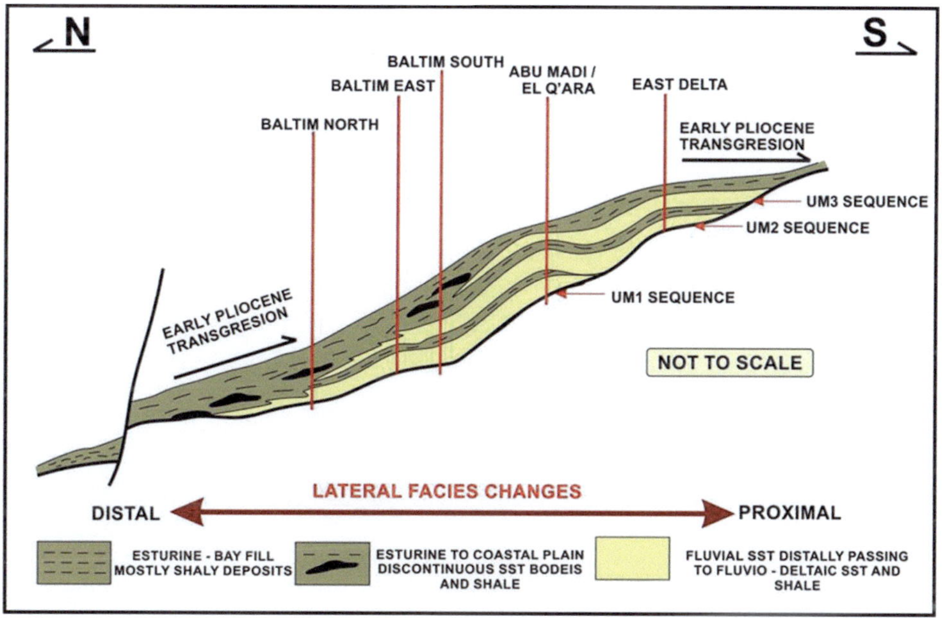

Fig. 4.11 Schematic cross-section showing the lithologic correlation of the Abu Madi formation from south (onshore) to north (offshore) direction. Notice the increased thickness of the sandstone channels and decreased thickness of mudstone in onshore and the vice versa in offshore northwards (after, El Heiny and Morsi 1992)

4.3 Depositional History of the Quaternary Nile Delta

The formations that make up the Quaternary Nile Delta, which unconformably lie over the Neogene Delta, include the Kafr El Shiekh Formation (Pliocene–Pleistocene), El Wastani Formation (Pleistocene), Mit Ghamr Formation (Pleistocene), and Bilqas Formation (Holocene) (Figs. 4.2, 4.3 and 4.12). The distribution of the Quaternary Nile Delta was associated with high wave energy, coastal subsidence, and an eastward littoral drift, as well as secular variations in the semi-arid climate and the input and influx of sediment from the Nile River (Barber 1981).

4.3.1 The Depositional History of Kafr El Sheikh Formation

After the Messinian period (Late Miocene), there were openings of the Atlantic Ocean with the Mediterranean Sea via the Jabel Tarek strait that occurred between Morocco and Spain, leading to the maximum transgression and flooding event. Such a transgression peak makes the Kafr El Shiekh Formation the only rock unit that has a wide geographic distribution in the Quaternary Delta. The initial transgression flooded and covered the whole delta and filled the ancient Nile Canyon as far south as Aswan, where its sediments are probably marine (Harms and Wray 1990; Issawi et al. 2009). This is delineated from the observed pure claystone facies enriched with planktonic foraminifera of the Kafr El Sheikh Formation (Fig. 4.13a, b). This formation has a monotonous and even constant lithological character that exclusively consists of thick claystone beds intercalated with thin fine-grained sandstone (Rizzini et al. 1978; Zaghloul et al. 1977), (Fig. 4.13a, b). This formation exhibits facies changes from onshore to offshore. In onshore (Kafr El Shiekh

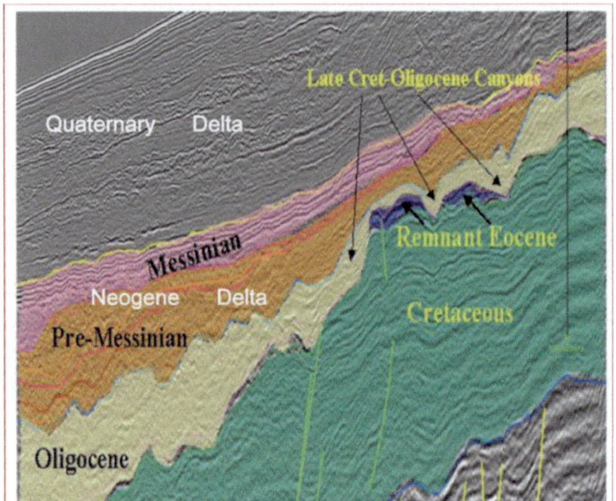

Fig. 4.12 Seismic cross-section along north–south direction showing the angular contact between the Neogene and Quaternary Deltas of the Mediterranean Nile Delta (after Elsheikh 2021)

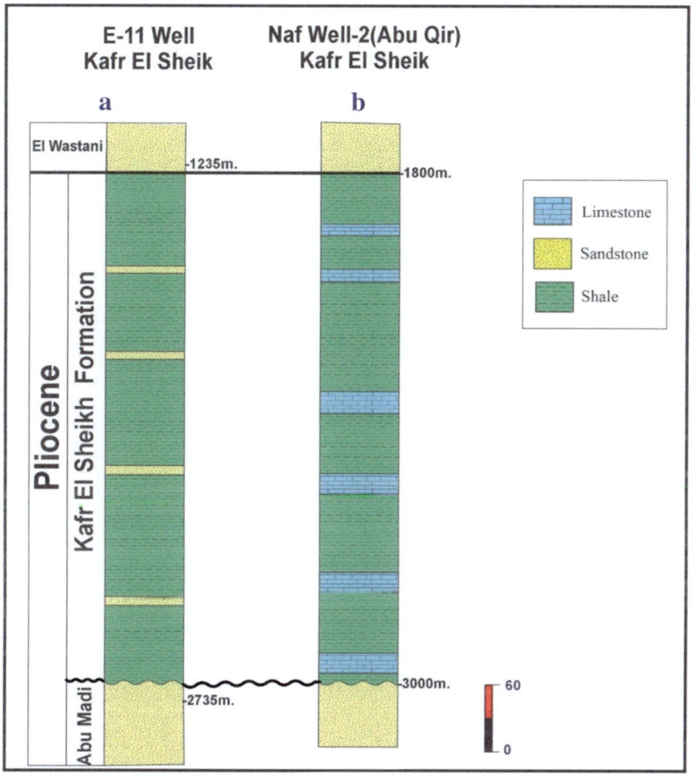

Fig. 4.13 **a** Lithologic log for the Kafr El Sheikh Formation at its type locality (Kafr El Sheikh well-1), which comprises pure shale and claystone in the central delta. **b** Lithologic log for the Kafr El Sheikh Formation at Naf well-2, which comprises pure shale and limestone intercalation in central delta

well-1, E-11 well), the formation includes two lithofacies, thick claystone and thin fine-grained sandstone lithofacies (Fig. 4.13a), while in offshore (Abu Qir wells), it comprises thick claystone and thin limestone interbeds (Fig. 4.13b). In this formation, the majority of the clays consist of kaolinite and montomorillonite, with very minor illite minerals. These minerals form a thick, voluminous section that has nearly identical characteristics throughout the entire Delta region (Khalid et al. 2011). The common presence of fine sand beds intercalated with claystone suggests that the sea level was dropping periodically throughout the interval. The sand supply most probably came from the coastal zone (Rizzini et al. 1978). Concerning the reduced transport capacity of the fluvial currents, the supply of sand also decreased quantitatively.

In general, the wide geographic distribution of the Pliocene facies all over the Quaternary Nile Delta suggests an overall transgression phase, in which the deeper water clayey sediments occur near the base of the Kafr El Shiekh Formation (Ayyad et al.

1987). The high transgression coincides with a Pliocene global rise in sea level (Sestini 1989). According to the structural contour maps created by Rizzini et al. (1978) and Zaghloul et al. (1977), the Pliocene Sea invasion over the Delta continued in a relatively wider V-shaped basin. According to Rizzini et al. (1978) and Zaghloul et al. (1977), the depositional environment of the Kafr El Sheikh sediments ranges from inner to outer neritic conditions. Such sediments were deposited in an outer-shelf environment as evidenced by the occurrence of certain benthonic and planktonic foraminifera, e.g., *Cassidulinabrocha, Ammonia beccarii, A. tepida, Pulleniaosloensis, Bolivinadilatata, Buliminaelongata, Uvigerinapygmoidea, and Vulvulineriacomplanata* (Rizzini et al. 1978). The faunal distribution in this formation reflects an increase of T.N.F. values and high P/B values mostly at the base and top of this formation, as compared to its middle part, indicating a fluctuating water depth, which is lower for the central section of the formation. Also, the same conditions prevailed in the upper part of the same formation as indicated by the occurrence of *Elphidium advenum, Quinqueloculinabosciana, Globocassidulina oblonga, and Fursenkoinaschreibersiana* (Ismail et al. 2010).

The Kafr El Sheikh Formation was mainly deposited in a slope-to-basin-floor environment. This was enhanced with salt and syn-depositional fault movement. These conditions include a graded slope that coexists along the strike, where unconfined turbidite deposition has occurred in slope channel systems. Sheet sands may be deposited along depressions caused by faults (Sestini 1989; Nini et al. 2010). In terms of the delta lithofacies setting, the Kafr El Shiekh Formation was deposited in pro-delta facies because it contains planktonic foraminifera more than benthonic foraminifera (Fig. 4.14). Dark grey siltstones and mudstones with centimeter-bedded levels that are rich in planktonic fauna make up the majority of the pro-delta sediments. Prodelta can also contain interbedded, extremely fine sandstones, which typically range in thickness from mm to cm and get thicker and more prevalent as they ascend (Plint 2003). The prodelta is the area where only suspension disturbance causes quiet sedimentation. According to Reading and Collinson (1996), the subaqueous delta's portion beneath the base of the fair-weather wave is known as the prodelta.

4.3.2 Depositional History of El Wastani Formation

After the deposition of the Kafr El Shiekh Formation, there was a decrease in sea level and contemporaneous regression of the sea water northwards. This gave rise to the deposition of coarse clastic facies of the El Wastani Formation. Therefore, the El Wastani Formation is regarded as a transitional facies between the underlying open marine shelf facies (prodelta) of the Kafr El Sheikh Formation and the overlying coastal and continental sands and delta plain of the overlaying Mit Ghamr Formation (Alfy et al. 1992). The thickness of the penetrated El Wastani section ranges from about 200 to 600 m. Isopach map of this formation in the Temsah area shows that thickness increases in the southwest

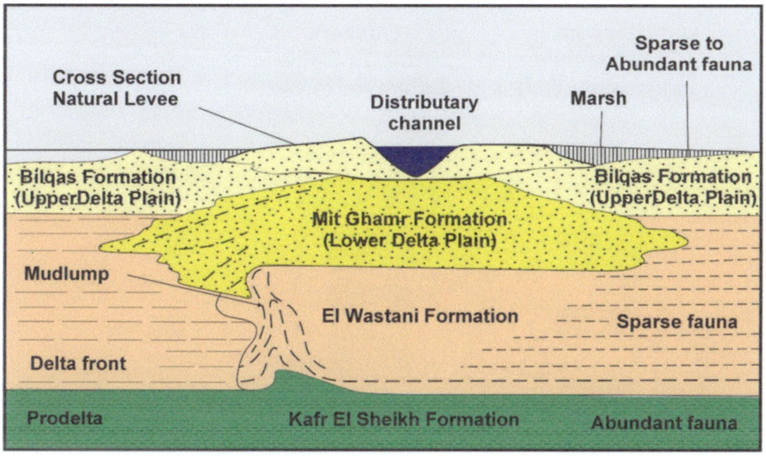

Fig. 4.14 Schematic cross-section of the depositional model of the Quaternary Nile Delta, showing the depositional setting of Kafr El Sheikh Formation (prodelta), Al Wastani Formation (Delta front), and Mit Ghamr and Bilqas formations (delta plain)

direction and decreases to the northeast direction, giving a reverse trend in comparison with trends of the older formation, especially the Kafr El Sheikh Formation (Rashid et al. 2018). This could be attributed to the regression of the channel belt in the Temsah area and the shifting of the depositional lobes toward the southwest direction with a low depositional rate at the deep marine areas to the northeast (Rashid et al. 2018). In onshore regions (El Wastani well-1), this formation consists of thick sand beds interbedded with a thin clay bed. Such a sequence comprises fining-upward cycles, each of which begins with thick sandstone, followed by thin claystone (Fig. 4.15a). The sandstone shows thinning toward the top of the formation (Fig. 4.15a). In offshore areas (Abu Qir and Baltim wells), the claystone exhibits an increase in thickness at the expense of sandstone. In such a case, the vertical arrangement of lithofacies exhibits fining-upward cycles. Each cycle includes sandstone at the base, followed by siltstone and limestone, capped by mudstone (Fig. 4.15b). This phenomenon is recorded in the Sequoia field area, where the El Wastani Formation comprises sandstones and mudstones arranged into a broad upward-fining succession (Abu Hashish and Ali 2021). The sands forming the basal cycles are laminated, massive, coarse- to medium-grained with little feldspar and rock fragments. The clays are soft, laminated, and massive with some plant remains.

The fining-upward cycles either onshore or offshore manifest the increased influence of tidal effects over time at the expense of fluvial processes (Walker and Cant 1984; Simpson et al. 2002). This transition is reflected in the upward decrease in the proportion of conglomerate, and concurrent increase in abundance of interlaminated sandstone–siltstone and mudstone at the top of fining-upward cycles. The upward transition within cycles from fluvial to tidal setting can be attributed to relative rises of sea level related either to

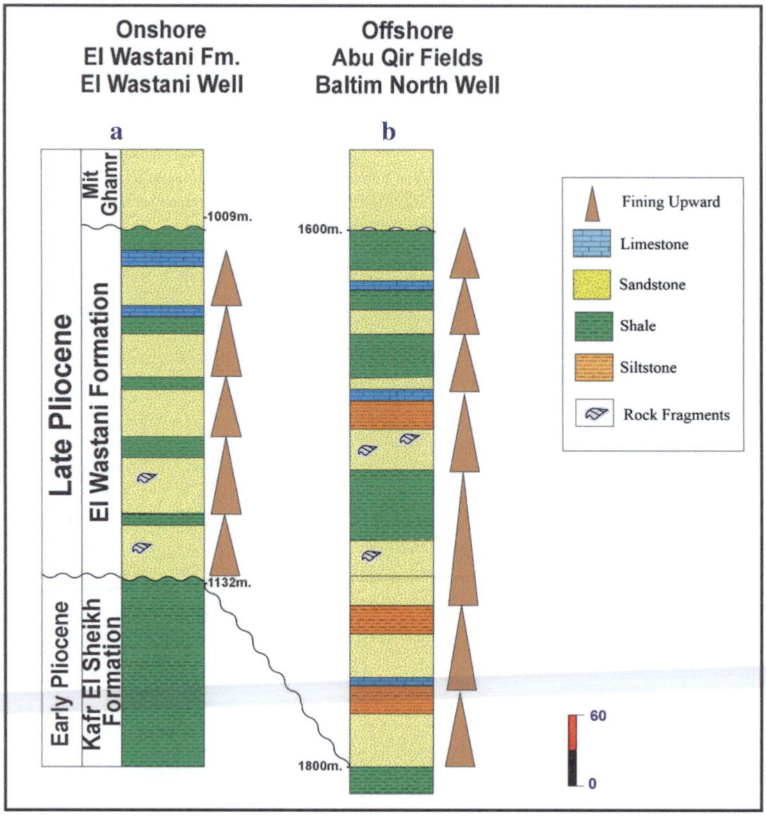

Fig. 4.15 Lithologic log for the Al Wastani formation that consists of fining-upward cycles in the Al Wastani well (**a**), and in the Abu Qir fields (**b**)

subsidence or, more likely, to absolute sea-level fluctuations (Simpson et al. 2002). The structure contour map made by Zaghloul et al. (1999) on top of the El Wastani Formation indicates a complex basin of general NE trend and is controlled by NE, NNE, and NW fault systems.

In three wells at the Scarab Gas Field using well logs data, declare two depositional sequences (sequence-1 and sequence-2) in the Pliocene–Pleistocene El Wastani Formation (Rashid et al. 2018). Each sequence includes lowstand, transgressive, and highstand systems tracts. This assumption is difficult to distinguish on seismic line due to the following: (1) because there is no adequate and demarcation of unconformable contact between sequences on the seismic images, (2) it is difficult to identify the systems tracts in each sequence. Hence, the adopted vertical stacking patterns of lithofacies are only described as fining-upward cycles as mentioned above. From the above facies sequence and its lithological component, it can be possible to set this formation in delta-front setting due to the absence of planktonic foraminifers and just presence of sparse fauna was recorded (Fig. 4.14). Delta-front deposits are characterized by fine- to medium-grained sandstone and the thickness of beds range from 1 to about 10 m (Plint and Nummedal 2000). The bases of sandstone bodies may be either gradational with underlying prodelta facies or sharp and erosive, commonly with spectacular gutter casts. Parallel lamination typifies fine sandstone, whereas trough cross-bedding is present in the upper fine- to lower medium-grained sandstones. The delta-front sandstone bodies are commonly capped by 1–2 m of planar-laminated to massive rooted sandstone interpreted as beach and backshore deposits (Plint and Nummedal 2000).

4.3.3 Depositional History of the Mit Ghamr Formation

After the sedimentation of the El Wastani Formation, further shallowing-upward and consequent fall in sea-level led to the deposition of coarse and very coarse clastic of the Mit Ghamr Formation. In the onshore, the Mit Ghamr well-1, the formation exhibits fining-upward cycles. Each cycle begins with pebbly conglomerate, followed by coarse-grained sandstone and capped by sandy siltstone (Fig. 4.16a). The pebbly conglomerate consists of quartzite, chert, and dolomitic lithoclasts, while the sandstone comprises medium- to coarse-grained quartz grains, chert grains, and shells of Pelecypods (Rizzini et al. 1978). The onshore, at the Qantara and Akhen fields, ranges in thickness from 400 to 600 m and shows the same manner of cyclic patterns encountered at Mit Ghamr well-1. It consists of thick pure coarse clastic (pebbly and coarse-grained sandstone) at cyclic base and capped by thin siltstone bed at the top (Fig. 4.16b). In offshore, the formation at Abu Qir field exhibits a drastic facies change, consisting solely of thick claystone intercalation and thin limestone beds (Fig. 4.16c). These lithofacies also show shallowing-upward cycles, where limestone caps each mudstone phase that starts (Fig. 4.16c). Zaghloul et al. (1989) drawn structure contour map of the upper Mit Ghamr Formation revealed a slightly different pattern than the El Wastani Formation below. It was visible that the NW Temsah and NE Rosetta fault systems controlled its structural orientation, with a gentle northward dip. There was a closed high structure in Abu Qir Bay and a closed low structure surrounding El Manzala lagoon and Tineh Bay. The sediments' average thickness was 700 m, which is thicker than the underlying rock unit (El Wastani Formation). In general, the Mit Ghamr Formation manifests facies changes from south to north (El Asmer 2023, Fig. 4.17). In the south (onshore) at Shiben El Kom, Qallin, and Hosh Isa, the sandstone constitutes the whole formation, while northwards (offshore), it comprises an intercalation of sandstone and claystone at Sindy-1, Tarif-1, Matariya-1, Temsah-1 and Baltime-1 (Fig. 4.17). At the Northwest Mediterranean Nile Delta specially and Abu Qir field, the formation exhibits radical facies changes, where the sandstone and siltstone are missed and substituted by limestone and claystone (Fig. 4.17). Lithofacies map of this formation declares the presence of three lithofacies (Serag El Din 1990), sandstone lithofacies covers the southern

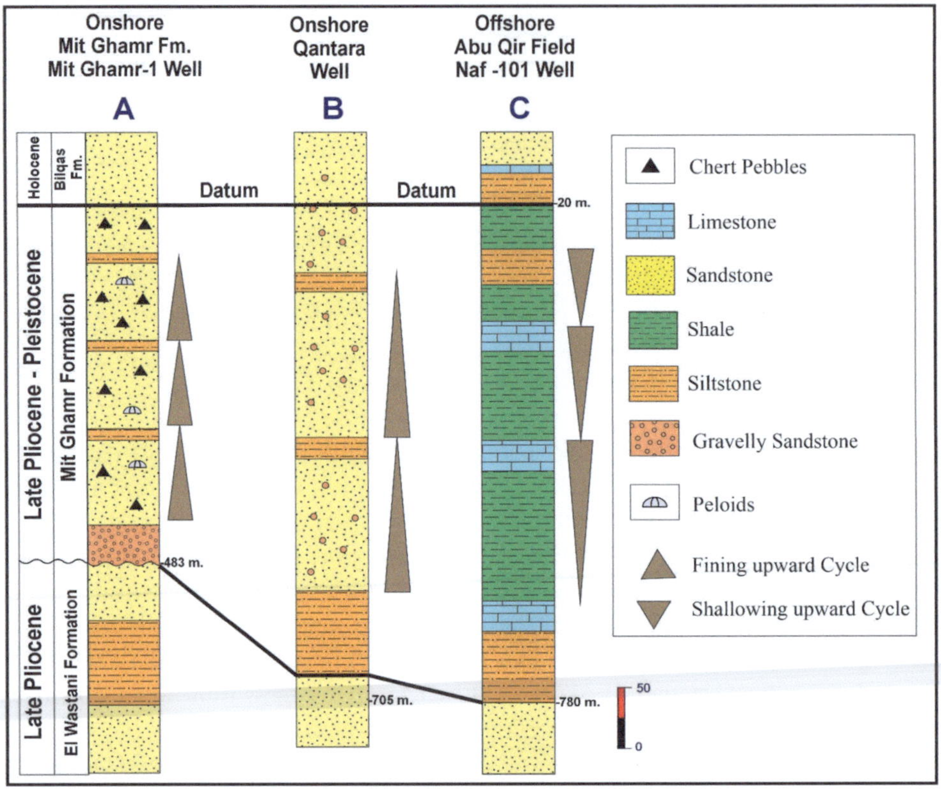

Fig. 4.16 Lithological characteristics and correlation of the lithology of Mit Ghamr Formation in onshore (Mit Ghamr well-1, **a**, Qantara well-1, **b**) and offshore (Naf well-101, **c**) localities in the Quaternary Nile Delta. Notice the facies changes between onshore and offshore

delta block and onshore, sandy clay lithofacies occurs on the flexure zone (shelf slope) and pure claystone lithofacies in northern delta and northern delta slope (Fig. 4.18).The formation constitutes the filling up of the basin by coastal sands or by deposits from Nile flooding that reflects shallow marine to fluvial environments and ranges from Late Pliocene to Pleistocene (El-Rawy 2020). On the other hand, Zaghloul et al. (1977) stated that the Mit Ghamr Formation was deposited in coastal and/or distributary mouth bar sand.

The facies type of the Mit Ghamr Formation at the south (onshore) that are represented by pure sandstone with sparse pebbles, indicating fluvial to coastal environments near the mouth of the Nile River. On the other hand, in the offshore region, the formation consists of mixed sandstone and claystone. This suggests that the substrate became more deepening northwards, where the amount of sandstone coming from the Nile River decreased in amount northwards. At the same time, the intercalated claystone with sandstone either

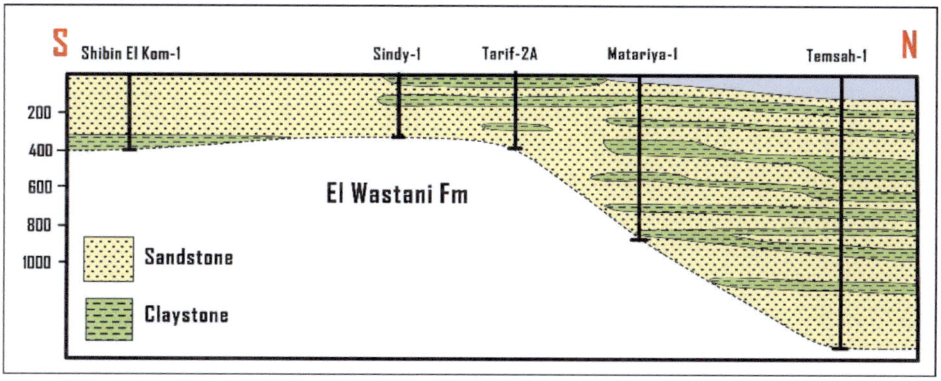

Fig. 4.17 Cross-section of the Mit Ghamr formation manifesting distinguished facies changes from south to north. Dominant sandstone facies occur in the south (onshore) at Shiben El Kom and Hosh Isa, while mixed sandstone and claystone occur northwards (offshore) at Sindy-1, Tarif-1, Matariya-1, Temsah-1, and Baltime-1 (after El Asmar 2023)

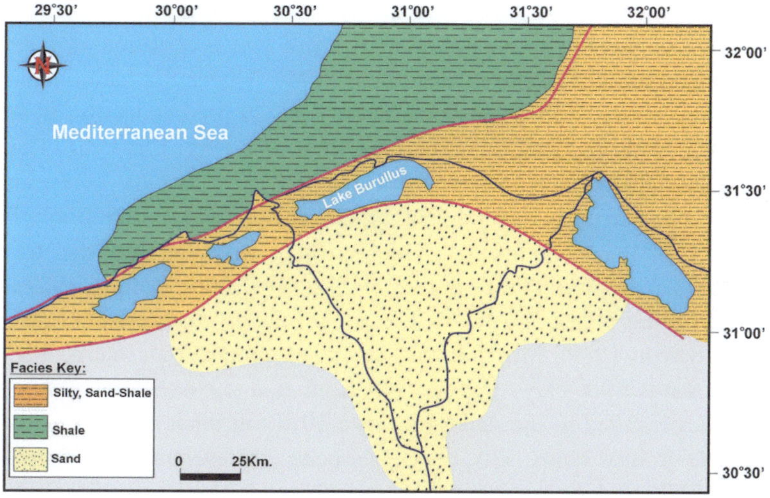

Fig. 4.18 Lithofacies map of the Mit Ghamr formation along the whole Mediterranean Delta. It manifests the gradual lithological changes, where it consists of pure sandstone on the onshore and southern delta block, sandy clays on the flexure one and shelf slope, and pure claystone on the northern delta block (after Serag El Din 1990)

came from inundation of marine sources or came from the Nile River, deposited in a deeper setting away from the coastal zone. Northwest at Abu Qir field, the formation was deposited in a slightly deeper setting or sheltered lagoon as evidenced by an increase in thickness (About 700 m) and consists entirely of mixed carbonate and claystone with the absence of coarse clastic influx. This area was sheltered from the open water waves, preventing the mixing of detrital sands coming from the Nile River. Moreover, this protected area was suitable for the living of micro-fauna, which in turn led to the formation of limestone. This interpretation is in harmony with Ismail et al. (2010), who stated that the Mit Ghamr Formation was probably deposited in a quiet shelf environment. However, these sediments were affected by post-depositional processes, forming some sort of canyon incised within the older and underlying sediments (Ismail et al. 2010). In addition, the depositional environment of this formation is probably coastal marine to fluvial. It is a typical fill-up basin with coastal sands, coquina beds, clay, and peat (Rizzini et al. 1978; Sestini 1995). On the subaerial Delta plain, silts and clays accumulate vertically in inter-distributary zones. At the mouths of deltaic rivers, marine processes rework fine-grained sediment, but more coarse deposits of sands and silts usually build forward while maintaining a steep seaward slope (Nini et al. 2010). In terms of the delta, this formation occupies the lower Delta plain setting (Fig. 4.14).

4.3.4 Depositional Environments of the Bilqas Formation

After the advent of the Mit Ghamr Formation, there was a cessation in sedimentation, producing an unconformity surface separating the Mit Ghamr Formation below and the Bilqas Formation above. The Bilqas Formation consists predominantly of very coarse sand interbedded with clay rich in molluscan fragments (Fig. 4.19a, b). The lithology of the Bilqas Formation is characterized by thin-bedded sediments, containing parallel bedding and small-scale cross-bedding. During the sedimentation of the Bilqas Formation, the development of the modern delta complex began with the Flandrian transgression following the Wirmian glacial stage during the last 6000 years (Sestini 1989). This was evidenced by drilling about 350 boreholes, which clearly manifested a post-Flandrian transgression (Sestini 1989). Most of such boreholes comprise an interbedded medium- to fine-grained sands, silts, clays, and peats (Rizzini et al. 1978; Sestini 1989). Such later transgression has resulted in the deposition of a 10–40 m thick complex of interbedded medium- to fine-grained sands, silts, clays, and peats, representing the Bilqas Formation (Rizzini et al. 1978) and is equivalent to the Neo-Nile 'delta' (Said 1981; Sestini 1989).

In the latest Pleistocene and Holocene times, the Bilqas Formation covers most, if not all the Nile Delta. During this time, the marine transgression covered most of the northern delta area and gave rise to deposits several meters (5–25 m thick) of marine sediments capped by agricultural soil (Azzam 1994). The faunal association reflects a lagoon or brackish swamps environment, intermittent with beach sand (Barakat 1982). The thickness of beds ranges from tens of centimeters to more than ten centimeters. This indicates a low-energy hydrodynamic condition of the deltaic facies (Pennington et al. 2017).

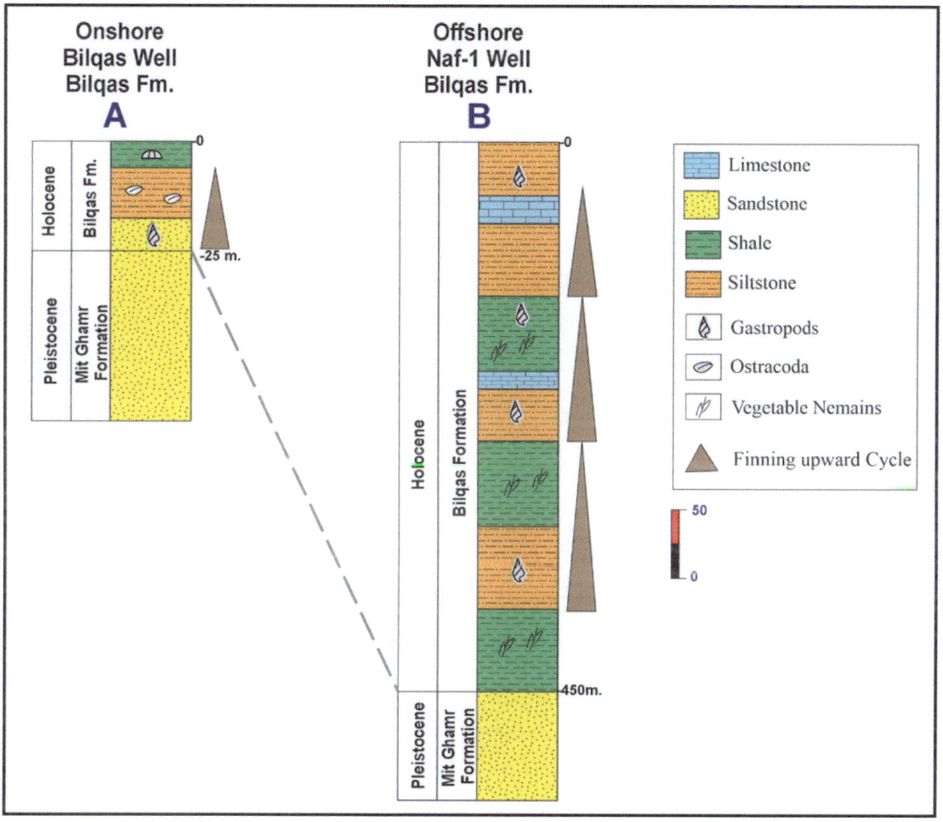

Fig. 4.19 Lithologic log for the Bilqas formation, notice the facies changes from onshore (mostly coarse clastics) and offshore (mainly fine clastics)

According to EGPC (1994), the probable depositional environment of the Bilqas formation ranges between coastal lagoons, swamps, and beaches. Frihy and Gamai (1991) studied the texture and mineralogy of 108 bottom samples from the Nile Delta, indicating a continental shelf environment. They classified such bottom samples into four facies: Facies I, (delta front), which includes fine and very fine sands, light minerals, heavy minerals, and "glauconite," extending from the shoreline to the inner shelf and decreasing away from the shoreline. Facies II (delta front) includes silty sand rich in biogenic components that occupies the entire outer shelf and its contiguous lower terraces. Facies III (prodelta) includes mud (silt plus clay), mica, and "glauconite," covering the middle shelf and the upper terraces. Facies IV (distributary mouth bar or progradational coastal sand) includes fine- to coarse-grained sediments and "glauconite," locally distributed along the coast off Lake Ldku, Burullus headland, and Damietta promontory. The configuration of the above facies pattern can help in identifying delta lobes related to former distributary branches of the Nile River (Canopic, Saitic, Sebennitic, Atribic, and Mendesian). Added to the above, the 4 facies-forming shelf sediments have resulted from sediment dynamics, sea-level fluctuation, and differences in sediment input and provenance (Frihy and Gamai 1991). Pennington et al. (2017) mentioned that the mid-Holocene environmental evolution of the Nile Delta is characterized by a transition from an earlier set of spatially varied landscapes dominated by swampy marshland to better-drained, more uniform floodplain environments. In the present work, we suggest that the Bilqas Formation was deposited in the upper delta plain that covers most of the Nile Delta either onshore or offshore (Fig. 4.14). This is in harmony with the interpretation of Orton and Reading (1993) who stated that the delta plains are extensive, lowland, marginal marine areas, that includes one or more distributary channels, separated by interdistributary zones of shallow water and emergent and near-emergent area. Such interdistributary zones include bay, lakes, lagoon, floodplain, marshes, tidal flat, beach ridges, and dunes (Fig. 4.14).

4.4 Depositional Model of the Mediterranean Nile Delta (Neogene and Quaternary Deltas)

The Mediterranean Delta was separated into two superimposed deltas, namely:

(1) Neogene Delta at the base (Late Oligocene-Late Miocene), and
(2) Quaternary Delta at the top (Pliocene, Pleistocene, and Holocene) (Figs. 4.12 and 4.20).

After the deposition of the Moghra Formation in the Fayoum Nile Delta (post Early Miocene), the Nile River shifted eastward during this period, precisely 25 km north of Cairo. This was most likely caused by geological forces reorienting the Nile River's route to the north. The geological forces were represented by the existence of down-faulting processes. One of the forces was the faulting and subsidence of the underlying basement, Cretaceous and Paleogene rocks (Fig. 4.20). The dominant structural elements, which enhanced the generation of the Mediterranean Delta (Neogene and Quaternary Deltas), took place during the Oligo-Miocene and continued up to the Holocene. Three structural trends dominated during the formation of the Mediterranean Delta (Abdel Aal et al. 1992, 1994). The first trend is called the Gulf of Suez trend (Oligo-Miocene), which is directed NNW parallel to the Gulf of Suez. This trend was observed in the western and central parts of the Delta. The second trend is NNE and is associated with the Gulf of Aqaba trend and is parallel to the Rosetta trend. The third trend developed in this phase is the NS Baltimore structural fault trend. It is believed that these mentioned trends were formed by rejuvenation and reactivation of pre-Tertiary structures during the Early Miocene (Abdel Aal et al. 1992, 1994). The above-mentioned structures were allowed for the generation

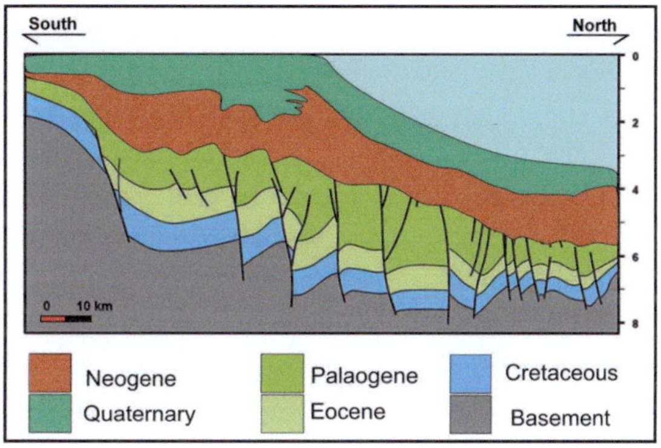

Fig. 4.20 Schematic block diagram showing the general occurrence of the Neogene and Quaternary Deltas of the Mediterranean Delta, which filled the rifted and subsidence of the Underlying rocks (basement, Cretaceous, and Paleogene)

of the rifted canyon on both sides of the delta, in which the deposits of the Neogene and the Quaternary Delta were deposited (Fig. 4.4).

Most of the drilled wells south of the hinge zone in onshore regions, specifically south of the Desouq area on the western side of the Nile Delta, the Cretaceous sediments and erosional remnants of Eocene strata directly unconformably overlain by the Oligocene continental sedimentary rocks (Barber 1981; Dolson et al. 2002) (Fig. 4.20). The Cretaceous and/or Eocene rocks underneath the Oligocene–Miocene rocks were eroded and incised creating an enormous accommodation spaces which subsequently filled with the Oligo-Miocene rocks (Fig. 4.12). The accommodation spaces or canyon valleys are often reached up to 6 km in width and 350 m depth (Barber 1981). These canyon valleys served as a main corridor for sediment to go across a shelf and feed a network of channels. At the end of the Eocene, a thick detrital series that was prograding northward was deposited offshore, marking the beginning of the development of the Nile Deep Sea Fan (Salem 1976; Hsü et al. 1973).

Above this significant canyon valley, the Haddadin Basalts (equivalent to the Oligocene Widan El Faras Basalts, north of the Fayoum depression) were erupted (Barber 1981; Dolson et al. 2002). According to Said (1962) and Barber (1981), these flood basalts were attributed to two significant igneous stages. One of them belongs to the Early rifting in the Gulf of Suez, and the second may belong to the Red Sea event. There was an interval period of around 2–3 million years in the Western Desert, where the Haddadin Basalt was worn by the impact of lower Miocene irregularities (El Heiny and Enani 1996). A comparable hiatus of around 1–2 M.yrs is recorded in certain offshore wells (Dolson et al. 2002). The primary reason of this hiatus was erosion, which may have been exacerbated by concurrent inversion brought on by the rifting of the Gulf of Suez and/or continent collision on the African plate's northern border (Aksu et al. 2005).

The Neogene Delta was deposited under two cycles; the first cycle represents the faulting process on both sides of the delta (east and west directions). This resulted in the rifting of the Lower Miocene Moghra Formation filling the basal parts of the rifted canyons (Fig. 4.3). This was followed by a transgression phase that resulted in the deposition of the claystone facies of the upper Oligocene Tineh and the Lower Miocene Qantara formations. They show angular unconformity with the underlying Cretaceous-Eocene rocks (Fig. 4.3). They were deposited on a narrow deep canyon on the shelf during the Oligo-Miocene, representing prodelta sediments (Fig. 4.20). This cycle was ended with the deposition of a conglomerate bed separating between the Qantara Formation below and the Sedi Salem Formation above. Upward, the Nile Delta Basin experienced a slight regression during the Middle Miocene, depositing the Sedi Salem Formation, which has a wider geographic distribution in the Neogene Delta as compared with the underlying Tineh and Qantara formations (Figs. 4.3 and 4.4). In general, the vertical staking pattern of the Sedi Salem Formation is which by the lithofacies arranged vertically into fining-upward cycles, each one starts with sandstone channels, followed by siltstone, and capped by massive claystone (Khalid et al. 2011). Such fining-upward cycles were mainly deposited on shelf slope or lower delta front, occurring above the pro-delta setting of the Tineh and Qantara formations (Fig. 4.5). Further upward regression gave rise to the deposition of the Qawasim Formation that represents the Late Miocene regression, after an early Late Miocene drop of sea level, and the Serravallian-Turonian tectonic event (Sestini 1995). This formation has a thick, irregular complex of sandstones and conglomerates interbedded with dark grey shale (fluvial-deltaic with littoral and lagoonal episodes in the upper part). In the northeastern, deeper sub-basin, it includes a thick (1–3 km) complex with deltaic clinoform facies (Sestini 1995). The lower part of this formation represents the distal front delta, while its upper part represents the proximal delta front (Fig. 4.5).

After the deposition of the Qawasim Formation, the Messinian crisis of the Mediterranean Sea took place, creating desiccation on the continental slope and leading to the formation of incised canyons or gulfs, which are known as the Pliocene Gulf (Ryan 1978; Issawi et al. 2009). Progradation and shallowing continued in the Messinian, subsidence being more prominent both east and west of the central delta uplift. The Messinian Abu Madi Formation sands were accumulated in deltaic, especially in the lower part, and littoral-marine in the (more shaly) upper part (Alfy et al. 1992). Further regression of the sea happened concurrent with the closure of Jabel Tarek Strait. This gave rise to the deposition of evaporites of the Rosetta Formation. Meanwhile, the equivalent fluvial to marginal marine facies of the Abu Madi Formation was deposited in deep incised valleys (Fig. 4.5).

The Quaternary Delta, which exhibits unconformity contact with the underlying Neogene Delta, is located in both the onshore and offshore delta. It occurs offshore where it fills the northern deeper sub-basin and down-dip of the hinge zone. It comprises the Pliocene Kafr El Sheikh Formation, Plio-Pleistocene Qawasim Formation, Pleistocene Mit Ghamr Formation, and Holocene Bilqas Formation (Fig. 4.14). It is a thick (1–3 km) complex with deltaic clinoform facies (Sestini 1995). It crosses over the Neogene Delta and extends southerly as far as Aswan (Said 1981; Issawi et al. 2009; Ryan 1978).

The Quaternary Delta was deposited during high transgression of the seawater. The reason for such transgression was formed due to the opening of the Atlantic Ocean to the Mediterranean Sea via the Jabel Tariq strait. This happened during the Pliocene period, causing a major sea transgression that filled the carved canyon formed on top of the Messinian unconformity with the Kafr El Sheikh Formation (Fig. 4.12) (Sestini 1984; Abdel Aal et al. 2001; Samuel et al. 2003). This formation began along the current Mediterranean Sea shoreline and moved southward, continuing or progressing as far south as Aswan and most likely north of Sudan (Issawi et al. 2009). At Lat. 29 30 N, it reaches its widest and deepest point, which is approximately 60 km away. In the Nile Delta's subsurface, a massive claystone thickness (1458 m) of the Kafr El Sheikh Formation was deposited at this location (Issawi et al. 2009). Chumakov (1967, 1973) identified the clayey facies of Kafr El Sheikh in the drilled holes in Aswan High Dam at depths of 170 and 180 feet below sea level. Furthermore, 50 km south of Cairo, near the western edge of the Nile Valley, between Beni Yosssef and Tahma, Issawi et al. (2009) have identified the shaly facies of the Kafr El Shiekh formation. This formation was also found south of the Suhaj Governorate and west of Al Minshah. It is characterized by limestone at the top that is 12 m thick, and has an exposed base, with shale at the base (Issawi et al. 2009). Numerous indicators point to the occurrence of a distinct pre-Pliocene marine regression phase in the Mediterranean Basin. The Nile River, for example, cut a deeper channel upstream up to Aswan, which was filled with Pliocene sediments (Chumakov 1967; Rizzini et al. 1978). Another example has been recorded along the whole western Mediterranean coast, suggesting erosional valleys. These valleys are now measured hundreds of meters below sea level, excavated in pre-Pliocene sediments. These troughs are filled with clastic sediments of Pliocene age (Rizzini et al. 1978; Peyton et al. 1998). Thus, it has become confirmed and documented that all these erosional phases could be due to a certain lowering of the sea level in correspondence with the emplacement of the Messinian sediments present in the Mediterranean Basin (Sestini 1984; Samuel et al. 2003). Contrary to the above, Sestini (1989) mentioned that the Early Pliocene transgression migrated inland up to the down-faulted Nile Valley without necessarily extending as far as Aswan, as stated by Ryan (1978), Said (1981).

After the deposition of the Kafr El Sheikh Formation (Early Pliocene), there was a gradual decrease in sea level that resulted in the deposition of coarse clastic facies of the El Wastani Formation (Late Pliocene-Early Pleistocene). These facies are characterized by large forest progradation (at least as far as the inner continental shelf in submarine fan (Sestini 1989; Loncker et al. 2006)). In Mit Ghamr Well-1, the formation exhibits fining-upward cycles. Each cycle begins with a pebbly conglomerate, followed by coarse-grained sandstone and capped by sandy siltstone (Fig. 4.16a). This indicates a delta-front setting (Fig. 4.14). Further shallowing-upward and regression and/or sea-level fall led to sedimentation of the Mit Ghamr and the Holocene Bilqas Formation in delta plain. The delta plain includes lowland, marginal marine areas that include one or more distributary channels, separated by interdistributary zones of shallow water and emergent and near-emergent areas. Such interdistributary zones include bay, lakes, lagoons, floodplains, marshes, tidal flats, beach ridges, and dunes (Fig. 4.14).

In general, the Quaternary Delta was deposited under a transgressive–regressive cycle. The transgressive was started during Early Pliocene Kafr El Sheikh Formation that comprises clays and silts with some sand representing pro-delta setting (Fig. 4.14). Meanwhile, the regressive part of the cycle comprises coarse-grained sandstone and pebbly sandstone of the Mit Ghamr and Bilqas formations which were deposited in delta plain, covering most of the Nile Delta either onshore or offshore (Fig. 4.14).

References

Abd El Aal A, Day RA, Lelek JJ (1992) Structural evolution and styles of the Northern Sinai, Egypt, In: EGPC, 7th exploration and production conference, Cairo, pp 546–562

Abdel Aal A, Price RJ, Vaitl JD, Shrallow JA (1994) Tectonic evaluation of the Nile Delta, its impact on sedimentation and hydrocarbon potential. In: Egyptian Geological Petroleum Company, 12th exploration and production conference, vol 1, pp 19–34

Abdel Aal AA, el Barkooky A, Gerrits M, Meyer H, Schwander M, Zaki H (2001) Tectonic evolution of the Eastern Mediterranean Basin and its significance for hydrocarbon prospectivity in ultradeep water of the Nile Delta. Geo Arabia 6:363–384

Abdel Fattah ML (2014) Petrophysical characteristics of the Messinian Abu Madi formation in the Baltim east and north fields. Offshore Nile Delta, Egypt. J Petrol Geol 37:183–195

Abdel Fattah ML, Slatt RM (2013) Sequence stratigraphic controls on reservoir characterization and architecture: case study of the Messinian Abu Madi incised-valley fill, Egypt. Cent Eur J Geosci 5:497–507

Abu El-Ella R (1990) The Neogene-Quaternary section in the Nile delta, Egypt; geology and hydrocarbon potential. J Petrol Geol 13:329–340

Abu Hashish ME, Ali SB (2021) Integration of well logs and seismic data for facies analysis and modeling of the El Wastani formation in the Sequoia field, offshore Nile Delta, Egypt. J Afr Earth Sci 104343

Abu Hashish MF, Al Shareif AW, Hassan NM (2022) Hydraulic flow unit and reservoir characterization of the Messinian Abu Madi Formation in west El Manzala development Lease, Onshore Nile Delta, Egypt. J Afr Earth Sci 189:104498

Abu Khadra AM, Selim SS, Abdel Baset AA (2018) Sedimentology and seismic geomorphology of the tidal-dominated estuarine incised valley fill. Late Miocene Qawasim reservoir, onshore east, east Nile Delta. Egypt J Geol 62:103–118

Aksu AE, Calon TJ, Hall J, Yaşar D (2005) Origin and evolution of the Neogene Iskenderun Basin, northeastern Mediterranean Sea. Mar Geol 221:161–187

Alfy M, Polo F, Shash M (1992) In the geology of the Abu Madi gas field. Paper presented at the 11th exploration and production conference, Egyptian General Petroleum Corporation, Cairo, vol 2, pp 485–513

Ali A, Gabr M, Hashem H, Ashraf MT, Elewa AMT, Ramadan SA, Mohamed RSA (2022) Sandstone reservoir quality in light of depositional and diagenetic processes of the Messinian Qawasim formation, onshore Nile Delta, Egypt. J Afr Earth Sci 223:104992

Aqrawi AAM (1993) Miocene evaporitic sequence of the southern Mesopotamian Basin. Mar Pet Geol 10:172–179

Aref MA, Attia OE, Wali AM (1997) Facies and depositional environment of the Holocene evaporites in the Ras Shukeir area, Gulf of Suez, Egypt. Sediment Geol 110:123–145

Ayyad S, El-Bassiouni AE, Ziko A (1987) Benthonic foraminifera, Bryozoan and the paleoenvironments of the Quaternary succession in north west Rosetta of shore well No.1, Nile Delta, Egypt. In: 6th symposium on quaternary and development, Mansoura University, Egypt
Azzam EI (1994) Adaptive responses to ionizing radiation in genesis to consideration of dose-rate. Mathematical normal human skin fibroblasts. Dissertation, University Biosciences 115:123–144
Barakat MG (1982) General review of the petroliferous provinces of Egypt with special emphasis on their geological setting and oil Potentialities. Geology Department, Faculty of Science, Cairo University, Cairo, Egypt, 87 p
Barber PM (1981) Messinian subaerial erosion of the Proto-Nile delta. Mar Geol 44:253–272
Burke K, Wells GL (1989) Trans-African drainage system of the Sahara: was it the Nile? Geology 17:743–747
Cant DJ, Walker RG (1978) Fluvial processes and facies sequences in the sandy braided Saskatchewan River, Canada. Sedimentology 25:625–648
Casshyap SM, Tewai RC (1984) Fluvial models of the Lower Permian coal measures of the Son-Mahanadi and Koel-Damodar valley basins, India. In: Rahamani RA, Flores RM (eds) Coal and coal-bearing sequences. Spec. Pub. Intl.. Assoc. Sediment. No 7, pp 121–147
Chen Z, Warne AG, Stanley DJ (1992) Late Quaternary evolution of the North western Nile Delta between the Rosetta, Promontory and Alexandria, Egypt. J Coast Res 18(3):527–561
Chumakov IS (1967) Pliocene ingression in the Nile Valley according to new data. In: Butzer KW, Hansen CL (eds) Desert and River in Nubia, geomorphology and prehistoric environments at the Aswan Reservoir. The University of Wisconsin Press, Madison, pp 521–522
Chumakove IS (1973) Pliocene and Pleistocene deposits of the Nile Valley in Nubia and upper Egypt. In: Ryan W, Hsu K (eds) Initial report of deep Sea drilling Project. US, Government Printing Office, Washington D.C. 13: 1242 p
Cita MB, Vergnaud-Grazzini C, Robert C, Chamley H, Ciaranfi N, d'Onofrio S (1977) Palaeoclimatic record of a long deep-sea core from the eastern Mediterranean. Quatern Res 8:205–235
Coutellier V, Stanley DJ (1987) Late Quaternary stratigraphyand paleogeography of the eastern Nile Delta shelf. Mar Geol 77:257–275
Dolson JC, Boucher PJ, Dodd T, Ismail J (2002) The petroleum potential of the emerging mediterranean offshore gas plays, Egypt. Oil Gas J 32–37
Dolson JC, Boucher PJ, Siok J (2005) Key challenges to realizing full potential in an emerging giant gas province: Nile Delta/Mediterranean offshore, deep water, Egypt. In: Petroleum geology conference series, vol 6, pp 607–624
EGPC (Egyptian General Petroleum Corporation) (1994) Nile Delta and North Sinai: fields, discoveries and hydrocarbon potentials (a comprehensive overview), Cairo, Egypt
El-Asmer HM (1986) Lithological and mineralogical investigation of the subsurface Quaternary succession in the Nile Delta region, Egypt. MSc thesis. Faculty of Science, El Mansoura University, Egypt
El Asmer HM (2023) Quaternary environmental and climatic changes in Egypt. In: Hemimi et al (eds) The phanerozoic geology and natural resources of Egypt. Springer, 719 pp
El-Asmer HM, Gheith AM (1995) Stable isotope stratigraphy and paleoclimatology of the offshore subsurface Quaternary sediments, Nile Delta, Egypt. Sedimentol Egypt 3:39–49
El-Barkooky AN, Helal MA (2002a) Some neogene stratigraphic aspects of the Nile Delta. In: Proceedings of the third mediterranean offshore conference, 35p
El-Barkooky AN, Helal MA (2002b) Sequence stratigraphy and sedimentary history of the neogene Nile Delta. In: AAPG annual international meeting, Cairo, Egypt
El-Bassiouni AE, Ayyad SN (1986) Quaternary Foraminifera of El-Temsah well-2, offshore Nile Delta, Egypt. El Mansoura Sci Bull 13(1):133–188
El Heiny I (1982) Neogene stratigraphy of Egypt. Newslett Stratigr 2:41–52

EI-Heiny E, Morsi S (1992) Stratigraphic correlation of Neogene sediments in the eastern Nile Delta and Gulf of Suez. In: 11th exploration and production conference, Cairo, pp 166–193
EI-Heiny E, Enani N (1996) Regional stratigraphic interpretation of Neogene sediments, northern Nile Delta. In: 13th EGPC petroleum conference exploration, Cairo (1996), pp 270–290
El Heiny I, Rizk R, Hassan M (1990) Sedimentological model for abumadireservoir sands, abumadi field, Nile delta, Egypt, vol 2, 515e551
El-Rawy AMA (2020) Visualization and extraction of sand reservoir units in Mit Ghamr formation using seismic frequency based a tributes, Tarut field east Nile Delta, Egypt. Egypt J Geol 64P:1–1
El-Shahat A, Kora MA (1986) Size analysis and paleoenvironmentsof the subsurface sediments, west of Nile Delta. El Mansoura Sci Bull 13(2):185–207
El sheikh M (2021) Seismic stratigraphic analysis and hydrocarbon potential of miocene-quaternary deposits in the western Nile Delta Basin. Iraqi Gel Ournal 5:1–12
El Sisi ZA, Sharaf LM, Dawood HY, Hassouba BA (1996) Source of clastic supply to the Sidi Salem and Abu Madi Reservoirs of the Nile Delta, Egypt, E.G.P.C. 13th Petrol Expl Prod Conf 1:291–296
Eriksson KA (1978) Alluvial and destructive beach facies from the Archean Moodies Group, Barberton Mountain Land, South Africa, and Swaziland. In: Miall AD (ed) Fluvial sedimentology. Canadian Society of Petroleum Geologists, Memoir, vol 5, pp 287–312
Farouk S, Ziko A, Eweda SA, Ali E, Said AE (2014) Subsurface Miocene sequence stratigraphic framework in the Nile Delta, Egypt. J Afr Earth Sci 91:89–109
Faris M, Samir AM, Shabaan M (2007) Calcareous nannofossil biostratigraphy of the subsurface Miocene sequence, northeast Nile Delta, Egypt. In: The fifth International Conference of the Geology of, Africa, 1, pp VII-1–VII-31
Frihy OE, Gamai IH (1991) Facies analysis of Nile delta continental shelf sediments off Egypt. Neth J Sea Res 27:165–171
Garrison R, Schreiber B, Bernoulli D, Fabricius F, Kidd R, Melieres F (1978) Sedimentary petrology and structures of Messinian evaporitic sediments in the Mediterranean Sea, Leg 42A, Deep Sea Drilling Project. In: Hsu K, Montadert L (eds) Initial Reports of the Deep Sea Drilling Project. Government Printing Office, Washington, U.S, pp 571–612
Gheith AM (1985) Sedimentological and mineralogical investigations of the subsurface Quaternary succession of Temsah offshore well-2, Nile Delta Coast. Egypt J Geol 29(1–2):)20–50)
Hampsons GJ, Gani MR, Sharman KF, Irfpan N, Bracken B (2011) Alongstrike and down-dip variations in shallow-marine sequence stratigraphic architecture: upper cretaceous star point sandstone, Wasatch Plateau, Central Utah, USA. J Sediment Res 81:159–184
Hardie LA, Eugster HP (1971) The depositional environment of marine evaporites: a case for shallow, clastic accumulation. Sedimentology 16:187–220
Harms JC, Wray JL (1990) Nile Delta. In: Said R (ed) Geology of Egypt. The Netherlands: AA Balkema, pp 329–343
Hota RN, Maejima W (2004) Comparative study of cyclicity of lithofacies in lower Gondwana formation of Talchir basin, Orissa, India: a statistical analysis of subsurface logs. Gondowana Res 7:353–362
Howell J, Schwarz E, Spalleitl L, Andvelga G (2005) The Neuque'n Basin: an overview. In: Veiga G, Spalletti L, Howell J, Schwarz E (eds) The Neuque'n Basin: a case study in sequence stratigraphy and basin dynamics: geological society of London, Special Publication 252, p 1–14
Hsü KJ, Ryan WBF, Cita MB (1973) Late Miocene desiccation of the Mediterranean. Nature 242:239–243
Hsü KJ, Montadert L, Bernoulli D, Cita MB, Erickson A, Garrison RE, Kidd RB, Melieres F, Muller C, Wright R (1978) History of the Messinian salinity crisis. In: Hsu KJ, Montadert L et al (eds)

Initial reports of the Deep Sea Drilling Project 42(1). U.S. Govt. Printing Office, Washington, D.C., pp 1053–1078
Ismail AA, Boukhary M, Abdel Naby AI (2010) Subsurface stratigraphy and micropaleontology of the Neogene rocks, Nile Delta Egypt. Geol Croat 63:1–26
Issawi B, McCauley JF (1992) The Cenozoic rivers of Egypt: The Nile problem. In: Adams B, Friedman R (eds) The followers of horus. Oxbow Press, Oxford, pp 1–18
Issawi B, Francies HM, Yussef EAA, Osman R (2009) The phanerozic geology of Egypt. A Geodynamic Approach, Special Pub. No.81, Geol. Surv., Egypt
Kassab MA, Tema MA, Cheadle BA, El Din ES, Mhamed IF, Meshah MA (2015) Reservoir characterization of lower Abu Madi zone using core analysis data, El Wastani gas field, Egypt. J Afr Earth Sci 110:116–130
Khalid A, Sharaf E, Abushady AI (2011) Reservoir characterization of Sidi Salem formation, Akhen fields, offshore Nile Delta, Egypt. Egypt J Geol 55
Kushnir J (1981) Formation and early diagenesis of varved evaporite sediments in a coastal hypersaline pool. J Sediment Res 51:1193–1203
Leila M, Moscariello A (2018) Depositional and petrophysical controls on the volumes of hydrocarbons trapped in the Messinian reservoirs, onshore Nile Delta, Egypt. Petroleum 4(250):267
Leila M, Moscariello A (2019) Seismic stratigraphy and sedimentary facies analysis of the pre- and syn-Messinain salinity crisis sequences, onshore Nile Delta, Egypt: implications for reservoir quality prediction. Mar Petrol Geol 101:303–321
Leila M, Kora MA, Ahmed MA, Ghanem A (2016) Sedimentology and reservoir characterization of the Upper Miocene Qawasim formation, El-Tamad Oil Field onshore Nile Delta, Egypt. Arab J Geosci 9:1–13
Leila M, Moscariello A, Kora M, Mohamed A, Samankassou E (2020) Sedimentology and reservoir quality of a Messinian mixed siliciclastic-carbonate succession, onshore Nile Delta, Egypt. Mar Petrol Geol 112:104076
Loncker V, Mascle J, Vendeville B, Camera L (2006) The Nile deep-sea fan: An example of interacting sedimentation, salt tectonics, and inherited subsalt paleotopographic features. Mar Petrol Geol 23:297–315
Maceachern JA, Bann KL (2008) The role of ichnology in refining shallow marine facies models. In: Hampson GJ, Steel RJ, Burgess PM, Dalrymple RW (eds) Recent advances in models of siliciclastic shallow-marine stratigraphy: SEPM, Special Publication 90, pp 73–116
Maceachern JA, Bann KL, Bhattacharya JP, Howell CD, JR (2005) Ichnology of deltas: organism responses to the dynamic interplay of rivers, waves, storms, and tides. In: Bhattacharya JP, Giosan L (eds) River Deltas—concepts, models and examples: SEPM, Special Publication 83, pp 49–85
Maelmni W, Taewari ARC, Hota RN (2008) Origin of fining-upward cycles in the early Permian Barakar formation Gondwana basins of Peninsular India. J Geosc. Osaka City University 51:21–26
McDougall I, Morton WH, Williams MAJ (1975) Age and rates of denudation of Trap Series basalts at Blue Nile Gorge, Ethiopia. Nature 254(5497):207–209
Miall AD (1978) A review of the braided-river depositional environment. Earth SciRev 13:1–62
Miall AD (1988) Reservoir heterogeneities in fluvial sandstones: lessons from outcrop studies. Am Assoc Pet Geol Bull 72:682–697
Middleton GV, Hampton MA (1973) Sediment gravity flows: mechanics of flow and deposition. In: Turbidites and deep-water sedimentation. Pacific section of the society of economic paleontologists and mineralogists, short course lecture notes, pp 1–38
Nini C, Checchi F, El Blasy A, Talaat A (2010) Depositional evolution of the Plio-Pleistocene succession as a key for unraveling the exploration potential of the post-Messinian play in the Central Nile Delta, MOC

Orton GJ, Reading HG (1993) Variabilty of deltaic processes in terms of sedimentary supply, with particular emphases on grain size. Sedimentology 40:475–512
Pennington BT, Sturtb F, Wilsonc P, Rowlandd J, Browna AG (2017) The uvial evolution of the Holocene Nile Delta. Quatern Sci Rev 1–22
Peyton L, Bottjer R, Partyka G (1998) Interpretation of incised valleys using new 3-D seismic techniques: a case history using spectral decomposition and coherency. Lead Edge 17:1294–1298
Plint AG, Nummedal G (2000) The falling stage systems tract: recognition and importance in sequence stratigraphic analysis. Geol Soc Lond Spec Publ 172:1–17
Plint AG (2003) Clastic sediment partitioning in a cretaceous delta system. Geol Croat 56:39–68
Rashid A, El-Gharabawy S, AbouShagar S (2018) Geological and structural evaluation on temsah gas field, NE-offshore Nile Delta, Egyp. Int J Sci Eng Appl Sci (IJSEAS) 4:2395–3470
Reading HG, Collinson JJD (1996) Clastic coasts. In: Reading HG (ed) Seidimentary environments: processes, facies and stratigraphy, 3rd edn. Blackwell Scientific Publication, Oxford, pp 231–254
Rizzini A, Vezzani F, Cococcetta V, Milod G (1978) Stratigraphy and sedimentation of the Neogene-Quaternary section in the Nile Delta area. Mar Geol 27:327–348
Rust BR, Koster EH (1984) Coarse alluvial deposits. In: Walker RG (ed) Facies models. Geoscience Canada, Reprint Series 1, 2nd ed, pp 53–69
Ryan WFB (1978) Messinian badlands on the SE margin of the Mediterranean Sea. Mar Geol 27:389–363
Said R (1962) The geology of Egypt: New York, Elsevier, 377 pp
Said R (1981) The geologic evolution of the River Nile. Springer-Verlag, New York, p 151
Said R (1993) The River Nile: geology, hydrology and utilization. Pergamon Press, p 320
Salem RM (1976) Evolution of Eocene-Miocene sedimentation patterns in parts of northern Egypt. Am Asso Petrol Geol Bull 60:34–64
Salem AM, Ketzer JM, Morad S, Rizk RR, Al-Aasm IS (2005) Diagenesis and reservoir-quality evolution of incised valley sandstones: evidence from the Abu Madi gas reservoirs (Upper Miocene), the Nile Delta Basim, Egypt. Sediment Res 75:572–584
Sambrook Smith GH (1996) Bimodal fluvial bed sediments: origin, spatial extent and processes. Prog Phys Geogr 20(4):402–417
Sambrook Smith GH, Ferguson RI (1995) The gravel–sand transition along river channels. J Sediment Res A 65(2):423–430
Samuel A, KnellerB, Raslan S, Sharp A, Parsons C (2003) Prolithic deep-marine slope channels of the Nile Delta, Egypt. AAPG Bull 87:541–560
Sarhan MA (2015) High resolution sequence stratigraphic analysis of the Late Miocene Abu Madi Formation, northern Nile Delta Basin. NRIAG J Astron Geophys 4(2):298–306. https://doi.org/10.1016/j.nrjag
Sarhan M, Hemdan K (1994) North Nile delta structural setting, trapping, and mechanism. Egypt. In: Proceedings of the 12th EGPC exploration and production conference, Cairo, Egypt 1, pp 1–18
Serag El Din HM (1990) Geological, hydrogeological, and hydrological studies on the Nile Delta aquifer. PhD thesis, Mansoura Univ. 300 pp
Sestini G (1984) Tectonic and sedimentary history of the NE African margin (Egypt—Libya). Geol Soc Lond Spec Publ 17:161–175
Sestini G (1989) Nile Delta: a review of depositional environments and geological history. In: Whateley MK, Pickering KT (eds) Deltas: sites and traps for Fossil fuels: geological society of America, Special Publication, 41, pp 99–127
Sestini G (1995) Egypt. In: Kulke H (ed) Regional petroleum geology of the world, part II: Africa, America, Australia and Antarctica, Beiträgezur regionalen Geologie der Erde, Band 22: Stuttgart, Gebrüder BornträgerVerlags-buchhandlung, pp 66–87

Simpson EL, Eriksson KA, Eriksson PG, Bumpy A (2002) Eolian Dune degradation and generation of massive sandstone bodies in the paleoproterozoic Makgabeng formation, Waterberg Group, South Africa. J Sediment Res 72(1):40–45

Sohn YK, Rhee CW, Kim BC (1999) Debris flow and hyperconcentrated flood-flow deposits in an alluvial fan, northwestern part of the Cretaceous Yongdon Basin, central Korea. J Geol 107:111–132

Stanley DJ (1988) Subsidence in the northeastern Nile delta: Rapid rates, possible causes and consequences. Science 240(497):500

Stanley DJ (1990) Recent subsidence and northeast tilting of the Nile delta, Egypt. Mar Geol 94:147–154

Stanley DJ (2005) Submergence and burial of ancient coastal sites on the subsiding Nile delta margin, Egypt. Méditerranée 1(2):65–73

Stanley DJ, Clemente PL (2014) Clay distributions, grain sizes, sediment thicknesses, and compaction rates to interpret subsidence in Egypt's Northern Nile Delta. J Coastal Res 30:88–102

Stanley DJ, Clemente PL (2017) Increased land subsidence and sea-level rise are submerging Egypt's Nile Delta coastal margin. Geol Soc Am Today 27(5):411–464

Stanley DJ, Corwin KA (2013) Measuring strata thicknesses incores to assess recent sediment compaction and subsidence ofEgypt's Nile Delta coastal margin. J Coastal Res 29(3):657–670

Stanley DJ, Warne AG (1993) Nile Delta: recent geological evolution and human impact. Science 260:628–634

Stanley DJ, Warne AG (1998) Nile Delta in its destruction phase. J Coastal Res 14(3):794–825

Summerhayes CP, Sestini G, Misdorp R, Marks N (1978) Nile Delta: nature and evolution of continental shelf sediments. Mar Geol 27:43–46

Tewari RC (1997) Numerical classification of coal-bearing cycles of early Permian Baraka!' coal measures of eastern-central India, Gondwana basins using Q-mode cluster analysis. J Geol Soc India 50:593–599

Walker RG, Can't DJ (1984) Sandy Fluvial systems. In: Walker RG (ed) Facies models, 2nd ed. Geoscience, Canada, Reprint Series, vol 1, pp 71−89

Williams MAJ, Adamson D, Cock B, Evedy R (2000) Later Quaternary environments in the White Nile region, Sudan. Global Planet Change 26:305–316

Zaghloul ZM, Andrawis SF, Ayyad SN (1977) New contribution to the stratigraphy of the Tertiary sediments of Kafr El-Dawar well no. 1, North West Nile Delta, Egypt. Ann Geol Surv Egypt 4:292–307

Zaghloul ZM, Taha AA, Hegab O, E1-Fawal F (1979) The Plio-Pleistocene Nile subenvironments stratigraphic section and genetic class. In: The fifth conference on African geology, geological survey, Egypt, Cairo

Zaghloul ZM, Gheith AM, Abdel-Daiem AA, El Asmar HM (1989) The pattern of sedimentation of the Quaternary Nile Delta, Egypt. El Mansoura Sci Bull 16(2):393–413

Zaghloul ZM, El Gamal MM, Shaaban FF, Yossef AFA (1999) In Zaghloul Z, El Gamal M (eds) Plates interaction and petroleum potentials in the Nile Delta, ancient and modern. Mansoura Univ., Delta, 14−53 pp

Quaternary Diatom Records of Environmental and Climate Changes in the Nile Delta, Egypt

5

Abdelfattah A. Zalat

Abstract

This study evaluates changes in the environmental and climatic conditions in the Nile Delta during the late Pleistocene-Holocene based on diatom analyses extracted from the sediment of four cores obtained from the central portion of the Nile Delta. Diatom taxa were abundant to rare and poorly to well-preserved throughout the core's sediments. A total of 216 species and varieties representing 49 genera were identified from the investigated 196 samples. The most dominant planktonic genera were *Aulacoseira,* followed by infrequent distribution of *Cyclostephanos, Cyclotella, Lindavia, Pantocsekiella, Stephanocyclus,* and *Stephanodiscus* species. The distribution pattern of benthic diatoms represented by a frequent abundance of *Staurosira, Staurosirella*, and *Ulnaria* taxa associated with low frequencies of *Craticula, Diploneis, Fragilaria, Navicula, Anomoeoneis, Caloneis, Diploneis, Gyrosigma, Pinnularia, Pseudostaurosira, Stauroneis, Nitzschia, Tryblionella, Cymatopleura, Surirella,* and *Campylodiscus*. The epiphytic diatoms were recorded by the frequent occurrence of *Encyonema* and *Gomphonema* species besides a low ratio of *Eunotia, Cocconeis, Amphora, Cymbella, Epithemia,* and *Rhopalodia* taxa. The planktonic forms were dominant in the Mit Ghamr formation and decreased gradually in their abundance, simultaneous with the abundance of the periphytic taxa in the Bilqas formation. Most recorded diatoms preferring slightly alkaline to alkaline water occurred in the sediments, which is documented by the most abundance of the alkaliphilous diatom group. Concerning the salinity, the recognized diatom assemblage from the sediments of Mit Ghamr formation

A. A. Zalat (✉)
Geology Department, Faculty of Science, Tanta University, Tanta, Egypt
e-mail: abzalat@science.tanta.edu.eg

M. A. Khalifa and K. S. Gemail (eds.), *Geology of the Nile Deltas and Natural Resources*, Deltas of the World, https://doi.org/10.1007/978-3-031-96594-4_5

is composed mainly of a freshwater oligohalobous group associated with a low occurrence of the mesohalobous forms. In contrast, the relative abundance of oligohalobous diatoms decreased relatively, accompanied by a marked increase in the abundance of both mesohalobous and polyhalobous diatom taxa in the Bilqas formation. Based on the relative abundance of dominant and subdominant taxa and the cluster-constrained analysis, the investigated core sections are divided into several diatom zones, which revealed an interesting history of changes in paleosalinity, pH, and water level of the Nile Delta basin as the water body filled in. The variation in the relative abundance of the recognized diatom taxa revealed five phases of Nile Delta development during the Quaternary time. The first two phases are characteristic of the Mit Ghamr formation, which are distinguished by the dominance of planktonic, oligohalobous, and alkaliphilous diatoms that reflect eutrophic, slightly alkaline, moderate stands of the freshwater level during humid, warm climate. Three phases represent the Bilqas formation and are marked by an increased abundance of the periphytic taxa over the planktonic forms, with the conspicuous occurrence of brackish and coastal marine taxa. The recognized diatom assemblages revealed marked lowering of water levels with increases in salinity and alkalinity during warm-arid conditions of Holocene time.

5.1 Introduction

The Nile Delta is one of the largest river deltas in the world. It is considered one of the oldest known ancient deltas and the most important depositional complex in the Mediterranean sedimentary basin (Shaaban et al. 2016). It is a part of the unstable shelf of Egypt and is located between the Western Desert and the Levant Basin along the northern margin of the African Plate (Morgan 1990), where the Nile River spreads out and drains into the Mediterranean Sea. It is an alluvial plain in the north of Egypt that occupied the area between latitudes 30°10′–31°30′ N and longitudes 30°20′–31°50′ E and has a widespread flat area of about 12,500 km^2 with very low relief and very flat at the north and reaches up to 18 m above sea level at the south close to Cairo (Fig. 5.1). It covers approximately 230 km of the Mediterranean coastline from Alexandria in the west to Port Said in the east (Fishar 2016) and a maximum of 170 km from north to south (El-Kady et al. 2000). The southern apex of the Delta is located approximately 30 km north of Cairo. Today, the Nile Delta covers an area of about 24,000 km^2 and is densely populated and intensively used for agriculture (Pennington et al. 2017).

The Nile Delta has attracted the attention of many geologists, sedimentologists, paleontologists, and archeologists for a long time. Earlier research on the Delta has been focused on the influence of sea level changes and subsidence rate as the basis for reconstructing the Nile Delta paleogeography and coastal geomorphology (e.g., Stanley 2005; Becker and Sultan 2009; Aly et al. 2012; Stanley and Corwin 2013; Tassy et al. 2015).

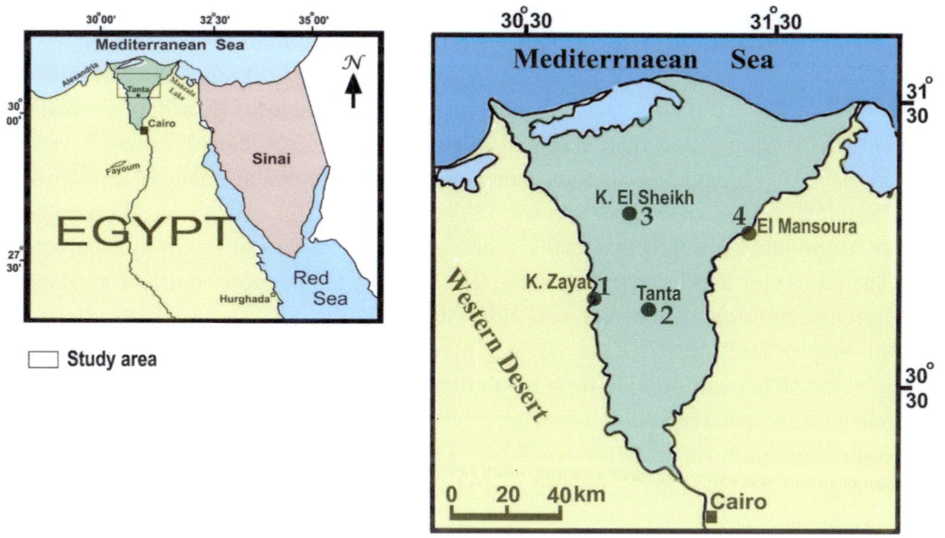

Fig. 5.1 Location map showing the position of the studied core sites in the central Nile Delta, Egypt

Other significant researches in the last six decades have addressed the Quaternary stratigraphy of the Delta and its depositional environments (e.g., Rizzini et al. 1978; Zaghloul et al. 1977, 1989; Said 1981, 1983, 1990; Sestini 1989; El-Asmar and Gheith 1995; El-Asmar 1998 and others); the Late Quaternary succession in response to paleoenvironment and paleoclimatic (e.g., Stanley 2005; Stanley and Clemente 2017; El-Asmar et al. 2015; Stanley and Wedl 2021) and the paleontological characters of the delta sediments (e.g., Saad et al. 1987; Zalat 1995; Kholeif and Mudie 2009; Ismail et al. 2010; Farouk et al. 2021). The paleontological studies were mainly concerned with the planktonic and benthic foraminifera, pollen, and calcareous nannoplankton. However, none of the previous studies have specifically addressed Quaternary fossil diatoms, except for Zalat's (1995) work.

Archeologically, the Nile Delta has been established and cultivated since the Neolithic period. The earliest known archeological remains within the Delta were documented from ca. 4530 BCE (Butzer 2002). The connection of the earliest settlements to preceding Nile branches is considered a vital issue for the vitality of a settlement to supply people and animals with fresh water, to irrigate fields, or as a connection to the transport system (Lange et al. 2016; Ginau et al. 2019).

Diatoms are the most diverse unicellular microalgae that represent an important group in any aquatic ecosystem. They are distributed in almost all aquatic habitats and abundant in the phytoplankton and phytobenthic of marine, brackish, and freshwater at all latitudes (Round et al. 1990), as well as in moist soil, on wet surfaces, in hot springs, or acidic environments (Mann 1999). In the Quaternary, diatoms are dominant in most sedimentary environments, including marine, lacustrine, and lagoonal sediments. They have been commonly used as bioindicators for a large range of applications in the environmental and earth sciences (Stoermer and Smol 1999) because of their robust siliceous wall that preserves well in the sediments, their abundance, wide geographical distribution, relatively well-known taxonomy, and their essential role in aquatic ecosystems. Diatoms have been used in paleolimnological studies as indicators of environmental changes through time (e.g., Stoermer 1998; Zalat and Servant Vildary 2005, 2007; Julius and Theriot 2010). Diatom community composition reflects and responds to several environmental conditions such as water depth, type of habitat, pH, salinity, temperature, nutrient availability, and current conditions (Stoermer and Smol 1999; Wolin and Duthie 1999; Zalat and Servant Vildary 2005, 2007; Zalat 2015). Therefore, the fossil diatoms in lacustrine and fluviatile sediments can provide high-resolution records of salinity, alkalinity, and water-level changes, which are mainly associated with fluctuations in climatic conditions such as precipitation and evaporation rates. Also, diatoms have been used as proxy indicators to reconstruct the paleoenvironmental and climatic changes through the Quaternary on every continent.

Until now, no detailed data on the diatom composition of the Quaternary Nile Delta sediments have been published except the work of Zalat (1995), and our knowledge of the diatom flora is limited to qualitative analysis of a few phytoplankton samples from some water habitats and the river Nile. Therefore, the present research aims to record the occurrence of fossil diatom assemblages in the Quaternary sediment of the Nile Delta and to discuss the grade of the response of diatom distribution to the environmental and climate changes during the Quaternary.

5.2 The Geological History of the Nile Delta

The geology of the Nile Delta has been extensively discussed by several researchers over several decades. Its geological history can be traced back to the Late Eocene–Oligocene time (Kamel et al. 1998). The delta basin was influenced by several tectonic events since the beginning of the Mesozoic (Zaghloul et al. 1999; Tassy et al. 2015), where the oldest known stratigraphic record is assigned to the Middle Triassic (Shaaban et al. 2016). This basin has been tectonically controlled by fault systems running mainly NW–SE and ENE-WSW (E1-Shazly et al. 1975). Six major structural trends delineate the present Nile Delta and have affected the distribution of its Neogene successions (Abdel Aal et al. 1994). These are the East–West Neogene Hinge Zone, the NW Temsah structural trend, the NE Rosetta fault trend, the Pelusim shear Zone, the minor NS Baltim fault trend, and the NW. Red Sea-Gulf of Suez fault trend (Farouk et al. 2021). The Neogene hinge line is one of the main tectonic features that subdivides the Nile Delta into two structural sub-basins: the North Nile Delta basin and the South Nile Delta sub-basin (Ismail et al. 2010).

Despite the delta basin being submerged during the Pliocene Sea transgression, the continuous accumulation of deltaic facies began with the retreat of the sea during the Late Pliocene (Zaghloul et al.1977). The delta basin received the Nile load from the Upper Pliocene onward, where no river deposits were deposited below the marine Pliocene strata (Hanter 1975). According to Zaghloul (1976), the Nile Delta occupies the place of an old bay with recent alluvium of several meters in thickness, which is underlain by coarse sand and gravels with occasional clay lenses of Pleistocene age. These deposits taper to the east and west and increase in thickness toward the north.

The Nile Delta basin has the complete record of Quaternary sediments (Said 1981), which reached a thickness of about 1000 m, with thinning southward and toward the fringes of the Delta, and reaches about 500 m in the central part of the Nile Delta (Rizzini et al. 1978; Shaaban et al. 2016). This great Quaternary sequence is comprised of Nile silt, clay, sandy clay, sands, and gravels and is divided into several units, which are unconformably stratified (Said 1981). According to Zaghloul et al. (1977) and Rizzini et al. (1978), the Quaternary Nile Delta is classified into the Pleistocene and Holocene cycles related to the transgression/regression of the sea and equivalent to the Mit Ghamr and Bilqas Formations. Zaghloul et al. (1977) interpreted the Mit Ghamr Formation as a coastal and/or distributary mouth bar sand, while the Bilqas Formation deposited in a lagoonal to the beach environment.

Said (1981) correlated the subsurface Quaternary sediments with the exposed Nile Valley sections and recognized four main fluvial intervals: Palaeo-, Proto -, Pre-, and Neo-Nile, which are separated by unconformities due to long periods of erosion accompanied by very reduced or no fluvial activity. The Palaeo- and Proto-Nile sediments belong, respectively, to the Late Pliocene and Early Pleistocene. The Pre-Nile system includes the fluviatile sediments of the river, which covers most of the Middle Pleistocene and terminates around 200,000 B.P. In the Nile Delta, the Pre-Nile sediments are mainly composed of sands and gravels (Mit Ghamr Formation) that lie unconformably under the famed agricultural silt layer of the succeeding Holocene Neo-Nile (Bilqas Formation). The Pre-Nile/Neo-Nile interval embraces the period elapsed between the cessation of the flow of the Pre-Nile in late Acheulian time and the beginning of the Neo-Nile at about 120,000 B.P. During this interval, a great pluvial episode, which was characterized by intense rains, prevailed over Egypt. There are three aggradational phases (β, γ, and δ) characteristic of the deposits of the Neo-Nile, which are interrupted by minor recessional episodes (Wendorf and Schild 1986; Paulissen and Vermeersch 1987; Said 1990). The first one encompasses the late-middle Paleolithic artifacts of the Khormusan tradition that terminated about 40,000 years B.P. The subsequent γ Neo-Nile phase contained the late Paleolithic artifacts and is dated between 21,000 and 12,000 B.P. These two phases were assigned to the Late Pleistocene (Said 1981) and recorded in various areas of the Nile Valley. The last (δ) Neo-Nile phase includes the deposits of the youngest of the aggradational cycles of the Nile that started after an episode of strong erosion at about 10,000 B.P. and belong to Holocene times. This phase is represented in the Delta by the

Bilqas Formation, which constructs the top layer of the floodplain of the modern Nile (Rizzini et al. 1978).

Stanley and Warne (1993) recognized three distinctive Quaternary sequences in the Nile Delta. Sequence I is of Late Pleistocene and includes the Mit Ghamr Formation. Sequence II started with transgressive olive-gray of medium-grained sand with shallow marine fauna at the beginning of the Holocene (cal. 11,000 and 7000 BP), which is a characteristic transgressive phase of the sea and described as Gezira cover formation of mixed coastal and aeolian sands. Sequence III comprised the sediments of the Bilqas Formation that were deposited during cal. 7–3 ka BP (Pennington et al. 2017). However, a marked decrease in sea level rise was detected at cal. 7–6 ka BP (Pennington et al. 2016). Tectonically, the Delta region can be subdivided into three zones. The first zone encompasses the southern area, which establishes as a part of the upward zone that is distinguished by a relatively thin sedimentary cover and by the development of the graben-like structure. The second zone occupied the central area, which coincides with the central part of the down-warp zone. The third zone covers the northern part of the Delta and is characterized by the presence of a thrust fault oriented in the northeast direction (Fergany and Omar 2017).

5.3 Lithostratigraphy

The Quaternary lithostratigraphy of the Nile Delta has been studied by many authors (e.g., Zaghloul et al. 1977, 1979; El-Fayoumi 1987; Rizzini et al. 1978; Serag El-Din 1983, 1989; Said 1981, 1990). The Quaternary Nile sediments lie over Pliocene or older sediments (Said 1990) and are classified into two rock units based on their lithological composition, including from base to top: Mit Ghamr and Bilqas Formations (Rizzini et al. 1978).

5.3.1 Mit Ghamr Formation

Mit Ghamr Formation is an unconsolidated sandy-gravelly unit of Pleistocene age and occasionally contains a thin clay interlayer. (Zaghloul et al. 1977; Serag El-Din 1983). The sand and gravel are more common in the southern part of the Delta. The clay is dominant in the north with occasional discontinuity and interaction of sand and gravel. The thickness of this formation gradually increases toward the north and northeast. It ranges from 100 to 400 m in the south, from 500 to 700 m in the middle, and reaches a maximum thickness of about 970 m in the north (Serag El-Din 1989). Hydrogeologically, this rock unit constitutes the main aquifer of the Nile Delta (Serag El-Din 1983). The lithofacies distribution of this formation displays a marked lateral lithologic variation fluctuating from sandy facies covering most of the Delta to sandy shale in the northern marginal zones of the Delta and containing between them a narrow belt of shaly sand facies (Serag El-Din 1990).

The formation is subdivided into two members; older and younger (the Zagazig and Minuf Members, respectively). The Zagazig Member mainly consists of coarse massive to laminated or cross-stratified, fine to medium-grained sands with scattered coarse clasts. The grains of sand deposits have very thin coatings of iron oxide, sometimes hornblende, epidote, augite, and carbonate, which are correlated within the Prenile at Qena (Said 1981; El-Asmar 1986; Hamdan 2003; Pennington et al. 2017). The subsequent Minuf Member unconformably the Zagazig Member and comprises fine-grained micaceous sand with stiff compacted clayey lenses. This member expanded over a long time in the Late Pleistocene and deposited under an earlier, braided Nilotic regime with a humid climate (Adamson et al. 1980; Said 1981).

Said (1981) considered that the Mit Ghamr Formation is coeval with the Prenile Qena Sands. It consists of quartzose sands and pebbles. The sands are medium to coarse-grained. The pebbles are mainly quartzose, but occasionally silicified limestone and chert pebbles are found. Thin beds of clay, silt, and peat containing coastal or lagoonal fauna mark the upper levels of this formation. Zaghloul et al. (1977) indicated that the sediments of the Mit Ghamr Formation were probably deposited under oscillatory deltaic conditions in a slightly subsiding embayment. However, Salem (2009) explained that the sediments of the Mit Ghamr Formation were probably deposited under shallow marine to fluvial conditions.

5.3.2 Bilqas Formation

Bilqas Formation consists mostly of clay and silt of Holocene age with a few sands of riverine, sometimes peat layer, and flood basin deposits. It constitutes the muddy cap of the Holocene sequence that covers the whole Delta and varies in thickness from 10 m in the southern part to more than 30 m in the northern part of the Nile Delta and has a maximum thickness of about 77 m (Zaghloul et al. 1977). The peat layer has been encountered in the subsurface Holocene sediments of the eastern part of the Nile Delta at depths ranging from 5 to 13 m (Hegab and Bahloul 1987). The maximum thickness of peat was 3.5 m that recorded around Mansoura city and decreases toward the south. Bilqas formation shows distinct lateral variation that exhibits silt facies in the south to clay facies in the north (El-Fayoumi 1987).

Bilqas formation may be included within the Neonile sediments, which are dated from the Late Pleistocene to the Holocene (Said 1981). Its lithofacies changes from clay and silt in the eastern and north-central parts of the Nile Delta, sand dunes in the northern coastal areas, to sandy facies to the south. These facies distributions may be attributed to the irregularity of the basin bottom during deposition and the density of the old Nile distributaries and hence their energy (El-Fayoumi 1987). The depositional environment of the Bilqas formation sediments varies from continental, fluviatile, and lagoonal to beach environments (Zaghloul et al. 1977) to lagoons and brackish swamps interrupted by beach sands in the northern parts of the Nile (Said 1981). Azzam (1994) clarified that during the Holocene, a marine transgression covered most of the Northern Delta area and gave rise to a few meters of marine sediments capped by agricultural soil.

5.4 Materials and Methods

A total of 196 subsurface sediment samples were obtained from four boreholes drilled in the central part of the Nile Delta, between Rosetta and Dumyat Branches (Fig. 5.1). The first borehole was obtained from the Kafr El-Zayat area with a total depth of about 55 m. The second borehole is taken from an area close to Tanta City with a depth of about 47 m. The third borehole was from Kafr El Sheikh City with a depth of about 44 m. The fourth borehole is taken from the eastern part of the Nile Delta at El-Mansoura City, and it has a thickness of about 50 m. The studied area is mainly covered with cultivated lands, ranging in thickness from 1 to 2 m. In each borehole, the samples were collected regularly at one-meter intervals (Fig. 5.2).

The recovered sediments were represented by two distinct units:

(i) Bilqas Formation: upper deltaic clay, silts, clayey sand, fine sand, and flood basin deposits with somewhat small thickness (1–2.5 m) of peat deposits. The section belongs to Holocene Neo-Nile sediments.
(ii) Mit Ghamr Formation: lower deltaic, medium to coarse sand and pebbles embedded in a coarse sand matrix. The section belongs to the Middle to Late Pleistocene Pre-Nile sediments.

Sediment samples were dried overnight at 80 °C, and fossil diatoms were extracted from 20 g of dry sediment using the disintegration method in HCl and H_2O_2, according to the technique proposed by Zalat and Servant-Vildary (2007). The sediments are boiled in 10% HCl to remove the carbonate fraction, rinsed in distilled water, and then treated with 30% H_2O_2 to eliminate the organic matter. Each sample was rinsed and decanted in distilled water. The coarse particles were removed by further decantation (settling for 5 s). Lastly, 0.1 ml of the final suspension was spread and dried onto coverslips 22 × 50 ml at low temperature, then mounted onto slides using Canada Balsam (R.I = 1.74).

The diatom assemblages were examined quantitatively using an Optika light photomicroscope combined with a digital camera and under oil immersion with differential interference contrast at 1000× magnification. Identification and ecological information of diatom species were acquired based mainly on previous works and comparisons with the most up-to-date literature (e.g., Hustedt 1930–1966, 1957; Simonsen 1962, 1979; Ehrlich

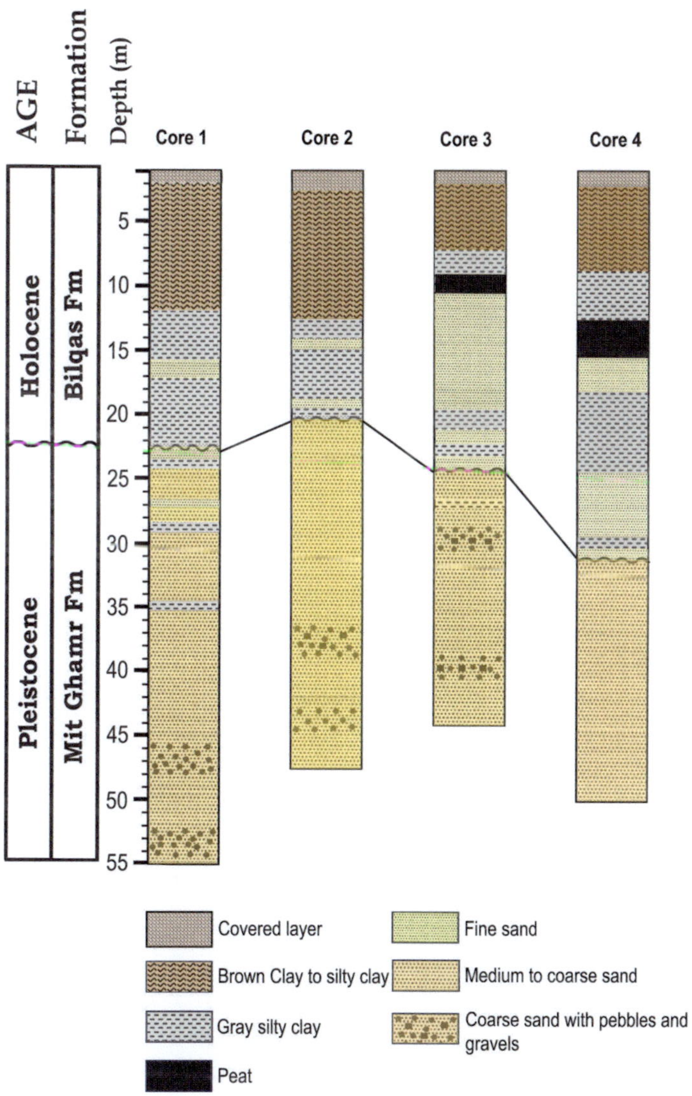

Fig. 5.2 Lithological profiles of the studied core sections in the central Nile Delta, Egypt

1973, 1975; Stoermer et al. 1974; Foged 1980, 1993; Gasse 1986; Krammer and Lange-Bertalot 1986–1991; Williams and Round 1987; Zalat 1991; Denys 1991–1992; van Dam et al. 1994; Witkowski et al. 2000; Wolfe et al. 2000; Zalat and Servant-Vidary 2005, 2007; Hofmann et al. 2011, Zalat et al. 2022).

For statistical analysis, in every slide, the diatoms were counted along 10 random transects, and the number of diatoms encountered was taken as a rough approximation of the total frequency. In samples of low diatom concentrations, the lower limit to allow percentage calculation was 300 diatom valves per sample. The relative frequencies of every species were calculated as a percentage of total diatom valves (%TDV) in each sample. The percentage contents of the species were used for the estimation of ecological parameters such as life-form groups, pH, and salinity. The grouping of diatoms into salinity preference categories has been used to estimate the salinity history during the time of deposition. Diatoms were classified as freshwater (oligohalobous), brackish water (mesohalobous), and marine water (polyhalobous forms). The ratio of planktonic to benthic and epiphytic forms was calculated for each sample to estimate the type of habitats and water-level changes. The pH grouping of diatoms included four categories according to Hustedt (1957) and Ehrlich (1973): acidophilous (pH 5.5 to < 7), circumneutral (pH around 7), alkaliphilous (pH over 7 to 8.5), and alkalibiontic (pH over 8.5). The most dominant species that have relative abundance over 4% of the total diatom valves in the three examined samples were used to construct diatom stratigraphic diagrams (Figs. 5.3, 5.4, 5.5, 5.6 and 5.7).

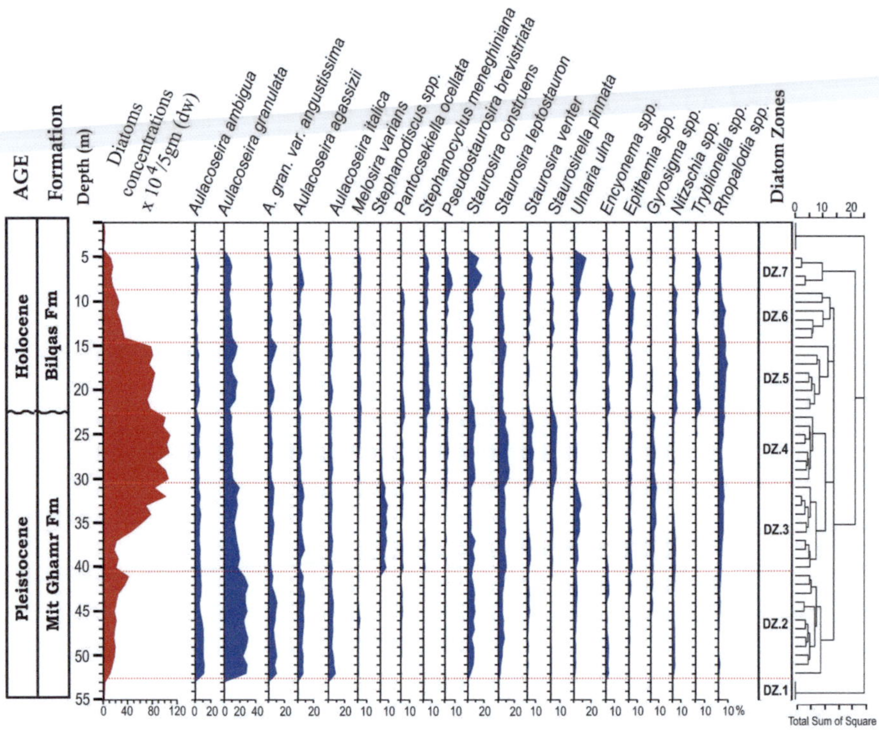

Fig. 5.3 Percentage diatom diagram of core 1, Kafr El-Zayat area, showing the relative abundances of the dominant and subdominant taxa and the diatom zones based on the cluster analysis

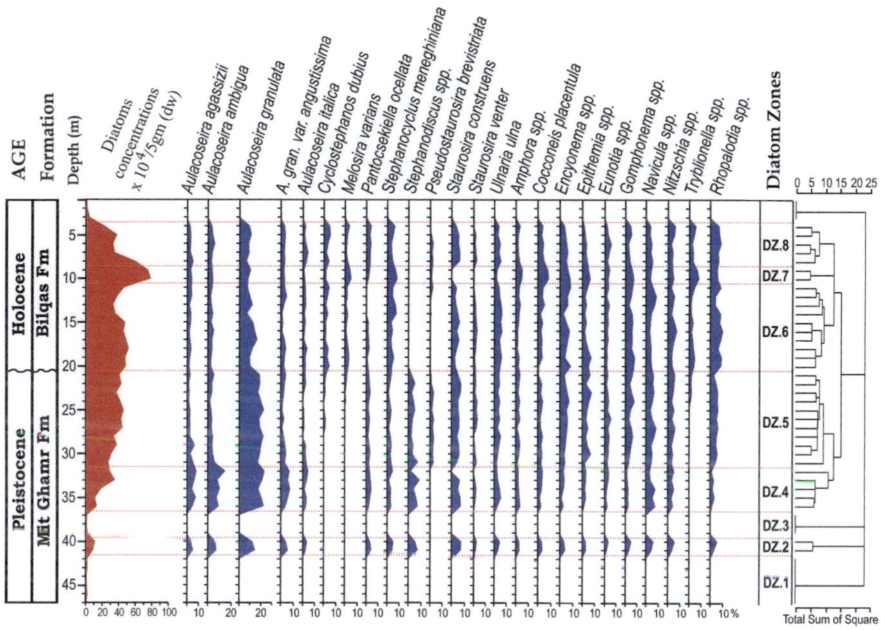

Fig. 5.4 Percentage diatom diagram of core 2, Tanta area, showing the relative abundances of the dominant and subdominant taxa and the diatom zones based on the cluster analysis

Multivariate statistical analysis is achieved by the computer program PAST (Hammer et al. 2001). Our calibration model included the diatom taxa that have a relative abundance over 3% of the total diatom valves. Cluster analysis was carried out using Euclidean distance measure and the average clustering method of the investigated samples to distinguish the diatom zones.

5.5 Results

5.5.1 Relative Diatom Frequency and Ecological Groups

Diatoms communities during the Quaternary in the sediments of the Nile Delta fluctuated between rich to rare and very well to poorly preserved. The diatom concentrations along the core's samples vary from 1.2×10^4 to 110×10^4. A total of 216 species and varieties representing 49 genera were identified from the investigated samples. Of the recorded species, 36 diatom taxa were distributed regularly, and 14 species were either common and/or abundant. The remaining taxa are infrequently distributed throughout the core's samples. *Aulacoseira* Thwaites, 1848 is the most dominant planktic genus

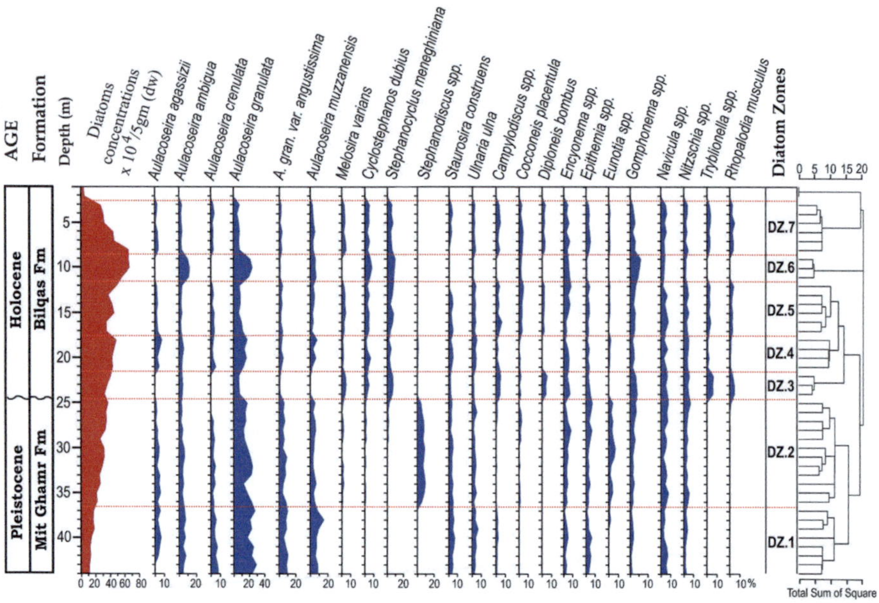

Fig. 5.5 Percentage diatom diagram of core 3, Kafr El-Sheikh area, showing the relative abundances of the dominant and subdominant taxa and the diatom zones based on the cluster analysis

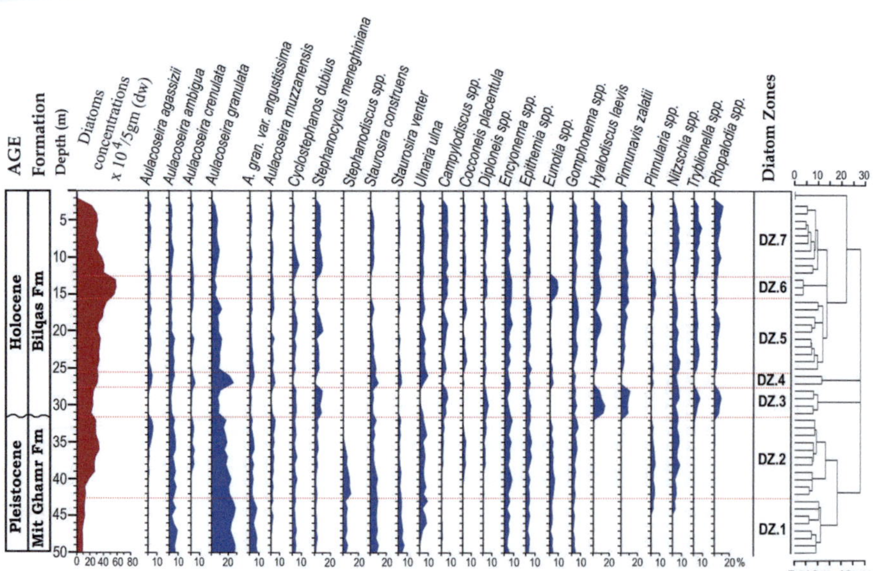

Fig. 5.6 Percentage diatom diagram of core 4, El Mansoura area, showing the relative abundances of the dominant and subdominant taxa and the diatom zones based on the cluster analysis

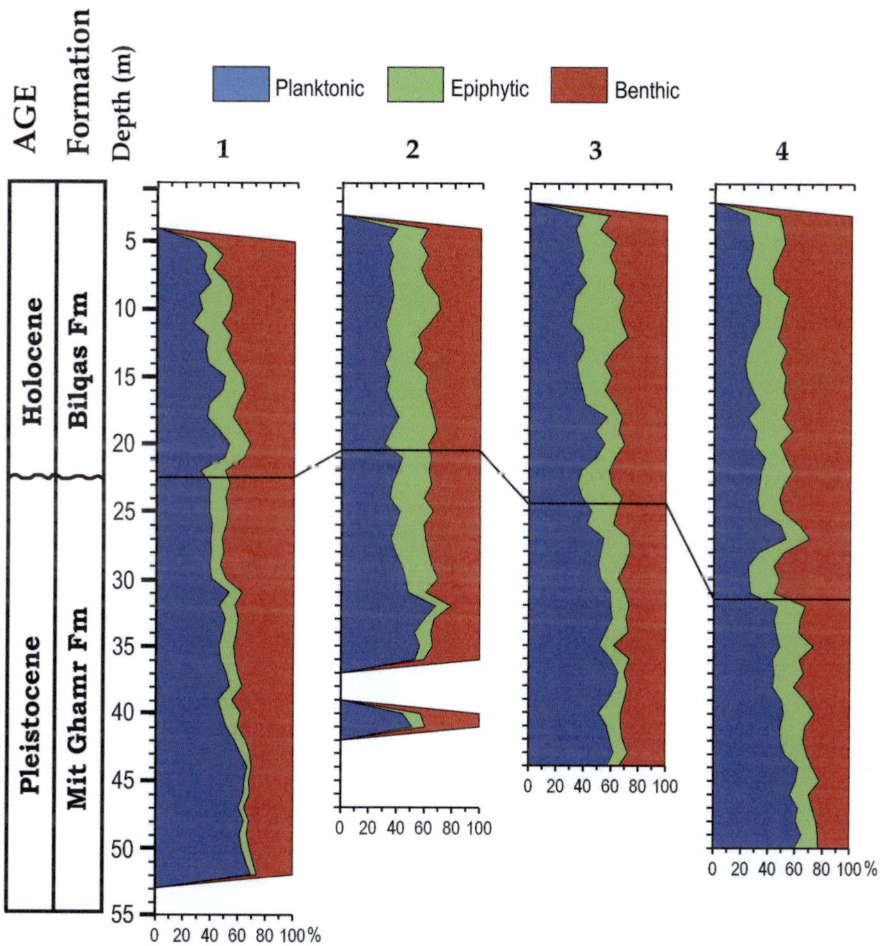

Fig. 5.7 Percentage diagram of the diatom life-form grouping (Planktonic/periphytic ratio) in the studied cores, central Nile Delta, Egypt

and accounted for an average of 40% of the total diatom assemblage. Other planktonic forms of the genera *Cyclostephanos* Round, 1987, *Cyclotella* (Kützing) Brébisson, *Lindavia* (Schütt) De Toni and Forti, 1900, *Pantocsekiella* Kiss and Ács, 2016, *Stephanodiscus* Ehrenberg (1845), and *Thalassiosira* Cleve, 1873 are reported in low frequencies. The most frequent epiphytic diatoms are *Encyonema* Kützing, 1834 and *Gomphonema* Ehrenberg, 1832, while the other forms, *Eunotia* Ehrenberg, 1837, *Cocconeis* Ehrenberg, 1836, *Amphora* Kützing, 1844, *Cymbella* Agardh, 1830, *Epithemia* Kützing, 1844, and *Rhopalodia* O.Müller, 1895 are found in small quantities, and *Tabellaria* Kützing, 1844, *Achnanthes* Bory 1822, *Rhoicosphenia* Grunow, 1860 are restricted to a few samples. The stratigraphical distribution of benthic diatoms includes a frequent abundance of species assigned to the genera *Staurosira, Staurosirella*, and *Ulnaria* species associated with low frequencies of species belonging to the genera *Craticula* Grunow, 1868, *Diploneis* Ehrenberg ex Cleve, 1894, *Fragilaria* Lyngbye, 1819, *Anomoeoneis* Pfitzer, 1871, *Caloneis* Cleve, 1894, *Gyrosigma* Hassall, 1845, *Navicula* Bory, 1822, *Pinnularia* Ehrenberg, 1843, *Pseudostaurosira* Williams and Round, 1988, *Stauroneis* Ehrenberg, 1843, *Nitzschia* Hassall, 1845, *Tryblionella* W. Smith, 1853, *Cymatopleura* W. Smith, 1853, *Surirella* Turpin, 1828 and *Campylodiscus* Ehrenberg ex Kützing, 1844. The relative abundance of the most dominant and subdominant diatom species is represented in Figs. 5.3, 5.4, 5.5 and 5.6.

Regarding the ecological habitat of the recognized diatom taxa, the planktonic forms were dominated by *Aulacoseira* spp., followed by common to frequent *Cyclostephanos dubius* (Fricke) Round 1987 and *Stephanocyclus meneghiniana* (Kützing) Skabichevskii 1975 and infrequently abundance of *Stephanodiscus* spp. in Mit Ghamr formation. The planktonic taxa decreased gradually in their abundance, with a marked fluctuation in the Bilqas formation. The epiphytic diatoms were relatively infrequent in the samples of Mit Ghamr formation in cores 1 and 2, with significant abundance in cores 3–5, and increased gradually in their abundance in the samples of Bilqas formation, in particular the peat layers of core 3–5. The benthic diatoms were less abundant in the Mit Ghamr formation and increased relatively in the Bilqas formation (Fig. 5.7). Most of the recorded diatoms preferring slightly alkaline to alkaline water occurred in the sediments. The most abundant taxa belonged to the alkaliphilous group with a relative frequency of 70–97% of the total assemblage. Alkalibiontic diatoms ranged from 2.5 to 26%, while the frequency of acidophilous forms varied between 1.3 and 12.6% (Fig. 5.8). Most of the recognized diatom taxa in the samples of Mit Ghamr formation belonging to oligohalobous salinity group (78 -97%) associated with infrequent mesohalobous forms and very limited taxa less than 10% were considered as polyhalobous at the top part of the formation. In contrast, the relative abundance of oligohalobous diatoms decreased relatively, accompanied by a marked increase in the abundance of both mesohalobous and polyhalobous diatom taxa in the Bilqas formation (Fig. 5.9).

5.5.2 Diatom Stratigraphy

The documented diatom assemblages displayed marked floristic changes along the studied core samples, which reflect marked environmental and climatic changes during the Quaternary. The pivotal change in diatom composition comprises a shift from an assemblage dominated by planktonic, eutrophic indicators of *Aulacoseira* taxa through marked intervals to periphytic taxa in distinct zones. A high proportion of periphytic (epiphytic and benthic) to plankton has been reported as indicative of a shallow environment during warm, dry climate, and a change from periphytic to planktonic *Aulacoseira* diatom taxa, reflecting a trend to a humid warm climate with a relative rise in water level during the Nile floods time. Multivariate statistical analysis included cluster-constrained analysis,

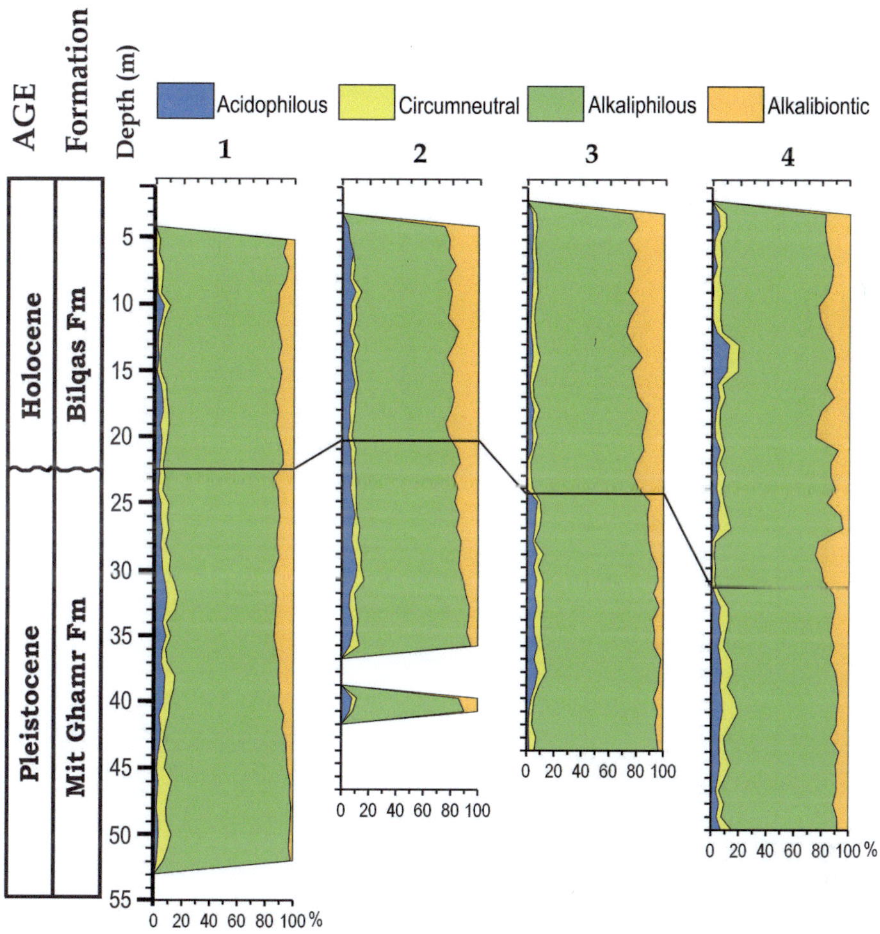

Fig. 5.8 Percentage diagram of the diatom pH grouping in the studied cores, central Nile Delta, Egypt

and the observed changes in the relative abundance of dominant and subdominant taxa explained several diatom zones for each core section, which reflect marked environmental changes during the time of sediment deposition.

Core 1 (Kafr El-Zayat site)

DZ.1 (53–55 m): Scarce diatom taxa

DZ.2 (41–52 m): Predominant of *Aulacoseira* taxa associated with a frequent abundance of *Fragilaria* sensu lato taxa such as *Staurosira construens* Ehrenberg, 1843 and *Staurosira leptostauron* (Ehrenberg) Kulikovskiy and Genkal, 2011. *Aulacoseira granulata*

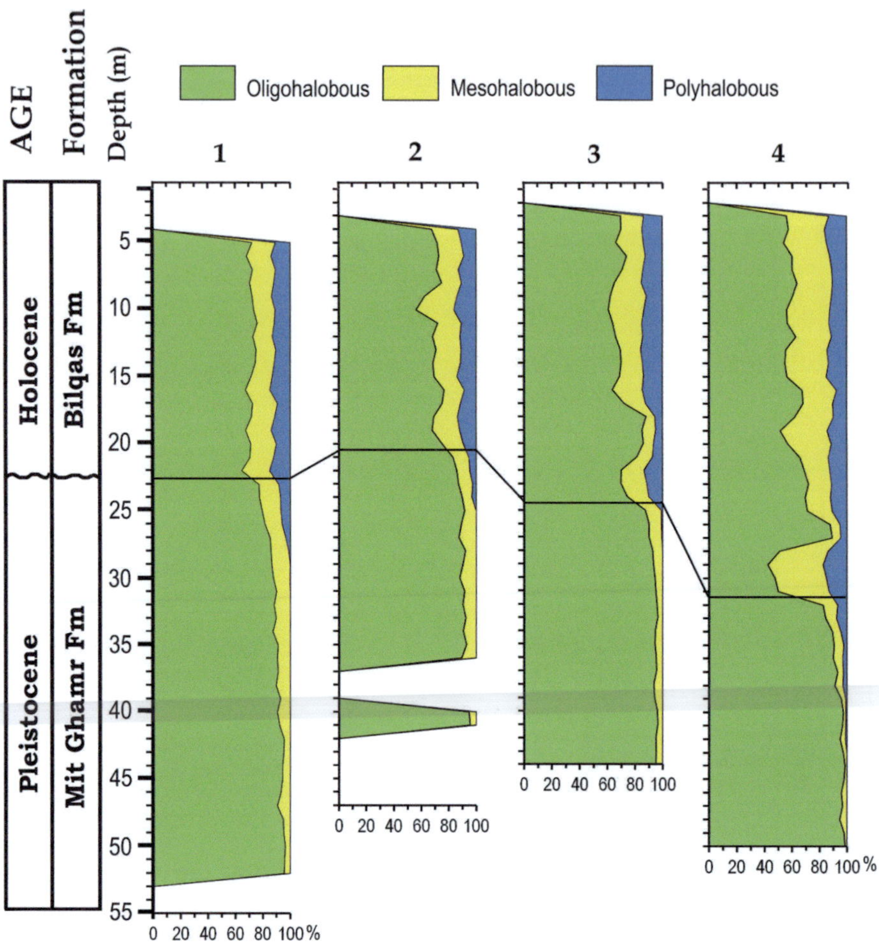

Fig. 5.9 Percentage diagram of the diatom halobien-groups (salinity spectrum) in the studied cores, central Nile Delta, Egypt

(Ehrenberg) Simonsen, 1979 has maximum abundance accompanied by a common occurrence of *A. granulata* var *angustissima* (O.Müller) Simonen 1979, *A. ambigua* (Grunov) Simonen 1979, *A. agassizii* (Ostenfeld) Simonen 1979, and *A. italica* (Ehrenberg) Simonen 1979. The average abundance of the planktonic taxa ranges from 52 to 69.3%, the benthic form varies from 26 to 35.6%, and the epiphytic taxa are less abundant by 3.5 to 8.7%. The diatom concentration was relatively low ($11.4–42.1 \times 10^4$).

DZ.3 (31–40 m): First appearance with a frequent abundance of *Stephanodiscus* taxa associated with a relative increase in abundance of *Staurosira construens, Staurosira leptostauron, Ulnaria ulna* (Nitzsch) Compère, 2001, and *Gyrosigma* spp. The abundance of *Aulacoseira* taxa decreased relatively than the prior zone with a common occurrence of

Aulacoseira granulata. The diatom concentration increased to reach about 103 × 10^4 at a depth of 32 m. The planktonic diatoms attained 45.8 to 53.6%, and the benthic forms 37.2 to 43.2%.

DZ.4 (23–30 m): There was a marked decrease in the abundance of *Aulacoseira* taxa with a relatively sparse occurrence of *Stephanodiscus* spp. The relative abundance of *Staurosira construens, S. leptostauron, S. venter* (Ehrenberg) Cleve & Möller, 1879, and *Staurosirella pinnata* (Ehrenberg) Williams & Round, 1987 increased relatively and characteristic of this zone. High diatom concentration 90 to 110 × 10^4. A relative decrease in abundance of planktonic taxa 38–40% accompanied by an increase in the rate frequency of the benthic forms (47–53%). The epiphytic taxa are less abundant.

DZ.5 (15–22 m): Marked oscillation in the abundance of freshwater *Aulacoseira* taxa associated with a significant abundance of mesohalobous and polyhalobous diatom taxa such as *Melosira varians* Agardh, 1827, *Diploneis bombus* (Ehrenberg) Ehrenberg, 1853, *Rhopalodia musculus* (Kützing) O. Müller, 1900, and *Tryblionella granulata* (Grunow) Mann, 1990. High diatom concentration 72.3 to 85.2 × 10^4. The diatom composition consists of a mixed assemblage of planktonic taxa (32–52.8%), the benthic forms (36–49.6%), and the epiphytic taxa increased relatively in this zone (13–19%).

DZ.6 (9–14 m): There was a marked decrease in diatom concentration, and the diatom composition contains a mixed assemblage of freshwater taxa accompanied by the significant occurrence of mesohalobous and polyhalobous taxa. The relative abundance of *Aulacoseira* taxa decreases relatively with the common occurrence of *Aulacoseira granulata,* associated with a frequent abundance of *Staurosira construens, S. leptostauron,* and *an* infrequent abundance of *Rhopalodia musculus* and *Tryblionella granulata.*

DZ.7 (5–8 m): There was a distinct increase in the abundance of *Staurosira construens* and *Ulnaria ulna* associated with the common occurrence of *Aulacoseira granulata,* and infrequent abundance of *Stephanocyclus meneghiniana, Pseudostaurosira brevistriata* (Grunow) Williams et Round, 1987, *Staurosira venter,* and *Tryblionella granulata.* Low diatom concentration 12.7–16.3 × 10^4. The periphytic taxa (61–72%) are more abundant than the planktonic forms (28–39%).

Core 2 (Tanta site)

DZ.1 (42–47 m): Scarce diatom taxa

DZ.2 (40–41 m): Sudden abundance of the planktonic *Aulacoseira* taxa accompanied by common to frequent abundance of *Stephanodiscus* spp., *Pantocsekiella ocellata* (Pantocsek) Kiss and Ács, 2016, *Stephanocyclus meneghiniana.* The benthic and epiphytic taxa were represented by the frequent occurrence of St*aurosira construens, Ulnaria ulna, Cocconeis placentula* Ehrenberg, 1838 associated with irregular distribution of the *Encyonema, Epithemia, Gomphonema, Eunotia, Nitzschia, Navicula,* and *Rhopalodia* species. The diatom concentration was low, and the diatom composition contains relatively equal frequency rates of planktonic and benthic forms.

DZ.3. (37–39 m): Marked disappearance of diatom taxa except for some sporadic *Aulacoseira* taxa.

DZ.4 (32–36 m): Predominance of the planktonic *Aulacoseira* taxa. *Aulacoseira granulata* has a maximum abundance (20–23.5%) associated with the common occurrence of *Aulacoseira ambigua, Aulacoseira agassizii,* and *Stephanodiscus* spp. The periphytic taxa are represented by a common occurrence of *Staurosira construens* and infrequently abundance of *Ulnaria ulna, Cocconeis placentula.* Other periphytic species are distributed irregularly. The planktonic species attain 53–68%, and the periphytic forms include 21–40% of benthic taxa and 7–11.8 epiphytic forms.

DZ.5 (21–31 m): *Aulacoseira granulata* continued in its high abundance with a marked decrease of *Aulacoseira ambigua* and *Aulacoseira agassizii* and infrequent distribution of *Stephanodiscus* spp. *Staurosira construens* is relatively common, accompanied by the frequent occurrence of *Ulnaria ulna.* Other periphytic taxa are represented by irregular distribution of the *Encyonema, Epithemia, Gomphonema, Eunotia, Nitzschia, Navicula,* and *Rhopalodia* species. The diatom concentration was moderate, and the diatom composition contained a relatively equal frequency rate of planktonic and benthic forms associated with a significant abundance of the epiphytic taxa (21–25%).

DZ. 6 (11–20): There was a marked decline in the abundance of *Aulacoseira* taxa with continuous relatively common *Aulacoseira granulata* and decreased abundance of *Staurosira construens.* The brackish water taxa have considerable occurrence, represented by a frequent abundance of *Stephanocyclus meneghiniana, Rhopalodia musculus,* and *Nitzschia sigma* (Kützing) W. Smith, 1853. The first appearance of some polyhalobous taxa included an irregular abundance of *Tryblionella granulata* and *T. punctata* W. Smith, 1853. The diatom concentration was moderate, and the diatom composition was composed of a relatively equal frequency rate of planktonic (31–41.2%) and the benthic forms (31.5–47.4%) associated with a substantial abundance of the epiphytic taxa (22–32.7%).

DZ.7 (9–10 m): Short interval characterized by increased abundance of the epiphytic and mesohalobous taxa. *Melosira varians, Stephanocyclus meneghiniana, Rhopalodia musculus, Nitzschia sigma,* and *Cocconeis placentula* have significance abundance. This zone is characterized by high diatom concentration with relatively equal occurrence of the life-form groups.

DZ.8 (4–8 m): The mesohalobous taxa of the previous zone are continuous in their abundance associated with re-abundance of *Staurosira construens, Ulnaria ulna,* and decreased abundance of *Epithemia* and *Amphora* taxa. *Aulacoseira granulata* is continuous in its frequent occurrence with an irregular abundance of other *Aulacoseira* taxa. The diatom concentration decreased gradually. The benthic taxa are common over the planktonic taxa, and the epiphytic forms are still common.

Core 3 (Kafr El Sheikh site)

DZ.1 (37–44 m): Predominance of the planktonic *Aulacoseira* taxa that included a great abundance of *A. granulata* associated with common to frequent occurrence of *A. granulata* var. *angustissima, A. agassizii, A. ambigua, A. crenulata* (Ehrenberg) Thwaites, 1848, and *A. muzzanensis* (Meister) Krammer, 1991. The benthic taxa are represented by a frequent abundance of *Staurosira construens* and *Ulnaria ulna.* The remaining taxa are distributed sparsely. The diatom concentration was relatively low with a high frequency of planktonic taxa (52–65%), the benthic (27.6–33.8%), and a limited occurrence of the epiphytic taxa (4.5–16%).

DZ.2 (25–36 m): Continuous abundance of *Aulacoseira* taxa (32–50.2%) with obvious oscillation in their profusion. First appearance of *Stephanodiscus* spp. that is frequently distributed, associated with the irregular occurrence of *Staurosira construens, Ulnaria ulna,* and other periphytic taxa. The diatom concentration was relatively moderate with a high frequency of planktonic taxa (42.3–60%), the benthic (26.3–38.1%), and a considerable frequency rate of the epiphytic taxa (10–19.6%).

DZ.3 (22–24 m): Marked reduction in *Aulacoseira* taxa (13.5–15%) and disappearance of *Stephanodiscus* spp. The first frequent appearance of *Stephanocyclus meneghiniana* and *Cyclostephanos dubius* is associated with a significant occurrence of mesohalobous and polyhalobous taxa, which are recorded by infrequent occurrence of *Diploneis bombus, Anomoeoneis sphaerophora* Pfitzer, 1871, *Rhopalodia musculus, Tryblionella granulata,* and *Campylodiscus* spp. The diatom concentration was moderate with a relatively equal frequency rate of planktonic (35.8–39.1%), and the benthic taxa (32.3–41.2%), and a substantial percentage of the epiphytic taxa (20.3–28.6%).

DZ.4 (18–21 m): Re-abundance and dominance of the freshwater planktonic *Aulacoseira* taxa (34–46%) with common occurrence of *Aulacoseira granulata.* The mesohalobous and polyhalobous taxa decreased with a relatively increased occurrence of the freshwater periphytic taxa comprising an infrequent abundance of *Staurosira construens, Ulnaria ulna, Encyonema* spp., *Epithemia adnata* (Kützing) Brébisson, 1838, and *Nitzschia frustulum* (Kützing) Grunow, 1880. The diatom concentration was relatively high with an increased frequency rate of planktonic (48.7–56.9%), the benthic taxa (30.4–36.8%), and a declining percentage of the epiphytic taxa (10.9–16.6%).

DZ.5 (12–17 m): Obvious reduction in abundance of *Aulacoseira* taxa (22–29%) with frequent occurrence of *Aulacoseira granulata.* The warm brackish water taxa have significant occurrence in the zone and are represented by the infrequent abundance of *Melosira varians, Stephanocyclus meneghiniana, Cyclostephanos dubius, Diploneis bombus, Rhopalodia musculus,* and *Campylodiscus* spp. The diatom concentration was moderate with an increased frequency rate of epiphytic taxa (17–33%), associated with a relatively equal abundance of the planktonic (36–41%) and the benthic taxa (28.2–44.8%).

DZ.6 (9–11 m): A sudden great abundance of *Aulacoseira* taxa (41–46%), including the common occurrence of *Aulacoseira granulata, A. ambigua,* and frequent *Aulacoseira muzzanensis. Stephanocyclus meneghiniana* and *Cyclostephanos dubius* have considerable abundance and marked decline with the sparse distribution of *Ulnaria ulna, Campylodiscus* spp., and *Rhopalodia* spp. However, the *Gomphonema* species have significant occurrences in this zone. This zone is characterized by high diatom concentration (58.9–70.4 $\times$

10^4), and the diatom composition contains a relatively equal abundance of the life-form groups.

DZ.7 (3–8 m): Marked reduction in the abundance of *Aulacoseira* taxa (19.8–27.5%) associated with continuous, frequent occurrence of *Stephanocyclus meneghiniana* and *Cyclostephanos dubius.* The mesohalobous and polyhalobous diatoms have considerable occurrence in this zone and are signified by the irregular abundance of *Diploneis bombus, Campylodiscus* spp. *Tryblionella* spp. *Rhopalodia musculus* and *Rhopalodia gibberula* (Ehrenberg) O.Müller, 1895. The diatom concentration decreased gradually upward in this zone with a relatively equal frequency rate of the planktonic (33.6–41.4%) and the benthic taxa (37–48.4%). The epiphytic taxa are distributed infrequently, with a proportion range of 18 to 27.9%.

Core 4 (El Mansoura site)

DZ.1 (43–50 m): Great abundance of the planktonic *Aulacoseira granulata* (23–28.5%) associated with frequent occurrence of *Aulacoseira ambigua* and *A. granulata* var. *angustissima* and *Stephanodiscus* species. The periphytic taxa are represented by a frequent abundance of *Staurosira construens, S. venter*, and *Ulnaria ulna* accompanied by an irregular distribution of *Encyonema, Gomphonema, Epithemia*, and *Eunotia* species. The diatom concentration was low, and the diatom composition contains a great percentage of the planktonic taxa (52.5–65%), and the benthic forms constitute 22 to 31.5%, with limited numbers of the epiphytic taxa.

DZ.2 (32–42 m): Continuous abundance of *Aulacoseira* taxa with dominance of *Aulacoseira granulata* and relative increase in abundance of *Stephanodiscus* spp. in the lower part and decrease in *Staurosira venter*. Other periphytic taxa such as *Staurosira construens* and *Ulnaria ulna* frequently occurred with irregular distribution of *Encyonema, Gomphonema, Epithemia, Eunotia, Pinnularia,* and *Nitzschia* species. The diatom concentration was relatively moderate, and the proportion of the planktonic forms (43–51.5%) exceeded the benthic taxa (27–41%). The epiphytic taxa are well developed in this zone with a frequency rate of 16 to 24.3%.

DZ.3 (28–31 m): Marked decline of freshwater planktonic *Aulacoseira* taxa associated with an obvious increase in abundance of mesohalobous diatoms signified by the frequent occurrence of *Stephanocyclus meneghiniana* and *Campylodiscus* spp. with the first appearance of *Hyalodiscus laevis* Ehrenberg, 1845 and *Pinnunavis zalatii* Saleh & Edgar, 2015. The polyhalobous taxa are documented by the first irregular occurrence of *Tryblionella* spp., *Rhopalodia musculus*, and *Rhopalodia gibberula.* The diatom concentration was relatively moderate, and the percentage of the benthic forms (48.5–55.8%) increased relatively over the planktonic taxa (25.7–33%). The epiphytic taxa attain a frequency rate of 18 to 22.6%.

DZ.4 (26–27 m): Short interval characterized by peak abundance of *Aulacoseira granulata* accompanied by a frequent occurrence of *Staurosira construens* and *Ulnaria ulna.* The mesohalobous taxa of the prior zone decreased markedly in this zone. The diatom concentration was moderate, and the ratio of the planktonic diatoms increased (48.9–52.6%) with a decline in the benthic taxa (29–37%). The epiphytic forms reach a frequency rate of 14 to 17.5%.

DZ.5 (16–25 m): Obvious reduction in the abundance of the *Aulacoseira* taxa with frequent occurrence of *Aulacoseira granulata.* The mesohalobous and polyhalobous taxa of the previous DZ.3 reappeared again with the frequent distribution. The diatom concentration was moderate, and the percentage of the planktonic diatoms decreased (24.8–37.5%) with an increase in the abundance of the benthic taxa (43–52%). The epiphytic forms have a considerable occurrence by a frequency rate of 16 to 26.5%.

DZ.6 (13–15 m): This zone is characterized by a marked reduction of the planktonic *Aulacoseira* taxa simultaneously with a common abundance of the mesohalobous periphytic taxa *Campylodiscus* spp., *Hyalodiscus laevis,* and *Pinnunavis zalatii.* The benthic acidophilous *Eunotia* species have significant abundance in this zone. The diatom concentration was high, and the proportion of the planktonic diatoms decreased (22.3–24%) with a continuous abundance of the benthic taxa (47–51.7%). The epiphytic forms distinguish this zone by a frequency rate of 26.3 to 28.8%.

DZ.7 (3–12 m): It is characterized by a mixed assemblage of the frequent occurrence of the *Aulacoseira* taxa, *Stephanocyclus meneghiniana, Staurosira construens, Ulnaria ulna,* associated with several mesohalobous and polyhalobous diatoms which distributed irregularly by *Diploneis bombus, Campylodiscus* spp. *Tryblionella* spp., *Nitzschia sigma, Rhopalodia musculus, Hyalodiscus laevis* and *Pinnunavis zalatii.* The diatom concentration was moderate, and the diatom composition contains a great percentage of the benthic taxa (46–53.8%), concurrent with a lower frequency rate of the planktonic (22.8–33.4) and the epiphytic forms (17–24.5%).

5.6 Discussion

The historical record of the moderate to well-preserved diatom assemblages in the Quaternary sediments of the Nile indicates marked environmental changes linked to climate variability during the late Pleistocene-Holocene. The diatom distribution pattern through the examined stratigraphic sections led to the recognition of several diatom zones. These zones reflected several phases of Nile Delta development and a series of alternations between fresh and slightly saline to saline episodes associated with rising and falling sea levels and Nile floods, which may relate to climatic changes during the late Pleistocene-Holocene time.

5.6.1 Quaternary Palaeohydrological Changes of the Nile Delta Ecosystem

Diatom analysis of the studied cores sediments explained marked changes in the environmental history of the Nile Delta linked by climate change during the Quaternary. Clear fluctuation in the relative abundance of the diatom taxa and the changes in the diatom assemblage composition can be seen along the core's sections over time, permitting the recognition of distinctive diatom zones corresponding to variations in the paleosalinity, alkalinity, and water level. The documented diatom assemblages are dominated by riverine planktonic taxa, which were generally similar to those reported from the waters of the modern Nile and the headwaters of the Nile such as the Atbara River in Sudan (Foged 1980), lake Mobuto Sese Seko and Lake Victoria in East Africa (Gasse 1986), Holocene sediments of the northern Egyptian lakes (Zalat and Servant Vildary 2005, 2007) and the Holocene sediments of the Faiyum depression (Zalat 2015; Zalat et al. 2017).

The recognized assemblages are composed mainly of planktonic *Aulacoseira* taxa, *Cyclotella* sensu lato species (*Cyclotella atomus, Lindavia glomerata, L. radiosa, Pantocsekiella kuetzingiana, P. ocellata,* and *Stephanocyclus meneghiniana*), *Cyclostephanos dubius, Melosira varians, Stephanodiscus* spp., accompanied by the infrequently distribution of many littoral and epiphytic taxa that assigned to genera *Fragilaria, Staurosira, Ulnaria, Eunotia, Cocconeis, Achnanthes, Amphora, Cymbella, Encyonema, Gomphonema, Epithemia, Planothidium, Pseudostaurosira, Rhopalodia, Tabellaria,* and benthic forms such as *Anomoeoneis, Campylodiscus, Caloneis, Craticula, Cymatopleura, Diploneis, Fallacia, Gyrosigma, Hantzschia, Hyalodiscus, Pinnularia, Placoneis, Mastogloia, Navicula, Nitzschia, Sellaphora, Stauroneis, Tryblionella,* and *Surirella.*

Numerous studies have sorted diatoms into groups based on the physicochemical conditions of where they are common. The most important ecological factors estimated from the study of diatoms are water level changes, pH, paleosalinity, trophic status, and dissolved silica concentration.

5.6.2 Water Level Changes

Water level changes result from a variety of geological, biological, and/or climatic processes (Wolin and Stone 2010). In aquatic environments, diatoms are highly sensitive indicator organisms for many ecological factors (Stone and Fritz 2004; Zalat and Servant Vildary 2007), and their fossils that are deposited in the sediments can be used to infer environmental changes (Smol 2008). Many paleolimnological studies have used the relative abundance of planktonic and periphytic taxa (epiphytic and benthic) ratio to qualitatively infer water level fluctuations (e.g., Wolin 1996; Xue et al. 2003; Bergner and Trauth 2004; Ekblom and Stabell 2008; Zalat et al. 2017). The fundamental idea of the planktic to periphytic ratio is that periphytic diatoms live within or on the sediments in the littoral zone or attach to any object in the milieu, such as stones and submerged vegetation, and are most common in shallower water where light penetration reaches the sediment–water interface. Thus, benthic diatoms are deposited primarily in sediments of shallow environment, in particular, nearer to shore, while planktic diatoms live drifting in open water and contribute cells to the sediment surface below all open-water areas (Wolin and Duthie 1999). Therefore, when water levels fall, more habitat is usually available for diatoms living attached to substrates (benthic) or aquatic vegetation (epiphytic) because a greater area of the bottom is exposed to sunlight. Moreover, a general reduction in diatom productivity can occur during low-water conditions, particularly in shallow lakes, owing to light limitation in benthic communities and nutrient limitation in plankton.

Changes in the relative abundance of planktonic and periphytic species can, therefore, reflect changes in the availability and location of habitats as a consequence of water level rise/fall (Wolin and Duthie 1999). Many benthic diatoms increase in abundance when declining water levels and shallow-water habitats are more abundant (Wolin and Duthie 1999). However, the adnate diatom species such as *Fragilaria* sensu lato (*Fragilaria, Pseudostaurosira, Staurosira,* and *Staurosirella*) are motile diatoms that can move along a substrate to adjust to changes in the photic zone (Hay et al. 2000). Consequently, the planktonic/periphytic ratio displays water depth because most planktonic and tychoplanktonic diatoms increase in abundance when water levels are high, whereas benthic species are more competitive under lowered water levels (Batterbee 1986; Zalat et al. 2017).

In the current study, essential fluctuations in the water levels took place in the Nile Delta during the late Pleistocene-Holocene time. This was documented by the variation in the ratio of planktic to epiphytic and benthic diatoms along the studied core sections over time (Fig. 5.7) that explained the water level fluctuation due to changes in water influx from the River Nile. During the late Pleistocene, the lower part of Mit Ghamr formation included the diatom zones (DZ.2 of core1, DZ.2, 4 of core 2, and DZ.1 of cores 3 and 4) that contained a high abundance of planktonic taxa over the periphytic forms which reflect relatively increased water level during the time of deposition. However, the most dominant planktonic taxa were *Aulacoseira* species, with a great occurrence of *Aulacoseira granulata* accompanied by a frequent abundance of *A. ambigua.* These taxa are heavily silicified diatoms with high sinking rates. Their ecology requires increased turbulent mixing and silica content to remain in the photic zone as well as low light conditions (Bradbury et al. 2002; Stager et al. 2005). They are often found in shallow eutrophic lakes in Europe (Rioual et al. 2007). Increased turbulence and corresponding nutrient increases during low-water stages can favor this genus over other planktonic species. However, in freshwater African lakes, *Aulacoseira* species are dominant with relatively high surface water levels during a humid warm climate (Gasse 1986; Bergner and Trauth 2004; Zalat 2015). The rise in the percentage of *Aulacoseira* taxa is associated with a lower concentration of diatom valves at the lower part of the Mit Ghmr Formation, which indicates relatively low-water conditions during a humid warm climate of the late Pleistocene. A combination of low-water conditions and high-wind exposure provides turbulent and high-nutrient conditions. The water level changed gradually upward of Mit Ghamr formation (DZ.3, 4 of core 1; DZ.5 of core 2; DZ.2 of cores 3 and 4) linked with gradual decrease in abundance of the planktonic taxa, concurrent with relative increase abundance of the periphytic forms (Fig. 5.7). The significant abundance of epiphytic taxa associated with many benthic forms reflect shallow-water environment with developed macrophytes.

An obvious water-level fluctuation occurred during the Holocene period and was documented by diatom zones of the Bilqas formation. An increase in the relative abundance of the periphytic taxa over the planktonic forms is demonstrated well, which reflects a marked lowering water level and shallow-water environment during the arid climate characteristic of the Bilqas formation. Increased abundance of the epiphytic taxa points to shallowness of the environment with developed macrophytes. Oscillations in the water level were owing to changes in the water supply from the river Nile into the delta and transgression of the Sea. This finding is estimated from the occurrence of the marine littoral diatoms associated with brackish and freshwater taxa in sediments of the Bilqas formation.

5.6.3 pH Reconstruction

The reconstruction of the pH history of the Nile Delta ecosystem during the Quaternary is based on a division of the recorded diatoms into four categories according to their pH tolerance as established by Hustedt (1957), Foged (1970, 1980), Ehrlich (1973), and Gasse (1975). These categories are: alkalibiontic: pH value of over 8; alkaliphilous: pH of over 7 to 8; circumneutral: pH of around 7; acidophilous: pH of below 7 to 6. The recognized pH diatom groups (Fig. 5.8) show little change from the late Pleistocene-Holocene and before the present. The majority of taxa encountered are alkaliphilous with a frequency rate of 67.5–94%. The alkalibiontic taxa have a significant occurrence along the core's sections, which attain a percentage range between 7 and 30%, except some samples in the lower part of the Mit Ghamr formation have a frequency rate of less than 5%. However, the relative abundance of circumneutral and acidophilous taxa suggests minor differences in paleoenvironmental conditions. The predominance of alkaliphilous species, accompanied by a considerable abundance of the alkalibiontic taxa in all the sections, clearly indicates that deposition mostly occurred in alkaline water with pH values ranging from 7 to 8.5, and the alkalinity increased slightly during the Holocene period by significant increase of the alkalibiontic diatoms in the sediments of Bilqas formation. The increasing abundance of the alkaliphilous and alkalibiontic taxa in the Nile delta sediments reflects the warm, dry climate conditions that prevailed during the late Pleistocene-Holocene period.

5.6.4 Paleosalinity Reconstruction

Variations in salinity in the water may directly reflect incoming water sources and evaporation. Diatoms are greatly influenced by water salinity, and marine and freshwater diatom floras are strikingly different (Potapova 2011). Various diatom assemblages from a stratigraphic sequence can be compared, and compositional differences thus detected are used to deduce salinity changes in the local aquatic environment. Changes in diatom-inferred salinity from short-term stratigraphic records can be used to determine the extent to which salinity change is related to lake water level and/or climatic shifts (Gasse et al. 1997). Numerous classifications of diatom species reflecting their relation to salinity were developed in the twentieth century, and the grouping of diatoms by halobien spectra is essentially based on the diatom classifications of Hustedt (1957), Simonsen (1962), Van der Werff and Huls (1957–1974), Ziemann (1971), Ehrlich (1975) and Ehrlich and Ortal (1979), according to their salinity preferences. Three main categories of diatoms concerning salinity were recognized (Fig. 5.9): oligohalobous (freshwater with salinity optima below 5 g/l), mesohalobous (brackish water with salinity optima in the 5–20 g/l range), and polyhalobous or euhalobien (coastal marine water with salinity optima in the 30–40 g/l range).

In the current study, the late Pleistocene sediments of Mit Ghamr formation are characterized by a great abundance of freshwater oligohalobous taxa, which attain a frequency rate of 77 to 98% associated with the limited occurrence of the mesohalobous taxa. The recognized diatom assemblages (Figs. 5.10, 5.11 and 5.12) indicate a freshwater environment due to freshwater influx from the river Nile and the salinity increased slightly at the end of the late Pleistocene, linked with the appearance of some mesohalobous taxa as a result of elevated evaporation during a warm, dry climate. Marked fluctuation in the salinity spectra has occurred during the deposition of the Holocene sediments of the Bilqas formation. A mixed diatom assemblage containing different dominant species assigned to oligohalobous, mesohalobous, and polyhalobous taxa was observed in the sediments, which reflects the coastal environment and transgression of the Sea on the delta area at the start of the Holocene period. At depth 18-21m of core 3 (DZ. 4) and 26-27m of core 4 (DZ.4), the conspicuous abundance of the oligohalobous species with a marked decline abundance of both mesohalobous and polyhalobous taxa suggests that deposition occurred in freshwater with a slight salinity of oligohalobien type during some phases of the Nile Delta development.

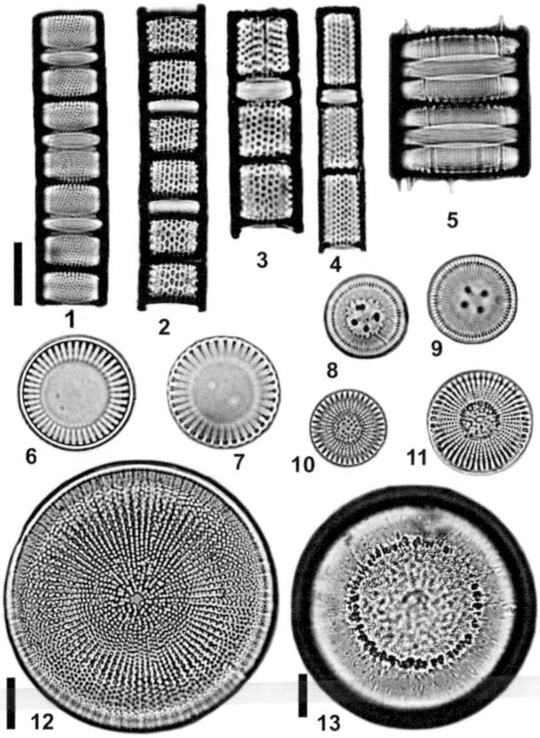

Fig. 5.10 LM micrographs of some selected diatom taxa at magnification ×1200, scale bar 10 μm. 1. *Aulacoseira ambigua* (Grunow) Simonsen, sample 34, core 2, Mit Ghamr Fm; 2–4. *Aulacoseira granulata* (Ehrenberg) Simonsen, 2. sample 32, core 2, Mit Ghamr Fm; 3–4. sample 40, core 4, Mit Ghamr Fm; 5. *Aulacoseira muzzanensis* (Meister) Krammer, sample 38, core 3, Mit Ghamr Fm; 6–7. *Stephanocyclus meneghiniana* (Kützing) Skabichevskii, sample 13, core 2, Bilqas Fm; 8–9. *Pantocsekiella ocellata* (Pantocsek) Kiss and Ács, 8. sample 33, 9. sample 41, core 2, Mit Ghamr Fm; 10–11. *Cyclostephanos dubius* (Fricke) Round, 10. Sample 10, core 3, Bilqas Fm, 11. Sample 19, core 4, Bilqas Fm; 12. *Stephanodiscus neoastraea* Håkansson and Hickel, sample 33, core 2, Mit Ghamr Fm; 13. *Hyalodiscus laevis* Ehrenberg, sample 30, core 4, Bilqas Fm

5.6.5 Development of the Nile Delta in the Late Pleistocene-Holocene

Diatom analysis from the sediments of the Nile Delta core's sections indicates five basic phases in the development of the Nile Delta ecosystem during the late Pleistocene-Holocene.

Phase 1

This phase covered the lower part section of Mit Ghamr formation and included DZ. 1 and 2 of core 1 (depth interval 41–55m); DZ.1–3 of core 2 (depth 37–47 m); DZ. 1 of core 3

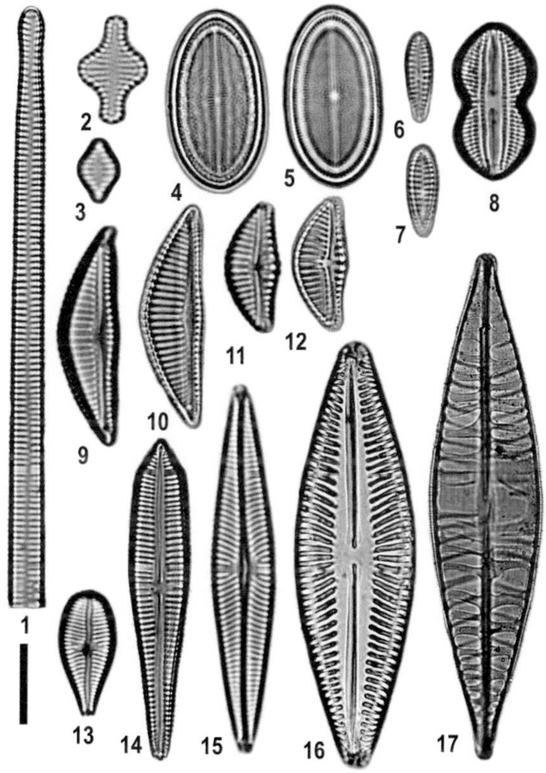

Fig. 5.11 LM micrographs of some selected diatom taxa at magnification ×1200, scale bar 10 μm. 1. *Ulnaria ulna* (Nitzsch) Compère, sample 33, core 1, Mit Ghamr Fm; 2. *Staurosira construens Ehrenberg*, sample 27, core 4, Bilqas Fm; 3. *Staurosira venter* (Ehrenberg) Cleve and Möller, sample 26, core 2, Mit Ghamr Fm; 4–5. *Cocconeis placentula* Ehrenberg, sample 10, core 2, Bilqas Fm; 6–7. *Rhoicosphenia abbreviata* (Agardh) Lange-Bertalot, sample 9, core 3, Bilqas Fm; 8. *Diploneis bombus* (Ehrenberg) Ehrenberg, sample 22, core 3, Bilqas Fm; 9–10. *Encyonema mesianum* (Cholnoky) Mann, sample 14, core 4, Bilqas Fm; 11–12. *Encyonema ventricosum* (C.Agardh) Grunow, sample 24, core 2, Mit Ghamr Fm; 13. *Gomphonema italicum* Kützing, sample 19, core 2, Bilqas Fm; 14. *Gomphonema turris* Ehrenberg, sample 9, core 3, Bilqas Fm; 15. *Navicula radiosa* Kützing, sample 27, core 3, Mit Ghamr Fm; 16. *Pinnunavis zalatii* Saleh and Edgar, sample 30, core 4, Bilqas Fm; 17. *Craticula cuspidata* (Kützing) Mann, sample 14, core 4, Bilqas Fm

(depth 37–44 m) and DZ.1 of core 4 (depth 43–50 m). There was a deposition of medium to coarse sand with an admixture of pebbles, gravel, and angular quartz grains, typical of a shallow environment. Diatom concentration was low, and the recognized diatom assemblages were dominated by the planktonic eutrophic *Aulacoseira* taxa, including the common occurrence of *A. granulata* and *A. ambigua* associated with a frequent abundance of other *Aulacoseira* species and the periphytic epipsammic *Staurosira* taxa and limited occurrence of other benthic forms.

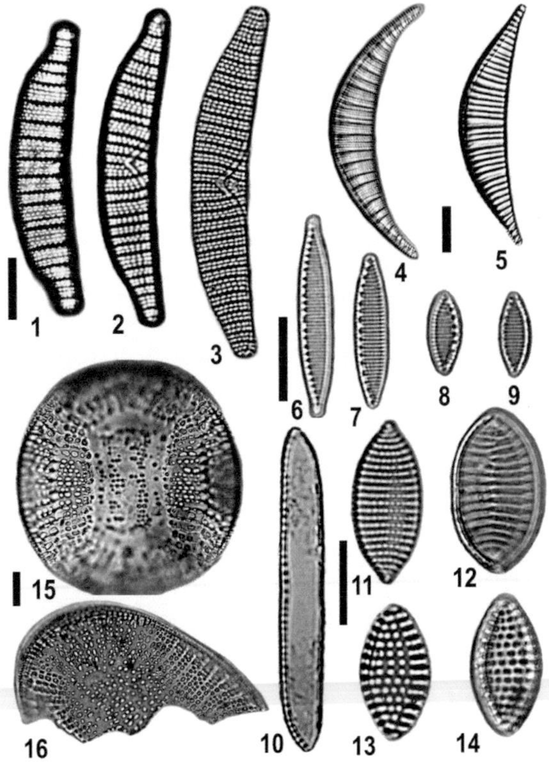

Fig. 5.12 LM micrographs of some selected diatom taxa at magnification ×1200, scale bar 10 μm. 1–2. *Epithemia adnata* (Kützing) Brébisson, sample 24, core 2, Mit Ghamr Fm; 3. *Epithemia turgida* (Ehrenberg) Kützing, sample 10, core 2, Bilqas Fm; 4. *Rhopalodia acuminata* Krammer, sample 16, core 2, Bilqas Fm; 5. *Rhopalodia gibberula* (Ehrenberg) O. Müller, sample 30, core 4, Bilqas Fm; 6. *Nitzschia frustulum* (Kützing) Grunow, sample 27, core 2, Mit Ghamr Fm; 7. *Nitzschia amphibia* Grunow, sample 16, core 2, Bilqas Fm; 8–9. *Nitzschia inconspicua* Grunow, sample 24, core 4, Bilqas Fm; 10. *Nitzschia obtusa* W. Smith, sample 8, core 4, Bilqas Fm; 11. *Tryblionella punctata* W.Smith, sample 10, core 2, Bilqas Fm; 12. *Tryblionella levidensis* W.Smith, sample 29, core 4, Bilqas Fm; 13–14. *Tryblionella granulata* (Grunow) Mann, sample 10, core 2, Bilqas Fm; 15. *Campylodiscus clypeus* Ehrenberg, sample 29, core 4, Bilqas Fm; 16. *Campylodiscus echeneis* Ehrenberg, sample 16, core 3, Bilqas Fm

A. granulata and *A. ambigua* are reported to thrive well in shallow water (Gasse 1986). *Aulacoseira* species respond to lake-level-induced changes in nutrient and light levels in African lakes (Gasse 1986; Marks et al. 2018). During deep lake and lower-nutrient conditions, *Aulacoseira ambigua* was dominant, while *Aulacoseira granulata* dominated during shallow lakes with higher-nutrient and lower-light stages (Verschuren et al. 2000). The dominance of planktonic diatoms and relatively high planktonic: periphytic ratio with higher proportions of *Aulacoseira granulata* and *A. ambigua* indicate a phase of moderate water levels and thorough mixing of the entire water column during the time of deposition. However, the low diatom content in the sediments was correlated with lower productivity in the Delta basin. A decline in diatom productivity can occur during low-water phases, particularly in shallow environments, due to light limitation in benthic communities and nutrient limitation in plankton. Moreover, the low abundances of all diatom species indicate stressful conditions during this period. The reduction in precipitation and moderate lake water levels may be attributed to drought climate conditions and low precipitation.

Phase 2

This phase enclosed the upper part section of the Mit Ghamr formation and encompassed DZ. 3 and 4 of core 1 (depth interval 23–40 m); DZ. 4 and 5 of core 2 (depth 21–36 m); DZ. 2 of core 3 (depth 25–36 m); and DZ.2 of core 4 (depth 32–42 m). During this phase, the diatom concentration was high and dominant with freshwater oligohalobous taxa (81–98%) that reflect a conspicuous connection with the Nile and the influx of freshwater into the Nile Delta basin. This interval was dominated by eutrophic freshwater planktonic taxa, with a common abundance of *Aulacoseira* spp. *Aulacoseira granulata* has great abundance accompanied by the frequent occurrence of *A. granulata* var. *angustissima, A. ambigua,* and *A. agassizii.* A significant occurrence of *Stephanodiscus* species distinguished this phase of delta development. Several periphytic taxa including *Staurosira, Ulnaria, Gomphonema, Encyonema*, and *Nitzschia* species, were distributed infrequently. Most of the *Aulacoseira* species indicate high growth requirements for silicon and demand high silica content in water (Kilham and Kilham 1971), presumably in different combinations of phosphorus and light (Kilham et al. 1986). *Aulacoseira* species were non-competitive, so their wide distribution normally coincided with low concentrations of other diatoms (Wolfe et al. 2000). These taxa are also indicators of a warmer climate, which may have led to wind-induced mixing in the water, higher input of humic substances, and increased precipitation (Laing and Smol 2003; Marks et al. 2018). *Aulacoseira granulata* was recognized as a freshwater planktonic and alkaliphilous species, common in eutrophic water of higher silica and temperature (Hustedt 1957; Ehrlich 1973; Stoermer et al. 1974; Kilham et al. 1986). The dominance of *A. granulata* requires high levels of silicon and high loadings of phosphorus and nutrient fluxes (Kilham et al. 1986). This species is most abundant in shallow, more eutrophic regions of the modern Great Lakes, which are not permanently stratified during summer (Stoermer et al. 1996). On the other hand, *Aulacoseira ambigua* required increased turbulence to remain in a photic zone and reported a high concentration of total phosphorus (TP) (Kilham et al. 1986; Fritz et al. 1993). *Aulacoseira granulata* and *A. ambigua* had summer blooms, which documented increasing moisture, greater flow through, and higher lake levels (Bradbury et al. 2004). However, *Stephanodiscus* species are known to occupy slightly alkaline and eutrophic freshwater with low silica content (Gasse 1986; Kilham et al. 1986; Zalat and Servant-Vildary 2007). They were dominant in winter and spring when increased turbulence suspended these relatively heavy diatoms;

therefore, they may denote moist winters and springs with active circulation (Bradbury 1992; Bradbury et al. 2004).

Great abundances of *Aulacoseira granulata* associated with other *Aulacoseira* species and decreased abundance of *Stephanodiscus* could reflect summer with high silica concentration, a freshwater environment with relatively moderate water level due to nutrient-rich influx from the Nile, and increased discharge of Nile water into the delta reservoir during a wet, warm period. Moreover, the maximum abundance of *Aulacoseira* species suggests stabilized conditions, remaining wet and windy with increased turbulence and upwelling in the water (Zalat 1995, 2015; Marks et al. 2018). The significant occurrence of small and intermediate-sized *Stephanodiscus* species (*S. minutulus, S. hantzschii,* and *S. agassizensis*), along with *Aulacoseira granulata* in this phase of delta development, reflect moist winters and springs with high-nutrient loading by runoff, increased turbulence and active circulation (Bradbury 1992; Bradbury et al. 2004; Zalat 2015).

During most of this time interval of the late Pleistocene, the water body of the Nile Delta reservoir was freshwater, slightly alkaline (pH = 7–8), and eutrophic, with moderate water level and wind-induced water mixing in spring and summer. This is documented by a great abundance of the alkaliphilous taxa (75–90%) of oligohalobien characters. The occurrence of *Aulacoseira* and *Stephanodiscus* assemblages indicates intensive seasonal water circulation, a large increase in trophy, and a seasonal influx of Nile water. Moreover, increasing relative frequencies of periphytic as well as of aerophilous or terrestrial species suggest more intensive erosion and wind action during this phase. The water body was colonized by some macrophytes as shown by the occurrence of epiphytic forms.

Phase 3

This phase comprised the lower part section of the Bilqas formation that included DZ.5 of core 1 (depth 15–22 m), subzone DZ.6a of core 2 (depth 18–20 m), DZ.3 of core 3 (depth 22–24 m), and DZ.3 of core 4 (depth 28–31 m). The deposition in the delta varied considerably in this phase at first, with a significant input of fine sand followed by gray silty clay. The reservoir was cut off from the Nile, but the deposition of clayey silt contained a significant abundance of freshwater oligohalobous taxa, suggesting that sporadic inflows of Nile water were possible, supplying the reservoir with nutrients as suggested by the frequent occurrence of planktonic *Aulacoseira* diatoms. This phase was generally characterized by dry conditions, with progressing desiccation of shallower parts of the reservoir and slight transgression of the Sea on the delta plain as indicated by rising contents of mesohalobous and polyhalobous taxa in cores 3 and 4 (Figs. 5.5, 5.6). The brackish water taxa were represented by the common occurrence of *Stephanocyclus meneghiniana*, *Campylodiscus* spp., and *Pinnunavis zalatii* associated with the common occurrence of the polyhalobous coastal marine diatoms such as *Hyalodiscus laevis*, *Tryblionella granulata,* and *Tryblionella punctata. Stephanocyclus meneghiniana,* which is known in many literatures as *Cyclotella meneghiniana,* is a facultative planktonic taxon whose salinity range extends from freshwater well into saline conditions; it is typically found in moderately alkaline conditions (Hecky and Kilham 1973; Richardson et al. 1978) in coastal and estuarine locations with water of varied chemistry (Trigueros and Orive 2000; Tibby and Reid 2004). Its most favorable development occurs at ~ 20°C, but it is eurythermal (Gasse 1986). This species is an important component of the phytoplankton flora in the modern eutrophic, brackish water of coastal Egyptian lakes, being dominant in spring and at the beginning of summer at water temperatures of 29–31 °C (Zalat and Servant-Vildary 2005, 2007). *Pinnunavis zalatii,* which is reported as *Navicula aegyptiaca,* is a common species in the brackish water of coastal Egyptian lakes (Zalat 2000). *Hyalodiscus laevis* is a coastal, euryhaline species, common in marine and brackish waters, polyhalobous and mesohalobous (Hendey 1964; Edsbagge 1968; Zalat and Servant Vildary 2005, 2007). *Tryblionella granulata* is considered a marine diatom, found in coastal lagoons and estuaries, a polyhalobous species (Foged 1980; Zalat and Servant Vildary 2005). *Tryblionella punctata* is a marine diatom, common in the coastal environment and estuaries, a polyhalobous species (Foged 1979; Zalat and Servant Vildary 2005).

An increased abundance of those mesohalobous and polyhalobous periphytic taxa accompanied by the conspicuous occurrence of oligohalobous species reflect coastal environments with increased salinity due to seawater transgression. The water level may have been dramatically low at that time, which led to the flourish of the periphytic diatoms over the planktonic forms. The benthic taxa dominated with a considerable abundance of the epiphytic taxa (Fig. 5.7), which reflects a lower water level of a shallow environment with the growth of some macrophytes. In addition, regular inflows of Nile water in late spring and early summer are evidenced by the abundance of the *Aulacoseira* species with many oligohalobous riverine taxa (Figs. 5.5 and 5.6). The predominance of the alkaliphilous diatom species (75–90%) with a considerable abundance of alkalibiontic taxa (10–27%) demonstrates alkaline conditions and high mineral input into a nutrient-rich environment. The common occurrence of *Stephanocyclus meneghiniana* with an abundance of *Aulacoseira* species indicates a swarm eutrophic water with increased salinity and alkalinity.

Phase 4

This phase is characterized by a short interval of clayey silts represented by the topmost part of DZ.5 of core 1 (at depth 15 m), subzone DZ.6b of core 2 (depth 15–17 m), DZ.4 of core 3 (depth 18–21 m), and DZ.4 of core 4 (depth 26–27 m). Re-abundance of the freshwater planktonic eutrophic *Aulacoseira* species comprising great abundance of *Aulacoseira granulata* accompanied by frequent occurrence of the periphytic riverine taxa such as *Staurosira construens, S. leptostauron, Ulnaria ulna* and simultaneous with reduction in abundance of mesohalobous and polyhalobous diatoms reflecting a hydrological connection with the Nile and the freshwater input into the delta basin during floods time. As it was mentioned, *Aulacoseira* is a good environmental indicator of eutrophic conditions, turbulence, high nutrients, and waters rich in silica. Predominance of *Aulacoseira* species is indicative of continuing rising temperature (Rühland et al. 2015), associated with weak stratification of a slightly alkaline, eutrophic open freshwater environment during a warm-humid climate (Pidek et al. 2021; Zalat et al. 2021). The relatively equal frequency rate of both planktonic and the periphytic with a dominance of *Aulacoseira granulata* indicates a relatively dropping water level with the occurrence of some macrophytes in a shallow environment. The alkaliphilous taxa were most dominant in this phase, which reflects slightly alkaline characters of the environment. The frequent occurrence of *Staurosira construens* and *Staurosira leptostauron,* which thrive in reduced light conditions, is an indication of reduced light penetration.

Phase 5

This phase is distinguished by the deposition of massive gray-brown clayey silt and fine sand interrupted by a peat layer in cores 3 and 4 with the admixture of diatom assemblages typical of a shallow and near-shore environment. It is covered the top part of the Bilqas formation and comprises the DZ. 6–7 of core 1 (depth 5–14 m), subzone DZ. 6c and DZ. 7–8 of core 2 (depth 4–14 m), DZ. 5–7 of core 3 (depth 3–17 m) and DZ. 5–7 of core 4 (depth 3–25 m). This phase indicates a restricted hydrological connection with the Nile. The delta basin was periodically brackish (Fig. 5.9), and the water level dropped significantly. The reservoir became shallower with a common occurrence of the brackish water diatoms such as *Stephanocyclus meneghiniana*, *Campylodiscus* spp., and *Pinnunavis zalatii*, associated with a frequent abundance of some polyhalobous diatoms *Hyalodiscus laevis*, *Tryblionella granulata*, *Tryblionella punctata* and a frequent occurrence of freshwater oligohalobous forms. The recognized diatom assemblage reflects deposition close to the shore in a shallow environment. The increase in the less silicified periphytic diatom species such as *Encyonema, Gomphonema, Navicula,* and *Nitzschia*, probably attributed to low Si:P ratio and hence reducing the production of diatoms that need high Si: P ratios like *Aulacoseira* species. The dominance of benthic species and a significant reduction in planktonic to benthic ratio indicate a low precipitation-to-evaporation ratio and warm, dry climate.

During this phase, a peat layer appears through soft to medium silty clay at a depth of 9–10 in core 3, Kafr El Sheikh area, and at a depth of 13–14 m in core 4, El Mansoura area. This layer has high diatom concentration with a dominance of the periphytic mesohalobous taxa *Campylodiscus* spp., *Pinnunavis zalatii*, and *Rhopalodia acuminata* associated with common occurrence of freshwater oligohalobous forms *Aulacoseira granulata, A. crenulata, A. ambigua, Cyclostephanos dubius, Ulnaria ulna, Encyonema* spp. and limited occurrence of the polyhalobous taxa. The epiphytic taxa belonging to *Encyonema, Epithemia, Gomphonema*, and *Eunotia* genera are relatively common, which points to well-developed macrophytes with marked lowering of water levels. The peat layer indicates the development of vegetation in the water body and the catchment area. The alkaliphilous diatoms were predominant with frequent abundance of alkalibiontic forms and significant occurrence of acidophilous taxa. The recognized diatom assemblage (DZ.

6 in both cores 3 and 4) mainly reflects deposition in shallow, slightly alkaline, fresh to brackish water conditions with reduced water levels like a marsh-swamp environment. The peat layer is not recorded in both the Kafr El-Zayat and Tanat regions (Cores 1 and 2). The recognized diatom assemblage is distinguished by a great abundance of freshwater oligohalobous taxa accompanied by the infrequent occurrence of mesohalobous and polyhalobous forms, which reflect a fresh to slightly brackish water environment. The great frequency of oligohalobous diatoms may suggest a limited hydrological connection to the river and an irregular freshwater inflow from the Nile accompanied by terrigenous sediments into the delta basin. The dominance of the alkaliphilous, benthic, and epiphytic diatoms over the planktonic signifies a prominently weak stratification of a slightly alkaline, relatively shallow environment associated with a lowering water level during a warm-arid climate.

5.7 Conclusions

The main goals of this study were to establish the presence of diatom fossils in the Nile Delta sediments and to reconstruct the environmental history and climate changes during the Quaternary. A total of 216 diatom species and varieties representing 49 genera were identified, and their abundances were abundant to rare and poorly to well-preserved throughout the four cores sediments that were investigated. The constrained-cluster analysis and relative abundance of dominant and subdominant taxa explained the local diatom zones for each core section. These zones showed an interesting history of changes in salinity, pH, and water level of the Nile Delta basin. The results of the diatom analysis described five phases of Nile Delta development during the Quaternary time. The first two phases are characteristic of the Mit Ghamr formation, and three phases distinguish the Bilqas formation. Summarized in terms of diatom analysis, it can be concluded that:

1. The deposition of the deltaic sediments of Mit Ghamr formation occurred in the slightly alkaline, fresh water of oligohalobien type (salinity ranged between 0.0‰ and 0.5‰) with occasional slight rises in salinity during the time of deposition, as shown by the predominance of both alkaliphilous and oligohalobous diatoms that dominant by *Aulacoseira* taxa. The water body was characterized by a moderate water level, moderate to high productivity, and a high concentration of dissolved silica. The recognized diatom assemblage reflects a conspicuous hydrological connection with the Nile and freshwater Nile influx into the delta during the warm climate in the Pleistocene time. The dominance of *Aulacoseira* taxa indicates increases in spring circulation and the degree of summer turbulences that are associated with drier and windier conditions when the annual Nile flood reached the delta basin.

2. The low abundance of diatoms in the lower part of Mit Ghamr formation that is characteristic of phase 1 of the development of the Nile Delta may be attributed to unfavorable conditions in the delta ecosystem, associated with a lowering water level during the period of deposition, or their dissolution due to greater alkalinity associated with warm-dry climate.
3. Marked variation in diatom composition and abundance happened in the Holocene time during the deposition of the Bilqas formation. The Holocene started with a marked lowering of the water level accompanied by a decreased abundance of oligohalobous taxa with a pronounced appearance of the mesohalobous and polyhalobous of coastal marine affinities, which reflect shallow brackish water to the coastal environment with increased salinity due to sea water transgression. The diatom assemblages were characterized by a great abundance of alkaliphilous taxa associated with a significant occurrence of alkalibiontic forms and a very limited abundance of acidophilous taxa, which reflect alkaline water of meso to eutrophic conditions with relatively high silica concentration during the warm-dry climate. The distinct variations visible in the ratio of planktic to epiphytic and benthic diatom species mainly reflect shallow environments and water level changes related to climatic fluctuations. The pronounced occurrence of the epiphytic taxa reflects shallow water with developed macrophytes. The presence of large diatom species with well-silicified valves during extreme water level fluctuations may also reflect the potential influence of increasing nutrient levels with decreasing depth.
4. The occurrence of peat layer dominant with alkaliphilous, oligohalobous diatom taxa associated with a considerable occurrence of mesohalobous forms intercalated silty clay beds in the Kafr El-Sheikh and El-Mansoura regions explained shallow, slightly alkaline brackish water conditions of the marsh-swamp environment.

References

Abdel Aal A, Price RJ, Vaiti JD, Shrallow JA (1994) Tectonic evolution of the Nile Delta, its impact on sedimentation and hydrocarbon potential. In: 12th Egyptian general petroleum corporation conference (EGPC), vol 1. Cairo, pp 19–34

Adamson DA, Gasse F, Street FA, Williams MAJ (1980) Late Quaternary history of the Nile. Nature 88:50–55

Aly MH, Klein AG, Zebker HA, Giardino JR (2012) Land subsidence in the Nile Delta of Egypt observed by persistent scattered interferometry. Remote Sens Lett 3(7):621–630

Azzam AS (1994) An integrated seismo-facies and seismo-tectonic study of the Nile Delta of Egypt, utilizing common-depth seismic reflection data. Ph.D. Thesis, Geology Department, Faculty of Science, Ein Shams University, Cairo, Egypt, p 240

Battarbee RW (1986) Diatom analysis. In: Berglund BE (ed) Handbook of holocene palaeoecology and palaeohydrologypy. John Wiley & Sons, Chichester, pp 527–570

Becker RH, Sultan M (2009) Land subsidence in the Nile Delta: inferences from radar interferometry. The Holocene 19(6):949–954
Bergner AGN, Trauth MH (2004) Comparison of the hydrological and hydrochemical evolution of Lake Naivasha (Kenya) during three highstands between 175 and 60 kyr BP. Palaeogeogr Palaeoclimatol Palaeoecol 215:17–36
Bradbury JP (1992) Late cenozoic lacustrine and climatic environments at Tule Lake, northern Great Basin, USA. Clim Dyn 6:275–284
Bradbury JP, Cumming B, Laird K (2002) A 1500-year record of climatic and environmental change in Elk Lake, Minnesota III: measures of past primary production. J Paleolimnol 27:321–340
Bradbury JP, Colman SM, Dean WE (2004) Limnological and climatic environments at Upper Klamath Lake, Oregon during the past 45 000 years. J Paleolimnol 31:167–188
Butzer KW (2002) Geoarchaeological implications of recent research in the Nile Delta. In: van den Brink ECM, Levy TE (eds) Egypt and the Levant: interrelations from the 4th through the early 3rd Millennium BCE. Leicester University Press, London, pp 83–97
Denys L (1991–1992) A check-list of the diatoms in the Holocene deposits of the western Belgian coastal plain with a survey of their apparent ecological requirements. I. Introduction, ecological code and complete list. Professional paper 1991/2 No. 246, Ministerie van Economische Zaken, Jennerstraat, Brussel, 13, p 1040
Edsbagge H (1968) Zur Ökologie der Marinen Ange-hefteten Diatomeen. Bot Gothoburgensia 6:1–137
Ehrlich A (1973) Quaternary diatoms of the Hula basin (Northern Israel). Bull Geol Surv Israel 58:1–39
Ehrlich A (1975) The diatoms from the surface sediments of the Bardawil Lagoon (Northern Sinai)—paleoecological significance. Nova Hedwigia 53:253–277
Ehrlich A, Ortal R (1979) The influence of salinity on the benthic diatoms communities of the lower Jordan River. Nova Hedwigia 64:325–336
Ekblom A, Stabell B (2008) Paleohydrology of Lake Nhaucati (southern Mozambique), c 400 AD to present. J Paleolimnol 40:1127–1141. https://doi.org/10.1007/s10933-008-9218-2
El-Asmar HM (1986) Lithological and mineralogical investigation of the subsurface Quaternary succession in the Nile Delta region, Egypt (M.Sc. Thesis). Faculty of Science, El Mansoura University Egypt
El-Asmar HM (1998) Middle and Late Quaternary palaeoclimatic evolution, northern Mediterranean coast of Egypt. In: Al-Shrhan AS, Glennie KW, Whittle GL, Kendall CGSC (eds) Quaternary deserts and climatic changes. Balkema, pp 261–271
El-Asmar HM, Gheith AM (1995) Stable isotope stratigraphy and paleoclimatology of the offshore subsurface Quaternary sediments, Nile Delta, Egypt. Sedimentol Egypt 3:39–49
El-Asmar HM, Taha MMN, El-Kafrawy SB, El-Sorogy AS (2015) Control of Late Holocene geoprocesses on the sustainable development plans of the Tineh Plain, NW Sinai coast, Egypt. J Coast Conserv 19:141–156. https://doi.org/10.1007/s11852-015-0377-9
El-Fayoumi IF (1987) Geology of the Quaternary succession and its impact on the groundwater reservoir in the Nile Delta region. Faculty of Science, Mansoura University, Egypt. Bull
El-Kady GA, Shoukry A, Reda LA, El-Badri YS (2000) Survey and population dynamics of freshwater snails in newly settled areas of the Sinai Peninsula. Egypt J Biol 2:42–48
El-Shazly EM, Abdel Hady MA, El Shazly MM, El Ghawabby MA, El Kassas IA, Salman AB, Morsi MA (1975) Geological and groundwater potential studies of El Ismailia master plan study area. Remote Sensing Research Project, Academy of scientific Research and Technology, Cairo, Egypt

Farouk S, Jain S, Abd-Elazez M, El Shennawy T, Shaker F (2021) High- resolution Pliocene-Pleistocene calcareous nannofossil distribution and palaeoenvironmental changes in the northwest Nile Delta, Egypt. Lethaia 54:845–870

Fergany E, Omar K (2017) Liquefaction potential of Nile delta, Egypt. NRIAG J Astron Geophys. https://doi.org/10.1016/j.nrjag.2017.01.004

Fishar MR (2016) Nile Delta (Egypt). In: Finlayson C, Milton G, Prentice R, Davidson N (eds) The wetland book. Springer, Dordrecht. https://doi.org/10.1007/978-94-007-6173-5

Foged N (1970) The diatomaceous flora in a Postglacial Kieselgur deposit in Southwestern Norway. Nova Hedwigia, Beih 31:169–202

Foged N (1979) Diatoms in New Zealand, the North Island. Bibl Phycol 47:1–225

Foged N (1980) Diatoms in Egypt. Nova Hedwigia 33:629–707

Foged N (1993) Some diatoms from Siberia, especially from Lake Baikal. Diatom Res 8:231–279

Fritz SC, Juggins S, Battarbee RW (1993) Diatom assemblages and ionic characterization of lakes of the northern Great Plains, North America: a tool for reconstructing past salinity and climate fluctuations. Can J Fish Aquat Sci 50(9):1844–1856

Gasse F (1975) L'évolution des lacs de l'Afar Central (Ethiopie et T.F.A.I.) du Plio-Pléistocène à l'actuel: reconstitution des paléomilieux à partir de l'étude des diatomées. D. Sc. Thesis, Univ. of Paris. Vol I : 390 p, Vol II : 103 p, Vol.III : 59 pls

Gasse F (1986) East African diatoms, taxonomy, ecological distribution. Bibl Diatomol 2:1–201

Gasse F, Barker P, Gell PA, Fritz SC, Chalie F (1997) Diatom-inferred salinity in palaeolakes: an 375 indirect tracer of climate change. Quat Sci Rev 16:547–563

Ginau A, Schiestl R, Wunderlich J (2019) Integrative geoarchaeological research on settlement patterns in the dynamic landscape of the northwestern Nile Delta. Quat Int 511:51–67

Hamdan MA (2003) Late Quaternary geology and geoarchaeology of Kafr Hassan Dawood, east delta. In: Hawass Z, Pinch Brock L (eds) Egyptology at the dawn of the twenty-first century: proceedings of the eighth international congress of Egyptologists, Cairo 2000, vol 1. The American University in Cairo Press, Cairo, pp 221–228

Hammer Ø, Harper DA, Ryan PD (2001) PAST: paleontological statistics software package for education and data analysis. Palaeontol Electron 4(1):1–9

Hanter G (1975) Contribution to the origin of the Nile delta. (B-3) In: 8th Arab petroleum congress, paper 117

Hay MB, Michelutti N, Smol JP (2000) Ecological patterns of diatom assemblages from Mackenzie Delta lakes, Northwest Territories. Canada Can J Bot 78:19–33

Hecky RE, Kilham P (1973) Diatoms in alkaline, saline lakes: ecology and geochemical significance. Limnol Oceanogr 18:53–71

Hegab O, Bahloul M (1987) On the occurrence of Peat in the subsurface Holocene sediments of the Nile Delta and its geotechnical implications. Egypt J Geol 31:73–83

Hendey NI (1964) An introductory account of the smaller algae of british coastal waters VI bacillariophyceae (diatoms). Fishery investigations. Her Majesty's Stationary Office, London, p 317, 45

Hofmann G, Werum M, Lange-Bertalot H (2011) Diatomeen im Süßwasser-Benthos von Mitteleuropa. A.R.G. Gantner Verlag, Rugell, Liechtenstein, p 908

Hustedt F (1957) Die Diatomeenflora des Fluss-systems der Weser im Gebiet der Hansestadt Bremen. Abh Naturw Ver Brem 34:181–440

Hustedt F (1930–1966) Die Kieselalgen Deutschlands, Österreichs und der Schweiz mit Berücksichtigung der übrigen Länder Europas sowie der angrenzenden Meeresgebiete. Rabenhorst's Kryptogamenflora. Band 7. Teil 1: 1–920 (1927–1930); Teil 2: 1–845; Teil 3:1–816 (1961–1966), Leipzig

Ismail AI, Boukhary M, Abdel Naby AI (2010) Subsurface stratigraphy and micropaleontology of the Neogene rocks, Nile Delta, Egypt. Geol Croatica 63(1). https://doi.org/10.4154/GC.2010.01

Julius ML, Theriot EC (2010) The diatoms: a primer. In: Smol JP, Stoermer EF (eds) The diatoms: application for the environmental and earth sciences, 2nd ed. Cambridge University Press, United Kingdom, pp 8–22

Kamel H, Eita T, Sarhan M (1998) Nile delta hydrocarbon potentiality, Egypt. In: Proceding 14th EGPC exploration and production conference, Cairo, Egypt, vol 1, pp 485– 503

Kholeif SEA, Mudie PJ (2009) Palynological records of climate and oceanic conditions in the Late Pleistocene and Holocene of the Nile cone, southeastern Mediterranean, Egypt. Palynology 33(1):1–24. https://doi.org/10.2113/gspalynol.33.1.1

Kilham P, Kilham SS (1971) *Melosira granulata* (Ehr.) Ralfs: morphology and ecology of a cosmopolitan freshwater diatom. Verhandlungen der Internationalen Vereinigung Limnologie 19:2716–2721

Kilham P, Kilham SS, Hecky RE (1986) Hypothesized resource relationships among African planktonic diatoms. Limnol Oceanogr 31:1169–1181

Krammer K, Lange-Bertalot H (1986–1991) Bacillariaceae. In: Ettl H, Gerloff J, Heynig H, Mollenhauer D (eds) Susswasserflora von Mitteleuropa, vol 2 (1–4). Gustav Fischer Verlag, Stuttgart/Jena, pp 1–2458

Laing TE, Smol JP (2003) Late Holocene environmental changes inferred from diatoms in a lake on the western Taimyr Peninsula, northern Russia. J Paleolimnol 30:231–247

Lange E, Ullmann T, Baumhauer R (2016) Remote sensing in the Nile Delta: spatio-temporal analysis of Bubastis/Tell Basta. Agypten Levante 26:377–392

Mann DG (1999) The species concept in diatoms. Phycologia 38(6):437–495

Marks L, Salem A, Welc F, Nitychoruk J, Chen Z, Blaauw M, Zalat A, Majecka A, Szymanek M, Chodyka M, Tołoczko-Pasek A, Sun Q, Zhao X, Jiang J (2018) Holocene lake sediments from the Faiyum Oasis in Egypt: a record of environmental and climate change. Boreas 47:62–79

Morgan P (1990) Egypt in the framework of global tectonics. In: Said R (ed) The geology of Egypt. A.A., Balkema, Rotterdam, Brookfield, pp 91–111

Paulissen E, Vermeersch PM (1987) Earth, man and climate in the Egyptian Nile Valley during the Pleistocene. In: Close AE (ed) Prehistory of arid North Africa. Southern Methodist University Press, pp 29–67

Pennington BT, Bunbury J, Hovius N (2016) Emergence of civilization, changes in fluvio-deltaic style, and nutrient redistribution forced by Holocene sea-level rise. Geoarchaeol Int J 31:194–210

Pennington BT, Sturt F, Wilson P, Rowland J, Brown AG (2017) The fluvial evolution of the Holocene Nile Delta. Quat Sci Rev 170:212–231

Pidek IA, Zalat AA, Hrynowiecka A, Żarski M (2021) A high-resolution pollen and diatom record of midto late-Eemian at Kozłów (Central Poland) reveals no drastic climate changes in the hornbeam phase of this interglacial. Quat Int 583:14–30

Potapova M (2011) Patterns of diatom distribution in relation to salinity. In: Seckbach J, Kociolek JP (eds) The diatom world. Cellular origin, life in extreme habitats and astrobiology (COLE) book series, vol 19, Springer, Dordrecht, pp 313–332

Richardson JL, Harvey TJ, Holdship SA (1978) Diatoms in the history of shallow East African lakes. Polskie Arch Hydrobiol 25:341–343

Rizzini A, Vezzani F, Cococcetta V, Milad G (1978) Stratigraphy and sedimentation of Neogene-Quaternary section in the Nile Delta area (A.R.E.). Mar Geol 27:327–348

Round FE, Crawford RM, Mann DG (1990) The diatoms. In: Biology and morphology of the genera. Cambridge University Press, Cambridge, p 747

Rühland KM, Paterson AM, Smol JP (2015) Lake diatom responses to warming: reviewing the evidence. J Paleolimnol 54:1–35

Saad S, Zaghloul Z, El-Beialy S (1987) Late Tertiary-Quaternary palynomorphs from the northern coast of the Nile Delta, Egypt. Neues Jb Geol Paläontol Monat 5:314–320
Said R (1981) The geological evolution of the River Nile. Springer Verlag, New York, p 151
Said R (1983) Proposed classification of the Quaternary of Egypt. J Afr Earth Sci 1:41–45. https://doi.org/10.1016/0899-5362(83)90030-1
Said R (1990) The geology of Egypt. Balkema Publication, Rotterdam, Netherlands, p 734p
Salem ZE (2009) Natural and human impacts on the groundwater under an Egyptian village, central Nile Delta e a case study of Mehallet Menouf. In: 13th International water technology conference (IWTC), vol 3, 12–15 Mar 2009, Hurghada, pp 1397–1414
Serag El-Din HM (1983) Geological and geochemical studies on the Nile Delta aquifer, Egypt. M.Sc. Thesis, Faculty of Science, Mansoura University, Egypt, p 235
Serag El-Din HM (1989) Geological and hydrogeological and hydrological studies on the quaternary aquifer. Ph.D. dissertation. Faculty of Science, Mansoura University, Mansoura, Egypt
Serag El-Din HM (1990) Geological, hydrogeological and hydrological studies on the Nile Delta quaternary aquifer. Ph.D. Thesis, Faculty of Science, Mansoura University, Egypt, p 300
Sestini G (1989) Nile Delta: a review of depositional environments and geological history. In: Whateley MKG, Picering KT (eds) Deltas: sites and traps for fossil fuels. Geological Society Sneciai Publication, vol 41(1), pp 99–127
Shaaban H, El-Qady G, Al-Sayed E, Ghazala H, Taha AI (2016) Shallow groundwater investigation using time-domain electromagnetic (TEM) method at Itay El-Baroud, Nile Delta, Egypt. NRIAG J Astron Geophys. https://doi.org/10.1016/j.nrjag.2016.05.004
Simonsen R (1962) Untersuchungen zur Systematik und Ökologie der Bodendiatomeen der westlichen Ostsee. Int Rev Gesamten 34:181–440
Simonsen R (1979) The diatom system: ideas on phylogeny. Bacillaria 2:9–72
Smol JP (2008) Pollution of lakes and rivers: a paleoenvironmental perspective, 2nd edn. John Wiley & Sons, Hoboken
Stager JC, Ryves D, Cumming BF, Meeker LD, Beer J (2005) Solar variability and the levels of Lake Victoria, East Africa, during the last millennium. J Paleolimnol 33:243–251
Stanley DJ (2005) Submergence and burial of ancient coastal sites on the subsiding Nile delta margin, Egypt. Méditerranée 1(2):65–73
Stanley DJ, Clemente PL (2017) Increased land subsidence and sea-level rise are submerging Egypt's Nile Delta coastal margin. Geol Soc Am Today 27(5):411–464
Stanley DJ, Corwin KA (2013) Measuring strata thicknesses in cores to assess recent sediment compaction and subsidence of Egypt's Nile Delta coastal margin. J Coast Res 29(3):657–670
Stanley DJ, Warne AG (1993) Nile Delta: recent geological evolution and human impact. Science 260:628–634. https://doi.org/10.1126/science.260.5108.628
Stanley DJ, Wedl SE (2021) Significant depositional changes offshore the Nile Delta in late third millennium BCE: relevance for egyptology. E&G Quat Sci J 70:83–92. https://doi.org/10.5194/egqsj-70-83-2021
Stoermer EF (1998) Thirty years of diatom studies on the Great Lakes. J Great Lakes Res 24:518–530
Stoermer EF, Smol JP (eds) (1999) The diatoms: applications for the environmental and earth sciences. Cambridge University Press, Cambridge, New York, Melbourne, xii + 469 pp
Stoermer EF, Emmert G, Julius ML, Schelske CL (1996) Paleolimnologic evidence of rapid recent change in Lake Erie's trophic status. Can J Fish Aquat Sci 53:1451–1458
Stoermer EF, Bowman M, Kingston JC, Schaedel AL (1974) Phytoplankton composition and abundance in Lake Ontario during IFYGL: Special Report 53. Great Lakes Research Division, Ann Arbor

Stone JR, Fritz SC (2004) Three-dimensional modeling of lacustrine diatom habitat areas: improving paleolimnological interpretation of planktic: benthic ratios. Limnol Oceanogr 49:1540–1548
Tassy A, Emmanuel C, Gorini C, Rubino JL, Bouroullec JL, François S (2015) Egyptian Tethyan margin in the Mesozoic: evolution of a mixed carbonate-siliciclastic shelf edge (from Western Desert to Sinai). Mar Pet Geol 68:565–581
Tibby J, Reid M (2004) A model for inferring past conductivity in low salinity waters derived from Murray River diatom plankton. Mar Freshw Res 55:587–607
Trigueros M, Orive E (2000) Tidally driven distribution of phytoplankton blooms in a shallow, macrotidal estuary. J Plankton Res 22:969–986
Van Dam H, Mertens A, Sinkeldam J (1994) A coded checklist and ecological indicator values of freshwater diatoms from the Netherlands. Neth J Aquat Ecol 28(1):117–133
Van der Werff A, Huls H (1957–1974) Diatomeeënflora van Nederland. Sprey Abcoude, Den Haag
Verschuren D, Laird KR, Cumming BF (2000) Rainfall and drought in equatorial east Africa during the past 1,100 years. Nature 403:410–414
Wendorf F, Schild R (1986) The prehistory of Wadi Kubbaniya. Southern Methodist University Press, p 85
Williams DM, Round FE (1987) Revision of the genus *Synedra* Ehrenb. Diatom Res 1(2):313–339
Witkowski A, Lange-Bertalot H, Metzeltin D (2000) Diatom flora of marine coasts I iconographia diatomologica 7:1–925
Wolfe BB, Edwards TWD, Duthie HC (2000) A 6000- year record of interaction between Hamilton Harbour and Lake Ontario: quantitative assessment of recent hydrologic disturbance using13 Cinlake sediment cellulose. Aquat Ecosyst Health Manag 3:47–54
Wolin JA (1996) Late Holocene lake-level fluctuations in Lower Herring Lake, Michigan, U.S.A. J Paleolimnol 15:19–45
Wolin JA, Duthie HC (1999) Diatoms as indicators of water level change in freshwater lakes. In: Stoermer EF, Smol JP (eds) The diatoms: applications for environmental and earth sciences. Cambridge University Press, Cambridge, pp 183–202
Wolin JA, Stone JR (2010) Diatoms as indicators of water-level change in freshwater lakes. In: Stoermer EF, Smol JP (eds) The diatoms applications to the environmental and earth sciences. Cambridge University Press, pp 174–185
Xue B, Qu W, Wang S, Ma Y, Dickman MD (2003) Lake level changes documented by sediment properties and diatom of Hulum Lake, China since the late Glacial. Hydrobiologia 498:133–141
Zaghloul ZM (1976) Stratigraphy of Nile Delta. In: Seminar Nile Delta Sediment, UNSCO, ARST, UNDP, Alexandria, Egypt, pp 40–49
Zaghloul ZM, Taha AA, Hegab OA, El-Fawal FM (1977) The Neogene-Quaternary sedimentary basins of the Nile Delta, Egypt. J Geol 21:1–19
Zaghloul ZM, Taha AA, Hegab OA, El-Fawal FM (1979) The Plio-Pleistocene Nile Delta sub-environments, stratigraphic section and genetic class. Ann Meet Geol Surv Egypt 23:282–291
Zaghloul ZM, Gheith AM, Abdel-Daiem AA, El Asmar HM (1989) The pattern of sedimentation of the Quaternary Nile Delta, Egypt. El Mansoura Sci Bull 16(2):393–413
Zaghloul ZA, Elgamal MM, El Araby H, Abdel Wahab W (1999) Evidences of neotectonics and ground motions in the northern Nile Delta. In: Zaghloul ZM, Elgamal MM (eds) Deltas-modern and ancient, Egypt, proceedings of Mansoura University, first international symposium on the deltas, pp 285–314
Zalat AA (1991) Paleontological studies on the Quaternary diatomite of the Fayoum Depression, Western Desert, Egypt. Ph.D. thesis, Tanta University, p 329
Zalat AA (1995) Diatoms from the quaternary sediments of the Nile Delta, Egypt, and their palaeoecological significance. J Afr Earth Sc 20(2):133–150

Zalat AA (2000) A new naviculoid diatom species *Navicula aegyptiaca* with two new varieties from Holocene sediments of brackish lakes in Egypt. Neues Jahrbuch Für Geol Palaeont MH 8:485–498

Zalat AA (2015) Holocene diatom assemblages and their palaeoenvironmental interpretations in Fayoum Depression, Western Desert, Egypt. Quat Int 369:86–98

Zalat AA, Servant-Vildary S (2005) Distribution of diatom assemblages and their relationship to environmental variables in the surface sediments of three northern Egyptian lakes. J Paleolimn 34:159–174

Zalat AA, Servant-Vildary S (2007) Environmental change in northern Egyptian Delta lakes during the Late Holocene, based on diatom analysis. J Paleolimn 37:273–299

Zalat AA, Bober A, Pidek IA, Żarski M (2021) Environmental and climate change during the Late Saalian-Eemian Interglacial at the Struga site (Central Poland): a diatom record against the background of palynostratigraphy. Rev Palaeobot Palynol 288:104386

Zalat A, Marks L, Welc F, Salem A, Nitychoruk J, Chen Z, Majecka A, Szymanek M, Chodyka M, Tołoczko-Pasek A, Sun Q, Zhao X, Jiang J (2017) Diatomstratigraphy of FA-1 core, Qarun Lake, records of Holocene environmental and climatic change in Faiyum Oasis, Egypt. Studia Quat 34(1):61–69. https://doi.org/10.1515/squa-017-0005

Zalat AA, Nitychoruk J, Chodyka M, Pidek IA, Welc F (2022) Recent and fossil freshwater diatoms of Poland: taxonomy, distribution, and their significance in the environmental reconstruction: part 1 coscinodiscophyceae, mediophyceae and fragilariophycidae. John Paul II University of Applied Science; Biala Podlaska, p 304

Ziemann H (1971) Die Wirkung des Salzgehaltes auf die Diatomeenflora als Grundlage für eine biologische Analyse und Klassifikation der Binnengewässer. Limnologica 8:505–525

Part II

Water Resources Challenges and Management

A Roadmap for Groundwater Sustainability in the Nile Delta: Assessing Challenges and Solutions Under the Impact of Climate Change and Anthropogenic Activities

6

Khaled S. Gemail, Youssef M. Youssef, Mohamed Mahdy, Hakeem Musaed, and Hafsa M. Atia

Abstract

On a global scale, climatic changes are widely expected to affect water resources and, therefore, the sustainable development in arid and hyper-arid regions. Thus, a full understanding of the role of groundwater in global water resources is essential for managing water deficiency in worldwide arid regions and for vastly improving water

K. S. Gemail (✉)
Environmental Geophysics Lab (ZEGL), Faculty of Science, Department of Geology, Zagazig University, Zagazig, Egypt
e-mail: khaledgemail@zu.edu.eg

Y. M. Youssef
Geological and Geophysical Engineering Department, Faculty of Petroleum and Mining Engineering, Suez University, Suez, Egypt
e-mail: youssef.ibrahim@pme.suezuni.edu.eg

M. Mahdy
Department of Geography, Faculty of Arts, Zagazig University, Zagazig, Egypt
e-mail: mmmahdy@arts.zu.edu.eg

H. Musaed
Department of Geological Engineering, Faculty of Oil and Minerals, University of Shabwah, Shabwah, Yemen

H. M. Atia
Geology Department, Faculty of Science, Mansoura University, New Mansoura, Egypt
e-mail: hafsa.atia@nmu.edu.eg

Geology Department, Faculty of Basic Science, New Mansoura University, New Mansoura, Egypt

M. A. Khalifa and K. S. Gemail (eds.), *Geology of the Nile Deltas and Natural Resources*, Deltas of the World, https://doi.org/10.1007/978-3-031-96594-4_6

resources management for generations to come. In this context, the Nile Delta represents a typical example of the complexity of land development in a region facing water scarcity due to rapid population growth. In the Nile Delta region, groundwater quality is controlled mainly by saltwater intrusion from the north due to Sea-Level Rise (SLR) and surface contaminants from the common wastewater networks and anthropogenic activities in heterogeneous topmost clay and silt soils. Thus, the Delta faces many environmental problems and challenges now and in the future due to unplanned water resource pumping and saltwater intrusion due to SLR. The groundwater in the area is considered the second most important water resource, after the surface Nile water network. It represents a key component for any sustainable development strategy in agricultural activities and urban expansion. To elaborate on these challenges and the future development of groundwater aquifers, in-depth analyses of vulnerability conditions related to water quality and recharge systems under the impact of SLR, surface wastewater pollution, and intensive use were addressed in the present chapter. The present work aims to draft a roadmap for groundwater management in the Nile Delta. Therefore, we present the status of the main aquifer systems in the Nile Delta according to their quality, change in water salinization, and future forecasting for the aquifer characteristics. The results of multidisciplinary hydrochemical, groundwater vulnerability modeling, and hydrogeophysical investigations were discussed in-depth to assess the water quality and validate the inferred transport models of contaminants and saltwater.

Keywords

Nile Delta • Groundwater aquifer • Saltwater intrusion • Sea-level rise (SLR) • Aquifer vulnerability • Hydrogeophysical modelling

6.1 Introduction

As a global research problem, the sea-level rise (SLR) and wastewater pollution have led to a reduction of freshwater resources in quantity and quality and may have effects on coastal regions in the short and long term, such as an increase in coastal erosion, population migration, and seawater intrusion (SWI). Water quality degradation problems are widespread, and the shortage of water resources has become a global problem and an issue that requires integrated efforts to minimize the negative impact of climate change. Also, the exchange of national and international experiences could help to initiate relevant solutions for these critical challenges.

The Nile Delta is a region in Egypt (see Fig. 6.1) that is highly dependent on groundwater from the Nile River for agriculture, industry, and domestic use. However, the Nile Delta is facing the challenges of climate change, with changes in precipitation patterns and rising sea levels. These changes affect the quantity and quality of groundwater resources

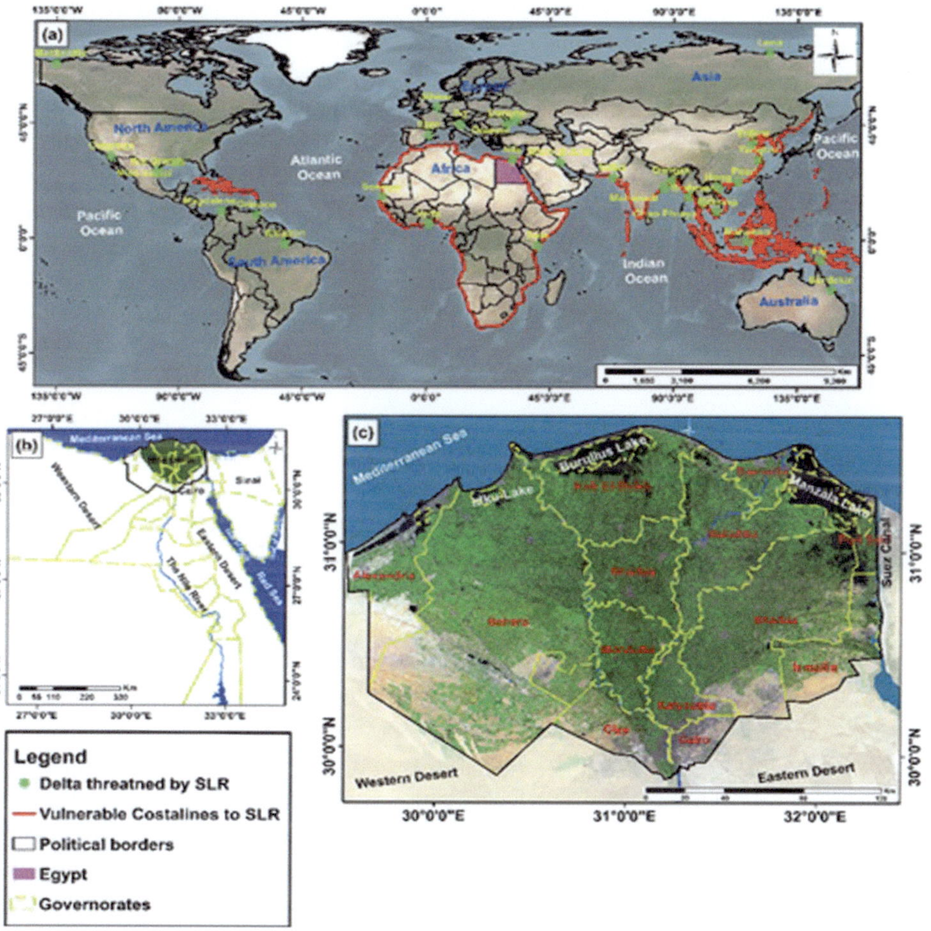

Fig. 6.1 Base of the Nile Delta. **a** Global map displays the vulnerability of coastal regions to flooding and saltwater intrusion menaces from Sea-Level Rise (SLR) where Egypt is one of these areas in North Africa (Oude Essink et al. 2010; Nicholls and Cazenave 2010), **b** Administrative image of Egypt showing the governorate boundaries (Yellow lines) and the location of the Nile Delta (green polygon) in the North of Egypt, and **c** Landsat-8 (RGB; 753)) of the Nile Delta region

in the region, exacerbating the existing challenges. Therefore, the climatic changes, SLR, and the overuse due to the high population, in addition to the mismanagement of groundwater resources, have led to several challenges in the region. The demand for groundwater in the Nile Delta has exceeded the recharge rate of the Nile Delta aquifer, resulting in groundwater depletion. This has led to a decline in water levels, which can lead to land subsidence, saltwater intrusion, and other environmental problems (Abotalib et al. 2023; Youssef et al. 2024; Salem et al. 2025a, b). On the other hand, the over-pumping of groundwater in the Nile Delta has led to saltwater intrusion into the shallow aquifer in the northern parts, which has resulted in the contamination of freshwater resources in the middle parts (Saqr and Abd-Elmaboud 2024). This has made it difficult and expensive to treat the water for domestic and agricultural use. Anthropogenic activities and excessive groundwater extraction have resulted in declining water quality, characterized by high levels of pollutants and contaminants from wastewater percolating into the shallow aquifer zones (Abu Salem et al. 2021; Gemail et al. 2024). This has led to health problems for people who rely on groundwater for drinking and other domestic uses. In addition to these challenges, there is a lack of regulation and management of groundwater resources in the Nile Delta, which has led to over-pumping, illegal drilling, and other unsustainable practices. The lack of effective management has made it difficult to address the challenges facing the region.

Several studies have investigated the potential impact of SLR and groundwater challenges on the Nile Delta aquifer using modeling and assessment techniques (i.e., Mabrouk et al. 2013, 2019; Attwa et al. 2016; Abd-Elhamid et al. 2016; Gemail et al. 2017a; Negm et al. 2019; Abd-Elaty et al. 2019, 2023; Abd-Elaty et al. 2021a, b; Mogren, and Shehata 2021; Abu Salem et al. 2022; Arafa et al. 2024; Salem et al. 2025a, b). These studies suggest that SLR could lead to increased saltwater intrusion into the aquifer, particularly in the northern parts of the Nile Delta region. The studies also highlight the need for effective management strategies to mitigate the impact of SLR on the Nile Delta aquifer, such as reducing groundwater pumping rates, increasing the use of treated wastewater for irrigation, and implementing coastal protection measures.

To address these challenges, there is a need for improved regulation and management of groundwater resources in the Nile Delta. This includes measures such as the enforcement of regulations on groundwater use, the promotion of sustainable practices, and the development of alternative water sources. Additionally, there is a need for research to better understand the groundwater system in the region and to develop effective strategies for managing this critical resource.

In the present chapter, the results of hydrochemical, hydrogeological, and geophysical datasets were reviewed and analyzed to assess and monitor the groundwater degradation and challenges, which are considered critical environmental problems in arid regions, considering the Egyptian Nile Delta as a typical example. In addition, index-based models, hydrogeological and geophysical models were employed as an assessment tool to locate potential hotspots and horizons of the lateral flow of groundwater under the impact of SLR and intensive pumping. Additionally, it may guide us in understanding the groundwater recharge systems to supply the growing population with safe water in arid and semi-arid regions elsewhere. The work should be of interest in the areas of groundwater exploration and water quality assessment under the impact of natural and anthropogenic stressors in the Nile Delta, one of the most vulnerable regions to water scarcity due to population growth. Therefore, the applied research approach in the present work can be considered an interesting case study for the mitigation of water quality degradation in coastal aquifers all over the world.

6.2 Geological and Hydrogeological Characterization of the Nile Delta Aquifer

Several researchers have studied the geological and hydrogeological characterization of the Nile Delta Quaternary aquifer and the challenges associated with groundwater, including Gheith (1976), Zaghloul et al. (1977, 1979), El-Fawal (1979), Sestini (1989), Sherif (1999), Sherif and Singh (2002), Sherif et al. (2012), Abd-Elhamid et al. (2016), Hegazy et al. (2020). These studies have highlighted the importance of the Nile Delta aquifer as a renewable water resource for domestic, agricultural, and industrial uses in northern Egypt. Most of these investigations have extensively studied the various sedimentary cycles in the Nile Delta to understand the composition and formation of the onshore sediments and their role in hydrocarbons and water resources in the area. These studies have identified three main sedimentary cycles in the region. The first cycle, known as the Miocene cycle, comprises the Sidi Salim Formation, the Qawasim Formation, and the Rosseta Formation, which are dated to the Early-Mid Miocene, Mid-Late Miocene, and Late Miocene periods, respectively. The second cycle, referred to as the Pliocene–Pleistocene cycle, includes the Abu Madi Formation from the Early Pliocene, the Kafr El-Sheikh Formation from the Middle Pliocene, the El-Wastani Formation from the Late Pliocene, and the Mit Ghamr Formation, which represents the transition from the Pliocene to the Pleistocene. Finally, the third cycle, known as the Holocene cycle, is characterized by the Bilqas Formation, which is found at the top of the sedimentary sequence. Figure 6.2 illustrates the geological map and sections in the Nile Delta.

Among these sedimentary cycles, the Mit Ghamr Formation of Quaternary age stands out due to its significance as the main aquifer in the Nile Delta region (Zaghloul et al. 1977; Said 1981; Sherif 1999; Gemail et al. 2017a). This formation is primarily of fluviatile origin, indicating its formation in an alluvial environment, and is composed mainly of gravelly sands and with a thickness of approximately 700 m. Sand and gravel deposits are abundant in the southern region of the study area, while they are relatively scarce in the eastern and western parts. In contrast, clays dominate the northern portion, with intermittent and discontinuous occurrences of sand and gravel. Notably, the northern and northeastern areas exhibit a distinct pattern of alternating layers of clays, sands, and gravel, which is indicative of the deltaic advancement and sedimentation processes.

The presence of such a significant aquifer underscores the importance of the Nile Delta as a vital water resource. However, the Nile Delta aquifer faces numerous challenges that threaten its sustainability. Overexploitation and contamination are two major issues that have led to declining water levels and deteriorating water quality. The ever-increasing demand for water, coupled with inadequate management practices, has resulted in the

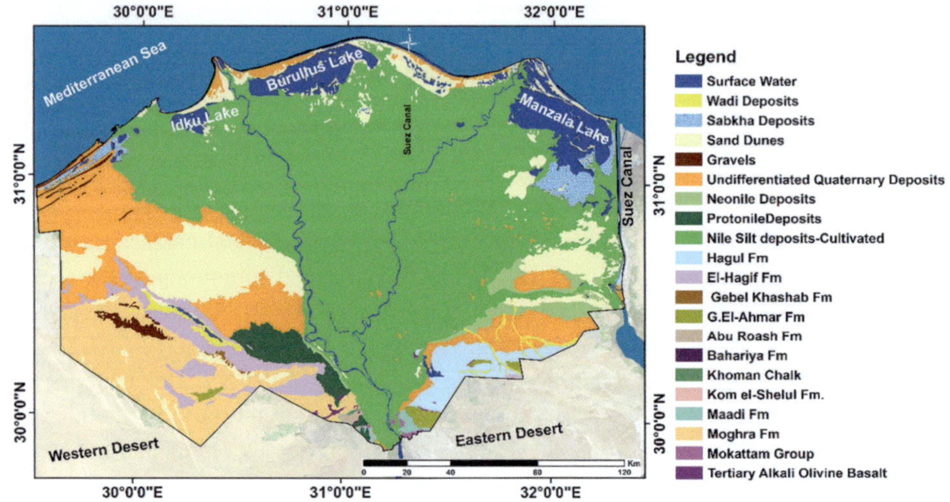

Fig. 6.2 Geological map showing the surficial lithological units in the Nile Delta region (modified after EGPC/CONOCO, 1987)

depletion of this crucial resource. Furthermore, contamination from various sources further exacerbates the problem, making it imperative to address these issues. Accordingly, the geological characterization of the Nile Delta aquifer is important for understanding the distribution and quality of the groundwater resource and for developing strategies to manage and protect this valuable resource.

The Bilqas Formation, consisting of clay and silty clay layers, overlies the extensive sequence of unconsolidated sand and gravel of the Mit Ghamr Formation. These two geological units form the primary components of the stratigraphic succession within the Nile Delta. Although the clay and silty clay layer and the underlying sand and gravel layer are hydraulically connected, each unit possesses its distinct water level. The surface clay cap, composed of silty and sandy deposits, is of particular significance as it plays a crucial role in the hydrological dynamics of the region. This layer serves as a hydraulic connection between the Nile Delta aquifer and the surface-cultivated lands. It functions as a regulatory mechanism, influencing the amount of water capable of infiltrating and subsequently recharging the Nile Delta aquifer.

Understanding the hydraulic relationship between these geological units is vital for comprehending the overall hydrogeology of the region. The interaction between the Bilqas Formation and the underlying Pleistocene aquifer shapes the movement and availability of water resources within the Nile Delta. By examining the characteristics and behavior of these interconnected hydrological units, researchers and water resource managers can gain insights into the dynamics of water infiltration and recharge, thereby contributing to the effective management and preservation of the Nile Delta aquifer.

In the Nile Delta region, the recharge of groundwater primarily occurs through vertical leakage, whereby water from the Nile River infiltrates the Bilqas aquitard and subsequently flows into the Nile Delta aquifer. The contribution of rainfall as a recharge source is minimal in this context. Additionally, recharge takes place in areas where the delta branches or canals indirectly interact with the aquifer sands and exhibit influent behavior. However, the most significant recharge originates from the vertical leakage of the Bilqas aquitard. This mechanism is considered a vital component in the dynamics of the groundwater flow system in the Nile Delta. It is worth noting that Barber and Carr (1981) emphasized the importance of this particular recharge process.

Accordingly, the water level in the Nile Delta aquifer is affected by some factors, including natural recharge from surface water infiltration, including Nile branches, irrigation and drainage networks, and pumping for irrigation and domestic use. In general, water levels in the aquifer have been declining in recent years due to overdraft, which occurs when more water is pumped out of the aquifer than is replenished through natural recharge, as indicated by GRACE data analysis in the area (Youssef et al. 2024).

6.3 Challenges of Groundwater Sustainability in the Nile Delta

The water resources in the Nile Delta have been experiencing a continuous decline in quality and quantity. This deterioration is primarily attributed to multiple challenges, including the diminishing flow of water from the Nile River, inadequate management of groundwater extraction, and various anthropogenic activities occurring along the Nile branches and wastewater drainage networks. Additionally, these anthropogenic activities have played a significant role in the deterioration of water quality. Factors such as industrial discharges, agricultural runoff, and inadequate sanitation systems in urban regions have contributed to water pollution, including contaminants such as heavy metals, nutrients, and salinization of the shallow aquifer zone.

These challenges negatively impact the quality and quantity of groundwater and have significantly impacted the availability of freshwater resources in the delta region. In addition, factors such as climate change, SLR, upstream water use, and altered flow patterns have contributed to this decrease in water flux. Furthermore, the mismanagement of groundwater extraction practices has further exacerbated the water resource challenges in the region. Over-pumping and inefficient use of groundwater resources have led to declining water levels, increased vulnerability to saltwater intrusion, and compromised overall water availability (El Osta et al. 2018; Abd-Elaty et al. 2019; Salem et al. 2019; Morsy et al. 2020; Gemail and Abd-Elaty 2023; Ali and Negm 2024; Gemail et al. 2024; Elshaarawy and Armanuos 2025).

Addressing these issues comprehensively and implementing sustainable water management strategies to safeguard the Nile Delta's water resources are crucial. Effective measures should include improved water governance, integrated water resources management practices, efficient irrigation techniques, and enhanced wastewater treatment systems. By addressing these challenges, stakeholders can work toward preserving and restoring the quality and quantity of water resources in the Nile Delta.

Certainly! Here is a more detailed review of the groundwater challenges in the Nile Delta, including saltwater intrusion, wastewater pollution, population growth, and water demands:

6.3.1 Urbanization and Water Demand

The Nile Delta is a vital ecosystem that crucially supports the livelihoods of millions while providing essential freshwater and food resources. This region is characterized by its rich biodiversity, fertile soils, and proximity to the Nile River and the Mediterranean Sea, making it an ideal area for agricultural production (Syvitski and Saito 2007; Mehrim and Refaey 2023). However, the rapid growth of the population and increasing land use pressures pose significant challenges to water demands in the Delta.

As the population in the Nile Delta continues to rise, the demand for freshwater intensifies. Urbanization, driven by economic growth, has led to extensive land use changes, including the conversion of wetlands, agricultural land, and natural habitats into urban areas. This shift not only reduces the availability of freshwater resources but also exacerbates environmental issues such as soil salinization and coastal erosion (Gemail et al. 2004; Youssef et al. 2020, Youssef et al. 2024).

The impacts of climate change, including rising sea levels and changing precipitation patterns, further threaten the Delta's water resources (Bennett et al. 2016; Gemail et al. 2017a; Rateb and Abotalib, 2020; Youssef et al. 2021; Darwish and Smith 2021; Ghosh et al. 2023). Studies indicate that even moderate SLR can have severe consequences for agriculture and aquaculture, which are heavily reliant on the Delta's freshwater systems. The Nile Delta's ancient, low-relief topography, coupled with urban expansion and dam construction, increases its susceptibility to these transformations (El-Raey et al. 1999; Hzami et al. 2021).

Egypt's dependence on the Nile Delta, which supports approximately 63% of its agricultural land and substantial aquaculture production (Fishar 2018; Bakr and Bahnassy 2019), underscores the growing significance of interactions between population growth, land use, and water demands. Ensuring water and food security in the region requires addressing these challenges through a comprehensive assessment of both natural and anthropogenic influences on the Delta's hydrogeological dynamics (Dasgupta et al. 2009; Hzami et al. 2021) (Fig. 6.3).

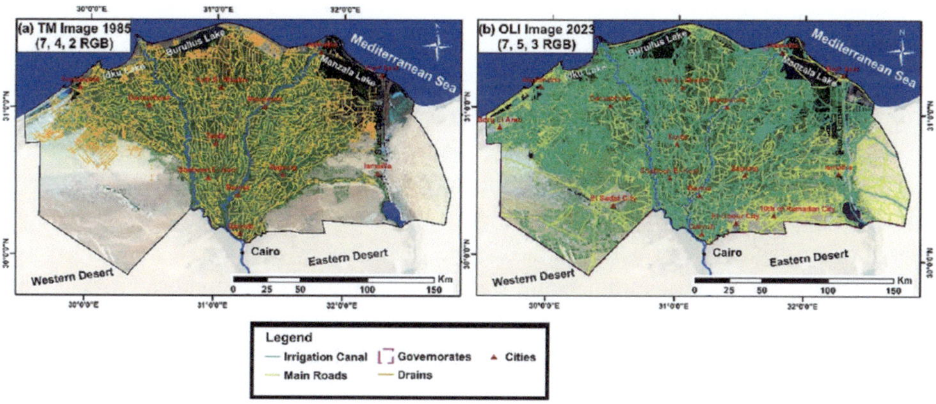

Fig. 6.3 Multi-temporal Landsat images of the studied period (1985 to 2023), which represent the false-color composite (FCC) after the preprocessing step, **a** Landsat 5 1985 and **b** Landsat 8 2023

6.3.2 Land Use Dynamics and Water Demands

In Vulnerable deltas, the Nile Delta in Egypt is a typical example, the existing critical situation is further intensified by a range of changes in land use patterns, which include the impact of SLR on coastal zones (Iskander et al. 2007), soil degradation due to the presence of near-surface saline water, and the conversion of cultivated land into urban areas (Shalaby 2012). These changes inevitably contribute to the acceleration of surface water pollution and the disruption of hydrological connectivity, ultimately leading to the degradation of both soil and groundwater quality (Gemail et al. 2017b). The drive for economic growth has prompted uncontrolled development, speeding up land use changes at an unprecedented rate and disrupting the natural landscape in deltas (Fouad et al. 2023). These changes, encompassing urbanization and agricultural expansion, have profound implications for the water resources and hydro-environmental aspects. For instance, the escalating trend of coastal urbanization has led to the transformation of farming lands, surface water bodies, wetlands, and natural vegetation into urban areas (Gemail et al. 2017b; Youssef et al. 2021). Moreover, unplanned urbanization has not only impaired surface water resources in riverine cities (Abu Salem et al. 2021) but has also escalated the demand for groundwater (Bennett et al. 2016), leading to overexploitation of deltaic aquifers beyond their safe yield (Magnan et al. 2022). The intensive pumping of groundwater can lead to soil degradation and land subsidence in arid regions. As the water table lowers, soils become drier, which can negatively impact agricultural productivity. Dry soils are more prone to compaction and salinization, reducing their fertility and suitability for farming, leading to reduced crop yields and financial losses for farmers.

To assess the impact of Land Use and Land Cover (LULC) changes on water demands and groundwater resources in the Nile Delta, the Maximum Likelihood Classifier (MLC) was employed to generate LULC maps spanning from 1985 to 2023. The MLC, a robust supervised classification method, relies on extensive pre-determined reference data for accurate classification. For the 2023 imagery, regions of interest (ROIs) were delineated through visual inspection, supported by high-resolution Google Earth maps and field surveys. Integration of False-Color Composite (FCC) imagery and ancillary data enabled the identification and classification of eight distinct LULC categories: urban areas, agricultural land, aquatic/natural vegetation, fish farms, drylands, Sabkha (salt flats), waterlogged areas, and surface water (including lakes and sea). These categories were selected to reflect the diverse land uses influencing water consumption and groundwater recharge.

A random sample of 70% of the ROI database was used to train the MLC for each LULC map (Fig. 6.4), while the remaining 30% was reserved for accuracy assessment (Lunetta and Lyon 2004). The classification accuracies were evaluated independently for each map to ensure reliability. The results revealed significant shifts in LULC over the study period, particularly the expansion of urban areas and agricultural land at the expense of natural vegetation and drylands. These changes have profound implications for water demands and groundwater resources in the Nile Delta. For instance, the increase in urban and agricultural areas has likely escalated water consumption, while the reduction of natural vegetation and drylands may have diminished groundwater recharge rates. Such transformations underscore the need for sustainable land and water management practices to mitigate the adverse effects on the region's hydrological balance.

Nile Delta faces persistent socio-economic challenges, including food insecurity, malnutrition, water scarcity, and groundwater degradation. These challenges are exacerbated by the ongoing transitions between different LULC categories, highlighting a critical trade-off between urban expansion and agricultural productivity. Such shifts have profound long-term implications for food security and the sustainability of water resources in the Nile Delta, a region already under significant hydrological stress.

One of the most notable changes observed between 1985 and 2023 is the decline in surface water bodies, which decreased from 4.06 to 2.16% of the area (Fig. 6.5). This

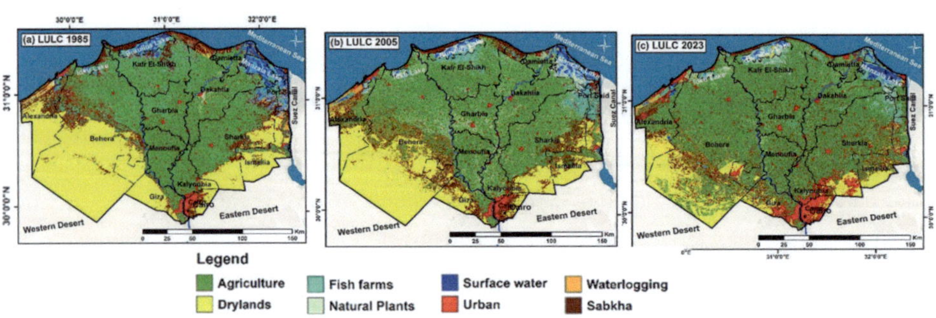

Fig. 6.4 Land use/land cover (LULC) changes in the Nile Delta area: **a** LULC map 1985, **b** LULC map 2005, and **c** LULC map 2023

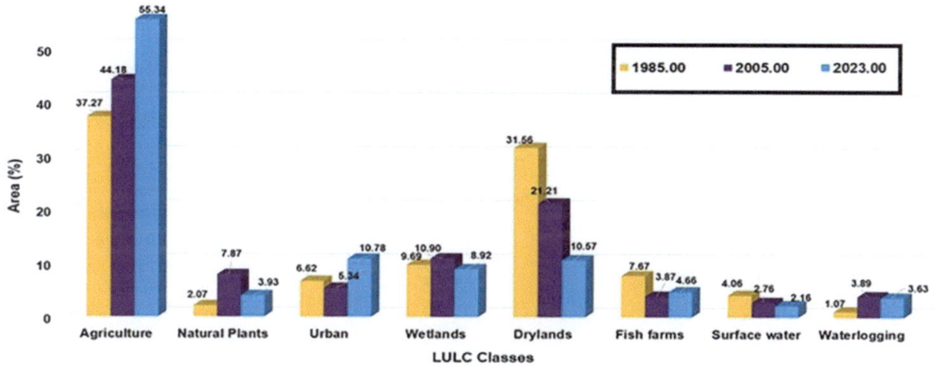

Fig. 6.5 LULC categories changes obtained from the classified maps in the Nile Delta area over 38 years

reduction is particularly evident in Manzala Lake (Fig. 6.4), where the implementation of national strategies aimed at lake preservation has paradoxically led to reduce water coverage in the area (El-Zeiny et al. 2022). While these efforts have contributed to environmental restoration, the resulting changes have also altered the hydrological dynamics of the region, potentially affecting groundwater recharge and local water availability.

Fish farms, another critical component of the Delta's water-dependent ecosystems, experienced a significant decline between 1985 and 2005, followed by a gradual recovery from 2002 to 2023. This trend is linked to the expansion of agricultural and aquacultural projects in the wetlands surrounding Manzala, Burullus, and Idku lakes (Fig. 6.4b, c). Agricultural development, particularly in the eastern and western parts of the Delta after 2000 (Othman et al. 2012), has intensified land use pressures. However, the proliferation of intensive agricultural and industrial activities in the Manzala region has raised concerns about soil salinity and saltwater intrusion, particularly in the Eastern Nile Delta. These phenomena threaten the quality and quantity of groundwater resources, exacerbating the region's water scarcity challenges.

Youssef et al. (2024) propose an integrative approach that combines LULC analysis with field measurements to comprehensively assess the dynamic interactions between land use changes, natural processes, and anthropogenic activities, and the resulting impact on groundwater resources in the Eastern Nile Delta (END). Accordingly, the combined effects of climatic changes and anthropogenic activities are exerting significant pressure on groundwater resources, soil quality, and food security in the END region (Fig. 6.6). Rapid population growth in the eastern governorates—increasing from 12,422,935 to 17,150,000 inhabitants between 1985 and 2018—has driven substantial land cover changes, particularly the conversion of natural and agricultural lands into urban areas (CAPMAS 2000). Sharkia, Dakahlia, Damietta, Ismailia, and Port Said have experienced the highest population densities, leading to unplanned urban expansion and the

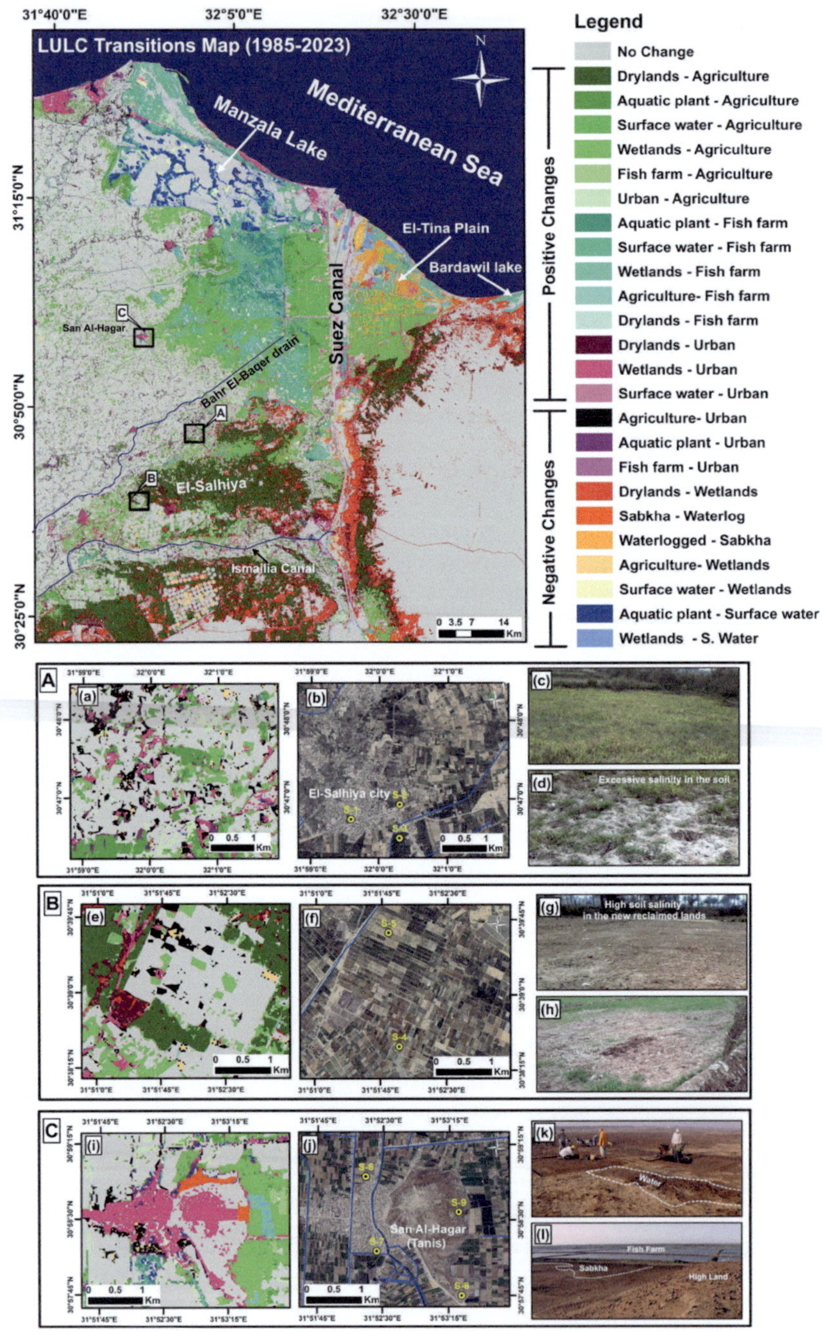
31°40'0"E
32°5'0"E
32°30'0"E
31°15'0"N
30°50'0"N
30°25'0"N
LULC Transitions Map (1985-2023)
N
Mediterranean Sea
Manzala Lake
El-Tina Plain
Bardawil lake
Suez Canal
San Al-Hagar
Bahr El-Bager drain
El-Salhiya
Ismailia Canal
0 3.5 7 14
Km
A
B
C
Legend
No Change
Drylands - Agriculture
Aquatic plant - Agriculture
Surface water - Agriculture
Wetlands - Agriculture
Fish farm - Agriculture
Urban - Agriculture
Aquatic plant - Fish farm
Surface water - Fish farm
Wetlands - Fish farm
Agriculture- Fish farm
Drylands - Fish farm
Drylands - Urban
Wetlands - Urban
Surface water - Urban
Agriculture- Urban
Aquatic plant - Urban
Fish farm - Urban
Drylands - Wetlands
Sabkha - Waterlog
Waterlogged - Sabkha
Agriculture- Wetlands
Surface water - Wetlands
Aquatic plant - Surface water
Wetlands - S. Water
Positive Changes
Negative Changes
(a)
(b)
(c)
(d)
Excessive salinity in the soil
El-Salhiya city
S-1
S-2
S-3
0 0.5 1
Km
(e)
(f)
(g)
High soil salinity
in the new reclaimed lands
(h)
S-4
S-5
(i)
(j)
(k)
Water
(l)
Fish Farm
Sabkha
High Land
San Al-Hagar
(Tanis)
S-6
S-7
S-8
S-9

◄**Fig. 6.6** LULC transition map in the END (1985–2023) (Youssef et al. 2024), verified using Google Earth and field data. Site A (El-Salhiya city): **a** Urban sprawl impacts; **c**, **d** phosphorus deficiency and high soil salinity reducing agricultural productivity. Site B (New El-Salhiya area): **e** Expansion of reclamation projects with minor soil degradation; **g**, **h** salt scald areas with barren, vegetation-free patches. Site C (San Al-Hagar area): **i** Urban and agricultural expansion near the Tanis archaeological site; **k**, **l** waterlogging and muddy soils near Tell Dibgou, indicating shallow saline groundwater. *Notes* **b**, **f**, **j** Google Earth imagery with yellow points marking soil sample locations. "Wetlands" includes waterlogged and sabkha classes

loss of fertile cultivated lands in the western parts of the region (Fig. 6.6a, e, I). Over the study period, illegal urban sprawl has resulted in the loss of approximately 183.17 km^2 of productive agricultural land, further exacerbating water scarcity and soil degradation (Youssef et al. 2024).

Climate change, characterized by rising temperatures and erratic rainfall patterns, has compounded these challenges by reducing surface water availability and increasing reliance on groundwater for irrigation and domestic use. This over-extraction of groundwater with reduced recharge rates due to urbanization and land cover changes has led to declining groundwater levels and increased vulnerability to saltwater intrusion. Additionally, the influx of migrants from rural areas in southern Egypt is expected to intensify land use pressures, further altering LULC patterns and straining already limited water resources.

These impacts are corroborated by the study of Youssef et al. (2024), which analyzed the temporal variability of long-term Terrestrial Water Storage (ΔTWS) changes to evaluate water challenges and storage dynamics between 2002 and 2017 (Fig. 6.7). The findings revealed significant decline in ΔTWS, with an average reduction of 77.89 mm over the study period and an annual depletion rate of −5.821 mm/year (Fig. 6.7b). Notably, the most severe decline occurred between 2005 and 2015, with a drop of −86.1 mm. These trends are closely linked to the expansion of land use and land cover (LULC), particularly the proliferation of aquaculture projects, unauthorized surface water use, and excessive groundwater extraction (Morsy 2009).

Groundwater extraction in the Nile Delta has risen steadily, increasing by an average of 0.1 billion cubic meters (Bm^3)/year since 1981, with a sharper rise of 0.2 Bm^3/year between 2003 and 2010 (Mabrouk et al. 2013). This over-extraction, combined with insufficient recharge from rainfall and limited contributions from the Nile River and irrigation, has led to significant groundwater depletion in the Nile Delta Aquifer, particularly during 2003–2012 (Mohamed 2020). The expansion of urban and agricultural areas has further strained groundwater resources, exacerbating water scarcity. These interconnected stressors highlight the urgent need for sustainable land and water management practices to mitigate the adverse impacts on groundwater resources. Without effective interventions, the continued degradation of groundwater quality and quantity will undermine the region's ability to support its growing population and agricultural demands, exacerbating food insecurity and water scarcity in the Nile Delta regions. The interplay between LULC

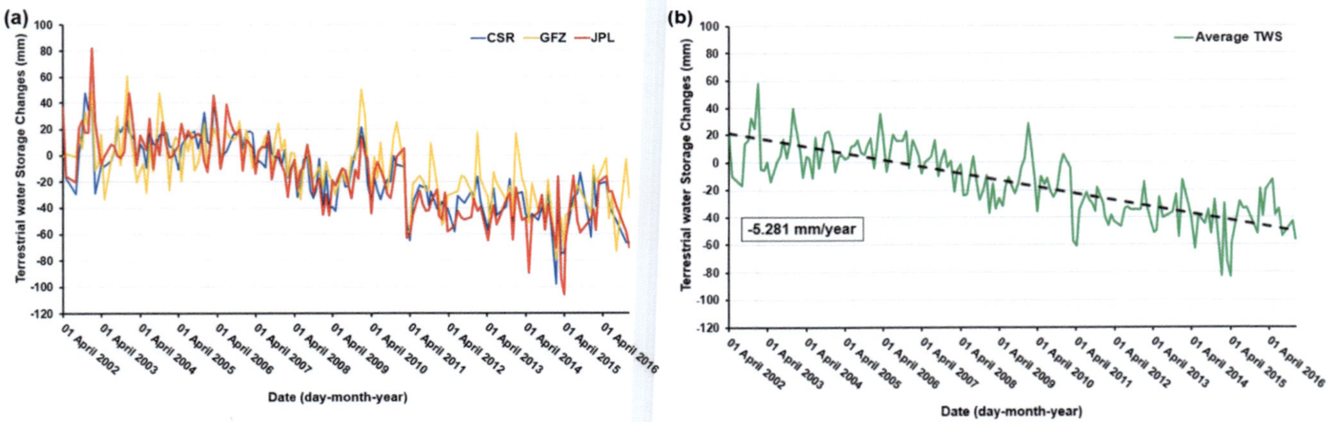

Fig. 6.7 The drawdown of the groundwater surface in the Eastern Nile Delta region (Youssef et al. 2024) **a** Time series of Terrestrial water storage changes (ΔTWS) in the Eastern Nile Delta from JPL, GFZ, and CSR GRACE satellite data. **b** Average of JPL, GFZ, and CSR datasets showing ΔTWS (mm) from April 2002 to January 2017

changes and water resources underscores the need for integrated land and water management strategies. Urbanization and agricultural expansion, while essential for economic development, must be balanced with measures to protect and sustainably manage the Nile Delta's fragile water ecosystems. Without such interventions, the continued degradation of surface and groundwater resources could undermine the region's ability to meet its growing water demands, further compounding socio-economic vulnerabilities.

6.3.3 Impact of Saltwater Intrusion on Groundwater Resources

Coastal aquifers like those in the Nile Delta are inherently hydraulically connected to saline water toward the sea. In such conditions, the hydraulic gradient maintains a delicate balance, which drives freshwater flow toward the sea, counteracting seawater intrusion. However, this gradient is often small and highly sensitive to disruptions (Fienen and Arshad 2016). Excessive groundwater extraction, driven by growing population demands, can reverse this gradient, allowing seawater to infiltrate and replace the extracted freshwater (Werner et al. 2013). This process, known as seawater intrusion, severely degrades groundwater quality, rendering it unsuitable for drinking or irrigation (Fig. 6.8).

The Nile Delta aquifer is particularly vulnerable due to the low natural recharge rates and high dependency on groundwater to meet the water needs of densely populated regions. Compounding the issue, climate change-induced sea-level rise accelerates saltwater encroachment into freshwater systems. While Ferguson and Gleeson (2012) argue that

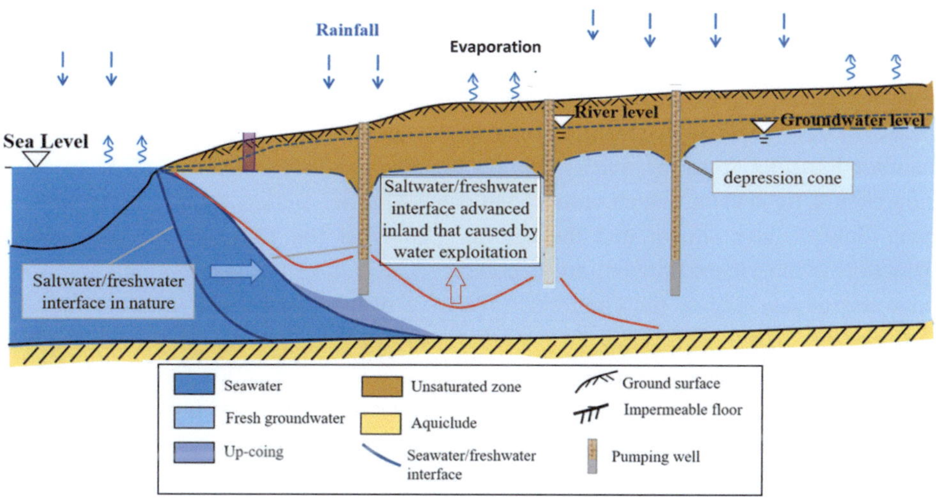

Fig. 6.8 A simplified diagram illustrates seawater intrusion and the impact of pumping in coastal aquifers such as the Nile Delta (Ma et al. 2019)

groundwater extraction has a more immediate impact on coastal aquifers than sea-level rise or recharge variations, the combined effects of these factors exacerbate the problem.

The extent of seawater intrusion depends on the interplay between groundwater extraction rates and natural recharge. Poorly positioned well fields or unplanned pumping can further accelerate intrusion, particularly in areas with high population densities and water demand. The consequences are dire in the Nile Delta regions, where much of the population resides along the coastal zone. Over-extraction has caused significant declines in groundwater levels, making it increasingly difficult to control seawater intrusion. As population growth and agricultural demands continue to strain the Nile Delta's groundwater resources, the aquifer's deficit is expected to worsen. This unsustainable exploitation, rising sea levels, and reduced recharge will likely intensify seawater intrusion, threatening water security for millions. Addressing this challenge requires integrated water management strategies that balance extraction with recharge, protect well fields, and mitigate the impacts of climate change.

The Nile Delta aquifer is a leaky aquifer system characterized by a direct hydraulic connection to the Mediterranean Sea through the northern coastal zone (spanning 220 km) and the lakes of the north and lagoons (Fig. 6.1). The aquifer is bounded by an upper semi-permeable layer of the Bilqas Formation and a lower impermeable clay layer of Pliocene origin (Gemail et al. 2017a). This geological configuration makes the aquifer highly vulnerable to external influences, particularly SLR and changes in Nile River flows, which significantly impact groundwater quality. Rising sea levels, combined with reduced freshwater inflow from the Nile, have led to increased seawater intrusion, elevating total dissolved solids (TDS) concentrations in the northern zone of the aquifer (El-Raey et al. 1999; Dawoud, 2004; Mabrouk et al. 2013; Engelen et al. 2018). This salinization process threatens the usability of groundwater for agricultural, drinking, and industrial purposes.

In addition to natural stressors, intensive and unplanned groundwater pumping in the central parts of the Delta has exacerbated the degradation of groundwater resources. Excessive abstraction lowers groundwater levels, creating hydraulic gradients that favor seawater intrusion from the north. This phenomenon is further intensified by the over-extraction of freshwater, which reduces the natural barrier against saline water encroachment. Studies have shown that the central regions of the Delta, where groundwater extraction is most concentrated, are experiencing declining water tables and deteriorating water quality due to the combined effects of over-pumping and seawater intrusion (Attwa et al. 2016; Abu Salem et al. 2022). Accordingly, the interplay between SLR, reduced Nile River flows, and unsustainable groundwater extraction poses a significant threat to the long-term sustainability of the Nile Delta aquifer. Without effective management strategies, such as regulated pumping, artificial recharge, and the development of alternative water sources, the aquifer will continue to face increasing salinization and depletion, jeopardizing water security for millions of people in the region.

Numerous studies have employed advanced simulation techniques to investigate seawater intrusion in the Nile Delta aquifer, providing critical insights into its dynamics and informing strategies for mitigation and adaptation. Among these, pioneering work by Sheriff et al. (1988, 1990), Sherif and Singh (1997), Sherif et al. (2012), Sefelnasr and Sherif (2014), and Nofal et al. (2015) utilized variable-density modeling approaches to delineate the freshwater-seawater interface in both horizontal and vertical cross-sections. These studies introduced the concept of equivalent freshwater heads employing sophisticated numerical models, such as FEFLOW—a 2D finite element variable-density model (FFD)—to simulate seawater intrusion under various scenarios, including the impacts of SLR and groundwater abstraction. Their findings revealed that the transition zone between freshwater and seawater shifts landward and deepens with increasing depth, highlighting the aquifer's vulnerability to salinization.

Modern computing (PC) technology can solve the complex equations related to variable-density flow and transport, which is beneficial for specific studies, including those on aquifers in the Nile Delta (Mazi et al. 2013, 2014). However, the high computational demands and issues with data scarcity limit the reliability of these detailed simulations. Insufficient data on hydraulic parameters, concentration fields, and boundary conditions (such as recharge and pumping rates) lead to uncertainties and simplifications in modeling. Consequently, the effectiveness of complex numerical models for studying seawater intrusion is often diminished due to guesswork in parameter estimation. In contrast, analytical approaches (Koussis et al. 2012; Lu et al. 2016; Idowu et al. 2022) and advanced machine learning (ML) models (i.e., Elshaarawy and Rajoub 2020; Tahmasebi et al. 2020; Bertrand et al. 2022; Narvaez-Montoya et al. 2023; Armanuos 2025) may offer more reliable insights under these complex hydrological conditions.

Although numerical models have been extensively used to monitor and forecast seawater intrusion in the Nile Delta aquifer using MODFLOW, FEFLOW, SEAWAT, and MT3DMS codes, their reliability is significantly hampered by outdated input datasets and an incomplete understanding of the aquifer's evolving hydrogeological dynamics. Recent studies (i.e., Armanuos and Negm 2018; Abd-Elaty et al. 2021a, b; Armanuos et al. 2022; Abd-Elaty and Zelenakova 2022; Abd-Elaty et al. 2024) rely heavily on abstraction and recharge rates from the Research Institute for Groundwater (RIGW 2008), which do not reflect the changes in LULC or account for anthropogenic activities over the past decade. These activities include increased groundwater extraction (Fig. 6.7), reduced recharge due to urbanization, and direct hydraulic connections with saline water sources, such as the Miocene in the southeast and the Moghra in the western Nile Delta. This dependence on obsolete data introduces substantial uncertainties in model predictions, undermining the accuracy of proposed mitigation strategies.

Moreover, many models treat the Nile Delta aquifer as a homogeneous system, overlooking its inherent geological complexity. Critical heterogeneities—such as discontinuous clay layers in the unsaturated upper zone, variations in layer thickness, and spatially diverse aquifer lithology—are often ignored. These oversights distort our understanding of saltwater intrusion and surface wastewater pathways, leading to flawed assessments of vulnerability hotspots. Consequently, outdated inputs and oversimplified aquifer geometry propagate errors in salinity distribution forecasts, diminishing confidence in long-term management plans.

Addressing these gaps is crucial for reducing uncertainties, enhancing predictive capabilities, and aligning mitigation strategies with the dynamic conditions of the Nile Delta. Failure to do so impedes impactful research and knowledge transfer between stakeholders and national institutions, raising concerns about the sustainability of groundwater supplies essential for supporting ecosystem services in rapidly growing coastal populations and agricultural systems. Geophysical, hydrochemical, and field observations can greatly enhance the accuracy of numerical models for the Nile Delta aquifer by addressing the shortcomings of outdated datasets and oversimplified geometries. Geophysical techniques, such as resistivity surveys, can identify subsurface heterogeneities and map salinity distribution while monitoring changes over time. Hydrochemical analyses provide insights into water quality and the impacts of anthropogenic activities, helping to track salinity sources and identify recharge areas. Field observations offer up-to-date data on groundwater levels and local conditions, allowing for the identification of vulnerability hotspots.

In this context, Abu Salem et al. (2022) adopted an integrative approach that combined holistic hydrogeochemical analysis, statistical methods, and DC resistivity measurements to investigate lateral and vertical variations in groundwater characteristics affected by salinization stress due to seawater intrusion in the Eastern Nile Delta. Their findings are consistent with the SWI numerical model developed by Mazi et al. (2014), as depicted in Fig. 6.9. Consequently, it is essential to calibrate numerical and SWI simulation models using geophysical and hydrogeochemical data. This integrated approach enables a more accurate representation of the aquifer's complexities and dynamics, improving predictions of salinity distribution and providing a robust foundation for effective management strategies. A proposed strategy for mitigating seawater intrusion involves controlled groundwater pumping in the northern and central regions of the Eastern Nile Delta, aimed at reducing aquifer degradation (Fig. 6.10). This approach underscores the importance of regularly updating models with field observations and integrating geophysical and hydrogeochemical insights to support sustainable water resource management practices.

To assess seawater intrusion in the Nile Delta aquifer, 69 DC resistivity soundings from prior geophysical studies (Gemail et al. 2011, 2017a; Attwa et al. 2016) were synthesized to map the spatial resistivity distribution of the freshwater-saltwater transition zone in the central Nile Delta (Fig. 6.11). This map reveals critical insights into groundwater salinization where the resistivity values (>20 Ω·m) align with the Nile River's and southern parts, indicating freshwater recharge from surface water infiltration. This pattern underscores the Nile's role in replenishing the aquifer and temporarily mitigating seawater intrusion in these zones. In contrast, resistivity values drop sharply (<5 Ω·m) in the central Delta south of Latitude 31°00′ (Fig. 6.11b), reflecting extensive groundwater salinization. This anomaly correlates with two key factors: Intensive groundwater pumping, driven by population growth and agricultural demands, has lowered hydraulic

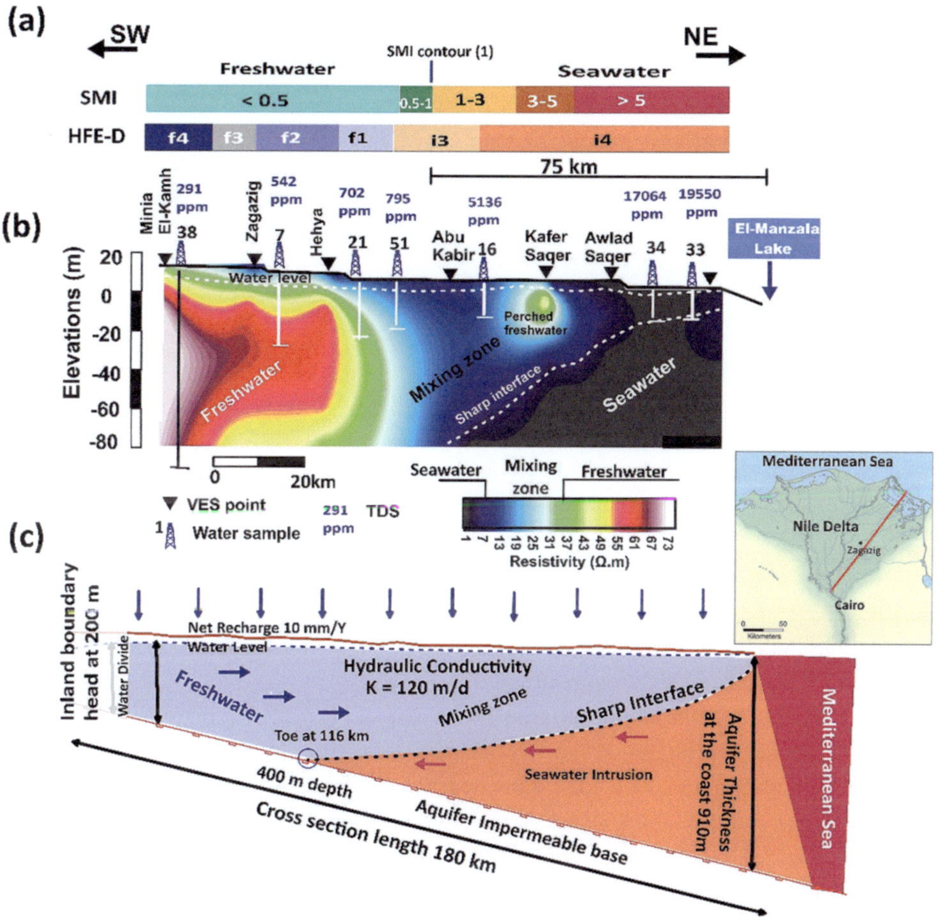

Fig. 6.9 Seawater Intrusion in the axis of Eastern Nile Delta: **a** Hydrochemical facies from HFE-D from groundwater samples and the calculated SMI of the coastal groundwater along the axis of Eastern Nile Delta (Abu Salem et al. 2022) *(the SMI values greater than 1 indicate significant seawater mixing)*, **b** Vertical resistivity cross-section showing the characteristic wedge shape of seawater intrusion inland. and **c** Adapted schematic cross-section from Sherif (1999), utilizing numerical solutions by Koussis et al. (2012), illustrating the extent of seawater intrusion in the Eastern Nile Delta (Fig. 3b in Mazi et al. 2014)

heads, reversed the natural seaward gradient, and drawn seawater inland, and urbanization and land use changes have diminished natural recharge rates, further destabilizing the freshwater-saltwater equilibrium. In addition, the resistivity map highlights a southward progression of saline water into the aquifer, particularly in areas with dense populations and high irrigation needs. The central Nile Delta's rapid population growth has amplified groundwater extraction, exceeding the aquifer's sustainable yield. Combined with

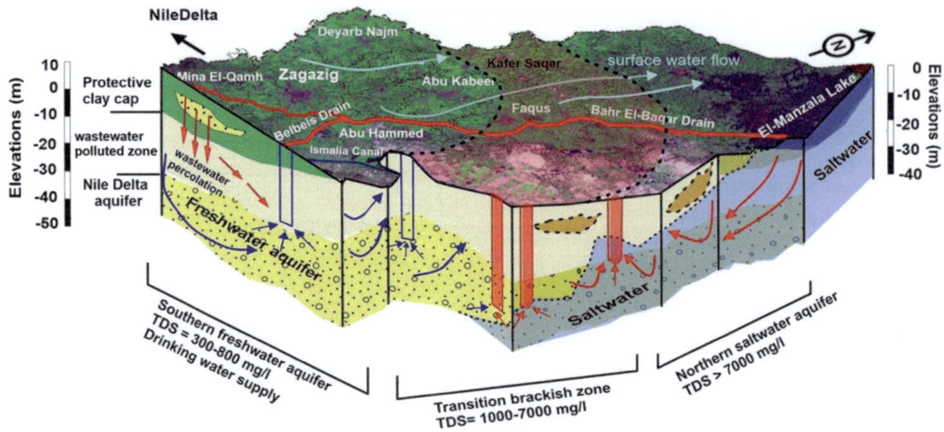

Fig. 6.10 A schematic 3D model shows the impact of pumping in the freshwater/saltwater mixing zone (in red) and wastewater percolation conditions due to clay heterogeneities in the Eastern Nile Delta

climate-driven and surface water reductions in the upper Nile flow and recharge, this overexploitation has accelerated seawater intrusion, created a feedback loop of declining water quality, and increased treatment costs.

The resistivity section along the Nile Delta axis (Fig. 6.11b) illustrates a southward-thickening wedge of higher resistivity values, reflecting freshwater dominance over underlying saline water. This wedge thins near the Mediterranean coast, where low resistivity (<2 Ω·m) confirms pervasive shallow seawater intrusion. A transitional zone (5–15 Ω·m) at Latitude 31°00′ marks the inland migration of the freshwater-saltwater interface. Over-pumping lowers hydraulic heads, destabilizing the natural equilibrium and drawing saline water southward. Vertical resistivity contrasts reveal deeper saline intrusion (>50 m) in the transition zone south of Latitude 31°00′, highlighting preferential pathways for saltwater encroachment. Studies by Sherif et al. (2012) and Mazi et al. (2014) document seawater intrusion advancing 98–116 km (1000 ppm, TDS) inland into the central and eastern Nile Delta, respectively. This salinization has necessitated the widespread shutdown of irrigation wells in the mixing zone.

These multidisciplinary studies with updated numerical models have played a pivotal role in characterizing the Nile Delta aquifer, offering a comprehensive understanding of its hydrogeological behavior under natural and anthropogenic stressors. By mapping the spatial and temporal evolution of seawater intrusion, they have identified high-risk zones and quantified the impacts of SLR and over-pumping on groundwater quality. This knowledge is indispensable for developing adaptive management strategies, such as regulated groundwater extraction, artificial recharge, and constructing hydraulic barriers to control saltwater encroachment.

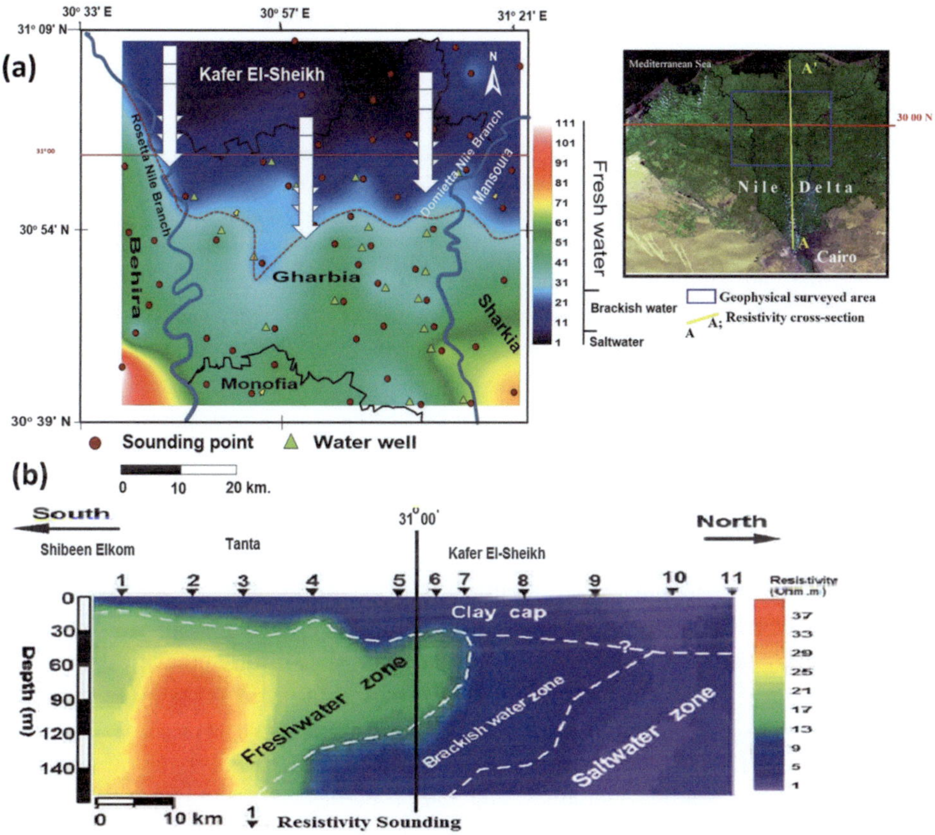

Fig. 6.11 Resistivity distribution in the Nile Delta aquifer illustrating the freshwater/saltwater interface located south of Latitude 31°00′, indicative of seawater intrusion. **a** The resistivity map highlights salinity patterns in the central region, with white arrows indicating the direction of seawater flow resulting from groundwater abstraction, and **b** The 2D resistivity section along the Nile Delta axis reveals a wedge-shaped high resistivity zone, representing the freshwater/saltwater interface in the mixing zone around Latitude 31°00′

Moreover, the methodologies and findings from these studies provide a robust scientific foundation for designing mitigation plans. For instance, using advanced numerical models allows policymakers to simulate the long-term effects of climate change and human activities on the aquifer, enabling the evaluation of various intervention strategies. By integrating these insights into water resource management frameworks, stakeholders can prioritize actions that enhance the resilience of the Nile Delta aquifer, ensuring sustainable water availability for agriculture, industry, and domestic use in the face of rising sea levels and increasing water demands.

6.3.4 Groundwater Pollution Under Anthropogenic Influences

Challenges in sustainable development are often linked to the availability and suitability of freshwater resources, particularly in regions where surface water is scarce or contaminated (Rehman et al. 2024). In Egypt, especially within semi-arid zones such as the Nile Delta, freshwater scarcity has intensified reliance on groundwater to meet diverse demands (Youssef et al. 2024). While groundwater generally offers greater accessibility, reliability, and resilience to contamination compared to surface water, it remains vulnerable to pollutants that can compromise its quality (Patrick et al. 2019; Salem and Osman 2017; Salem et al. 2022a, b). Moreover, traditional flood irrigation practices present a significant risk of groundwater contamination due to excessive percolation rates (Abdel-Fattah and Helmy 2015; Arnous et al. 2015; Mabrouk et al. 2013; Salem et al. 2021). Several studies have investigated groundwater contamination by chemical pollutants (Salem et al. 2021; Salem et al. 2022a, b), highlighting the substantial impact of anthropogenic activities, including agricultural runoff, industrial discharge, improper waste disposal, and insufficient wastewater treatment (Poshtegal and Mirbagheri 2019). Consequently, groundwater contamination remains a pressing concern in this rapidly urbanizing mega-delta aquifer, where economic development often conflicts with environmental sustainability, posing risks to water quality and availability (Arafa et al. 2024; Youssef et al. 2024).

Assessing groundwater contamination risk (GwCR) is therefore crucial for managing the vulnerable mega-delta aquifer, as it provides insights into contamination susceptibility based on hydrogeological conditions and overlying strata (Salem et al. 2025a, b). These assessments quantify the likelihood of pollutants reaching the water table by integrating geological structure, aquifer hydraulics, and soil properties (Oke and Fourie 2017). Such evaluations play a vital role in guiding policymakers and urban planners in land use planning, industrial development, and sustainable water management strategies (Nahin et al. 2020). While hydrogeochemical analyses (Salem and Osman 2017; Salem and Hassan 2021), ionic ratio comparisons (Salem et al. 2022a), simulation models (Sefelnasr and Sherif 2014; Eslamian et al. 2023), and geophysical techniques (Attwa et al. 2016; Salem and Osman 2017; Gemail et al. 2017a; Abu Salem et al. 2022) have advanced groundwater pollution studies in the region, these approaches are often limited in their ability to capture the spatial variability of GwCR across two dimensions (Arafa et al. 2022).

Index-based models are widely utilized to assess GwCR, including DRASTIC (Aller et al. 1987), GOD (Foster 1987), SINTACS (Civita and De Maio 1997), and GALDIT (Chachadi et al. 2002). Among these, GALDIT and DRASTIC are the most commonly applied frameworks for evaluating intrinsic groundwater vulnerability in the Nile Delta (Arafa et al. 2022; Salem et al. 2025a, b). Numerous studies have further emphasized the growing influence of land use on water pollution (Nahin et al. 2020; Arafa et al. 2022), demonstrating how various land use practices and human activities introduce contaminants that percolate into groundwater systems (Maqsoom et al. 2020). To address these concerns, Arafa et al. (2022) conducted a GwCR assessment across the Nile Delta aquifer

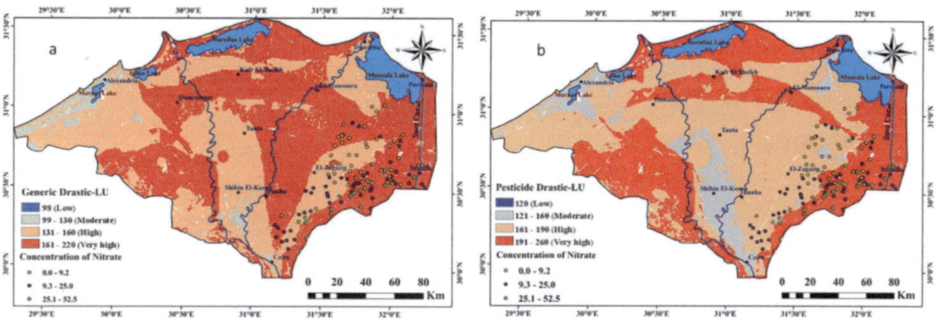

Fig. 6.12 GwCR vulnerability map in the whole Nile Delta generated by **a** generic DRASTIC-LU and **b** pesticide DRASTIC-LU model

using modified DRASTIC-LU models—generic and pesticide-based (Fig. 6.12)—by integrating eight key variables: Depth to Groundwater (D), Net Recharge (R), Aquifer Media (A), Soil Media (S), Topography (T), Impact of the Vadose Zone (I), Hydraulic Conductivity (C), and Land Use (LU). The study employed Multi-Criteria Decision Analysis (MCDA) within a Geographic Information System (GIS) framework, revealing that the optimized pesticide DRASTIC-LU model classified 50.68% of the Nile Delta as highly vulnerable and 39.53% as very highly vulnerable (Fig. 6.12b).

Despite the widespread application of MCDA techniques such as the Analytical Hierarchy Process (AHP) and fuzzy logic for weighting DRASTIC and GALDIT parameters, these approaches have been criticized for their subjectivity (Kazakis et al. 2019; Pautz et al. 2023). To enhance groundwater vulnerability assessments, Salem et al. (2025a, b) employed ML-based GALDIT models, including Support Vector Machine (SVM), Generalized Linear Model (GLM), and eXtreme Gradient Boosting (XGBoost), optimized using grid search hyperparameter tuning. This methodology refined the assessment of delta aquifer vulnerability to SWI by incorporating GALDIT factors—groundwater occurrence, hydraulic conductivity, groundwater level above sea level, proximity to the shoreline, impact of existing SWI, and aquifer thickness—while integrating a Conditioned Vulnerability Index (CVI) based on Total Dissolved Solids (TDS) as an input variable. The $GALDIT_{XGBoost}$ map detected previously undetected high-vulnerability areas west of Alexandria and localized pockets within southern Port Said along the Mediterranean coast (Fig. 6.13). Piper diagrams also confirmed SWI risks, with dominant Na-Cl and Ca–Mg–Cl facies indicating elevated Cl^- and SO_4^{2-} levels. A shift from HCO_3^- to Cl^- further validated salinization, while Ca–HCO_3 facies represented freshwater in the southern parts.

While these advancements enhance spatial prediction in the Nile Delta, the standalone application of GALDIT and DRASTIC models often lacks accuracy in identifying salinization risks driven by natural and anthropogenic factors (Taghavi et al. 2022). To address this limitation, Arfa et al. (2025) introduced an integrated GALDIT-NUTS framework,

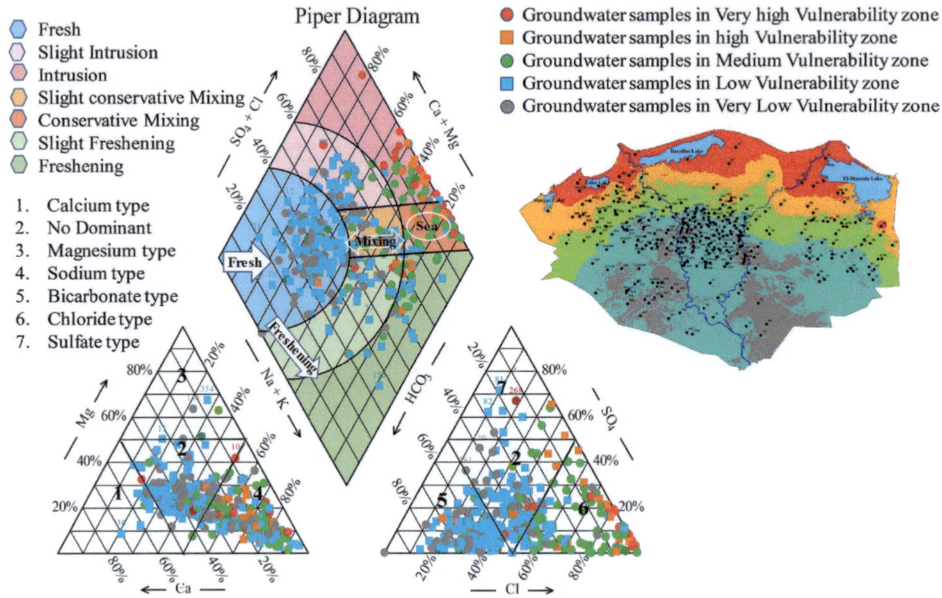

Fig. 6.13 Piper trilinear diagram illustrating the hydrochemical facies of groundwater samples, overlaid with corresponding zones on the GALDIT-XGBoost vulnerability map of the Nile Delta aquifer

combining ML models—Random Forest Regression (RFR) and GLM—with hydrogeochemical analyses to assess GwCR in the Sharqia aquifer of Egypt's eastern Nile Delta. The refined GALDIT-NUTS-RFR model effectively delineated high-vulnerability zones characterized by elevated TDS, Na^+, Cl^-, and Sr concentrations, where Na-Cl-type water indicated seawater intrusion (Fig. 6.14). Moderately vulnerable regions exhibited higher Na/Cl ratios, suggesting anthropogenic influences, whereas southern areas with Ca–HCO_3-type water displayed minimal vulnerability (Figs. 6.13 and 6.14). This study highlights the complex interactions among physiographical, hydrogeological, and anthropogenic factors in determining groundwater susceptibility to salinization (Fig. 6.15).

These studies underscore the critical need for sustainable water management strategies to mitigate the detrimental effects of human activities on groundwater resources in the Nile Delta. Future research should focus on enhancing the GALDIT and DRASTIC models by integrating additional salinization drivers, including excessive groundwater extraction (Gorgij and Moghaddam 2016), surface saline water intrusion (Kazakis et al. 2019), saltwater up-coning (Motevalli et al. 2018), and electrical conductivity (Gemail et al. 2017a). Incorporating these factors will improve groundwater vulnerability assessments related to SWI and anthropogenic influences.

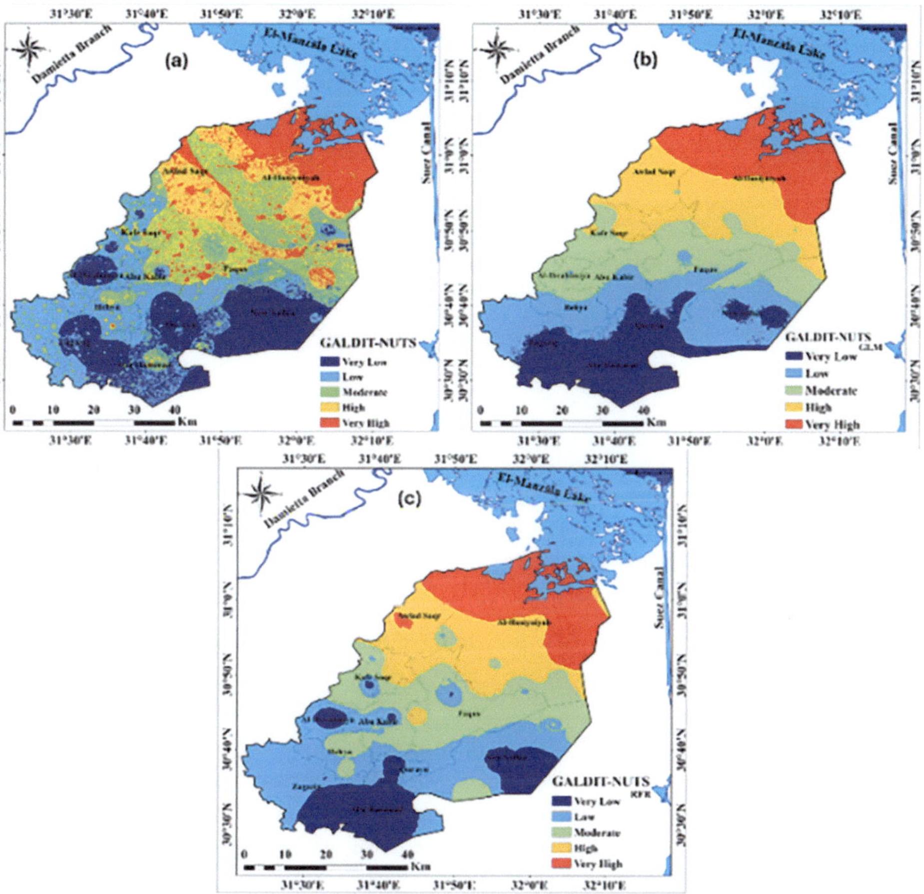

Fig. 6.14 GwCR vulnerability map in the eastern Nile Delta: **a** Basic GALDIT-NUTS framework, **b** Generalized linear models (GLM), **b** and **c** Random Forest Regression (RFR)

6.3.5 Integrated Groundwater Resources Management Roadmap

The Nile Delta is grappling with significant challenges, particularly from seawater intrusion, which severely impacts both the quality and quantity of groundwater. This updated roadmap incorporates strategies designed to mitigate groundwater degradation while addressing other pressing water resource issues. Central to this plan is the implementation of Integrated Water Resources Management (IWRM), which promotes sustainable groundwater extraction and enhanced recharge efforts. Through adaptive management practices, this roadmap aims to ensure long-term water security in the face of groundwater pollution and increasing population pressures. This holistic approach seeks to balance the

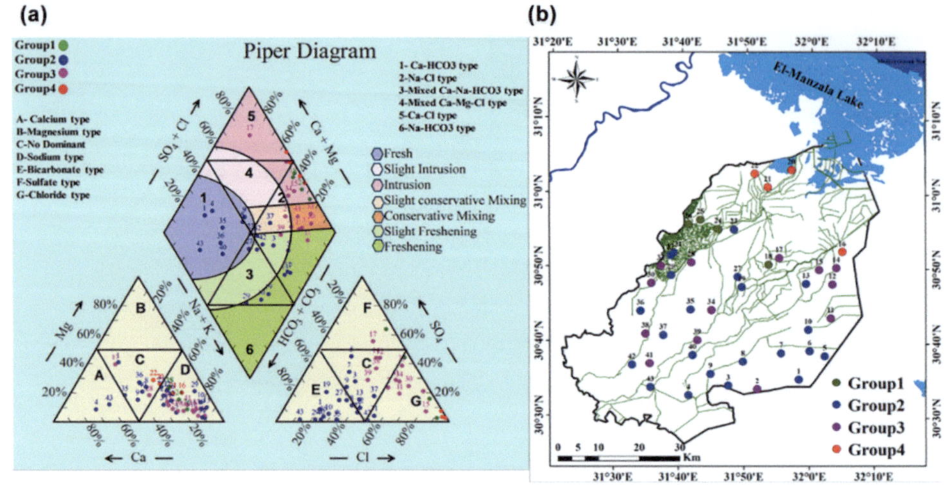

Fig. 6.15 **a** Piper trilinear diagram of the groundwater samples and **b** hydrochemical facies representing groundwater classification in the aquifer

needs of diverse stakeholders while protecting the fragile groundwater resources of the Nile Delta.

To tackle the interconnected challenges of the Nile Delta, a framework for Integrated Water Resource Management (IWRM) is proposed, engaging government authorities, agricultural representatives, and academic experts. This initiative aims to balance groundwater extraction with natural recharge rates, establishing an annual extraction limit that aligns with aquifer replenishment. A key focus of the IWRM strategy is the integration of non-conventional water resources, including treated wastewater, desalinated seawater, and recycled agricultural drainage water, to enhance water use efficiency and alleviate the pressure on overexploited aquifers. For example, treated wastewater from urban communities can significantly contribute to agricultural demand in the Delta, while small-scale desalination plants in coastal areas can serve as freshwater buffers against saline intrusion. The IWRM will utilize participatory GIS platforms to co-develop policies that map regional water needs, vulnerabilities, and opportunities for deploying non-conventional resources. By fostering collaboration among stakeholders, this strategy aims to ensure equitable water access, mitigate saltwater intrusion, and support long-term sustainability goals. Additionally, the framework will incorporate pilot projects for rainwater harvesting and wastewater treatment in peri-urban areas and solar-powered desalination units in the coastal zone, diversifying supply sources and enhancing resilience to climate variability.

To construct an Integrated Water Resources Management (IWRM) framework for groundwater in the Nile Delta, several key steps should be undertaken based on the findings in the current chapter (Fig. 6.16).

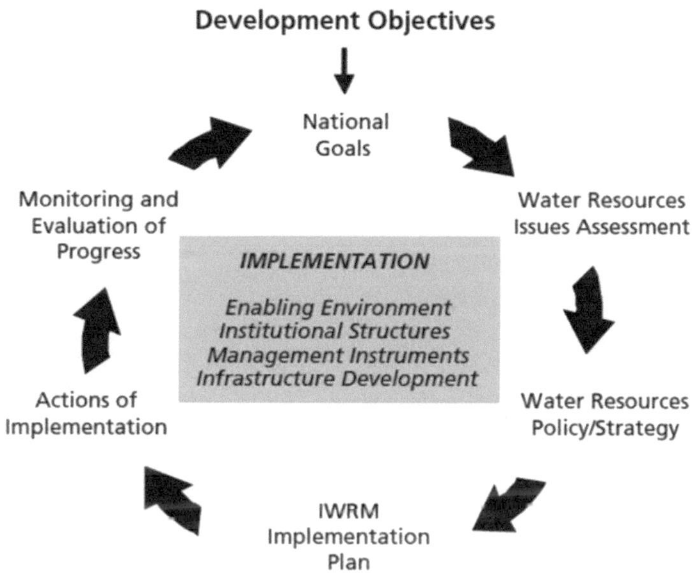

Fig. 6.16 Schematic showing the steps and objectives of IWRM development and implementation (UNEP 2012)

(1) A comprehensive assessment of the current groundwater resources, including quantity and quality, is essential to understand the existing conditions and challenges using multifilamentary hydrogeological, hydrogeophysical, and hydrogeochemical modeling with GIS-based and numerical models.
(2) Next, stakeholder engagement is crucial; this involves collaborating with local communities, government agencies, and agricultural sectors to gather input and ensure inclusive decision-making.
(3) Developing a robust data management system, such as a Geographic Information System (GIS) and Machine Learning, will facilitate effective monitoring and evaluation of groundwater levels and usage patterns.
(4) Additionally, promoting sustainable practices, such as controlled groundwater extraction and artificial recharge initiatives, will help mitigate issues like over-extraction and saltwater intrusion. Implementing educational programs to raise awareness about groundwater conservation among users is also vital.
(5) Finally, establishing clear policies and regulations that support IWRM principles will ensure long-term sustainability and equitable access to groundwater resources, ultimately enhancing the resilience of the Nile Delta against climate change impacts.

The proposed IWRM framework plays a vital role in achieving sustainable development in the Nile Delta by addressing the multifaceted challenges posed by water scarcity, pollution, and climate change. Recognizing the critical role of water in the Sustainable Development Goals (SDGs), the IWRM framework supports not only SDG 6 (Clean Water and Sanitation) but also contributes to several other goals that are interconnected with water management.

For instance, under SDG 2 (Zero Hunger), the IWRM framework promotes sustainable water use to enhance agricultural productivity and food security. By optimizing irrigation practices and integrating non-conventional water resources, the framework ensures that agricultural needs are met without depleting natural water sources. This is particularly crucial in the Nile Delta, where agricultural activities are a significant part of the economy and livelihood.

Moreover, the IWRM framework directly impacts SDG 3 (Good Health and Well-being) by facilitating access to clean water. Improved water quality and management reduce the incidence of waterborne diseases, thereby enhancing public health outcomes. Through initiatives such as wastewater treatment and community-led water testing programs, the framework contributes to ensuring safe drinking water for all.

In terms of urban development, the IWRM framework aligns with SDG 11 (Sustainable Cities and Communities) by supporting urban resilience against water-related disasters. Proper water management practices help mitigate flooding and ensure that urban populations have reliable access to water, thus fostering sustainable urban growth.

Furthermore, the IWRM framework addresses SDG 13 (Climate Action) by implementing water conservation strategies that help mitigate the impacts of climate change. By promoting practices such as rainwater harvesting and the use of treated wastewater, the framework enhances the resilience of water resources against climate variability. Additionally, this farmwork focuses on sustainable agricultural practices, water quality improvement programs, and innovative water management technologies, exemplifying effective strategies that can be replicated globally.

Despite these advancements, challenges such as climate change, increasing water demand, and governance issues persist. To overcome these obstacles, a multi-stakeholder approach is essential. Strengthening policy frameworks, enhancing investments in water conservation technologies, and fostering public–private partnerships are critical steps toward sustainable financing and effective water governance. This approach will help ensure groundwater resources sustainability, mitigate the impacts of seawater intrusion, and secure water security for future generations.

In addition to this farmwork, the Enhanced Water Resources Management Project (World Bank 2016), supported by the World Bank (2013–2016), aimed to address Egypt's critical water scarcity issues exacerbated by socio-economic demands, climate change, and pollution. Relying on the Nile River for 98% of its renewable water resources, Egypt faces challenges in managing water quality, particularly in agriculture, which consumes 85% of freshwater. The project focused on building the capacity of water sector institutions for Integrated Water Resource Management (IWRM), enhancing pollution control, and improving the ecosystem health of the Mediterranean Sea. Key achievements included the development of a groundwater strategy, significant water savings through innovative agricultural practices, training for over 4,400 individuals, and the establishment of a Geographic Information System (GIS) for better data management. Additionally, the project improved rural water quality through the installation of monitoring stations and the construction of wetland facilities to reduce pollutants. The initiative fostered better-informed water management practices, benefiting rural communities and improving health outcomes while enhancing Egypt's resilience to climate impacts.

6.4 Conclusions

The Nile Delta, a crucial ecosystem supporting the livelihoods of millions, faces significant challenges due to the dynamics of land use and land cover (LULC) changes, compounded by rapid urbanization, climate change, and the impacts of sea-level rise (SLR) and saltwater intrusion. The increasing demand for freshwater, driven by population growth and economic development, has led to extensive alterations in the landscape, including the conversion of agricultural land and wetlands into urban areas. In the northern part of the Nile Delta, these transformations have severe implications for groundwater quality and quantity, exacerbating issues such as soil salinization, coastal erosion, and the overexploitation of aquifers. In addition, the intrusion of saltwater into freshwater aquifers, driven by rising sea levels, threatens the quality of groundwater resources in the middle parts of the Delta, making them unsuitable for agricultural and drinking purposes. In the mixing zone, the saltwater intrusion not only diminishes the availability of freshwater but also accelerates soil salinization, which reduces agricultural productivity and compromises food security in the region. Additionally, the decline of natural vegetation and surface water bodies complicates the hydrological balance, further reducing groundwater recharge rates and leading to land subsidence. As the water table drops due to excessive groundwater extraction, soils become drier and more prone to compaction, resulting in decreased fertility and increased vulnerability to erosion.

As the Nile Delta continues to grapple with these interconnected challenges, it is essential to implement effective solutions to safeguard its water resources and ensure the sustainability of this vital ecosystem.

Continuous research and effective management efforts are essential to ensure the long-term viability of the Nile Delta aquifer. Robust studies should be conducted to further understand the hydrogeological dynamics of the region and devise sustainable strategies for water resource management. Additionally, measures to mitigate contamination and promote responsible water use must be implemented. Only through concerted efforts can the challenges facing the Nile Delta aquifer be overcome and its sustainability secured for future generations.

These governmental initiatives demonstrate a proactive approach toward addressing the challenges faced by the Nile Delta aquifer. However, ongoing monitoring and assessment of water levels, along with the implementation of effective management strategies, will be crucial for maintaining the aquifer's health and mitigating further decline. By adopting a multifaceted approach that combines conservation, efficient water use, and alternative water sources, the sustainability of the Nile Delta aquifer can be safeguarded for future generations.

6.5 Recommendations for Sustainable Water Management in the Nile Delta

1. **Integrated Water Resource Management (IWRM)**:

Develop and implement IWRM strategies that recognize the interconnectedness of water, land, and ecosystems. Engage stakeholders at all levels, including local communities, to promote sustainable water use and equitable access. Emphasize balancing groundwater extraction with natural recharge rates to mitigate saltwater intrusion.

2. **Regulation of Agricultural Practices**:

Promote sustainable agricultural methods, including crop rotation, and efficient irrigation techniques like drip and smart sensing irrigation. Educate stockholders about the impacts of saltwater intrusion on crop viability and strategies to minimize soil salinization.

3. **Monitoring and Assessment**:

Establish a comprehensive monitoring system to track land use, groundwater levels, and water quality. Utilize this data to inform policy decisions and adaptive management strategies. Integrate geophysical and hydrogeochemical data to enhance the reliability of assessments, particularly regarding seawater intrusion dynamics.

4. **Public Awareness and Education**:

Conduct awareness campaigns to inform communities about groundwater conservation, the impacts of urbanization, and the risks of saltwater intrusion. Implement educational programs that empower local populations to adopt sustainable practices and actively participate in groundwater management.

5. **Policy Frameworks and Incentives**:

Develop and enforce policies that incentivize sustainable land use while penalizing practices that exacerbate groundwater depletion and pollution. Consider zoning laws that restrict urban expansion into vulnerable areas and establish regulations to control groundwater extraction rates.

6. **Controlled Groundwater Pumping**:

Implement controlled groundwater pumping strategies in regions vulnerable to seawater intrusion in the mixing zone. Regulating extraction, especially in northern and central delta areas, will help maintain groundwater levels and reduce saline encroachment.

7. **Artificial Recharge Initiatives**:

Riverbank infiltration (RBF) and artificial recharge along Egypt's Kotichanr and Baher Elbaqir drains could mitigate seawater intrusion and groundwater pollution by filtering agricultural runoff via aquifer sediments and replenishing aquifers with treated water. However, large-scale implementation requires further research to assess aquifer suitability, economic viability (e.g., infrastructure costs versus reduced desalination needs), and ecosystem impacts (e.g., wetland health). Pilot projects and modeling are critical to optimize site-specific designs, address risks like clogging, and validate benefits. Collaboration between researchers, policymakers, and communities is key to scaling these solutions sustainably in the Nile Delta.

8. **Advanced Modeling Techniques**:

It is a device to utilize advanced numerical and machine learning models to simulate groundwater dynamics and seawater intrusion under various scenarios. Regularly update these models with current data to enhance predictive capabilities and inform effective management strategies.

9. **Research and Development**:

Support research initiatives aimed at understanding the dynamics of the Nile Delta aquifer, particularly concerning saltwater intrusion and groundwater salinization, and water quality degradation due to natural and anthropogenic stressors. Collaborate with academic institutions and research organizations to develop innovative solutions and best practices.

Acknowledgements This work was funded as part of a comprehensive review for the research project titled "Roadmap for Sustainable Groundwater Resources in the Nile Delta Using Integrated Approaches Under the Impact of Climate Change and Wastewater Pollution." We gratefully acknowledge the Science, Technology, and Innovation Funding Authority (STDF), Egypt, for their generous support under grant number 46209, Applied Science 1st Call, which made this study possible.

The authors also extend their heartfelt thanks to the Environmental Geophysics Research Center of Excellence team members at Zagazig University, Egypt, for their invaluable support through discussions and case studies throughout the research process. Their contributions have significantly enriched this work.

References

Abd-Elaty I, Zelenakova M (2022) Saltwater intrusion management in shallow and deep coastal aquifers for high aridity regions. J Hydrol Reg Stud 40:101026. https://doi.org/10.1016/j.ejrh.2022.101026

Abd-Elaty I, Sallam GA, Straface S, Scozzari A (2019) Effects of climate change on the design of subsurface drainage systems in coastal aquifers in arid/semi-arid regions: case study of the Nile Delta. Sci Total Environ 672:283–295

Abd-Elaty I, Javadi AA, Abd-Elhamid H (2021a) Management of saltwater intrusion in coastal aquifers using different well systems: a case study of the Nile Delta aquifer in Egypt

Abd-Elaty I, Zeleňáková M, Krajníková K, Abd-Elhamid HF (2021b) Analytical solution of saltwater intrusion in coastal aquifers considering climate changes and different boundary conditions. Water 13(7):995. https://doi.org/10.3390/w13070995

Abd-Elaty I, Abd-Elmoneem SM, Abdelaal GM, Vrána J, Vranayová Z, Abd-Elhamid HF (2022) Groundwater quality modeling and mitigation from wastewater used in irrigation: a case study of the Nile Delta aquifer in Egypt. Int J Environ Res Public Health 19(22):14929

Abd-Elaty I, Abdoulhalik A, Ahmed A (2023) The impact of future hydrology stresses and climate change on submarine groundwater discharge in arid regions: a case study of the Nile Delta aquifer, Egypt. J Hydrol Reg Stud 47:101395

Abd-Elaty I, Kuriqi A, Ramadan EM, Ahmed AA (2024) Hazards of sea level rise and dams built on the River Nile on the water budget and salinity of the Nile Delta aquifer. J Hydrol Reg Stud 51:101600

Abdel-Fattah MK, Helmy AM (2015) Assessment of water quality of wastewaters of Bahr El-Baqar, Bilbies, and El-Qalyubia drains in the east delta, Egypt for irrigation purposes. Egypt J Soil Sci 55(3):287–302

Abd-Elhamid HF, Abd-Elaty I, Ramadan EM (2016) The impact of climate change and population growth on groundwater resources in the Eastern Nile Delta Aquifer, Egypt. In: The international conference of engineering sciences and applications, Aswan, Egypt, pp 29–31

Abotalib AZ, Abdelhady AA, Heggy E, Salem SG, Ismail E, Ali A, Khalil MM (2023) Irreversible and large-scale heavy metal pollution arising from increased damming and untreated water reuse in the Nile Delta. Earth's Future 11(3):e2022EF002987

Abu Salem HA, Gemail KS, Nosair A (2021) A multidisciplinary approach for delineating wastewater flow paths in shallow groundwater aquifers: a case study in the southeastern part of the Nile Delta, Egypt. J Contam Hydrol 236:103701. https://doi.org/10.1016/j.jconhyd.2020.103701

Abu Salem HS, Gemail KS, Junakova N, Ibrahim A, Nosair AM (2022) An integrated approach for deciphering hydrogeochemical processes during seawater intrusion in coastal aquifers. Water 14(7):1165. https://doi.org/10.3390/w14071165

Ali S, Negm A (eds) (2024) Groundwater quality and geochemistry in arid and semi-arid regions. Springer

Aller L, Bennet T, Leher JH, Petty RJ, Hackett G (1987) DRASTIC: a standardized system for evaluating groundwater pollution potential using hydro-geological settings, EPA 600/2-87-035:622

Arafa NA, El-Said Salem Z, Abdeldayem AL, Ghorab MA, Moustafa YM, Soliman SA, Youssef YM et al (2024) Advancing deltaic aquifer vulnerability mapping to seawater intrusion and human impacts in eastern Nile Delta: insights from machine learning and hydrochemical perspective. Earth Syst Environ:1–26

Arafa NA, Salem ZES, Ghorab MA, Soliman SA, Abdeldayem AL, Moustafa YM, Ghazala HH (2022) Evaluation of groundwater sensitivity to pollution using GIS-based modified DRASTIC-LU model for sustainable development in the Nile Delta region. Sustainability 14(22):14699

Armanuos AM, Eldin Ibrahim MG, Negm A, Takemura J, Yoshimura C, Mahmod WE (2022) Investigation of seawater intrusion in the Nile Delta aquifer, Egypt. J Eng Res 6(1):10–28

Armanuos AM, Negm A (2018) Integrated groundwater modeling for simulation of saltwater intrusion in the Nile Delta aquifer, Egypt. In: Negm A (eds) Groundwater in the Nile Delta. The Handbook of Environmental Chemistry, vol 73. Springer, Cham. https://doi.org/10.1007/698_2017_184

Arnous MO, El-Rayes AE, Green DR (2015) Hydrosalinity and environmental land degradation assessment of the East Nile Delta region, Egypt. J Coast Conserv 19:491–513

Attwa M, Gemail KS, Eleraki M (2016) Use of salinity and resistivity measurements to study the coastal aquifer salinization in a semi-arid region: a case study in northeast Nile Delta, Egypt. Environ Earth Sci 75:1–18

Bakr N, Bahnassy MH (2019) Land use/land cover and vegetation status. In: The soils of Egypt, pp 51–67

Barber W, Carr DP (1981) Water management capabilities of the alluvial aquifer system of the Nile Valley, Upper Egypt. Technical Report No. 11, Water Master Plan, Ministry of Irrigation, Cairo, Egypt

Bennett NJ, Blythe J, Tyler S, Ban NC (2016) Communities and change in the anthropocene: understanding social-ecological vulnerability and planning adaptations to multiple interacting exposures. Reg Environ Change 16:907–926

Bertrand G, Petelet-Giraud E, Cary L, Hirata R, Montenegro S, Paiva A, Almeida C et al (2022) Delineating groundwater contamination risks in southern coastal metropoles through the implementation of geochemical and socio-environmental data in decision-tree and geographical information systems. Water Res 209:117877

CAPMAS (2000) A.R.E - Annual Bulletin of Health Services 2000, CAPMAS - capmas - Data collection and processing. Ministry of Planning. Accessed at: https://censusinfo.capmas.gov.eg/metadata-env4.2/index.php/catalog/604. Last update 10 Aug 2022

Chachadi AG, Lobo Ferreira JP, Noronha L, Choudri BS (2002) Assessing the impact of sea-level rise on salt water intrusion in coastal aquifers using the GALDIT model. Coastin Coast Policy Res Newsl 7:27–32

Civita M, De Maio M (1997) SINTACS: Un sistema parametrico per la valutazione e la cartografia delle vulnerabilità degli acquiferi all'inquinamento. Metodologia e automatizzazione

Darwish K, Smith S (2021) A comparison of Landsat-8 OLI, Sentinel-2 MSI, and PlanetScope satellite imagery for assessing coastline change in El-Alamein, Egypt. Eng Proc 10(1):23

Dasgupta S, Laplante B, Murray S, Wheeler D (2009) Climate change and the future impacts of storm-surge disasters in developing countries. Center for Global Development Working Paper 182

Dawoud AM (2004) Design of national groundwater quality monitoring network in Egypt. J Environ Monitor Assess 96:99–118

El Osta M, Hussein H, Tomas K (2018) Numerical simulation of groundwater flow and vulnerability in Wadi El-Natrun depression and vicinities, West Nile Delta, Egypt. J Geol Soc India 92:235–247

El-Fawal FM (1979) Geological and hydrogeological studies on the Quaternary aquifer in the Nile Delta. Unpublished doctoral dissertation, Faculty of Science, Cairo University, Egypt

El-Raey M, Frihy O, Nasr SM, Dewidar KH (1999) Vulnerability assessment of sea level rise over Port Said governorate, Egypt. Environ Monit Assess 56:113–128

Elshaarawy MK, Armanuos AM (2025) Predicting seawater intrusion wedge length in coastal aquifers using hybrid gradient boosting techniques. Earth Sci Inf 18(2):243

El-Zeiny A, Effat H, Mansour K, Shahin A, Elwan K (2022) Geo-environmental monitoring of coastal and land resources of Port Said Governorate, Egypt. Egypt J Remote Sens Space Sci 25(1):157–172

Eslamian S, Harooni Y, Sabzevari Y (2023) Simulation of nitrate pollution and vulnerability of groundwater resources using MODFLOW and DRASTIC models. Sci Rep 13(1):8211

Ferguson G, Gleeson T (2012) Vulnerability of coastal aquifers to groundwater use and climate change. Nat Clim Chang 2(5):342–345

Fienen MN, Arshad M (2016) The international scale of the groundwater issue. In: Jakeman AJ, Barreteau O, Hunt RJ, Rinaudo JD, Ross A (eds) Integrated groundwater management. Springer, Cham. https://doi.org/10.1007/978-3-319-23576-9_2

Fishar MR (2018) Nile Delta (Egypt). In: Finlayson CM, Milton GR, Prentice RC, Davidson NC (eds) The wetland book: II: distribution, description, and conservation, pp 1251–1260

Foster SSD (1987) Fundamental concepts in aquifer vulnerability, pollution risk, and protection strategy

Fouad SS, Heggy E, Ramah M, Abotalib AZ, Palmer EM, Jomaa S, Weilacher U (2023) Egypt's waterways conservation campaigns under growing intrinsic demand and Nile upstream damming. J Hydrol Reg Stud 50:101537

Gemail KS, Abd-Elaty I (2023) Unveiling the hidden depths: a review for understanding and managing groundwater contamination in arid regions. In: Groundwater quality and geochemistry in arid and semi-arid regions, pp 3–35

Gemail KS, El-Shishtawy AM, El-Alfy M, Ghoneim MF, Abd El-Bary MH (2011) Assessment of aquifer vulnerability to industrial wastewater using resistivity measurements: a case study, along El-Gharbyia main Drain, Nile Delta, Egypt. J Appl Geophys 75(1):140–150

Gemail KS, Ghoneimi A, Shebl S, Soliman S, El-Sayed K, Saleh AH (2024) Flow pathways in fractured basalt under challenging environmental conditions: a synergistic approach of electrical resistivity tomography and seismic refraction imaging. Environ Process 11(2):24

Gemail K, Samir A, Oelsner C, Mousa SE, Ibrahim S (2004) Study of saltwater intrusion using 1D, 2D, and 3D resistivity surveys in the coastal depressions at the eastern part of the Matruh area, Egypt. Near Surf Geophys 2(2):103–109

Gemail KS, Alfy ME, Ghoneim MF et al (2017a) Comparison of DRASTIC and DC resistivity modeling for assessing aquifer vulnerability in the central Nile Delta, Egypt. Environ Earth Sci 76:350. https://doi.org/10.1007/s12665-017-6688-4

Gemail KS, Attwa M, Eleraki M et al (2017b) Imaging of wastewater percolation in heterogeneous soil using electrical resistivity tomography (ERT): a case study at east of tenth of Ramadan City, Egypt. Environ Earth Sci 76:666. https://doi.org/10.1007/s12665-017-7013-y

Gheith H (1976) Hydrogeological studies on the Nile Delta. Unpublished doctoral dissertation, Faculty of Science, Cairo University, Egypt

Ghosh T, Loc HH, Park E, Mortreux C, Okoh R, Nesterenko M, Syvitski J et al (2023) Adaptation for changing deltas. One Earth 6(3):185–189

Gorgij AD, Moghaddam AA (2016) Vulnerability assessment of saltwater intrusion using simplified GAPDIT method: a case study of Azarshahr Plain Aquifer, East Azerbaijan, Iran. Arab J Geosci 9:1–13

Hegazy D, Abotalib AZ, El-Bastaweesy M, El-Said MA, Melegy A, Garamoon H (2020) Geo-environmental impacts of hydrogeological setting and anthropogenic activities on water quality in the quaternary aquifer southeast of the Nile Delta, Egypt. J Afr Earth Sci 172:103947

Hzami A, Heggy E, Amrouni O, Mahé G, Maanan M, Abdeljaouad S (2021) Alarming coastal vulnerability of the deltaic and sandy beaches of North Africa. Sci Rep 11(1):2320

Idowu TE, Jepkosgei C, Nyadawa M, Korowe MO, Waswa RM, Lasisi KH, Ajibade FO et al (2022) Integrated seawater intrusion and groundwater quality assessment of a coastal aquifer: GALDIT, geospatial and analytical approaches. Environ Sci Pollut Res 29(24):36699–36720

Iskander MM, El-Anssary AE, El-Mooty MA, Nagy HM (2007) Investigating the value of monitoring the implemented coastal structures along El Agami Beach, Alexandria, Egypt: case study. J Coast Res 23(6):1483–1490

Kazakis N, Busico G, Colombani N, Mastrocicco M, Pavlou A, Voudouris K (2019) GALDIT-SUSI is a modified method to account for surface water bodies in the assessment of aquifer vulnerability to seawater intrusion. J Environ Manage 235:257–265

Koussis AD, Mazi K, Destouni G (2012) Analytical single-potential, sharp-interface solutions for regional seawater intrusion in sloping unconfined coastal aquifers, with pumping and recharge. J Hydrol 416:1–11

Lu C, Xin P, Kong J, Li L, Luo J (2016) Analytical solutions of seawater intrusion in sloping confined and unconfined coastal aquifers. Water Resour Res 52(9):6989–7004

Lunetta RS, Lyon JG (2004) Remote sensing and GIS accuracy assessment. CRC Press

Ma J, Zhou Z, Guo Q, Zhu S, Dai Y, Shen Q (2019) Spatial characterization of seawater intrusion in a coastal aquifer of northeast Liaodong Bay, China. Sustainability 11(24):7013. https://doi.org/10.3390/su11247013

Mabrouk MB, Jonoski A, Solomatine D, Uhlenbrook S (2013) A review of seawater intrusion in the Nile Delta groundwater system–the basis for assessing impacts due to climate changes and water resources development. Hydrol Earth Syst Sci Discuss 10(8):10873–10911

Mabrouk M, Jonoski A, Oude Essink GH, Uhlenbrook S (2019) Assessing the fresh–saline groundwater distribution in the Nile Delta aquifer using a 3D variable-density groundwater flow model. Water 11(9):1946

Magnan AK, Oppenheimer M, Garschagen M, Buchanan MK, Duvat VK, Forbes DL, Pörtner HO et al (2022) Sea level rise risks and societal adaptation benefits in low-lying coastal areas. Sci Rep 12(1):10677

Maqsoom A, Aslam B, Khalil U, Ghorbanzadeh O, Ashraf H, Faisal Tufail R, Blaschke T et al (2020) A GIS-based DRASTIC model and an adjusted DRASTIC model (DRASTICA) for groundwater susceptibility assessment along the China-Pakistan economic corridor (CPEC) route. ISPRS Int J Geo Inf 9(5):332

Mazi K, Koussis AD, Destouni G (2014) Intensively exploited Mediterranean aquifers: resilience to seawater intrusion and proximity to critical thresholds. Hydrol Earth Syst Sci 18(5):1663–1677

Mazi K, Koussis AD, Destouni G (2013) Tipping points for seawater intrusion in coastal aquifers under rising sea level. Environ Res Lett 8:014001. https://doi.org/10.1088/1748-9326/8/1/014001

Mehrim AI, Refaey MM (2023) An overview of the implications of climate change on fish farming in Egypt. Sustainability 2023(15):1679

Mogren S, Shehata M (2021) Research on groundwater vulnerability and risk mapping of the quaternary aquifer system in the Northeastern part of the Nile Delta, Egypt. Afr J Geogr Reg Plann ISSN 8(1):001–013

Mohamed A (2020) Gravity applications to groundwater storage variations of the Nile Delta aquifer. J Appl Geophys 182:104177

Morsy KM, Mishra AK, Galal MM (2020) Water quality assessment of the Nile Delta lagoons. Air Soil Water Res 13:1178622120963072

Morsy WS (2009) Environmental management of groundwater resources for the Nile Delta region. Doctoral dissertation, Faculty of Engineering, Cairo University

Motevalli A, Moradi HR, Javadi S (2018) A comprehensive evaluation of groundwater vulnerability to saltwater up-coning and seawater intrusion in a coastal aquifer (case study: Ghaemshahr-Juybar aquifer). J Hydrol 557:753–773

Nahin KTK, Basak R, Alam R (2020) Groundwater vulnerability assessment with DRASTIC index method in the salinity-affected southwest coastal region of Bangladesh: a case study in Bagerhat Sadar, Fakirhat and Rampal. Earth Syst Environ 4:183–195

Narvaez-Montoya C, Mahlknecht J, Torres-Martínez JA, Mora A, Bertrand G (2023) Seawater intrusion pattern recognition supported by unsupervised learning: a systematic review and application. Sci Total Environ 864:160933

Negm AM, Sakr S, Abd-Elaty I, Abd-Elhamid HF (2019) An overview of groundwater resources in the Nile Delta aquifer. In: Groundwater in the Nile Delta, pp 3–44

Nicholls RJ, Cazenave A (2010) Sea-level rise and its impact on coastal zones. Science 328(5985):1517–1520

Nofal ER, Amer MA, El-Didy SM, Fekry AM (2015) Delineation and modeling of seawater intrusion into the Nile Delta aquifer: a new perspective. Water Sci 29(2):156–166

Oke SA, Fourie F (2017) Guidelines to groundwater vulnerability mapping for Sub-Saharan Africa. Groundw Sustain Dev 5:168–177

Othman AA, Rabeh SA, Fayez M, Monib M, Hegazi NA (2012) El-Salam canal is a potential project reusing the Nile Delta drainage water for Sinai desert agriculture: microbial and chemical water quality. J Adv Res 3(2):99–108

Oude Essink GH, Van Baaren ES, De Louw PG (2010) Effects of climate change on coastal groundwater systems: a modeling study in the Netherlands. Water Resour Res 46(10)

Patrick RJ, Grant K, Bharadwaj L (2019) Reclaiming indigenous planning as a pathway to local water security. Water 11(5):936

Pautz C, Santos ARD, Ferrari JL, Guerra Filho PA, Kunz SH, Dias HM, Marin Scaramussa L et al (2023) Mapping the environmental vulnerability of a lagoon using fuzzy logic and the AHP method. Water 15(11):2102

Poshtegal MK, Mirbagheri SA (2019) The heavy metals pollution index and water quality monitoring of the Zarrineh river, Iran. Environ Eng Geosci 25(2):179–188

Rajoub B (2020) Supervised and unsupervised learning. In: Biomedical signal processing and artificial intelligence in healthcare. Academic Press, pp 51–89

Rateb A, Abotalib AZ (2020) Inferencing the land subsidence in the Nile Delta using Sentinel-1 satellites and GPS between 2015 and 2019. Sci Total Environ 729:138868

Rehman A, Xue L, Islam F, Ahmed N, Qaysi S, Liu S, Alarifi N, Youssef YM, Abd-Elmaboud ME (2024) Unveiling groundwater potential in Hangu district, Pakistan: A GIS-driven bivariate modeling and remote sensing approach for achieving SDGs. Water 16(22):3317. https://doi.org/10.3390/w16223317

Research Institute for Groundwater (RIGW) (2008) Annual technical report for the year 2008. Kanater El-Khairia, Egypt

Said R (1981) The geological evolution of the river Nile. Springer-Verlag, New York

Salem EZ, Arafa NA, Abdeldayem AL, Youssef YM (2025a) Machine learning-enhanced GALDIT modeling for the Nile Delta aquifer vulnerability assessment in the Mediterranean region. Groundw Sustain Dev 28:101403

Salem ZES, Arafa NA, Abdeldayem AL, Youssef YM (2025b) Machine learning-enhanced GALDIT modelling for the Nile Delta aquifer vulnerability assessment in the Mediterranean region. Groundw Sustain Dev 28:101403

Salem ZES, Kováčiková S, Badran OM (2021) Use of various statistical techniques to assess the vertical and lateral change in the groundwater chemistry of the Quaternary aquifer in an irrigated highly populated area. J King Saud Univ Sci 33(7):101556

Salem ZE, Hasan SS (2021) Use of GALDIT model and HFE-diagram to assess seawater intrusion vulnerability in West Nile Delta, Egypt. Arab J Geosci 14(14):1318

Salem ZE, Osman OM (2017) Use of geoelectrical resistivity to delineate the seawater intrusion in the northwestern part of the Nile Delta, Egypt. In: Groundwater in the Nile Delta, pp 425–459

Salem ZE, ElNahrawy A, Attiah AM, Edokpayi JN (2022a) Vertical and spatial evaluation of the groundwater chemistry in the Central Nile Delta quaternary aquifer to assess the effects of human activities and seawater intrusion. Front Environ Sci 10:961059

Salem ZE, Elsaiedy G, ElNahrawy A (2019) Hydrogeochemistry and quality assessment of groundwater under some central Nile Delta villages, Egypt. In: Groundwater in the Nile Delta, pp 625–645

Salem ZE, Sefelnasr A, Hasan SS (2022b) As a water resources management tool, groundwater quality assessment for irrigation in the Young Alluvial Plain of Western Nile Delta, Egypt. Sustainability of groundwater in the Nile valley, Egypt. Springer International Publishing, Cham, pp 179–199

Saqr AM, Abd-Elmaboud ME (2024) Saltwater intrusion in coastal aquifers: a comprehensive review and case studies from Egypt. Qeios, CC-BY 4.0 Article. https://doi.org/10.32388/XLIS2F

Sefelnasr A, Sherif M (2014) Impacts of seawater rise on seawater intrusion in the Nile Delta aquifer, Egypt. Groundwater 52(2):264–276

Sestini G (1989) Nile Delta: a review of depositional environments and geological history. Geol Soc Lond Spec Publ 41(1):99–127

Shalaby A (2012) Assessment of urban sprawl impact on the agricultural land in the Nile Delta of Egypt using remote sensing and digital soil map. Int J Environ Sci 1(4):253–262

Sherif MM (1999) Hydrogeology and hydrochemistry of the Nile Delta aquifer, Egypt. Ph.D. Dissertation, Faculty of Science, Cairo University

Sherif MM, Singh VP (1997) Groundwater development and sustainability in the Nile Delta aquifer. Final report submitted to Binational Fulbright Commission, Egypt

Sherif MM, Singh VP, Amer AM (1988) A two-dimensional finite element model for dispersion (2D-FED) in coastal aquifers. J Hydrol 103(1–2):11–36

Sherif MM, Singh VP, Amer AM (1990) A note on saltwater intrusion in coastal aquifers. J Water Resour Manage 4:113–123

Sherif M, Sefelnasr A, Javadi A (2012) Incorporating the concept of equivalent freshwater head in successive horizontal simulations of seawater intrusion in the Nile Delta aquifer, Egypt. J Hydrol 464:186–198

Syvitski JP, Saito Y (2007) Morphodynamics of deltas under the influence of humans. Global Planet Change 57(3–4):261–282

Taghavi N, Niven RK, Paull DJ, Kramer M (2022) Groundwater vulnerability assessment: a review including new statistical and hybrid methods. Sci Total Environ 822:153486

Tahmasebi P, Kamrava S, Bai T, Sahimi M (2020) Machine learning in geo and environmental sciences: from small to large scale. Adv Water Resour 142:103619

United Nations Environmental Programme (UNEP)(2012) Status report on the application of integrated approaches to water resources management. http://www.un.org/waterforlifedecade/pdf/unwater_status_report_2012.pdf

Van Engelen J, Essink GHO, Kooi H, Bierkens MF (2018) On the origins of hypersaline groundwater in the Nile Delta aquifer. J Hydrol 560:301–317

Werner AD, Bakker M, Post VE, Vandenbohede A, Lu C, Ataie-Ashtiani B, Barry DA et al (2013) Seawater intrusion processes, investigation and management: recent advances and future challenges. Adv Water Resour 51:3–26

World Bank Group (2016) EG-enhanced water resources management project. https://projects.worldbank.org/en/projects-operations/project-detail/P118090?lang=en&tab=overview. Last update 28 Feb 2025

Youssef YM, Gemail KS, Atia HM, Mahdy M (2024) Insight into land cover dynamics and water challenges under anthropogenic and climatic changes in the eastern Nile Delta: inference from remote sensing and GIS data. Sci Total Environ 913:169690. https://doi.org/10.1016/j.scitotenv.2023.169690

Youssef YM, Gemail KS, Sugita M, AlBarqawy M, Teama MA, Koch M, Saada SA (2021) Natural and anthropogenic coastal environmental hazards: an integrated remote sensing, GIS, and geophysical-based approach. Surv Geophys:1–33

Zaghloul ZM, Taha AA, Hegab O, El Fawal F (1977) The Neogene-quaternary sedimentary basins of the Nile Delta. Egypt J Geol 21(1):1–19

Zaghloul ZM, Taha AA, Hegab O, El-Fawal FM (1979) The Plio-Pleistocene Nile Delta subenvironments, stratigraphic section, and genetic classification. Ann Geol Surv Egypt 9:282–291

Nile Delta Groundwater and Impacts of Environmental Changes

7

Zenhom E. Salem and Mohamed K. Zidan

Abstract

Egypt struggles to meet its freshwater needs due to its location in a semi-arid and arid region, where rainfall is scarce and irregular. The Nile River, its primary surface water source, faces growing challenges, further compounding the issue. Groundwater is significant in the Nile Delta, as it is the second water source in an area that is the most densely populated and humanly active in Egypt. There are many challenges to the Nile Delta and its groundwater, including climate change and the subsequent rising sea level that occupies land areas and increases groundwater salinization. The decrease in recharge sources of the Nile Delta Quaternary aquifer is considered another important challenge, as the construction of dams on the Nile River affects the flow of surface water, and the subsequent excessive pumping of groundwater negatively affects the freshwater reserve in the aquifer and stimulates the intrusion of salty seawater. For all of the above, this chapter intended to compile a comprehensive review of the Nile Delta Quaternary aquifer in terms of (1) the surface environmental factors that affect the aquifer, including climate, geomorphology, land use, and assessment of the amounts of recharge to the aquifer and the surface waterways network, (2) the Nile Delta geological nature and tectonics and the impact on the aquifer system, (3) sedimentological, hydrogeological and aquifer, and Holocene clay cap layer, petrophysical characteristics, as well as the groundwater flow system, (4) evaluation of the origin and source of groundwater using stable isotopes, (5) evaluating the quality of surface

Z. E. Salem (✉) · M. K. Zidan
Geology Department, Faculty of Science, Tanta University, Tanta, Egypt
e-mail: zenhom.salem@science.tanta.edu.eg

M. K. Zidan
e-mail: mohamed.kandeel@science.tanta.edu.eg

M. A. Khalifa and K. S. Gemail (eds.), *Geology of the Nile Deltas and Natural Resources*, Deltas of the World, https://doi.org/10.1007/978-3-031-96594-4_7

and groundwater for different uses, as well as assessing the extent of groundwater's susceptibility to pollution in different areas of the delta, (6) evaluating the impact of agricultural reclamation operations on groundwater in the east and west of the Nile Delta, and (7) evaluating the impact of seawater intrusion in the Quaternary aquifer and its spatial extension in the aquifer based on hydrogeochemical, hydrogeophysical, and GALDIT methods.

Keywords

Nile Delta groundwater • Quaternary aquifer • Groundwater quality • Seawater intrusion • Groundwater pollution • Environmental impacts on groundwater

7.1 Introduction

Around the world, more than a billion people live in coastal regions, and coastal aquifers where seawater interacts with the continental fresh groundwater are crucial water supply areas (Post 2005). According to Befus et al. (2017, 2023), groundwater flow from coastal aquifers to marine seas comprises the release of both fresh and saltwater. Due to carrying heat, nutrients, pollutants, and dissolved ions, undersea groundwater flow affects the quality of coastal waters and geochemical cycles (Sawyer et al. 2016). Sea-level rise brought on by climate change, excessive groundwater pumping, and land use changes are major threats to coastal aquifers. The landward intrusion of seawater will deteriorate the fresh groundwater quality (Werner et al. 2013).

It has been demonstrated that coastal aquifers are more vulnerable to groundwater extraction than projected sea level rise (SLR) estimated by global climate models for a variety of hydrogeological situations (Ferguson and Gleeson 2012). The majority of studies for understanding seawater intrusion mechanisms have been done on sand-tank experiments and numerical modeling experiments (Werner et al. 2013). Future work should concentrate on in-depth field research and measurement.

Egypt is situated in a semi-arid zone and faces significant challenges as a result of its limited water resources, which are primarily reflected by its fixed share of the Nile water and its general aridity (ENCID 1996). The requirement for exploitation of the existing groundwater resources to meet the water demand has increased due to attention being focused on the possibility of groundwater utilization due to the dearth of suitable surface freshwater sources (Neshat et al. 2014; Patrick et al. 2019).

The groundwater of the coastal aquifer systems is greatly relied upon by populations living along the Mediterranean coastline region of Egypt. The greatest depo-center in the Mediterranean region, the Nile Delta, is essentially a coastal plain that has been transformed by man, has stopped expanding into the Mediterranean, and is locally receding. The Nile Delta region, which is about 22,000 km^2 in size and makes up two-thirds of

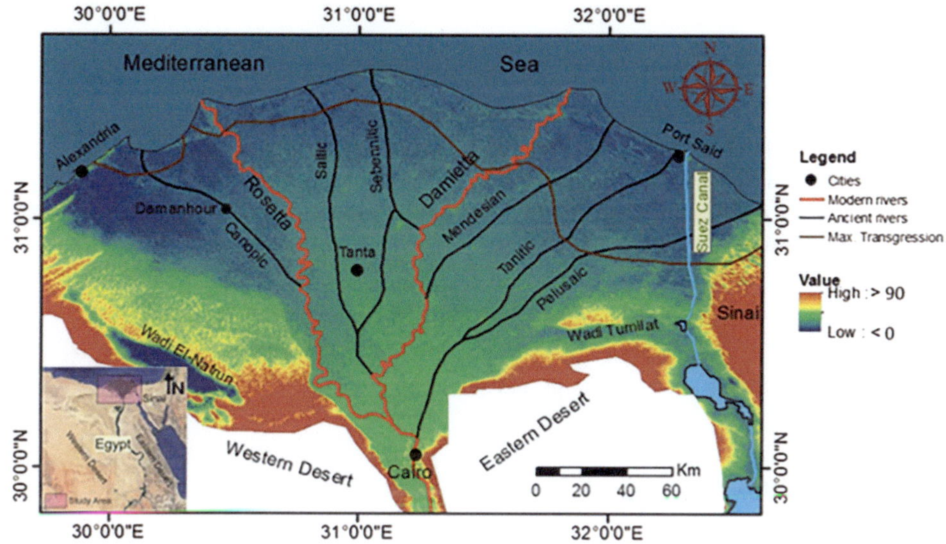

Fig. 7.1 Map shows the location and topography of the Nile Delta as well as the ancient branches of the Nile River

Egypt's agricultural land, is located 160 km north of Cairo along a 225 km long, accurate, and smooth coastline (Fig. 7.1). One of the world's largest groundwater aquifer systems, the aquifer of the Nile Delta has an estimated capacity of 500 billion m^3. Egypt, nevertheless, experiences severe water scarcity. The coastal aquifers are mostly used for agriculture, although significant amounts of groundwater are collected for domestic purposes, including drinking water.

The effects of climate change and water pollution are the main issues facing the Nile Delta. Four significant dangers caused by the climate were identified, along with their effects in Egypt. These threats of the effects of climate change were concluded by MWRI (2013a) and Negm et al. (2017). An overview of the Nile Delta groundwater resources is provided in this chapter. The Nile Delta aquifer's groundwater has been the subject of a thorough examination.

7.2 Climate

The Nile Delta is situated in an area with arid weather; on average, the delta receives only 25 mm of rain per year in the south and center of the delta and 200 mm per year in the northern section of the delta bordering the Mediterranean Sea (RIGW 1992). The delta region experiences maximum temperatures between 26 and 34 °C and lowest temperatures between 6 and 13 °C. According to SNC (2010), the daily average temperature is between

17 and 20 °C. The average maximum air temperature climbed by 0.34 °C each decade between 1961 and 2000, whereas the average lowest air temperature grew by 0.31 °C per decade (SNC 2010).

Morning and evening relative humidity averages range from 60 to 80% per year, respectively. By 2050 and 2100, respectively, the Met Office Hadley Center (2011) predicted a temperature increase over Egypt of roughly 1–1.5 °C and 3–3.5 °C, with generally high consistency between the models over the Middle East region. According to WMRI-NWRC (2002), the current evaporation rates in Egypt range from 7 mm per day in Upper Egypt to roughly 4 mm per day along the Northern Mediterranean coast. According to Eid (2001) and IPCC (2007), an increase in temperature of 1 °C may increase evapotranspiration by 4–5%, while an increase of 3 °C may increase evapotranspiration by 15%. This suggests that the aquifer recharge may be impacted by the rise in air temperature.

7.3 Geologic Setting of the Nile Delta

Over several decades, numerous researchers have extensively explored the geology of the Nile Delta basin (Hurst 1952; Said 1962, 1993; Kashef 1981a, b). The Quaternary sediments blanket the Nile Delta area. There is a very thick sedimentary sequence in the investigated region, as revealed by deep drilling. The principal geological units and lithostratigraphic cross-sections of the Nile Delta are shown in Fig. 7.2 (Schlumberger 1984; Kamel et al. 1998; Abd El-Bary 2004).

In the Nile Delta region, there are two major geological units, which are:

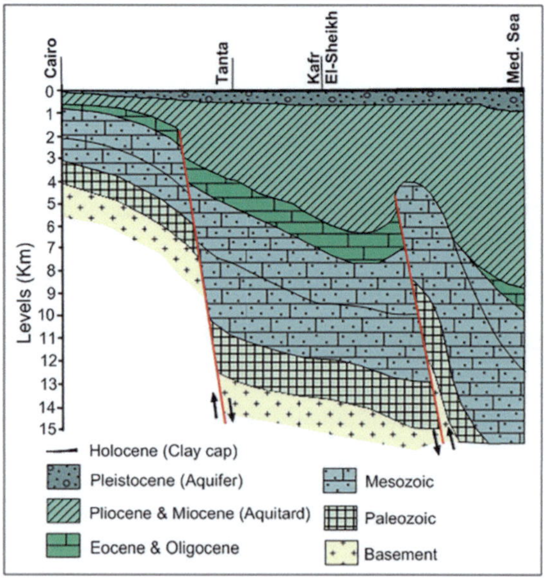

Fig. 7.2 Cross-section of the subsurface geologic succession of the Nile Delta (Said 1962; RIGW 1980)

- The Holocene and Pleistocene sediments of the Quaternary deposits. Dunes, coastal deposits, sabkha deposits, and silty clay sediments that cover the floodplain are all products of the Holocene. Desert crusts, kurkar ridges, and graded sand and gravel make up the principal water-bearing formation in the Pleistocene. The Quaternary aquifer's thickness is around 100 m at Cairo, 1,000 m along the shore, and zero meters at the eastern and western edges.
- The Pliocene, Miocene, Oligocene, Eocene, and Paleocene sediments are tertiary deposits. The primary water-bearing formation (bottom limit) is formed by the Pliocene deposits. The Miocene deposits, like the Moghra, are present below the surface and can be up to 2,000 m thick. Due to their negligible impact on groundwater, hydrogeologically, the Oligocene and Eocene deposits are only somewhat interesting.

7.4 Tectonic and Structural Settings of the Nile Delta

It is believed that the Nile Delta is located within the scope of the unstable shield, which is characterized by complex structural models affected by multiple groups of faults and layers. The Nile Delta is located within the large lower torsion zone, which was affected by several faults of different intensities (RIGW 1980). It was possible to distinguish two groups of these faults, with the first heading northwest, resulting in the large basin structure in the delta. As for the second group, it heads to the northeast. Opinions differed on how the delta was formed, and it is believed that its formation was structurally linked to the events that passed through the Gulf of Suez (Said 1962).

It was possible to distinguish 3 regions in the Delta area (Shatta and El-Fayoumey 1970), as shown in Fig. 7.3. They are distinguished as follows: The Southern Region is the one that occupies part of the upper torsion range and is characterized by the presence of a large basin structure, and this region is bordered to the north by a fault in the northeast direction. This region is characterized by the presence of fine sedimentary formations. The Central Region is located in the large lower torsion range and is bordered from the north by a large thrust fault in the northeastern direction, and represents the central delta region. The Northern Region is bordered to the north by the edge of another upper torsion range, which is characterized by the presence of several large convex folds. It becomes clear that there was a continuous struggle between the river and the sea, and due to the thick sedimentation of the Nile during the continuous floods, it was able to capture part of the land of the delta and separate it from the sea, forming lagoons whose size continued to decrease continuously with the increase in sedimentation. Therefore, the Nile River has great merit in the formation of the sediments of the river delta, which were able to stop the progress of the sea and prevent it from submerging the lands of the entire delta.

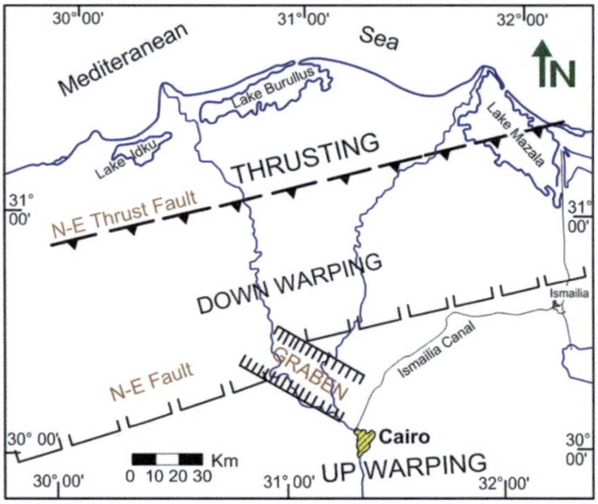

Fig. 7.3 Map shows the tectonic setting of the Nile Delta (Shatta and El-Fayoumey 1970)

7.5 Nile Delta Geomorphology

The Nile Delta is considered the ideal model for deltas in the world and has a wide base with a length of about 220 km at the Mediterranean Sea (Fig. 7.1). The length of its right branch (Damietta) is about 245 km, and its left branch (Rosetta) is about 239 km. The shape and size of the Nile Delta area are indicators of the high degree of physiographic maturity, e.g., arcuate shoreline, silting of old branches, shallow coastal lagoons, and the smoothening of the coast. Rosetta and Damietta are the only remaining branches of the river, while several others have been completely or partially silted up (Fig. 7.1). The two active branches today are the Fatimid, which corresponds to the Damietta branch, and the Al-Balbti, which aligns with the Rosetta branch. It is believed that most of the eastern estuaries are in the era of the Ptolemaids; it was removed as a result of movements in the earth's crust that caused the land to rise east of Egypt. This is also attributed to the small size of the Damietta branch compared to the Rosetta branch (RIGW 1980).

The area of the Nile Delta is characterized by low relief. Depression like Wadi El-Farigh and Wadi El-Natrun are present on the southwestern side of the Nile Delta and Wadi El-Tumylat on the southeastern side (Figs. 7.1 and 7.4). Along the northern coast, lakes of Idku, Al-Burullus, and El-Manzala occur to the north of a one-meter contour line. They are characterized by an elongated shape aligned with the direction of the coast. The ground elevation ranges about 18 m above the mean sea level (AMSL) in the south at El Qanater El Khayreya and about 5 m (AMSL) near Tanta and sloping very gently in the northward direction by the average value of 1 m/m/10 km (Saleh 1980).

Figure 7.4 depicts the Nile Delta's geomorphologic features. The area defined by the coastal lakes is occupied by the foreshore plain and its inland extension into brackish water lagoons. The sabkhas and wetland regions of the major lakes are among the landforms present on that coastal plain. The offshore plain, which was submerged during the

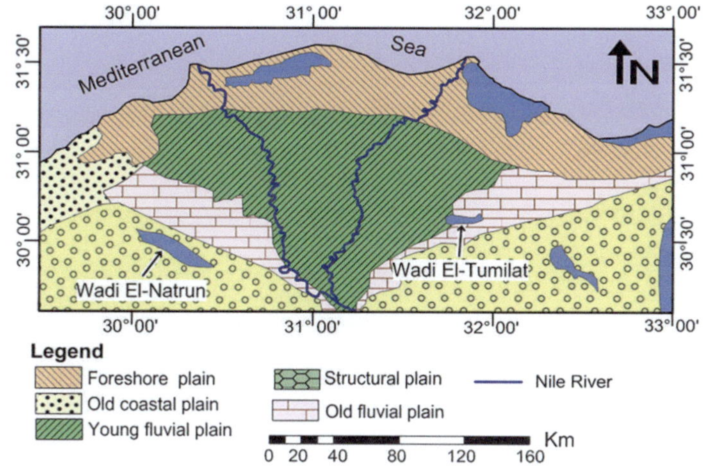

Fig. 7.4 Geomorphologic units for the Nile Delta region (Zeidan and Elshinnawy 2015)

early Holocene, may still be present in the Nile Delta's northwest corner, where it is known as the old coastal plain. The agricultural areas that border the Nile River's course and its two branches are dominated by the young alluvial plains. The majority of the Nile Delta is covered by these plains (Shata and El-Fayoumey 1970). The irrigation canals and drains break through these plains. Damietta Branch is about 3 m higher in elevation than Rosetta Branch due to the Nile Delta's east-to-west slope. The area is flat and covered by recent and Quaternary sediments that were brought from igneous and metamorphic rock at the Nile watersheds in Ethiopia and Sudan. These regions are dominated by ancient, traditionally farmed farmland.

The eastern and western sides of the Nile Delta have old alluvial plains, which have varied elevations above the young alluvial plains. This plain's gently undulating surface has well-known instances of wind-eroded landforms. A significant number of dry channels (wadis), which functioned as active rivers during the Middle Pleistocene, also cut through the surface.

7.6 Irrigation-Drainage System

The Nile River represents an outstanding feature that has shaped not only the physical characteristics of the study area but also its history and the nature of its settlers. The surface water sources on the Nile Delta are formed of complicated irrigation channels and drainage network systems that are connected with the underlying aquifer. These hydraulic features are directed to the natural discharging areas, namely, the northern lakes and the Mediterranean Sea. In the Nile River and its irrigation canals, the delta barrages and other

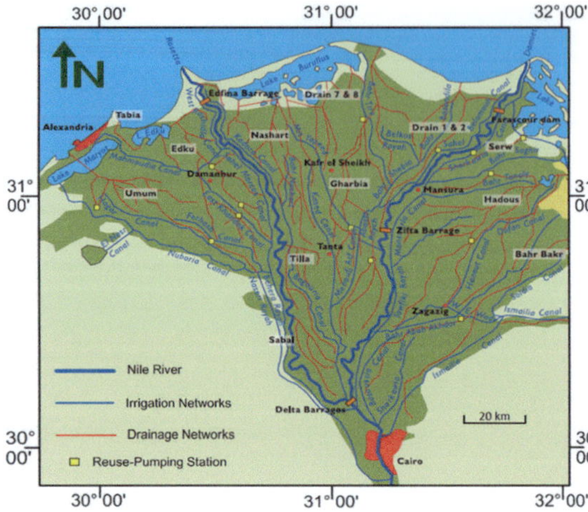

Fig. 7.5 Distribution of the Nile Delta's main canals and surface drains (Negm et al. 2018 and http://entroportal.nilebasin.org)

hydraulic control structures are used to regulate the levels of the surface water. To stop the saline water from invading from the Mediterranean Sea to the Nile River, hydraulic structures were built at the endpoints of the two branches. Figure 7.5 shows the irrigation network in the Nile Delta. The water pathway network is complicated, and a large portion of the agricultural drainage water is reused to supplement the shortage of freshwater, especially in the low reaches of the canals (MWRI 2013b). Since the construction of the Aswan High Dam (AHD), the flow rates in the irrigation system vary according to the variable water demand. This water demand does not change significantly from year to year unless new expansion projects are implemented. During the low flood periods, there is a chance of less flow in the Nile River, and hence, the water levels in all the irrigation-drainage systems are affected. Water levels in the Nile branches and irrigation channels are controlled by several barrages.

The lands of the delta were converted to perennial irrigation at the beginning of the last century. Irrigation-drainage channels in the Nile Delta area alternate with each other. The alternate drainage system stretches regionally toward the north, debouching in the Mediterranean Sea and coastal lakes, while the irrigation system begins in the south and moves generally northward. In the Nile Delta, open drains are widely installed and primarily utilized to remove surplus water used for irrigation. For improving agricultural drainage, tile drains are currently installed on the majority of Nile Delta fields. The depth of these tile drains is about 1.5 m below the land surface. The majority of the irrigation canals take their water from the Damietta branch, especially to the south due to the difference in elevation between the Damietta and Rosetta branches as a result of the general slope, e.g., El Raiah El Monoufy and Bahr Sheibin which travel northwest in the central delta near the Rosetta branch.

7.7 Land Use/Land Cover Change

The Nile Delta's various land use/land cover (LULC) variations between 1987 and 2015 are shown in the Fig. 7.6 (Elagouza et al. 2020). As shown, most classes exhibited a consistent trend, and the change in various land use/land cover classifications varied across areas of rise and reduction. The area has been detected to have ongoing growth in agricultural, urban, fish farming, and natural vegetation areas, while there is an ongoing decline in water bodies and sand areas. The agricultural area rose the most (305,296.1 ha) between 1987 and 2000, and it increased the least (170,578.1 ha) between 2000 and 2015. The interval from 2000 to 2015 showed the greatest expansion in urban areas, followed by the interval from 1987 to 2000, with mean values, respectively, 97,940.8 and 53,112.6 ha. It is revealed that most of the governorates in the Egyptian Delta have been considerably impacted by the various types of land use/land cover change brought on by agricultural operations and the dynamic effects of urbanization, especially in the eastern and western parts of the Nile Delta.

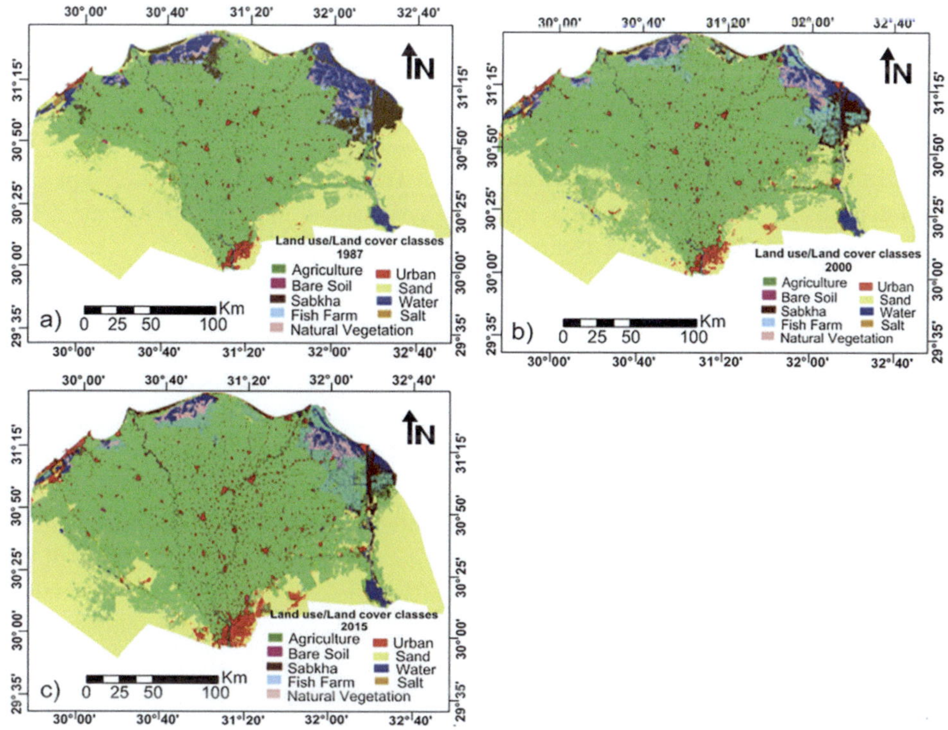

Fig. 7.6 Maps show the spatial and temporal change in land use (Elagouza et al. 2020)

7.8 Nile Delta Aquifer System

The Nile Delta groundwater is considered one of the most important aquifers in Egypt and the world. The groundwater in the reservoir is present in a geological formation called Mit Ghamr Formation, which belongs to the Quaternary period. This aquifer is covered by a layer of clay called the Bilqas Formation. In the leading sections, we will present the characteristics of each layer separately.

7.8.1 Bilqas Formation

Geologic Characteristics

The surface mud layer in the delta region acts as the upper, semi-permeable boundary of the aquifer. Composed of Nile alluvium deposits from the Holocene Epoch, this layer represents the earliest millennia of the recent era (RIGW 1980). It is irregular and heterogeneous in composition, consisting of clay, silt, and sandy mud. The thickness of this layer varies across the delta, generally increasing toward the Mediterranean coast while thinning out at the eastern and western edges. This formation plays a critical role in defining the type and degree of confinement of the groundwater-bearing layer in the delta and its adjacent areas, as well as influencing the interaction between saline seawater and the freshwater aquifer. The surface mud layer also contains the delta's surface water mass, with its upper level marking the free water surface. This water mass functions as a perched aquifer, primarily fed by leakage from irrigation systems and filtration through the delta's dense irrigation network (RIGW 1980).

The surface clay layer in the delta (Bilqas Formation) can be divided into two layers: The first is represented by the surface clay layer (Fig. 7.7), which represents the semi-permeable layer and is found with a thickness of more than 20 m toward the north with low permeability. The second layer, the clayey sand layer, is located below the first layer, and its thickness does not exceed 10 m in most of the region (Fig. 7.8), with much higher permeability than the first layer. Generally, the thickness of the mud layer reaches 70 m or more at the northern coast of the Delta.

Figures 7.9 and 7.10 show the shape of this mud cover along the northern coast (I–I′) and perpendicular to it (II–II′), respectively, and it can be seen that the thickness of this mud cover increases irregularly toward the sea. There is an important fact that imposes itself when studying the problem of the interference of seawater and the salinization of the aquifer in the northern coast of the delta, is that the seawater creeps into the layers carrying fresh water in the delta irregularly through streams that may be separate and at different depths, and therefore the interference does not occur in all vertical sector of the northern coast to the same degree (Salem et al. 2016a).

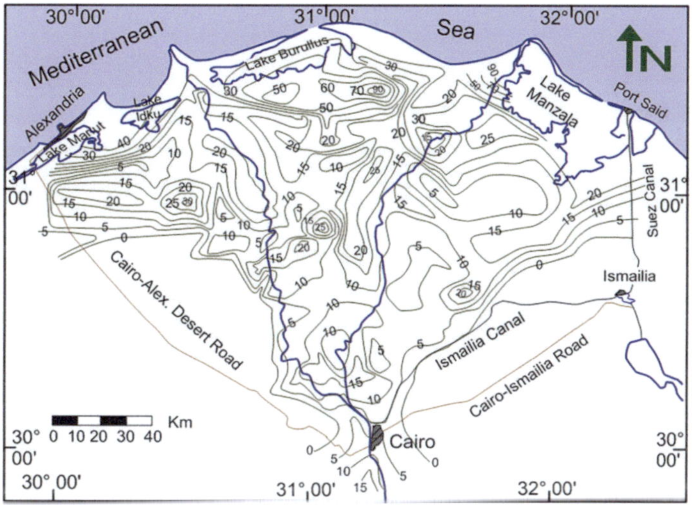

Fig. 7.7 Thickness contour map of clay cap aquitard (RIGW 1980)

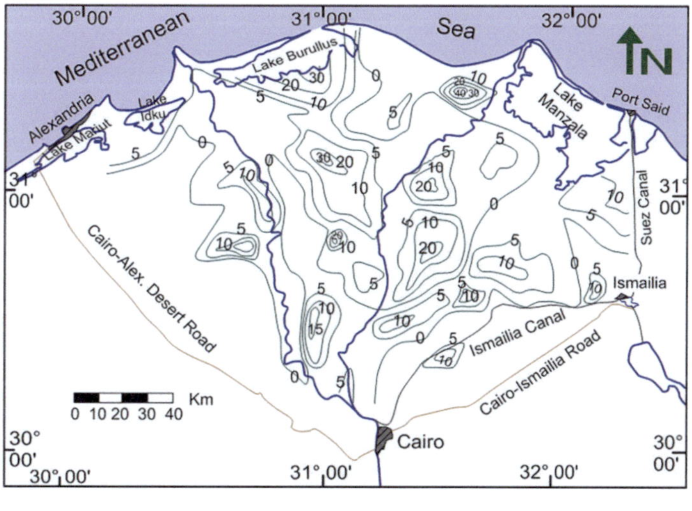

Fig. 7.8 Thickness contour map of the clay sand layer underlying clay cap aquitard (RIGW 1980)

These communication channels consist of either sand, loamy sand, or sandy clay. It should be mentioned that some of these sewers might be sea-connected but not to the aquifer, or they might be connected to the aquifer but not to the sea. This means that in some places along the northern shore, there might not be a connection between the aquifer water and seawater. The construction of the High Dam led to the elimination

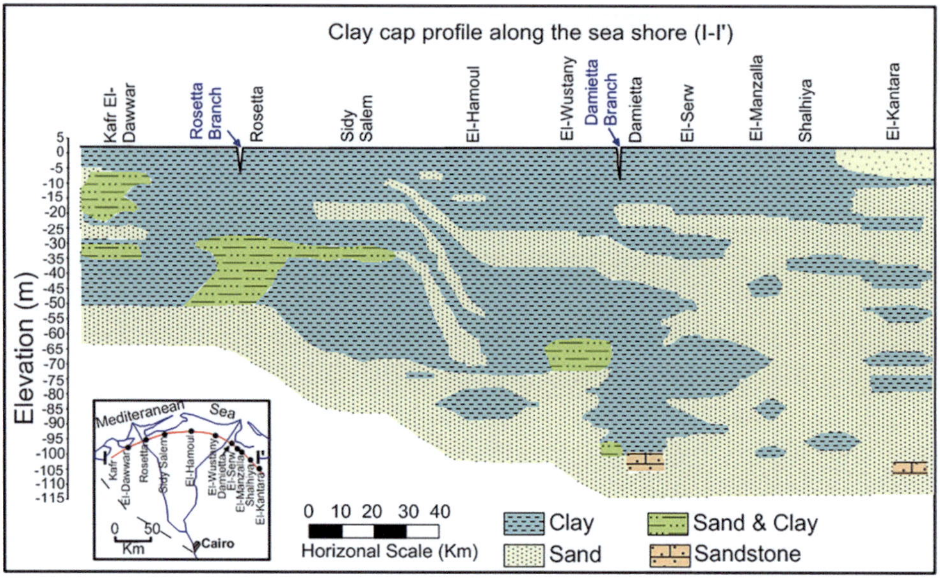

Fig. 7.9 East–West cross-section in the northern part of the Nile Delta shows the change in thickness and sedimentary facies of the Bilqas Formation (RIGW 1980)

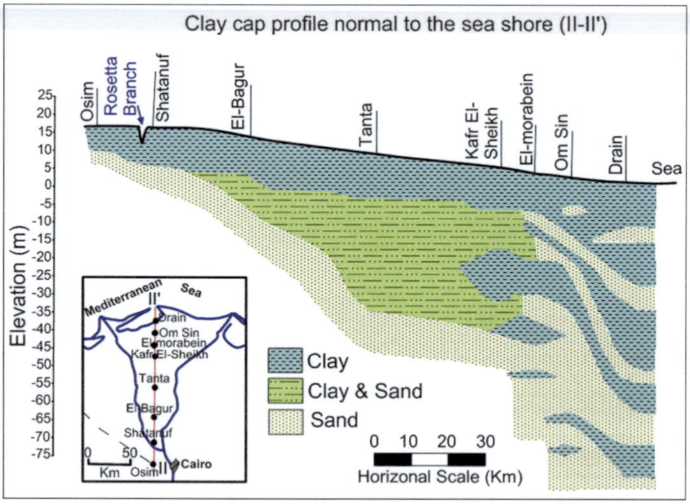

Fig. 7.10 South-North cross-section in the middle Nile Delta shows the change in thickness and sedimentary facies of the Bilqas Formation (RIGW 1980)

of floods and the sediments they carried, which were functioning to counteract or resist erosion. It is inconceivable that the erosion on the Mediterranean coast would reduce the thickness of this mud cover and increase the amount of seawater that seeps into the aquifer (RIGW 1980).

Hydrogeologic Characteristics

Figure 7.11 shows the contour lines of the vertical permeability coefficient derived from the surface muddy layer of the aquifer in the delta region. The value of this coefficient ranges between 1 and 5 mm/day, that is, with a mean average of 2.5 mm/day (RIGW 1980). In the middle of the Nile Delta, Salem et al. (2017a) calculated the petrophysical characteristics of the Bilqas Formation using geophysical logs (Fig. 7.12). They concluded that the north and northeastern directions saw an increase in thickness, porosity, and water salinity. Values for hydraulic conductivity and permeability both decline in the same way. The thickness of the Bilqas Formation varies from 3 m in the southwest to 31 m in the northeast. The total porosity ranges from 21 to 55%, and the shale content ranges from 54 to 97%. However, this layer has poor hydraulic conductivity (2×10^{-9} cm/s) and permeability (16×10^{-9} to 78×10^{-9} mD). The water in this layer has a salinity range of 200 mg to 1,600 mg/l. According to Taha et al. (2022), the areal change in the calibrated vertical hydraulic conductivity of the upper clay layer (Fig. 7.13) ranges from 0.00365 to 0.024 m/day.

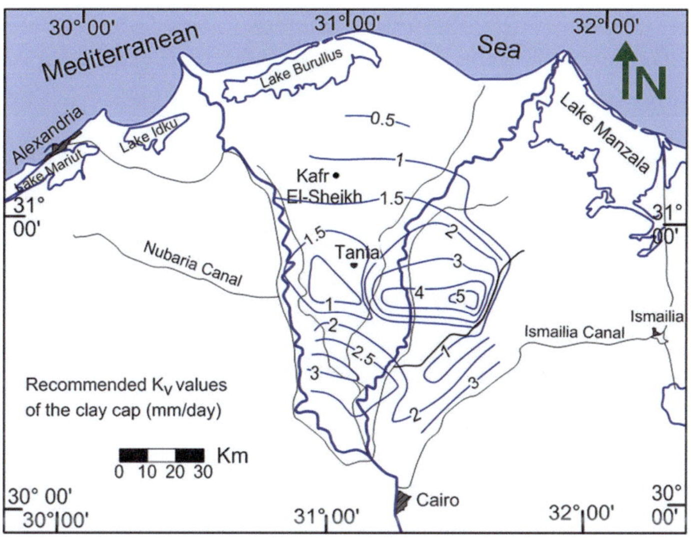

Fig. 7.11 Vertical permeability values of clay cap (RIGW 1980)

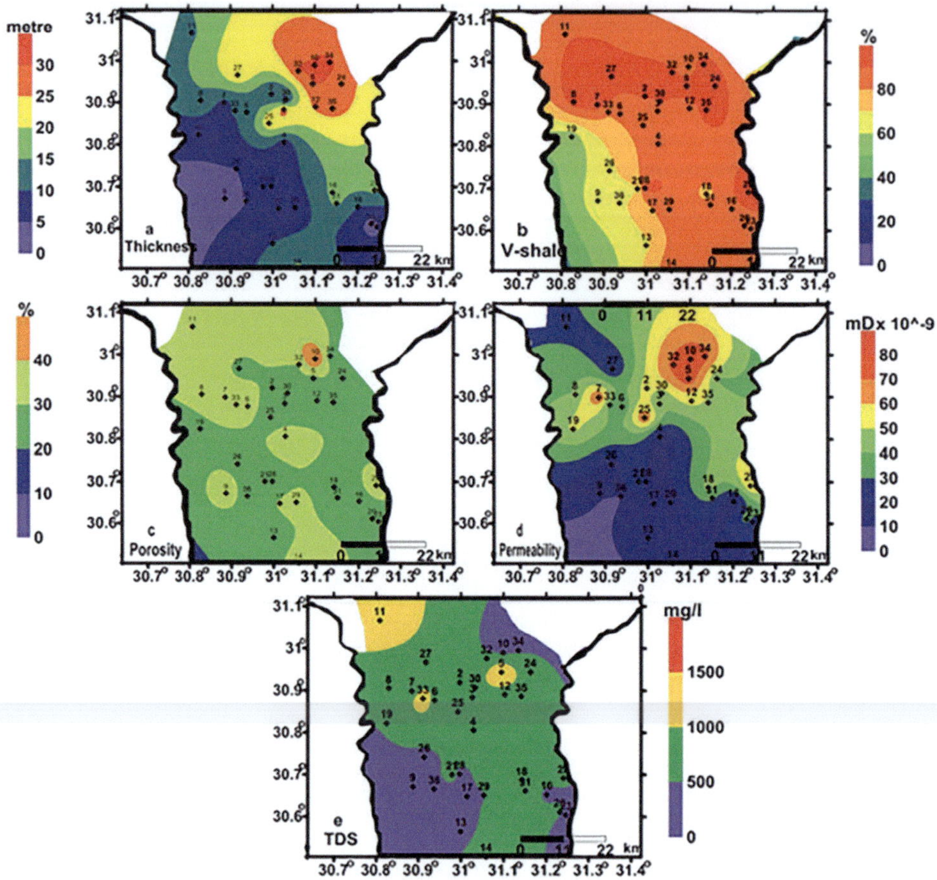

Fig. 7.12 Spatial distributions maps of average values of different petrophysical parameters for Bilqas Formation in Central Nile Delta (Salem et al. 2017a)

7.8.2 Mit Ghamr Formation (Quaternary Aquifer)

Aquifer types

The groundwater-bearing layer in the Nile Delta dates back to the Pleistocene and generally consists of permeable sand and gravel layers, sometimes interspersed with clay lenses of different thicknesses from one location to another. The upper limit of these layers is the clay cap aquitard, whose formation dates back to the Holocene. As for the lower boundary, they are marine sediments dating back to the Neogene and consist of thick layers of poorly permeable clay (RIGW 1980; Abdelhameed et al. 2017).

Thus, the Pleistocene layers in the delta are considered a semi-confined aquifer due to the semi-permeability of its surface clay cover. Figure 7.14 shows the extension of these

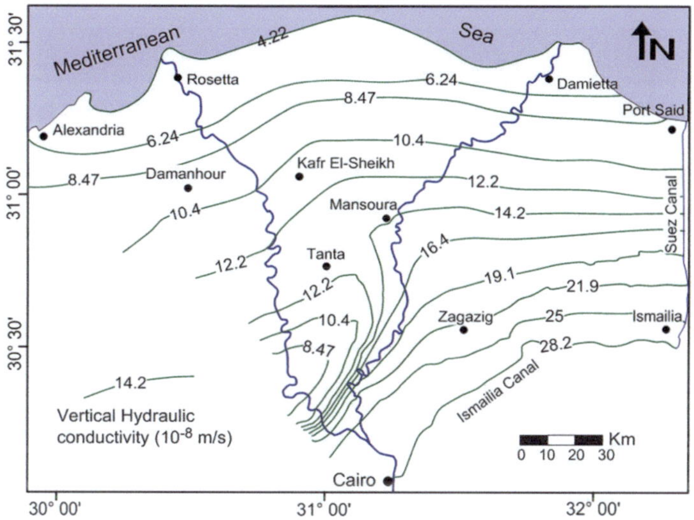

Fig. 7.13 Calibrated vertical hydraulic conductivity of the upper clay layer (Sherif et al. 2012)

layers below the area located between the two branches of the Nile, Damietta and Rosetta, and in the direction of the west until the Nubariya Canal, and in the east until the line that connects Cairo-Belbeis-Abu Hammad Abu Shalaby and Qantara. As for the rest of the delta area, the Pleistocene layers are considered to have a free aquifer (Phreatic aquifer), as there is no surface mud cover (RIGW 1980).

On the south side, the aquifer decreases in thickness due to the height of the muddy bottom and its proximity to the surface of the earth near Cairo, which suggests that there is no good hydraulic connection between the aquifers in the Delta and those in Upper Egypt (Fig. 7.15). As for the western side, these layers open up to the Western Desert, and many studies showed the decrease in thickness of these layers in this direction.

The isopach map of the Mit Ghamr Formation (Fig. 7.16) shows thickening toward the northeastern direction with a rather regular thickness gradient. The contours are not closed but open into the sea. The basin topography is rather uniform, sloping generally to the north except for a part in the northwest with a marked irregularity. The thickness of the sediments reaches a maximum value of over 900 m near the southwest Bilqas area (RIGW 1980). The average slope of the bottom of the aquifer in the delta in the north direction is about 4 m/km, which is approximately 40 times the average slope of the Earth's surface or the piezometric pressure of the reservoir water.

The general lithofacies map of Mit Ghamr Formation (Fig. 7.17) shows distinct lateral variation from sandy facies (S/Sh. ratio 11:1) covering most of the delta to sandy shale (S/Sh. Ratio 2:3) in a marginal zone to the north. The latter is termed the Baltimore Formation in the offshore wells. Between these two facies types occurs a narrow belt of

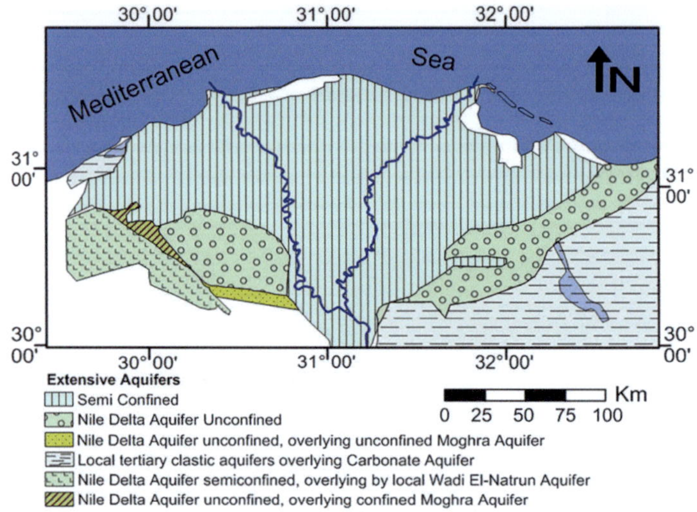

Fig. 7.14 Aquifer types of the Nile Delta region (RIGW 1980)

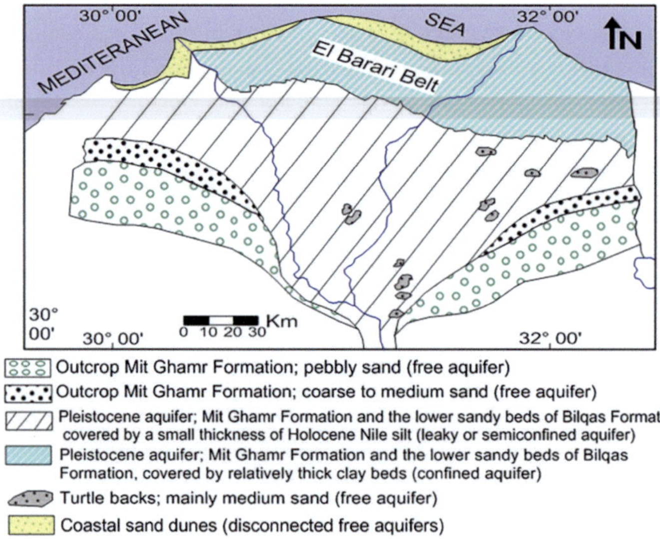

Fig. 7.15 Illustrates the Nile Delta's Pleistocene aquifer hydrogeologic boundary conditions (RIGW 1980)

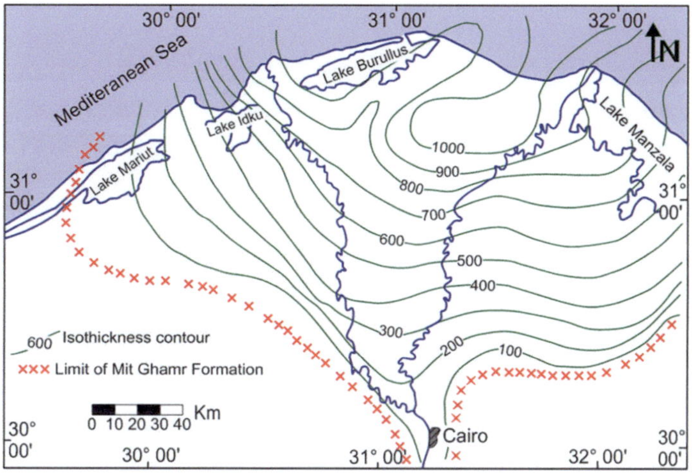

Fig. 7.16 Isopach map of Mit Ghamr Formation in the Nile Delta area (El-Fayoumey 1987)

shaly sand facies (S/Sh. Ratio 2:1) widening toward the east. The western arm of this belt almost coincides with the present Nile Delta shore (RIGW 1980).

Six longitudinal and transverse lithological sectors were drawn in the Nile Delta (RIGW 1980). Thus, it was possible to describe the geometry of the Pleistocene water-bearing layers and their surroundings, as shown in Figs. 7.18 and 7.19. Inspection of these figures may lead to the following observations:

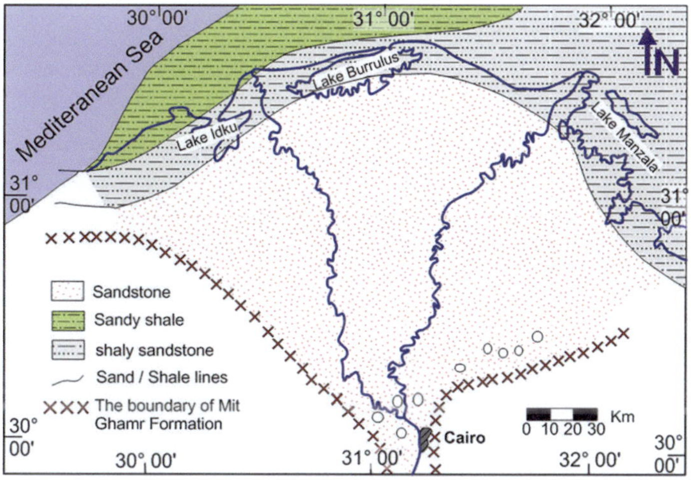

Fig. 7.17 Lithofacies map of Mit Ghamr Formation (RIGW 1980)

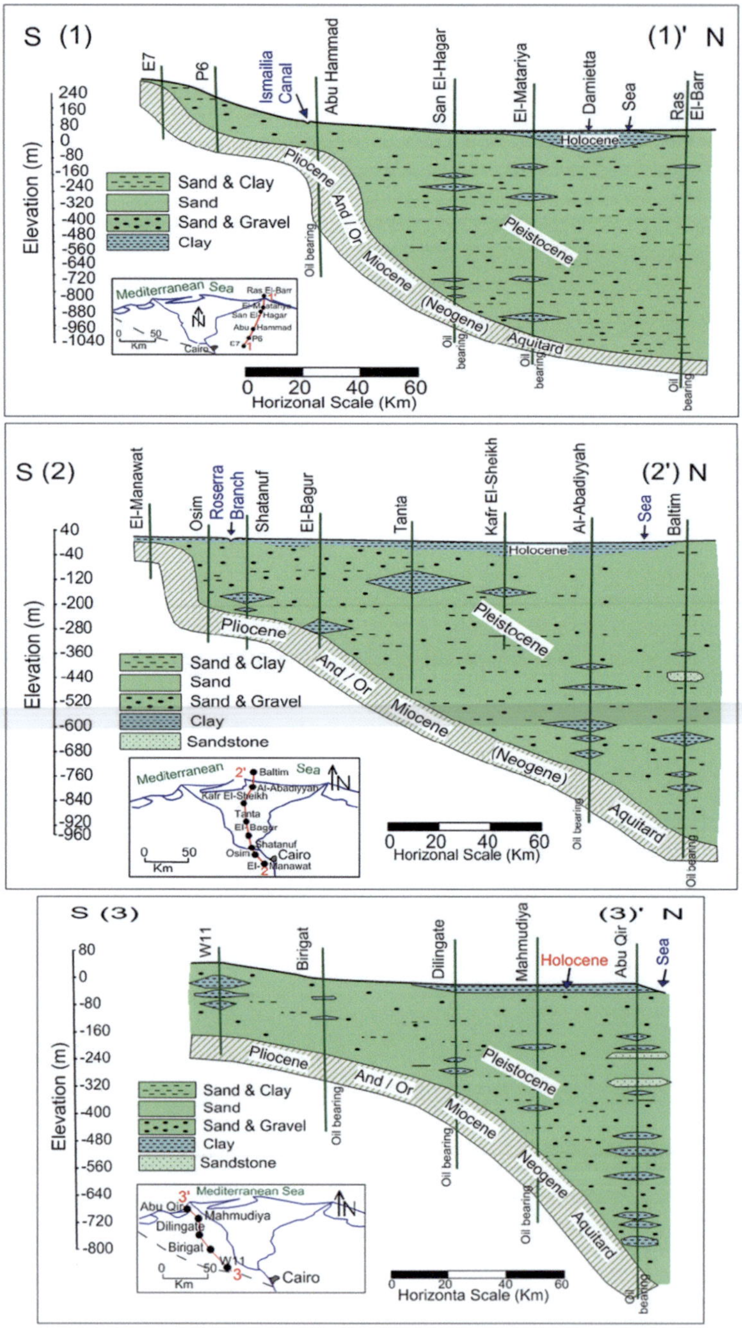

Fig. 7.18 North–South cross-Sects. 7.1, 7.2, and 7.3 show the change in lithology and thickness of Mit Ghamr Formation (RIGW 1980)

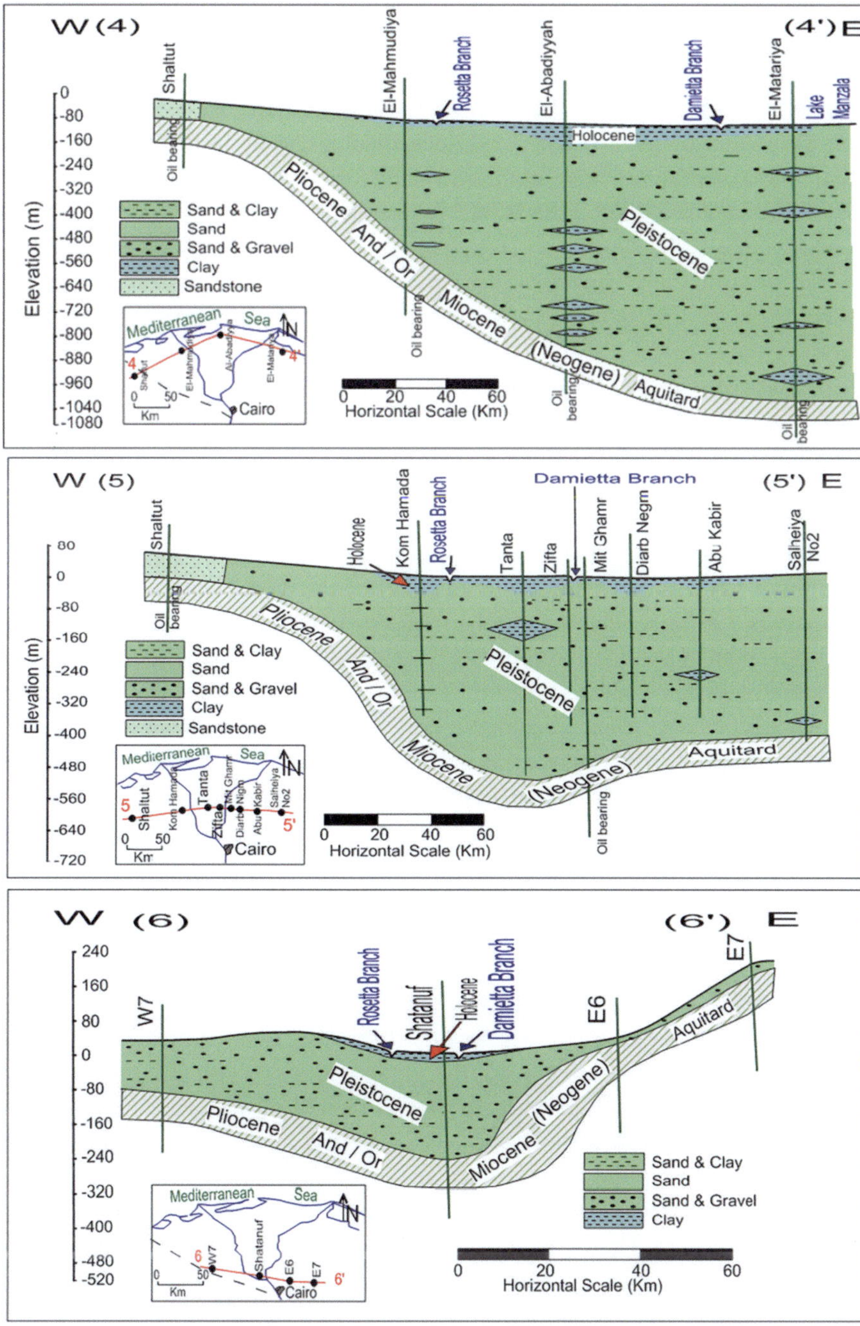

Fig. 7.19 East–West cross-Sects. 7.4, 7.5, and 7.6 show the change in lithology and thickness of Mit Ghamr Formation (RIGW 1980)

- The Pleistocene layers' thickness increases in the north in toward the Mediterranean Sea, while it decreases toward the south until it almost fades near the Manawat area, as this indicates a separation between the Pleistocene in the Delta and its counterparts in Upper Egypt (cross-Sects. 7.1, 7.2 and 7.3 in Fig. 7.18).
- Cross-Sections 7.5 and 7.6 (Fig. 7.19), located south of the Delta and extending from Tanta, reveal that the reservoir's bottom exhibits an almost concave shape, with thickness decreasing toward the east and west while retaining its greatest value in the central region of the delta. In contrast, transverse sector 4 (Fig. 7.19), situated north of the delta and starting from Tanta, indicates that the maximum reservoir thickness occurs in the eastern direction. The thickest part of the reservoir is likely found in the Al-Mataria sector.
- The decrease in the thickness of the reservoir in the southeast direction is due to the faulty structure of this region.

Hydrogeologic Characteristics

It is clear that to the south the permeability coefficient between Bagour and Cairo is less than 50 m per day, while for the middle part of the delta between Bagour and Kafr El-Sheikh, it is 100 m per day on average (RIGW 1980, Fig. 7.20). It can be seen that generally the coefficient of permeability increases in the north and east direction, while it decreases in the west and south directions. 100 m/day can be considered an average value of the permeability coefficient K, and the values 10^{-4} and 10^{-3} as an average value of the storage coefficient (S). According to Taha et al. (2022), the spatial distribution of the calibrated horizontal hydraulic conductivity of the Mit Ghamr aquifer (Fig. 7.21) ranges from 40 to 243 m/day.

7.9 Recharge-Discharge System

The Nile Delta groundwater recharge has three sources: infiltration of rainfall, excess irrigation, vertical leakage, and water seepage from the water pathways (irrigation and drains). The downward leakage toward the aquifer ranges from 0.25 to 0.80 mm/day (Morsy 2009, Fig. 7.22). The southern and central regions of the aquifer are continuously recharged by irrigation water. The water vertically moves into the clay layer in two successive steps. The first is the movement of irrigation water infiltration from the surface vertically to the free water surface. This movement occurs through the unsaturated area of the surface clay layer. The water velocity in this layer is considered as the vertical percolation velocity. The second movement is the vertical leakage of surface water, starting from its free surface to the groundwater. This movement occurs through the waterlogged zone of the surface mud layer.

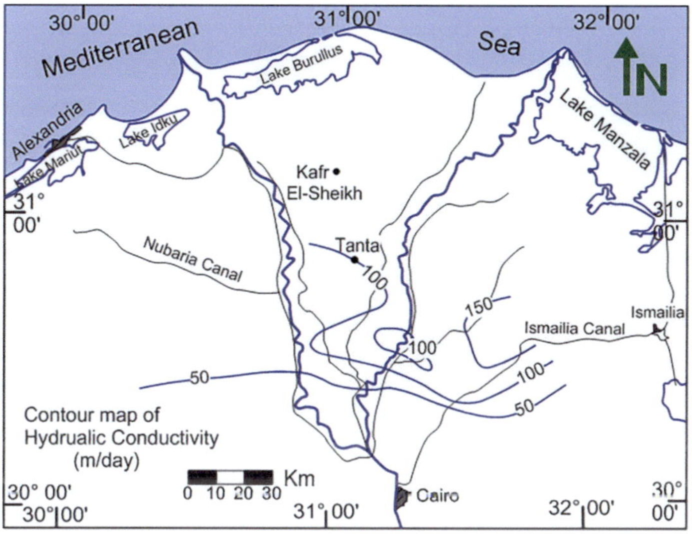

Fig. 7.20 Spatial distribution of the hydraulic conductivity of Nile Delta Aquifer (RIGW 1980)

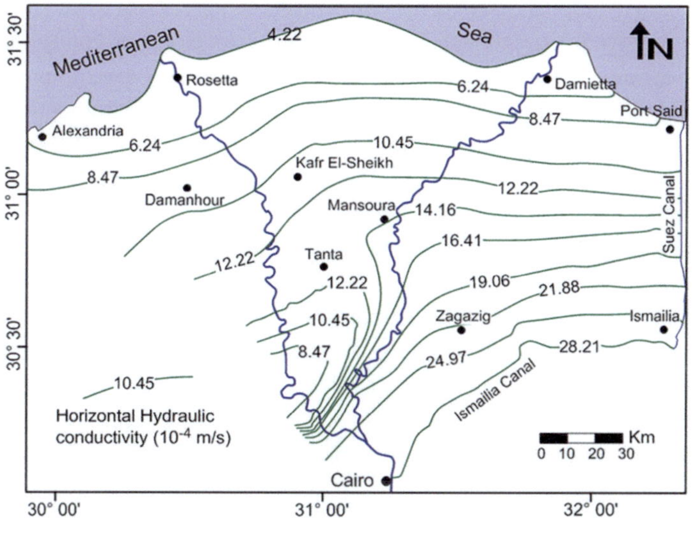

Fig. 7.21 Calibrated horizontal hydraulic conductivity of the Quaternary Nile Delta aquifer (Sherif et al. 2012)

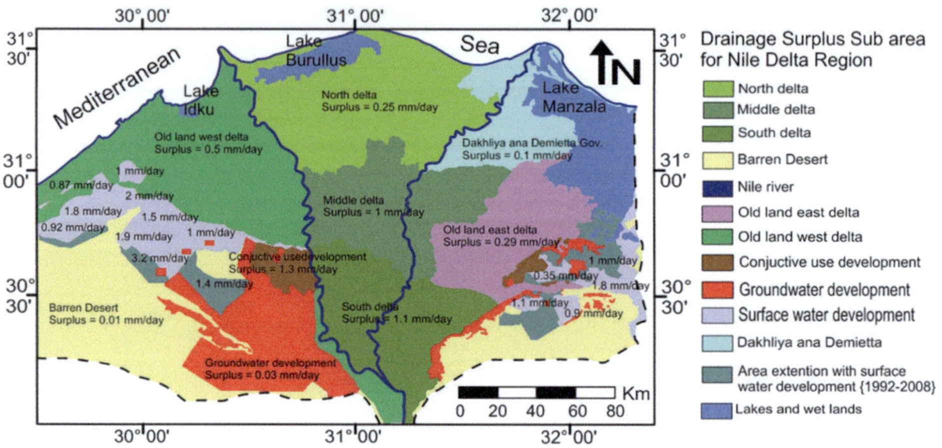

Fig. 7.22 Drainage surplus distribution in the Nile Delta during 2008 (Morsy 2009)

The WetSpass model was used to predict the groundwater recharge in the Nile Delta from rainfall for the years 1972, 1984, 1990, 2000, and 2009 (Fig. 7.23, Armanuos et al. 2016a). The model outputs showed a good degree of agreement with the amounts that were manually derived using the equation of water balance. According to the findings for 1991, the lowest and greatest recharge values were 0.0 and 304 mm, and they changed to 0.0 and 134 mm for the year 2000, respectively. The groundwater recharge results revealed that varying amounts of rainfall generate a considerable difference in recharge. It has been demonstrated that the Nile Delta's various soil classes significantly affect groundwater recharge. Due to the combination of the sand soil type and the considerable amount of rainfall, high levels of recharge were seen to the north.

The discharge of groundwater takes place by four components: outflow into the drainage system, direct evaporation, extraction, and inter-aquifer flow. In the northern region of the delta, upward leakage causes groundwater discharge to the drainage system at a daily rate of 0.2–0.9 mm. Shallow groundwater evaporation occurs in the low-lying regions. Groundwater discharges naturally in the Wadi El Natrun depression, northern lakes, and sabkhas. An insignificant aspect of discharge is inter-aquifer flow, which takes place between the Wadi El Natrun and Moghra aquifers and the Nile Delta aquifer. The groundwater flow amount from the Nile Delta Quaternary aquifer to the western borders where the Moghra aquifer is located is determined to range between 50 and 106 million m^3/year (Sakr 2005).

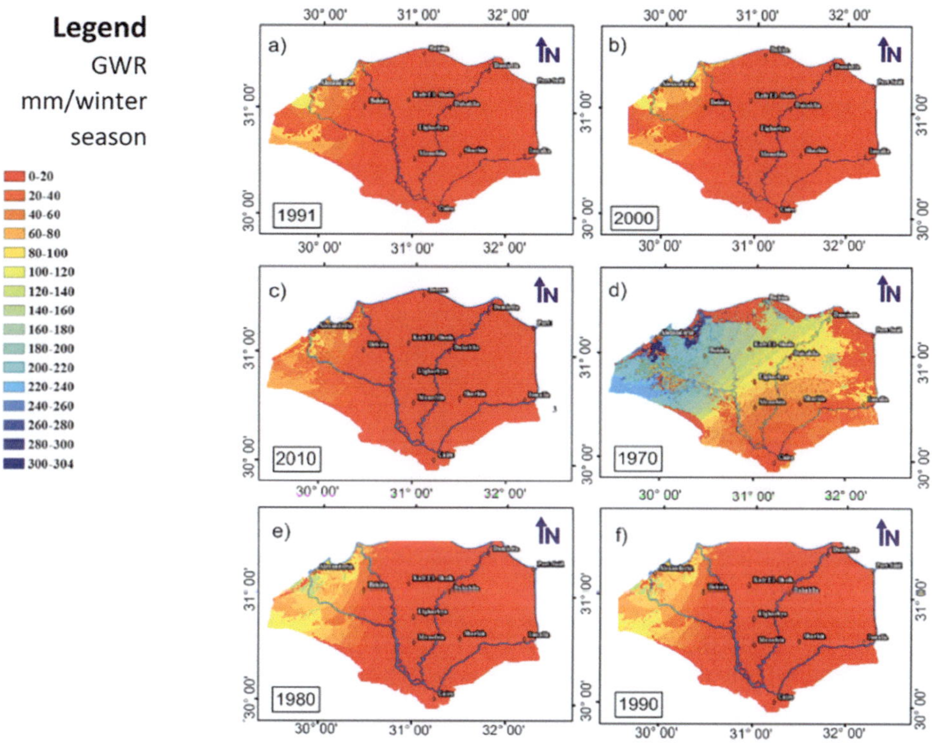

Fig. 7.23 Maps show the spatial and temporal change in groundwater recharge based on precipitation (Armanuos et al. 2016a)

7.10 Groundwater Flow System

7.10.1 Vertical Movement

There is a mutual connection in the form of vertical movement between the surface water in the semi-permeable surface reservoir and the deep water in the aquifer in the Delta. This vertical movement is either to the bottom or to the top, according to the distinction between the levels of each free surface of water in the superficial muddy layer and the piezometric pressure in the water-bearing layer. The vertical movement in the clay cap layer is downward to the south, where the free water surface is higher than the piezometric pressure, and vice versa to the north (see Fig. 7.24) and RIGW (1980). The vertical movement of water through the surface mud layer greatly affects the movement of groundwater and the drainage conditions of agricultural lands in the delta. This phenomenon also suggests that the Bilqas mud layer has a major role in controlling the water discharges to and from the aquifer as a result of this vertical movement of water.

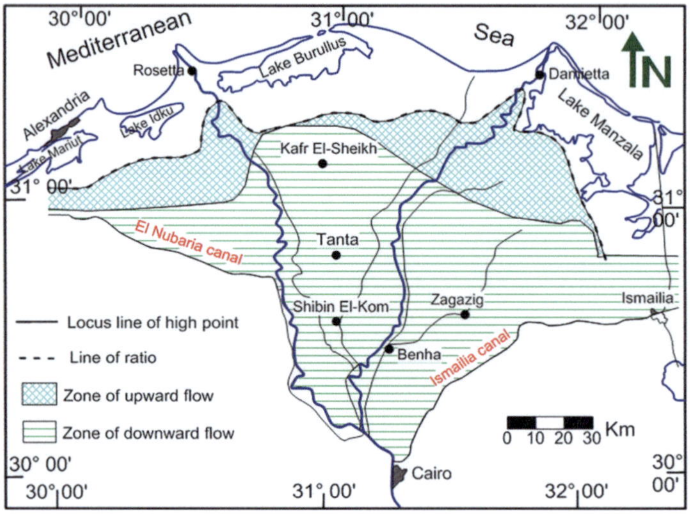

Fig. 7.24 Map shows the vertical movement of water in the upper clay layer, RIGW (1980)

Salem and others used a shallow subsurface thermal regime to trace the groundwater flow system and calculate the groundwater vertical flow velocity and flux in the Nile Delta Quaternary aquifer in several areas. Salem et al. (2008, 2018a) applied subsurface temperature to map the system of groundwater flow in the middle of the Nile Delta (Fig. 7.25) and estimated that the Tala area has a downward freshwater flux of 0.8 m/year. From south ElGharbia Governorate to south Kafr Elsheikh Governorate, fresh groundwater started to flow upward; the recorded upward fluxes from south to north were 0.1–0.5, 0.35, 0.23, 0.6, 1.2, and 2.8 m/year for the Kafelarab, Nawag, Elkarada, Kafr Mesaaed, Elhadady, and Motobes wells. The fresh water outflow uphill was explained by seawater intrusion.

The research area in the central Nile Delta had vertical downward groundwater flow velocities between 11.53 cm and 69.20 cm per year and vertical upward groundwater flow velocities between 498.26 cm and 10.65 cm per year. The upward groundwater movement in the northern section and the downward flow of freshwater in the southern part were used to estimate the seawater intrusion effect (Salem and El-Nahrawy 2020).

Salem et al. (2020) noticed the presence of shallow subsurface warm zones around the Ismailia Canal, which shows the canal's increasing influence in the area where the majority of the investigated boreholes around the Ismailia Canal had upward groundwater flow with velocities ranging from 13 to 162 cm/year. According to Salem and Osman (2016, 2018), the groundwater in the El-Nubaria area (the northwest portion of the Nile Delta) demonstrated downward flow (recharge type), with groundwater flux values ranging from 0.3 to 1.5 m/year, while the groundwater in the older agricultural land has a vertical upward flow (discharge type) with a flux range of 0.1 to 1.5 m/year. The groundwater

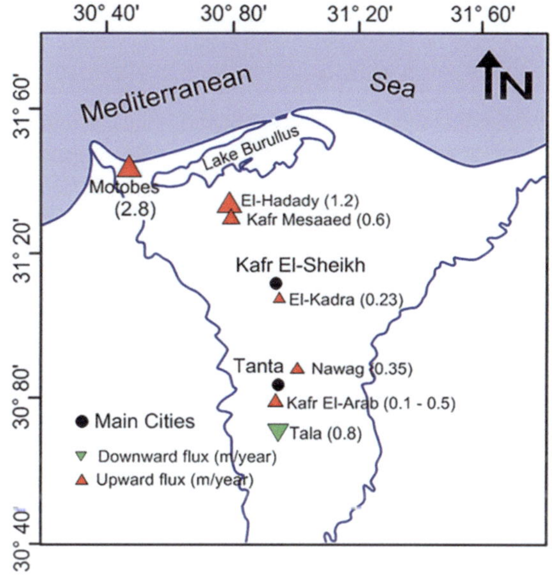

Fig. 7.25 Location map of the measured borehole temperature in the Middle part of the Nile Delta. The south-north cross-section illustrating the vertical and spatial change in temperature and groundwater flow is presented in Fig. 7.44a (Salem et al. 2008, 2018a)

flowing from the west and southwest discharges to the El-Nubaria irrigation canal, which functions as a gaining stream.

According to 2D groundwater flow and heat transport models conducted by Salem and Bayumy (2016a), the sealing faults are the main causes disrupting the regional subsurface thermal regime in the area to the east of Wadi El-Natrun. The limited permeability of the faults results in local discharge regions on their upthrown side and local recharge areas on the downthrown side in addition to the main recharge and discharge areas. The estimated average linear groundwater velocity in the recharge region is 0.9 mm/day to the eastern side and 14 mm/day to the direction of the northwest. The normal linear groundwater discharge rates in the southern region vary between 0.4 and 0.9 mm/day.

7.10.2 Horizontal Groundwater Flow System

According to RIGW (2002, Fig. 7.26), the depth of the groundwater table in the majority of the Nile Delta area is between less than 1 and 2 m in the north, 3 to 4 m in the middle, and 5 m in the south. The depth of groundwater reaches 50 m in the Nile Delta's reclaimed areas in the southeast and southwest. The groundwater level steadily drops toward the north and northeast, with an average gradient of 11 cm/km (Fig. 7.27). Near the Mediterranean coast, the groundwater head drops from 15 m in Cairo to less than one meter.

In the Wadi El Natrun Valley, groundwater is naturally discharged. An insignificant aspect of the discharge is inter-aquifer movement, which occurs among the Wadi El

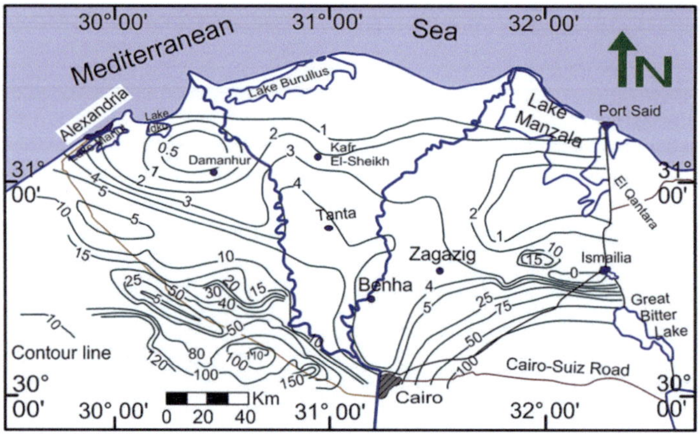

Fig. 7.26 Average depth to ground water spatial distribution (RIGW 2002)

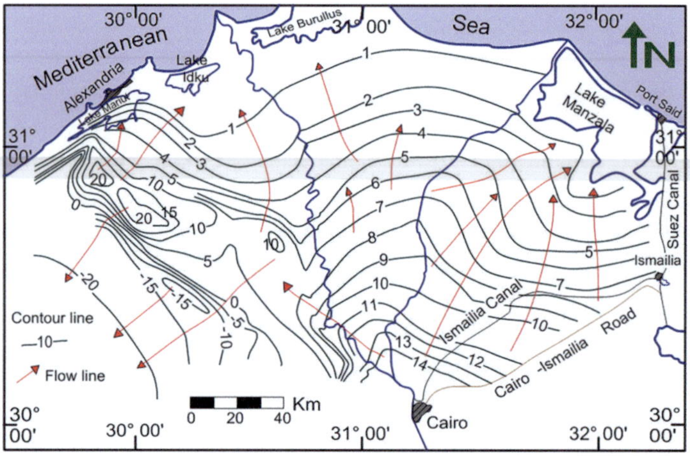

Fig. 7.27 Flow net and flow directions of Nile Delta aquifer (RIGW 2002)

Natrun, Moghra, and Nile Delta aquifers. Between 50 and 106 million m^3/year of groundwater is thought to flow from the Nile Delta Quaternary aquifer to the western borders, where the Moghra aquifer is located (Sakr 2005).

7.11 Groundwater Origin

7.11.1 West Nile Delta

Groundwater supply sustainability and environmental conservation are crucial considerations in the integrated environmental development of the Nile Basin's arid to sub-arid areas. Hussein et al. (1994) investigated the origin of groundwater in the west Nile Delta using stable isotopes. They concluded that there are two main groundwater components in the West Nile Delta. The paleo-groundwater (in the range 10–80% with means of 39% and 52% for Tahrir and Khatatbah areas, respectively) as well as the recent recharge from irrigation conveyance canals in the desert and sub-recent recharge from the delta aquifer. Along 75 km of the western strip of the Rosetta Nile branch, it is suggested that there is low northward groundwater recharge and around 0.30‰ $\delta^{18}O$ depletion per 1 km downstream. Khattbah is thought to have seen more recent recharge from the main delta aquifer than Tahrir, but the latter is thought to be now being refilled from its drainage and irrigation network and irrigated areas with increased pollution hazards for the groundwater system. Groundwater stable isotopes also proved that in the northern parts of the Behira area, the Quaternary aquifer is subjected to sea incursion.

El-Gamal et al. (2004) investigated the aquifers that feed water to new reclamation zones in the southwest of the Nile Delta by stable and radioactive isotopes. Although at a slow rate, the Nile River is the main source of groundwater recharge in this region. Most of the groundwater was recharged before the Aswan High Dam was finished in 1969 and had ages of >50 years; just a few wells that have shallow depths give recently recharged groundwater.

Using oxygen and hydrogen stable isotopes, tritium, and major ions, Geirnaert and Laeven (1992) sought to gain further understanding of the historical groundwater flow pattern and to distinguish between fossil groundwater and contemporaneous replenishment by Nile water in the Western portion of the Nile Delta. Regardless of total salinity, higher concentrations of sodium were found in the majority of studies compared to the concentration obtained from the direct mixing between seawater and Nile water. The extreme arid period between 14,000 and 10,000 years BC is thought to have given rise to the Na-rich water. During that period, rain evaporation created trona minerals (non-marine evaporite minerals) at the surface, which gradually increased the unconfined aquifer's salinity. Because the young delta floodplain was overlaid by clay, the subsequent humid phase, which occurred from 8000 to 5000 years B.P. when the sea level was higher than the current situation, groundwater freshening occurred only in the phreatic parts of the aquifer on the desert edges. Around 3500 years ago, the arid era began, and the introduction of widespread floodplain irrigation began. Groundwater tables on the desert's edges dropped as a result of climate changes and aquifer depletion from leakage to the Qattara Depression in the west. As a result, the Delta's Moghra aquifer and Quaternary underwent a freshening process. Since that time, Nile water has traveled 55 km to the west of the western floodplain border, to Wadi Natrun.

7.11.2 Middle Nile Delta

Geriesh et al. (2015) examined the impact of Nile Delta groundwater variations caused by paleoclimatic sea-level changes using water-stable isotopes. Four climate-ocean level variations during the past 22,000 years are believed to affect the Nile Delta groundwater. The sea level decreased by a maximum of 125 m during these significant sea-level fluctuations in comparison to its current level, favoring groundwater movement in a seaward direction.

Salem (2009) investigated and assessed the issue of groundwater logging at Mehallet Menouf village in the center of the Nile Delta using an integrated approach, including stable isotopes. It was indicated that groundwater logging had happened due to the surface clayey layer's low hydraulic load relative to the amount of recharged water. It was also found that the shallow groundwater was contaminated by sewage water.

The stable isotopic and hydrogeochemistry data were used to investigate the origin of the unusual presence of brackish groundwater in the Northern Nile Delta's Abu Madi shoreline region (Salem et al. 2016). It was found that the Abu Madi area's deep groundwater is predominantly freshwater mixed with a small amount of seawater. During the first and third humid Holocene climatic cycles, when the sea level was approximately 125 and 25 m below the present sea level, respectively, this groundwater was refilled. The existence of a vertical low-salinity seawater zone in front of the research area, the brackish character of higher groundwater piezometric levels, and other factors all support the potential of subsea groundwater discharge.

Salem et al. (2022a) used stable isotopes to trace the sources of groundwater recharge in the central Nile Delta. The isotopic results (Fig. 7.28) showed that surface water has a greater isotopic composition than recent Nile River water, which indicates the influence of evaporation during flow. The evaporation mechanism also affected the soil water via infiltration. The investigated groundwater is a mixture of groundwater recharged before and after the building of the High Dam. A few groundwater samples had low isotopic composition, suggesting that precipitation and Nile River water had been mixed before the construction of the High Dam. The departure from the global meteoric line (GMWL) indicates that the shallow groundwater is subject to evaporation. Deep groundwater showed more deviation from the GMWL, which is a result of freshwater and seawater mixing.

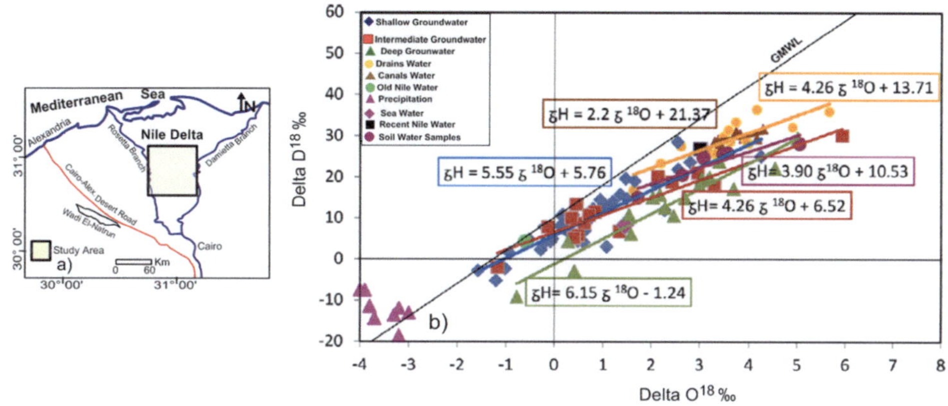

Fig. 7.28 Relationship between delta ^{18}O and delta D for the surface and Groundwater in the central Nile Delta, Salem et al. (2022c)

7.11.3 East Nile Delta

In East Nile Delta, Al-gamal et al. (2018) suggested that the most common recharging mechanisms for the Pleistocene aquifer, including Nile water leakage, excess water from irrigation, waste water seepage, and old Nile water (before High Dam construction), are thought to be explained by mixing from four poles of waters. Several hydrochemical and environmental evidence allowed for the mixing and dissolving processes to be proven.

Hamza et al. (1997) identified four distinct groupings of groundwater samples in the East Nile Delta based on water-stable isotopes. With average values of 17.6‰ (δD) and 1.7‰ (δ^{18}O), the first group is attributed to varied mixing between old Nile water and current Nile water. δD and δ^{18}O concentrations in the second group vary from 26.3‰ to 39.2‰ and 3.49‰ to.57‰, respectively. Concerning the current Nile water, this group is the richest. The third group of samples is situated geographically close to the Mediterranean Sea and Lake Manzala, and the isotopic composition of this group shows a mixing of irrigation water return and seawater intrusion. The stable isotope content in the fourth group is depleted due to the mixing of the rainwater and old Nile water in the northern area. According to the contribution of each recharge source, the stable isotopic compositions of groundwater for these aquifers are widely varied. It has been previously discussed that paleo-water may have contributed to some localized locations.

To identify the sources of groundwater contamination adjacent to the Bahr El-Baqar outflow in the eastern Nile Delta, Eissa et al. (2019) used hydrochemical and stable isotopes as markers. According to the relationship between δ^{18}O and δD, the groundwater's primary sources in the targeted area are direct surface water infiltration and the mixing of groundwater with surface water (irrigation canals, agricultural drains, and the Bahr-El Baqar sewage drain). The groundwater samples' ^{15}N isotopic composition shows that the region is polluted by sewage and fertilizer substances, including the Bahr El-Baqar sewage drain.

Hydrogeochemistry and stable isotopes were used by Gomaah et al. (2016) to investigate the groundwater beneath the East Nile Delta's Ismailia and El Kassara canals. There are three sources of water and dissolved minerals in the aquifer that were looked into: recently recharged Nile water enriched in heavy isotopes; old recharged Nile water, which is characterized by depleted isotopic composition; and a small contribution from the deeper Miocene aquifer due to severe withdrawing, which is richest water in ^{34}S and ^{18}O coming from the sulfate.

7.12 Water Quality

7.12.1 Surface Water Quality

The main canals (Fig. 7.5) that collect water from the Nile River at Cairo are used to distribute irrigation water among places all over the Nile Delta. This main waterway split into a network of auxiliary canals as it moved downstream. The issue of how much water should be delivered to each channel is decided according to the anticipated water demands for various uses, including farming, businesses, and so on. In the Nile Delta, 22 catchment areas make up the drainage system (Fig. 7.29), where the drainage water salinity increases northward. To increase the availability of freshwater, effluents are either fed into irrigation canals at 21 locations along the main channels, depending on their quality, or dumped into the northern lakes (DRI 1998).

Abukila (2015) evaluated twelve years (2002–2013) of pollution abatement activities in all drain estuaries in the Nile Delta that discharge to the Egyptian Northern Lakes and the Mediterranean Sea. Four drains (Bahr El-Baqar, Drain No. 1, Tabia, and No. 11) had significant increases in their water quality index (WQI) scores, one drain (Ramsis Drain) had a significant decline in its WQI score over the tested period, and Burullus Drain had insufficient data to analyze for trends. The remaining 14 drains, however, displayed negligible trends.

The water quality in the Damietta Nile Branch was evaluated by Badr et al. (2013). Data gathered showed that the Nile in the Damietta district had a higher concentration of BOD, COD, ammonia, nitrite, nitrate, and phosphate than in the Dakahlia district. Fe, Mn, and Pb, which are classified as heavy metals, were found to be higher than allowed by Egyptian and WHO guidelines. As a result, the water suitability showed that concentrations of contaminants in the Nile water were connected with the downstream away from the origin of the river, with water in the upstream portions of acceptable quality but being relatively contaminated toward the estuary.

Abdo et al. (2022) investigated the water quality along the Rossetta Nile Branch and found that, at stations exposed to wastewater released from drains into the water branch,

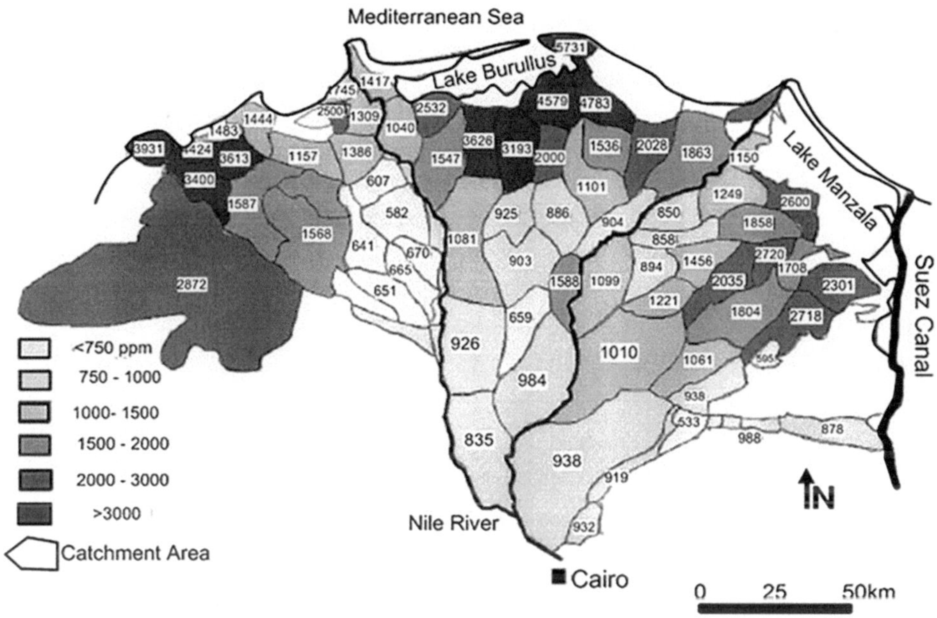

Fig. 7.29 A 1992/1993 drainage water average salinity (mg/l) in the Nile Delta, DRI (1994) and Salem et al. (2018b)

the mean annual averages of the Water Quality Index (WQI) values ranged for irrigation, from good to bad, and for aquatic life, from poor to extremely poor. The obtained results showed that the analyzed variables' values were greater than the advised levels of WHO (2008), FAO (1994), and Canadian Water Quality Guidelines, which have an impact on the local environment and cause health issues. Physicochemical parameter concentrations were higher in the winter than in the summer. El Bouraie et al. (2011) concluded that considerable transport of NH_3, TDS, and low COD, BOD, and DO values by the El-Rahawy drain significantly affected the water suitability in the downstream of the Rosetta branch, particularly during the low flow conditions in winter.

In the Middle Nile Delta, Salem et al. (2018b) examined the spatiotemporal hydrochemical properties and water quality of some canals and drains. They observed a general increase in salinity and the principal ions along the channel flow from south to north. The salinity of streams is higher in the winter (270–590 mg/l with an average of 370 mg/l) than it is in the summer (180–200 mg/l with an average of 190 mg/l), when the water level in the streams is high due to the high input into the stream. In drains, the salinity of the water exceeds 3190 mg/l in the winter (average: 1484) and 1880 mg/l in the summer (average: 900 mg/l). The appropriateness of drain water for irrigation is moderate to low. P, N, Mn, Zn, and Fe levels in the drain water used for the irrigation of fish farms and other reasons were higher than allowed.

According to Al-Agha et al. (2015), the electrical conductivity of drainage water in the central Delta's boundary drain typically ranges between 2000 and 4000 μS/cm. Lower values are observed during the summer when significant amounts of relatively good-quality drainage water from the Delta's rice fields accumulate in the system. In contrast, higher values occur in the winter, as saltier groundwater is intercepted by drains and canals during off-periods in the northern region. Although drain salinity is fortunately at it's lowest during the summer—when water demand and drain abstraction peak—the salinity levels remain harmful to agricultural productivity. Farmers using water with conductivity exceeding 2000 μS/cm are likely to experience a noticeable decline in crop yields.

In the northern region of the Nile Delta, Gad et al. (2020) looked into the water suitability of the surface water network. According to the drinking water guidelines, 67% of the tested samples showed poor to unfit water for human consumption, whereas 33% of the samples showed clean water. The pollution index results showed that surface water samples were significantly impacted by Pb and Mn, while Fe and Cr had a less significant impact.

The suitability of water in the main canals (El-Nubaria, El Mahmoudia, El Hager, El Nassr, El Khandak, and Abo Diab canals) in the Western side of the Nile Delta was evaluated by Agrama and El-Sayed (2013). The quality of water in the examined channels is evaluated as marginal for of them; accordingly, the water quality index averages for various places at each canal. This is explained by the fact that freshwater from canals in the investigated region is mixed with low-quality water from agricultural drains. Therefore, it is suggested that the quantity of water utilized in the western delta be limited to what is necessary to maintain its quality or to enhance it by discharging more fresh water or lowering the water quantity mixed with drainage.

The Palmer index indicates that organic pollution affects the Nile Delta's surface water (Ghobara and Salem 2017; Salem et al. 2017b). The organic pollution was lesser in winter than it was in summer because of the shallow water and little water that either canals or drains received. The Palmer index rose from south to north along the waterways. Elshobary et al. (2020) found that most sampling sites had high relative Palmer's Index scores, ranging from 13 to 21, especially during the summer. This finding is consistent with the Shannon-Weiner index findings, which show that the Ismailia Canal is contaminated with organic matter in most parts along the flow direction.

7.12.2 Groundwater Salinity Distribution

Several authors have indicated the presence of highly saline water in the north of the Nile Delta at 45000 mg/l (Farid 1985) and 120,000 mg/l (Saleh 1980), which could be related to the subsurface intrusion of seawater (Leaven 1991). Contour maps of groundwater salinity for the years 1960 to 2000 were created by RIGW in 2002 (Fig. 7.30). It was estimated that the recent human and natural activities impacted the salinity of the

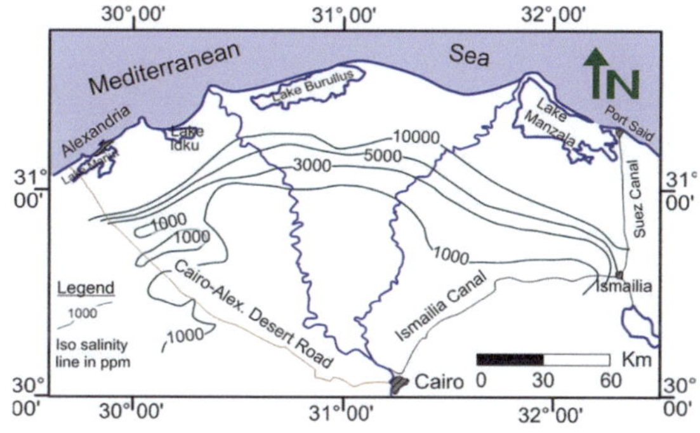

Fig. 7.30 Groundwater salinity, RIGW (2002)

groundwater in the Nile Delta. The contour line representing 1000 ppm salinity migrated southward, suggesting greater saline water incursion, which had the evident impact of raising groundwater salinity after 1980. As groundwater salinity is the main factor controlling the use of groundwater for both drinking and irrigation, the groundwater located to south of the isolines 1000 mg/l and 3000 mg/l (Fig. 7.30) is suitable to permissible portion of Nile Delta groundwater for domestic and agriculture purposes, respectively.

7.12.3 Freshwater Mass in the Nile Delta Aquifer

An attempt was made to determine the thickness of the freshwater layer in the reservoir to calculate its volume. The volume of dead storage of fresh groundwater in the aquifers of the Nile Delta was estimated at 500 billion cubic meters based on the assumption of the average porosity of the water-bearing formations of 30% (RIGW 1980).

From Fig. 7.31, it is clear that the mass of fresh water in the aquifer in the delta takes the form of a large convex lens. The largest thickness of it is situated to the south of the delta at Al-Bajour, while its thickness decreases in all directions. To the north and the east, respectively, the thickness decreases due to the advance of the salty sea waterfront from the Mediterranean Sea and the Suez Canal into the freshwater delta below. In the west direction, the fresh water thickness decreases due to the decrease in the thickness of the Pleistocene layers carrying fresh water in this direction.

7.12.4 General Groundwater Quality

Based on several studies (e.g., El Ramly 1997; Taha et al. 2003; Khalil et al. 2013; Armanuos el al. 2016b; Negm and Armanuos 2016; El-Fakharany et al. 2017; Eltarabily and Negm 2017; Embaby et al. 2017; Abu El Ella 2018; Eltarabily and Negm 2018; Hegazy et al. 2018; Eltarabily et al. 2018; Salem et al. 2017c, d, 2018b, c, 2021a, 2022, 2025a, b), the groundwater located south to the line 1000 mg/l in Fig. 7.30 and line zero in

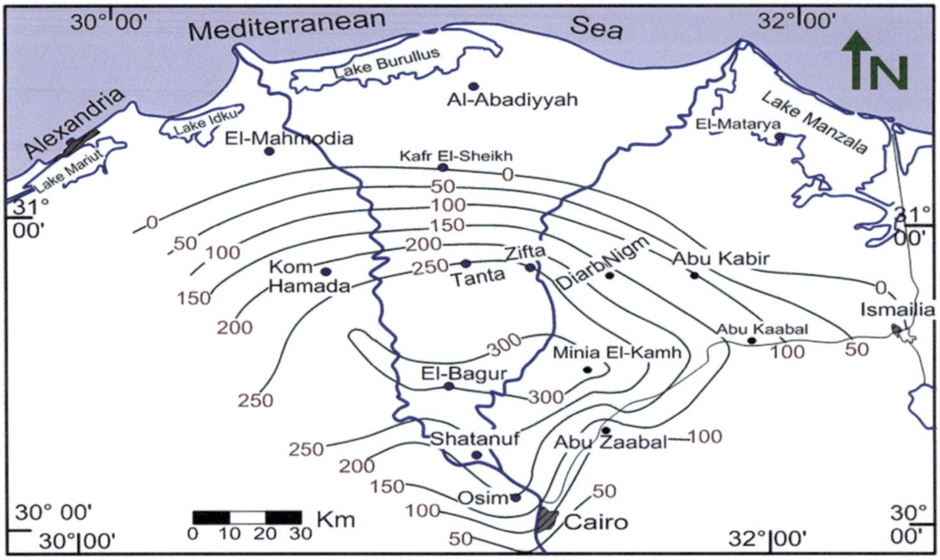

Fig. 7.31 Contour map of freshwater thickness in meters, RIGW (1980)

Fig. 7.31, in the southern and central regions of the Nile Delta is suitable to permissible for drinking purposes. Still, several localities have high unacceptable Fe and Mn concentrations, especially in the middle regions of the Nile Delta. High nitrate concentrations in the shallow zone of groundwater, especially when the groundwater depth is close to the land surface, lead to a deterioration in the groundwater suitability of this zone, but the groundwater in the deep range of the reservoir does not contain harmful concentrations of nitrates.

Groundwater, which lies south of isoline 3000 for total salt concentrations, ranges in quality for irrigation from suitable to permissible as salinity increases in the north. As for the groundwater located in the area between the Mediterranean coast and the line that represents a salinity of 3000 mg/l (Fig. 7.30) from east to west, it is unsuitable for agriculture because its use will lead to salinization of the soil and also to a significant decrease in crop productivity. Generally, in the Nile Delta, the problems of groundwater use in the field of irrigation are related to the high salinity and the risks of sodium as we head north, especially in the presence of poorly drained clay soil.

7.12.5 Groundwater Quality Under Villages

The shallow groundwater contamination beneath the south Ismailia villages (Egypt) was investigated by Geriesh and EL-Rayes (2001). They concluded that seepage from inadequately built septic tanks was the main cause of groundwater contamination. As a result, human activity was the main cause of greater nitrate and heavy metal concentrations, as well as a high bacterial count in the shallow groundwater.

According to Emara et al. (2007), the primary causes of groundwater pollution in the water-bearing formation of the Quaternary aquifer beneath some rural regions in the Giza Governorate include industrial operations, wastewater drains, and fertilizers. Since most of the settlements in the center portion of the Nile Delta lack sewers, domestic water consumption is stored in septic tanks, which artificially recharges the groundwater and results in waterlogging and groundwater pollution. Salem (2009) conducted an investigation and evaluation of the issue by integrating data gathered from various hydrogeological and hydrogeochemical tools.

Salem et al. (2017d) assessed the groundwater quality beneath sixteen villages in the central Nile Delta in 2016. They concluded that the seepage from the majority of the villages' septic tanks and/or drains was having an impact on the groundwater. High salin ity, K, Mg, Na, Ca, SO_4, Cl, HCO_3, Mn, Zn, P, NH_4, Ba, and unsuitable WQI values were found in the groundwater impacted by drains. The mean ion concentrations of the samples that were directly impacted by seepage from the septic tanks were mild, while Fe, Sr, and Si showed higher values. Additionally, the WQI ratings ranged from unfit to extremely poor. The samples with the lowest TDS concentrations and the best WQI were those that were unaffected by septic tanks or sewers. Unwanted quantities of NH_4, as in most groundwater samples, are signs of leakage from drains and septic tanks. These two substances pose a serious risk to human health.

7.13 Groundwater Pollution Sensitivity

The Nile Delta is one of the most severely water-challenged regions in the world. As urban areas grow and populations increase, water resources are depleted quickly. Therefore, the strategic use of groundwater is essential for maintaining human health, economic development, and environmental diversity. The analysis of groundwater's vulnerability to contamination has proven to be extremely useful for developing effective policies and strategies for the sustainable management of groundwater sources. This prompted the present review, which aimed to identify areas of increased groundwater vulnerability through the Nile Delta.

Figure 7.32 depicts how Abu El Ella (2018) categorized the Eastern Nile Delta into four categories according to the DRASTIC index: low, moderate, high, and extremely high. The areas with extremely high and high susceptibility are north of the Ismalia Canal and

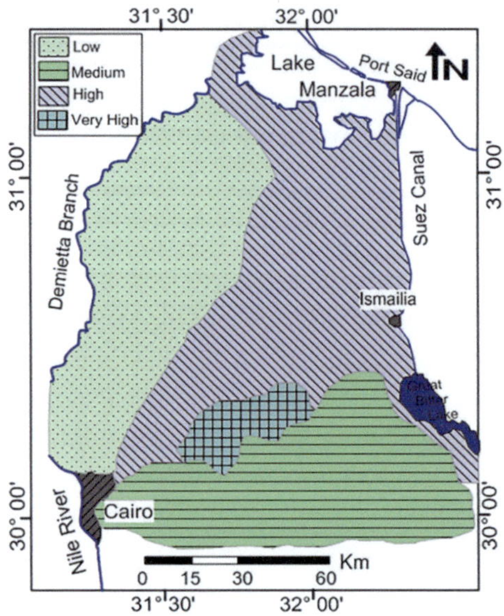

Fig. 7.32 Groundwater Vulnerability Map for the Study Area, Abu El Ella (2018)

west of the Suez Canal, respectively, whereas the areas with low and medium sensitivity are located south of the Ismalia Canal and east of the Damietta Nile Branch.

Gemail et al. (2017) evaluated the groundwater vulnerability in El-Gharbia Governorate (Middle Nile Delta) using the DRASTIC model and electrical resistivity measurements. Very high vulnerability zones, high vulnerability zones, moderate vulnerability zones, and low vulnerability zones, respectively making up 19.89%, 41.0%, 27.0%, and 12%, were found. Using DRASTIC methodologies applied to the middle Nile Delta, Metwally et al. (2023) conducted comparative research to assess groundwater sensitivity to contamination. Most of the Qotur region is classified as moderately sensitive to pollution based on all DRASTIC models and highly sensitive according to AHP-pesticide DRASTIC. South of Zefta is classified as a very highly susceptible zone. Most of the Samanod region and the area to the north of Al-Mahalla Al-Kubra are considered to be in a very low vulnerability zone. Sensitivity analyses showed that the groundwater depth characteristic has the biggest impact on determining susceptibility.

In the Western Nile Delta, Armanuos et al. (2020) evaluated groundwater susceptibility to pollution based on the integration of several hydrogeological characteristics (Fig. 7.33). 31% of the region was classified by the Modified-DRASTIC model as being extremely sensitive to pollution. Additionally, 25% of the region was assessed as being extremely highly sensitive to pollution by Pesticide-DRASTIC-LU. With the help of nitrate concentrations, the model was validated. The sensitivity analysis also revealed that the vadose zone is the most crucial variable for the DRASTIC model.

Salem et al. (2019) used the DRASTIC Index to study the Western Nile Delta's young alluvial plain to evaluate the groundwater pollution vulnerability (Fig. 7.34). The research

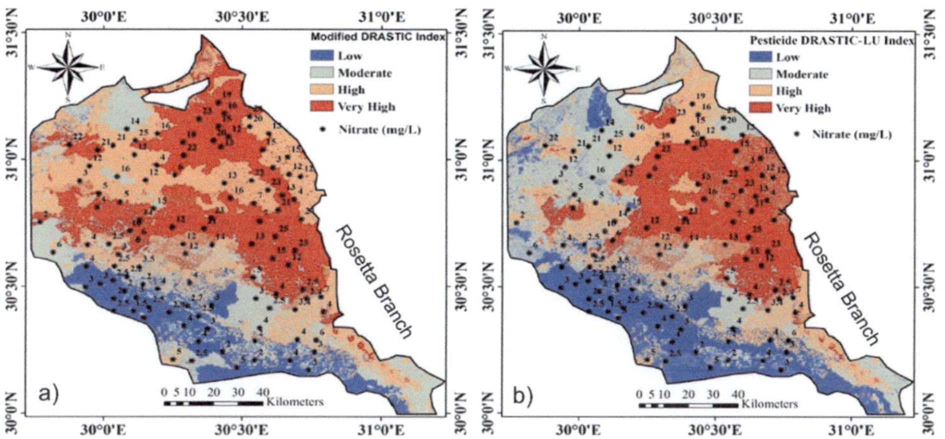

Fig. 7.33 Vulnerability index maps: **a** modified DRASTIC, **b** pesticide DRASTIC-LU, Armanuos et al. (2020)

area was divided into three susceptible zones, according to the general DRASTIC model vulnerability map. The study area's percentage of intermediate zones is 28.4%, whereas the area's overall percentage of highly vulnerable zones is 58.9%. The center and southern regions are both included in them. Additionally, 12.7% of the northern region demonstrated a very high capacity for contamination. About 22.4% of the research region has a very high vulnerability to pollution, according to the groundwater vulnerability map created using the pesticide DRASTIC model. The coastal region and the southwesterly region are included in these areas. Also, the coastline region and the region in the southwest along the El-Nubaryia canal are included. Additionally, 77.6% of the research area demonstrated moderate to high pollution susceptibility. The southern and central regions are included in these areas.

Drastic-LU was used by Arafa et al. (2022) to evaluate groundwater pollution sensitivity in the entire Nile Delta area (Fig. 7.35). The generic prevalence of pollution in the study area is so high that, apart from a moderate vulnerability in the northwestern regions (3.38%), pollution vulnerability is very high (42.69%) and correspondingly high (53.91%) across the entire Nile Delta. On the other hand, the pesticide findings of the Drastic-LU model showed that, apart from moderate vulnerability in the northwest and southern areas (9.78%), the predominant area in the Nile Delta had very high and very low vulnerability (39.53%) and very high vulnerability (50.68%), respectively. The credibility of the groundwater nitrate model was confirmed when compared to nitrate values, and a sensitivity analysis was conducted. The output map showed a highly consistent relationship between the pesticide model and the groundwater model. Going into the model's sensitivity assessment, the importance of groundwater depth and net recharge became clear.

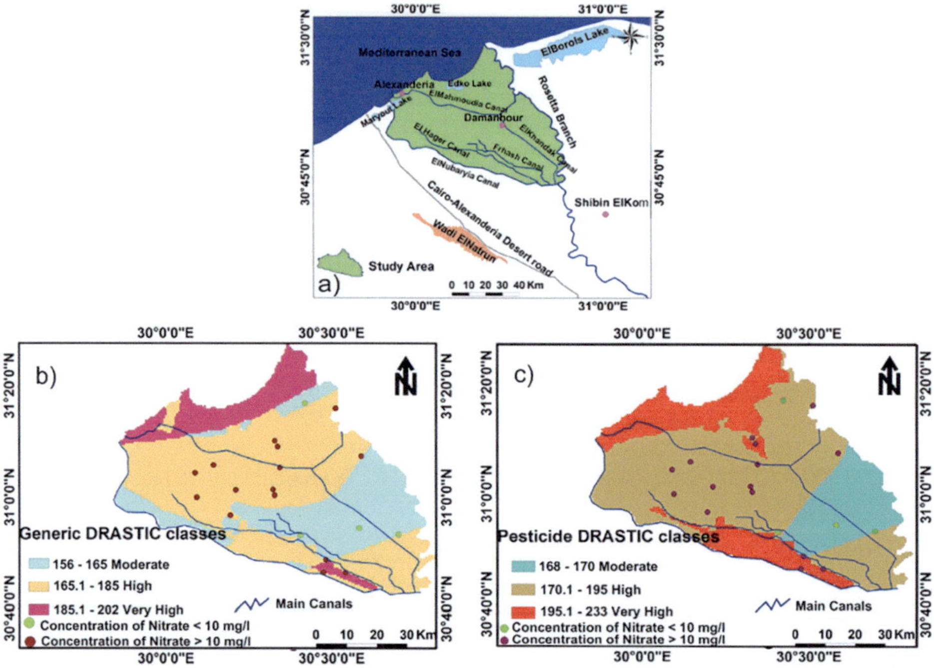

Fig. 7.34 Location map of the young alluvial plain **a** and vulnerability maps generated by **b** Generic DRASTIC Model and **c** pesticide DRASTIC model, Salem et al. (2019)

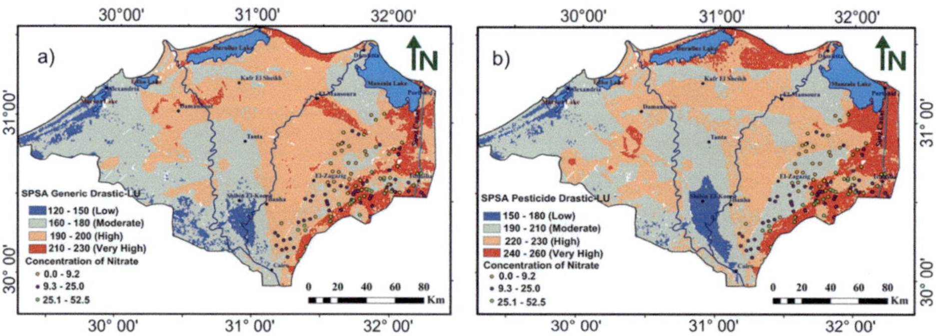

Fig. 7.35 Validation of sensitivity map constructed by **a** SPSA Generic DRASTIC-LU and **b** pesticide DRASTIC-LU model (Arafa et al. 2022)

7.14 Sea-Level Rise

The rise in sea level is regarded as a prominent indicator of changes in global climate and is thought to be closely tied to the phenomenon of global warming. It could be ascribed to several fundamental factors, such as ocean thermal expansion, melting of polar ice caps and mountain glaciers, and adjustments to terrestrial storage (Dawod et al. 2019). It is a fact that, as a global average, the sea level has dramatically increased in the 20th century due to climate change. The observed worldwide mean sea level rises for the periods 1901–1990, 1971–2010, and 1993–2010, respectively, were 1.5, 2.0, and 3.2 mm/year. Sea-level changes follow a clear regional pattern, and in some locations, the local and regional sea-level rise rates deviate significantly from the rate of global sea-level increase. According to Stanley and Warne (1994), the Mediterranean experienced a dramatic rise in sea level from −120 m below mean sea level at 18,000 years B.P. to around −16 m at 8,000 years B.P. According to Gornitz (1991), the consequences of sea-level rise are typically spatially uneven, with some places experiencing uplift and others experiencing subsidence.

Dawod et al. (2019) stated that Alexandria, Damietta, Burullus, and Port Said, respectively, have secret relative sea-level rise estimates in the region of the Nile delta of 4.3, 7.4, 10.3, and 2.6 mm/year, respectively (Fig. 7.36). Due to their unique geographic locations near the mouths of the Nile and the Lake of Burullus, Damietta and Burullus have high values. The projected vertical land movement rates in Alexandria, Damietta, Rosetta, Port Said, and Burullus are −0.7, −1.8, −1.0, −4.3, and +0.09 mm/year, respectively. In the Nile Delta, tectonic movements, groundwater extraction, and soil compaction are the causes of the vertical land deformations. The Nile Delta region should be addressed individually and not as a single entity when dealing with sea-level rise projection and other related scenarios because there is significant diversity in subsidence or uplift and diverse geomorphologic and morphodynamic elements for each coastal sector. Due to the considerable topographical diversity, not all of the Nile Delta's coastal plains are vulnerable to accelerated sea-level rise at the same rate.

The surface elevations of the Nile Delta coastal plain range from high topographic features to low-lying regions, each of which reflects a different degree of sensitivity to erosion and, consequently, sea intrusion. In Fig. 7.36, the main topographic features and their levels are spread spatially. It is also shown as a bar chart representing the respective proportions of these units up to the 4-m contour about the Nile Delta area's coastal plain.

Sea-level rise threatens vulnerable areas of the sand belt, which is necessary to safeguard lagoons and low-lying reclaimed land in Egypt's Nile Delta. The effects would be quite negative as the lagoons produce a third of Egypt's fish catches. Most freshwater fish would be affected by water suitability changes brought on by sea-level rise. It would flood valuable agricultural land. Furthermore, the consequences of SLR impacts are made worse by land subsidence in the Delta coastal region. According to El-Raey et al. (1995) and Hassanin (2010), it would cause an environmental disaster such as storm surges and coastal floods that would result in shoreline erosion, saltwater intrusion, land inundation,

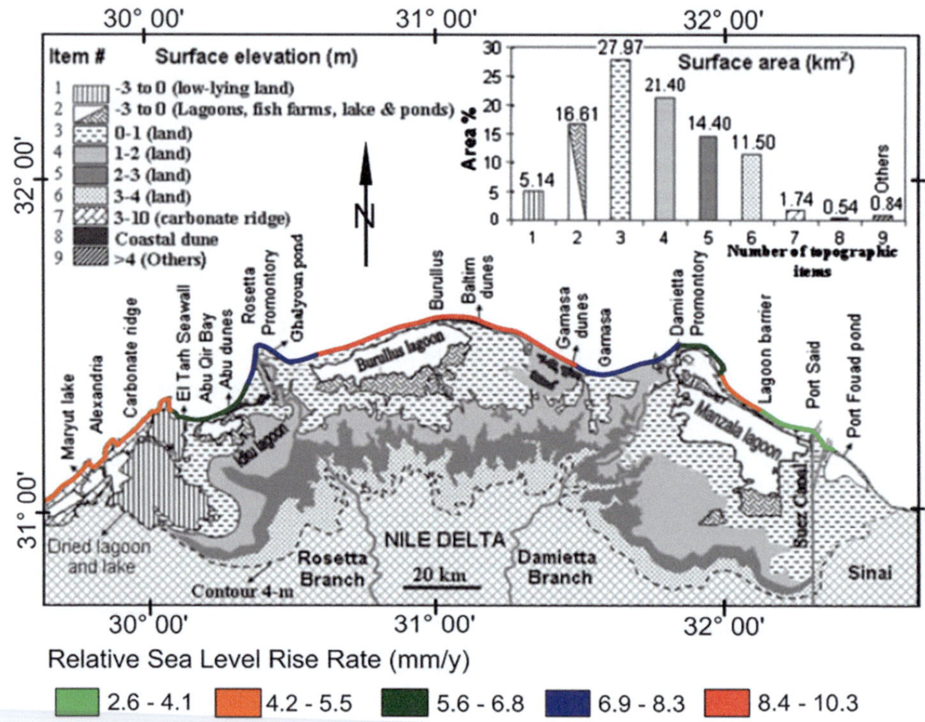

Fig. 7.36 Map shows SLR and its effect on the Nile Delta (Frihy 2003; Hassanin 2010; Dawod et al. 2019)

and soil salinity. 6.1 million people who live along the seaside may move away, and 4500 km^2 of agricultural land would be lost (World Bank 2007). This would cause Egypt's gross domestic product (GDP) to decline by 6% for 1 m SLR and 16% for 5 m SLR.

7.15 Change in Groundwater Level and Waterlogging

Out of all significant land use changes, expanding irrigated agriculture in desert areas using newly constructed channel water sources led to the biggest impact on groundwater. Near land surface water tables and waterlogging in soils are obtained. According to numerous studies (Bromley et al. 2001; Foster et al. 2004; Liu and Xia 2004; Salem et al. 2015; Salem and El-Horiny 2014; Yang et al. 2015), if irrigation activities, in such a development project, depend on groundwater pumping, a lowering in the water table is created and increases with time. In semiarid areas, the sediments of the unsaturated zones hold high amounts of salts that have formed over thousands of years. Therefore, groundwater recharge changes due to land use/land cover change can have a deleterious impact on groundwater quality (Allison et al. 1990; Phillips 1994; Walvoord et al. 2003; Javi et al. 2015; Pulido-Velazquez et al. 2015; Salem et al. 2015; Chavez et al. 2016).

Fekry (2004) studied the groundwater level and salinity changes in the Western part of the Nile Delta due to agricultural reclamation projects. He found an increase in the water level in the El-Nubaria area and a lowering in the water level to the south of the West Nile Delta around Cairo-Alexandria Desert Road. He estimated that the water level rise (5–25 m) in the El-Nubaria area might be due to excessive irrigation from the El-Nasr and El-Nubaria canals and their tributaries. The undulation of the topography and aquifer heterogeneity is responsible for the appearance of waterlogging in some parts of the area. In the El-Nubaria area, Salem and Osman (2017a) monitored the flow pattern of groundwater and water level changes between 1966 and 2010. They noticed that in 1966, groundwater was flowing from El-Nubaria Canal to the west and the east (Fig. 7.37a) but after constructing the El-Nasr Canal and its tributaries in 1975, groundwater in 2010 was noticed flowing from west and southwest to northeast crossing El-Nubaria Canal (Fig. 7.37b). The water table in the southwest of the region increased by up to 30 m between 1966 and 2010, according to the generated piezometric difference map (Fig. 7.37c). It is possible that this divergence developed due to the canal networks and irrigation techniques. In the west Nile Delta's southern region, excessive groundwater pumping has resulted in a noticeable drawdown in groundwater levels ranging between 0.5 m to more than 10 m due to lack of aquifer recharge. Accordingly, pumping wells may get dry with time. Groundwater pumping management is urgently needed, and managed surface aquifer recharge is important Salem and Bayumy (2016b).

The West El-Nubaria region (Fig. 7.38) has been the focus of reclamation efforts since the 1980s. An artificial salty lake (water EC > 20 dS/m) has been created by the return flow of agricultural drainage water from nearby cultivated land due to poor management (Abd El-Kawy et al. 2011). This lake began to take shape in the early 2000s and expanded till it achieved its current size of about 883 acres. There are several land use classes in the investigated area in addition to the free water class, including cultivated fields, bare soil, and urban areas. Based on the result of Bakr and Abd El-kawy's study (2020), the lake's surface water area (Fig. 7.38) increased from 593 to 883 ha between 2013 and 2017, and it is predicted to continue growing to 1068, 1711, and 2267 ha in 2020, 2030, and 2040, respectively. Land degradation will result from the lake extending across the adjacent agricultural and uncultivated regions.

El-Salhiya project, East Nile Delta (Fig. 7.39), reclamation operations began around 23,000 feddan in 1981/82 (Meyer 1998). At that time, the depth from the land surface to subsurface water varied from 1 to 20 m north of El-Salhiya and near Ismailia Canal, respectively. At New El-Salhiya, the water level varied from 2.19 to 9.31 m, while the depths ranged from 1.3 to 18 m (Rizk 1997). By that time, a waterlogging issue had arisen due to rising shallow groundwater getting to or above the earth's surface because farmland reclamation had taken place. Awad and El Fakharany (2020) identified that elevated groundwater levels in regions north of El-Salhiya canal and surrounding the El-Heseneya

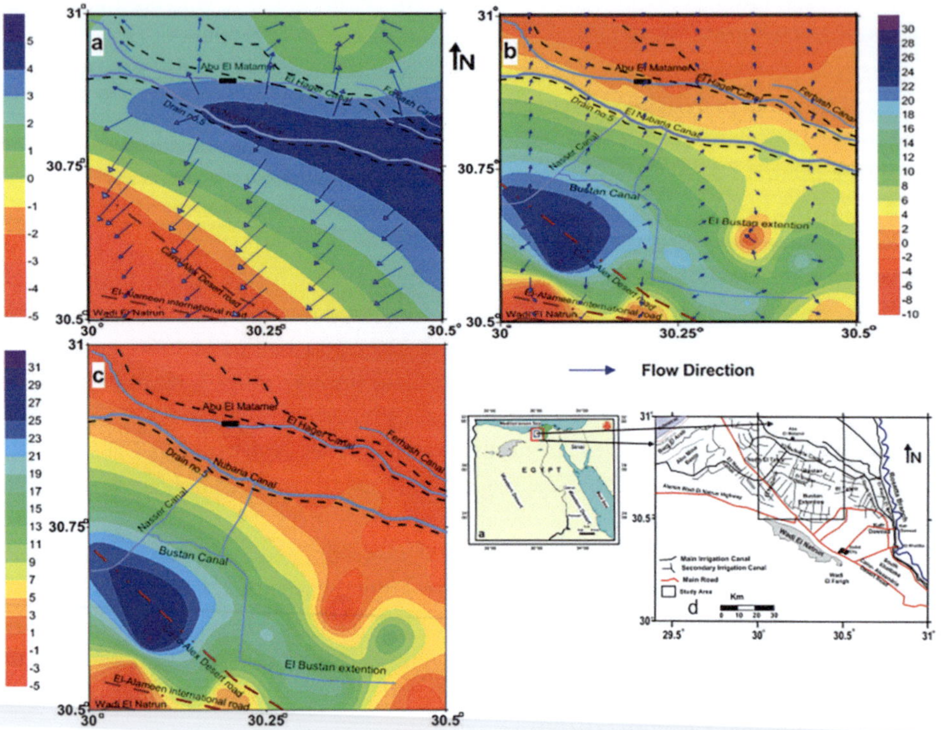

Fig. 7.37 Maps show the flow patterns of the groundwater in 1966 (**a**) and 2010 (**b**). The difference in meters between **a** and **b** of the water table is constructed in **c**, Salem and Osman (2017a)

canal contributed significantly to waterlogging conditions observed in the El-Salhiya area (Fig. 7.39). Their research, corroborated by additional studies, indicates that while excessive irrigation practices exacerbate the problem, the primary contributors to waterlogging are seepage from irrigation canals and inefficiencies in the drainage infrastructure.

During the years 1984–2008, water-logged sites were identified and assessed in the Wadi El Tumilat basin (East Nile Delta) by Kaiser et al. (2013). It became known that water logging surface area increased by 0.7 km^2 per year, from 9.1 km^2 in 1984 to 18.8 km^2 in 2000 and to 25.3 km^2 in 2008. The vegetation cover increased from 453 km^2 in 1984 to 719 km^2 in 2008. This illustrates a strong relationship between the vegetation increase in the desert area and the surface area growth of the waterlogging. Between 3 and 8 cm/year of groundwater is increasing around the flooded locations. According to estimates, the topography, surface elevation, drainage pattern, groundwater flow system, excessive use of water for irrigation, inadequate drainage system, the existence of subsurface impervious shallow clay lenses (at or below 1 m depth), and lineaments direction, all contribute to the waterlogging issue. On average, 25.3 km^2 out of 350 km^2 (7.3%) of the former agricultural areas are suffering from waterlogging (Fig. 7.40).

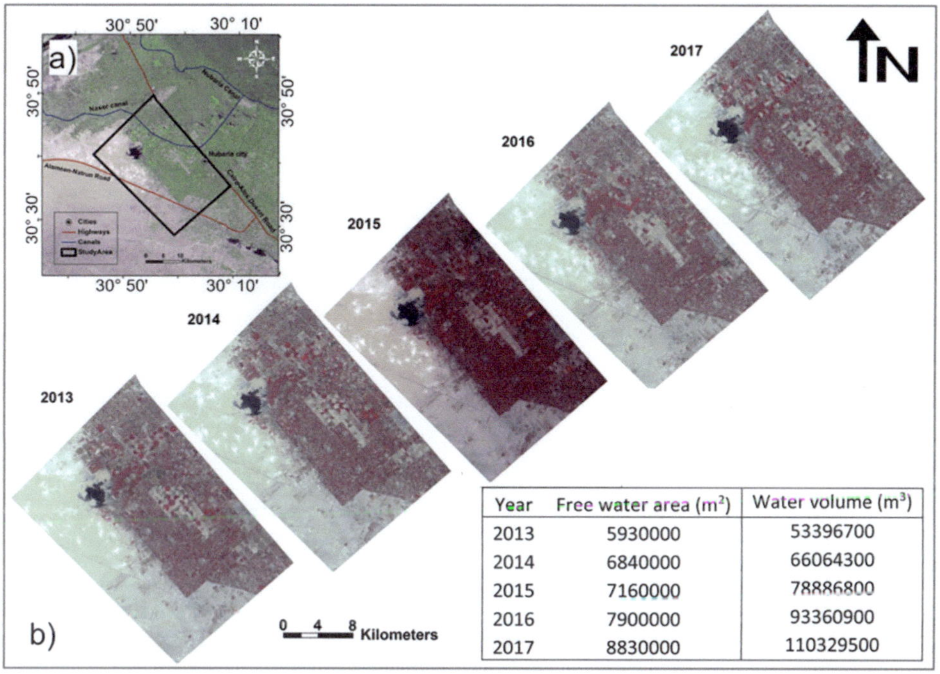

Year	Free water area (m²)	Water volume (m³)
2013	5930000	53396700
2014	6840000	66064300
2015	7160000	78886800
2016	7900000	93360900
2017	8830000	110329500

Fig. 7.38 Artificial lake location map in West El-Nubaria area (**a**) and the temporal change in lake surface area in (**b**) (Bakr and Abd El-kawy 2020)

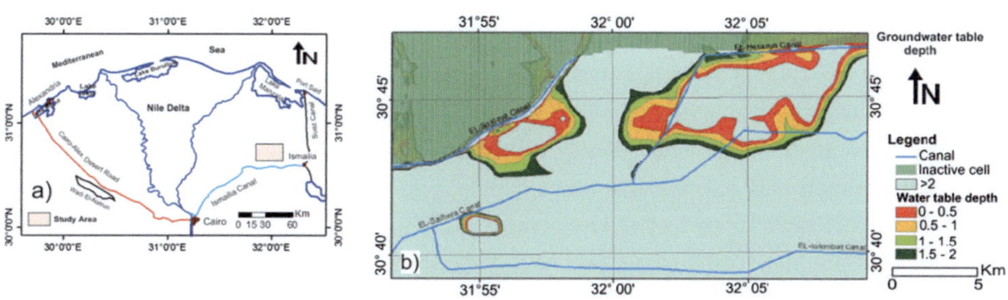

Fig. 7.39 Groundwater table depth map in El-Salhiya area (Awad and El Fakharany 2020)

7.16 Seawater Intrusion (Hydrogeochemical and Hydrogeophysical Criteria)

Seawater intrusion is a significant issue in the semiarid and arid coastal zones, where groundwater is the main freshwater supply (Nhu et al. 2020). The complexity of their geological environments also makes the maintenance of water supplies in semiarid coastal

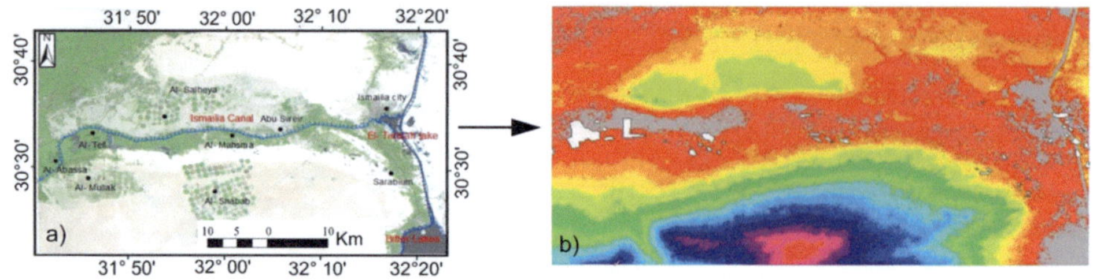

Fig. 7.40 Maps indicate the land use of the area (**a**) and waterlogged sites showing white color (**b**) that are closely distributed at the low topographic areas shown by grey color (Kaiser et al. 2013)

zones a major problem (Arafa et al. 2024; Salem et al. 2025a, b). High population expansion, rapid urbanization, and ineffective water resource management put the developing countries of the Mediterranean Basin under environmental stress (UN 2003; WHO 2006). Numerous variables, including anthropogenic activity, agricultural practices, biogeochemical processes, and seawater intrusion, cause the salinity of groundwater to rise (Kanagaraj et al. 2018).

Groundwater that has a salinity higher than 35 g/l can be called hypersaline groundwater. Diab and Saleh (1982) detected hypersaline groundwater close to the shore (Fig. 7.41) but did not mention the sampling depth. Shata (1999) recorded a hot brine in oil wells at 1000 m depth. Groundwater salinity at 200 m, 400 m, and 600 m depths was measured by RIGW 2013 (Nofal et al. 2015), where high values of TDS were recorded starting from 400 m depth. Engelen et al. (2018) estimated that hypersaline groundwater can come from either Quaternary Free Convective systems or by compaction-induced upward salt transport that occurred during the Messinian salinity crisis.

Sharaky et al. (2007) reported that the salinity of the groundwater in the West Nile Delta ranged from 430 mg/l fresh water to 24.407 mg/l saline water, with an average of 2,609 mg/l brackish water. The salinity minimum value was observed in the vicinity of the Rosetta branch and various irrigation canals, and maximum salinity was detected in the northern part near Edku Lake. The spatial distribution map of salinity in the western Nile Delta shows that freshwater (<1,000 mg/L) predominates in the central and eastern parts of the area. High salinity zones are located to the north due to seeping saltwater from the Mediterranean Sea.

In their study, Hasan et al. (2023) attempted to identify the geochemical mechanisms and effects of seawater intrusions in the aquifers on the western region of the Nile Delta in the young alluvial plains (Fig. 7.42). To carry out this study, they used hydrogeochemical, multivariate statistics, and graphical tools. As shown in Fig. 7.42a, seawater intrusion increases salinity in the northern part of the site, where the coast is over 30 km away. They also calculated the seawater intrusion index; the values ranged from 17.8 to 82.3 (see Fig. 7.42b). The groundwater in the site is classified as saline, mixed or fresh water.

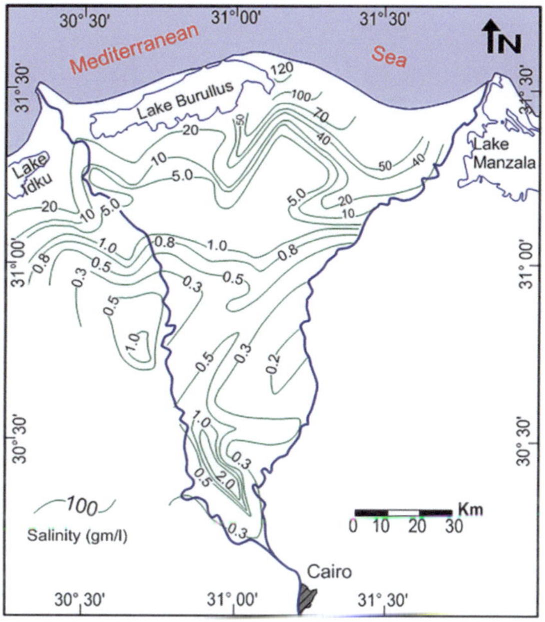

Fig. 7.41 Map showing the areal change in salinity (g/l) of the Nile Delta groundwater (Diab and Saleh 1982)

The northern coastal areas have low values, reflecting groundwater salinization due to seawater intrusion.

Using the GALDIT model, Salem and Hassan (2021) sought to assess the Pleistocene aquifer of the west Nile Delta's groundwater vulnerability to seawater intrusion (Fig. 7.42c). The coastal and central regions, which make up 37.7% of the total land, are categorized as having a high susceptibility to seawater intrusion. The investigation area's southern regions, which make up 62.3% of the entire area, are moderately susceptible to intrusion hazards. Based on the findings of the sensitivity study, the aquifer's hydraulic conductivity and thickness are the two elements that have the most impact on the updated vulnerability model. The modified GALDIT map demonstrated that the northern regions continue to have a high vulnerability percentage that is 15% higher than the traditional GALDIT index.

Salem and Osman (2017a) investigated the salinity origin and the governing geochemical processes of the Northwest Nile Delta Quaternary aquifer as an illustration of groundwater in an arid area affected by reclamation and seepage from seawater intrusion. The main determinants of groundwater chemistry are soil salinity, seepage from irrigation canals, and seawater intrusion. Due to salt water's higher electrical conductivity than fresh water, Salem and Osman (2017b) used surface resistivity surveys to map the boundary between the two types of water. 40–150 m BSL is the range of seawater intrusion depth (Fig. 7.43). Due to seawater invasion, the minimum depth in the area's northwest reaches roughly 40 m. Maximum depth (150 m) was measured at the area's center.

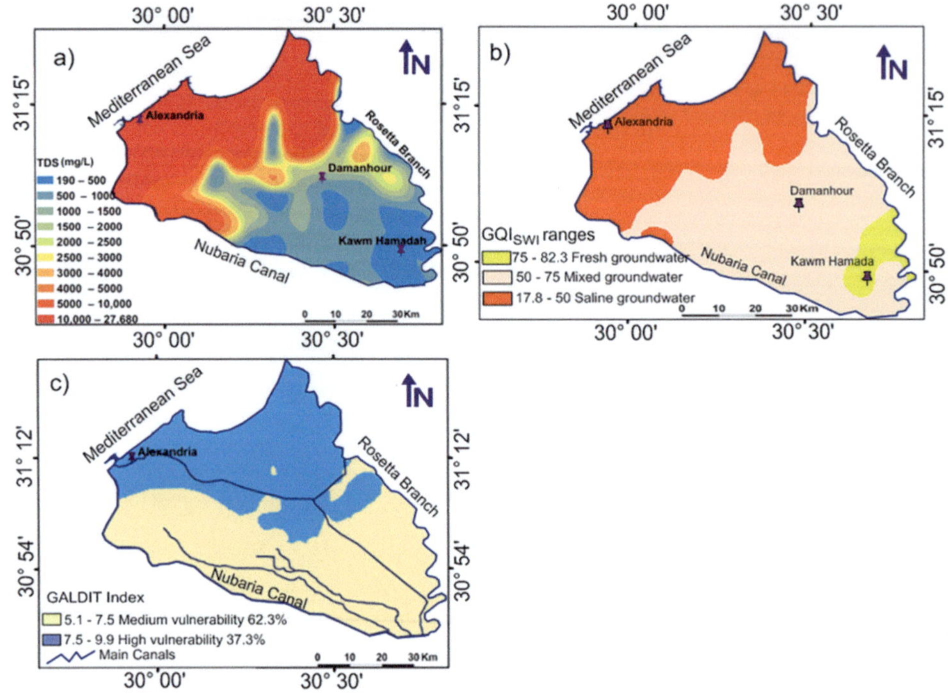

Fig. 7.42 Maps show the groundwater salinity distribution (**a**), groundwater classes relating to seawater intrusion index (**b**), and seawater effect according to GALDIT Model (**c**) in the young alluvia plain, West Nile Delta (Salem and Hasan 2021; Hasan et al. 2023). The geographic location is presented in Fig. 7.34a

The Middle Nile Delta's hydrogeochemical data and shallow thermal regime (Fig. 7.44) provided information regarding the extent of seawater intrusion. The following are the hypothesised criteria for seawater intrusion based on the subsurface temperature (Fig. 7.44a): (1) Despite their great distance from one another, the Kafrelarab and Motobes deep wells in the deeper zones showed only a small temperature difference, which suggests that the seawater in the mentioned area is nearly static; (2) The deeper zones also revealed almost horizontal isotherms beneath the majority of the area under study; (3) The Motobes wells have a modest thermal gradient in comparison to other wells, despite their proximity to the Mediterranean Sea. The aforementioned thermal criteria point to the existence of a water body is practically static deeper in the south and shallower in the north of Kafr Elsheikh City, which may constitute seawater intrusion.

The following recognized hydrogeochemical indicators (Fig. 7.44b) point to seawater invasion in the research area: (1) In the shallow and deeper zones, respectively, the groundwater in Qutor wells contains both $CaCl_2$ and $MgCl_2$ water types. This suggests that under the Qutor area, two zones of freshwater and seawater are interconnected. The

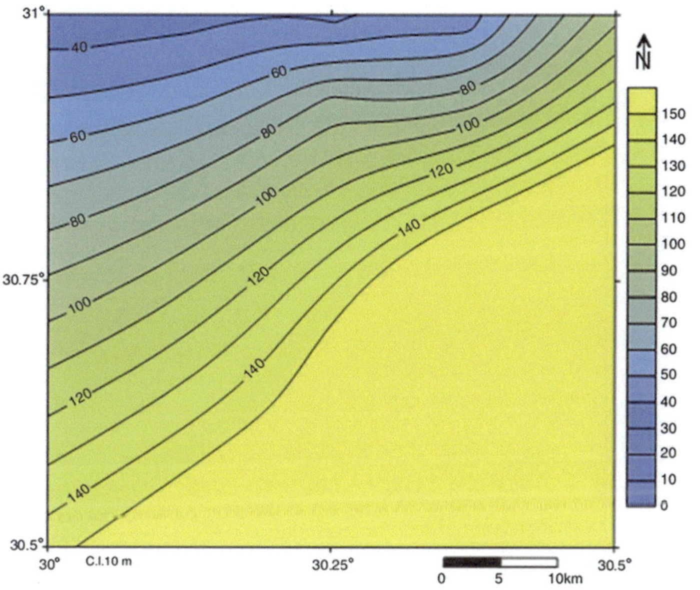

Fig. 7.43 Freshwater zone thickness of the area as deduced from the electrical resistivity method (Salem and Osman 2017b)

maximum depth of the Qutor well is 100 m, hence, this estimation is restricted to that depth. (2) Partial ion exchange results in a relative increase in Ca and a decrease in Na rather than Cl in Kafr Mesaaed and Elhadady wells, which is another indication of seawater intrusion. (3) Elkarada samples exhibit a relative rise in HCO_3 and higher Na than Cl, in contrast to the regional groundwater flow system of the southern section and the local flow system in the northern part. Such a characteristic suggests that a partial ion exchange occurred during the propagation of freshwater into seawater. In a different sense, the groundwater beneath El-karada was originally seawater that had been somewhat contaminated with freshwater. The most perilous revelation from this research is that the Nile Delta's groundwater is impacted not just by seawater intrusion but also by the inland spread of hypersaline brine water. The upper 200–250 m is affected by seawater, but the 420 m-deep Motobes well has hypersaline water. The fresh groundwater recharge flux in Tala wells (0.8 m/year) is substantially lower than the upward seawater intrusion flux rates in Motobes wells, which were 2.8 m/year.

To identify groundwater contamination brought on by saltwater intrusion, Ebraheem et al. (1997) undertook a resistivity investigation in the Middle Nile Delta region from south to north (Fig. 7.45). To measure the areal fluctuations in water salinity, existing monitoring wells were employed. An empirical relation between the concentration of all solids dissolved and the predicted earth resistivity was proved using data from vertical electrical sounding (VES) for sounding stations near monitoring wells and the water salinity

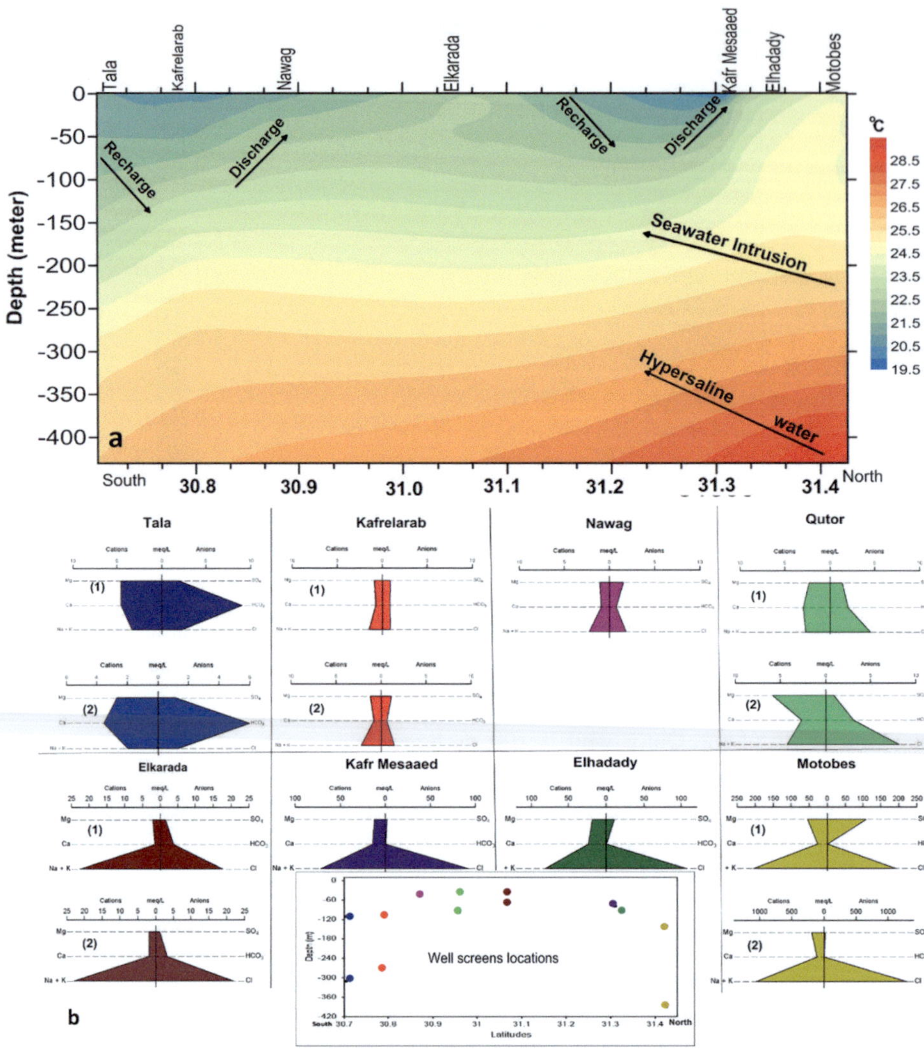

Fig. 7.44 South-north cross-section **a** shows the subsurface shallow thermal system of the investigated region. The groundwater flow system is also indicated by the white arrows, **b** stiff diagram shows the difference in the chemical types along the same cross-section

of these wells. This relationship was utilized to distinguish between three water-bearing formation zones (fresh-, brackish-, and salt-water zones) using the calculated apparent resistivity depth and geoelectric longitudinal sections. The fresh-brackish contact depth along the examined profile is greater than 150 m near Tanta City and falls to 40 m or less far north between Qutor and Kafr EI-Sheikh. The depth to the brackish-saline interface

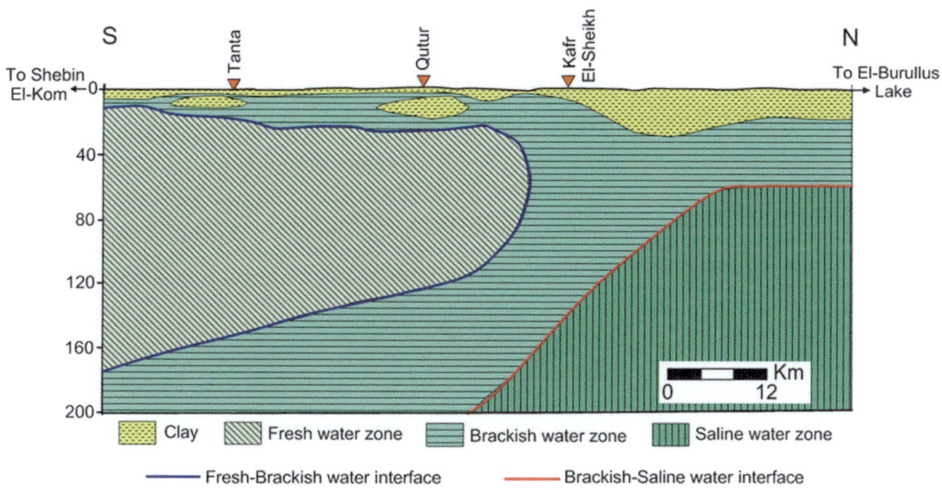

Fig. 7.45 South-north cross-section in the middle part of the Nile Delta aquifer using geoelectrical resistivity shows the effect of seawater intrusion (Ebraheem et al. 1997)

is more than 180 m south of Kafr-El-Sheikh and drops to 70 m in the north, close to the village of Hadadi.

To evaluate the effects of human activity and seawater intrusion, Salem et al. (2021b, 2022b) described the vertical and Geographical assessment of the Central Nile Delta's Quaternary aquifer's groundwater chemistry (Fig. 7.46). In the three-depth zones of shallow, middle, and deep, the salinity of the groundwater varied from fresh to saline. The groundwater salinity rose in a northerly direction, which may have been caused by surface conditions and/or seawater intrusion. The shallow groundwater was detected to possess a higher salinity than the intermediate and deep zones, indicating that human activity, soil salinity, and surface activities are to blame for the high salinity of this zone. The presence of meteoric and marine water types suggested that recent recharging from the Nile branches, canals, and drains as well as seawater intrusion and evaporation, affected the groundwater. The main elements affecting the chemistry of groundwater in the three zones were evaporation, dissolution, ion exchange, seawater intrusion, leakage, recent recharge, and human activities. As shown in the cross-section A–A′ (Fig. 7.47), the vertical groundwater salinity increases from south to north, with the highest values found on the shallow zones.

In the eastern region of the Nile Delta (Fig. 7.48), Mogren and Shehata (2020) tested the salinity of the groundwater. They showed that the TDS values in the northern zone, which is categorized as being salinized, are over 10,000 mg/l. However, the southern portion of the research area has recorded comparatively low TDS values (less than 1500 mg/

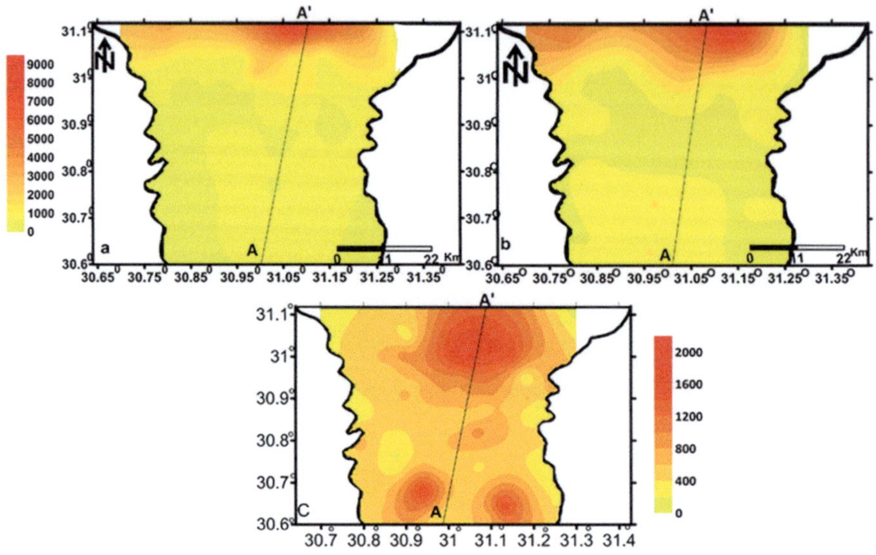

Fig. 7.46 Spatial distribution maps of TDS (mg/l) in the aquifer of the Nile Delta. Central part. **a**, **b** and **c** are related to the shallow, intermediate, and deep zones. A–A′ is the location of the cross-section shown in Fig. 7.47

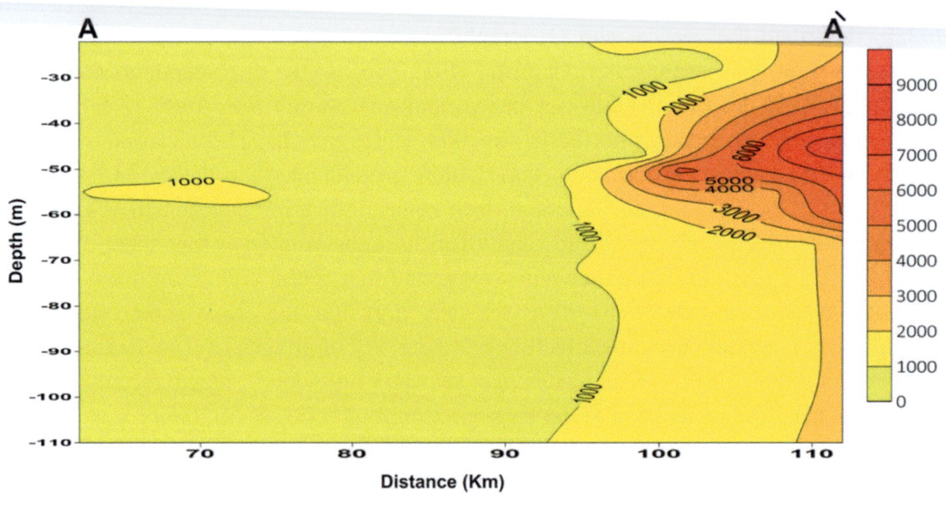

Fig. 7.47 Vertical distribution of the groundwater total dissolved solids (TDS, mg/l) along the cross-section A–A′ from south to north

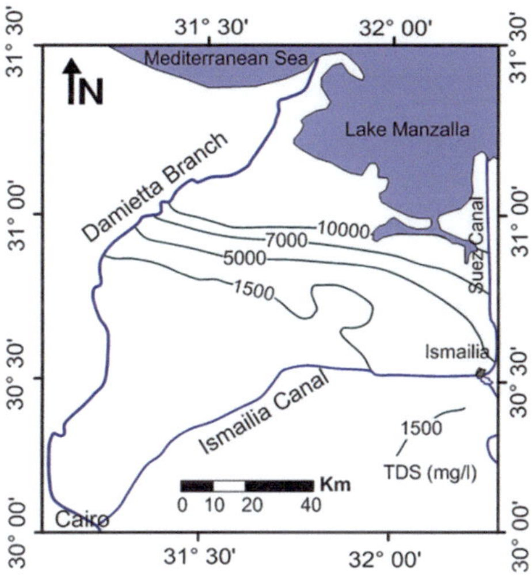

Fig. 7.48 Groundwater total dissolved solids spatial distribution map showing the effect of seawater invasion to the north

l). The TDS map's salinization trend foresaw the effects of saltwater intrusion on freshwater deterioration because the area's northern regions are much more salinized than its center and southern regions.

An extensive hydrogeochemical, statistical, and geophysical study was carried out by Abu Salem et al. (2022) in the East Nile Delta's coastal shallow aquifer, which is well recognized for its high susceptibility to seawater salinization. The groundwater was divided into fresh, mixed, and saline categories (Fig. 7.49). The geophysical findings show that due to substantial seawater intrusion, seawater largely invaded inland up to 75 km at 100 m depth from the shoreline. These results, which are based on geophysical and hydrogeochemical modeling, are in line with earlier simulation studies in the eastern and middle regions of the Nile Delta and show that the investigated aquifer is currently significantly threatened by the sensitivity to seawater intrusion. They advised that a strategy for preventing seawater intrusion is to raise groundwater levels by at least 5 m and keep them there by carefully monitoring pumping to prevent aquifer damage. This might be accomplished by artificially replenishing several moderately salinity wells in the mixing zone in the center of the study area within a 12–75 km radius. Desalinated groundwater or properly treated wastewater are potential sources of recharge.

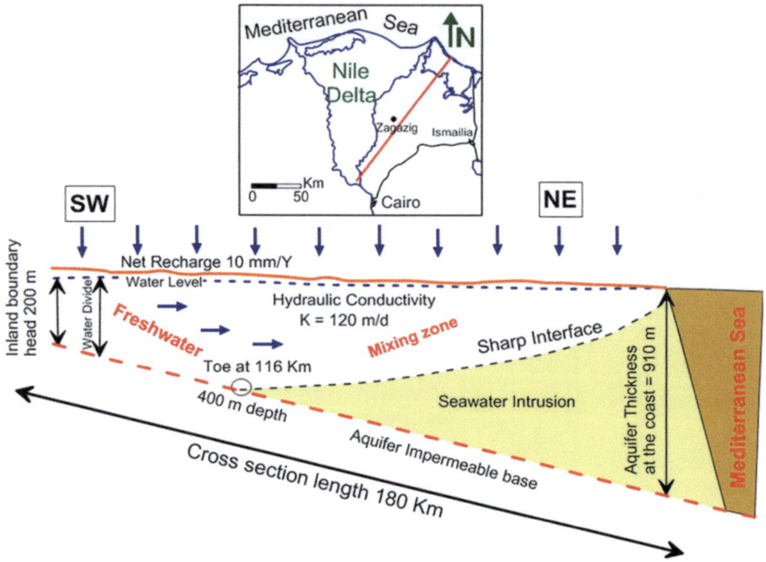

Fig. 7.49 Schematized modeled cross-section shows the seawater intrusion in the eastern Nile Delta aquifer (Abu Salem et al. 2022)

Nosair et al. (2022) created a model of predictive regression depending on an original method employing seawater instruction indices and artificial intelligence methodologies to monitor groundwater salinization owing to seawater intrusion in the eastern Nile Delta's aquifer (Fig. 7.50). Data from groundwater wells collected during three time periods (1996, 2007, and 2018) were utilized as input for artificial intelligence techniques. All of the examined indicators were included in the feature extraction procedure, which determined the study year, the seashore distance, the kind of aquifer, and the water table as the most important inputs. For the years 2020 and 2030, an electrical conductivity prediction model was created. The findings indicate that between 2018 and 2030, the 1000 S/cm border is predicted to shift inland by 9.5 km eastward, 10 km westward, and 12.4 km in the central section. If no steps are taken, this encroachment could be dangerous for agricultural and water resources.

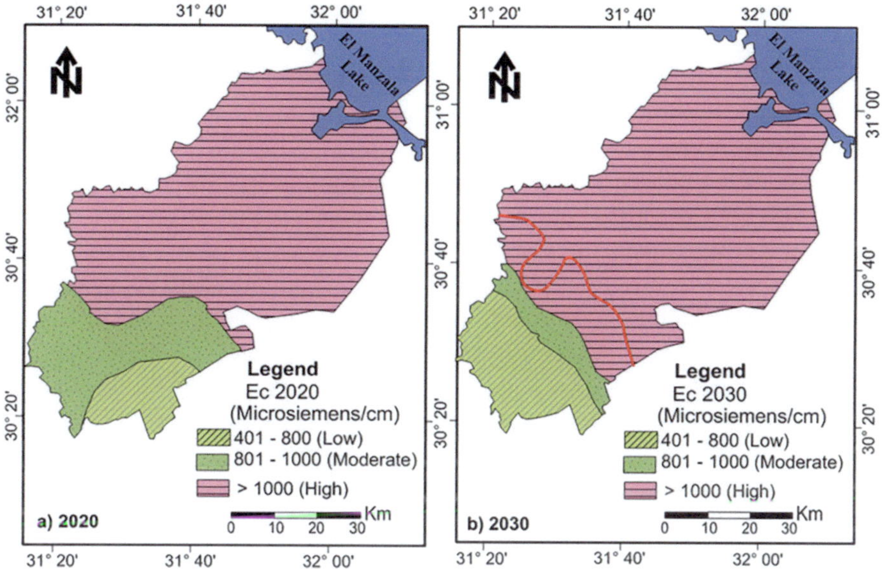

Fig. 7.50 Sharkya Governorate (East Nile Delta) maps show the groundwater salinity in 2020 and the predicted groundwater salinity in 2030 (Nosair et al. 2022)

7.17 Conclusion

The Nile Delta Quaternary aquifer is considered one of the most important coastal aquifers in the arid regions of the world, as it is a source of water for a large portion of Egypt's population, where human activities are diverse. This aquifer suffers from many problems, including climate change and the resulting increase in seawater intrusion, as well as increased reclamation operations in the east and west of the delta, a significant increase in pumping from the reservoir in the last decades, and a change in the amount of water feeding the reservoir as a result of the construction of the High Dam. The construction of the Renaissance Dam may lead to a significant decrease in the amount of feeding.

The Nile Delta extends over an area of about 22,000 km^2, extending about 225 km along the Mediterranean coast and reaching a depth of about 160 km, reaching Cairo in the south. The delta has a thick sequence of sediments ranging from Quaternary to Tertiary sediments. Tectonically, the delta sediments were affected by a group of complex geological structures with many successive faults. Two groups of faults have been distinguished, the first toward the northwest and the second toward the northeast. The delta contains many coastal lakes (Idku, Mariout, and Manzala lakes) and many ancient buried Nile branches, of which only the Rosetta and Damietta branches remain. The surface of the delta slopes gently from south to north, with the elevation in the south reaching 18 m above sea level, in the middle 5 m above sea level, and the elevation in some areas in the north reaching sea level or less.

The surface of the delta is covered by a complex network of waterways used for irrigation and drainage, which in turn are directly connected to the aquifer, and this network discharges excess water into the Mediterranean Sea or the northern lakes. Over the last four decades, there has been a radical change in land use in the Nile Delta, with the area of agricultural land, residential areas, and fish farms increasing. In contrast, the area of water bodies and sandy areas has decreased.

The groundwater aquifer in the Nile Delta is considered one of the most important in the world and is located in a geological formation known as the Mit Ghamr Formation, which is of Quaternary age. This formation is covered by a layer of clay dating back to the Holocene era called the Belqas Formation. The Belqas Formation is a semi-permeable layer consisting of silt and clay deposits with some sand. This layer increases in thickness to the north to reach 70 m and disappears to the east and west of the delta. This layer plays a very important role in the interaction between surface water and the aquifer, as well as in the intrusion of seawater into the delta groundwater. This layer is characterized by poor hydrogeological properties, with an average vertical permeability of 2.5 mm/day, although porosity may reach 50%.

The main aquifer of the Nile Delta dates back to the Pleistocene era and consists of permeable layers of sand and gravel interspersed with clay lenses. The number and density of these clay lenses increase toward the north. This aquifer is considered semi-confined, and it is considered unconfined to the east and west of the delta, where the low-permeability Belqas Formation layer is absent. The thickness of the aquifer increases from the south, east, and west toward the northeast, reaching 1000 m, and the slope of the lower surface of the aquifer is 40 times the slope of the ground surface. The hydrogeological characteristics of the reservoir vary from place to place, with hydraulic conductivity in the south being around 50 m/day and in the center, 100 m/day. In general, its value ranges from 40 to 243 m/day.

The main water source feeding this aquifer is irrigation water and small amounts of rainwater. Irrigation water feeds the aquifer through direct seepage from irrigation canals and drains or the infiltration of crop irrigation water through the surface clay layer into the aquifer. The rate of feeding with Nile water ranges from 0.25 to 0.80 mm/day. As for feeding from rainwater, it is low and is restricted to the north and ranges from 0.0 to 134 mm in the year 2000. Water is drained from the reservoir through several processes, including evaporation, pumping groundwater for various human uses, and feeding the adjacent aquifers, such as the Pliocene and Moghra aquifers to the west.

Several studies have been conducted using stable isotopes to identify the origin of groundwater in the Nile Delta. They concluded that the groundwater in this aquifer throughout the delta can be traced back to three main sources: the Nile River water after the construction of the High Dam, which forms groundwater at shallow depths; the second source is the Nile River water before the construction of the High Dam, which represents the deep groundwater in the reservoir; and the third source is the intrusion of seawater, which dominates the aquifer toward the north. In the western Delta, traces of reasonable proportions of ancient deep water have been found.

The Nile Delta is divided according to the quality of surface water into 22 areas where the salinity of the drains increases to the north. These drains feed irrigation canals in 21 places, and the rest of the drains lead to the northern lakes or the Mediterranean Sea. The Rosetta and Damietta branches suffer from some water quality problems, especially at their ends toward the sea. It was also noted in the middle of the delta that the concentrations of P, N, Mn, Zn, and Fe exceeded the permissible limit for irrigation. It was also noted that the use of algae in the so-called Palmer Index provides strong evidence of the extent of pollution in the water of canals and drains in the Nile Delta.

As for the quality of groundwater for various uses, it was noted that the northern part of the Delta, which contains the thickest part of the reservoir and is adjacent to the Mediterranean Sea, has water with salinity higher than 1000 mg/L and 3000 mg/L, making it unsuitable for drinking or irrigation. The thickness of the freshwater from the reservoir is about 50 m to the south, which is the thickness of the aquifer, and zero to the north at Kafr El-Sheikh City and its equivalents to the east and west due to the intrusion of seawater. The maximum thickness of the freshwater in this aquifer is 300 m, mostly below Menoufia Governorate.

Based on studies conducted on groundwater in the Nile Delta villages, it became clear that the lack of a sewage system hurt the quality of groundwater, as sewage water seeped into it and increased the percentage of ammonia, heavy elements, and other elements that reduce the quality of the water, especially in the shallow range.

Many studies have been conducted in different areas of the Nile Delta to identify the susceptibility of this reservoir to pollution as a result of various human activities. All these studies concluded that the thickness of the surface clay layer and the depth of groundwater, as well as the amount of recharge to the reservoir, play an important role in the possibility of groundwater pollution in the region.

The Nile Delta suffers from a continuous rise in sea level due to climate change. In the Mediterranean Sea, the water surface increased from −120 m below the current sea level in 18,000 BC to −16 m around 8,000 BC. Recently, several studies have shown that the sea level has increased at different rates for the Nile Delta, ranging from 2.6 to 10.3 mm/year, which has led to a vertical movement of the earth's surface ranging from 0.7 to 4.3 mm/year. This rise in sea level also leads to negative changes in coastal areas, such as the erosion of beaches and the erosion of the sand belts protecting them. Coastal lakes also suffer negative changes in water quality, which affects fish wealth in Egypt.

As a result of agricultural reclamation operations in the eastern and western Delta in desert lands and the use of groundwater and surface water in irrigation, qualitative changes have occurred in the depth and quality of the groundwater. Areas that depend on surface water for irrigation have experienced a noticeable increase in water level in recent years, which has led to some of their low-lying areas turning into lakes, causing severe damage to agricultural soil. These areas include the Abu Mena Basin and the El-Nubaria area in the western Delta, and the Al-Salihiya area in the eastern Delta. As for areas that depend primarily on groundwater for irrigation, such as the Al-Khattaba and Wadi El-Natrun areas in the western Delta, they suffer from water level decrease as well as an increase in their salinity.

One of the most important environmental problems that groundwater in the Nile Delta suffers from, and which is becoming more harmful over time, is the intrusion of seawater into the aquifer, causing high salinity in the largest part of the Nile Delta and even more than 70 km inland. Many hydrogeochemical and geophysical studies have been conducted to determine the extent of the impact of this phenomenon on the reservoir in different areas of the Delta. It was observed that the regions where reservoir water interacts with intruding seawater are marked by a variety of abundant geochemical processes, largely due to the presence of clay sediments in the reservoir. The degree of intrusion of saltwater also varies from one place to another in the Delta, according to what was monitored by geophysical methods.

7.18 Recommendation

- Addressing the hydrogeological complexities of the Nile Delta requires a multi-faceted, sustainable approach. Collaboration among government agencies, scientific research institutions, and local communities is vital to develop effective water management policies. Incorporating modern technologies, such as satellite imaging and geospatial analysis, can aid in better understanding groundwater dynamics and managing resources effectively.
- Investments in infrastructure to facilitate water conservation, such as rainwater harvesting and the construction of artificial recharge basins, can provide additional avenues for replenishing aquifers. Moreover, implementing integrated water resource management strategies that consider the ecological, hydrological, and socio-economic frameworks will be essential for achieving sustainable development in the region.
- In conclusion, the hydrogeology of the Nile Delta embodies a delicate balance of natural processes and human activities. As challenges mount due to climate change, urbanization, and pollution, a concerted effort is necessary to protect and preserve this vital landscape. The interplay of water and land in the Nile Delta serves as a reminder of the intricate connections between the environment and human existence, emphasizing the importance of sustainable management for future generations.

References

Abd El-bary MH (2004) Sedimentological, geochemical and hydrogeological characteristic of the Central Nile Aquifer, Egypt. MSc thesis, Geology Department, Fac of Sci Tanta Univ

Abd El-Kawy OA, Rød JK, Ismail HA, Suliman AS (2011) Land use and land cover change detection in the western Nile delta of Egypt using remote sensing data. Appl Geogr 31:483–494

Abdelhameed AT, Salem ZE, Osman OM (2017) Sedimentological characteristics of the quaternary groundwater aquifer, northwestern Nile Delta, Egypt. In: The handbook of environmental chemistry (groundwater in the Nile Delta, 2019). Springer, Berlin, Heidelberg. ISBN 978-3-319-94283-4

Abdo MH, Ahmed HB, Helal MH, Fekry MM, Abdelhamid AE (2022) Water quality index and environmental assessment of Rosetta Branch aquatic system, Nile River, Egypt. Egypt J Chem 65:321–331

Abu El Ella EM (2018) Environmental management of groundwater in the eastern Nile Delta, Egypt. Int J Water Resour Arid. Environ 7:130–143

Abu Salem HS, Gemail KS, Junakova N, Ibrahim A, Nosair AM (2022) An integrated approach for deciphering hydrogeochemical processes during seawater intrusion in coastal aquifers. Water 14:1165. https://doi.org/10.3390/w14071165

Abukila AF (2015) Assessing the drain estuaries' water quality in response to pollution abatement. Water Sci. https://doi.org/10.1016/j.wsj.2015.02.002

Agrama AA, El-Sayed EA (2013) Assessing and mapping water quality (case study: Western Delta-Egypt). Int Water Technol J (IWTJ) 3:158–169

Al-Agha DE, Molle F, El Baily M, El Desouqy E, Abou El-Hassan W (2015) Spatial and temporal variability of water quality in the Nile delta. Report by IWMI/WMRI, 25pp

Al-Gamal SA, El-Sayed SA, Atta EA (2018) Exploring surface and groundwater interactions in East Delta aquifer using conventional and non-conventional techniques. IOSR J Appl Geol Geophys (IOSR-JAGG) 6:48–58. https://doi.org/10.9790/0990-0602024858

Allison GB, Cook PG, Barnett SR (1990) Land clearance and river salinization in the western Murray Basin, Australia. J Hydrol 119:1–20

Arafa NA, Salem ZE, Abdeldayem AL, Ghorab MA, Moustafa YM, Soliman SA, Farag MH, Purohit S, Elhag M, Youssef MY (2024) Advancing deltaic aquifer vulnerability mapping to seawater intrusion and human impacts in Eastern Nile Delta: insights from machine learning and hydrochemical perspective. Earth Syst Environ 1–26. https://doi.org/10.1007/s41748-024-00518-6

Arafa NA, Salem ZE, Ghorab MA, Soliman SA, Abdeldayem AL, Moustafa YM, Ghazala HH (2022) Evaluation of groundwater sensitivity to pollution using GIS-based modified DRASTIC-LU model for sustainable development in the Nile Delta region. Sustainability 14:14699. https://doi.org/10.3390/su142214699

Armanuos AM, Allam A, Negm AM (2020) Assessment of groundwater vulnerability to pollution in Western Nile Delta aquifer, Egypt. Int Water Technol J (IWTJ) 10:1

Armanuos AM, Negm A, Valeriano OCS (2016a) Groundwater quality investigation using multivariate analysis—case study: western Nile Delta Aquifer, Egypt. Int J Environ Sci Develop 7(1):1–9. https://doi.org/10.7763/IJESD.2016.V7.732

Armanuos AM, Negm A, Yoshimura C, Saavedra Valeriano OC (2016b) Application of Wet Spass model to estimate groundwater recharge variability in the Nile Delta aquifer. Arab J Geosci 9:553. https://doi.org/10.1007/s12517-016-2580-x

Awad SR, El Fakharany ZM (2020) Mitigation of waterlogging problem in El-Salhiya area. Egypt, Water Sci 34(1):1–12. https://doi.org/10.1080/11104929.2019.1709298

Badr EA, El-Sonbati MA, Nassef HM (2013) Water quality assessment in the Nile River, Damietta Branch, Egypt. CATRINA 8:41–50

Bakr N, Abd El-kawy OR (2020) Modeling the artificial lake-surface area change in arid agro-ecosystem: a case study in the newly reclaimed area, Egypt. J Environ Manag 271:110950. https://doi.org/10.1016/j.jenvman.2020.110950

Befus KM, Barnard PL, Hoover DJ, Finzi Hart JA, Voss CI (2023) Increasing threat of coastal groundwater hazards from sea-level rise in California. Nat Clim Change 10:946–952. www.nature.com/natureclimatechange

Befus KM, Kroeger KD, Smith CG, Swarzenski PW (2017) The magnitude and origin of groundwater discharge to eastern U.S. and Gulf of Mexico coastal waters. Geophys Res Lett 44:396–10,406. https://doi.org/10.1002/2017GL075238

Bromley J, Cruces J, Acreman M, Martinez L, Llamas MR (2001) Problems of sustainable groundwater management in an area of over-exploitation: the upper Guadiana catchment, central Spain. Int J Water Resour Dev 17:379–396

Chavez RO, Clevers J, Decuyper M, de Bruin S, Herold M (2016) 50 years of water extraction in the Pampa del Tamarugal basin: can Prosopis tamarugo trees survive in the hyper-arid Atacama Desert (Northern Chile)? J Arid Environ 124:292–303

Dawod MG, Mohamed HF, Haggag GG (2019) Relative and absolute sea level rise based on recent heterogeneous geospatial data: a case study in the Nile Delta, Egypt. J Sci Eng Res 6(6):55–64

Diab MS, Saleh MF (1982) The hydrogeochemistry of the Pleistocene aquifer in the Nile Delta area, Egypt. Environ Int 7:221–230. https://doi.org/10.1016/0160-4120(82)90109-X

DRI (Drainage Research Institute) (1994) Drainage water in the Nile Delta, yearbook of 1992/1993. Reuse Monitoring Programme, National Water Research Centre, Cairo

DRI (Drainage Research Institute) (1998) Monitoring and analysis of drainage water quality (MADWQ) in Egypt. Interim Report, Cairo

Ebraheem AM, Senosy MM, Dahab KA (1997) Geoelectrical and hydrogeochemical studies for delineating groundwater contamination due to salt water intrusion in the northern part of the Nile Delta, Egypt. Ground Water 35:216–222

Eid HM (2001) Climate change studies on Egyptian agriculture. Soils, Water and Environment Research Institute (SWERI) ARC, Ministry of Agriculture, Giza

El-Fayoumey I (1987) Geology of the Quaternary succession and its impact on the groundwater reservoir in the Nile Delta region, Egypt. Bulletin Faculty of Science, Menoufia University, Egypt

Eissa M, Ali M, Zaghlool E, Stash OS (2019) Hydrochemical and stable isotopes indicators for detecting sources of groundwater contamination close to Bahr El-Baqar drain, eastern Nile Delta, Egypt. Water Sci 33:54–64. https://doi.org/10.1080/11104929.2019.1635346

El Bouraie MM, Motawea EA, Mohamed GG, Yehia MM (2011) Water quality of Rosetta branch in Nile delta, Egypt. Suo 62:31–37

El Ramly IM (1997) Hydrogeological and water quality characteristics of the saturated zone beneath the various land uses in the Nile Delta region, Egypt. In: Freshwater contamination in proceedings of Rabat symposium S4. IAHS Publ. no. 243. pp 255–261

Elagouza MH, Abou-Shleela SM, Belalb AA, El-Mohandesa MAO (2020) Detection of land use/cover change in Egyptian Nile Delta using remote sensing. Egypt J Remote Sens Space Sci 23:57–62

El-Fakharany MA, Mansour NM, Yehia MM, Monem M (2017) Evaluation of groundwater quality of the quaternary aquifer through multivariate statistical techniques at the southeastern part of the Nile Delta, Egypt. Sustain Water Resour Manag. https://doi.org/10.1007/s40899-017-0087-6

El-Gamal H, Dahab K, Werner A (2004) A multi-tracer study of groundwater in reclamation areas southwest of the Nile Delta, Egypt. International workshop on the application of isotope techniques in hydrological and environmental studies UNESCO, September 6–8, 2004, Paris, France

El-Raey M, Nasr S, Frihy O, Desouki S, Kh, Dewidar (1995) Potential impacts of accelerated sea-level rise on Alexandria Governorate, Egypt. J Coast Res 14:190–204

Elshobary ME, Essa DI, Attiah AM, Salem ZE, Qi X (2020) Algal community and pollution indicators for the assessment of water quality of Ismailia canal, Egypt. Stochastic Environmental Research and Risk Assessment. https://doi.org/10.1007/s00477-020-01809-w

Eltarabily MG, Negm AM, Yoshimura C, Abdel-Fattah S, Saavedra OC (2018) Quality assessment of southeast Nile Delta groundwater for irrigation. Water Resour 45(6):975–991. https://doi.org/10.1134/S0097807818060118

Eltarabily MGA, Negm AM (2017) Groundwater management for sustainable development east of the Nile Delta aquifer. In: Negm AM (ed) Groundwater in the Nile Delta, Hdb Env Chem. https://doi.org/10.1007/698_2017_102

Eltarabily MGA, Negm AM (2018) Groundwater management for sustainable development plans for the Western Nile Delta. In: Negm AM (ed) Groundwater in the Nile Delta, Hdb Env Chem. https://doi.org/10.1007/698_2018_247

Emara MM, El Sabagh I, Kotb A, Turkey AS, Husseen D (2007) Evaluation of drinking groundwater for the rural areas adjacent to the nearby desert of Giza Governorate of greater Cairo, Egypt. In: Linkov I, Kiker GA, Wenning RJ (eds) Environmental security in harbors and coastal areas. NATO security through science series (series C: Environmental Security). Springer, Dordrecht

Embaby AAA, Beheary MS, Rizk SM (2017) Groundwater quality assessment for drinking and irrigation purposes in El-Salhia plain East Nile Delta Egypt. Int J Innov Eng Technol (Special Issue on EGE) 51–73

ENCID (1996) Egyptian national committee on irrigation and drainage. https://www.icid.org/v_egypt.pdf

Engelen J, Essink GHP, Kooi H, Bierkens MFP (2018) On the origins of hypersaline groundwater in the Nile Delta aquifer. J Hydrol 560:301–317. https://doi.org/10.1016/j.jhydrol.2018.03.029

FAO (1994) Water quality for agriculture. Irrigation and drainage paper. 29 Rev.2, FAO, Rome

Farid MSM (1985) Nile Delta groundwater study. MSc thesis, Fac Sci

Fekry AM (2004) Assessment of impacts of groundwater development in the west Nile Delta. In: Proceeding of the 3rd international symposium on geophysics, Tanta (2004), pp 262–270

Ferguson G, Gleeson T (2012) Vulnerability of coastal aquifers to groundwater use and climate change. Nat Clim Change 2:342–345. https://doi.org/10.1038/nclimate1413

Foster S, Garduno H, Evans R, Olson D, Tian Y, ZhangW HZ (2004) Quaternary aquifer of the North China plain—assessing and achieving groundwater resource sustainability. Hydrogeol J 12:81–93

Frihy OE (2003) The Nile Delta-Alexandria coast: vulnerability to sea-level rise, consequences and adaptation. J Mitigation Adapt Strateg Glob Change 8:115–138

Gad M, Elsayed S, Moghanm FS, Almarshadi MH, Alshammari AS, Khedher KM, Ebrahim EM, Hussein H (2020) Combining water quality indices and multivariate modeling to assess surface water quality in the Northern Nile Delta, Egypt. Water 12:2142. https://doi.org/10.3390/w12082142

Geirnaert W, Laeven MP (1992) Composition and history of ground water in the western Nile Delta. J Hydrol 138:169–189

Gemail KS, El Alfy M, Ghoneim MF, Shishtawy AM, Abd El-Bary M (2017) Comparison of DRASTIC and DC resistivity modeling for assessing aquifer vulnerability in the central Nile Delta, Egypt. Environ Earth Sci 76:350. https://doi.org/10.1007/s12665-017-6688-4

Geriesh MH, Balke K, El-Rayes AE, Mansour BM (2015) Implications of climate change on the groundwater flow regime and geochemistry of the Nile Delta, Egypt. J Coast Conserv 19:589–608

Geriesh MH, EL-Rayes AE (2001) Municipal contamination of shallow groundwater beneath south Ismailia villages. 5th International conference on geochemistry. Alex. Univ., Egypt, pp 241–253. Accessed 12–13 Sept 7
Ghobara M, Salem ZE (2017) spatiotemporal fluctuations in phytoplankton communities and their potential indications for the pollution status of the irrigation and drainage water in the Middle Nile Delta Area, Egypt. In: The handbook of environmental chemistry (conventional water resources and agricultural in Egypt, 2019). Springer, Berlin, Heidelberg. ISBN: 978-3-319-95065-5
Gomaah M, Meixner T, Korany EA, Garamoon H, Gomaa MA (2016) Identifying the sources and geochemical evolution of groundwater using stable isotopes and hydrogeochemistry in the Quaternary aquifer in the area between Ismailia and El Kassara canals. Northeastern Egypt. Arab J Geosci 9:437. https://doi.org/10.1007/s12517-016-2444-4
Gornitz V (1991) Global coastal hazards from future sea level rise, paleogeography, paleoclimatology, palaeoecology. Glob Planet Change Sect 89:379–398
Hamza MS, Awad MA, Salem WM, Aly AIM (1997) Isotope techniques in the study of groundwater in the Eastern Nile Delta and the transport of pollution from drainage water into Lake Manzala, Egypt. Isotope Techniques in the Study of Environmental Change Proceedings of a Symposium, April 1997, Vienna, pp 14–18
Hasan SS, Salem ZE, Sefelnasr A (2023) Assessment of hydrogeochemical characteristics and seawater intrusion in coastal aquifers by integrating statistical and graphical techniques: quaternary aquifer, West Nile Delta, Egypt. Water 15:1803. https://doi.org/10.3390/w15101803
Hassanin L (2010) Egypt GIS Watch. http://www.giswatch.org/sites/default/files/gisw2010country egypten.pdf. Accessed at 20 Feb 2018
Hegazy D, El-Bastaweesy M, Garamoon H, Melegy A, Elsaid MA (2018) The impacts of human and agricultural activities on the groundwater quality in Mostorod Abu Zabal Area, East Nile Delta, Egypt. Al Azhar Bullet Sci 29(2):59–71
Hurst HE (1952) Long-term storage capacity of reservoirs. Trans Am Soc Civ Eng 116:770–808
Hussein MF, Nada AA, Awad MA (1994) Groundwater geochemistry of Nile Delta-Desert interface. In: Isotope hydrology (5–9 Nov, 1994) 2nd Arab conference on the peaceful uses of atomic energy, IAEA, Cairo
IPCC (2007) An assessment of the intergovernmental panel on climate change. In: Adopted section by section at IPCC Plenary XXVII, represents the formally agreed statement of the IPCC concerning key findings and uncertainties contained in the working group contributions to the fourth assessment report, Valencia, Spain
Javi ST, Mokhtari H, Rashidi A, Javi HT (2015) Analysis of spatiotemporal relationships between irrigation water quality and geoenvironmental variables in the Khanmirza Agricultural Plain, Iran. J Biodiv Environ Sci 6:240–252
Kaiser MF, El Rayes A, Ghodeif K, Geriesh B (2013) GIS data integration to manage waterlogging problem on the Eastern Nile Delta of Egypt. Int J Geosci 4:680–687. https://doi.org/10.4236/ijg.2013.44063
Kamel H, Eita T, Sarhan M (1998) Nile Delta hydrocarbon potentialities. In: 14th exploration and production conference, EGPC, Cairo, pp 485–503
Kanagaraj G, Elango L, Sridhar SGD, Gowrisankar G (2018) Hydrogeochemical processes and influence of seawater intrusion in coastal aquifers south of Chennai, Tamil Nadu, India. Environ Sci Pollut Res 25:8989–9011. https://doi.org/10.1007/s11356-017-0910-5
Kashef AI (1981a) The Nile-one River and nine countries. J Hydrol 53:53–71
Kashef AI (1981b) Technical and ecological impacts of the high Aswan Dam. J Hydrol 53:73–84
Khalil MA, Salem ZE, Gheda SF, El-Sheekh MM (2013) Quality assessment of drinking water in Tanta City, Egypt. J Environ Sci Eng B 2:257–275

Leaven MT (1991) Hydrogeological study of the Nile Delta and adjacent desert areas, Egypt, with emphasis on hydrochemistry and isotope hydrology, thesis, Free University, Amsterdam, also published by RIGW/IWACO as Technical note TN 77.01300-91-01

Liu CM, Xia J (2004) Water problems and hydrological research in the Yellow River and the Huai and Hai River basins of China. Hydrol Process 18:2197–2210

Met Office Hadley Centre (2011) Climate: observations, projections and impacts: Egypt. Produced by the Met Office

Metwally MI, Armanuos AM, Zeida BA (2023) Comparative study for assessment of groundwater vulnerability to pollution using DRASTIC methods applied to central Nile Delta, Egypt. Int J Energy Water Resour 7:175–190. https://doi.org/10.1007/s42108-022-00198-w

Meyer G (1998) Economic changes in the newly reclaimed land: from state farming to smallholding and private agricultural enterprises. In: Hopkins NS, Westergaard K (eds) Direction of change in rural Egypt. American University of Cairo Press, pp 334–356

Mogren S, Shehata M (2020) Groundwater vulnerability and risk mapping of the quaternary aquifer system in the northeastern part of the Nile Delta, Egypt. Afr J Geogr Reg Plann 7:01–13

Morsy WS (2009) Environmental management to groundwater resources for Nile Delta Region. PhD thesis, Faculty of Engineering, Cairo University, Cairo

MWRI (2013a) Adaptation to climate change in the Nile Delta through integrated coastal zone management. Ministry of Water Resources and Irrigation, Giza

MWRI (2013b) National water resources plan for Egypt 2017, Ministry of Water Resources and Irrigation

Negm AM, Armanuos AM (2016) GIS-based spatial distribution of groundwater quality in the Western Nile Delta, Egypt. In: Negm AM (ed) The Nile Delta, Hdb Env Chem. https://doi.org/10.1007/698_2016_66

Negm AM, Saavedra O, El-Adawy A (2017) Nile Delta biography: challenges and opportunities. In: Negm AM (ed) The Nile Delta, Hdb Env Chem 55:3–18. https://doi.org/10.1007/698_2016_62

Negm AM, Sakr S, Abd-Elaty I, Abd-Elhamid HF (2018) An overview of groundwater resources in Nile Delta aquifer. In: Negm (ed) Groundwater in the Nile Delta, Hdb Env Chem. https://doi.org/10.1007/698_2017_193

Neshat A, Pradhan B, Dadras M (2014) Groundwater vulnerability assessment using an improved DRASTIC method in GIS. Resour Conserv Recycl 86:74–86. https://doi.org/10.1016/j.resconrec.2014.02.008

Nhu VH, Rahmati O, Falah F, Shojaei S, Al-Ansari N, Shahabi H (2020) Mapping of groundwater spring potential in karst aquifer system using novel ensemble bivariate and multivariate models. Water 12:985. https://doi.org/10.3390/w12040985

Nofal ER, Amer MA, El-Didy SM, Akram MF (2015) Sea water intrusion in Nile Delta in perspective of new configuration of the aquifer heterogeneity using the recent stratigraphy data J. Am Sci 11:567–570

Nosair AM, Shams MY, AbouElmagd LM, Hassanein A, Fryar AE, Abu Salem HS (2022) Predictive model for progressive salinization in a coastal aquifer using artificial intelligence and hydrogeochemical techniques: a case study of the Nile Delta aquifer, Egypt. Environ Sci Pollut Res 29:9318–9340

Patrick RJ, Grant K, Bharadwaj L (2019) Reclaiming Indigenous planning as a pathway to local water security. Water 11:936. https://doi.org/10.3390/w11050936

Phillips FM (1994) Environmental tracers for water movement in desert soils of the American southwest. Soil Sci Soc Am J 58:15–24

Post VEA (2005) Fresh and saline groundwater interaction in coastal aquifers: is our technology ready for the problems ahead. Hydrogeol J 13:120–123. https://doi.org/10.1007/s10040-004-0417-2

Pulido-Velazquez M, Peña-Haro S, García-Prats A, Mocholi-Almudever AF, Henriquez-Dole L, Macian-Sorribes H, Lopez-NicolasA (2015) Integrated assessment of the impact of climate and land use changes on groundwater quantity and quality in the Mancha Oriental system (Spain). Hydrol Earth Syst Sci 19:1677–1693

RIGW (1980) Project of groundwater study in the Nile Delta. Ministry of Irrigation, 108pp

RIGW (1992) Hydrogeological map of Egypt. Nile Delta; scale 1:500,000, 1st edn. El-Qanatir, Egypt

RIGW (2002) Nile Delta Groundwater modeling report, Research Inst. for Groundwater, Kanater El-Khairia, Egypt

Rizk, TM (1997) Hydrogeological studies on El-Salhia district, East Nile Delta, Egypt. M.Sc Thesis, Fac. Sci., Mansoura Univ., 125pp

Said R (1962) The geology of Egypt. Elsevier, Amsterdam, the Netherlands

Said R (1993) The Nile River: geology, hydrology, and utilization. Pergamon, New York

Sakr SA (2005) Enhancing groundwater availability in the Nile Delta Aquifer of Egypt. Ain Shams University-Faculty of Engineering. Sci Bull 40(3)

Saleh MF (1980) Some hydrogeological and hydrochemical studies on the Nile Delta. MSc thesis, Fac Sci Ain Shams Univ

Salem ZE (2009) Natural and human impacts on the groundwater under an Egyptian village, central Nile Delta—a case study of Mehallet Menouf. In: Thirteenth international water technology conference (IWTC 13), March 12–15, 2009. Hurghada, Egypt, 3:1397–1414

Salem ZE, El-horiny MM (2014) Hydrogeochemical evaluation of calcareous eolianite aquifer with saline soil in a semiarid area. Environ Sci Pollut Res 21:8294–8314

Salem ZE, Osman OM (2016) Shallow subsurface temperature in the environs of El-Nubaria canal, northwestern Nile Delta of Egypt: implications for monitoring groundwater flow system. Environ Earth Sci 75:1241. https://doi.org/10.1007/s12665-016-6046-y

Salem ZE, Abdelrahman k, Kováčiková S, Badran OM (2021b) Use of various statistical techniques to assess the vertical and lateral change in the groundwater chemistry of quaternary aquifer in an irrigated highly populated area. J King Saud Univ Sci 33:1015562. https://doi.org/10.1016/j.jksus.2021.101556

Salem ZE, Al Temamy AM, Salah MK, Kassab M (2016) Origin and characteristics of brackish groundwater in Abu Madi coastal area, Northern Nile Delta, Egypt. Estuar Coast Shelf Sci 178:21–35

Salem ZE, Arafa NA, Abdeldayem AL, Youssef YM (2025a) Machine learning-enhanced GALDIT modeling for the Nile Delta aquifer vulnerability assessment in the Mediterranean region. Groundw Sustain Dev 28:101403. https://doi.org/10.1016/j.gsd.2024.101403

Salem ZE, Atwia MG, El-Horiny MM (2015) Hydrogeochemical analysis and evaluation of groundwater in the reclaimed small basin of Abu Mina, Egypt. Hydrogeol J 23(8):1781–1797

Salem ZE, El Bayumy DA (2016a) Use of the subsurface thermal regime as a groundwater-flow tracer in the semi-arid western Nile Delta, Egypt. Hydrogeol J 24:1001–1014. https://doi.org/10.1007/s10040-016-1377-z

Salem ZE, El Bayumy DA (2016b) Hydrogeological, petrophysical and hydrogeochemical characteristics of the groundwater aquifers east of Wadi El-Natrun, Egypt. NRIAG J Astron Geophys 5:106–123

Salem ZE, El-Nahrawy A (2020) Use of 1-D subsurface thermal profiles to characterize the groundwater flow in the Central Nile Delta region. Arab J Geosci 13:12p

Salem ZE, ElNahrawy A, Attiah AM, Edokpayi JN (2022b) Vertical and spatial evaluation of the groundwater chemistry in the Central Nile Delta quaternary aquifer to assess the effects of human activities and seawater intrusion. Front Environ Sci 10:961059. https://doi.org/10.3389/fenvs.2022.961059

Salem ZE, ElNahrawy A, Ghobara M (2018b) Spatiotemporal hydrochemical evaluation and quality assessment of drainage water compared to canal surface water in the Middle Nile Delta, Egypt. In: The handbook of environmental chemistry (unconventional water resources and agricultural in Egypt, 2019). Springer, Berlin, Heidelberg. ISBN 978-3-319-95071-6

Salem ZE, Elnahrawy AM, Abdelrahman K, Fnais MM, Abu-Alam T (2022a) Oxygen and hydrogen stable isotopes as recharge indicators, Central Nile Delta Quaternary aquifer, Egypt. J King Saud Univ Sci 34:101834

Salem ZE, Elsaiedy G, and ElNahrawy A (2017c) Assessment of the groundwater quality for drinking and irrigation purposes in the Central Nile Delta Region, Egypt. In: Negm AM (ed) Groundwater in the Nile Delta, Hdb Env Chem. https://doi.org/10.1007/698_2017_137_19

Salem ZE, Elsaiedy G, ElNahrawy A (2017d) Hydrogeochemistry and quality assessment of groundwater under some central Nile delta villages, Egypt. In: Negm AM (ed) Groundwater in the Nile Delta, Hdb Env Chem. https://doi.org/10.1007/698_2017_111_18

Salem ZE, Fathy MS, Helal AI, Afifi SY, Attiah AM (2020) Use of subsurface temperature as a groundwater flow tracer in the environs of Ismailia Canal, Eastern Nile Delta, Egypt. Arab J Geosci (2020) 13:13p

Salem ZE, Fathy MS, Helal AI, Afifi SY, Attiah AM (2021a) Groundwater quality for irrigation as ANA of sustainable development approaches: a case study of semi-arid area around Ismailia Canal, Eastern Nile Delta, Egypt. In: Negm A, Elkhouly A (eds) Groundwater in Egypt's deserts. Springer Water, https://doi.org/10.1007/978-3-030-77622-0_12

Salem ZE, Gaame OM and Hassan TM (2008) Using temperature logs and hydrochemistry as indicators for seawater intrusion and flow lines of groundwater in quaternary aquifer, Nile Delta, Egypt. In: Fifth international symposium on geophysics (ISG 5), pp 25–38

Salem ZE, Gaame OM, Hassan TM (2018a) Integrated subsurface thermal regime and hydrogeochemical data to delineate the groundwater flow system and seawater intrusion in the Middle Nile Delta, Egypt. In: Negm AM (ed) Groundwater in the Nile Delta, Hdb Env Chem 73:461–486. https://doi.org/10.1007/698_2018_253

Salem ZE, Ghobara M, ElNahrawy AA (2017b) Spatio-temporal evaluation of the surface water quality in the middle Nile Delta using Palmer's algal pollution index. Egypt J Basic Appl Sci 4(3):219–226

Salem ZE, Hasan SS (2021) Use of GALDIT model and HFE-diagram to assess seawater intrusion vulnerability in West Nile Delta, Egypt. Arab J Geosci 14:1318. https://doi.org/10.1007/s12517-021-07678-z

Salem ZE, Hasan SS, Sefelnasr AM (2025b) Health risk assessment of groundwater use for drinking in West Nile Delta, Egypt. Sci Rep 15:7414. https://doi.org/10.1038/s41598-025-90477-3

Salem ZE, Negm AM, Nahrawy AM (2017a) Hydrogeophysical characteristics of the Central Nile Delta. In: Negm (ed) Groundwater in the Nile Delta, Hdb Env Chem (2019) 73:187–210. https://doi.org/10.1007/698_2017_75

Salem ZE, Osman OM (2017a) Use of major ions to evaluate the hydrogeochemistry of groundwater influenced by reclamation and seawater intrusion, West Nile Delta, Egypt. Environmental Science and Pollution Research, https://doi.org/10.1007/s11356-016-8056-4

Salem ZE, Osman OM (2017b) Use of geoelectrical resistivity to delineate the seawater intrusion in the northwestern part of the Nile Delta, Egypt. In: Negm AM (ed) Groundwater in the Nile Delta, Hdb Env Chem 73:425–460. https://doi.org/10.1007/698_2017_175

Salem ZE, Osman OM (2018) Use of one-dimensional subsurface temperature profiles to characterize the groundwater flow system in the northwestern part of the Nile Delta, Egypt. In: The handbook of environmental chemistry (Groundwater in the Nile Delta, 2019). Springer, Berlin, Heidelberg. ISBN: 978-3-319-94283-4

Salem ZE, Sefelnasr A, Hasan SS (2022c) As a water resources management tool, groundwater quality assessment for irrigation in the Young Alluvial Plain of Western Nile Delta, Egypt. In: Negm AM, El-Rawy M (eds) Sustainability of groundwater in the Nile Valley, Egypt, Earth and Environmental Sciences Library. https://doi.org/10.1007/978-3-031-12676-5_8

Salem ZE, Sefelnasr AM, Hasan SS (2019) Assessment of groundwater vulnerability for pollution using DRASTIC index, young alluvial plain, Western Nile Delta, Egypt. Arab J Geosci 12:727. https://doi.org/10.1007/s12517-019-4883-1

Salem ZE, Temamy AMA, Salah MK, Kassab M (2018c) Evaluation of water resources qualities for agriculture irrigation in Abu Madi area. Northern Middle Nile Delta 74:277–316. https://doi.org/10.1007/698_2018_27320

Sawyer AH, David CH, Famiglietti JS (2016) Continental patterns of submarine groundwater discharge reveal coastal vulnerabilities. Science 353:705–707. https://doi.org/10.1126/science.aag1058

Schlumberger (1984) Well evaluation. In: Conference Egypt. Schlumberger Middle East S.A. Chapter 1. Geology of Egypt, pp 1–64

Sharaky AM, Atta SA, El Hassanein AS, Khallaf KMA (2007) Hydrogeochemistry of groundwater in the Western Nile Delta aquifers, Egypt. In: 2nd international conference on the geology of Tethys, 19–21 March 2007, Cairo University, Egypt

Shata A, El-Fayoumey I (1970) Remarks on the regional geological structure of the Nile Delta.Bucharest symposium, UNESCO publication chapter, Catalog No 14859:189–197

Shata AA (1999) Contribution to the hydrogeology of the Nile delta region, Egypt. In Deltas, Modern and Ancient, p 19, Cairo

Sherif MM, Sefelnasr A, Javad A (2012) Incorporating the concept of equivalent freshwater head in successive horizontal simulations of seawater intrusion in the Nile Delta aquifer, Egypt. J Hydrol 464:186–198

SNC (2010) Egypt's second national communication. Egyptian Environmental Affairs Agency (EEAA-May 2010) under the United Nations framework convention on climate change

Stanley DJ, Warne AG (1994) Worldwide initiation of Holocene marine deltas by deceleration of sea-level rise. Science 265:228–232

Taha AA, El Mahmoudi AS, El-Haddad IM (2003) Evaluation of the water quality in new communities southeast the Nile Delta, Egypt. Emirates J Eng Res 8(2):51–67

Taha MS, Armanuos AM, Zeidan BA (2022) Creep of seawater intrusion in the Nile Delta Aquifer. Hydrol Current Res 13:425

Un-United Nations (2003) Water for people, water for life; UN world development report (WWDR). New York, NY, USA, UN

Walvoord MA, Phillips FM, Stonestrom DA, Evans RD, Hartsough PC, Newman BD, Striegl RG (2003) Areservoir of nitrate beneath desert soils. Science 302:1021–1024

Werner AD, Bakker M, Post VEA, Vandenbohede A, Lu C, Ataie-Ashtiani B, Simmons CT, Barry DA (2013) Seawater intrusion processes, investigation and management: recent advances and future challenges. Adv Water Res 51:3–26. https://doi.org/10.1016/j.advwatres.2012.03.004

WHO (2008) Guidelines for drinking- water quality. http://www.who.int/water_sanitation_health/dwq/fu

WHO (World Health Organization) (2006) Guidelines for drinking-water quality. In: First Addendum to Third Edition, vol. 1. Recommendations. WHO, Geneva, Switzerland

WMRI-NWRC (2002) Unpublished report under the matching supply and demand project. Water Management Research Institute, National Water Research Center, Ministry of Water Resources and Irrigation, Cairo

World Bank (2007) World development indicators. Washington, D.C.: World Bank. http://databank.world bank.org/ddp/home.do

Yang X, Chen Y, Pacenka S, Gao W, Zhang M, Sui P (2015) Recharge and groundwater use in the North China Plain for six irrigated crops for eleven years. PLoS ONE 10:e0115269. https://doi.org/10.1371/journal.pone.0115269
Zeidan BA, Elshinnawy H (2015) Impact of hydraulic conductivity uniformity on seawater intrusion in Nile Delta aquifer, Egypt. J Al-Azhar Univ Eng Sector 37(4)

Seawater Intrusion Challenges in the Central Nile Delta Aquifer: Assessment and Monitoring

8

Moheb H. Abdelbary, Youssef M. Youssef, Samir Z. Kamh, and Khaled S. Gemail

Abstract

Coastal aquifers worldwide confront significant challenges from salinization, with seawater intrusion (SWI) particularly severe in urbanized coastal areas. The Nile Delta Aquifer (NDA), a crucial coastal resource, is notably affected by SWI, especially in its central regions, underscoring the need for comprehensive assessment and monitoring frameworks. This chapter reviews the impacts of sea level rise (SLR) and SWI on the NDA, offering an overview of salinity sources in coastal aquifers and evaluating previous numerical modeling studies related to SWI in this area. It also examines the potential impacts of climate change. A central focus is integrating hydrogeochemical and geophysical methods to characterize the freshwater-saltwater interface accurately. This study employs high-resolution vertical electrical resistivity surveys alongside extensive hydrogeochemical analysis, including major ion chemistry, stable isotope analysis, and geochemical modeling. The findings from this interdisciplinary approach provide valuable insights for developing sustainable groundwater management strategies aimed at mitigating the impacts of SWI and ensuring the long-term availability of freshwater resources in the Nile Delta region. The chapter concludes

M. H. Abdelbary (✉) · S. Z. Kamh
Geology Department, Faculty of Science, Tanta University, Tanta, Egypt
e-mail: mhabib1959@yahoo.com

Y. M. Youssef
Geological and Geophysical Engineering Department, Faculty of Petroleum and Mining Engineering, Suez University, Suez, Egypt

K. S. Gemail
Environmental Geophysics Lab (ZEGL), Geology Department, Faculty of Science, Zagazig University, Zagazig, Egypt

M. A. Khalifa and K. S. Gemail (eds.), *Geology of the Nile Deltas and Natural Resources*, Deltas of the World, https://doi.org/10.1007/978-3-031-96594-4_8

with a discussion of the research implications and outlines potential future directions for addressing this critical environmental challenge.

Keywords

Nile delta aquifer (NDA) • Seawater index • Climatic changes and SLR • Vertical electrical sounding • Groundwater management • Hydrogeochemical modeling

8.1 Introduction

Coastal aquifer salinization poses a significant challenge for cities situated along coastlines worldwide. Various sources contribute to the salinity of these freshwater aquifers, including seawater intrusion (SWI) (Van Pham and Lee 2015), mixing with highly saline fossil groundwater (Tijani 2004), and contamination from wastewater (Gemail et al. 2024). Among these factors, SWI remains the predominant cause of salinization. The intrusion of saltwater into groundwater can severely degrade water quality; even a minor influx of saltwater, just 2%, can render freshwater supplies unsuitable for drinking, leading to the abandonment of vital water resources (Abd-Elhamid and Abdelaty 2017; Armanous et al. 2022; Abd-Elaty et al. 2024; Hagage et al. 2025).

The Nile Delta Aquifer (NDA), covering an area of approximately 40,000 km^2, is a critical freshwater resource supporting over 50 million people and irrigating 33% of Egypt's agricultural land. However, the NDA is highly susceptible to SWI, particularly in its northern regions adjacent to the Mediterranean Sea. Excessive groundwater extraction, estimated at 6.1 billion m^3/year, has exacerbated SWI, with intrusion rates ranging from 50 to 100 m/year in some areas (Sherif 1999; Mazi et al. 2014; Arafa et al. 2024a, b). This aquifer requires thoughtful management strategies to limit further freshwater degradation. Groundwater modeling can be valuable for simulating future saltwater intrusion scenarios and developing long-term adaptation plans.

Numerous studies have explored seawater intrusion in the Nile Delta Aquifer (NDA) using various numerical modeling techniques. These studies have employed a range of numerical models, including finite difference (MODFLOW), finite element (FEFLOW), and coupled variable-density flow and solute transport models (SEAWAT), to simulate SWI dynamics under various scenarios. Notable contributions include the works of El-Fayoumi (1968), Wilson et al. (1979), and Kashef (1983), among others, which collectively demonstrate significant inland seawater intrusion, extending over 100 km from the Mediterranean coast. Researchers (e.g., Mabrouk et al. 2018; Abd-Elaty et al. 2024; Armanuos et al. 2024) have simulated factors influencing seawater intrusions (SWI), such as groundwater pumping rates and SLR. While increased pumping and rising sea levels exacerbate saltwater intrusion, some studies suggest that rising groundwater discharge poses an even greater threat to freshwater salinization than projected sea level rise.

While numerous studies have investigated SWI in the NDA, the complex interplay between climate change factors and SWI remains underexplored. Climate projections for the Nile Delta region indicate a potential sea level rise of 0.5–1.0 m by 2100, coupled with changes in precipitation patterns and evapotranspiration rates, which could significantly alter groundwater recharge dynamics and SWI progression (IPCC 2021). For instance, Tarik (2013) utilized the groundwater modeling system (GMS) to assess expected sea level rise, finding that the seawater-freshwater interface would stabilize after approximately 400 years, with a maximum freshwater loss of around 2.2%. This study underscored the necessity for adequate recharge from the Nile River to mitigate saltwater intrusion. Similarly, Abd-Elhamid et al. (2016), Abd-Elalty et al. (2024), and Saqr and Abd-Elmaboud (2024) identified the NDA as vulnerable to climate change, highlighting substantial water losses attributable to rising sea levels.

While there is consensus regarding the importance of increasing freshwater recharge to combat SWI in the NDA—primarily through absorption from the Nile River, runoff from irrigation canals, and surplus irrigation water (Kashef 1983; Sherif et al. 2012; Taha and Armanous 2023) the advancement of seawater into the aquifer has been correlated with declining water levels in irrigation canals (Abd-Elaty et al. 2014; Saqr and Abd-Elmaboud 2024).

Given the projected impacts of climate change and continued anthropogenic pressures on the NDA, there is an urgent need for effective, scientifically informed management strategies to mitigate SWI and ensure long-term aquifer sustainability, few studies have proposed effective strategies to manage SWI. Nasar (2004) evaluated the environmental impacts of extracting brackish water for desalination and reinjecting brine into deep wells, while Sherif and Kacimov (2008) suggested using brackish water pumping to mitigate SWI. Recently, Abd-Elhamid and Abdelaty (2017) introduced the treatment and recharge of wastewater, and abstraction and desalination of brackish water (TRAD) methodology as a sustainable approach to managing SWI.

In addition to modeling studies, various methods have been employed to assess SWI in the NDA, including hydrogeochemical analysis (Arlt 1995; Nosair et al. 2021), hydrogeological investigations (Sherif 1999), geophysical surveys, Gemail et al. (2004), isotope tracer studies (Carreira et al. 2018; Campillo et al. 2021), and integrated geological, hydrogeochemical and electrical resistivity approaches (Ebraheem et al. 1997; Abd-Elbary 2012; Attwa et al. 2016; Gemail et al. 2017a; Abu Salem et al. 2022). These diverse techniques have delineated the extent of SWI and mapped aquifer vulnerability.

However, a significant research gap exists between traditional numerical and hydrochemical modeling approaches and comprehensive assessments of SWI mechanisms. Existing models often fail to capture the complex, nonlinear interactions between hydrogeological, geochemical, and anthropogenic factors influencing SWI dynamics. Moreover, the spatial and temporal heterogeneity of the NDA system necessitates a more integrated, multi-scale approach to accurately characterize and predict SWI processes. The current study aims to bridge this gap by integrating water sampling and geophysical mapping to provide a holistic understanding of the processes driving seawater intrusion in the Nile Delta. Additionally, this study emphasizes the role of an assessment and monitoring framework to support integrated groundwater management and sustainability efforts, ultimately contributing to more effective strategies for mitigating the impacts of SWI in coastal aquifers.

The current chapter integrates hydrogeochemical analysis of water samples from wells with geophysical data to assess water quality impacted by seawater intrusion. Samples were collected from wells at varying depths, which complicates the comprehensive understanding of the freshwater-saltwater interface and hinders insight into the mechanisms of intrusion in this coastal aquifer. To address this challenge, the vertical electrical resistivity (VES) method, a cost-effective surface geophysical tool, is employed, as it effectively maps the seawater interface due to the significant contrast in resistivity values between freshwater and seawater. The primary objective of this research is to elucidate the state of seawater intrusion in the study area and to define the complex hydrogeochemical processes that occur during the mixing of groundwater and saline water. This includes investigating the variables that govern salinity sources and the development of brackish groundwater in coastal aquifers. The study utilizes a range of hydrogeochemical analyses, including the Revelle index, Stuyfzand indicator based on the base exchange index, Piper diagram, hydrochemical facies evolution diagram (HFE-D), and the seawater mixing index (SMI), alongside statistical techniques such as cluster analysis.

Geophysical measurements will complement the analysis to validate the hydrogeochemical findings and delineate the vertical boundary of saltwater/freshwater intrusion. Additionally, spatial mapping of the various factors using GIS will facilitate visual analysis and interpretation. By integrating these findings, the study aims to provide a deeper understanding of seawater intrusion mechanisms in the Nile Delta aquifer system. The resulting model is anticipated to contribute to the development of improved management strategies for sustainable groundwater resources in coastal aquifers.

8.2 Nile Delta and Groundwater Degradation

The Nile Delta is a major agricultural and the most densely populated part of Egypt. The study areas occupy the central part of the Nile Delta and include Gharbiyyah governorate from the middle to northern parts and Kafr Al-Shikh governorate from the contact with Gharbiyyah governorate in the south to the south of Burllus Lake and the part between the Damietta branch and Mansoura governorate to the east and Rosetta branch to the west. The area falls between latitudes 30°50' and 31° 18' N and longitudes 30° 4' and 31°17' E (Fig. 8.1).

The Nile Delta aquifer has been exposed to extensive groundwater pumping during the past 30 years (Mabrouk et al. 2013). According to the General Authority for Drinking Water and Sanitation (2009) and the Authority of Irrigation in Gharbiyyah Governorate

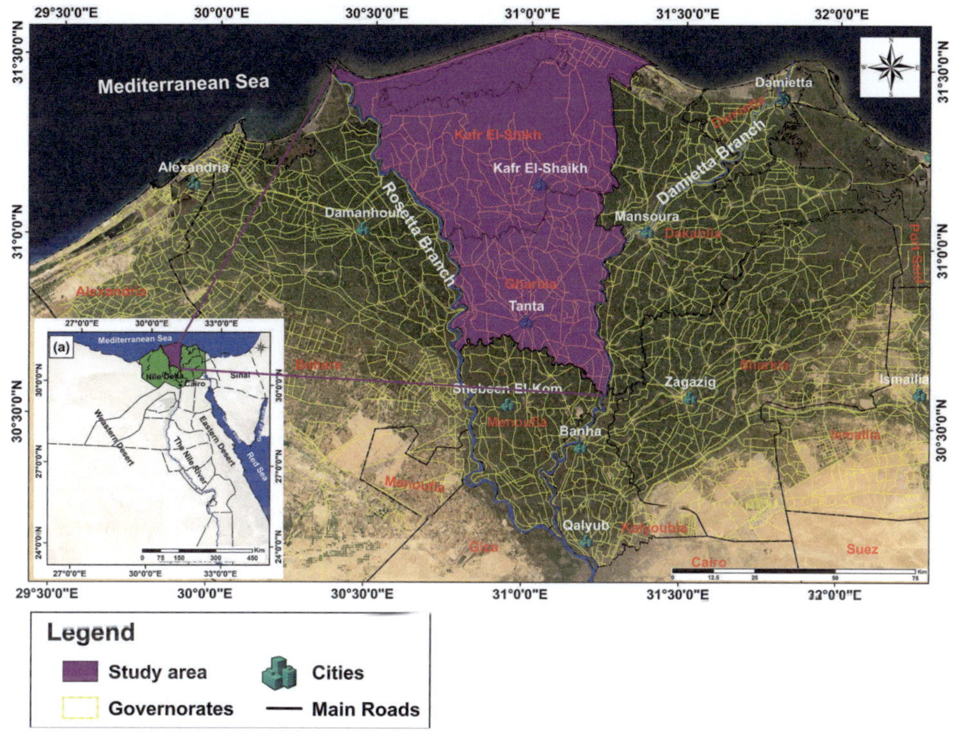

Fig. 8.1 Research region is located in the center of the Nile Delta, as shown in this Landsat image

for the period 2003–2005, the groundwater is considered the main source of drinking and domestic water, where extensive pumping took place through more than 100 groundwater wells distributed over the area of the governorate. In the southern region, groundwater salinity is less than 220 ppm, whereas in the northern region, it exceeds the permissible threshold of 1000 ppm. In the penultimate decade, as the northern part of the area was affected by increasing salinity as a result of extensive groundwater pumping, most of the groundwater wells in Gharbiyyah Governorate were contaminated by iron and manganese. For this reason, more than 10 large treatment stations for the Nile water were constructed using the two main branches of the Nile Delta and its sub-branches for better drinking and domestic water.

The General Authority for Drinking Water and Sanitation in Kafr Al-Sheikh Governorate has highlighted a concerning trend of escalating salinity levels from the southern to northern regions. Consequently, most groundwater wells have been decommissioned, with only a handful of observation wells managed by the Research Institute for Groundwater strategically placed throughout the governorate (Arlt 1995).

In a similar vein, up until 2015, the Qellen region relied exclusively on groundwater sources. Consequently, the Kafr El-Sheikh governorate and its coastal cities, spanning 120 km, now heavily rely on Nile water for consumption after treatment at specialized facilities. Notably, collaborative efforts between the Ministries of Irrigation and Agriculture have been instrumental in diverting significant volumes of Nile water for rice cultivation, aimed at safeguarding soil integrity against salinization. Moreover, recent years have witnessed a surge in the drilling of deep underground wells, exceeding 60 m in depth, to support aquaculture ventures, particularly in areas like Hamul, Riyadh, Sidi Salem, and Baltim. This shift toward increased groundwater extraction for fish farming purposes is anticipated to exacerbate seawater intrusion into the Northern Delta Aquifer (NDA), raising concerns about its sustainability and the potential impact on soil heterogeneities within the region (Gemail et al. 2017b).

8.3 Geological and Hydrogeological Settings

Several studies have been interested in studying the geology and hydrogeology of the Nile Delta (e.g., Attia 1975; Gheith 1976; Zaghloul et al.1977, 1979; El-Fawal 1979, 1981; Sestini 1989; El-Fayoumi et al. 1991; Bahr 1992; Sherif 1999; Sherif and Singh 2002; Sherif et al.2012; Abd-Elhamid et al.2016). The Nile Delta basin contains a thick sequence of Neogene–Quaternary deposits. These deposits include Sidi Salim, Qawasim, Rosetta (Miocene), and Abou Madi formations, Kafer El-Sheikh, El-Wastani, and Mit Ghamer formations (Pliocene–Pleistocene cycle), and Bilqas Formation (Holocene cycle) at the top.

The Bilqas Formation of the Holocene, a coastal lagoonal deposit, is predominantly comprised of silts and sandy mud, constituting the agricultural soil in the region. The lower section of the Bilqas Formation is characterized by silt and fine sand, with a thickness of a few meters, while the upper half features a clay cap, primarily composed of montmorillonite, exhibiting lower hydraulic conductivity (Sherif 1999; Troger et al. 1993) reported varying thicknesses of the Bilqas Formation, ranging from a few meters in the southern part of the study area to potentially exceeding 40 m and reaching up to 70 m in the northern portion, situated northeast of the Nile Delta. This clay cap acts as a protective layer for the underlying sandy aquifer in the Nile Delta.

The Nile Delta aquifer comprises the Mit Ghamr and El Wastani formations (Pleistocene/Pliocene), characterized by coarse sand and gravel with interspersed silt–clay lenses (Fig. 8.2). The dissolution of halite within the Quaternary aquifer is attributed to the presence of numerous clay lenses, as noted by Sallouma (1983), serving as a potential source of salinity, a phenomenon under scrutiny in this study. The Quaternary aquifer within the Nile Delta region typically lies within 50 m of the ground surface (Sherif et al. 2012), with a thickness ranging from 100 to 400 m in the south, increasing to an average of 800 m in the north (Fig. 8.2).

The hydraulic connection between the Bilqas Aquitard and the Quaternary aquifer is acknowledged, albeit with distinct water levels. Water levels range from 6 to 12 m

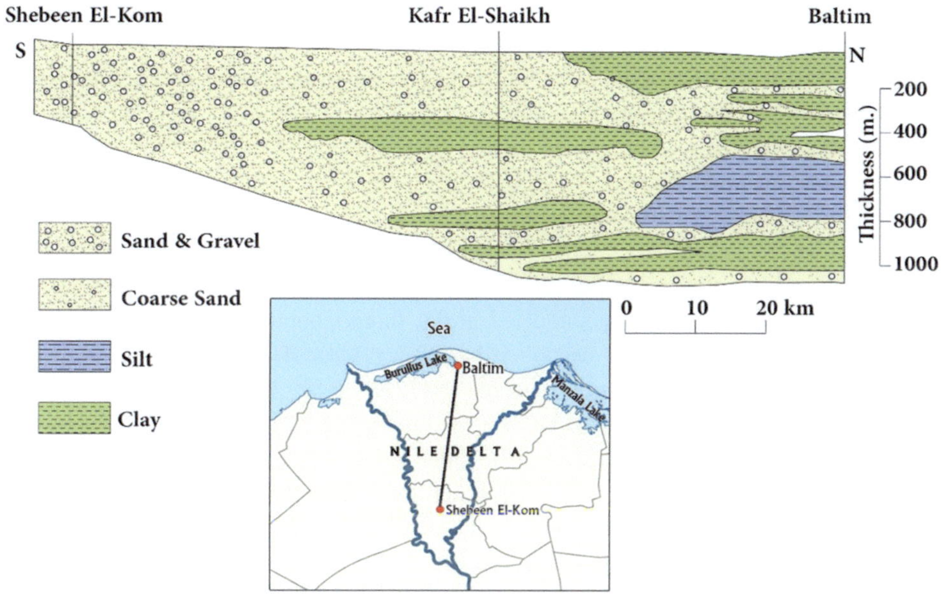

Fig. 8.2 Geological longitudinal section showing the change in the lithological characteristics of the Mit Ghamr Formation in the Nile Delta from south to north (Serage El-Din 1989)

above sea level in the central and southern regions of the Nile Delta, contrasting with levels between 0 and 5 m in the northern areas, resulting in a general groundwater flow toward the north. Recharge sources for the Nile Delta aquifer include Nile branches, canals, open drainage networks, and surplus irrigation water. The influence of wastewater from drainage networks, downward leakage of irrigation return flow, and canals through the clay aquitard impact the water balance of the aquifer systems in the region. The average hydraulic conductivity of the clay protection zone in the Nile Delta is reported to be 2.9×10^{-8} m/s (Sherif 1999). Dahab (1993) has shown variations in the hydraulic conductivity of the Nile Delta aquifer, ranging between 6.8×10^{-4} and 8.2×10^{-4} m/s in the southwestern and northern regions, with higher values (1.21×10^{-3} to 2.12×10^{-3} m/s) in the central and southeastern parts. Past pumping tests in the study area indicated an average hydraulic conductivity of 1.2×10^{-3} m/s, an average transmissivity of 10.26 m2/s, and an average storage coefficient of 1.5×10^{-4}.

The salinity of groundwater shows a large variation with depth in the northern part of the Nile Delta, and in the upper zone, salinity values range between 2500 and 77,000 ppm (Ebraheem et al. 1997). The thickness of this zone increases toward the north to reach 50 m in the area north of Kafer Shiekh. The area's northern region has shallow brackish water that was created by the mixing of freshwater from the south and saltwater from the north. The depth of this zone decreases toward the north of the studied area.

8.4 Methods and Materials

Samples of groundwater were taken in 2008 from 51 groundwater wells with varying total depths ranging from 15 to 100 m to evaluate the research area's incursion of seawater. Forty-seven groundwater samples were collected from the southern part of the study area (Gharbiyyah governorate), but due to the increase in salinity and the scarcity of groundwater wells, 4 groundwater samples were collected from the northern part (Kafer Al-Shikh governorate). Chemical analyses were carried out in the laboratory of the Geology Department, Tanta University, and the laboratory of the Agriculture Authority in Tanta for major elements. Tests conducted in the Tanta laboratory were performed using the standard analytical methods described by Hach (1990). These samples were used to determine TDS, EC, and the major elements (Na^+, K^+, Ca^{+2}, Mg^{+2}, Cl^-, NO^{-3}, $SO_4{}^{-2}$, and HCO^{-3}). The concentration of HCO^{-3} and pH values was measured in the field. A pH Meter (P314) was used to measure pH (Table 8.1).

To investigate SWI, a group of hydrogeochemical processes were performed using the Revelle index, Stuyfzand intrusion indicator, which is famed as (BEX), index of seawater mixing (SMI), diagram of Piper, hydrochemical facies evolution diagram (HFE-D), saturation index (SI) of a mineral, multivariate statistical techniques, including cluster analysis. The outcome results are validated using an electric resistivity tool.

8.4.1 Revelle Index

The Revelle ratio ($Cl^-/HCO^{-3} + CO_3{}^-$) (Revelle 1941) was used to determine the origin of the groundwater based on the fact that chloride is found mainly in seawater but relatively seldom in fresh groundwater. The reverse case is true for bicarbonate and carbonate, commonly found in groundwater but relatively low in concentrations in seawater. A Revelle ratio>200 represents seawater, while a ratio<1 represents freshwater.

Table 8.1 Classification of water type according to the chloride concentration (Stuyfzand 1986)

Main type	Cl–mg/l	Code	Samples
Very oligohaline	<5	G	
Oligohaline	5–30	g	
Fresh	30–150	F	1, 3, 4, 5, 6, 7, 8, 9, 13, 14, 16, 30
Fresh-brackish	150–300	Fb	2, 10, 11, 12, 15, 17, 18, 19, 20, 21, 22, 31, 32, 33
Brackish	$300–10^3$	B	23, 24, 25, 26, 27, 28, 29, 34, 35, 36, 37
Brackish-saline	$10^3–10^4$	Bs	38, 39, 40, 41, 42, 43, 44, 45, 46, 47, 48, 49, 50, 51
Saline	$10^4–2*10^4$	S	
Hyper-saline	$>2*10^4$	H	

8.4.2 Stuyfzand Water Classifications

The type of water samples has been classified according to a modified version of the Stuyfzand method (Stuyfzand 1986). The main type is determined by the chloride content of samples, starting from < 5 mg/l of Cl^- as code (G) represents under freshwater type to > 20,000 mg/l of Cl^- as code (H) represents hyper–saline type water while the value < 150 mg/l as code (F) represents freshwater. Stuyfzand index was defined as BEX = Na+ + K+ + Mg2+ − 1.0716Cl^- (meq/L). A positive BEX number indicates freshening salinization by a negative BEX value, and no base exchange is shown by a zero BEX value.

8.4.3 Piper Diagram

The trilinear diagram developed by Hill (1940) and later modified by Piper (1944) does not show the various concentrations of particular solutions, but instead, it helps to determine whether a particular water may be a simple mixture of different waters or whether it is influenced by solution or precipitation. The Piper trilinear diagram can display a large number of analyses and does show the product of mixed waters from different sources. The Piper diagram displays the outcome of a chemical analysis as meq/1%. In the central diamond diagram, average freshwater and seawater can be plotted. The line connecting these two points or any water sample found on or near this line represents the conservative mixing of freshwater and seawater (Fig. 8.3). Freshwater and seawater intrusion can be represented on the Piper diagram as compositional changes that take place on account of cation exchange. Simply, a sodium surplus is indicative of freshwater encroachment; a calcium surplus indicates saltwater intrusion, according to Appelo and Geirnaert (1983). In the SWI phase, a reverse exchange reaction takes place as Eq. (8.1) liberates Ca^{2+} and adsorbs Na^+, and the solution becomes supersaturated by Ca^{2+}. Calcite can precipitate as a result of Eq. (8.2). This process drives the results to the upper right of the maximum line in the diagram (Fig. 8.3). During freshening, direct ion exchange takes place (or "refreshing"), when a salinized aquifer is flushed by fresh calcium bicarbonate water (Appelo and Geirnaert 1983) Eq. (8.3), then Ca^{2+} ion is removed from the solution, and the solution becomes undersaturated concerning calcite, so that calcite dissolution then lies to the lower left of the conservative mixing line in the diagram (Fig. 8.3).

$$\mathrm{Na^+ + Ca - [clay] \rightarrow Ca^{2+} + [clay] - Na} \tag{8.1}$$

$$\mathrm{Ca^{2+} + 2HCO_3 \longleftrightarrow CaCO_3 + CO_2 + H_2O} \tag{8.2}$$

$$\mathrm{Ca^{2+} + Na[clay] \rightarrow Na^+ + [clay] - Ca} \tag{8.3}$$

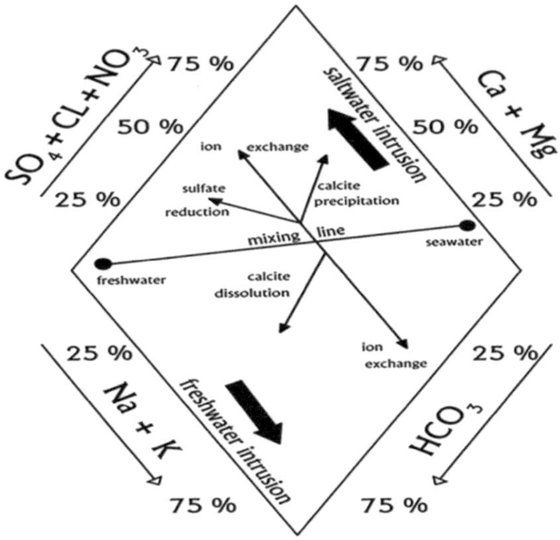

Fig. 8.3 Piper diagram with average freshwater and seawater compositions and the line of cautious mixing. Chemical analysis plotted away from this line may indicate intrusion (after Arlt 1995)

8.4.4 Hydrochemical Facies Evolution Diagram (HFE-D)

The hydrochemical facies evolution diagram (HFE-D) gives an easy way to recognize the state of a coastal aquifer concerning intrusion/freshening phases that take place over time (Giménez-Forcada 2014). When it comes to hydrochemical facies identification, HFE-D is a more valuable diagram than Durov and Piper diagrams, which have portions that represent two or more distinctions. The X-axis represents the line of evolution with two end members (%Ca and %Na, respectively), which can be interpreted as the evolution of the Na\Ca and Ca\Na exchanges as the calcium and sodium waters mix where the ordinates show the mixing procedure and the recharging water's primary anion is denoted by $\%HCO_3$ or $\%SO_4$.

The HFE-Diagram identifies four principal heteropic facies: Na-Cl representing seawater, Ca-HCO_3 for natural freshwater, Ca–Cl for salinized water resulting from reverse exchange, and Na-HCO_3 for salinized water resulting from direct exchange. Each of these heteropic facies is associated with four closely resembling sub-facies. The intrusion sub-stages are labeled i1 through i4, while the freshening sub-stages are labeled f1 through f4.

The distal facies, indicated by the i1 and f1 sub-stages, represent the separation of the two intrusive flows; this separation should not be confused with the cautious mixing line depicted in the HFE graphic. In contrast, the intermediate levels are represented by the sub-stages f2, f3, i2, and i3, where cation exchange is typically more pronounced. The proximal facies, defined by f4 and i4, are closely related to the final members of freshwater (FW) and seawater (SW).

Within the HFE-Diagram, a dividing line distinguishes the freshwater field on the left from the seawater intrusion field on the right, known as the conservative mixing line (CML). Samples categorized within the saltwater field may indicate an incursion stage, even if they reside in the freshening field. Although these samples are positioned in the incursion field, they are still classified as part of the freshening phase in the freshwater field (Giménez-Forcada 2014). The end members of freshwater (FW) and seawater (SW) can be either real or hypothetical (Turekian 1968). Since the saltwater-freshwater mixing line serves as the starting point for all process evaluations, with recharge water as the only variable, given that seawater is considered stable, it is crucial to select the FW end member with care (Giménez-Forcada 2019).

8.4.5 Seawater Mixing Index (SMI)

The degree of seawater mixing can be estimated using the seawater mixing index (SMI) (Park et al. 2005; Armanuos et al. 2024) where the (Cl^-, $SO_4{}^{2-}$, Na^+, and Mg^{2+}) are considered as the primary ionic components of seawater (Eq. 8.4)

$$SMI = a \times \frac{CNa}{Tna} + b \times \frac{CMg}{Tmg} + c \times \frac{Ccl}{Tcl} + d \times \frac{Cso4}{Tso4} \tag{8.4}$$

where a, b, c, and d are constants that show the fraction of relative concentration of Na^+, Cl^-, Mg^{2+}, and $SO_4{}^{2-}$, their values are 0.31, 0.04, 0.57, and 0.08 in seawater, respectively. The measured concentration of these ions is denoted by C in mg/L. The T value is computed using the cumulative probability curves, where T denotes the detected ions' regional threshold values, and for the concentrations, the curves' inflection points are utilized for Cl^-, $SO_4{}^{2-}$, Mg^{2+}, and Na^+ in groundwater samples (Sinclair 1974). SMI values less than 1 indicate freshwater, whereas values greater than 1 suggest groundwater that has been affected by the mixing of saltwater and freshwater (Park et al. 2005; Arafa et al. 2024a, b).

8.4.6 Resistivity Measurements

The resistivity method is the best geophysical method for assessing SWI, founded on the idea that as water's salinity (TDS) increases, so does its electrical conductivity (the inverse of electrical resistivity. Low-resistivity anomalies in coastal aquifers, such as those found in the Nile Delta, indicate areas of seawater intrusion, typically characterized by resistivity values of less than 2 Ω m (Gemail et al.2004). The integration of hydrochemical data and DC resistivity surveys is widely used to map subsurface issues related to groundwater quality and describe the spatial distribution of freshwater and saltwater in coastal zones (Gemail et al. 2017a).

To investigate saltwater intrusion and validate the freshwater/seawater interface boundary outcomes using SMI calculations at varying depths and hydrochemical facies modeling, 29 Schlumberger resistivity sounding points were strategically positioned across the study area. The subsurface electrical resistivity distributions were estimated by driving four steel electrodes into the ground surface in a linear arrangement, symmetrically positioned around the central points (A and B). Voltage measurements were then taken between the remaining two electrodes (M and N). The spacing between the primary electrodes (AB) ranged from 400 m for four vertical electrical soundings (VES) in the southern region to 600 m across most of the surveyed area, particularly in the northern section south of Brules Lake. Some soundings were conducted near wells to ascertain the resistivities of the underlying layers by correlating resistivity measurements with lithological and hydrogeological data.

Utilizing the IPI2WIN 3.01 software, the apparent resistivities were inverted, and layer resistivities and thicknesses were determined by incorporating measurements from AB/2. This scheme, as described by Bobachev et al. (2003), enabled the characterization of vertical variations in lithology and water quality, as illustrated in Fig. 8.4.

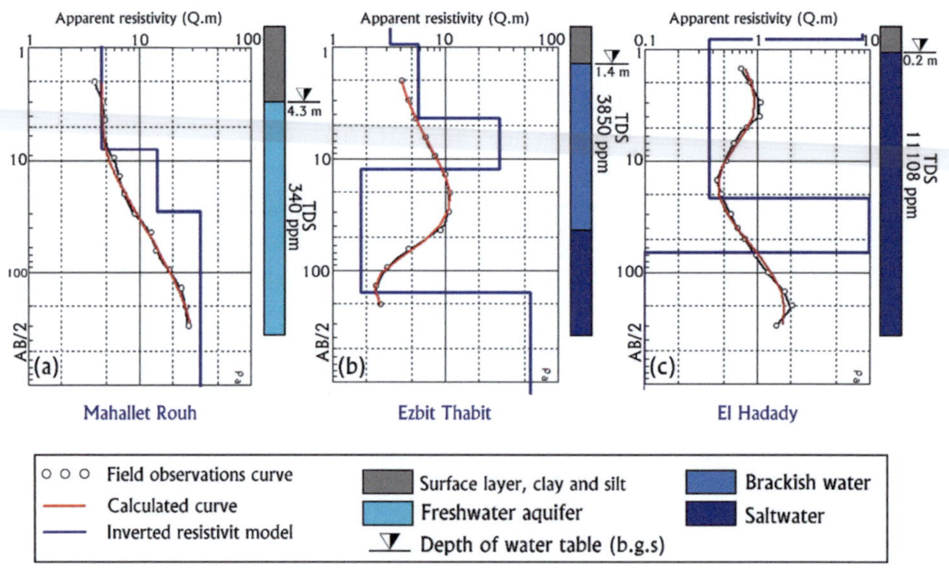

Fig. 8.4 VES survey's inversion data demonstrate the vertical fluctuations brought about by the effects of seawater intrusion in the area: **a** zone of freshwater in the southern; **b** zone where freshwater and saltwater mixing in the center; **c** northern portion's severely affected seawater

8.4.7 GIS Spatial Mapping

Nearly fifteen maps were created using the geostatistical interpolation analysis tools of the inverse distance weighting (IDW) method in the ArcMap (10.8) software developed by ESRI (2020). The IDW method calculates the values at the unknown (non-surveyed) locations using a linear combination of known points weighted by the inverse function of the distance between them (Chen and Liu 2012).

8.5 Results and Discussions

8.5.1 Monitoring of Seawater Intrusion

Assessing and predicting groundwater deterioration in coastal aquifers is crucial for monitoring seawater intrusion. Effective monitoring involves selecting the most reliable methods for data collection. There are both direct and indirect methods for monitoring seawater intrusion. Direct methods include measuring groundwater salinity and sampling groundwater from observation wells, while indirect methods often utilize geophysical techniques, specifically the time-domain electromagnetic method (TDEM).

Various researchers and organizations are engaged in monitoring seawater intrusion. For instance, the Sustainable Groundwater Management Act (SGMA 2014) has adopted chloride concentration measurements from hydrochemical analyses to monitor seawater intrusion in its projects. Tal et al. (2019) has introduced a new downhole geophysical tool known as the Subsurface Monitoring Device (SMD), which offers high vertical resolution for monitoring seawater intrusion in boreholes. Melloul (1997) advocated for using groundwater salinity measurements as a direct method, complemented by TDEM as an indirect method for monitoring seawater intrusion.

In this study, hydrochemical analyses of groundwater samples were employed to monitor seawater intrusion over three decades: the 1980s, 1990s, and the early 2000s. The 1980s data is represented by a modified iso-salinity map (Fig. 8.5a) based on the work of Diab and Saleh (1982), while a salinity map for the 1990s was constructed using data from Arlt (1995) (Fig. 8.5b). The salinity map for the 2000s was created from data collected in this study (Fig. 8.5c). Latitude 31°00’ was established as a reference line to track changes in salinity during these three decades. Variations in chloride content, closely linked to groundwater salinity, serve as a critical factor for monitoring seawater movement. This is illustrated by the maps for the 1990s and 2000s (Fig. 8.6a, b, respectively). Unfortunately, analyses of chloride content from the 1980s were not available.

A comparative analysis of the three salinity maps reveals an increase in salinity from north to south over time. In the 1980s map, salinity levels north of latitude 31° were below 1000 mg/L. By the 1990s, salinity levels approached 1000 mg/L near Latitude 31°, and by 2008, salinity levels exceeded 1000 mg/L in areas south of latitude 31°,

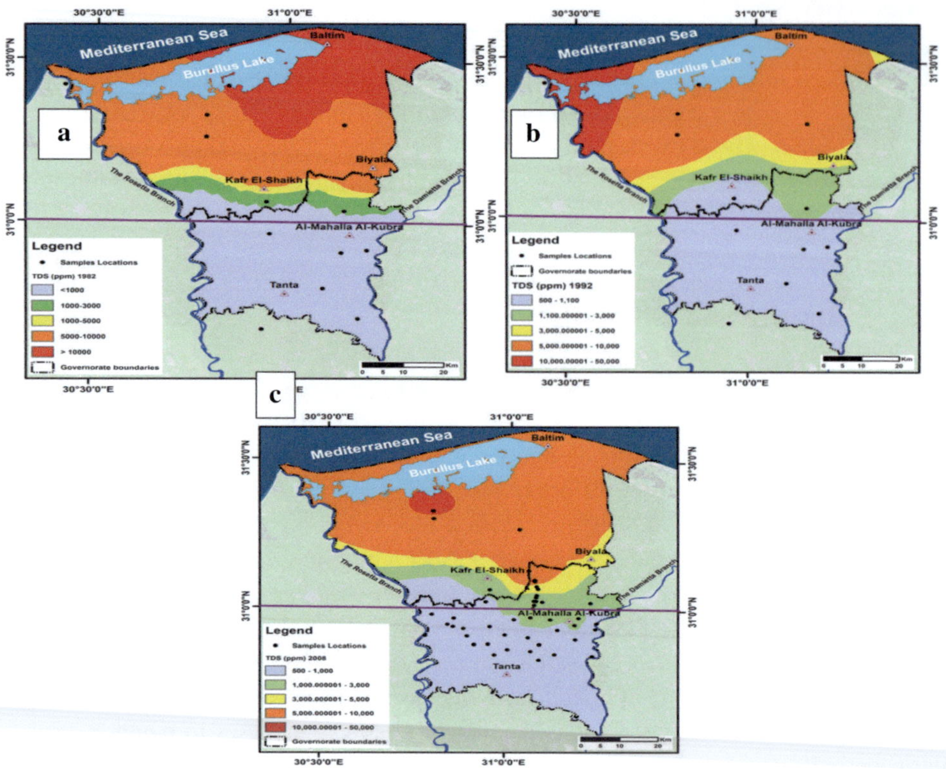

Fig. 8.5 **a** Groundwater iso-salinity map (mg/l) of the central part of NDA (modified after Diab and Saleh 1982), **b** constructed salinity map (mg/l) of the central part of NDA (1992) (collected data from Arlt 1995), **c** groundwater iso-salinity map of the central part of NDA (2008)

particularly around samples No. 24, 25, 27, 28, 29, 34, and 35 in the El Mahallah El Kobra and Samannod regions. Comparison between the chloride map of 1992 and the chloride map of 2008 indicates that the chloride content in groundwater follows the same path of salinity and increases in the southward direction with time, confirming that seawater intrusion increases with time, leading to the progression of contamination of the study area.

8.5.2 Hydrochemical Results

8.5.2.1 Stuyfzand Water Classifications

The groundwater samples are classified according to the classification system of Stuyfzand (1986), shown in Table 8.1. According to the classification code for the water type, which refers to the Cl contents, four types were distinguished: F (Fresh), Fb (Fresh-Brackish),

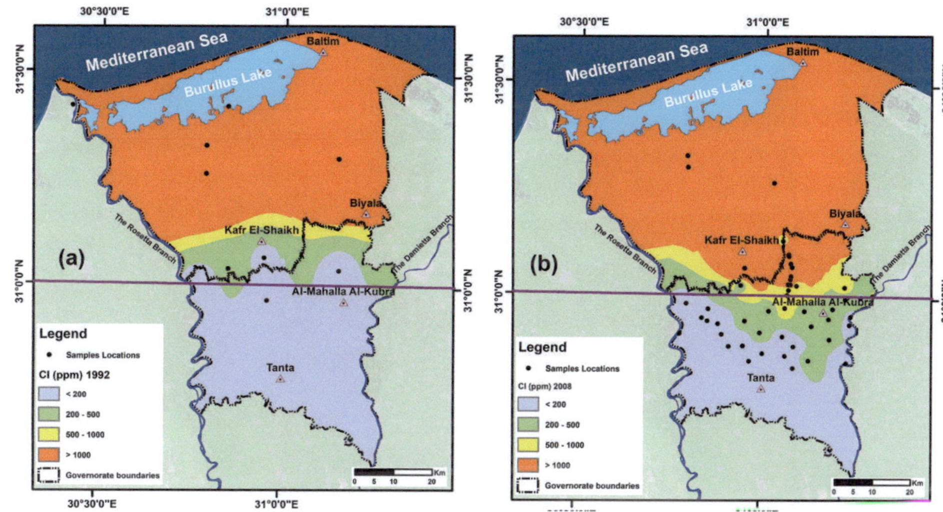

Fig. 8.6 **a** Distribution of chloride (Cl) concentration in the central part of NDA (1992) data from Arlt 1995), **b** distribution of chloride (Cl) concentration in the central part of NDA (2008)

B (Brackish), and Bs (Brackish-Saline). Respectively, it is located in the south, middle, middle north, and northern parts of the study area.

The water samples are numbered sequentially from south to north, with sample no. 1 located in the southern region and sample no. 51 in the northern part of Burullus Lake. According to Stuyfzand's intrusion indicator (BEX), most groundwater samples with positive BEX values indicate a mixing phase of freshwater and saltwater, categorized as F, Fb, and B types. However, samples no. 18, 19, 20, 21, 24, 44, 45, 46, 47, 48, 49, 50, and 51 exhibit negative BEX values, predominantly representing the salinization or seawater intrusion (SWI) phase, classified as Bs type water. Notably, sample no. 23 (Amut) has a BEX value of 0, indicating no base exchange (see Fig. 8.7 and Table 8.1).

The accompanying map illustrates the impact of seawater intrusion, showing it reaching the boundary between two governorates and extending south of Latitude 31°. This intrusion has created three saline pockets within the freshwater zone to the south, likely due to halite dissolution resulting from the numerous clay lenses present in the Quaternary aquifer (Sallouma 1983).

8.5.2.2 Revelle Index

Some authors using ionic ratio by Revelle (1941) ($Cl-1/CO_3-2 + HCO_3-1$) to study the seawater intrusion into the Nile Delta Aquifer; for example, El Ghandour et al. (1985) indicated that the southern parts of that less than 1.0, near to the ionic ratio of freshwater (River Nile give an ionic ratio of 0.22), and the northern parts give ionic ratio above 200 (i.e., prevalence of seawater).

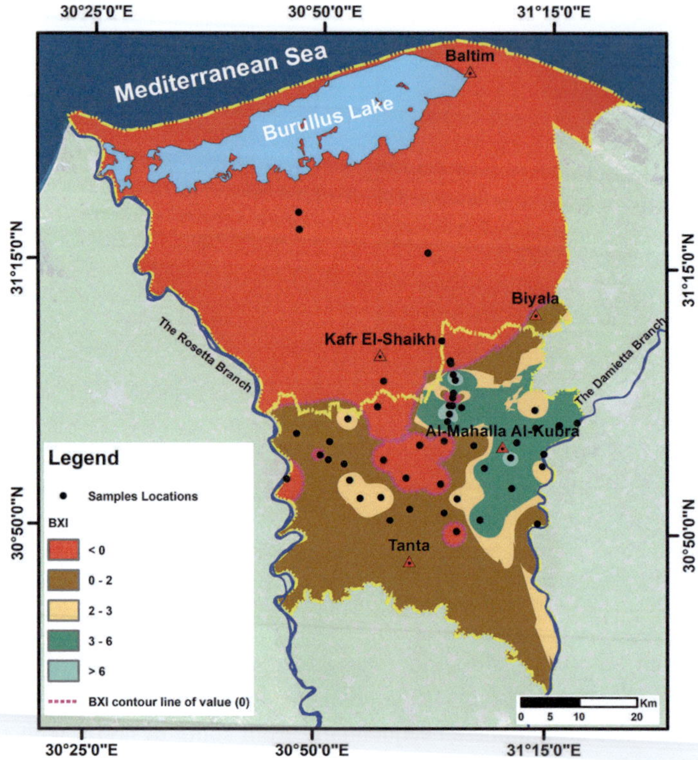

Fig. 8.7 Groundwater samples' spatial distribution of the computed Stuyfzand indicator (base exchange index, BEX)

Habib (1996) studied the seawater intrusion into the middle aquifer of the Nile Delta according to the Revelle ionic ratio (the highest ionic ratio value of 5.42 was found in the north part of Gharbiyyah governorate at Mahalla El-kubra city).

In the present study, according to Revelle ionic ratio values listed in Table 8.2, the groundwater samples (Table 8.1) show that the south part of the study area represents a freshwater zone (<1) except sample no. 2 (Metyzied Village) well whose ionic ratio is more than one. While the samples of groundwater from no. 22–29 and from no. 34–48 are represented by a diffusion water zone (1–50), which is located in the north of Gharbiyyah governorate with contact of Kafer Al-Shikh governorate, indicating intermixing of Nile River water and saltwater intrusion. But two groundwater samples, no. 49 (Sidi Salem region) and 51 (El-Hamol region) are represented by an interface water zone (50–150) reflecting more encroachment of seawater intrusion. Only one groundwater sample (well no. 50 El-Hadady well) has an ionic ratio greater than 200 and was discovered around 25 km from the coastline, indicating the seawater intrusion.

Table 8.2 Classifications of the type of water according to ionic ratio Revelle (1941)

Ionic ratio ($Cl^{-1}/CO_3{}^{-2} + HCO_3{}^{-1}$)	Water class or zone
>200	Saltwater zone, i.e., seawater
200–50	Interface water zone
50–1	Diffusion water zone
<1	Freshwater

8.5.2.3 Piper Diagram

The composition of groundwater samples was meticulously analyzed using a Piper diagram and its conservative mixing line. The Mahallet Monof sample (no. 8) was identified as the freshwater end member, while the Mediterranean seawater sample from Arlt (1995) served as the saltwater end member (Fig. 8.8). According to the Piper diagram analysis, water samples classified as type (F) and located within zone (A) indicate freshwater intrusion. These samples, characterized by Ca–HCO_3 and Na-HCO_3 water types, predominantly originate from the southern region and display significantly low total dissolved solids (TDS) and chloride (Cl) values. The positions of these samples suggest active $CaCO_3$ dissolution, as evidenced by the declining calcium concentrations resulting from adsorption (Fig. 8.3).

Conversely, the B zone on the diagram, located in proximity to the south, central (bordering two governorates), and northern sectors of the study area, illustrates a mix of type (F), (Fb), (B), and a small portion of (Bs) water types. This zone signifies the amalgamation of various water types, with prevalent Na–Cl and mixed Mg–Na–Ca–Cl–HCO_3 water types characterizing this region. Notably, the presence of Na–Cl and Mg–Na–Cl water types underscores the pronounced influence of seawater intrusion on coastal aquifers (Appelo et al. 2005).

Additionally, the high salinity Bs-type water samples, located along a trajectory toward the Mediterranean seawater sample, are found within the C zone. In the northern region, groundwater experiences various cation exchange processes, which contribute to the precipitation of $CaCO_3$ due to increasing calcium concentrations from desorption. Samples situated above the conservative mixing line indicate salinization effects, whereas those below it reflect freshwater processes.

8.5.2.4 Hydrochemical Facies Evolution Diagram (HFE-D)

To evaluate the extent of seawater intrusion in the study area, the HFE diagram was employed. Sample 16, from the Mallet Monof well, was designated as the freshwater (FW) end member, while a Mediterranean seawater sample from Arlt (1995) represented the saltwater (SW) end member (Bear et al. 1999). In the hydrochemical facies displayed in the HFE diagram, samples affected by seawater intrusion (SWI) are highlighted in red, whereas freshwater samples are indicated in blue (Fig. 8.9a).

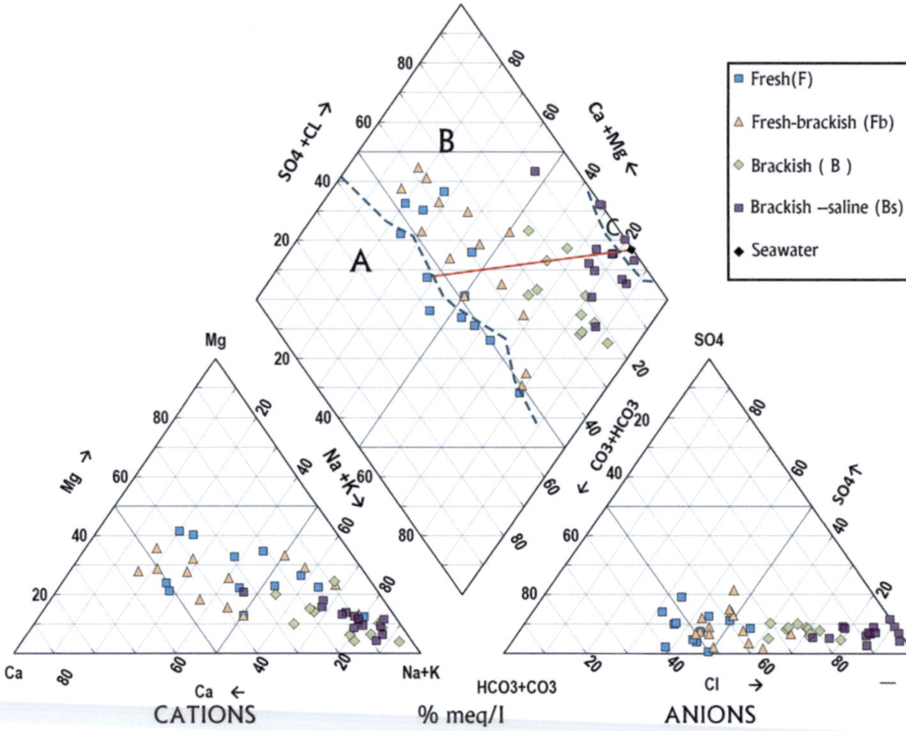

Fig. 8.8 Plotting hydrochemical data of the Nile Delta coastal aquifer on a piper diagram, the conservative mixing of fresh groundwater (sample 16) and seawater is shown by the red line. The three zones—A, B, and C—representing the southern, middle, and northern regions of the research area are divided by the dashed blue lines. The mixed water affinity is represented by the B zone

Ten hydrochemical sub-stages correspond to the freshening stage (20 samples, or 39.3% of all the samples, were obtained) and the intrusion stage (31 samples, or 60.7.3%) of the samples. Logically, we studied seawater intrusion from coast of sea to land; thus, let us start with the incursion stage, when the aquifer is replenished with new water Ca–HCO_3 water, the water evolved increasingly through sub-stage (i1) representing Ca–MixHCO_3 facies the distant incursion represented by the F and Fb, samples 5, 11 (Table 8.3 and Fig. 8.9a). The water then progresses to sub-stage (i2) Ca-MixCl and MixCa-MixCl represented by the F and Fb, samples 15, 17, 18.

The water undergoes sub-stage evolution (i3) MixNa–Cl, MixCa–Cl, Ca–Cl_2, representing F and Fb, (samples 3, 19, 20, 21). At the same time, two samples from the (i3) sub-stage are located diverging from the cautious mixing line, which symbolizes the active cation exchange processes that take place when aquifers get salinized. Sub-stage (i4) is the last one that represents the intrusion facies Na-Cl, and more samples of this phase are in Table 8.3. Ultimately, during this stage, the aquifer is dominated by the major heteropic

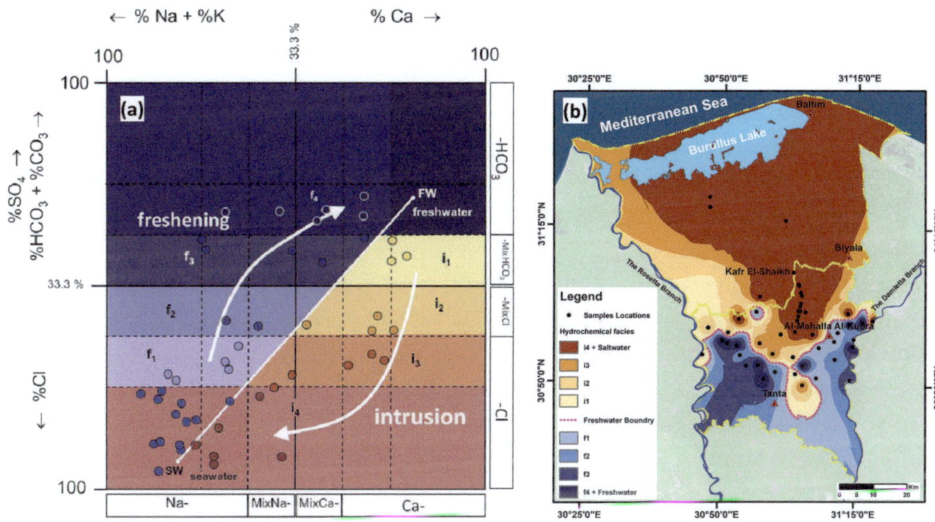

Fig. 8.9 a HFE-D of the groundwater samples being examined; b heatmap showing the hydrochemical facies distribution that corresponds to the aquifer's sub-stages

Table 8.3 Each HEF-D sub-stage's percentage, facies, and related samples

Fresh sub-stages	Samples	% of sub-stages	Code of type water	Type of facies
f1	2, 23, 26, 27, 28	11.8	Fb. B	NaCl, NaMixCl
f2	6, 10, 12, 31	3.9	Fb	NaMixCl
f3	1, 7, 32	7.8	F. Fb	$NaMixHCO_3$, $MixCaMixHCO_3$
f4	4, 8, 9, 13, 14, 16, 30, 33	13.8	F	$CaHCO_3$, $MixCaHCO_3$, $MixNaHCO_3$
Intrusion sub-stages				
i1	5, 11	5.9	F. Fb	$CaMixHCO_3$
i2	15, 17, 18	7.8	Fb. F	MixCaMixCl, Ca-MixCl
i3	3, 19, 20, 21	7.8	Fb. F	MixNaCl, MixCaCl, $CaCl_2$,
i4	22, 24, 25, 29, 34, 35, 36, 37, 38, 39, 40, 41, 42, 43, 44, 45, 46, 47, 48, 49, 50, 51	41.2	B. Bs	NaCl

Na-Cl facies, which are symbolized by B and Bs (samples 22, 24, 25, 34, 35, 36), and from 37 to 51 included 22 samples (Table 8.3 and Fig. 8.9a).

Conversely, the freshening phase observed in the studied samples originated from an initial condition where the aquifer had been entirely salinized due to the intrusion of seawater. The aquifer represented by the distant freshening facies (f1) Na-Cl, MixNa-Cl, which the Na-Cl facies is in harmony with, begins to receive a flush from surface canal replenishment water, representing F and B; samples 2,23,26,27, 28, see (Table 8.3 and Fig. 8.9a). The water then progresses to sub-stages (f2) Na-MixCl, Mix Na-MixCl representing Fb; samples 6, 10, 12, 31 and (f3) MixCa–MixHCO_3, Na-MixHCO_3, representing F and Fb it is demonstrated by samples 1, 7, 32. Lastly, we arrive at sub-stage (f4), which is the closest freshening face Ca–HCO_3, Na-MixHCO_3, MixCa-HCO_3, Na-HCO_3 representing F and Fb of samples 4, 8, 9, 13, 14, 16, 30, 33 (Table 8.3 and Fig. 8.9a). Finally, the Ca-HCO_3 facies, which correspond to F samples 14 and 16, indicate the full recovery of the aquifer (Table 8.3). A heat map illustrating the spatial distribution of each phase's several sub-stages was created, revealing that the intrusion phase dominated almost two-thirds of the research area, (Fig. 8.9b).

8.5.2.5 Seawater Mixing Index (SMI)

Plot the logarithmic concentration values of specific hydrochemical parameters alongside their cumulative probability percentages (Fig. 8.10). The threshold values (T) for each ion are determined by the crossing points of the plots (Sinclair 1974; Park et al. 2005). Accordingly, calculated the T values for Cl, Na, Mg, and SO_4 for the investigated samples are 550, 60,950, and 180 mg/L, respectively, the relative concentration fractions and seawater's ion threshold values ($a = 0.31$, $b = 0.04$, $c = 0.57$, $d = 0.08$) were substituted in Eq. 8.4 to determine each groundwater sample's SMI values. The SMI values indicated that the SMI values of all groundwater samples belonging to the F, Fb, and B water types are less than 1, representing two zones F type water zone in the south and the dynamic zone where different levels of seawater mixing have an impact in the middle, while groundwater samples of Bs possess values more than one represents SWI zone in north (Fig. 8.11 and Table 8.4).

The assessment of inflection moments is based on Sinclair (1974), where the background concentration is represented by the trend line passing through the horizontal dots. The vertical trend line of the samples indicates atypical concentrations. The regional threshold for the element in question is identified at the intersection of these two trend lines.

Accordingly, 38 groundwater samples (74.5%) are not significantly affected by the mixing of components with seawater, while 13 groundwater samples (25.4%) are affected (Table 8.4).

The study area's coastal groundwater SMI value spatial distribution is comparatively similar to the BEX map and the heat map, where SWI has varying degrees of impact

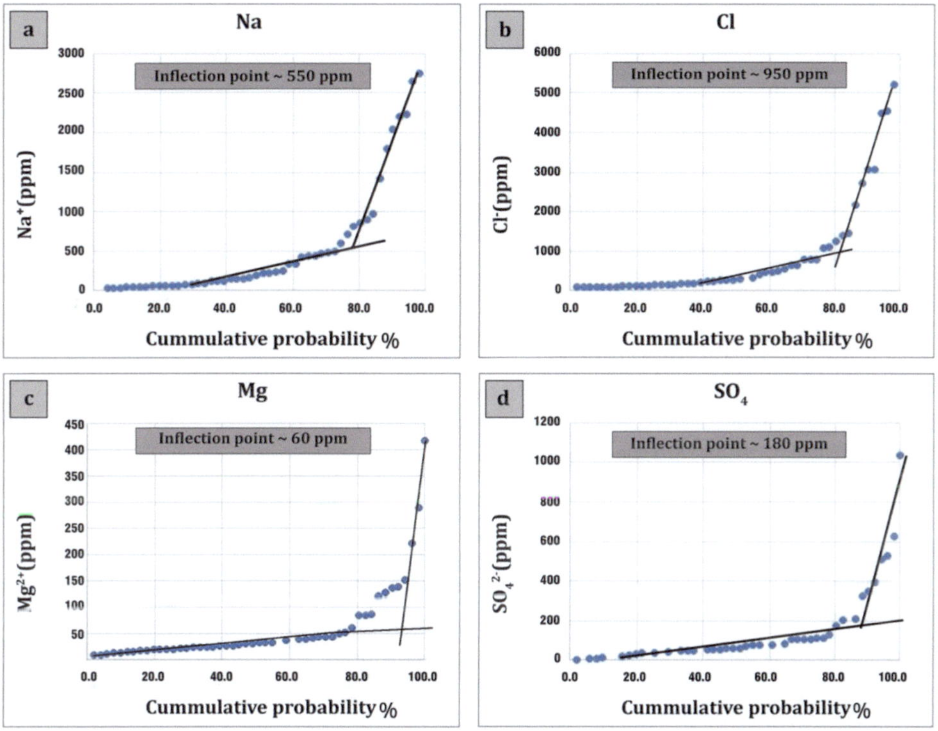

Fig. 8.10 Curves of cumulative probability for **a** Na, **b** Cl, **c** Mg, and **d** SO_4 concentrations in groundwaters

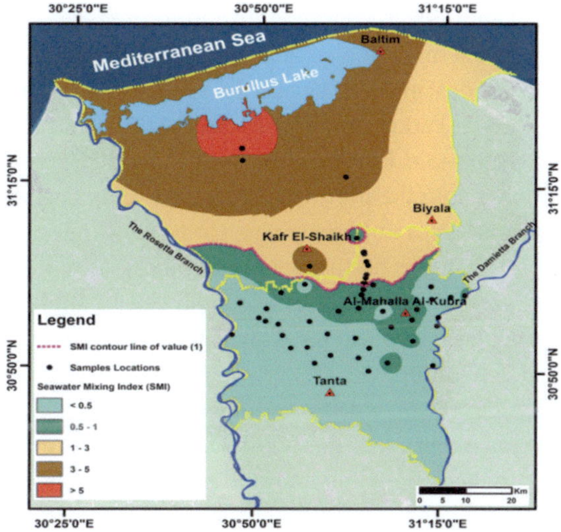

Fig. 8.11 Geographic distribution of the groundwater computed SMI (significant seawater mixing is indicated by SMI values larger than 1)

Table 8.4 SMI values of study groundwater samples

SMI values	Samples	Type of water	Code	% SMX of samples
<1	1, 3, 4, 5, 6, 7, 8, 9, 13, 14, 16, 30	Fresh	F	74.5
	2, 10, 11, 12, 15, 17, 18, 19, 20, 21, 22, 31, 32, 33	Fresh-brackish	Fb	
	23, 24, 25, 26, 27, 28, 29, 34, 35, 36, 37, 47	Brackish	B	
>1	38, 39, 40, 41, 42, 43, 44, 45, 46, 48, 49, 50, 51	Brackish-saline	Bs	25.4

on the research area's northern parts. The central area illustrates the dynamic zone influenced by varying degrees of seawater mixing, while the southern portion represents the freshwater groundwater zone.

8.5.2.6 Chemical Equilibrium and Saturation Indices SI

Geochemists have focused on quantitatively interpreting the chemical evolution within groundwater systems, particularly through the advancement of ion association models. Using numerical tools like PHREEQE and PHREEQCI, researchers simulate hypothetical reactions by defining initial conditions, proposing sets of reactions, and utilizing a comprehensive thermodynamic database (Glynn and Plummer 2005). As groundwater flows through aquifers, it interacts with minerals, leading to two primary geochemical processes: dissolution and precipitation. Negative saturation index (SI) values indicate under-saturation conditions, causing mineral phases to dissolve, while positive SI values signify over-saturation, resulting in mineral phase precipitation. Equilibrium between water and mineral phases is reached when the SI of the mineral is zero. Saturation indices for various mineral phases in groundwater samples were calculated using PHREEQC software (Parkhurst and Appelo 2001). The positive SI values of calcite (−0.08 in the south to −3.6 in the north), aragonite, and dolomite mineral phases suggest their deposition within the pore spaces of sand sediments, forming calcareous sandstone. However, dolomite does not precipitate. Conversely, negative SI values for halite (−7.1 in the south to −0.3 in the north), anhydrite, and gypsum mineral phases indicate their dissolution. Regions recharged by surface freshwater exhibit a high propensity for mineral dissolution, particularly of minerals with higher solubility, like calcite, aragonite, gypsum, dolomite, and anhydrite.

Analysis of calcite and gypsum saturation index maps reveals that the discharge zone demonstrates the ability to precipitate minerals (Fig. 8.12), with negative saturation indices predominantly observed in the east and west near the Nile River branches, indicating under-saturation conditions and mineral dissolution tendencies. In contrast, positive saturation indices in the high-salinity regions in the north and northern parts indicate

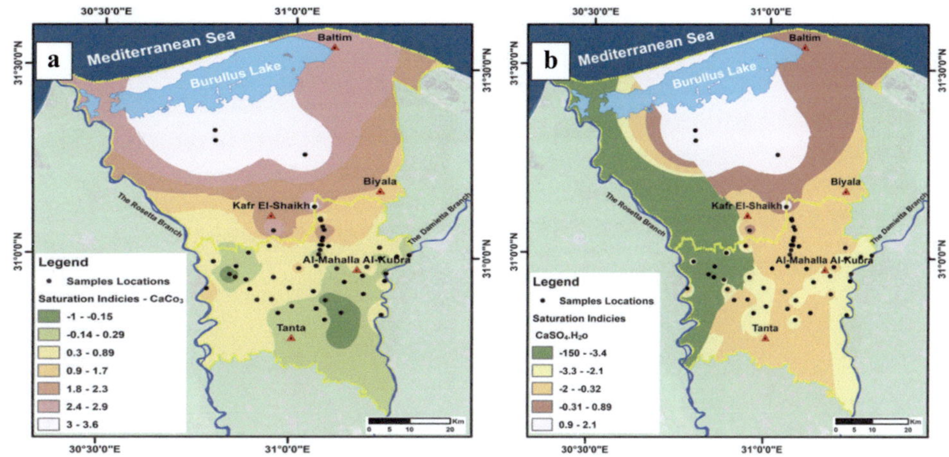

Fig. 8.12 Saturation indices maps of **a** calcite and **b** gypsum mineral phase in groundwater samples of the Nile Delta aquifer (2008)

saturation conditions, with minerals tending to precipitate. This suggests that groundwater in the northern areas is saturated with these minerals, depositing them within the aquifer sediments as disseminated grains of calcareous sandstone. The precipitation of calcite supports the saturation water index (SWI) in the northern study area, as illustrated in Figs. 8.3 and 8.12a, while gypsum dissolution, particularly in the north and northern regions, as depicted in Fig. 8.12b, contributes to increased salinization (El Ghandour et al. 1985).

8.5.3 Statistical Analysis

8.5.3.1 Cluster Analysis

Cluster analysis was successfully used to classify water chemistry data (Jaquet et al. 1975; Alther 1979; Williams 1982; Farnham et al. 2000; El Alfy 2003; El Alfy and Merkel 2006), where it can be used to test water quality data. Cluster analysis assumptions include homoscedasticity (equal variance) and normal distribution of the variables (Alther 1979). The entire cluster analysis was conducted with the result of the main ions of the water samples collected in 2008 using the SPSS and STATISTICA software. The variables assessed in all samples (depth, pH, EC, TDS, K^+, Na^+, Mg^{++}, Ca^{++}, Cl^-, SO_4^-, and HCO_3^-) were taken into account for the calculations. The hydrochemical data were equally weighted using the z-standardization; the samples were grouped according to their similarity, where the classification of samples according to their parameters is termed Q-mode classification cluster analysis of the z-standardized. Input data matrix of 51 cases

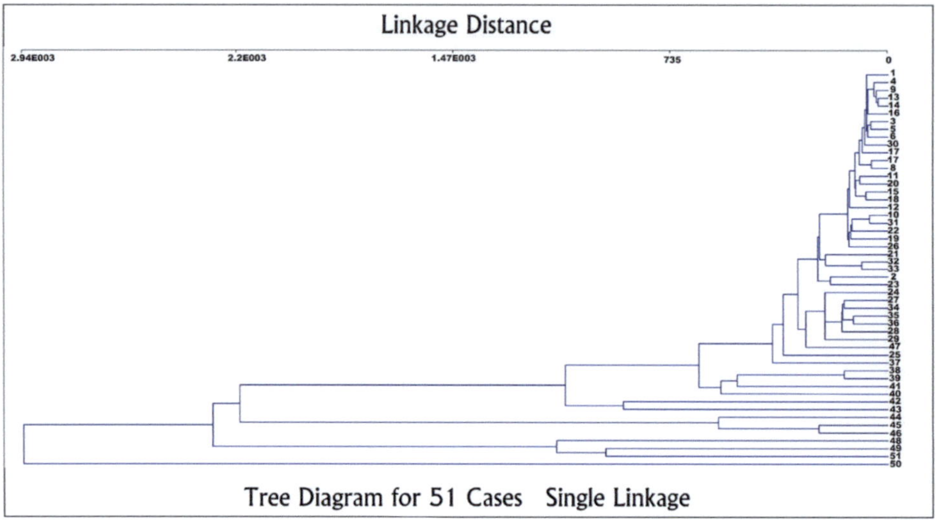

Fig. 8.13 Q-mode dendrogram of the Nile Delta aquifer using the Wards method linkage rule

using the hydrochemical composition of 11 variables (see Table 8.1) gives the Q-mode dendrogram (Fig. 8.13).

The Q-mode dendrogram is interpreted at a similarity linkage distance of 0.32, where there are two clusters. The first cluster contains wells no. 38, 39, 40, 41, 42, 43, 44, 45, 46, 48, 49, 50, and 51, representing Bs types of water (Fig. 8.13). It has salinity values varying between 2270 and 11,109 mg /1 with an average value of 5412 mg/1 (Table 8.1). NaCl is prevalent in hydrochemical facies where the samples are located from the contact of two governorates (Azbit Elshok in Gharbiyyah drain) to El-Hadady village south of Burullus Lake. This shows the gradual encroachment of seawater into the freshwater aquifer from the north to the south. This cluster is spatially distributed in the northern part of the Gharbiyyah Governorate and extends to the north of Kafer-Shikh governorate (Fig. 8.14). The second cluster contains a mixing of three types of water samples F, Fb, and B arrangement from south to middle of the area (north of Gharbiyyah Governorate) F type consists of samples no. 1, 3, 4, 5, 6.7, 8, 9, 13, 14, 16 and 30, while Fb type represents by samples 2, 10, 11, 12, 15, 17, 18, 19, 20, 21, 22, 31, 32, and 33. The final B type consists of samples no. 23, 24, 25, 26, 27, 28, 29, 34, 35, 36, 37, and 47 (Fig. 8.13).

This cluster is characterized by a gradual increase in salinity, with values ranging from 270 mg/L in the southern region (F type) to 1740 mg/L in the central area (B type). The dominant hydrochemical facies in this cluster include $CaCO_3$, $MgHCO_3$, $NaHCO_3$, and NaCl, transitioning from freshwater in the south to fresh-brackish and brackish water in the north as water quality changes over time. This spatial distribution is illustrated in

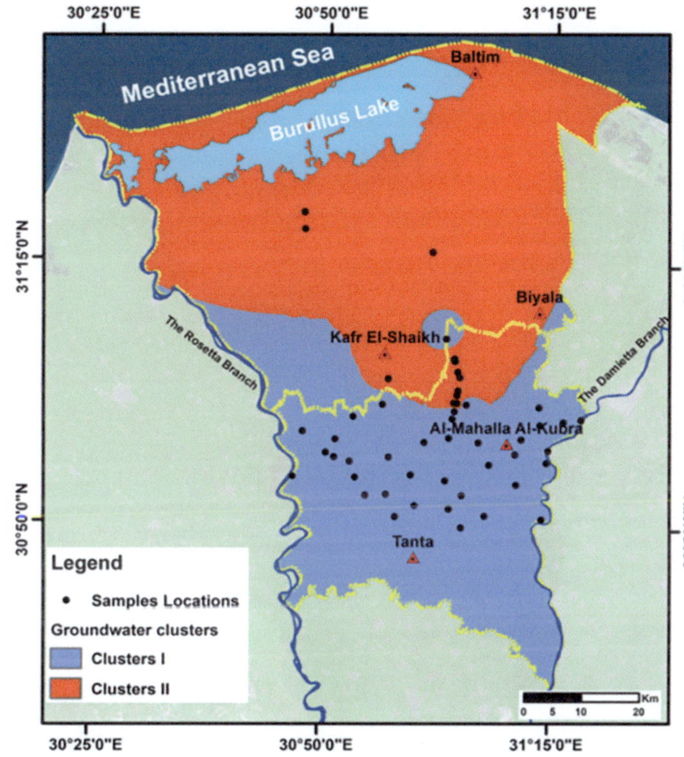

Fig. 8.14 Areal distribution of the groundwater clusters of the center part of the Nile Delta aquifer

Fig. 8.14. The findings from the cluster analysis are further supported by hydrochemical analyses, specifically the Piper diagram and SMI results.

8.5.4 Distribution of Spatial Resistivity and Hydrogeochemical Facies

The spatial distribution of interpreted resistivity derived from geophysical surveys was cross-referenced with both maps to validate the accuracy of the computed saturation water index (SMI) and the resulting hydrochemical facies, as depicted in (Fig. 8.15). Upon this comparison, it was observed that the resistivity distribution bears a strong resemblance to the hydrochemical facies evolution (HFE) and SMI maps, aligning well within the framework of the saturation water index (SWI). Subsequently, by analyzing the resistivity data alongside the distribution of hydrochemical facies along the axis of the Nile Delta (refer to Fig. 8.15b), the region can be delineated into three distinct zones with clear boundaries. The resistivity profile highlights the influence of SWI, notably interconnected

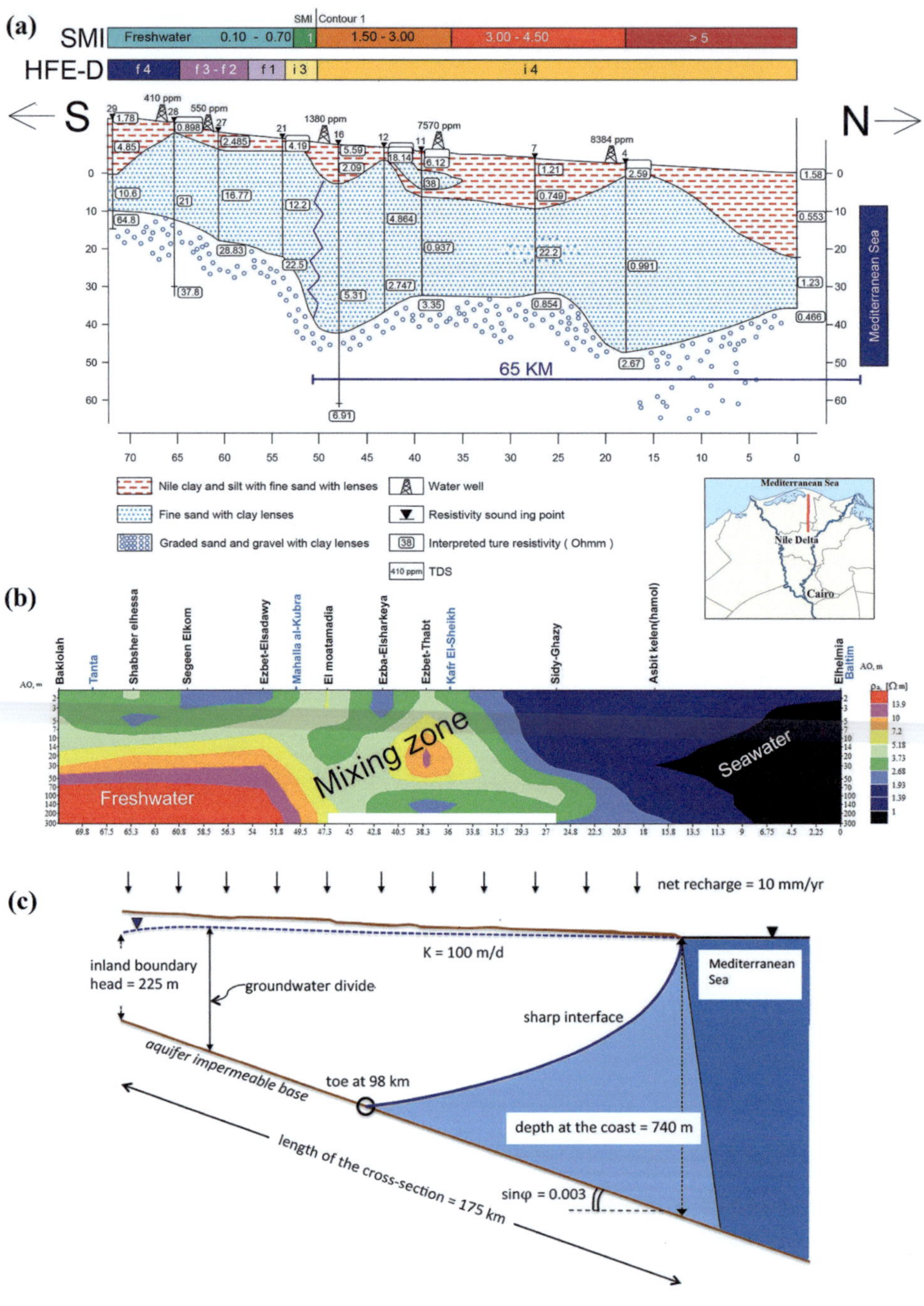

Fig. 8.15 Encroachment of seawater on the central Nile Delta axis: **a** geoelectric cross-section along profile E–E' above it, the HFE-D and SWI hydrochemical facies; **b** apparent resistivity in the vertical cross-section displays the usual seawater wedge shape; **c** seawater intrusion is visible in the schematized modeling cross-section in middle Nile Delta aquifer (Mazi 2014)

with El Burls Lake and the Mediterranean Sea, exerting a notable impact on the northern segment of the aquifer characterized by lower resistivity levels (<1Ωm).

Furthermore, the high chloride concentration (>1000 ppm) and the i4- i4-hydrogeochemical facies emerge as defining features of this zone, with chloride levels exceeding 66% and sodium surpassing 50% (as depicted in Fig. 8.9a), attributed to the influence of SWI in this region.

The integration of the resistivity method and hydrogeochemical modeling offers a comprehensive approach to understanding the mechanisms and spatial distribution of saltwater intrusion (SWI) in the Nile Delta, particularly under the influence of sea level rise (SLR) and intensive groundwater pumping. In the southern part of the area, freshwater facies (Ca-HCO_3) dominate, as indicated by high resistivity values (22–64 Ωm) and low chloride concentrations, which suggests minimal SWI in this region. This is further supported by the analysis of resistivity and hydrochemical maps (Figs. 8.9 and 8.16), which reveal that the middle area is moderately impacted by SWI. Here, resistivity values ranging from 3 to 22 Ωm delineate a well-defined mixing zone, confirming the presence of mixed geochemical facies (i3, i2, f2, and f3; Fig. 8.9a), consisting of both freshwater and seawater.

To further investigate the mechanisms of seawater incursion, a vertical resistivity section (Fig. 8.15) was constructed perpendicular to the shoreline, extending from north to south. This section provides valuable insights into the subsurface resistivity distribution, demonstrating how resistivity methods, when integrated with hydrogeochemical modeling, can enhance the understanding of saltwater intrusion dynamics. This combined approach enables a more detailed assessment of the spatial variations in SWI and helps monitor the evolving impacts of SLR and intensive groundwater pumping in the Nile Delta's mixing zone.

The resistivity map (Fig. 8.16) also highlights the role of intensive groundwater pumping, driven by SWI, in contributing to localized low-resistivity anomalies in the central region. These anomalies suggest the upward movement of seawater from deeper zones, exacerbated by the pumping activities. In contrast, elevated resistivity readings in the southern portion, as well as in areas primarily to the east and west of the two Nile Delta branches, indicate substantial recharge from both the delta branches and surrounding irrigation networks, effectively mitigating SWI.

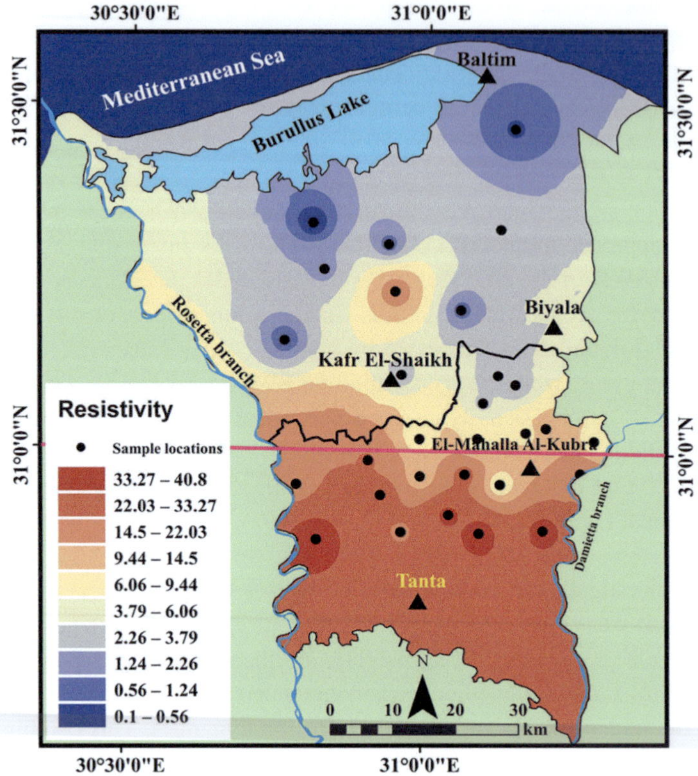

Fig. 8.16 Resistivity values (Ohm. m) spatial distribution of the Nile Delta aquifer in the area. The low-resistivity values of less than 2 Ohm.m. indicate seawater intrusion

8.6 Conclusion

The challenge of seawater intrusion in coastal groundwater aquifers is an ongoing concern, exacerbated by the proximity of seas and the pressures of global climate change, population growth, and increased development activities. This study highlights the critical role of integrated hydrogeochemical and geophysical modeling in understanding the dynamics of seawater intrusion (SWI) in the Nile Delta aquifer.

The findings indicate a clear consensus across statistical, geophysical, and hydrogeochemical analyses that the study area is significantly affected by seawater intrusion. The northern regions experience varying degrees of influence, extending to the border of Gharbiya and Kafr El-Sheikh governorates. The central region, situated between these two governorates, is identified as a dynamic zone characterized by fluctuating seawater mixing, while the southern zone remains predominantly fresh groundwater.

Particularly, the saturation indices (SI) underscore the importance of Nile River recharge and high aquifer pumping rates as primary factors driving the movement of seawater. This integrated approach not only elucidates the nature and mechanisms of seawater intrusion in urban environments but also provides actionable insights for the management and conservation of groundwater resources.

By employing these findings, stakeholders can develop targeted strategies to mitigate the impacts of anthropogenic activities and climate change, particularly sea level rise, ensuring the long-term sustainability of freshwater resources in the Nile Delta region. The integration of hydrogeochemical and geophysical data thus emerges as a vital tool in addressing the pressing challenges posed by seawater intrusion.

8.7 Recommendations for Groundwater Sustainability

1. **Establishment of a Continuous Monitoring System**: Implement a robust monitoring network that combines hydrogeochemical and geophysical data collection. This system should regularly assess the salinity levels, groundwater levels, and the freshwater-saltwater interface to track changes over time.
2. **Adaptive Water Management Strategies**: Develop adaptive management strategies that respond to real-time data on groundwater levels and quality. This approach should include flexible pumping regulations in the mixing zone that can be adjusted based on the current findings from TDS and geophysical data and the degree of seawater intrusion.
3. **Reduction of Pumping in the Mixing Zone**: Limit groundwater extraction in the middle parts of the Nile Delta aquifer, where seawater intrusion is most pronounced. This reduction in pumping will help maintain pressure in the freshwater zone and slow the advancement of saltwater.
4. **Recharge Enhancement Initiatives**: Promote artificial recharge projects that utilize surplus surface water, such as Nile River water, to replenish the aquifer. This should include the construction of recharge basins and the restoration of natural recharge areas.
5. **Community Engagement for Risk Assessment**: Involve local communities in water management decisions and educate them about the impacts of over-extraction and seawater intrusion. Awareness programs can foster responsible water use and support conservation efforts.
6. **Developing an Integrated Water Resource Management (IWRM)**: Adopt an integrated approach that considers the interconnections between surface water, groundwater, and land use. This strategy should align with regional development plans to ensure sustainable water resource use and water demands in the Nile Delta.

7. **Policy and Regulation Review**: Regularly review and update water management policies to address emerging challenges related to climate change and population growth. Regulations should support sustainable withdrawal limits and incentivize conservation practices.

Acknowledgements The authors express their sincere gratitude to all those who contributed to this work with their expertise and support. Special thanks are extended to Prof. Zinhom El-Saeed, Professor of Hydrogeology, Geology Department, Faculty of Science, Tanta University, for his invaluable insights. Prof. Mohammed Ezzat, Professor of Statistical Analysis, Mathematics Department, Faculty of Science, Tanta University, for his guidance in statistical methodologies. Dr. Hend S. Abu Salem, Faculty of Science, Cairo University, for her expertise in geology and thoughtful contributions. Mohammed Moheb Abd-Elbary, Urban Designer, for his assistance in producing the final illustrations for this chapter.

References

Abd El-Bary M (2012) Integration of hydrogeological, hydrochemical, and geophysical investigations to assess groundwater contamination in central Nile Delta, Egypt. Ph.D. thesis, Geology Department, Tanta University, Egypt

Abd-Elaty I, Kuriqi A, Ramadan EM, Ahmed AA (2024) Hazards of sea level rise and dams built on the River Nile on the water budget and salinity of the Nile Delta aquifer. J Hydrol Reg Stud 51:101600

Abd-Elaty IM, Abd-Elhamid HF, Fahmy MR, Abdelaal GM (2014) Investigation of some potential parameters and its impacts on saltwater intrusion in Nile Delta aquifer. J Eng Sci, Assiut University, Faculty of Engineering 42(4):931–955

Abd-Elhamid H, Javadi A, Abdelaty I, Sherif M (2016) Simulation of seawater intrusion in the Nile Delta aquifer under the conditions of climate change. Hydrol Res 47(6):1198–1210

Abd-Elhamid H, Abdelaty I (2017) Application of a new methodology (TRAD) to control seawater intrusion in the Nile Delta aquifer, Egypt. Conference Paper Res 2016 47/5, pp 1–14

Abu Salem HS, Gemail KS, Junakova N, Ibrahim A, Nosair AM (2022) An integrated approach for deciphering hydrogeochemical processes during seawater intrusion in coastal aquifers. Water 2022(14):1165. https://doi.org/10.3390/W14071165

Appelo CAJ, Geirnaert W (1983) Processes accompanying the intrusion of salt water. Geologia Applicata e Idrogeologia, Bari Italy 18(II):29–40

Arafa NA, El-Said Salem Z, Abdeldayem AL, Ghorab MA, Moustafa YM, Soliman SA, Youssef YM et al (2024) Advancing deltaic aquifer vulnerability mapping to seawater intrusion and human impacts in eastern Nile Delta: insights from machine learning and hydrochemical perspective. Earth Syst Environ:1–26

Arafa NA, El-Said Salem Z, Abdeldayem AL, Ghorab MA, Moustafa YM, Soliman SA, Youssef YM et al (2024) Advancing deltaic aquifer vulnerability mapping to seawater intrusion and human impacts in eastern Nile Delta: insights from machine learning and hydrochemical perspective. Earth Syst Environ:1–26

Arlt DH (1995) A Hydrogeological study of the Nile Delta Aquifer with emphasis on saltwater intrusion in the Northern Area. Ph.D. Thesis Technical University, Berlin, p 182

Armanuos AM, Eldin Ibrahim MG, Negm A, Takemura J, Yoshimura C, Mahmod WE (2022) Investigation of seawater intrusion in the Nile Delta aquifer, Egypt. J Eng Res 6(1):10–28

Armanuos AM, Taha MS, Zeidan BA (2024) Impact of climate change on seawater intrusion in the Nile Delta Aquifer (Egypt). In: Groundwater quality and geochemistry in arid and semi-arid regions. Springer Nature Switzerland, Cham, pp 97–163

Attia SM (1975) Pedology and soil genesis of the Quaternary deposits in the region west of the Nile Delta (north and east Wadi El-Natrun). Ph.D. thesis, Faculty of Science, Ain Shams University, p 213

Attwa M, Gemail KS, Eleraki M (2016) Use of salinity and resistivity measurements to study the coastal aquifer salinization in a semi-arid region: a case study in northeast Nile Delta, Egypt. Environ Earth Sci 75:1–18

Authority of Environmental Health in Gharbiyyah Governorate (2009) Regular report about the environmental health in the Gharbiyyah governorate

Bahr C (1992) Ein geologisches Strukturmodell des Nildeltas als Basis der Klalibrierung eines Grundwasser-stromungsmodlells. Zeitschrift der deutschen geologischen Gesellschaft 143:303–315

Bear J, Cheng AHD, Sorek S, Ouazar D, Herrera I (1999) Seawater intrusion in coastal aquifers: concepts, methods and practices. In: Theory and applications of transport in porous media, vol 14, p 640

Bobachev A, Modin I, Shevinin V (2003) IPI2Win V2.0: user's guide. Moscow State 452 University, Geological Faculty, Department of Geophysics, Moscow, Russia

Campillo JD, Taupin T, Betancur N, Patris V, Vergnaud VP, Villegas P (2021) A multi-tracer approach for understanding the functioning of heterogeneous phreatic coastal aquifers in humid tropical zones. Hydrol Sci J 2021(66):600–621

Carreira PM, Bahir M, Salah O, Fernandes PG, Nunes D (2018) Tracing salinization processes in coastal aquifers using an isotopic and geochemical approach: comparative studies in western Morocco and southwest Portugal. Appl Hydrogeol 2018(26):2595–2615

Chen FW, Liu CW (2012) Estimation of the spatial rainfall distribution using inverse distance weighting (IDW)in the middle of Taiwan. Paddy Water Environ 10:209–222

Dahab KA (1993) Hydrogeological evolution of the Nile Delta after the high dam construction. Ph.D dissertation, Geology Dept, Faculty of Science, Menofia University, Shibin El-Kom, Egypt, p 256

Diab MSh, Saleh MF (1982) The hydrochemistry of pleistocene aquifer in the Nile Delta area, vol 7. Environment International, Egypt, pp 221–230

Ebraheem AM, Senosy MM, Dahab KA (1997) Geoelectrical and hydrogeochemical studies for delineating groundwater contamination due to saltwater intrusion in the northern part of the Nile Dela, Egypt. Ground Water 35:216–222

EL Alfy M (2003) Environmental impact of the geomorphological and hydrogeological aspects of Rafah area, North Sinai, Egypt. Ph.D. Thesis, Masoura University, Faculty of Science, p 216

El Alfy M, Merkel B (2006) Hydrochemical relationships and geochemical modeling of groundwater in Al Arish area, north Sinai, Egypt. Hydrol Sci Technol 22(1–4):47–62. ISSN 0887-686X

El–El-Ghandour MFM, Hefny K, Khalil JB, Atta SA (1985) Sea water intrusion into the Nile Delta aquifer, Egypt. J Middle East Water Sew:265–268

El-Fawal FM (1979) Neogene Quaternary succession of the northern part of the Nile Delta area. M.Sc. thesis, Faculty of Science, Mansoura University, p 233

El–Fayoumi IF, Hegab OA, El Nahas HA (1991) Holocene sedimentary pattern and evolution of the northern coastal plain of the Nile Delta, Egypt. Mansoura Sci Bull Spec:133–149

El-Fayoumi LF (1968) Geology of groundwater supplies in the eastern region of the Nile Delta and its extension in the north Sinai. PhD Thesis, Cairo University, Egypt

ESRI (2020) ArcGIS desktop 10.8 help (includes Arc globe and Arcscene software). https://desktop.arcgis.com/en/ArcMap/. Accessed 1 Apr 2020

Farnham IM, Klaus JS, Ashok KS, Johannesson KH (2000) Deciphering groundwater flow systems in Oasis Valley, Nevada, using trace element chemistry, multivariate statistics, and geographical information system. Math Geol 32(8):943–968

Gemail K, Samir A, Oelsner C, Mousa SE, Ibrahim S (2004) Study of saltwater intrusion using 1D, 2D, and 3D resistivity surveys in the coastal depressions at the eastern part of the Matruh area, Egypt. Near Surf Geophys 2(2):103–109

Gemail KS, El Alfy M, Ghoneim MF, Shishtawy AM, Abd El-Bary MH (2017a) Comparison of DRASTIC and DC resistivity modeling for assessing aquifer vulnerability in the central Nile Delta, Egypt. Environ Earth Sci 2017(76):350

Gemail KS, Attwa M, Eleraki M, Zamzam S (2017b) Imaging of wastewater percolation in heterogeneous soil using electrical resistivity tomography (ERT): a case study at east of tenth of Ramadan City, Egypt. Environ Earth Sci 76:1–17

Gemail KS, Ghoneimi A, Shebl S, Soliman S, El-Sayed K, Saleh AH (2024) Flow pathways in fractured basalt under challenging environmental conditions: a synergistic approach of electrical resistivity tomography and seismic refraction imaging. Environ Process 11(2):24

Gheith AM (1976) Geological studies on the subsurface sediments of Kafr El Sheikh well no. 1, Nile Delta. M.Sc. thesis, Faculty of Science, Mansoura University, p 156

Giménez-Forcada E (2014) Space/time development of seawater intrusion: a case study in Vinaroz coastal plain (eastern Spain) using HFE-diagram, and spatial distribution of hydrochemical facies. J Hydrol 2014(517):617–627

Giménez-Forcada E (2019) Use of the Hydrochemical Facies Diagram (HFE-D) for the evaluation of salinization by seawater intrusion in the coastal Oropesa Plain: comparative analysis with the coastal Vinaroz Plain, Spain. J Hydro-Environ Res 2019(2):76–84

Glynn PD, Plummer LN (2005) Geochemistry and the understanding of groundwater systems. Hydrogeol J 13:263–287

Habib MM (1996) Hydrogeological studies on El-Gharbiyyah, central Nile Delta, Egypt. M.Sc. thesis, Faculty Science, Tanta University, p 136

Hach (1990) Chemical procedure explained. Hach Technical Center for Applied Analytical Chemistry, Colorado, USA, p 48

Hagage M, Hewaidy AGA, Abdulaziz AM (2025) Saltwater intrusion and land subsidence destroy northern Nile Delta archaeological sites: an assessment using hydrochemical indices, SAR satellite imagery, and analytic hierarchy process (AHP). Mar Pollut Bull 212:117460

Hill (1940) Geochemical patterns in Coachella Valley, vol 21. Transactions/American Geophysical Union, California

Jaquet JM, Froidevoux R, Verned JP (1975) Comparison of automatic classification methods applied to lake geochemical samples. Math Geol 7(3):237–265

Kashef AI (1983) Saltwater intrusion in the Nile Delta. Ground Water 21(2):160–167

Mabrouk MB, Jonoski A, Solomatine D, Uhlenbrook S (2013) A review of seawater intrusion in the Nile Delta groundwater system—the basis for assessing impacts due to climate changes and water resources development. Hydrol Earth Syst Sci Discuss 10(10873–10911):2013. https://doi.org/10.5194/hessd-10-10873-2013

Mabrouk M, Jonoski A, HP Oude Essink G, Uhlenbrook S (2018) Impacts of sea level rise and groundwater extraction scenarios on fresh groundwater resources in the Nile Delta governorates, Egypt. Water 10:1690. https://doi.org/10.3390/w10111690

Melloul (1997) Monitoring of seawater intrusion in coastal aquifers: basics and local concerns. J Environ Manag. Accepted 27 May 1997, 25 May 2002

Nasar MKK (2004) Design and management of discharge and recharge well field for coastal desalination plants. M.Sc. thesis, Cairo University, Egypt

Nosair AM, Shams MY, Abou Elmagd LM, Hassanein AE, Frayer A, Abu Salem HS (2021) A predictive model for progressive salinization in a coastal aquifer using artificial intelligence and hydrogeochemical techniques: a case study of the Nile Delta aquifer, Egypt. Environ Sci Pollut Res 2021(29):9318–9340

Park S-C, Yun S-T, Chae G-T, Yoo I-S, Shin K-S, Heo C-H, Lee S-K (2005) Regional hydrochemical study on salinization of coastal aquifers, the western coastal area of South Korea. J Hydrol 2005(313):182–194

Piper AM (1944) A graphic procedure in the geochemical interpretation of water analysis. Eos Trans Am Geophys Union 1944(25):914–928

Revelle R (1941) Criteria for recognition of seawater in groundwater. Trans Am Geophys Union 22:293

Sallouma MKM (1983) Hydrogeological and hydrochemical studies east of Nile Delta, Egypt. Ph.D. Thesis, Faculty of Science, Ain Shams University, Cairo, Egypt

Saqr AM, Abd-Elmaboud ME (2024) Saltwater intrusion in coastal aquifers: a comprehensive review and case studies from Egypt. https://doi.org/10.32388/XLIS2F

Serage El-Din HM (1989) Geological, hydrogeological and hydrological studies on the Nile Delta quaternary aquifer: Dissertation. Mansoura University, Egypt, p 289

Sestini G (1989) Nile Delta: a review of depositional environments and geological history. In: Whateley MKG, Pickering KT (eds), vol 40. Geological Society Special Publication, pp 99–127

SGMA (2014) An excerpt from the groundwater surface water interactions monitoring method [MM-07]. California Water Science Center. https://www.usgs.gov/centers/ca-water

Sherif MM, Singh VP (2002) Effect of groundwater pumping on seawater intrusion in coastal aquifers. Agric Sci 7(2):61–67

Sherif M, Sefelnasr A, Javadi A (2012) Incorporating the concept of equivalent freshwater head in successive horizontal simulations of seawater intrusion in the Nile Delta aquifer, Egypt. J Hydrol 2012(464–465):186–198

Sherif M, Kacimov A (2008) Pumping of brackish and saline water in coastal aquifers: an effective tool for alleviation of seawater intrusion. In: Proceedings of 20th SWIMS, Naples, Florida, USA, 23–27 June 2008

Sherif MM (1999) The Nile Delta aquifer in Egypt. In: Seawater intrusion in coastal aquifers: concepts, methods, and practices, vol 14. Theory and application of transport in porous media. Kluwer Academic Publishers, Alphen aan den Rijn, The Netherlands, pp 559–590

Sinclair A (1974) Selection of threshold values in geochemical data using probability graphs. J Geochem Explor 1974(3):129–149

Stuyfzand PJ (1986) A new hydrochemical classification of water types: Principles and application to the coastal dunes aquifer system of the Netherlands. In: Proceedings of the 9th salt water intrusion meeting, Delft, The Netherlands, pp 641–655, 12–16 May 1986

Tal A et al (2019) High-resolution monitoring of seawater intrusion in a multi-aquifer system-implementation of a new downhole geophysical tool. Water 11(9):1877. https://doi.org/10.3390/w11091877

Taha MS, Armanuos AM, Zeidan BA (2023) Impact of climate change on seawater intrusion, and shoreline advance in the Nile Delta, Egypt. Delta Univ Sci J 6(2):182–219

Tarik B (2013) Effect of climate change on sea water intrusion in the Nile Delta coastal aquifer. M.Sc. Thesis, Civil Engineering—Cairo University

Tijani MN (2004) Evolution of saline waters and brines in the Benue-Trough, Nigeria. Appl Geochem 2004(19):1355–1365
Troger U, Holtorff G, Arlt H, Asbrand M, Bahr C, Baumann R, Moser H (1993) Meerwasserintnision im Delta arider und senuaride:Gebieteam Beispiel des Nildeitas: Berlin. Germany. Sondenorschungsbereich 69 (SFB69—TUB): Geouisscnschaftlichc problem in anden und semiariden Gebieten, Berichisband 1990–1993. Tell C {internal report}, pp 1411–1470
Turekian KK (1968) Oceans. Foundations of earth science series. Prentice-Hall, Hoboken NJ, USA, p 120
Van Pham H, Lee S-I (2015) Assessment of seawater intrusion potential from sea-level rise and groundwater extraction in a coastal aquifer. Desalin Water Treat 2015(53):2324–2338
Williams R (1982) Statistical identification of hydraulic connections between the surface of a mountain and internal mineralized zones. Ground Water 20:466–478, 79–2, M.I.T., Cambridge, Massachusetts
Wilson J, Townley LR, Sa da Costa A (1979) Mathematical development and verification of a finite element aquifer flow model AQUIFEM-1. Technology Adaptation Program Report N
Zaghloul Z, Taha A, Hegab O, El Fawal F (1977) The Neogene-Quaternary sedimentary basins of the Nile Delta. Egypt J Geol 21(1):1–19
Zaghloul ZM, Taha HH, Hegab OA, El-Fawal FM (1979) The Plio-Pleistocene Nile Delta subenvironments, stratigraphic section, and genetic class. Geol Soc V. lx, Cairo

Physiochemical Characterization and Groundwater Quality Evaluation for Irrigation Purposes in Eastern Nile Delta Aquifer, Egypt

9

Mahmoud E. Abd-Elmaboud, Hossam A. Abdel-Gawad, Kassem S. El-Alfy, Mohsen M. Ezzeldin, and Ahmed M. Saqr

Abstract

Groundwater is a critical resource globally, supporting agriculture, domestic use, and industrial activities, especially in arid and semi-arid locations with inadequate surface water. However, groundwater quality is increasingly deteriorating due to over-extraction, pollution, and saltwater intrusion. The eastern Nile Delta, a region of strategic importance in Egypt for agricultural development, faces significant challenges in groundwater quality management. This study evaluates the groundwater suitability for irrigation purposes in the eastern Nile Delta by conducting a comprehensive hydro-geochemical analysis. Data from 488 groundwater wells and 94 surface water samples were utilized, employing robust methods including Piper diagrams, spatial mapping, and water quality metrics such as sodium adsorption ratio (SAR) and residual sodium carbonate (RSC). The findings indicated that 53.6% of groundwater samples were fresh, 24.3% were brackish, and 22.1% were saline, with fresh water predominantly located in the southern and southwestern regions. Piper diagram analysis identified key water types, including calcium bicarbonate ($Ca2^{+}-HCO_3^{-}$) in fresher zones and sodium chloride ($Na^{+}-Cl^{-}$) in saline-impacted areas. Spatial mapping (mention the type of map/s and the applied method, e.g., weighting criteria analysis in ArcGIS) revealed that approximately 30% of the region was highly suitable for irrigation, while

M. E. Abd-Elmaboud (✉) · H. A. Abdel-Gawad · K. S. El-Alfy · M. M. Ezzeldin · A. M. Saqr
Irrigation and Hydraulics Department, Faculty of Engineering, Mansoura University, Mansoura, Egypt
e-mail: mahmoud_elsayed@mans.edu.eg

A. M. Saqr
e-mail: ahmedsaqr@mans.edu.eg

M. A. Khalifa and K. S. Gemail (eds.), *Geology of the Nile Deltas and Natural Resources*, Deltas of the World, https://doi.org/10.1007/978-3-031-96594-4_9

severe restrictions existed in 15–20% of the area, particularly in the north and northeast. These findings provided actionable insights for optimizing irrigation practices and managing water resources effectively. This research offers a replicable framework for similar regions globally, particularly in water-stressed areas facing comparable challenges, enhancing agricultural sustainability.

Keywords

Physiochemical properties • Groundwater quality • Irrigation • Piper diagram • Nile Delta • Sustainability

9.1 Introduction

Groundwater is an essential element for human survival and development, playing a critical role in meeting water demands for drinking, industrial activities, and particularly agriculture (Alao et al. 2024). Groundwater is used for irrigation in arid and semi-arid locations when surface water is sparse or unreliable, supporting food production (Paudel et al. 2024). In Egypt, the role of groundwater is magnified due to the increasing pressure on Nile River resources, which are already overextended by rapid population growth and intensive agricultural activities (Elewa et al. 2013). In the eastern Nile Delta, groundwater is a vital resource for irrigation, especially in reclaimed desert lands and regions where expanding agricultural activities demand additional water supplies (Faid and Abdulaziz 2012). However, the quality of groundwater in these areas has deteriorated significantly over the past decades (Abdulaziz et al. 2009). This decline is attributed to factors such as over-extraction, agricultural return flows containing fertilizers and pesticides, industrial pollution, and the intrusion of saline water from the Mediterranean Sea and underlying aquifers (El Alfy et al. 2020; Saqr et al. 2023b). The preservation of groundwater quality is thus crucial for sustaining agricultural productivity, and this underscores the importance of employing efficient and systematic monitoring techniques to evaluate and manage this critical resource (Noman et al. 2024).

Various methods have been used to monitor groundwater quality for agricultural purposes, from physical assessments to advanced chemical and geospatial analyses (Abo-El-Fadl 2013). Among these, chemical characterization techniques are particularly valuable as they provide insights into groundwater's composition, origin, and evolution (Arnous and Green 2015). The Piper diagram, for example, has proven to be a powerful tool for identifying water sources and understanding the hydrogeochemical processes that influence groundwater quality (He and Li 2020). Additionally, a suite of water quality indicators, including total dissolved solids (TDS) and sodium adsorption ratio (SAR), is commonly employed to assess groundwater suitability for irrigation (Paudel et al. 2024). These indices help evaluate the potential impact of groundwater on soil health and crop productivity (Saqr et al. 2023a). In the context of the eastern Nile Delta aquifer, several earlier studies have highlighted the region's hydrogeochemical challenges, including saline intrusion and agricultural contamination (Elewa et al. 2013). For instance, research by Shata and El Fayoumy (1970) provided initial insights into the region's groundwater conditions, while Abdulaziz et al. (2009) highlighted the significance of multi-temporal remote sensing data for monitoring alterations in land cover and their impact on water quality. However, these studies often lack the depth and granularity required to address the complexities of groundwater quality in the region (Arnous and El-Rayes 2013). The rapid pace of land reclamation and agricultural expansion necessitates a more detailed and up-to-date assessment of groundwater resources along the east of the Nile Delta (Elewa et al. 2013).

This research seeks to address this critical gap by providing a comprehensive physicochemical characterization and groundwater quality evaluation in the east of the Nile Delta to determine its suitability for irrigation. By employing hydrogeochemical techniques, this study aims to identify the key physical and chemical parameters influencing groundwater quality, including using the Piper diagram and other analytical tools to delineate water types and trace contamination sources. Besides, the assessment involves calculating key irrigation water indices such as TDS, SAR, MH, and RSC, which are instrumental in determining the compatibility of groundwater with agricultural practices. This work not only sheds light on the current status of groundwater in the region but also provides actionable insights for sustainable water management. The findings are expected to contribute to the formulation of evidence-based policies that enhance the resilience of Egypt's agricultural sector against water scarcity and quality challenges.

9.2 Materials and Methods

9.2.1 Study Area

The investigation focuses on a distinctive region within the northeastern Nile Delta that encompasses approximately 12,284 km^2 (Fig. 9.1). This area is strategically positioned between 30°00′–31°30′N latitude and 31°00′–32°30′E longitude, with clear geographical boundaries defined by significant landmarks: the Suez Canal to the east, the Mediterranean Sea and El-Manzala Lake to the north, the Damietta Branch of the Nile River to the west, and the Cairo-Suez Desert Road marking its southern extent. The region's topographical characteristics reveal a subtle yet significant gradient, transitioning from a gentle northward slope to an expansive southern zone featuring an elevated plateau, with elevations varying between 2 and 24 m above sea level (Abd-Elmaboud et al. 2021).

The climatic conditions distinctly reflect an arid to semi-arid environment, characterized by intense, dry summers devoid of rainfall and temperate winters experiencing occasional precipitation (Mansour et al. 2025). Temperature patterns demonstrate notable

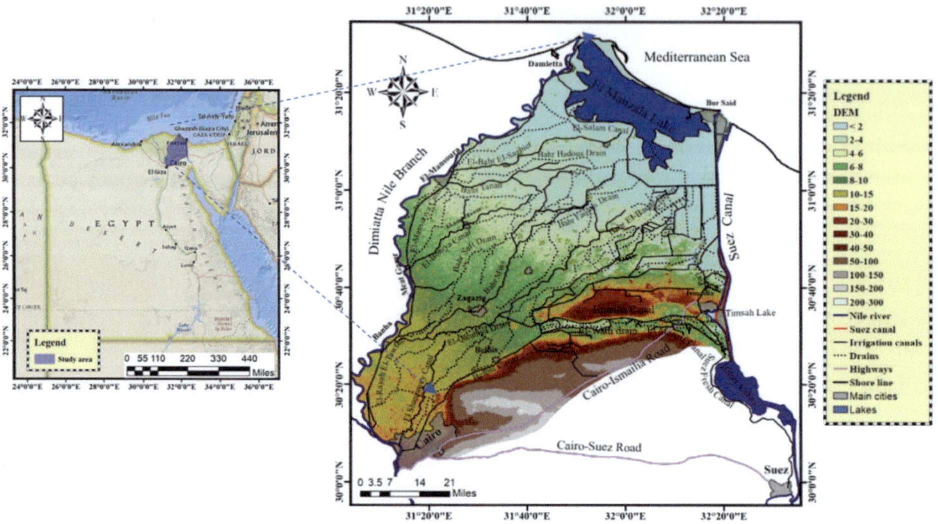

Fig. 9.1 Location map of the study area, eastern Nile Delta region, Egypt

seasonal variation, with monthly averages fluctuating from approximately 12 °C during winter to 27 °C in summer. Annual precipitation in the region remains minimal, typically ranging from 20 to 100 mm. The landscape exhibits remarkable diversity, incorporating a mosaic of land-use patterns including productive and dormant agricultural territories, urban developments, desert expanses, freshwater canal networks, marine bodies, and wetland ecosystems (Ezzeldin et al. 2018). Of particular interest is the eastern sector of the study area, which has undergone substantial agricultural and hydrogeological development. This section, primarily consisting of reclaimed land, covers an area of approximately 2,931 km^2 and is delineated by several significant hydrological features: the Bahr El-Bakar drain, El-Manzala Lake, the Suez Canal, and the Cairo-Ismailia Road. This area exemplifies the diverse conditions present within the Quaternary aquifer system (Abd-Elmaboud, 2022; Abdulaziz et al. 2009).

9.2.2 Research Methodology

A systematic approach was adopted to fulfill the study objectives, involving multiple steps to collect and analyze groundwater quality and its suitability for irrigation. Advanced hydrogeochemical techniques were employed to collect and interpret data, with the use of AQ.QA software to classify and interpret water chemistry and GIS software (ESRI ArcGIS 10.8) to generate spatial maps of groundwater quality parameters.

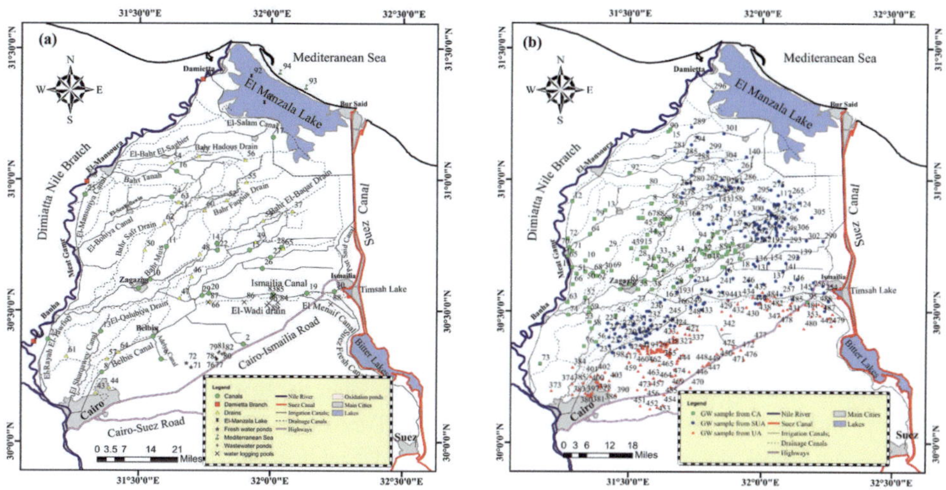

Fig. 9.2 Location maps of the collected samples: **a** surface samples and **b** groundwater samples

9.2.2.1 Data Collection

This study utilized hydrochemical data obtained from previous studies, e.g., Embaby et al. (2014) and Geriesh et al. (2015), in addition to a field survey. The Quaternary aquifer of the eastern Nile Delta region comprises three primary subsystems determined by the prevailing hydrogeological conditions: unconfined aquifer (UA), semi-confined aquifer (SUA), confined aquifer (CA) (Ismael et al. 2007). A total of 488 wells employed for groundwater sampling and water table monitoring were divided into three categories: 95 wells in CA, 212 wells in SUA, and 181 wells in UA (Fig. 9.2). This study employed 94 surface water samples, categorized as follows: 31 samples from main canals, 36 samples from main drains, 3 samples from the Damietta Branch of the Nile River, 2 samples from a freshwater pond, 10 samples from a wastewater pond adjacent to Tenth of Ramadan city, 8 samples from waterlogging pools south of Ismailia Canal, 2 samples from El-Manzala Lake, and 2 samples from the Mediterranean Sea.

Field measurements of physical parameters such as pH, TDS, electrical conductivity (EC), and sodium chloride percentage (NaCl%) were performed using portable, calibrated devices. (Mastrocicco et al. 2011). Water samples were stored in sterilized polyethylene bottles, filtered, and acidified as required to ensure the accuracy of subsequent chemical analyses.

9.2.2.2 Analysis of Data

A comprehensive chemical characterization of the groundwater samples was implemented through systematic analytical procedures to quantify the predominant ionic constituents. The analysis encompassed detailed measurements of principal cationic species, including calcium (Ca^{2+}), magnesium (Mg^{2+}), sodium (Na^{+}), and potassium (K^{+}), alongside essential anionic components such as chloride (Cl^{-}), sulfate ($SO_4{}^{2-}$), bicarbonate ($HCO_3{}^{-}$), and nitrate ($NO_3{}^{-}$). All analytical protocols strictly adhered to the standardized methodologies established by the American Public Health Association (APHA 2017). To ensure data reliability, quality control measures were implemented throughout the analytical process, including duplicate analyses (10% of samples), reagent blanks, and certified reference materials. The charge balance error (CBE) was calculated for each sample using the equation: $CBE(\%) = [(\sum cations - \sum anions)/(\sum cations + \sum anions)] \times 100$, with 92% of samples exhibiting CBE values within the acceptable range of $\pm 5\%$, confirming the analytical reliability of the dataset. Samples exceeding this threshold were re-analyzed or excluded from further interpretation. The investigation extended beyond basic ionic measurements to include the computation of crucial derivative parameters, particularly water hardness indices and carbonate system equilibria. This methodical analytical approach generated a robust dataset that served as the foundation for the detailed characterization of groundwater properties and the subsequent evaluation of its irrigation suitability parameters.

9.2.2.3 Identification of Hydrochemical Facies

To identify hydrochemical facies and understand the origin and evolution of groundwater, Piper diagrams were used. Piper diagrams are an essential tool for visualizing the chemical composition of groundwater and classifying water types based on ionic dominance (Piper 1944). By plotting the concentrations of cations and anions on trilinear diagrams, the study identified trends and mixing processes within the aquifer systems. This step was instrumental in distinguishing between different water types, such as $Ca^{2+}-HCO_3{}^{-}$ and $Na^{+}-Cl^{-}$, and identifying zones influenced by saline intrusion or agricultural return flows.

9.2.2.4 Assessment of Groundwater Quality

Groundwater quality for irrigation was assessed using six key indices, which provide a quantitative framework for evaluating the impact of water on soil and crop health. Salinity hazard (SH) categorizes water based on its salinity levels and its potential impact on crops and soil (Simmons 2000). The SAR measures the relative proportion of Na^{+} to Ca^{2+} and Mg^{2+} ions, which is critical for understanding water's effect on soil permeability (Thorne 1954). The exchangeable sodium ratio (ESR) evaluates the balance between exchangeable Na^{+} and other cations in the soil, influencing soil structure and infiltration capacity (Poonia and Talibudeen 1977). Residual sodium carbonate (RSC) assesses the carbonate and $HCO_3{}^{-}$ content relative to Ca^{2+} and Mg^{2+}, where high values indicate risks of carbonate precipitation that could degrade soil quality (Doneen 1964). Magnesium hazard (MH) examines the ratio of Mg^{2+} to Ca^{2+}, with higher values suggesting potential harm to crops due to Mg^{2+} dominance (Kaddah 1975). Lastly, sodium percentage (Na%) determines Na^{+}'s proportion relative to total cations, aiding in the assessment of sodicity

Table 9.1 Assessment parameters and quality indicators for irrigation groundwater suitability analysis

Index	Abbreviation	Formula	References
Salinity hazard	SH	$\begin{cases} Low,\ TDS < 200mg/L;\ Ec < 250\mu m/cm \\ Medium,\ 200 < TDS < 500;\ 250 < Ec < 750 \\ High,\ 500 < TDS < 1500;\ 750 < Ec < 2250 \\ Very\,high,\ 1500 < TDS;\ 2250 < Ec \end{cases}$	(Zaman et al. 2018)
Sodium adsorption ratio	SAR	$SAR = \frac{Na^{+}}{\sqrt{(Ca^{2+}+Mg^{2+})/2}}$	(Thorne 1954)
Exchangeable sodium ratio	ESR	$ESR = \frac{Na^{+}}{Ca^{2+}+Mg^{2+}}$	(Poonia and Talibudeen 1977)
Residual sodium carbonate	RSC	$C = (CO^{-} + HCO_3^{-}) - (Ca^{2+} + Mg^{2+})$	(Doneen 1964)
Magnesium hazard	MH	$MH\% = \frac{Mg^{2+}}{Ca^{2+}+Mg^{2+}} * 100$	(Kaddah 1975)
Sodium percentage	Na%	$Na\% = \frac{Na^{+}+K^{+}}{Na^{+}+K^{+}+Ca^{2+}+Mg^{2+}} * 100$	(Khodapanah et al. 2009)

Note All the ionic concentrations are in Milliequivalents per Liter (meq/l)

risks (Khodapanah et al. 2009). These indices were calculated for each water sample, enabling the classification of irrigation water into suitability categories and facilitating the creation of spatial maps to identify regions with high irrigation potential and areas requiring mitigation measures. Table 9.1 provides the formulas and references for these indices.

9.3 Results and Discussion

9.3.1 Physical Characteristics

Four Physical characteristics, i.e., pH, TDS, EC, and NaCl, were examined for the aquifer systems within the study area, along with the surface water in the following sections.

9.3.1.1 Total Dissolved Solids (TDS)

TDS distribution in the study area is shown in Fig. 9.3a. It was evident that groundwater samples from the CA ranged from 277.5 to 35,600 mg/l, with an average of 2773.7 mg/l. In contrast, samples from the UA ranged from 260 to 41,394 mg/l, with an average of 2643.1 mg/l. The TDS of groundwater from the SUA varied from 224 to 25,500 mg/l, with a mean of 3879.2 mg/l. The United States Environmental Protection Agency (EPA) secondary drinking water guidelines restrict water with TDS values exceeding 500 mg/

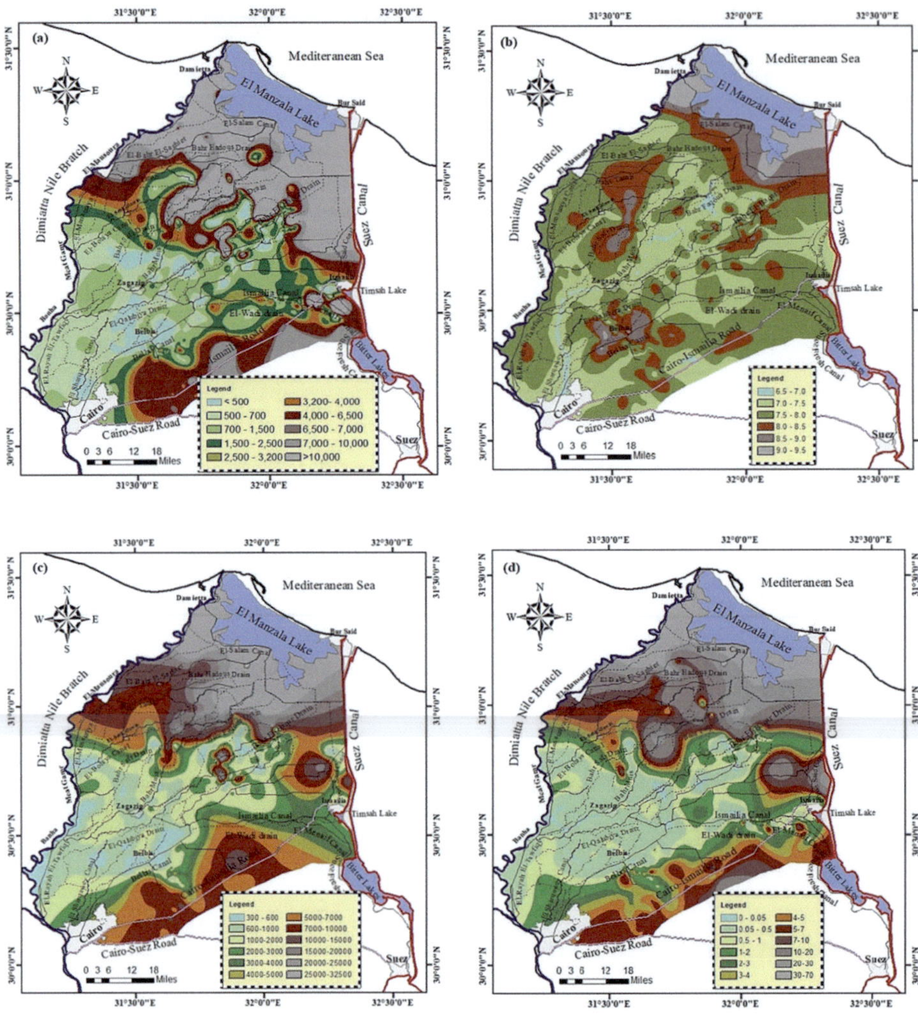

Fig. 9.3 Distribution of physical parameters within aquifer systems in the study area: **a** TDS, **b** pH, **c** EC, and **d** NaCl%

l. For crops with low to moderate salt tolerance, 1500–2600 mg/l TDS irrigation can be difficult (Harter 2003). Also, the results showed that 53.6% of groundwater samples were fresh, 24.3% were brackish, and 22.1% were saline.

The spatial distribution of freshwater resources exhibits distinct patterns, with significant concentrations of high-quality groundwater identified predominantly in the region north of Cairo. This distribution is particularly pronounced along major hydrological features, notably the Damietta branch and principal irrigation networks including the Ismailia Canal and El-Rayah El-Tawfiqy, where the hydrogeological conditions are characterized by deltaic sedimentary deposits and relatively shallow water table configurations. In contrast, the northern and eastern sectors of the study area demonstrate elevated salinity levels, a phenomenon attributed to the presence of fluviomarine deposits and saline water bodies. A notable transitional zone exists in the vicinity of the southern Ismailia canal, where fresh groundwater undergoes a progressive transformation into brackish conditions, primarily due to the vertical migration of brackish water from underlying aquifer systems. The observed spatial variations in groundwater quality effectively reflect the influence of depositional environments, with areas of reduced salinity corresponding to fluvial deltaic deposits, while zones of high salinity align with marine sedimentary formations, particularly evident in the Miocene aquifer system (Indugula and Venkata Vajjha 2023).

9.3.1.2 Power of Hydrogen (pH)

The study area had a low pH zone (<7.5) in the middle, spanning northeast-southwest parallel to Bahr El-Baqar (Fig. 9.3b). pH abnormality usually depends on the surface water canal type (Paudel et al. 2024). As drains dominated this region, the pH zone was low. The high pH zones located in the central northwest and southeast regions are primarily recharged by the Damietta branch and major irrigation canals, such as the Ismailia Canal. Notably, pH levels exhibit an inverse relationship with the salinity of surface water canals.

The EPA's secondary drinking water standard is 6.5–8.5 pH. For surface water samples, freshwater pH varied from 7.3 to 8.8, while brackish water pH was 6.9–8.7. Aquatic plants burning dissolved CO_2 gas might raise the pH (Abo-El-Fadl 2013), while organic matter oxidation lowered pH and increased HCO_3^- and H^+ species.

9.3.1.3 Electric Conductivity (EC)

Figure 9.3c displays groundwater EC distribution in the research area. The map indicated a gradual increase in EC from latitude 31° 00′ N to the north and northeast of the study area. Saltwater intrusion from the Mediterranean and El-Manzala Lake caused significant changes in groundwater EC measurements (Saqr and Abd-Elmaboud 2024). Saltwater penetration from the Suez Canal and El Temsah Lake created a narrow high EC zone to the east. The south and southeast saw a considerable increase in EC measurements of Quaternary aquifer groundwater due to inter-aquifer mixing with southern Miocene brackish water.

Additionally, the EC values ranged from 323–1020 μs/cm for the Damietta branch, irrigation canals, and freshwater ponds, 774–5130 μs for drains, El-Manzala Lake, and wastewater ponds, 6929.1 μs for waterlogging pools south of Ismailia Canal, and 46,282 μs for the Mediterranean Sea. Besides, the EC of groundwater from the Quaternary aquifer ranged from a few hundred μs/cm in the south to many tens of thousands in the north, with most samples below 5000 μs/cm. EC values represent a solution's ability to conduct electrical current and are connected to ions (Geriesh et al. 2015).

9.3.1.4 Sodium Chloride Percentage (NaCl%)

NaCl% represents water sample salinity along with TDS and EC (Abdulaziz et al. 2009). The NaCl% distribution map (Fig. 9.3d) resembled the TDS and EC distribution maps in Fig. 9.4a and b. It was evident from Fig. 9.4d that the south and middle of the research area (deltaic section) had NaCl typically less than 10%, indicating meteoric groundwater origin. The north and northeast had higher NaCl values, often above 10%, due to saltwater and fluviomarine sediment mixing (Moorthy et al. 2024).

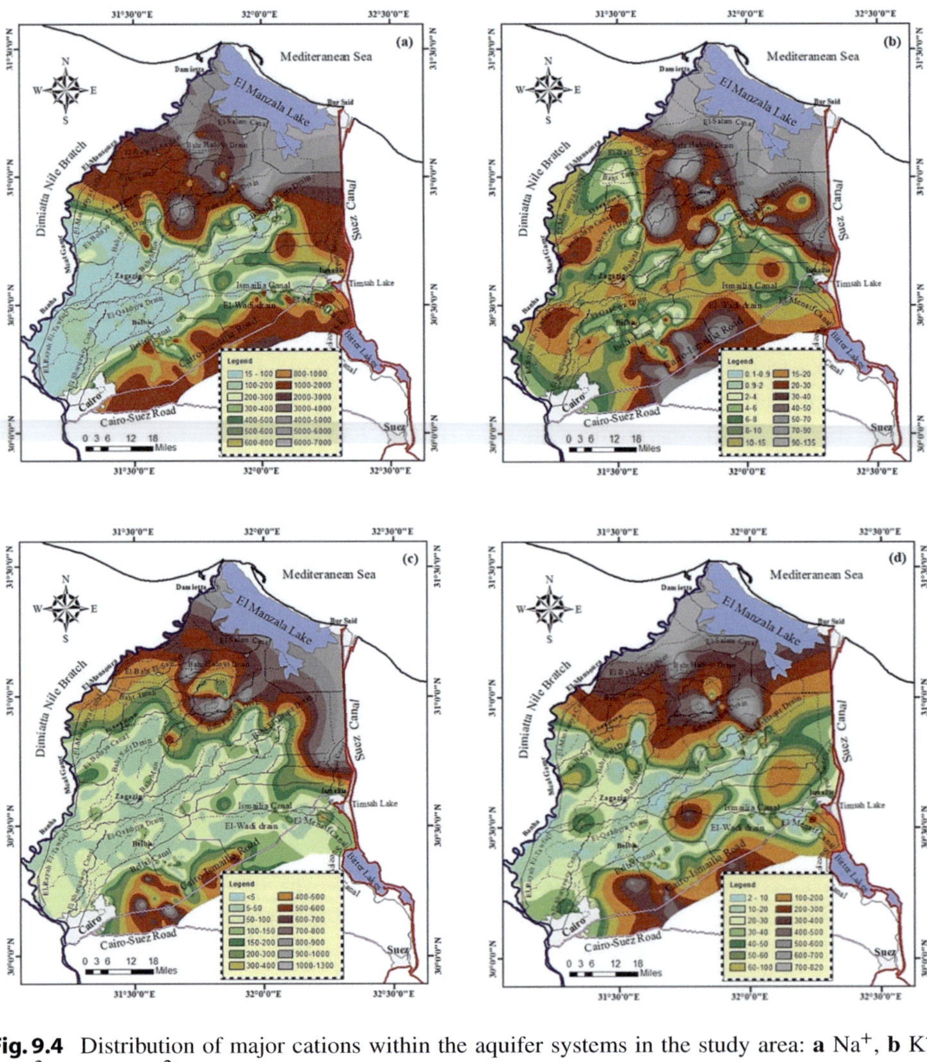

Fig. 9.4 Distribution of major cations within the aquifer systems in the study area: **a** Na^+, **b** K^+, **c** Ca^{2+}, and **d** Mg^{2+}

9.3.2 Chemical Properties

The description of the main chemical properties, i.e., cations and anions, was given in the following lines.

9.3.2.1 Major Cations

Sodium (Na^+)

The spatial distribution of sodium concentrations across the groundwater aquifer systems, as illustrated in Fig. 9.4a, reveals distinct hydrochemical patterns throughout the study area. In the western region, particularly at latitudes approximating 30º 50´ N, sodium concentrations maintain relatively low levels ranging from 15 to 200 mg/l, a phenomenon attributed to the presence of substantial clay deposits confirmed the ion exchange process where the Ca and Mg in the clay deposits have been replaced by sodium indicating groundwater with longer residence time, and sophisticated irrigation network systems (El-Amier et al. 2021). Conversely, the southeastern unconfined aquifer exhibits markedly elevated sodium content, ranging from 200 to 3000 mg/l, primarily resulting from the utilization of brackish irrigation water coupled with intensive evapotranspiration processes characteristic of the prevailing arid climatic conditions.

The southern region demonstrates notably high sodium concentrations, a condition primarily influenced by two factors: recharge from Miocene marine sedimentary facies and the vertical flow of saline water from deep aquifers through fault systems (Faid and Abdulaziz 2012). This situation is further complicated by anthropogenic factors, particularly the application of sodium-based fertilizers, which contribute to the elevated sodium levels. The most significant sodium concentrations, exceeding 2000 mg/l, are observed in the northern and northeastern sectors, where saltwater body infiltration substantially influences groundwater chemistry.

Potassium (K^+)

Figure 9.4b shows that the southern and middle parts of the study area aquifer had low K^+ concentrations (0.9–8) mg/l, with many low K^+ locations around irrigation canals. The southeast section of the research area, the new reclaimed zone, has high K^+ content (15–70) mg/l, likely due to inorganic K-fertilizer use (Embaby et al. 2014). Surface infiltrations from agricultural/municipal drains to groundwater also cause elevated K^+ concentrations near major drains. K^+ content steadily increases (>50 mg/l) in the north and northeast, indicating seawater intrusion, notably near El-Manzala Lake and the Suez Canal (Saqr and Abd-Elmaboud 2024).

Calcium (Ca^{2+})

Groundwater Ca^{2+} distribution for the study area aquifers is shown in Fig. 9.4c. Ca^{2+} values were often below 150 mg/l in areas south of latitude 30º50´ N, especially near major freshwater canals like the Damietta branch to the west and Ismailia Canal to the south. An increase in Ca^{2+} content along the southern border of the research region might be due to brackish groundwater upwelling from the Miocene aquifer. (Faid and Abdulaziz 2012). In the north and northeast near the sea bodies, Ca^{2+} concentrations were high.

Magnesium (Mg^{2+})

Figure 9.4d depicts the regional distribution of Mg^{2+} in the aquifer's groundwater, which was comparable to Na^{+}. Generally, Mg^{2+} concentrations below 40 mg/l were found in the western research region, notably near freshwater canals in the southern latitude of 30º 50' N. Mg^{2+} infiltration from the breakdown of a natural salt or mineral deposit left after water evaporation caused slight increases in Mg^{2+} concentration near the El Salhyia canal barrier and in the southeast of the research region. Active agricultural activity dominates the east of the research area near Ismailia city, where Mg^{2+} content rises to (40–200) mg/l. Saltwater incursion raises Mg^{2+} levels to around 200 mg/l in the north (Saqr and Abd-Elmaboud 2024).

9.3.2.2 Major Anions

Chloride (Cl^{-})

Cl^{-} distribution maps indicated low, middle, and high concentration zones (Fig. 5a). The western zone of the study region, dominated by constrained conditions, has low Cl^{-} concentrations ($\leq$300 mg/l). This might be due to the thick surface canal network (El-Amier et al. 2021). For 50 years, the freshly reclaimed area north and south Ismailia canal has had an intermediate concentration zone with Cl^{-} levels between 400 and 3000 mg/l. Evapotranspiration, salt richness in upper soil zones, and excessive inorganic fertilizer use enhance Cl^{-} content in this zone. A high Cl^{-} concentration zone (<3000 Mg/l) is seen in the northern study region near saltwater bodies.

Sulfate (SO_4^{-2})

Groundwater SO_4^{-2} distribution is depicted in Fig. 9.5b. The CA had a low SO_4^{-2} concentration zone ($\leq$ 200 ppm) from south of latitude 30º50´ N to north of Cairo city. Leaching SO_4^{-2} from topsoil zones and dissolving gypsum crystals from the aquifer component raise SO_4^{-2} concentrations to 1000 mg/l in the southeast SUA and UA (Faid and Abdulaziz 2012). SO_4^{-2}, like other major ions, grows steadily north and northeast.

Bicarbonate (HCO_3^{-})

Figure 9.5c shows HCO_3^{-} concentrations in water samples. It was evident that the study area's total alkalinity was unevenly distributed, indicating that lithology and local contamination from agricultural and industrial effluents have caused significant hydrogeological changes (Saqr et al. 2023a). The southern half of the aquifer had low to moderate total HCO_3^{-} concentrations, while the eastern part of the Ismailia canal, which might provide groundwater with HCO_3^{-} ions, had greater values. Alkalinity was higher in the north,

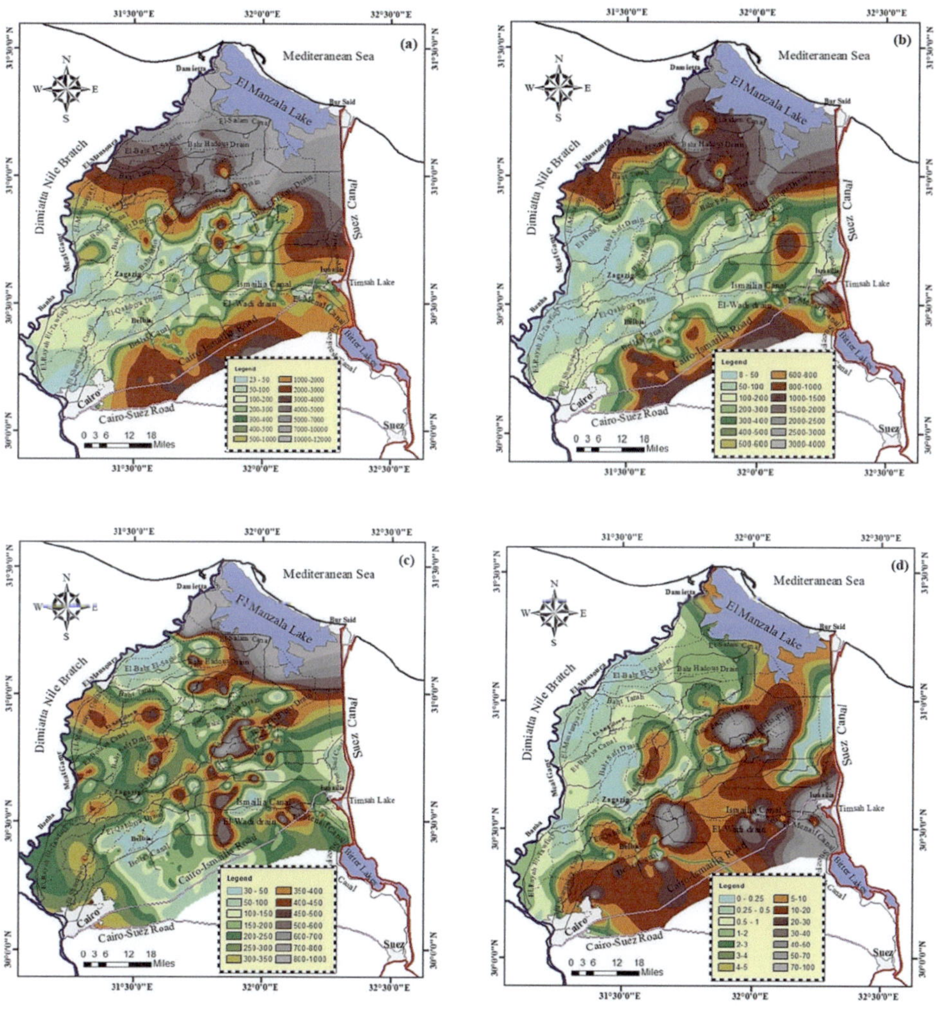

Fig. 9.5 Distribution of major anions within the aquifer systems in the study area: **a** Cl^-, **b** SO_4^{-2}, **c** HCO_3^-, and **d** NO_3^-

where groundwater mixes with the sea. Since the brackish water of the Miocene aquifer was low in HCO_3^-, HCO_3^- readings were low near its borders.

Nitrate (NO_3^-)

The NO_3^- distribution map (Fig. 9.5d) showed a significant rise in groundwater NO_3^- content (<10 mg/l) in newly reclaimed areas. There are significant NO_3^- concentration zones near Ismailia and Al Salhyia around the Bahr El-Baqar Drain. These high NO_3^- concentrations are due to the excessive use of inorganic NO_3^- fertilizers on poor sandy soil. Sandy soils with high-permeability help smooth NO_3^-leaching to groundwater. West of the study region had the lowest NO_3^- values. Old cultivated land overused NO_3^- fertilizers. Because of the deep clay layer under the surface soils, large-scale clay soil allows microorganisms to extract nitrogen or nitrogen compounds from water near the surface, trapping the NO_3^- in drainage systems and sending it to farm drains (Paudel et al. 2024).

9.3.3 Frequency Distribution of the Major Solute Concentrations

Correlation analysis of primary dissolved ions (Ca, K, Na, Mg, HCO_3, Cl, and SO_4) in the examined groundwater revealed significant relationships among various constituents (Table 9.2). The correlation matrix demonstrated particularly strong associations between sodium and chloride (r = 0.96), while sodium exhibited moderate correlations with magnesium and sulfate (both r = 0.79) and a weaker relationship with calcium (r = 0.64). Calcium displayed substantial correlation coefficients with both magnesium and chloride (r = 0.78), whereas sulfate showed notable correlation with magnesium (r = 0.72) and a moderate association with calcium (r = 0.64). These statistical relationships provide valuable insights that enhance our understanding of solute distribution patterns throughout the groundwater system within the investigated region.

Table 9.2 Correlation matrix of the major solutes in groundwater

	Ca^{2+}	Mg^{2+}	Na^+	K^+	HCO_3^-	SO_4^{2-}	CL^-
Ca^{2+}	1.00						
Mg^{2+}	0.78	1.00					
Na^+	0.64	0.79	1.00				
K^+	0.41	0.50	0.56	1.00			
HCO_3^-	−0.14	−0.09	0.01	0.04	1.00		
SO_4^{2-}	0.64	0.72	0.79	0.53	−0.07	1.00	
CL^-	0.78	0.89	0.96	0.55	−0.09	0.76	1.00

Figure 9.6 illustrates the frequency distribution of major ionic constituents across fresh, brackish, and saline groundwater samples. Freshwater samples, constituting 56.7% of the total analyzed dataset, primarily contained Na, HCO_3, and Cl as predominant solutes, while Mg and K occurred at comparatively lower concentrations. Approximately 70% of these samples exhibited concentration ranges of 20–100 mg/l for Ca, 40–100 mg/l for Mg, 0–200 mg/l for Na, and 4–16 mg/l for K. The majority of samples showed HCO_3 and Cl levels between 0–400 mg/l, with half the samples containing SO_4 concentrations between 0–100 mg/l.

Brackish water samples, representing 22.9% of the total dataset, demonstrated similar predominance of Na, HCO_3, and Cl as major constituents, with notable SO_4 presence. While K maintained relatively low concentrations, Mg levels were moderately elevated compared to freshwater samples. Between 60–70% of brackish samples contained Ca (15–150 mg/l), Mg (2–80 mg/l), Na (200–800 mg/l), and K (2–40 mg/l). The anions HCO_3, SO_4, and Cl typically ranged from 0–800 mg/l, 30–800 mg/l, and 200–1600 mg/l, respectively.

Saline water samples, accounting for 20.4% of the total analysis, were characterized by the dominance of Na and Cl ions, with substantial contributions from Ca, Mg, and SO_4. The majority of these samples exhibited Na concentrations between 300–2000 mg/l and Cl levels between 1000–5000 mg/l. Over 60% of saline samples showed concentration ranges of 200–800 mg/l for Ca, 200–600 mg/l for Mg, 0–50 mg/l for K, 0–400 mg/l for SO_4, and 500–1500 mg/l for HCO_3.

9.3.4 Investigation of Water Type

Figure 9.7a–d shows Piper's CA, SUA, and UA chemical analysis. Upper diamond-shaped field groundwater samples were gathered. Every SUA and UA brackish and saline groundwater sample on the top diamond-shaped field's (Ca^{2+} + Mg^{2+}) axis was part No. 6 (Ca^{2+}–Cl^-) and part No.7. Main influence of Miocene aquifer brack SUA and UA fresh groundwater samples contained No. 9 (Mixed type), and No. 5 (Mg^{2+}–HCO_3^-) on the (SO_4^{2-} + Cl^-)axis. CA groundwater samples declined while SUA and UA rose. The diamond-shaped field's bottom was predominantly CA and SUA meteoric groundwater. SUA and UA plotted in zone (B), while CA collected fresh groundwater samples in section (E) with the high alkalinity-low Cl^- anion triangle. Brackish, saline, SUA, and UA freshwater had more Cl^-–SO_4^{2-}. Freshness was controlled by underground Na^+ and K^+ (Elewa et al. 2013) (D) mainly plotted brackish/saline. According to the Piper diagram of groundwater samples, SUA and UA showed more meteoric origin modifications than CA. SUA/UA groundwater contained saline water (Na^+–Cl^-) and freshwater (Ca^{2+}–HCO_3^-).

Further, Fig. 9.8 shows the water type distribution map created from this data. It was obvious that the HCO_3^- water type (primarily Na^+–HCO_3^-, Ca^{2+}–HCO_3^-, and

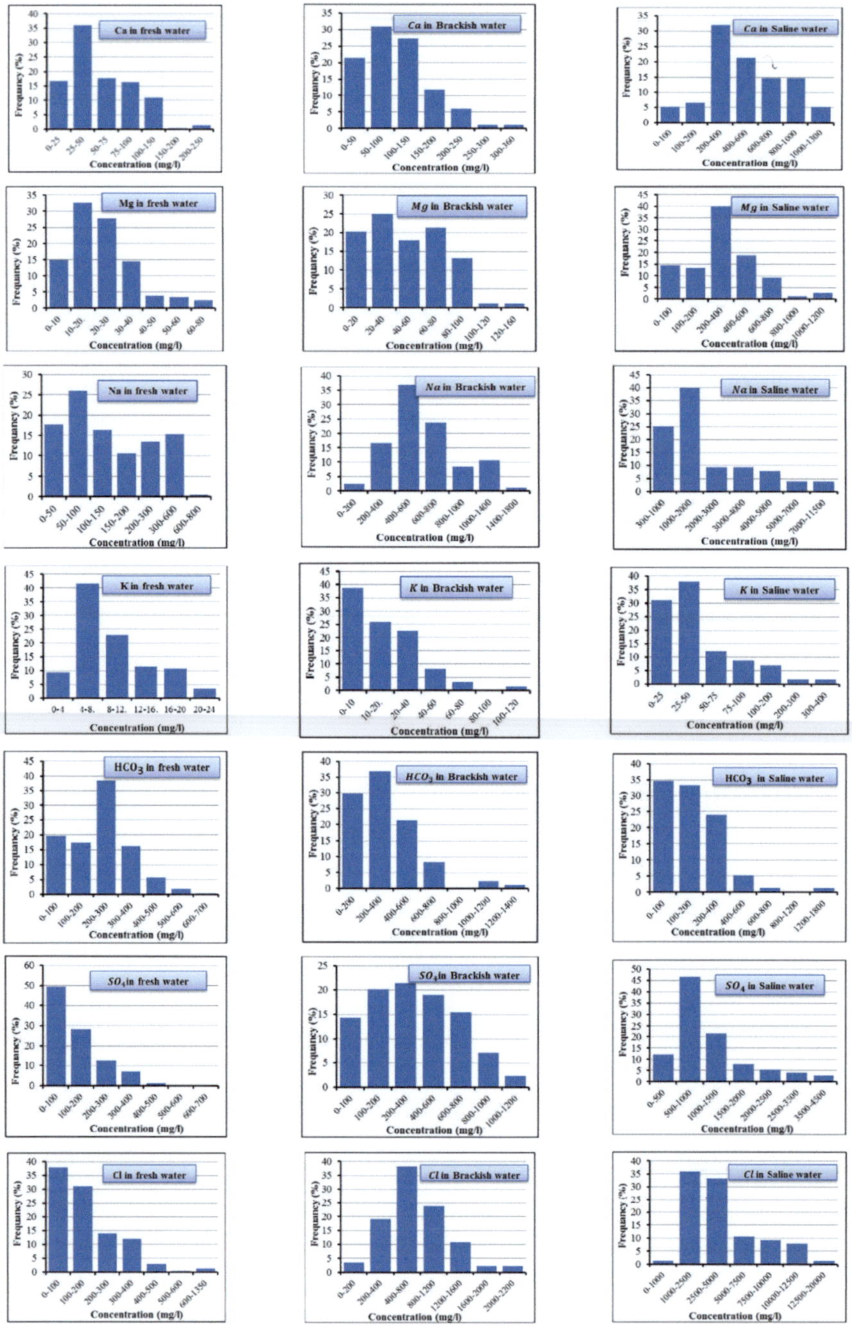

Fig. 9.6 Frequency distribution of the major solutes in groundwater in the QAEND

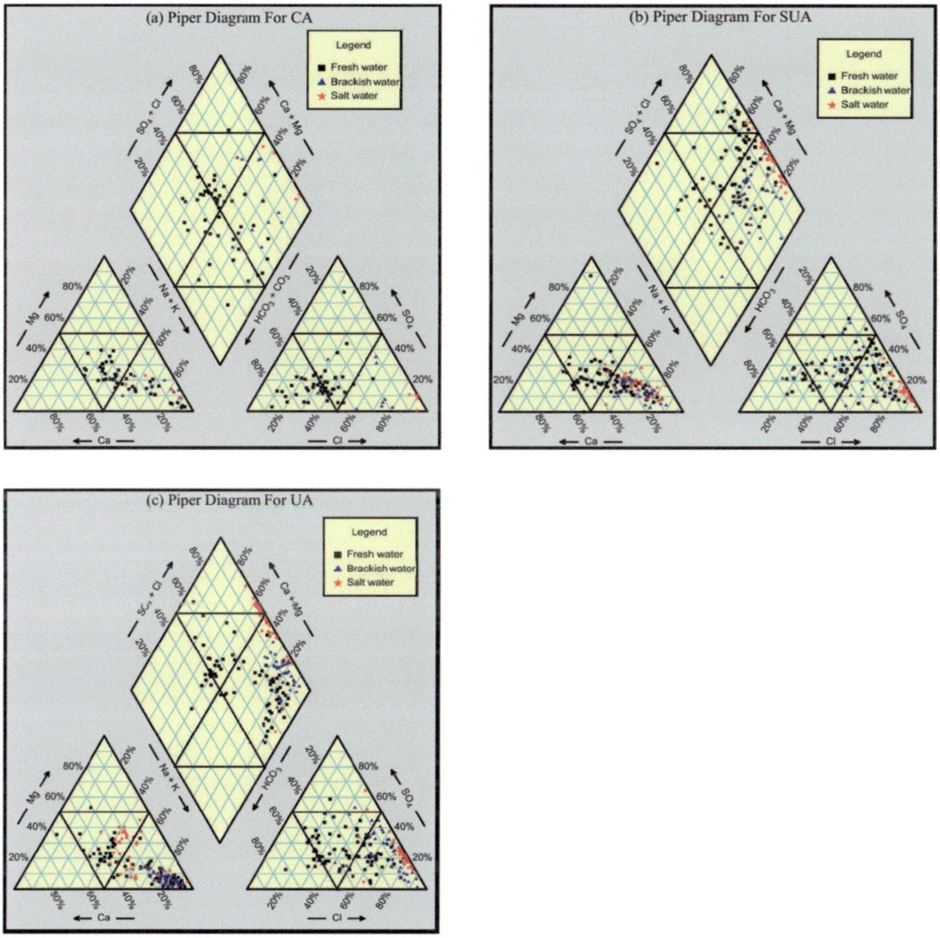

Fig. 9.7 Piper's trilinear diagram for the classification of the water types within the aquifer systems in the study area

less often $Mg-HCO_3^-$) dominated the CA with its meteoric signature. Due to saltwater intrusion (Saqr and Abd-Elmaboud 2024), (Na^+-Cl^-) water with a seawater impact dominates the north and northeast. Several significant areas of ($Ca^{2+}-Cl^-$), ($Na^+-SO_4^{2-}$), and ($Mg^{2+}-Cl^-$) water types were encountered within the (Na^+-Cl^-) type, especially near the northeast border of the mapped area around El-Manzala Lake.

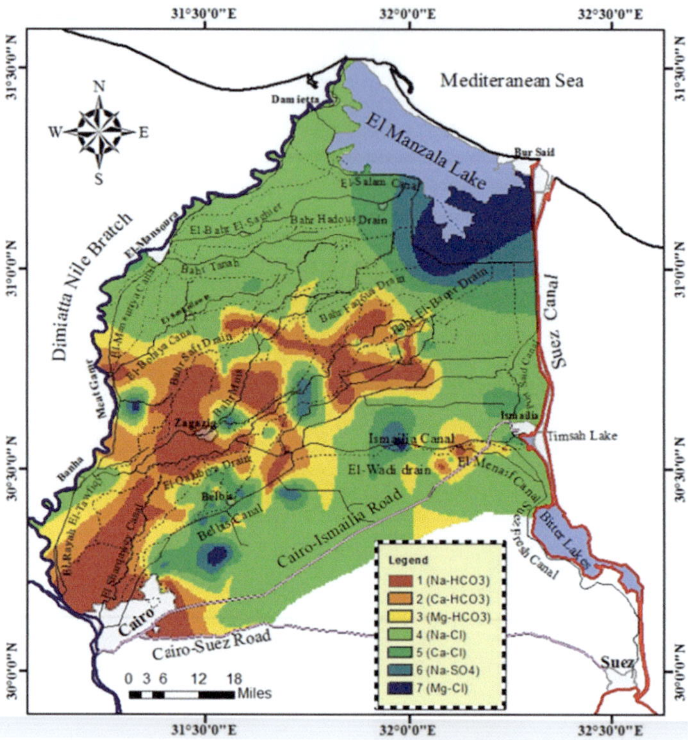

Fig. 9.8 Distribution of water types within the aquifer systems in the study area

9.3.5 Suitability of Groundwater for Irrigation

Six indices were adopted to test the suitability of groundwater for irrigation as described in the following sections.

9.3.5.1 Salinity Hazard (SH)

Figure 9.9a depicts the SH within the study area. The majority of the study area was high SH groundwater, which could harm many plants, and needs mixing with surface water before irrigation (Mohanavelu et al. 2021). The southern half of the CA, parallel to the southern part of the Damietta branch, had hard and medium salinity danger. Wadi El Tumiliat, near Ismailia City, had a medium salinity zone because freshwater from the canal infiltrated the high-permeability sand and gravel formation. Drains, wastewater ponds near the Tenth of Ramadan City, waterlogging pools south of Ismailia Canal, and El-Manzala Lake had high and very high salinity, but surface water, the Damietta branch, irrigation canals, and freshwater ponds had medium hazard.

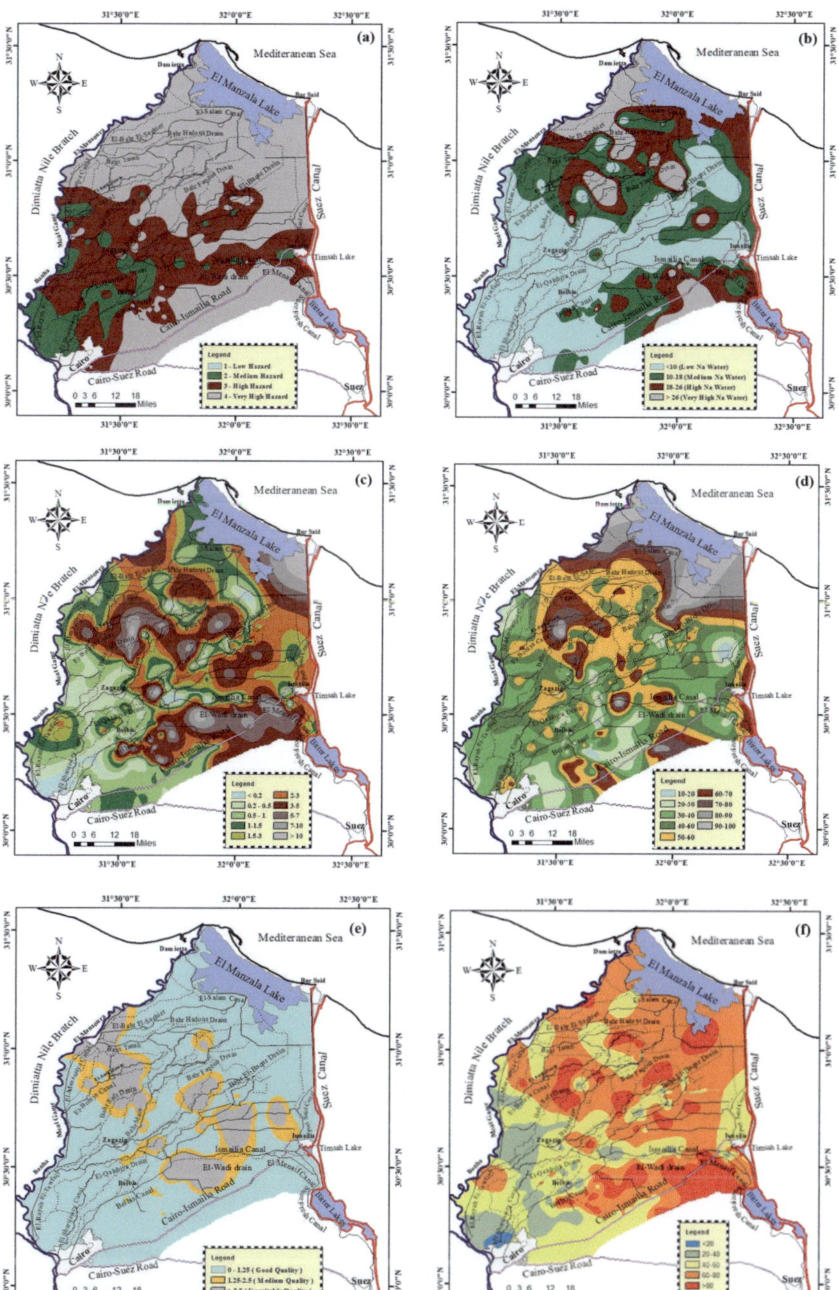

Fig. 9.9 Distribution of different water quality indices within aquifer systems in the study area: **a** SH, **b** SAR, **c** ESR, **d** RSC, **e** MH, and **d** Na%

9.3.5.2 Sodium Adsorption Ratio (SAR)

Figure 9.9b illustrates the SAR distribution map, which gradually increased from north and northeast to southeast, where the freshly reclaimed land existed. Generally, the CA south of 30º 50′ N had a low to medium Na^+ Hazard of Water. The eastern study area had a low SAR zone with a freshwater canal (Ismailia canal) and extremely permeable surface sediments. However, the newly recovered ground has significant SAR values.

9.3.5.3 Exchangeable Sodium Ratio (ESR)

The ESR distribution map is shown in Fig. 9.9c. In the research area, low ESR (<1.0) was found in the west and southwest, while higher ESR values (>3.0) rose toward the north, northeast, and south Ismailia canal in newly reclaimed lands, particularly near the Miocene aquifer.

9.3.5.4 Magnesium Hazard (MH)

The MH distribution map is given in Fig. 9.9d. South latitude 30º50′ N exhibited low to moderate MH, except for the SUA and UA regions in the east and south of the research area. Due to Ca^{+2} and Mg^{+2} interaction on clay minerals in CA, most useable groundwater samples had somewhat more MH than SUA and UA.

9.3.5.5 Residual Sodium Carbonate (RSC)

Figure 9.9e shows the RSC distribution in the research area. It was clear that the aquifer of the study area had mostly a low RSC of ~ 0.125. A few isolated high RSC values > 2.5 were detected around El-Mansoura city and extended in the northwestern-southeastern direction. Other high RSC locations were found in reclaimed land.

9.3.5.6 Sodium Percentage (Na%)

The Na% map is presented in Fig. 9.9f. The southern and southwest part of the CA was generally less than 60%, but most of the groundwater from the UA and north part above the latitude of 30º50′ N was greater than 60%. Therefore, groundwater from the UA was not recommended to be used independently for irrigation purposes but required mixing with surface water as a fundamental process to reduce the chances of soil salinity problems.

To evaluate the groundwater's suitability for irrigation in the East Nile Delta aquifer, I developed a comprehensive assessment framework utilizing the previously calculated indices. A spatially explicit groundwater quality zone map was generated through GIS-based overlay-weighted modeling techniques. As shown in Table 9.3, the weighting criteria were systematically assigned based on each parameter's relative importance: SAR and SH were weighted most heavily at 25% each, ESR and MH received moderate weights of 15% each, while RSC and Na% were allocated 10% each. These weighted values were integrated within the spatial analysis framework to produce a composite suitability index, with the resulting scores categorized into four distinct irrigation suitability classes as presented in Table 9.4: no restrictions, slight to moderate restrictions, high restrictions, and severe restrictions. The estimated values were utilized to map the research area's groundwater suitability for irrigation (Fig. 9.10). The distribution of groundwater suitability for irrigation in the eastern Nile Delta revealed distinct spatial variability, with approximately 30–35% of the total area classified as having no restrictions in use, predominantly located in the southern and southwestern regions where freshwater sources and favorable hydrogeological conditions enhance water quality. Slight to moderate restrictions cover about 25-30% of the area, forming transitional zones often found near agricultural drains and areas with moderate salinity levels. High restrictions in use accounted for roughly 20–25% of the region, particularly in the central and eastern parts of the delta, where salinity issues were aggravated by over-extraction and poor drainage. The remaining 15–20% of the area fell under severe restrictions, primarily concentrated in the northern and northeastern zones near El-Manzala Lake and the Mediterranean Sea, where saline intrusion and fluviomarine deposits heavily compromised water quality (Geriesh et al. 2015). This distribution emphasized the pressing need for regional water management strategies to mitigate salinity risks and ensure sustainable agricultural practices across the delta.

9.4 Conclusions

This study provided a comprehensive evaluation of groundwater quality in the eastern Nile Delta, focusing on its suitability for irrigation through advanced hydrogeochemical analyses. Using data from 488 groundwater wells and 94 surface water samples, the study employed Piper diagrams, water quality indices, and spatial mapping to assess key physical and chemical parameters. The results showed that 53.6% of groundwater samples were fresh, 24.3% were brackish, and 22.1% were saline. Freshwater zones were primarily located in the southern and southwestern parts, while severe restrictions were identified in 15–20% of the area, mainly in the north and northeast, due to saline intrusion and poor drainage. Piper diagram analysis highlighted dominant water types, with $Ca2^{+}-HCO_3^{-}$ characterizing fresher regions and $Na^{+}-Cl^{-}$ dominating saline zones. Key indices, including SAR and RSC, indicated significant variability in groundwater suitability, with approximately 30% of the region deemed highly suitable for irrigation and the remainder requiring management interventions such as blending with surface water or selective crop use. These findings have far-reaching implications, providing a robust methodological framework for assessing groundwater quality in other arid and semi-arid regions. The research contributes to evidence-based decision-making in water resource management, promoting sustainable agricultural practices and mitigating salinity risks. Future studies should explore the dynamic interactions between groundwater and surface water under changing climatic and land-use conditions to refine management strategies further.

Table 9.3 Categorization of water quality indices for groundwater irrigation suitability assessment

Index	Category	Range	Rank	Weight (%)
SH	Low	TDS < 200 mg/L; EC < 250 μS/cm	1	25
	Medium	200 < TDS < 500 mg/L; 250 < EC < 750 μS/cm	2	
	High	500 < TDS < 1500 mg/L; 750 < EC < 2250 μS/cm	3	
	Very high	TDS > 1500 mg/L; EC > 2250 μS/cm	4	
SAR	Excellent	<10	1	25
	Good	10–18	2	
	Fair	18–26	3	
	Poor	> 26	4	
ESR	Excellent	< 1	1	15
	Good	1–2	2	
	Permissible	2–3	3	
	Doubtful	> 3	4	
MH	Safe	<50%	1	15
	Moderate	50–65%	2	
	Severe	>65%	3	
RSC	Safe	<1.25	1	10
	Marginal	1.25–2.5	2	
	Unsuitable	>2.5	3	
Na%	Excellent	<20%	1	10
	Good	20–40%	2	
	Permissible	40–60%	3	
	Doubtful	60–80%	4	
	Unsuitable	>80%	5	

Table 9.4 Irrigation suitability classes based on weighted index

Weighted score range	Irrigation suitability class	Restrictions on use
1.0–1.5	Class I	No restrictions
1.5–2.5	Class II	Slight to moderate restrictions
2.5–3.5	Class III	High restrictions
>3.5	Class IV	Severe restrictions

Note The weighted score is calculated as: Σ(Rank × Weight/100) for all indices. All ionic concentrations used in calculations are in milliequivalents per liter (meq/L)

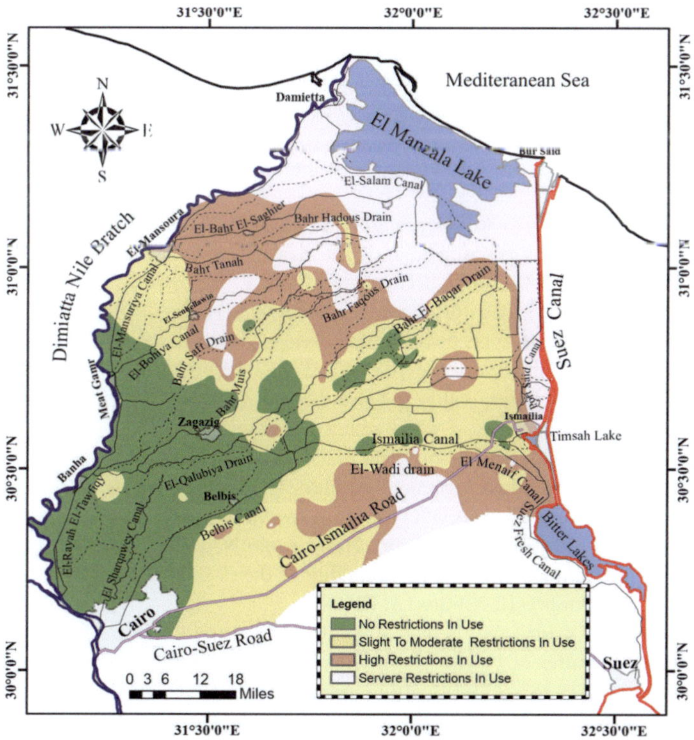

Fig. 9.10 Distribution of groundwater suitability for irrigation purposes within aquifer systems in the study area

Acknowledgements The authors thank the Faculty of Engineering, Mansoura University, Mansoura, Egypt, for its support. This research is a part of the Ph.D. thesis of the first author under the supervision of the second, third, and fourth authors. The last author prepared the first draft of the chapter. All authors reviewed the final draft and agreed to publish the current version of the chapter.

References

Abd-Elmaboud ME, Abdel-Gawad HA, El-Alfy KS, Ezzeldin MM (2021) Estimation of groundwater recharge using simulation-optimization model and cascade forward ANN at East Nile Delta aquifer, Egypt. J Hydrol: Reg Stud 34:100784. https://doi.org/10.1016/j.ejrh.2021.100784

Abd-Elmaboud ME (2022) Evaluation and management of groundwater using remote sensing techniques, geochemical analysis, and 3d groundwater flow modeling: case study quaternary aquifer east Nile Delta, Egypt. PhD Thesis, University of Mansoura, Mansoura, Egypt

Abdulaziz AM, Hurtado JM, Al-Douri R (2009) Application of multitemporal landsat data to monitor land cover changes in the Eastern Nile Delta region, Egypt. Int J Remote Sens 30(11):2977–2996. https://doi.org/10.1080/01431160802558675

Abo-El-Fadl (2013) Possibilities of groundwater pollution in some areas, East of Nile Delta, Egypt. Int J Environ 1(1):1–21

Alao JO, Ayejoto DA, Fahad A, Mohammed MAA, Saqr AM, Joy AO (2024) Environmental burden of waste generation and management in Nigeria. In: Souabi S, Anouzla A (eds) Technical landfills and waste management. Springer Water. Springer, Cham, pp 27–56. https://doi.org/10.1007/978-3-031-55665-4_2

APHA (2017) Standard methods for the examination of water and wastewater. Choice reviews online. https://doi.org/10.5860/choice.49-6910

Arnous MO, El-Rayes AE (2013) An integrated GIS and hydrochemical approach to assess groundwater contamination in West Ismailia area, Egypt. Arab J Geosci 6(8):2829–2842. https://doi.org/10.1007/s12517-012-0555-0

Arnous MO, Green DR (2015) Monitoring and assessing waterlogged and salt-affected areas in the Eastern Nile Delta region, Egypt, using remotely sensed multi-temporal data and GIS. J Coast Conserv 19(3):369–391. https://doi.org/10.1007/s11852-015-0397-5

Doneen LD (1964) Notes on water quality in agriculture, water science and engineering. University of California, Davis, Department of water sciences and engineering, p 400

El Alfy K, El-Ghandour H, Abd-Elmaboud M (2020) Tidal effect on groundwater fluctuations and saltwater intrusion in coastal heterogeneous aquifers. (Department C. (irrigation and hydraulics)). Mansoura University, Bull Fac Eng 39(1):1–22. https://doi.org/10.21608/bfemu.2020.103086

El-Amier YA, Kotb WK, Bonanomi G, Fakhry H, Marraiki NA, Abd-Elgawad AM (2021) Hydrochemical assessment of the irrigation water quality of the el-salam canal, Egypt. Water (Switzerland) 13(17). https://doi.org/10.3390/w13172428

Elewa HH, Shohaib RE, Qaddah AA, Nousir AM (2013) Determining groundwater protection zones for the quaternary aquifer of Northeastern Nile Delta using GIS-based vulnerability mapping. Environ Earth Sci 68(2):313–331. https://doi.org/10.1007/s12665-012-1740-x

Embaby AAA, Afifi SY, El Atabany NM (2014) Hydrochemical evaluation of Belbies district groundwater. Life Sci J 11(12s):1097–8135. http://www.lifesciencesite.com.228

Ezzeldin MM, El Alfy KS, Gawad HAA, Abd Elmaboud ME (2018) Comparison between structured and unstructured MODFLOW for simulating groundwater flow in three-dimensional multilayer quaternary aquifer of East Nile Delta, Egypt. Hydrol Curr Res 09(01). https://doi.org/10.4172/2157-7587.1000297

Faid AM, Abdulaziz AM (2012) Monitoring land-use change-associated land development using multitemporal Landsat data and geoinformatics in Kom Ombo area, South Egypt. Int J Remote Sens 33(22):7024–7046. https://doi.org/10.1080/01431161.2012.697207

Geriesh MH, Balke KD, El-Rayes AE, Mansour BM (2015) Implications of climate change on the groundwater flow regime and geochemistry of the Nile Delta, Egypt. J Coast Conserv 19(4):589–608. https://doi.org/10.1007/s11852-015-0409-5

Harter T (2003) Groundwater quality and groundwater pollution. Groundw Qual Groundw Pollut. https://doi.org/10.3733/ucanr.8084

He S, Li P (2020) A MATLAB based graphical user interface (GUI) for quickly producing widely used hydrogeochemical diagrams. Chemie Der Erde 80(4). https://doi.org/10.1016/j.chemer.2019.125550

Indugula R, Venkata Vajjha H (2023) Spatial variations of trace metals, salinization processes and multivariate analysis in the near surface aquifer systems of Krishna Delta, Andhra Pradesh, India. Environ Earth Sci 82(10). https://doi.org/10.1007/s12665-023-10910-6

Ismael AMAA, Doser DI, Chair C, Ellis AS, Hurtado JM, Cornell WC, Walton J, Arenaz P (2007) Applications of remote sensing, GIS, and groundwater flow modeling in evaluating groundwater resources: two case studies; East Nile Delta, Egypt and Gold Valley, California, USA. Approved

Kaddah MT (1975) Irrigation with saline water. Soil Sci Soc Am J 39(5). https://doi.org/10.2136/sssaj1975.03615995003900050009x

Khodapanah L, Sulaiman WNA, Khodapanah N (2009) Groundwater quality assessment for different purposes in Eshtehard district, Tehran, Iran. Eur J Sci Res 36(4):543–553

Mansour N, Sarhan T, El-Gamal M, Nassar K, Abd-Elmaboud M (2025) Geospatial time series analysis for coastal systems: AI-powered NARX neural networks integrating remote sensing for advanced shoreline change prediction. Int J Ocean 19(1):1–38. https://doi.org/10.37622/IJOO/19.1.2025.1-38

Mastrocicco M, Prommer H, Pasti L, Palpacelli S, Colombani N (2011) Evaluation of saline tracer performance during electrical conductivity groundwater monitoring. J Contam Hydrol 123(3–4):157–166. https://doi.org/10.1016/j.jconhyd.2011.01.001

Mohanavelu A, Naganna SR, Al-Ansari N (2021) Irrigation induced salinity and sodicity hazards on soil and groundwater: an overview of its causes, impacts and mitigation strategies. Agriculture (Switzerland) 11(10). https://doi.org/10.3390/agriculture11100983

Moorthy P, Sundaramoorthy S, Roy PD, Jonathan MP, Usha T, Gowrappan M, Chokkalingam L (2024) Identifying seawater interaction with coastal aquifers using hydrochemical and GIS for Nagapattinam district, southeast coast of India. Desalination Water Treat 320. https://doi.org/10.1016/j.dwt.2024.100703

Noman M, Ullah I, Khan MA, Qazi A, Farooq W, Saqr A, Elsheikh A (2024) Analysis of overcurrent protective relaying as minimum adopted fault protection for small-scale hydropower plants. Int J Environ Sci Technol 21(4):4457–4470. https://doi.org/10.1007/s13762-023-05284-y

Paudel G, Pant RR, Joshi TR, Saqr AM, Đurin B, Cetl V, Kamble PN, Bishwakarma K (2024) Hydrochemical dynamics and water quality assessment of the Ramsar-listed Ghodaghodi lake complex : unveiling the water-environment nexus. Water 16(3373):1–23. https://doi.org/10.3390/w16233373

Piper A (1944) A graphic procedure in the geochemical interpretation of water-analyses. EOS Trans Am Geophys Union 25(6):914–928

Poonia SR, Talibudeen O (1977) Sodium-calcium exchange equilibria in salt-affected and normal soils. J Soil Sci 28(2):276–288. https://doi.org/10.1111/j.1365-2389.1977.tb02236.x

Saqr AM, Abd-Elmaboud ME (2024) Management of saltwater intrusion in coastal aquifers: a review and case studies from Egypt. Online J Eng Sci 3(1):1–17. https://doi.org/10.31586/ojes.2024.982

Saqr AM, Nasr M, Fujii M, Yoshimura C, Ibrahim MG (2023a) Monitoring of agricultural expansion using hybrid classification method in southwestern fringes of Wadi El-Natrun, Egypt: an appraisal for sustainable development. Environ Sci Eng 349–362. https://doi.org/10.1007/978-981-99-4101-8_27

Saqr AM, Nasr M, Fujii M, Yoshimura C, Ibrahim MG (2023b) Optimal solution for increasing groundwater pumping by integrating MODFLOW-USG and particle swarm optimization algorithm: a case study of Wadi El-Natrun, Egypt. Environ Sci Eng 59–73. https://doi.org/10.1007/978-981-99-1381-7_6

Shata A, El Fayoumy I (1970) Remarks on the hydrogeology of the Nile Delta, UAR. Hydrol Deltas 385–396

Simmons HB (2000) Water quality for agriculture. Coast Eng Proc 1(2):7

Thorne DW (1954) Diagnosis and improvement of saline and alkali soils. Agron J 46(6):290–290. https://doi.org/10.2134/agronj1954.00021962004600060019x

Zaman M, Shahid SA, Heng L (2018) Irrigation water quality. In: Guideline for salinity assessment, mitigation and adaptation using nuclear and related techniques, pp 113–131. https://doi.org/10.1007/978-3-319-96190-3_5

Groundwater Modeling in the Nile Delta, Egypt: A Comprehensive Review of Challenges, Methodologies, and Research Trends

10

Mahmoud E. Abd-Elmaboud, Ahmed M. Saqr, and Youssef M. Youssef

Abstract

Groundwater resources constitute approximately 30% of the world's freshwater supply, providing essential support for agricultural production, industrial development, and domestic needs for billions of people globally. With increasing pressures from population growth, climate change, and intensifying land use, sophisticated groundwater modeling approaches have become critical tools for sustainable water resource management in vulnerable regions worldwide. This comprehensive review critically examines groundwater modeling approaches in the Nile Delta, Egypt, where intensive anthropogenic activities and climate change threaten this vital resource. The study evaluates methodological evolution from simple analytical techniques to sophisticated three-dimensional numerical models, highlighting advances in process representation across seawater intrusion, contaminant transport, and management optimization domains. Despite significant progress, persistent challenges include inadequate spatiotemporal monitoring networks, limited data integration, and insufficient representation of

M. E. Abd-Elmaboud (✉) · A. M. Saqr
Irrigation and Hydraulics Department, Faculty of Engineering, Mansoura University, Mansoura, Egypt
e-mail: mahmoud_elsayed@mans.edu.eg

A. M. Saqr
e-mail: ahmedsaqr@mans.edu.eg

Y. M. Youssef
Geological and Geophysical Engineering Department, Faculty of Petroleum and Mining Engineering, Suez University, Suez, Egypt
e-mail: youssef.ibrahim@pme.suezuni.edu.eg

M. A. Khalifa and K. S. Gemail (eds.), *Geology of the Nile Deltas and Natural Resources*, Deltas of the World, https://doi.org/10.1007/978-3-031-96594-4_10

human–natural system interactions. Seawater intrusion modeling reveals coastal intrusion extending 60–80 km inland, with projections suggesting 5–15 km additional intrusion by 2100. Contaminant transport studies document agricultural pollutants affecting 30% of shallow aquifers, while management models demonstrate spatially variable sustainable extraction limits. The review identifies critical research gaps, including inadequate uncertainty quantification, limited implementation pathway analysis, and insufficient integration of socioeconomic factors with physical models. Future research directions should prioritize integrated socio-hydrological frameworks, enhanced stakeholder engagement, and robust decision-making approaches under deep uncertainty. By synthesizing methodological advances, persistent limitations, and promising research directions, this review provides guidance for sustainable groundwater management in the Nile Delta and comparable deltaic systems worldwide facing similar challenges from climate change, population growth, and intensive resource exploitation.

Keywords

Groundwater modeling • Nile Delta aquifer • Seawater intrusion • Contaminant transport • Climate change impacts • Sustainable management • Socio-hydrological frameworks • Numerical simulation • Model uncertainty • Integrated water resources

10.1 Introduction

Groundwater represents a critical resource for sustainable development worldwide, constituting nearly 30% of global freshwater resources and supporting approximately 40% of the world's irrigated agriculture (Scanlon et al. 2012; Famiglietti 2014). As climate change intensifies hydrological extremes and population growth increases water demands, groundwater systems face unprecedented pressures globally, with over-extraction, pollution, and salinization threatening this vital resource across diverse geographical contexts (Taylor et al. 2013; Rodell et al. 2018). Accurate modeling of groundwater dynamics has consequently emerged as an essential tool for water resource planning in regions experiencing severe water stress (Anderson et al. 2015).

Across the Mediterranean Basin and North Africa, coastal aquifers face particularly acute challenges from the combined effects of climate change and intensive human activities (Sherif and Singh 1999a, b; Wassef and Schüttrumpf 2016a, b). Regional studies indicate that over 70% of coastal aquifers in this region show evidence of degradation, with seawater intrusion affecting freshwater availability for millions of people (Bear et al. 1999; Abd-Elhamid et al. 2016a, b). Countries sharing similar deltaic and coastal aquifer systems, including Turkey, Morocco, and Tunisia, have increasingly employed sophisticated groundwater modeling approaches to manage these complex hydrogeological systems under changing environmental conditions (Shaban et al. 2006; Ghouili et al. 2017).

The Nile Delta, a triangular alluvial plain covering approximately 30,000 km^2 at the northern terminus of the Nile River where it meets the Mediterranean Sea, represents one of the world's most historically significant agricultural and demographic regions (Fig. 10.1). With elevations typically below 10 m above sea level, this densely populated region exhibits particular vulnerability to environmental changes and anthropogenic pressures (Mabrouk et al. 2019a, b; Eltarabily and Negm 2018). Despite constituting only 2.5% of Egypt's total land area, the Delta houses approximately 60% of the country's population and produces over 60% of its agricultural output, creating substantial demands on its water resources (El-Raey 2010; Nofal et al. 2015).

The Nile Delta groundwater system faces several critical challenges. Seawater intrusion represents one of the most significant threats, with studies documenting intrusion extending 60–80 km inland in parts of the eastern and western delta, where total dissolved solids exceed 10,000 mg/L (Sefelnasr and Sherif 2014). This intrusion results from excessive groundwater extraction and sea-level rise, with the natural hydraulic gradient from land to sea reversed in many coastal areas. Water quality degradation extends

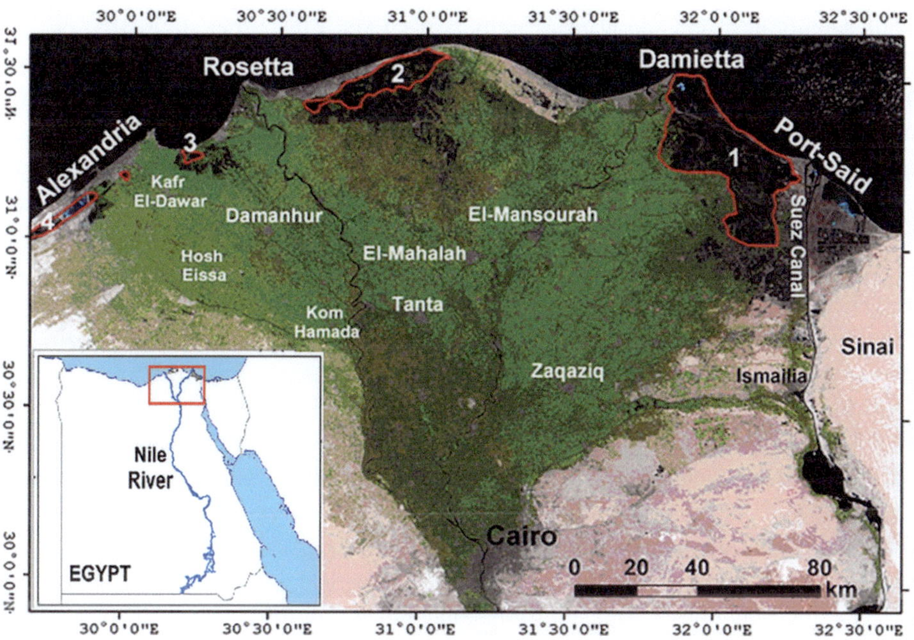

Fig. 10.1 Location map of the study area, eastern Nile Delta region, Egypt (Modified after Hereher 2010)

beyond salinization to include widespread nitrate contamination from agricultural activities, with concentrations exceeding 50 mg/L in many areas of the central and southern delta (Eltarabily et al. 2017). Industrial zones have introduced heavy metals and organic compounds, particularly in the eastern delta near the 10th of Ramadan City. At the same time, inadequate sewage infrastructure in densely populated areas has resulted in pathogen contamination (El-Nimer et al. 2020).

Groundwater extraction has increased dramatically, from approximately 2.3 billion cubic meters per year in the 1980s to nearly 6.1 billion cubic meters per year by 2015, resulting in declining piezometric levels of 5–15 m in parts of the eastern and western delta (Armanuos et al. 2016a, b; Nofal et al. 2015). Concurrently, natural recharge has diminished due to several factors, including urbanization, reducing permeable surfaces by 15–20% in newly developed areas, and modernization of irrigation infrastructure, decreasing seepage by an estimated 0.5–0.7 billion cubic meters annually (Eltarabily and Negm 2018; Nofal et al. 2022).

Climate change further compounds these challenges, with the Nile Delta considered among the world's most vulnerable regions to climate impacts. Rising temperatures are projected to increase evapotranspiration rates by 11–19% by 2070, correspondingly reducing groundwater recharge by 20–37% (Armanuos et al. 2023). Additionally, the Grand Ethiopian Renaissance Dam (GERD) introduces transboundary water management complexities that could potentially reduce freshwater delivery to the delta, exacerbating existing salinity intrusion and water scarcity issues (Abd-Elhamid et al. 2019).

The hydrogeological framework of the Nile Delta primarily consists of Quaternary deposits overlying Tertiary formations, with the main Pleistocene aquifer varying in thickness from 200 m in southern areas to over 1,000 m near the coast (Sefelnasr and Sherif 2014). Numerical modeling has emerged as an essential tool for understanding this complex system, evolving from early analytical approaches to sophisticated three-dimensional variable-density simulations incorporating multiple processes. Various modeling codes have been employed, including MODFLOW, SEAWAT, MT3DMS, and FEFLOW (Abd-Elhamid et al. 2016a, b; Mabrouk et al. 2019a, b).

Despite these advances, significant challenges persist in effective modeling. Data limitations are particularly acute for deep groundwater and offshore areas, with approximately 30% of the delta lacking sufficient monitoring data for reliable model calibration (Nofal et al. 2015). Fragmentation of data across multiple agencies creates integration challenges, while limited documentation of uncertainty constrains robust calibration and prediction (Mabrouk et al. 2018). Most studies have focused on specific aspects rather than adopting holistic approaches that consider multiple interacting processes (El-Agha et al. 2017).

This study conducts a comprehensive review of numerical groundwater modeling in the Nile Delta, focusing on seawater intrusion, contaminant transport, management optimization, and integrated climate–land-use models. Through critical analysis of methodological approaches and limitations across these domains, we identify best practices and propose future research directions that can better support sustainable groundwater management in this critical region, with potential applications to similar deltaic systems worldwide.

10.2 Groundwater Resources in the Nile Delta

The Nile Delta aquifer system represents one of Egypt's largest groundwater repositories, characterized by a complex hydrogeological structure that has evolved over geological time. This system primarily consists of Quaternary deposits overlying Tertiary formations, with Pleistocene-aged sandy and gravelly deposits forming the principal water-bearing formation (Sherif et al. 2012). Aquifer thickness exhibits significant spatial variability, ranging from approximately 200 m in southern regions to over 1,000 m near the Mediterranean coast, creating distinct hydrogeological zones with varying vulnerability characteristics (Sefelnasr and Sherif 2014).

The aquifer system's multilayered structure fundamentally influences groundwater flow dynamics and quality. The uppermost layer consists predominantly of Holocene clay and silty clay deposits, forming a semi-confining unit with thickness varying from 5 to 50 m, thinning southward and thickening northward. This layer exhibits relatively low vertical hydraulic conductivity (0.05–0.3 m/day), providing partial protection to the underlying aquifer in northern regions (Nofal et al. 2015). The main Pleistocene aquifer beneath this clay cap comprises coarse to fine sands and gravels with occasional clay lenses, characterized by high hydraulic conductivity (15–150 m/day) and transmissivity values ranging from 5,000 to 15,000 m2/day (Sefelnasr and Sherif 2014). A thick Pliocene clay layer underlies the system, acting as an aquiclude that separates the Quaternary aquifer from deeper Miocene formations and minimizes vertical flow at the aquifer base (Mabrouk et al. 2018).

From a hydrogeological perspective, the Nile Delta aquifer can be divided into three distinct zones with varying confinement conditions. The southern zone exhibits predominantly unconfined conditions with direct recharge from rainfall, irrigation return flow, and seepage from the Nile River and canals. The central zone displays semi-confined conditions as the clay cap becomes more continuous, while the coastal zone features fully confined conditions with a thick clay cap, making it particularly vulnerable to seawater intrusion. While natural groundwater flow generally follows the topographic gradient from south to north, intensive groundwater pumping has created local depressions in the piezometric surface, significantly altering these natural flow patterns in many areas (Eltarabily and Negm 2018).

Groundwater in the Nile Delta plays a critical role in supporting various sectors of the Egyptian economy and society. Agriculture, consuming approximately 85% of Egypt's water resources, depends significantly on groundwater for irrigation, especially in areas with limited access to canal irrigation (El-Agha et al. 2017). Total groundwater extraction for agriculture in the Nile Delta is estimated at 4.9 billion cubic meters annually, supporting the cultivation of strategic crops such as wheat, rice, cotton, and various fruits and vegetables (Nofal et al. 2015). The rapid urbanization and industrialization of the region have substantially increased the demand for reliable water sources, with groundwater serving essential functions for both domestic and industrial uses. Annual extraction for these purposes is estimated at 1.2 billion cubic meters, with approximately 15% of the domestic water supply in the delta region sourced from groundwater, serving an estimated 7–8 million people (Armanuos et al. 2016a, b; El Alfy et al. 2015).

Despite its critical importance, the Nile Delta aquifer faces multiple interconnected challenges that threaten its long-term sustainability. Seawater intrusion represents one of the most significant threats to groundwater quality, with documented intrusion extending up to 60–80 km inland in parts of the eastern and western delta (Sefelnasr and Sherif 2014). Intensive pumping has also caused upconing of the saltwater interface beneath wells, particularly in the northeastern and northwestern sectors where extraction rates exceed 3,000 m^3/day (Abd-Elhamid et al. 2016a, b). Groundwater quality degradation extends beyond salinization to include widespread nitrate contamination from intensive agriculture, with concentrations exceeding 50 mg/L in many areas (Eltarabily et al. 2017), and industrial pollution introducing heavy metals and organic chemicals into the aquifer system, especially in the eastern delta near 10th of Ramadan City (El-Nimer et al. 2020).

Groundwater extraction has increased substantially, from approximately 2.3 billion cubic meters per year in the 1980s to nearly 6.1 billion cubic meters per year by 2015, resulting in declining piezometric levels of 5–15 m in parts of the delta over the past three decades (Nofal et al. 2015; Armanuos et al. 2016a, b). Simultaneously, natural recharge has diminished due to urbanization reducing permeable surfaces, modernization of irrigation infrastructure reducing seepage, and climate change effects altering precipitation patterns and increasing evapotranspiration (Eltarabily and Negm 2018; Nofal et al. 2022; Armanuos et al. 2023). These challenges interact in complex ways, creating feedback loops that accelerate groundwater degradation and necessitate integrated management approaches based on comprehensive scientific understanding and robust modeling frameworks.

10.3 Groundwater Modeling Methodologies

Groundwater models represent essential analytical tools for hydrogeological investigations, providing mathematical approximations of complex subsurface flow systems (Saqr et al., 2025). These models describe physical processes through simplified mathematical equations that, while not perfectly replicating field conditions, enable prediction and comparative analysis of alternative hydrogeological scenarios (Kumar 2002). The three-dimensional saturated groundwater flow in porous media is governed by the equation:

$$\frac{\partial}{\partial x}\left(K_{xx}\frac{\partial h}{\partial x}\right)+\frac{\partial}{\partial y}\left(K_{yy}\frac{\partial h}{\partial y}\right)+\frac{\partial}{\partial z}\left(K_{zz}\frac{\partial h}{\partial z}\right)-\omega=S_S\frac{\partial h}{\partial t}$$

Here, K_{xx}, K_{yy}, and K_{zz} are hydraulic conductivity values along x–, y–, and z-axis; h is the hydraulic head; ω is a source and sink term and/or volumetric flux per unit volume per day; S_S is the specific storage, and (t) is time (Abdulaziz and Faid 2013; Panday et al. 2013). This equation, combined with appropriate boundary and initial conditions, constitutes the mathematical foundation for groundwater flow simulations, enabling the prediction of hydraulic potential changes in response to withdrawal and recharge variations (Ezzeldin and Abd-Elmaboud 2024).

Model development follows a systematic process beginning with a clear objective definition, followed by comprehensive hydrogeological characterization that informs conceptual model formulation—a critical step that systematizes field data to describe groundwater flow mechanisms (Karadacs 2009), as shown in Fig. 10.2. The selection of appropriate modeling software depends on site-specific conditions; simpler analytical solutions may suffice for straightforward flow patterns, while complex hydrogeological settings necessitate multidimensional numerical models. Model design encompasses parameter specification, including grid configuration, layer properties, boundary conditions, hydraulic conductivity distributions, and stress factors such as pumping and recharge.

Calibration, the systematic adjustment of input parameters to match observed field conditions, represents a cornerstone of model reliability. This inverse problem typically employs statistical metrics, including mean absolute error (MAE), root-mean-square error (RMSE), correlation coefficient (R), and mean relative error (MRE) to quantify calibration quality (Woessner and Anderson 1990). Sensitivity analysis subsequently evaluates model robustness by measuring output response to parameter variations across logical ranges, identifying parameters requiring more precise specification. Model verification further validates the calibrated model against independent datasets, ensuring its capability to reproduce measured field conditions without parameter modification.

The calibrated, verified model ultimately serves as a predictive tool for simulating potential groundwater flow scenarios and evaluating management alternatives. However, inherent uncertainties in input data and parameter values necessitate expressing predictions as a range of possible outcomes rather than deterministic results (Dawoud and Raouf 2009). This comprehensive modeling framework provides invaluable insights for groundwater management in complex systems such as the Nile Delta aquifer, where multiple stressors interact to influence groundwater dynamics.

10.3.1 Types of Groundwater Models

Groundwater modeling in the Nile Delta has evolved significantly over time, employing various methodologies to address the complex hydrogeological conditions and water

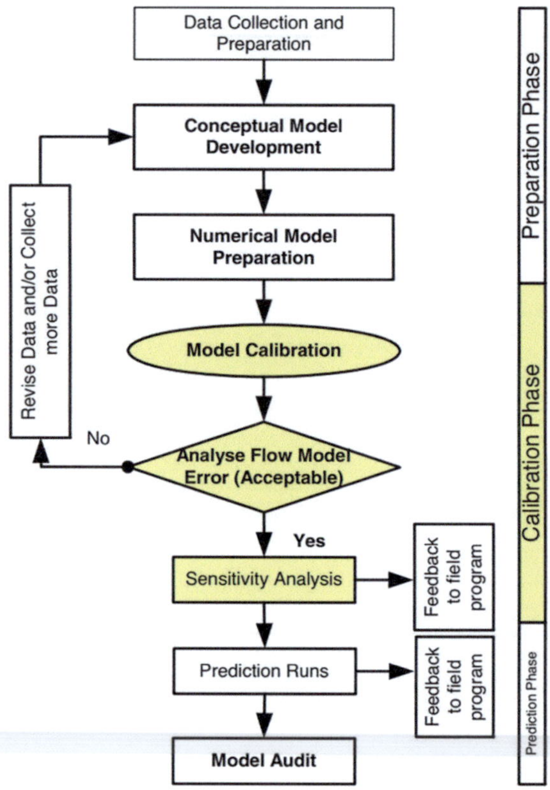

Fig. 10.2 The groundwater modeling process (Dawoud and Raouf 2009)

management challenges in the region. Mathematical models for solving the governing differential equations of groundwater flow can be broadly categorized as illustrated in Fig. 10.3, which delineates the primary solution methodologies. Both analytical and numerical modeling approaches have been utilized, each with distinct advantages, limitations, and applications.

10.3.1.1 Analytical Models

Analytical models, such as those using the Ghyben–Herzberg principle (Kashef 1983) and Theis's equation (El-Bihery 2009), provide explicit solutions for groundwater flow and transport. In the Nile Delta, these have been used for saltwater intrusion assessment, well hydraulics analysis, and one-dimensional transport studies (Sherif and Singh 1999a, b). Their advantages include computational efficiency and minimal data needs. However, analytical models are limited by simplifying assumptions, hindering their ability to represent spatial heterogeneity, complex boundary conditions, and three-dimensional processes, as shown in Fig. 10.4. Consequently, they are best suited for initial conceptualization and preliminary assessments, with numerical models being essential for detailed analysis of the complex Nile Delta aquifer.

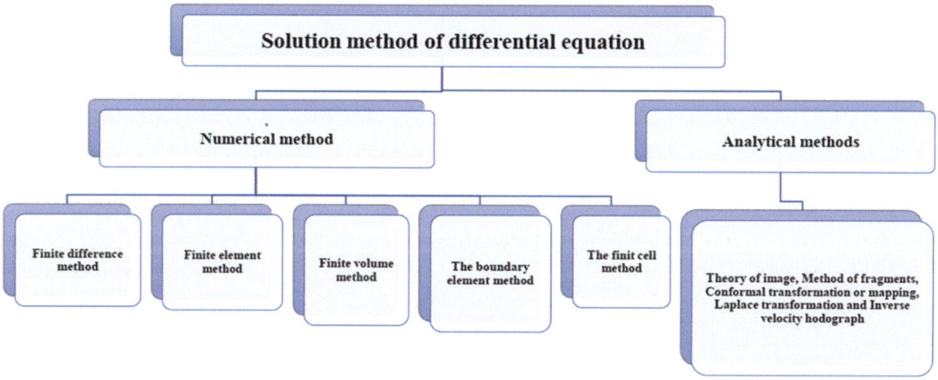

Fig. 10.3 Summary of the solution methods for differential equations

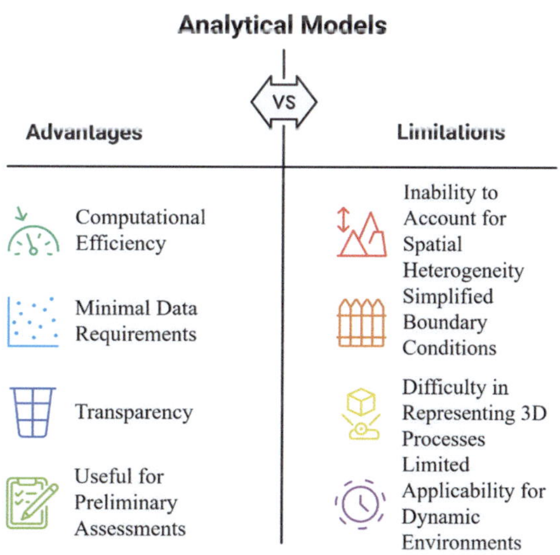

Fig. 10.4 Advantages and limitations of analytical models in groundwater modeling

10.3.1.2 Numerical Models

Numerical models solve groundwater flow equations through spatial and temporal discretization, enabling the representation of complex hydrogeological systems with heterogeneous properties and irregular boundaries (Fig. 10.5). In the Nile Delta, three primary numerical techniques are employed: finite-difference methods (FDM), which discretize the aquifer into grid cells with nodes at centers or corners, extensively applied through MODFLOW software (Armanuos et al. 2016a, b); finite element methods (FEM), which divide domains into variably shaped elements for better representation of complex boundaries, implemented through FEFLOW for coastal studies (Sefelnasr and Sherif

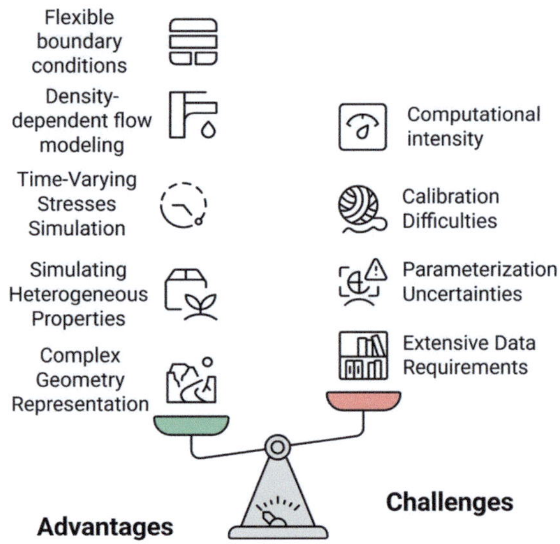

Fig. 10.5 Balancing benefits and drawbacks of numerical models

2014); and finite volume methods (FVM), which have gained prominence for computational fluid dynamics applications over the past two decades (Ezzeldin et al. 2018; Abd-Elmaboud et al. 2021). These approaches offer significant advantages in representing heterogeneity, simulating time-varying stresses, and incorporating density-dependent flow processes crucial for seawater intrusion analysis. However, they face challenges, including extensive data requirements, computational intensity, parameterization uncertainties, and calibration complexities when multiple processes interact simultaneously.

10.3.2 Common Software and Tools

The complexity of groundwater systems in the Nile Delta necessitates specialized modeling software capable of representing various hydrological, geological, and chemical processes (Fig. 10.6). MODFLOW, developed by the U.S. Geological Survey, dominates Nile Delta groundwater studies with its modular, three-dimensional finite-difference approach. Armanuos et al. (2017) employed MODFLOW-2005 to develop a comprehensive regional model evaluating the Grand Ethiopian Renaissance Dam's impacts on groundwater levels, while Embaby et al. (2017) utilized it to simulate artificial recharge scenarios that demonstrated potential 15% increases in local groundwater storage through strategic well placement. MODFLOW's flexibility has been extended through specialized packages, including the Saltwater Intrusion Package (SWI2) for computationally efficient seawater intrusion modeling (Nofal et al. 2015) and the Farm Process (FMP) for simulating irrigation–groundwater interactions (Eltarabily et al. 2018).

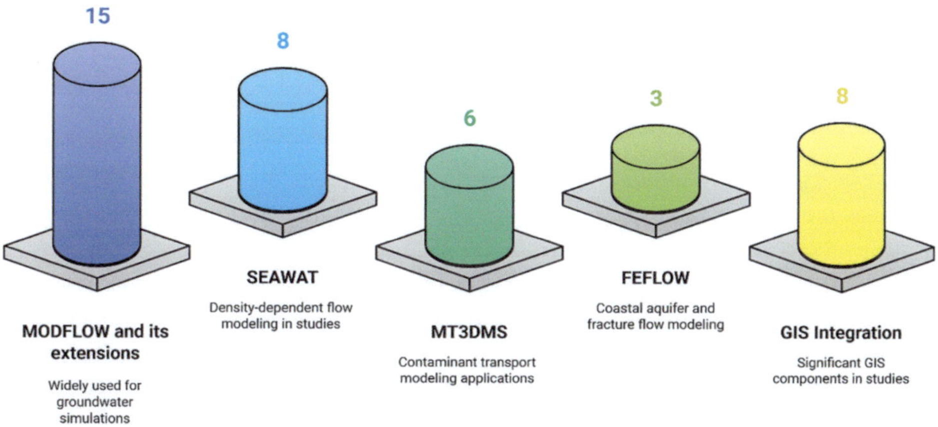

Fig. 10.6 Usage of modeling software in Nile Delta groundwater studies

SEAWAT, coupling MODFLOW with MT3DMS to simulate variable-density flow, has proven essential for seawater intrusion studies. Abd-Elhamid et al. (2016a, b) applied SEAWAT to project future intrusion patterns under various management strategies, while Mabrouk et al. (2019a, b) identified critical pumping thresholds that trigger saltwater upconing beneath extraction wells. For complex geological structures requiring flexible mesh discretization, FEFLOW has been employed by researchers like Sefelnasr and Sherif (2014), who leveraged its triangular mesh capabilities to accurately represent the irregular Nile Delta coastline when modeling sea-level rise impacts.

MT3DMS has been widely implemented for contaminant transport modeling, with Eltarabily et al. (2017) simulating nitrate transport from agricultural areas and El-Nimer et al. (2020) modeling heavy metal migration from industrial zones. Geographic Information Systems (GIS) have become integral to these modeling workflows, with Nofal et al. (2022) developing comprehensive databases integrating hydrogeological and meteorological data and El-Agha et al. (2017) creating thematic vulnerability maps from model outputs.

Specialized interfaces like Groundwater Modeling System (GMS) and automated calibration tools such as PEST (Abd-Elhamid et al. 2019) have further enhanced modeling efficiency. The integration of these software tools has enabled increasingly sophisticated representations of the Nile Delta groundwater system, though their effective application remains critically dependent on input data quality and the modeler's understanding of both software capabilities and physical system dynamics (Saqr and Abd-Elmaboud 2024).

10.3.3 Data Requirements: Hydrological, Geological, and Climatic Inputs

Successful groundwater modeling in the Nile Delta requires comprehensive datasets spanning multiple disciplines. The quality, quantity, and spatial–temporal coverage of these data significantly influence model reliability and predictive capability. This section examines the critical data requirements for groundwater modeling in the Nile Delta and discusses the challenges associated with data acquisition and management (Table 10.1).

Groundwater modeling in the Nile Delta confronts significant data limitations that constrain model reliability. Monitoring networks exhibit uneven spatial distribution, with approximately 30% of the delta lacking sufficient data for proper calibration (Nofal et al. 2015). Historical records suffer from inconsistent measurement protocols, particularly before 1990, complicating long-term trend analysis (El-Agha et al. 2017). Data fragmentation across multiple institutions creates integration challenges, with researchers spending considerable time harmonizing information from diverse sources (Armanuos et al. 2016a, b). Innovative approaches, including remote sensing, geophysical techniques, and machine learning algorithms, help address these limitations (Eltarabily et al. 2023; Mabrouk et al. 2019a, b), but model reliability remains fundamentally constrained by data quality and comprehensiveness, necessitating continued improvements in collection and management practices across this complex hydrogeological system (Noman et al. 2024).

10.4 Critical Review of Groundwater Modeling Studies in the Nile Delta

10.4.1 Studies Addressing Seawater Intrusion

Seawater intrusion represents one of the most significant threats to groundwater resources in the Nile Delta, prompting numerous modeling studies focused on understanding, predicting, and mitigating this phenomenon. The evolution of these studies traces the methodological advancement from simplified analytical approaches to sophisticated numerical models that incorporate multiple interacting processes.

The foundations for seawater intrusion modeling in the Nile Delta were established by several pioneering studies applying analytical approaches. Kashef (1983) conducted one of the earliest systematic assessments, applying the Ghyben–Herzberg principle to estimate seawater intrusion extending 30–40 km inland along the central delta coastline. While groundbreaking, this work was limited by assumptions of a sharp freshwater–saltwater interface and homogeneous aquifer properties. Farid (1985) extended this work with an early attempt at groundwater system management in the Nile Delta, providing a critical baseline understanding of saltwater dynamics. Sherif et al. (1988) advanced

Table 10.1 Review of data requirements for groundwater modeling in the Nile Delta

Category	Subcategory	Data description	Key studies	Key findings
Hydrogeological data	Aquifer geometry and boundaries	Borehole logs and well-cutting analyses	Ezzeldin et al. (2018)	Data from 400 + boreholes used to construct a 3D representation of the Quaternary aquifer
		Geophysical surveys (resistivity, seismic)	Mabrouk et al. (2018)	Integration of surface geophysics with borehole data to map aquifer base and clay lenses
		Geological and structural maps	El-Bihery (2009)	Identified structural controls on groundwater movement in the eastern delta
	Hydraulic properties	Hydraulic conductivity and transmissivity	Sefelnasr and Sherif (2014)	Compiled 150 + pumping tests; transmissivity ranges from 500 to 15,000 m^2/day
		Storage parameters	El-Agha et al. (2017)	Storage coefficient values of 0.0001–0.001 in confined portions
		Porosity measurements	Eltarabily et al. (2017)	Values between 25 and 40% in main aquifer materials
		Dispersivity parameters	Abd-Elhamid et al. (2016a, b)	Longitudinal dispersivity: 50–100 m; transverse: 5–10 m
	Hydrodynamic measurements	Piezometric head observations	Armanuos et al. (2017)	Data from 120 observation wells with monthly to quarterly measurements

(continued)

Table 10.1 (continued)

Category	Subcategory	Data description	Key studies	Key findings
		Time-series groundwater levels	Nofal et al. (2022)	30 years of records showing declining trends of 0.2–0.5 m/year
		Water quality parameters	Mabrouk et al. (2019a, b)	Data from 155 wells; TDS variations from 500 to 35,000 mg/L
		Aquifer response to stresses	Saleem et al. (2018)	Tidal response data used to estimate specific storage and hydraulic diffusivity
Hydrological inputs	Recharge components	Precipitation data	El-Agha et al. (2017)	25 rainfall stations; annual precipitation: 25 mm (south) to 200 mm (coast)
		Irrigation application rates	Eltarabily et al. (2018)	15–25% of applied irrigation water percolates to groundwater
		Canal seepage	Embaby et al. (2017)	Seepage rates of 3–7 mm/day from unlined irrigation canals
		Evapotranspiration data	Armanuos et al. (2023)	Remote sensing-based actual evapotranspiration estimates
	Surface water interaction	River stage and discharge	Masterson et al. (2007)	Time series from 35 gauging stations on main waterways

(continued)

Table 10.1 (continued)

Category	Subcategory	Data description	Key studies	Key findings
		Riverbed/canal bed conductivity	Sobeih et al. (2017)	Hydraulic conductivity values of 0.01–0.5 m/day
		Drainage network configuration	Eltarabily et al. (2016)	Water level data from 42 monitoring points on major drains
	Abstraction data	Well inventory information	Nofal et al. (2015)	Documented 12,000 + production wells across the delta
		Pumping rates and variations	El-Agha et al. (2017)	Actual extraction rates exceed permitted amounts by 20–40%
		Historical trends	Mabrouk et al. (2018)	Increase from 2.3 billion m^3/year (1980s) to 6.1 billion m^3/year (2015)
Geological and hydrogeochemical data	Stratigraphy and lithology	Lithological descriptions	Sefelnasr and Sherif (2014)	Data from 200 + deep boreholes showing complex interbedding
		Sedimentological analyses	El-Bihery (2009)	Analysis of 22 core samples to relate sediment characteristics to hydraulic conductivity
		3D distribution of clay lenses	Ezzeldin et al. (2018)	Mapping of clay layers using borehole data and geophysical surveys

(continued)

Table 10.1 (continued)

Category	Subcategory	Data description	Key studies	Key findings
	Geochemical parameters	Background water chemistry	Eltarabily et al. (2017)	Analysis of major ions, trace elements, and isotopes in 85 wells
		Reaction kinetics	El-Nimer et al. (2020)	Sorption coefficients and degradation rates for various contaminants
		Isotopic tracers	Mabrouk et al. (2019a, b)	Used to differentiate modern recharge from paleowaters
Climatic and sea-level data	Current climate parameters	Temperature, precipitation, humidity	Armanuos et al. (2016a, b)	Daily climatic data from 18 stations across the delta
		Spatial interpolation	El-Agha et al. (2017)	Geostatistical methods to develop continuous climate surfaces
		Climate variability	Eltarabily et al. (2023)	30 years of data characterizing wet and dry cycles
	Sea-Level and coastal parameters	Sea-level measurements	Abd-Elhamid et al. (2019)	Sea-level rise of 1.8–2.2 mm/year over the past century
		Coastal bathymetry	Sefelnasr and Sherif (2014)	Detailed surveys extending 5 km offshore
		Seawater density and salinity	Mabrouk et al. (2019a, b)	Mediterranean seawater density of 1,025 kg/m^3
	Climate change projections	Downscaled climate models	Armanuos et al. (2023)	Ensemble projections from five regional climate models

(continued)

Table 10.1 (continued)

Category	Subcategory	Data description	Key studies	Key findings
		Sea-level rise projections	Abd-Elhamid et al. (2019)	IPCC projections of 0.3–0.8 m by 2100
		Nile flow projections	El-Hamidi et al. (2018)	Scenarios with 10–30% reductions in Nile flow

analytical techniques by incorporating a two-dimensional finite element model for dispersion, indicating the 1,000 mg/L isochlor had penetrated up to 60 km inland, though their steady-state approach could not represent intrusion dynamics. Diab et al. (1997) further documented saltwater intrusion in the coastal aquifer, providing field evidence that validated and refined these early modeling efforts.

As computational capabilities improved, researchers developed increasingly sophisticated numerical models. Sherif and Singh (1999a, b) employed SUTRA for two-dimensional variable density flow modeling, estimating the 35,000 mg/L isochlor at 41 km inland and the 1,000 mg/L isochlor at 84 km inland. Sherif and Hamza (2001) explored mitigation strategies for coastal aquifers, while Sherif and Al-Rashed (2001) expanded previous work by developing separate horizontal and vertical models that estimated the 35,000 mg/L isochlor at 63 km inland. A significant methodological advancement came with Sefelnasr and Sherif's (2014) FEFLOW-based three-dimensional model, which projected substantial freshwater volume reductions to approximately 513 km^3 under combined sea-level rise and increased pumping scenarios. Nofal et al. (2015) presented a novel perspective on delineation and modeling of seawater intrusion, developing a comprehensive conceptual model that accounted for aquifer heterogeneity, representing an important methodological improvement in regional-scale modeling.

Climate change impact assessments have evolved toward increasingly integrated approaches. Early work by Sherif and Singh (1999b, a) estimated that a 0.5 m rise in the Mediterranean Sea level would cause an additional 9.0 km inland movement of the saltwater front. Abd-Elhamid et al. (2016a, b) developed a coupled transient finite element model (2D-FEST) suggesting that a 100 cm sea-level rise would advance the 1,000 mg/L isochlor by an additional 10 km inland. Wassef and Schüttrumpf (2016a, b) incorporated different climate scenarios (RCP 2.6 and 8.5), projecting saltwater interface extensions of up to 43 km and 57 km inland, respectively, by 2100. Abd-Elhamid et al. (2019) evaluated the combined impacts of sea-level rise and reduced Nile flow from the Grand Ethiopian Renaissance Dam, demonstrating potential inland advancement of the 35,000 mg/L isochlor to 110.2 km under worst-case scenarios. Recent work by Abd-Elaty et al. (2023, 2024) has further analyzed future hydrology stresses and climate change impacts on submarine groundwater discharge and water budget in the Nile Delta aquifer.

Detailed process studies have enhanced the understanding of specific mechanisms controlling seawater intrusion. Abd-Elaty et al. (2021a) found that in the central Nile Delta, the 35,000 mg/L isochlor had reached 63.75 km inland, while the 1,000 mg/L isochlor extended to 93.75 km. Hashish et al. (2014) simulated the effects of spatial layout and pumping/recharging rates on saltwater intrusion, demonstrating that the best well alignment is a regular one or two rows of wells subjected to uniform pumping rates. Mabrouk et al. (2019a, b) developed a sophisticated three-dimensional variable-density model with a comprehensive salinity classification system, revealing complex spatial patterns across the delta. El-Hamidi et al. (2018) incorporated tidal effects and seasonal recharge variations in their coastal zone analysis. Engelen et al. (2019) conducted a remarkable three-dimensional paleo-reconstruction of groundwater salinity distribution over 32,000 years, demonstrating that freshwater volumes declined by factors of 1.9 to 5.4 due to marine transgression, resulting in a steeper fresh–salt interface.

Management-oriented research has evaluated various mitigation strategies with varying degrees of success. Abd-Elhamid (2017) demonstrated that combined approaches could reduce seawater intrusion by up to 21.3%, while Abd-Elaty et al. (2021a, b) found that recharge enhancement with treated wastewater (19.5% intrusion reduction) outperformed pumping reduction (6.2%) and brackish water extraction (5.9%). Armanuos and Negm (2018) provided practical recommendations for reducing coastal extraction while maintaining agricultural productivity through improved irrigation efficiency.

Despite the considerable methodological advancements in modeling seawater intrusion in the Nile Delta, several critical research gaps and challenges persist that limit the practical application of these models for sustainable groundwater management (Fig. 10.7). A fundamental limitation across many studies is the reliance on outdated hydrogeological data, with some key parameters derived from field measurements conducted in the 1990s or early 2000s that may not reflect current aquifer conditions. The most comprehensive salinity measurement campaigns, such as those referenced by Mabrouk et al. (2019a, b), included only 155 observation points across the entire delta, providing insufficient spatial resolution for robust model calibration and validation. This data scarcity is particularly pronounced in deeper aquifer zones (below 200 m) and offshore extensions, creating substantial uncertainty in boundary condition specifications.

Most modeling approaches employ simplified representations of the delta's complex stratigraphy, often neglecting the heterogeneity and discontinuities in the confining clay layer that significantly influence local intrusion patterns. Sefelnasr and Sherif (2014) acknowledged this limitation, noting that their regional-scale model could not adequately represent preferential flow paths through clay layer breaches. Additionally, there exists a critical disconnect between model calibration periods (typically historical data) and projection horizons (often extending to 2100), with insufficient validation against intermediate observations to build confidence in long-term predictions.

Parameter uncertainty quantification remains inadequate in most studies. While sensitivity analyses are sometimes performed (e.g., Abd-Elaty et al. 2021a), few researchers

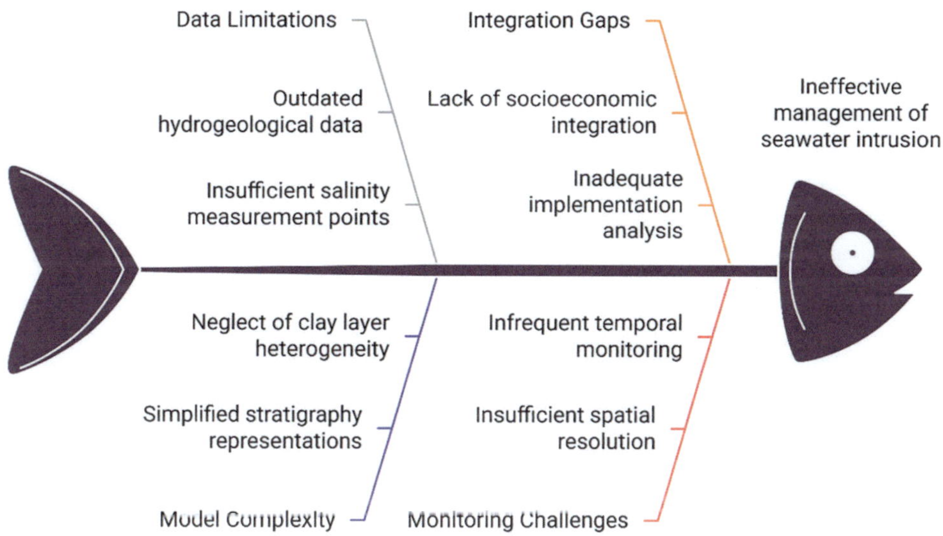

Fig. 10.7 Challenges in managing seawater intrusion in the Nile Delta

have systematically quantified the propagation of parameter uncertainties through model predictions to establish confidence intervals for projected intrusion rates. This limitation is particularly problematic for climate change impact assessments, where uncertainties in climate projections compound with hydrogeological parameter uncertainties. The downscaling of global climate models to regional hydrological impacts introduces additional uncertainty rarely acknowledged in existing literature (Subbarayan et al. 2025).

The spatial and temporal resolution of monitoring networks represents another significant constraint. Nofal et al. (2015) noted that approximately 30% of the delta area lacks sufficient monitoring data, with particular gaps in the eastern and western fringes where significant land-use changes are occurring. Temporal monitoring frequency (often quarterly or semiannual) is insufficient to capture seasonal dynamics and extreme events that may disproportionately influence intrusion processes. Furthermore, most observation wells sample only shallow portions of the aquifer, creating substantial uncertainty in the vertical distribution of salinity (Mumtaz et al. 2025).

Perhaps most critically, there exists a profound integration gap between technical modeling approaches and socioeconomic realities. Future extraction scenarios typically rely on simplified assumptions rather than incorporating demographic trends, agricultural policy changes, and economic development plans that will ultimately drive groundwater demand (Saqr et al., 2022). The economic feasibility and implementation challenges of proposed technical solutions receive inadequate attention, with limited consideration of institutional capacities, financial resources, and stakeholder acceptance. Abd-Elhamid (2017) and Armanuos and Negm (2018) identified potentially effective management approaches but provided insufficient analysis of implementation pathways within Egypt's complex water governance framework.

Addressing these research gaps requires not only methodological refinements but also fundamental shifts in approach: from discipline-specific studies toward truly interdisciplinary frameworks, from snapshot measurements toward continuous monitoring systems, and from technical optimization toward socially embedded management approaches that recognize the complex interplay between hydrogeological, economic, and institutional factors shaping groundwater use in this critically important region.

10.4.2 Research on Contaminant Transport and Pollution

The Nile Delta aquifer faces significant contamination threats beyond seawater intrusion, including agricultural, industrial, and urban pollution sources. Numerous modeling studies have investigated these contamination processes, employing various approaches to understand contaminant transport dynamics and develop mitigation strategies (Paudel et al. 2024).

Agricultural activities represent the most spatially extensive contamination source in the region. Eltarabily et al. (2017) developed a comprehensive groundwater flow and contaminant transport model using MODFLOW and MT3DMS to investigate nitrate pollution in the southern Nile Delta. Their research revealed nitrate concentrations exceeding WHO drinking water standards (50 mg/L) in approximately 30% of the shallow aquifer, with the highest concentrations in intensive vegetable cultivation areas. Similar findings emerged from Ghoraba et al. (2013), who conducted a broader study in the central delta (El-Gharbiya Governorate), documenting significant ammonium contamination and proposing a groundwater remediation scheme involving extraction wells for polluted water recovery. Abd-Elaty et al. (2020) further contributed to this understanding by developing a three-dimensional mass transport model examining soil type effects on contaminant attenuation, demonstrating that silty clay soils provided substantially better groundwater protection than sandy soils (Alao et al. 2024).

Industrial and urban pollution sources have received focused attention through several modeling studies. El-Nimer et al. (2020) simulated heavy metal contamination migration from industrial oxidation ponds in the eastern Nile Delta's 10th of Ramadan City area, projecting that contamination plumes containing aluminum, iron, and strontium would travel 1.8–2.3 km northeastward over 50 years. Hussein et al. (2019) confirmed these findings, documenting metal concentrations exceeding WHO guidelines. Addressing urban pollution sources, Abd-Elaty et al. (2021a, b) modeled groundwater contamination from wastewater irrigation, finding that untreated wastewater usage significantly increased chemical oxygen demand (COD) levels in groundwater. Their proposed solution—mixing treated wastewater with freshwater—demonstrated potential water quality improvements, with average COD concentrations decreasing from 53.15 mg/L to 32.83 mg/L.

Several researchers have developed regional vulnerability assessments to identify areas at greatest contamination risk. El Alfy et al. (2015) utilized a modified DRASTIC methodology with GIS integration to classify aquifer vulnerability zones, finding approximately 25% of the western delta highly vulnerable to pollution, particularly where protective clay layers thinned. Salem et al. (2019) expanded this approach, documenting that 22.4% of their study area qualified as "very highly vulnerable" to contamination. Metwally et al. (2022) further refined these methods through comparative analysis of multiple vulnerability models, consistently identifying areas south of Zefta as extremely vulnerable across all assessment approaches.

The most sophisticated research has integrated multiple contamination processes into unified modeling frameworks. Abd-Elaty et al. (2019) evaluated strategies to protect groundwater from polluted surface drains, demonstrating that cut-off walls along drain edges reduced contamination by approximately 28.5%. Nofal et al. (2022) developed a comprehensive contaminant transport model incorporating agricultural, industrial, and urban sources, while Abd-Elaty et al. (2022) specifically addressed the severe contamination from the Bahr El-Baqar drain system.

Despite these advances, significant research gaps persist in Nile Delta contaminant transport modeling (Fig. 10.8). Most studies have focused narrowly on nitrates and heavy metals, neglecting emerging contaminants such as pharmaceuticals, microplastics, and engineered nanomaterials that potentially threaten human and ecosystem health. Methodologically, research has predominantly employed standard advection–dispersion frameworks rather than incorporating advanced reactive transport capabilities that could better represent complex biogeochemical interactions. Spatiotemporal limitations are evident in the prevalence of localized, short-duration studies (typically 5–10 years) that inadequately capture long-term accumulation dynamics and regional-scale contaminant interactions.

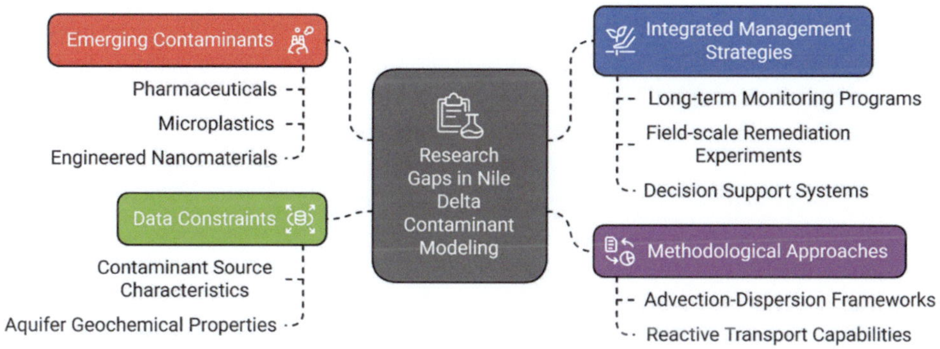

Fig. 10.8 Addressing research gaps in Nile Delta contaminant modeling

Data constraints represent perhaps the most significant challenge, with limited information available on contaminant source characteristics, aquifer geochemical properties, and field-measured concentration distributions for model calibration (Rehman et al. 2024). Additionally, most studies remain descriptive rather than prescriptive, offering limited analysis of remediation feasibility or implementation pathways. Research has inadequately addressed critical aspects, including multi-contaminant interactions, climate change effects on contaminant transport, remediation system modeling, and economic integration, which are necessary for policy formulation.

Advancing from scientific understanding to effective management will require more integrated, multidisciplinary approaches that incorporate socioeconomic factors, institutional frameworks, and implementation pathways alongside sophisticated contaminant transport modeling (Masria et al. 2024). Future research should prioritize long-term monitoring programs, field-scale remediation experiments, and decision support systems that connect technical models with practical management options.

10.4.3 Models for Sustainable Groundwater Management and Recharge Strategies

As groundwater challenges in the Nile Delta have intensified, researchers have developed increasingly sophisticated models to support sustainable management and explore artificial recharge strategies. Regional management models have provided critical insights into groundwater sustainability at the basin scale. Dawoud et al. (2005) created a GIS-based groundwater management model for the Western Nile Delta that evaluated scenarios including reduced surface water inflow with increased extraction, demonstrating a potential 91% decrease in aquifer potentiality. Similar concerns were echoed by Soliman et al. (2014), whose MODFLOW simulations identified scenarios with drawdowns of 11.5-15 m combined with increased canal leakage as representing worst-case outcomes for sustainability. More holistic approaches were developed by El Molla et al. (2005), who addressed water shortages through the evaluation of multiple water sources for supporting reclaimed land, and Sobeih et al. (2017), who quantified groundwater recharge at 28.1 million m3/day from surface water sources while highlighting data limitations in rapidly changing areas.

Optimization and decision support models have enhanced management capabilities through the integration of multiple factors. El-Agha et al. (2017) developed a comprehensive system that integrated hydrological, economic, and institutional factors to evaluate extraction patterns and regulatory approaches, though challenges remained in quantifying informal extraction. Nofal et al. (2016) further demonstrated that doubling extraction rates would significantly impact groundwater heads, with central and southwestern delta regions showing the highest vulnerability. Eltarabily et al. (2018) provided specific sustainability thresholds, finding the eastern Delta aquifer would remain stable through 2023 at current extraction rates but could not sustain a 50% extraction increase beyond 2020.

Artificial recharge has emerged as a promising management strategy for addressing extraction–recharge imbalances. Embaby et al. (2017) demonstrated that eight injection wells could increase local groundwater storage by approximately 15% over a decade, though questions remained regarding water quality and maintenance requirements. El Arabi (2012) pioneered innovative approaches through soil aquifer treatment systems that simultaneously enhanced storage and improved water quality. Abd-Elaty et al. (2021a, b) quantified comparative effectiveness, showing that artificial recharge could delay saltwater intrusion by 19.5%, outperforming pumping reduction (6.2%) and brackish water extraction (5.9%), though implementation challenges were not fully addressed.

Climate change adaptation has driven specialized modeling approaches. Akbarpour and Niksokhan (2018) conducted an integrated assessment of climate change impacts, highlighting significant effects on recharge rates and water demands despite projection uncertainties. Transboundary concerns were addressed by Armanuos et al. (2017) and Abd-Elhamid et al. (2019), who evaluated the Grand Ethiopian Renaissance Dam impacts, finding that groundwater pumping would need to be reduced by 40–60% to mitigate increased saltwater intrusion, though dam operation uncertainties remained significant.

Integrated water resources management frameworks have provided more comprehensive perspectives. Eltarabily and Negm (2018) developed holistic approaches demonstrating that annual aquifer potential could be improved through irrigation canal construction and extraction management. Mabrouk et al. (2019a, b) evaluated adaptation strategies, identifying crop pattern adjustments and irrigation practice improvements as more promising than artificial recharge. Nofal et al. (2018) emphasized regional sensitivity differences, highlighting central and southwestern areas as most vulnerable to extraction stresses.

Canal management studies have clarified critical surface–groundwater interactions. Nofal et al. (2022) demonstrated significant impacts of canal lining on groundwater recharge, with aquifer recharge effectiveness decreasing from 66 to 10% under maximum canal efficiency scenarios. Abd-Elziz et al. (2022) documented reduced groundwater tables from canal lining, with implications for crop water availability and saltwater intrusion. Armanuos et al. (2016a, b) highlighted the need for coordinated management of surface and subsurface water resources, though institutional coordination challenges remained.

Despite these advances, significant research gaps persist (Fig. 10.9). Most models identify technically feasible solutions but provide limited implementation guidance addressing institutional, financial, and social constraints. Few studies adequately represent the coevolution of human and natural systems, including feedback between groundwater conditions, human behavior, and policy interventions. Management models typically present deterministic predictions without adequately characterizing uncertainties in model structure and future conditions. Financial mechanisms for sustainable management, including appropriate pricing and cost recovery for artificial recharge, have received insufficient attention. Distributional impacts across socioeconomic groups and regions remain underexplored despite their importance for social acceptability. Additionally, most research lacks the adaptive management frameworks necessary for navigating uncertainties in climate, demographic, and economic conditions.

Addressing these gaps will require more integrated, participatory approaches incorporating stakeholder perspectives throughout model development. Future research should prioritize coupled socio-hydrological frameworks, robust decision-making under uncertainty, and implementation-oriented studies that bridge the gap between technical solutions and practical application. Improving monitoring networks and data sharing would strengthen model parameterization and validation, while transdisciplinary collaboration would better address the complex intersections of hydrological, economic, social, and institutional factors shaping groundwater sustainability in the Nile Delta.

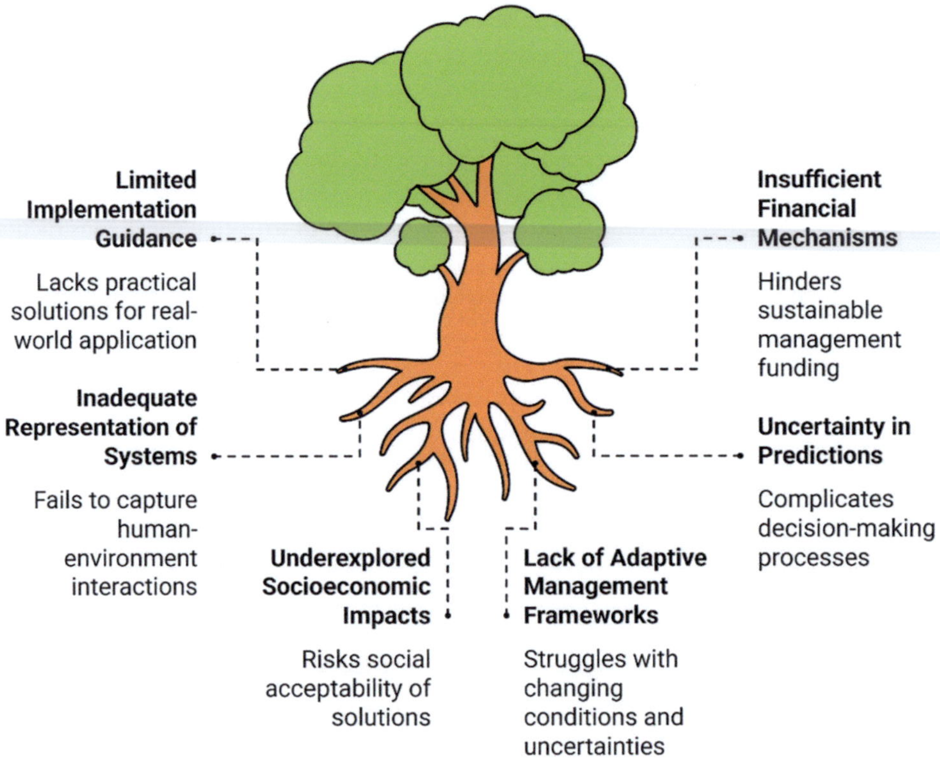

Fig. 10.9 Research gaps in groundwater management implementation

10.4.4 Integration of Climate Change and Land-Use Dynamics

The complex interactions between climate change, land-use transformations, and groundwater dynamics present significant modeling challenges in the Nile Delta. Recent research has increasingly focused on integrating these drivers to develop a more comprehensive understanding and management approaches. Climate-focused studies have provided critical quantitative insights into future groundwater conditions. Armanuos et al. (2023) projected progressive increases in crop evapotranspiration of 11.2%, 15.0%, and 19.0% by 2030, 2050, and 2070, respectively, leading to corresponding recharge decreases of 19.74%, 27.16%, and 36.84% and subsequent groundwater head declines. Complementing this research, Eltarabily et al. (2023) identified solar radiation, humidity, and vapor pressure as primary drivers of evapotranspiration changes while validating groundwater head predictions against observed data. Xiao et al. (2016) further demonstrated climate change's spatially variable impacts, with low-relief coastal areas showing particular vulnerability to sea-level rise and altered precipitation patterns.

Land-use transformation studies have documented significant changes affecting groundwater dynamics throughout the Nile Delta. El-Agha et al. (2015) revealed that land-use shifts between 1990 and 2015 reduced natural recharge areas by approximately 12% while simultaneously increasing water demands, creating growing imbalances in the groundwater budget. Investigating more localized impacts, Eltarabily et al. (2018) documented groundwater head decreases of 0.2–0.42 m in the southeastern Nile Delta resulting from agricultural expansion and urbanization. The rapid pace of these transformations was comprehensively documented by Shalaby and Tateishi (2007), who used remote sensing to identify the conversion of approximately 20% of agricultural land to urban uses between 1987 and 2005, though connections to specific policy drivers remained unclear.

The most advanced research has integrated both climate change and land-use dynamics into unified modeling frameworks. Akbarpour and Niksokhan (2018) evaluated the compound impacts of climate projections, urbanization trends, and sea-level rise, demonstrating that these combined stressors would reduce freshwater availability beyond single-factor predictions despite calibration challenges from limited historical multi-stressor data. Armanuos et al. (2016a, b) predicted that agricultural expansion in desert fringes would increase extraction by 15–30% while climate change simultaneously reduced recharge by 10–20%, resulting in groundwater level declines of 0.5–1.5 m by 2050. Extending this integrative approach, Aziz et al. (2019) examined how Grand Ethiopian Renaissance Dam operations would interact with changing land use and climate conditions, suggesting that reservoir filling within three years would necessitate 60% groundwater extraction reductions to maintain sustainable aquifer conditions.

Adaptation-focused research has evaluated management strategies under changing conditions. Mabrouk et al. (2019a, b) assessed various adaptation measures, finding that crop pattern modifications and irrigation practice improvements showed greater promise than technical interventions like artificial recharge or desalination. Hugman et al. (2017)

demonstrated that managed aquifer recharge could partially offset climate impacts but would require substantial infrastructure investment and careful site selection. Emphasizing integration, Eltarabily and Negm (2018) proposed coordinated surface and groundwater management strategies for the western Nile Delta, though implementation pathways under uncertainty remained underdeveloped.

Methodological innovations have enhanced modeling capabilities for addressing these complex challenges. Armanuos and Negm (2018) employed integrated 3D groundwater modeling with SEAWAT, coupling climate projections with land-use scenarios despite computational limitations for fine-scale processes. Gomaa et al. (2024) linked crop water demand models with groundwater flow simulations to provide holistic climate and land-use impact assessments. Wassef and Schüttrumpf (2016a, b) employed sophisticated 3D finite element models featuring enhanced coastal representations to more accurately assess the impacts of sea-level rise. However, they encountered challenges in parameterization, particularly in regions with limited data availability.

Despite these advances, significant research gaps and methodological challenges persist. Integration of climate and land-use factors substantially increases model complexity and computational demands, creating inevitable trade-offs between comprehensiveness and tractability. Scale mismatches between coarse climate projections and fine-scale groundwater processes remain difficult to reconcile despite various downscaling approaches. Models integrating multiple dynamic systems face compounding uncertainties from climate projections, land-use trajectories, and hydrogeological parameterization, yet few studies have systematically analyzed uncertainty propagation through model chains. The different temporal scales of climate trends and land-use conversions are inadequately captured by conventional time-stepping approaches, potentially missing threshold behaviors. Perhaps most significantly, feedback mechanisms between groundwater levels, irrigation demands, and subsequent land-use decisions remain underrepresented despite their critical importance for system dynamics.

Future research should prioritize several underdeveloped areas. Agent-based modeling approaches could better represent human decision-making processes driving land-use changes in response to environmental shifts (see Fig. 10.10). Ensemble modeling frameworks would provide more robust decision support by systematically exploring multiple plausible futures under high uncertainty. Integrated models could support early warning systems identifying emerging threats to groundwater sustainability based on monitored climate and land-use indicators. Greater attention to equity dimensions would strengthen the analysis of how climate and land-use impacts disproportionately affect vulnerable communities. Finally, more exploration of transformative adaptation strategies—rather than merely incremental approaches—would better prepare for potentially severe future changes requiring fundamental restructuring of water use systems in the Nile Delta.

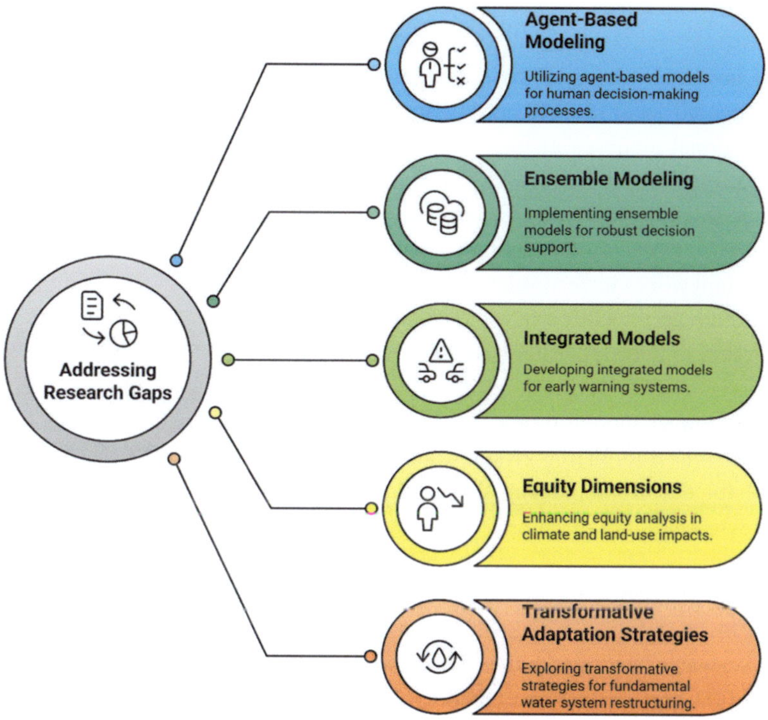

Fig. 10.10 Navigating research gaps in groundwater sustainability

10.5 Synthesis of Findings and Research Gaps

10.5.1 Comparative Analysis of Methodologies and Outcomes

The methodological approaches employed in Nile Delta groundwater modeling have evolved significantly over recent decades, reflecting both technological advances and deepening understanding of aquifer complexities. Earlier studies predominantly utilized one-dimensional analytical approaches or two-dimensional numerical models (Kashef 1983; Sherif et al. 1988), while contemporary research increasingly employs sophisticated three-dimensional representations (Sefelnasr and Sherif 2014; Mabrouk et al. 2019a, b). This dimensional progression coincides with the growing recognition of vertical flow processes, particularly for an accurate representation of seawater intrusion dynamics. Concurrently, process integration has advanced substantially, transitioning from single-process flow models toward comprehensive frameworks incorporating variable-density flow, solute transport, and occasionally heat transport (Abd-Elhamid et al. 2016a, b; Blanco-Coronas et al. 2021). This integration enhances the representation of complex physical interactions but necessarily increases computational demands and data requirements (Masria et al. 2023).

Spatial coverage and resolution exhibit considerable variation across studies, ranging from detailed local models investigating specific contamination sites (El-Nimer et al. 2020) to regional analyses encompassing the entire delta (Armanuos et al. 2017). Each approach offers distinct advantages: Local models effectively represent fine-scale processes and heterogeneity but may overlook regional context, while delta-wide assessments capture broader patterns but inevitably simplify local complexity (Timilsina et al. 2025). Temporal frameworks have similarly evolved, with early research predominantly employing steady-state approaches while recent studies increasingly utilize transient simulations spanning years to decades (Eltarabily et al. 2018; Nofal et al. 2022), reflecting heightened interest in dynamic responses to climate change and management interventions.

The selection of modeling codes demonstrates consistent patterns, with MODFLOW and its extensions dominating groundwater flow modeling, SEAWAT emerging as the preferred tool for density-dependent flow and transport studies, and FEFLOW occasionally employed when flexible mesh discretization is required (Sefelnasr and Sherif 2014). This selection typically reflects balanced consideration of process representation requirements, computational efficiency constraints, and researcher familiarity with specific platforms.

Despite methodological differences, consistent outcomes have emerged across research domains. Seawater intrusion studies generally agree that significant intrusion has occurred, with the 1,000 mg/L isochlor extending 80–110 km inland in the central delta (Abd-Elhamid et al. 2016a, b; Mabrouk et al. 2019a, b). The lateral extent varies substantially, with greater intrusion in eastern and western portions compared to the central region, reflecting differences in extraction patterns and hydrogeological conditions. Climate change and sea-level rise are projected to exacerbate intrusion by 5–15 km over the next century under most scenarios (Abd-Elhamid et al. 2019; Wassef and Schüttrumpf 2016a, b), though groundwater extraction remains the dominant driver in most areas.

Contaminant transport research indicates that agricultural contamination, particularly from nitrates, affects approximately 30% of the shallow aquifer with concentrations exceeding drinking water standards (Eltarabily et al., 2017). Industrial and urban pollution creates localized but severe contamination plumes typically extending 1–5 km from sources (El-Nimer et al. 2020). Vulnerability to contamination demonstrates high spatial heterogeneity, with areas lacking the protective clay cap showing significantly elevated risk (Salem et al. 2019).

Management and recharge studies reveal that sustainable extraction limits vary substantially across the delta, ranging from approximately 0.2 to 1.2 million m3/day per 100 km^2 depending on local hydrogeological conditions and recharge sources (Nofal et al. 2018). Artificial recharge using treated wastewater appears technically feasible and potentially effective, particularly for coastal areas threatened by seawater intrusion (Abd-Elaty et al.

2021a, b). Canal lining for irrigation efficiency creates a notable trade-off with groundwater recharge, potentially reducing aquifer storage by 15–40% in affected areas (Nofal et al. 2022).

Climate change research projects reductions in natural recharge of 15–35% by 2070 through increased evapotranspiration and altered precipitation patterns (Armanuos et al. 2023). Land-use changes, particularly urbanization, have reduced recharge areas by approximately 1% annually over recent decades (El-Agha et al. 2015). The combined effects could reduce groundwater availability by 30–50% in severely affected areas by 2100 if current trends continue (Akbarpour and Niksokhan 2018). Adaptation through modified crop patterns and irrigation practices appears more promising and cost-effective than engineering solutions in most regions (Mabrouk et al. 2019a, b).

These comparative findings highlight the necessity of spatially differentiated management approaches rather than uniform policies across the delta. Different stressors dominate in different regions, with seawater intrusion being most critical in coastal zones, agricultural contamination being predominant in the central delta, and water table decline being most severe in southern and desert-fringe areas. Temporal dynamics vary significantly by process, with seawater intrusion responding relatively slowly to interventions (decades), while contaminant transport and water table changes demonstrate more rapid responses (years).

10.5.2 Limitations in Data Availability and Model Validation

Despite significant advances in groundwater modeling for the Nile Delta, persistent challenges in data availability and model validation continue to constrain research quality and management applications. The spatial distribution of monitoring infrastructure across the Nile Delta is highly uneven, creating significant data gaps that hinder comprehensive analysis. Monitoring networks show substantial disparities in spatial coverage, with relatively dense networks in the central delta but significant gaps in the eastern and western fringes, coastal areas, and newly reclaimed lands. Nofal et al. (2015) noted that approximately 30% of the delta area lacks sufficient monitoring data for reliable model calibration, creating significant uncertainties in these regions. This limitation is particularly pronounced in vertical sampling, as most observation wells sample only shallow portions of the aquifer (typically the upper 50–100 m), with limited data available on deeper zones. Mabrouk et al. (2019a, b) highlighted how limited deep salinity observations constrained their ability to validate three-dimensional salinity distributions, especially in the central and eastern delta.

The coastal and offshore regions present particular challenges for data collection, with extremely limited information on submarine groundwater discharge, offshore aquifer extensions, and saltwater intrusion beneath the Mediterranean Sea. This scarcity creates significant uncertainty in coastal boundary conditions, which are essential for accurate seawater intrusion modeling. Sefelnasr and Sherif (2014) emphasized that limited offshore data constrained their ability to represent submarine groundwater discharge processes accurately. Further complicating the data landscape is the prevalence of unregistered private wells with limited documentation of location, construction details, or pumping rates. El-Agha et al. (2017) estimated that 20–30% of total groundwater abstraction in the delta is not captured in official records, creating substantial uncertainty in extraction patterns and rates that directly impact model accuracy.

Temporal data limitations further constrain modeling efforts in the Nile Delta. Long-term consistent monitoring records needed for trend analysis and model validation under varying conditions are limited in many areas. Many observation wells have discontinuous records or were established relatively recently, limiting historical baseline data. El-Agha et al. (2017) highlighted inconsistencies in groundwater level measurement protocols before 1990, complicating the establishment of long-term trends. Most monitoring programs collect data at quarterly or biannual intervals, missing short-term variations and event responses that could provide valuable insights into aquifer behavior. Armanuos et al. (2017) noted that the limited temporal resolution of available monitoring data constrained their ability to capture seasonal dynamics in their models. Additionally, some long-term monitoring locations have experienced changes in measurement methodology, reference points, or equipment over time, creating artificial discontinuities in records that complicate trend analysis and model validation.

Parameter estimation represents another significant challenge for groundwater modeling in the Nile Delta. The spatial heterogeneity of hydraulic conductivity is poorly characterized in many areas, with available measurements concentrated at specific test sites rather than systematically distributed across the delta. Ezzeldin et al. (2018) highlighted how limited field measurements of hydraulic conductivity, particularly in newly developed areas, constrained their ability to represent aquifer heterogeneity accurately. Storage parameters (specific yield and specific storage) are among the most challenging to estimate accurately yet significantly influence transient model behavior. Eltarabily et al. (2018) acknowledged substantial uncertainty in storage parameter values used in their transient simulations due to limited field data. Transport parameters such as longitudinal and transverse dispersivity are rarely measured directly in the field and are typically estimated through model calibration, introducing significant uncertainty in contaminant transport predictions. Abd-Elhamid et al. (2016a, b) determined dispersivity values primarily through model calibration against observed concentration patterns, acknowledging the limitations of this approach compared to field-based estimation.

Model validation faces several significant constraints in the Nile Delta context. The limited overall data availability often forces researchers to use all available data for model calibration, leaving few or no independent observations for validation. This approach risks overfitting and reduces confidence in the model's predictive capability. Nofal et al. (2022) acknowledged this limitation, noting that the scarcity of monitoring data forced them to use all available observations for calibration, preventing rigorous validation against independent measurements. Many models are calibrated against historical data and used to make future projections, but opportunities to validate these projections against subsequent observations are limited. While water level observations are relatively common, data on processes such as flow velocities, travel times, and geochemical transformations are extremely limited. Abd-Elaty et al. (2020) highlighted how limited field measurements of contaminant transformation processes constrained their ability to validate reaction module performance in their transport model.

Institutional and data management factors create additional constraints beyond the physical limitations of monitoring networks. Groundwater data are often scattered across multiple agencies and institutions with different monitoring objectives, methods, and data management systems. Armanuos et al. (2016a, b) reported spending approximately 30% of their research time on data compilation and harmonization from different sources, highlighting the inefficiency created by this fragmentation. Even existing data may be difficult to access due to institutional policies, proprietary concerns, or inadequate data-sharing mechanisms. Many datasets lack adequate documentation of measurement methods, quality control procedures, and uncertainty estimates. Mabrouk et al. (2018) emphasized the challenges of incorporating data uncertainty into model calibration and prediction due to inadequate uncertainty documentation in many datasets. Complete assessment of the Nile Delta groundwater system ideally requires data on the broader Nile Basin, which spans multiple countries with varying data-sharing policies, further complicating comprehensive analysis.

Researchers have increasingly turned to innovative approaches to address these data limitations. Satellite-based observations provide spatially continuous data on land cover, soil moisture, evapotranspiration, and land subsidence. Eltarabily et al. (2023) utilized GRACE satellite data to estimate regional-scale groundwater storage changes but acknowledged significant uncertainty in their satellite-based estimates compared to direct measurements. Noninvasive geophysical techniques help fill gaps in subsurface characterization, with Mabrouk et al. (2019a, b) employing airborne electromagnetic surveys to map the freshwater–saltwater interface along the Mediterranean coast. Engaging local communities in data collection expands monitoring networks, while data mining and machine learning techniques can extract patterns from existing datasets to fill gaps. Systematic efforts to combine datasets from different sources can expand data availability but require addressing differences in measurement methods, timing, and quality across sources.

These persistent data limitations have important implications for the reliability and application of groundwater models in the Nile Delta. Limited data for calibration and validation inevitably increase uncertainty in model predictions, particularly for future scenarios involving climate change or management interventions. Model reliability varies substantially across the delta, with greater confidence in well-monitored central areas and much higher uncertainty in poorly monitored fringes and coastal zones. Data limitations often force the simplification of process representations, potentially omitting important mechanisms or interactions. These challenges underscore the critical importance of expanding and improving monitoring networks, enhancing data sharing and integration, and developing modeling approaches that explicitly address uncertainty while providing a framework for decision-making under these constraints.

10.5.3 Future Directions for Integrated and Multidisciplinary Approaches

The escalating complexity of groundwater challenges in the Nile Delta demands a paradigm shift toward integrated, multidisciplinary research approaches. As shown in Table 10.2, six key research directions emerge as critical for advancing sustainable groundwater management in this vital region.

Socio-hydrological integration represents a fundamental advancement in modeling approaches, linking natural hydrogeological systems with human dimensions. As highlighted in Table 10.2, agent-based modeling offers significant potential for simulating stakeholder decisions and exploring policy scenarios (Timalsina et al. 2025a, b). This approach, coupled with economic-hydrological models, can evaluate market-based instruments and pricing policies while revealing governance mechanisms better suited to the Nile Delta's institutional landscape. These integrated frameworks provide essential insights into the reciprocal relationships between groundwater conditions and human adaptations that shape extraction patterns. Hydrological process integration constitutes another critical direction detailed in Table 10.2, advancing beyond current fragmented approaches. Future research must develop comprehensive models that simultaneously simulate surface water, groundwater, and unsaturated zone processes with dynamic feedback. Particular emphasis should be placed on reactive transport modeling for contaminant transformation processes and ecosystem linkages, offering more realistic representations of the complete hydrological system and its environmental interactions.

Emerging technologies, as identified in Table 10.2, present substantial opportunities for enhancing groundwater modeling capabilities. Machine learning applications can improve parameter estimation and uncertainty analysis, while advanced sensor networks and remote sensing provide critical data for real-time model updating. These technological advances are especially valuable for monitoring seawater intrusion and developing early warning systems in vulnerable coastal zones. Transdisciplinary collaboration, emphasized in Table 10.2, requires participatory modeling approaches that engage diverse stakeholders while maintaining scientific rigor (Sathiyamurthi et al. 2025). Complex groundwater challenges necessitate expertise beyond traditional hydrogeology, requiring research teams that integrate social scientists, economists, agronomists, and policy experts with groundwater specialists. Stronger science-policy interfaces can ensure research addresses relevant questions while informing policy development. As outlined in Table 10.2, specific applied research priorities include managed aquifer recharge systems,

Table 10.2 Future directions for integrated and multidisciplinary approaches in Nile Delta groundwater modeling

Research direction	Key methods and approaches	Expected benefits	Current challenges	Priority applications for Nile Delta
Socio-hydrological integration	• Agent-Based Modeling (ABM) • Economic-hydrological model coupling • Institutional framework analysis	• Better understanding of human–water resource interactions • Evaluation of stakeholder responses to policies • Improved effectiveness of management tools	• Complexity of representing human decision-making • Calibration and validation challenges • Lack of behavioral data	• Assessment of farmer responses to water pricing policies • Simulation of land-use change impacts • Identification of effective governance mechanisms
Hydrological process integration	• Integrated surface–groundwater models • Reactive transport modeling • Ecosystem-groundwater linkages	• More realistic representation of a hydrological system • Comprehensive assessment of interventions • Improved understanding of biogeochemical interactions	• Computational challenges for complex models • Lack of data on chemical interactions • Model efficiency constraints	• Management of irrigation canal–groundwater interactions • Assessment of emerging contaminant impacts • Study of seawater intrusion effects on coastal ecosystems
Emerging technologies	• Machine learning applications • Advanced sensors and remote sensing • Hybrid (Physics-Statistical) models	• Enhanced parameter estimation and uncertainty analysis • Real-time monitoring and model updating • Filling critical data gaps	• Challenges in integrating multisource data • Infrastructure requirements for continuous monitoring • Balancing accuracy with computational efficiency	• Monitoring seawater intrusion in coastal areas • Developing early warning systems • Dynamic mapping of groundwater level changes

(continued)

Table 10.2 (continued)

Research direction	Key methods and approaches	Expected benefits	Current challenges	Priority applications for Nile Delta
Transdisciplinary collaboration	• Participatory modeling • Multidisciplinary research teams • Science-policy interface strengthening	• Improved model quality and implementation potential • Integration of specialized and local knowledge • Increased research impact on policy	• Methodological differences between disciplines • Communication challenges • Institutional barriers	• Engaging local communities in data collection • Developing integrated water resource management strategies • Designing socially acceptable policies
Specific applied research directions	• Managed aquifer recharge systems • Integrated coastal zone management • Innovative agricultural water management • Water-energy nexus exploration	• Enhanced groundwater sustainability • Increased hydrological system resilience • Maximized resource use efficiency	• Large-scale implementation challenges • Substantial investment requirements • Technical and regulatory constraints	• Coastal aquifer recharge systems • Assessment of Ethiopian Renaissance Dam impacts • Safe and efficient wastewater reuse
Implementation pathways and action research	• Policy analysis and design • Transition management frameworks • Adaptive management methodologies	• Bridging technical modeling and practical solutions • Developing effective implementation strategies • Enhanced adaptation to changing conditions	• Complexities of decision-making contexts • Competing stakeholder priorities • Limited institutional resources	• Development of monitoring protocols and evaluation criteria • Detailed case studies of implementation experiences • Adjustment mechanisms for changing conditions

integrated coastal zone management, innovative agricultural water practices, and water-energy nexus exploration. Additionally, implementation pathways need greater emphasis, focusing on policy analysis, transition management frameworks, and adaptive approaches with clear monitoring protocols and evaluation criteria. By pursuing these integrated directions identified in Table 10.2, researchers can develop a more comprehensive understanding of groundwater dynamics and more effective management strategies for the Nile Delta's sustainable future, bridging the persistent gap between technical modeling and practical implementation.

10.6 Conclusion

Groundwater modeling research in the Nile Delta has evolved significantly over recent decades, transitioning from basic analytical approaches to sophisticated numerical simulations that incorporate multiple physical processes, environmental stressors, and management considerations. This evolution reflects technological advancements in computational capabilities and growing recognition of the complex challenges confronting this critical aquifer system. The Nile Delta aquifer faces converging pressures from seawater intrusion, agricultural and industrial contamination, excessive extraction, reduced recharge, and climate change impacts—challenges that interact synergistically, amplifying their individual effects. Research consistently demonstrates significant spatial heterogeneity in aquifer characteristics and vulnerability, with temporal dynamics varying substantially by process and location, underscoring the necessity for spatially and temporally tailored management approaches.

Despite methodological advances, groundwater modeling in the Nile Delta remains constrained by limitations in data availability, quality, and accessibility. These constraints generate substantial uncertainties in model predictions and management recommendations, particularly in areas with sparse monitoring networks. Translating technical understanding into effective management actions represents a persistent challenge, with institutional fragmentation, resource constraints, and competing priorities complicating the implementation of technically sound solutions. The most promising approaches for addressing these challenges involve integrated frameworks combining multiple processes, disciplines, and stakeholder perspectives—essential for developing a comprehensive understanding and implementing contextually appropriate management strategies.

Sustainable groundwater management in this region requires differentiated strategies that acknowledge spatial heterogeneity, with coastal areas focusing on seawater intrusion, agricultural zones emphasizing contamination control, and desert fringes prioritizing balanced development. The substantial uncertainties in future conditions necessitate adaptive management approaches capable of adjusting to emerging trends and unexpected developments. Strategic expansion of monitoring networks represents a high-priority investment, potentially enhancing model reliability and enabling more effective early warning systems for emerging threats. Additionally, management initiatives must integrate surface water considerations, land-use planning, agricultural policy, and coastal zone management through robust institutional mechanisms that coordinate across these domains.

Moving forward, the research community should prioritize not only refinement of technical modeling approaches but also enhanced cross-disciplinary integration, improved stakeholder engagement, and greater attention to implementation pathways. By combining sophisticated technical understanding with practical implementation strategies and inclusive stakeholder processes, researchers can contribute more effectively to sustainable groundwater management in the Nile Delta—one of the world's most significant and vulnerable aquifer systems. The future of this region and its population depends significantly on how effectively its groundwater resources are managed in the coming decades.

10.7 Recommendations

10.7.1 Future Research

Future groundwater research in the Nile Delta should prioritize methodological innovation through the development of coupled socio-hydrological frameworks that explicitly represent feedback between human activities and groundwater conditions. Researchers should emphasize comprehensive uncertainty quantification, moving beyond parameter sensitivity to embrace robust decision-making approaches that identify management strategies performing adequately across multiple plausible futures. Field-scale experimental studies examining managed aquifer recharge effectiveness under local hydrogeological conditions would provide essential implementation guidance beyond theoretical assessments. Real-time monitoring techniques integrating remote sensing, advanced sensors, and machine learning algorithms should be developed to enable early detection of emerging threats, particularly for seawater intrusion and contaminant transport. Process-based studies of emerging contaminants—including pharmaceuticals, personal care products, and agricultural chemicals—deserve particular attention given their potential long-term health implications. Critical knowledge gaps concerning offshore groundwater dynamics, submarine discharge processes, and coastal aquifer responses to sea-level rise require targeted research programs combining geophysical surveys with numerical modeling to better constrain boundary conditions.

10.7.2 Policy Planners

Policy planners must develop spatially differentiated regulatory frameworks acknowledging the Nile Delta's hydrogeological diversity, and implementing stricter extraction controls in vulnerable coastal zones while promoting managed aquifer recharge in suitable inland areas. A comprehensive groundwater monitoring strategy should be established with clear institutional responsibilities, standardized protocols, and sufficient funding for sustainable operation, addressing critical spatial gaps, particularly in eastern and western delta margins. Regulatory approaches should transition from static extraction permits toward adaptive allocation systems that respond to changing groundwater conditions, potentially incorporating economic instruments such as tiered pricing structures or tradable extraction rights. Coordination mechanisms must be formalized between water resource agencies, agricultural ministries, urban planning authorities, and environmental protection bodies to facilitate integrated management of surface and groundwater resources. Long-term national water security planning should incorporate scenario-based assessment of transboundary developments, particularly factoring potential Nile flow alterations from Ethiopian Renaissance Dam operations into groundwater management strategies. Policy frameworks should explicitly acknowledge climate change impacts, mandating periodic reassessment of management approaches against updated projections and providing institutional flexibility for adaptive responses.

10.7.3 Decision-Making

Decision-makers should adopt structured decision-making processes that explicitly acknowledge deep uncertainty in future conditions, employing adaptive management frameworks that establish clear monitoring triggers for management adjustments. Multi-criteria decision analysis techniques can help balance competing objectives, economic development, food security, environmental protection, and social equity while clarifying trade-offs among stakeholders with diverse interests in groundwater resources. Critical decisions about infrastructure investments, particularly irrigation system modernization and groundwater recharge facilities, should incorporate comprehensive cost–benefit analysis, including long-term sustainability values rather than focusing solely on immediate economic returns. Technical decision support systems should be developed that translate complex modeling outputs into accessible visualization tools, enabling policymakers to explore scenario outcomes and management alternatives interactively. Decisions affecting groundwater sustainability should integrate traditional hydrological criteria with broader socioeconomic considerations, including population vulnerabilities, agricultural livelihood dependencies, and gender-differentiated water access implications. Emergency response frameworks should be established for acute groundwater crises such as contamination events or saltwater breakthroughs, with clearly defined institutional responsibilities and decision protocols.

10.7.4 Stakeholders

Meaningful stakeholder engagement requires establishing multilevel participatory platforms where farmers, urban water authorities, industries, and environmental organizations can contribute meaningfully to groundwater management decisions through structured dialogue processes. Capacity-building programs should be developed for local water user associations, providing technical training in groundwater monitoring techniques, water quality testing, and basic hydrological principles to enable informed participation in management decisions. Knowledge translation initiatives should systematically communicate research findings to diverse audiences through appropriate channels, developing simplified models, visualization tools, and practical guidance documents tailored to stakeholder needs. Agricultural extension services should be strengthened to promote water-efficient farming practices, appropriate crop selection for different hydrogeological zones, and safe management of agricultural inputs to minimize groundwater contamination. Industrial stakeholders should be engaged through regulatory compliance assistance programs combined with collaborative pollution prevention initiatives focusing on cleaner production technologies and wastewater reuse opportunities. Community-based monitoring networks can complement official systems while building local ownership of groundwater management processes, particularly in underserved areas where institutional monitoring coverage remains limited.

References

Abd El Hamid A, Tahael Mahmoudi AS, El-Haddad UAEIM (2005) Geoelectrical exploration for groundwater around the new communities, the east Nile Delta. In: The 18th EEGS symposium on the application of geophysics to engineering and environmental problems

Abd El-Mogheeth SM, Sallouma MK (1988) Some hydrological and hydrochemical data for land reclamation activities east of the Nile Delta, Egypt. Bull, Fac Sci, Mansoura Univ 15(2):227–244

Abdel Baki AA, Atwa SM, Haiba MFA (1995) Water chemistry and pollution of groundwater in some new communities west and east Nile delta, Egypt. J Environ Sci 9:127–153

Abd-Elaty I, Abd-Elmoneem SM, Abdelaal GM, Vrana J, Vranayova Z, Abd-Elhamid HF (2022) Groundwater quality modeling and mitigation from wastewater used in irrigation, a case study of the Nile Delta aquifer in Egypt. Int J Environ Res Public Health 19(22):14929. https://doi.org/10.3390/ijerph192214929

Abd-Elaty I, Abdoulhalik A, Ahmed A (2023) The impact of future hydrology stresses and climate change on submarine groundwater discharge in arid regions: A case study of the Nile Delta aquifer, Egypt. J Hydrol: RegNal Stud 47:101395. https://doi.org/10.1016/j.ejrh.2023.101395

Abd-Elaty I, Javadi AA, Abd-Elhamid HF, Abd-Elhamid HF (2021a) Management of saltwater intrusion in coastal aquifers using different well systems: a case study of the Nile Delta aquifer in Egypt. Hydrogeol J 29(5):1767–1783. https://doi.org/10.1007/s10040-021-02344-w

Abd-Elaty I, Kuriqi A, Ramadan EM, Ahmed AA (2024) Hazards of sea level rise and dams built on the River Nile on the water budget and salinity of the Nile Delta aquifer. J Hydrol: RegNal Stud 51:101600. https://doi.org/10.1016/j.ejrh.2023.101600

Abd-Elaty I, Pugliese L, Zelenakova M, Mésároš P, El Shinawi A (2020) Simulation-based solutions reducing soil and groundwater contamination from fertilizers in arid and semi-arid regions: Case study the Eastern Nile Delta, Egypt. Int J Environ Res Public Health 17(24):9373. https://doi.org/10.3390/IJERPH17249373

Abd-Elaty I, Said AM, Abdelaal GM, Zeleňáková M, Jandora J, Abd-Elhamid HF (2021b) Assessing the impact of lining polluted streams on groundwater quality: A case study of the Eastern Nile Delta aquifer. Egypt. Water 13(12):1705. https://doi.org/10.3390/W13121705

Abd-Elhamid HF (2017) Investigation and control of seawater intrusion in the Eastern Nile Delta aquifer considering climate change. Water Sci & Technol: Water Supply 17(2):311–323. https://doi.org/10.2166/WS.2016.129

Abd-Elhamid HF, Javadi AA (2008) An investigation into control of saltwater intrusion considering the effects of climate change and sea level rise. In: Proceedings of 20th saltwater intrusion meeting, pp 4–7

Abd-Elhamid HF, Abd-Elaty I, Sherif M (2019) Evaluation of the potential impact of the Grand Ethiopian Renaissance Dam on seawater intrusion in the Nile Delta aquifer. Int J Environ Sci Technol 16(5):2321–2332. https://doi.org/10.1007/S13762-018-1851-3

Abd-Elhamid HF, Javadi AA, Abd-Elaty I, Sherif M (2016a) Simulation of seawater intrusion in the Nile Delta aquifer under the conditions of climate change. Hydrol Res 47(6):1198–1210. https://doi.org/10.2166/NH.2016.157

Abd-Elhamid HF, Javadi AA, Abd-Elaty I, Sherif M (2016b) Simulation of seawater intrusion in the Nile Delta aquifer under the conditions of climate change. Hydrol Res 47(6):1198–1210

Abd-Elmaboud ME, Abdel-Gawad HAA, El-Alfy KS, Ezzeldin MM (2021) Estimation of groundwater recharge using simulation-optimization model and cascade forward ANN at East Nile Delta aquifer, Egypt. J Hydrol: RegNal Stud 34:100784. https://doi.org/10.1016/J.EJRH.2021.100784

Abd-Elziz YM, Al Shenawy AS, El Tahlawi MR (2022) Effects of irrigation canal rehabilitation on land and crop yields in the Nile Delta using SEEP/W model. Water Science 36(1):85–97. https://doi.org/10.1080/23570008.2022.2025557

Abdulaziz AM, Faid AM (2013) Evaluation of the groundwater resources potential of Siwa Oasis using three-dimensional multilayer groundwater flow model, Mersa Matruh Governorate, Egypt. Arab J Geosci 8(2):1–17. https://doi.org/10.1007/s12517-013-1199-4

Abdulaziz AM, Hurtado JJM, Al-Douri R (2009) Application of multitemporal Landsat data to monitor land cover changes in the Eastern Nile Delta region, Egypt. Int J Remote Sens 30(11):2977–2996. https://doi.org/10.1080/01431160802558675

Abdulla FA, Al-Khatib MA, Al-Ghazzawi ZD (2000) Development of groundwater modeling for the Azraq Basin. Jordan. Environ Geol 40(1–2):11–18. https://doi.org/10.1007/s002540000154

Abdulla F, Al-Assa'd T (2006) Modeling of groundwater flow for the Mujib aquifer, Jordan. J Earth Syst Sci 115(3):289–297. https://doi.org/10.1007/BF02702043

Abo-El-Fadl MM (2013) Possibilities of groundwater pollution in some areas, East of Nile Delta, Egypt. Int J Environ 1(1):1–21

Akbarpour M, Niksokhan MH (2018) Integrated assessment of climate change, urbanization, and sea level rise impacts on groundwater in the coastal aquifer of the Nile delta. Environ Monit Assess 190(11):651. https://doi.org/10.1007/s10661-018-7026-3

Alao JO, Ayejoto DA, Fahad A, Mohammed MAA, Saqr AM, Joy AO (2024) Environmental burden of waste generation and management in Nigeria. In: Souabi S, Anouzla A (eds) Technical landfills and waste management. Springer Water. Springer, Cham, pp 27–56. https://doi.org/10.1007/978-3-031-55665-4_2

Al-Nimer AA, Gad MI, Hussein EE (2020) Numerical simulation of groundwater flow and contaminant transport in the Quaternary aquifer in the 10th of Ramadan city area, East Delta, Egypt. Bull Fac Eng, Mansoura Univ 40(5):83–99. https://doi.org/10.21608/BFEMU.2020.104971

Anderson MP, Woessner WW, Hunt RJ (2015) Applied groundwater modeling: simulation of flow and advective transport. Academic Press

Armanuos AM, Negm AM (2018) Integrated groundwater modeling for simulation of saltwater intrusion in the Nile Delta aquifer, Egypt. In: Negm AM (ed) Groundwater in the Nile Delta, pp 489–544. Springer, Cham. https://doi.org/10.1007/698_2017_184

Armanuos AM, Ahmed K, Shiru MS, Jamei M (2021) Impact of increasing pumping discharge on groundwater level in the Nile Delta aquifer, Egypt. Knowl Based Eng Sci 2(2):13–23. https://doi.org/10.51526/KBES.2021.2.2.13-23

Armanuos AM, Ibrahim MG, Mahmod WE, Negm AM, Yoshimura C, Takemura J, Zidan BA (2017) Evaluation of the potential impact of the Grand Ethiopian Renaissance Dam and pumping scenarios on groundwater level in the Nile Delta aquifer. Water Sci & Technol: Water Supply 17(5):1356–1367. https://doi.org/10.2166/WS.2017.037

Armanuos AM, Jonoski A, Oude Essink GHP, Saavedra OC (2016a) Estimation of bed and bank levels of an irrigation canal network towards accurate groundwater modeling of the Nile delta aquifer. J Hydrol: Reg Stud 6(1):74–83. https://doi.org/10.1016/j.ejrh.2016.03.001

Armanuos AM, Negm AM, Taha M, Zeidan BA (2023) Impact of climate change on seawater intrusion in the Nile Delta aquifer (Egypt). In: Negm AM (ed) Water-energy-food nexus in the Arab region, pp 97–163. Springer Nature. https://doi.org/10.1007/698_2023_1062

Armanuos AM, Negm AM, Yoshimura C, Saavedra Valeriano OC (2016b) Application of WetSpass model to estimate groundwater recharge variability in the Nile Delta aquifer. Arab J Geosci 9(10):553. https://doi.org/10.1007/S12517-016-2580-X

Aziz MA, Elmoustafa AM, El Alfy M (2019) Assessment of the potential impacts of the Grand Ethiopian Renaissance Dam on water resources and soil salinity in the Nile Delta, Egypt. Environ Socio-Econ Stud 7(1):1–13. https://doi.org/10.2478/environ-2019-0001

Bear J, Cheng AHD, Sorek S, Ouazar D, Herrera I (1999) Seawater intrusion in coastal aquifers: concepts, methods and practices. Springer Science & Business Media

Blanco-Coronas AM, Martinez-Moreno FJ, Martos-Rosillo S, Díaz-Peña Á, Rubio-Campos JC (2021) Contribution of hydrogeophysics to the study of the saline intrusion in coastal aquifers: A case study in the Nile Delta. J Afr Earth Sc 175:104107. https://doi.org/10.1016/j.jafrearsci.2021.104107

Dawoud MA, Raouf ARA (2009) Groundwater exploration and assessment in rural communities of Yobe State, Northern Nigeria. Water Resour Manag 23(3):581–601. https://doi.org/10.1007/s11269-008-9289-x

Dawoud MA, Darwish MM, El-Kady MM (2005) GIS-based groundwater management model for Western Nile Delta. Water Resour Manage 19(5):585–604. https://doi.org/10.1007/s11269-005-5603-z

Diab MS, Dahab K, El-Fakharany M (1997) Impact of the present drinking water consumption on the groundwater quantity in the Nile Delta aquifer system. In: The third conference of geology, mineral and water resources in the Arab world, Cairo University

El Alfy M, Lashin A, Abdalla F, Al-Bassam A (2015) Assessing the hydrogeochemical processes affecting groundwater pollution in arid areas using an integration of geochemical equilibrium and multivariate statistical techniques. Environ Pollut 198:144–156. https://doi.org/10.1016/j.envpol.2015.01.008

El Arabi N (2012) Environmental management of groundwater in Egypt via artificial recharge, extending the practice to soil aquifer treatment (SAT). Int J Environ Sustain 1(3):66–82. https://doi.org/10.24102/ijes.v1i3.96

El Molla AM, Dawoud MA, Hassan SS (2005) Management scenarios for increasing water use efficiency in the Western Nile Delta, Egypt. In: International commission on irrigation and drainage 19th congress, Beijing, China

El-Agha DE, Closas A, Molle F (2017) Below the radar: The boom of groundwater use in the central part of the Nile Delta in Egypt. Hydrogeol J 25(6):1621–1631. https://doi.org/10.1007/s10040-017-1570-8

El-Agha DE, Molle F, Rap E, El Hassan WA, Abou El Hassan W (2015) Modelling groundwater use in the central part of the Nile Delta. Int Water Manag Inst Res Rep 166:1–42. https://doi.org/10.5337/2016.203

El-Bihery MA (2009) Groundwater flow modeling of Quaternary aquifer Ras Sudr, Egypt. Environ Earth Sci 58(5):1095–1105. https://doi.org/10.1007/S00254-008-1589-1

El-Hamidi MJ, Subyani AM, Shephard WD (2018) Hydrologic and land cover changes in the Nile Delta using remote sensing techniques. J Geogr Geol 10(2):85–95. https://doi.org/10.5539/jgg.v10n2p85

El-Nimer AA, Gad MI, Hussein EE (2020) Numerical simulation of groundwater flow and contaminant transport in the Quaternary aquifer in 10th of Ramadan city area, East Delta, Egypt. Bull Fac Eng, Mansoura Univ 40(5):83–99. https://doi.org/10.21608/BFEMU.2020.104971

El-Raey M (2010) Impact of sea level rise on the Arab region. Arab Climate Initiative/UNDP Regional Bureau for Arab States. https://www.preventionweb.net/files/12218_ImpactsofSLRontheArabRegion.pdf

Eltarabily MG, Negm AM (2017) Groundwater management for sustainable development East of the Nile Delta aquifer. In: Negm AM (ed) The Nile Delta, pp 687–708. Springer, Cham. https://doi.org/10.1007/698_2017_102

Eltarabily MG, Negm AM (2018) Groundwater management for sustainable development plans for the Western Nile Delta. In: Negm AM (ed) Conventional water resources and agriculture in Egypt, pp 709–727. Springer, Cham. https://doi.org/10.1007/698_2018_247

Eltarabily MG, Abd-Elaty I, Elbeltagi A, Zeleňáková M, Fathy I (2023) Investigating climate change effects on evapotranspiration and groundwater recharge of the Nile Delta aquifer. Egypt. Water 15(3):572. https://doi.org/10.3390/w15030572

Eltarabily MG, Abd-Elaty I, Yoshimura C, Takemura J (2018) Groundwater modeling in agricultural watershed under different recharge and discharge scenarios for Quaternary aquifer Eastern Nile Delta, Egypt. Environ Model & Assess 23(3):289–308. https://doi.org/10.1007/S10666-017-9577-Z

Eltarabily MG, Eltarabily MG, Negm AM, Saavedra O (2017) Modeling the effect of nitrate fertilizers on groundwater quality in the Southern part of the Nile Delta, Egypt. In: Negm AM (ed) The Nile Delta, pp 265–293. Springer, Cham. https://doi.org/10.1007/698_2017_114

Embaby AA, El Sabri MASH, Rayan RA (2017) Using Modflow as a management tool for the groundwater system in the area Northeastern Cairo, Egypt. IOSR J Appl Geol Geophys 05(02):78–88. https://doi.org/10.9790/0990-0502017888

Engelen J, Verkaik J, King J, Nofal ER, Bierkens MFP, Oude Essink GHP (2019) A three-dimensional palaeo-reconstruction of the groundwater salinity distribution in the Nile Delta aquifer. Hydrol Earth Syst Sci 23:5175–5198. https://doi.org/10.5194/HESS-2019-151

Ezzeldin MM, El-Alfy KS, Abdel-Gawad HAA, Abd-Elmaboud ME (2018) Comparison between structured and unstructured MODFLOW for simulating groundwater flow in three-dimensional multilayer Quaternary aquifer of East Nile Delta, Egypt. J Hydrol & Hydromech 9(1):1–12. https://doi.org/10.4172/2157-7587.1000297

Ezzeldin R, Abd-Elmaboud M (2024) Modeling flow resistance and geometry of dunes bed form in alluvial channels using hybrid RANN–AHA and GEP models. Int J Sedim Res. https://doi.org/10.1016/j.ijsrc.2024.08.002

Famiglietti JS (2014) The global groundwater crisis. Nat Clim Chang 4(11):945–948

Farid MS (1985) Management of groundwater system in the Nile Delta. Ph.D. Thesis, Faculty of Engineering, Cairo University, Egypt

Ghoraba SM, Zyedan BA, Rashwan IMH (2013) Solute transport modeling of the groundwater for quaternary aquifer quality management in Middle Delta, Egypt. Alex Eng J 52(2):197–207. https://doi.org/10.1016/J.AEJ.2012.12.007

Ghouili N, Horriche FJ, Zammouri M, Benabdallah S, Farhat B (2017) Coupling WetSpass and MODFLOW for groundwater recharge assessment: case study of the Takelsa multilayer aquifer, northeastern Tunisia. Geosci J 21(5):791–805

Gomaa SS, Fleskens L, Nunes JP, Badr M (2024) An integrated modelling approach to support sustainable water resources management and climate change adaptation for irrigated agriculture in the Nile Delta, Egypt. EGUsphere, preprint. https://doi.org/10.5194/egusphere-egu24-693

Hashish HA, El-Ghandour HA, El-Nimr AA (2014) Simulating the effects of spatial layout and pumping/recharging rates of wells on saltwater intrusion. Ain Shams Eng J 5(2):581–592. https://doi.org/10.1016/j.asej.2013.11.006

Hereher ME (2010) Vulnerability of the Nile Delta to sea level rise: an assessment using remote sensing. Geomat, Nat Hazards Risk 1(4):315–321. Abd El Hamid A, Tahael Mahmoudi AS, El-Haddad UAEIM (2005) Geoelectrical exploration for groundwater around the new communities, the east Nile Delta. In: the 18th EEGS symposium on the application of geophysics to engineering and environmental problems

Hugman R, Stigter TY, Monteiro JP, Nunes L (2017) Influence of aquifer properties and the spatial and temporal distribution of recharge and abstraction on sustainable yields in semi-arid regions. Hydrol Process 31(7):1392–1404. https://doi.org/10.1002/hyp.11111

Hussein EE, Gad MI, El-Nimer AA (2019) Predicting solute transport parameters for Quaternary aquifer at 10th of Ramadan city area, using sand box experiments. Bull Fac Eng Mansoura Univ 17 (1):103–114. https://doi.org/10.21608/bfemu.2019.74765

Karadacs D (2009) The development of a regional groundwater flow model for the Tahtali watershed. DEÜ Fen Bilimleri Enstitüsü

Kashef AI (1983) Salt-water intrusion in the Nile Delta. Groundwater 21(2):160–167. https://doi.org/10.1111/j.1745-6584.1983.tb00713.x

Kumar CP (2002) Groundwater flow models. Scientist 'E1' National Institute of Hydrology Roorkee-247667 (Uttaranchal) Publication

Mabrouk M, Jonoski A, Oude Essink GHP (2019a) Regional groundwater modelling for determining adaptation strategies in the Nile delta aquifer. In: 38th IAHR world congress: Panama City, Panama, September 1–6, 2019. https://doi.org/10.3850/38WC092019-1397

Mabrouk M, Jonoski A, Oude Essink GHP, Uhlenbrook S (2018) Impacts of sea level rise and groundwater extraction scenarios on fresh groundwater resources in the Nile Delta governorates, Egypt. Water 10(11):1690. https://doi.org/10.3390/W10111690

Mabrouk M, Jonoski A, Oude Essink GHP, Uhlenbrook S (2019b) Assessing the fresh-saline groundwater distribution in the Nile delta aquifer using a 3D variable-density groundwater flow model. Water 11(9):1946. https://doi.org/10.3390/W11091946

Masria A, Alshammari TO, Ghareeb M, Seif AK, Abd-Elmaboud ME, Ammar AI (2024) 2D and 3D Modeling Of Resistivity And Chargeability To Identify The Type Of Saturated Groundwater For Complex Sedimentary Facies. Hydrology 11(8). https://doi.org/10.3390/hydrology11080120

Masria A, Seif AK, Ghareeb M, Abd-Elmaboud ME, Soliman K, Ammar AI (2023) Numerical modeling of vadose zone electrical resistivity to evaluate its hydraulic parameters. Appl Water Sci 13(12). https://doi.org/10.1007/s13201-023-02024-y

Masterson JP, Fienen MN, Thieler ER, Gesch DB, Gutierrez BT, Plant NG (2007) Effects of sea-level rise on barrier island groundwater flow and saltwater intrusion. USGS Sci Investig Rep 2007-5052. https://doi.org/10.3133/sir20075052

Metwally UAA, Al Saud WA, Hassan HH (2022) Groundwater vulnerability assessment using different DRASTIC vulnerability models: A case study from the central Nile delta, Egypt. Environmental Earth Sciences 81:177. https://doi.org/10.1007/s12665-022-10320-0

Mumtaz M, Jahanzaib SH, Hussain W, Khan S, Youssef YM, Qaysi S, Abdelnabi A, Alarifi N, Abd-Elmaboud ME (2025) Synergy of remote sensing and geospatial technologies to advance sustainable development goals for future coastal urbanization and environmental challenges in a riverine megacity. ISPRS Int J Geo-Inf, 14(1). https://doi.org/10.3390/ijgi14010030

Nofal ER, Amer M, El-Didy SM, Fekry AM (2015) Delineation and modeling of seawater intrusion into the Nile Delta aquifer: A new perspective. Water Science 29(2):156–166. https://doi.org/10.1016/J.WSJ.2015.11.003

Nofal ER, Eid AE, Morsy W, Al-Sittawy M (2022) Assessing the impacts of canal rehabilitation and lining on groundwater conditions. Water Science 38(1):348–358. https://doi.org/10.1080/23570008.2024.2361972

Nofal ER, Fekry A, Ahmed MH (2018) Decision support system for the groundwater extraction management in the Nile delta aquifer. J Groundw Sci Eng, 6(2):138–147. https://doi.org/10.19637/j.cnki.2305-7068.2018.02.006

Nofal ER, Mostafa HF, van Lanen HA (2016) Groundwater management options in Nile Delta aquifer in Egypt under different scenarios of sea level rise: a case study. J Afr Earth Sc 124:254–267. https://doi.org/10.1016/j.jafrearsci.2016.09.029

Noman M, Ullah I, Khan MA, Qazi A, Farooq W, Saqr A, Elsheikh A (2024) Analysis of overcurrent protective relaying as minimum adopted fault protection for small-scale hydropower plants. Int J Environ Sci Technol 21(4):4457–4470. https://doi.org/10.1007/s13762-023-05284-y

Panday S, Langevin CD, Niswonger RG, Ibaraki M, Hughes JD (2013) MODFLOW-USG version 1: An unstructured grid version of MODFLOW for simulating groundwater flow and tightly coupled processes using a control volume finite-difference formulation. U.S. Geological Survey

Paudel G, Pant RR, Joshi TR, Saqr AM, Đurin B, Cetl V, Kamble PN, Bishwakarma K (2024) Hydrochemical dynamics and water quality assessment of the ramsar-listed Ghodaghodi lake complex : Unveiling the water-environment nexus. Water 16(3373):1–23. https://doi.org/10.3390/w16233373

Rehman A, Xue L, Islam F, Ahmed N, Qaysi S, Liu S, Alarifi N, Youssef YM, Abd-Elmaboud ME (2024) Unveiling groundwater potential in Hangu District, Pakistan: A GIS-driven bivariate modeling and remote sensing approach for achieving SDGs. Water (Switz) 16(22). https://doi.org/10.3390/w16223317

Rodell M, Famiglietti JS, Wiese DN, Reager JT, Beaudoing HK, Landerer FW, Lo MH (2018) Emerging trends in global freshwater availability. Nature 557(7707):651–659

Salem ZES, Al Temamy AM, Salah MK, Kassab M (2019) Origin and characteristics of brackish groundwater in Abu Madi coastal area, Northern Nile Delta, Egypt. Estuar, Coast Shelf Sci 218:342–358. https://doi.org/10.1016/j.ecss.2019.01.005

Saqr AM, Abd-Elmaboud ME (2024) Management of saltwater intrusion in Coastal Aquifers: A review and case studies from Egypt. Online J Eng Sci 3(1):1–17. https://doi.org/10.31586/ojes.2024.982

Saqr AM, Ibrahim MG, Fujii M, Nasr M (2022) Simulation-optimization modeling techniques for groundwater management and sustainability: A critical review. Adv Eng Forum 47:89–100. https://doi.org/10.4028/p-5011j1

Saqr AM, Nasr M, Fujii M, Yoshimura C, Ibrahim MG (2025) Integrating MODFLOW-USG and Walrus optimizer for estimating sustainable groundwater pumping in arid regions subjected to severe drawdown. In: Ali S, Negm A (eds) Groundwater in developing countries. Springer Water. Springer, Cham. https://doi.org/10.1007/978-3-031-79122-2_5

Sathiyamurthi S, Youssef YM, Gobi R, Ravi A, Alarifi N, Sivasakthi M, Praveen Kumar S, Dąbrowska D, Saqr AM (2025) Optimal land selection for agricultural purposes using hybrid geographic information system–fuzzy analytic hierarchy process–geostatistical approach in attur Taluk, India: Synergies and trade-offs among sustainable development goals. Sustainability (Switzerland), 17(3). https://doi.org/10.3390/su17030809

Subbarayan S, Youssef YM, Singh L, Dąbrowska D, Alarifi N, Ramsankaran RAAJ, Visweshwaran R, Saqr AM (2025) Soil and water assessment tool-based prediction of runoff under scenarios of land use/land cover and climate change across Indian Agro-climatic zones: Implications for sustainable development goals. Water (Switzerland) 17(3). https://doi.org/10.3390/w17030458

Scanlon BR, Faunt CC, Longuevergne L, Reedy RC, Alley WM, McGuire VL, McMahon PB (2012) Groundwater depletion and sustainability of irrigation in the US High Plains and Central Valley. Proc Natl Acad Sci 109(24):9320–9325

Sefelnasr A, Sherif M (2014) Impacts of seawater rise on seawater intrusion in the Nile Delta aquifer, Egypt. Ground Water 52(2):264–276. https://doi.org/10.1111/GWAT.12058

Shaban A, Khawlie M, Abdallah C (2006) Use of remote sensing and GIS to determine recharge potential zones: the case of Occidental Lebanon. Hydrogeol J 14(4):433–443

Shalaby A, Tateishi R (2007) Remote sensing and GIS for mapping and monitoring land cover and land-use changes in the Northwestern coastal zone of Egypt. Appl Geogr 27(1):28–41. https://doi.org/10.1016/j.apgeog.2006.09.004

Sherif MM, Al-Rashed MF (2001) Vertical and horizontal simulation of seawater intrusion in the Nile Delta aquifer. In: Proceedings of the 1st international conference on saltwater intrusion and Coastal Aquifers, pp 23–25

Sherif MM, Hamza KI (2001) Mitigation of seawater intrusion by pumping brackish water. Transp Porous Media 43(1):29–44. https://doi.org/10.1023/A:1010601208708

Sherif MM, Singh VP (1999a) Effect of climate change on sea water intrusion in coastal aquifers. Hydrol Process 13(8):1277–1287. https://doi.org/10.1002/(SICI)1099-1085(19990615)13:8%3c1277::AID-HYP765%3e3.0.CO;2-W

Sherif MM, Singh VP (1999b) Effect of climate change on sea water intrusion in coastal aquifers. Hydrol Process 13(8):1277–1287

Sherif MM, Singh VP, Amer AM (1988) A two-dimensional finite element model for dispersion (2D-FED) in coastal aquifers. J Hydrol 103(1–2):11–36. https://doi.org/10.1016/0022-1694(88)90003-0

Sherif M, Sefelnasr A, Javadi A (2012) Incorporating the concept of equivalent freshwater head in successive horizontal simulations of seawater intrusion in the Nile Delta aquifer, Egypt. J Hydrol 464–465:186–198. https://doi.org/10.1016/j.jhydrol.2012.07.007

Sobeih MM, El-Arabi NE, Helal EEDY, Awad BS (2017) Management of water resources to control groundwater levels in the southern area of the western Nile delta. Egypt. Water Science 31(2):137–150. https://doi.org/10.1016/J.WSJ.2017.09.001

Soliman MM, LaMoreaux PE, Memon BA, Assaad FA (2014) Environmental hydrogeology. CRC Press. https://doi.org/10.1201/9781482293876

Stanley DJ, Warne AG (1998) Nile Delta in its destruction phase. J Coast Res 14(3):795–825. https://www.jstor.org/stable/4298830

Taylor RG, Scanlon B, Döll P, Rodell M, Van Beek R, Wada Y, Longuevergne L, Leblanc M, Famiglietti JS, Edmunds M, Konikow L (2013) Ground water and climate change. Nat Clim Chang 3(4):322–329

Timalsina R, Acharya S, Đurin B, Awasthi MP, Pant RR, Joshi GR, Byanju RM, Panthi KP, Joshi S, Kumar A, Thakur TK, Saqr AM (2025a) An assessment of seasonal water quality in Phewa Lake, Nepal, by integrating geochemical indices and statistical techniques: A sustainable approach. Water (Switzerland) 17(2). https://doi.org/10.3390/w17020238

Timilsina B, Kharel M, Shrestha R, Saqr AM, Pant RR (2025b) Assessment of soil quality along the elevation gradient of the Seti River watershed in Pokhara Metropolitan City, Nepal. Environ Monit Assess 197(3). https://doi.org/10.1007/s10661-025-13716-0

Wassef R, Schüttrumpf H (2016a) Impact of sea-level rise on groundwater salinity at the development area western delta, Egypt. Groundw Sustain Dev 2:85–103. https://doi.org/10.1016/J.GSD.2016.06.001

Wassef R, Schüttrumpf H (2016b) Impact of sea-level rise on groundwater salinity at the development area western delta. Egypt. Groundw Sustain Dev 2:85–103

Woessner WW, Anderson MP (1990) Setting calibration targets and assessing model calibration—room for improvement: an example from North America. ModelCARE'90:279–288

Xiao H, Wang D, Hagen SC, Medeiros SC, Hall CR (2016) Assessing the impacts of sea-level rise and precipitation change on the surficial aquifer in the low-lying coastal alluvial plains and barrier islands, east-central Florida (USA). Hydrogeol J 24(7):1791–1806. https://doi.org/10.1007/s10040-016-1437-4

11 Monitoring Irrigation Water Quality for Sustainable Agriculture in the Northeastern Nile Delta, Egypt

Manal A. Alnaimy, Khalid G. Soliman, Nabih A. Atia, El-Sayed A. El-Naka, and Marmar Mabrouk

Abstract

Ensuring sustainable irrigation water quality is vital for global food security, as agriculture relies on safe and efficient water management to support crop production and soil health. With increasing water scarcity and contamination risks worldwide, monitoring and optimizing irrigation water quality play a crucial role in sustaining agricultural productivity, protecting ecosystems, and adapting to climate change. The El-Salam Canal, a crucial irrigation project in Egypt, facilitates agricultural expansion by blending drainage water with Nile water to reclaim and cultivate new farmland while establishing a key developmental axis. This chapter adopts an integrative approach, combining Geographic Information Systems (GIS) to evaluate water quality in the canal and its associated drains—an essential factor in enhancing food security. The analysis revealed that pH levels in irrigation water samples ranged from 7.71 to 8.03, remaining within normal limits, while electrical conductivity (EC) varied from 0.46 to 2.62 dS/m, reflecting differences in salinity. Alkalinity indicators showed the following ranges: Sodium Soluble Percentage (SSP): 30.22–51.22%, Sodium Adsorption Ratio (SAR): 1.29–5.72 mmolc/L, adjusted SAR: 2.49–14.61 mmolc/L, adjusted R Na: 1.38–6.30 mmolc/L, and Exchangeable Sodium Percentage (ESP): 0.65–7.25%. Additionally, trace element concentrations in the water remained below recommended thresholds, and the permeability index averaged 62%, indicating suitable irrigation

M. A. Alnaimy (✉) · K. G. Soliman · N. A. Atia · E.-S. A. El-Naka
Soil Science Department, Faculty of Agriculture, Zagazig University, Zagazig, Egypt
e-mail: dr_manalalnaimy@yahoo.com

M. Mabrouk
Nile University, Giza, Egypt

M. A. Khalifa and K. S. Gemail (eds.), *Geology of the Nile Deltas and Natural Resources*, Deltas of the World, https://doi.org/10.1007/978-3-031-96594-4_11

conditions. These findings provide valuable insights into the water quality and soil suitability of the El-Salam Canal region, offering essential data for agricultural planning, irrigation management, and resource sustainability—all critical for bolstering Egypt's food security and optimizing land use.

Keywords

El-Salam Canal • Resource sustainability • Nile Delta • Alkalinity indicators • Irrigation management • Water quality

11.1 Introduction

Water quality is a fundamental factor in sustainable agriculture, directly impacting soil health, crop productivity, and global food security. With increasing water scarcity and contamination risks due to climate change and human activities, ensuring safe and efficient irrigation water management is critical for maintaining agricultural sustainability worldwide (Gupta et al. 2021). In Egypt, where water resources are limited and agricultural expansion is essential for food security, the quality of irrigation water plays a crucial role in land reclamation and sustainable farming. The El-Salam Canal project, which blends Nile water with agricultural drainage water, represents a significant effort to maximize water use efficiency and expand cultivable land in the Northeastern Nile Delta (El-Saidi 2002). However, continuous monitoring of water quality is necessary to ensure long-term soil fertility, prevent salinization, and support sustainable agricultural practices in the region (Cipolla et al. 2019).

The Government of Egypt is undertaking a major initiative with the El-Salam Canal project, aimed at reusing drainage water, developing new communities along the canal, and redistributing the population to support sustainable development. The canal is designed to irrigate 600,000 feddans (1 feddan = 1.04 acres, 4,200 m^2), with 200,000 feddans located west of the Suez Canal, spanning the governorates of Damietta, Dakahlia, El-Sharkia, El-Ismailia, and Port Said. The remaining 400,000 feddans lie east of the Suez Canal, covering El-Tena Plain, the Northern Coast, Ser, and Kawarer (Hafez 2005). The El-Salam Canal's water supply is a carefully controlled blend of fresh Nile water and drainage water, ensuring that total dissolved solids (TDS) remain within acceptable limits for irrigation. The canal receives 2.11 billion cubic meters (Bcm) of freshwater per year from the Damietta branch of the Nile, combined with 1.905 Bcm from the Bahr Hadous drain and 0.435 Bcm from the El-Serw drain, totaling 4.45 Bcm annually. To maintain water quality suitable for cultivated crops, the blending ratio of Nile water to drainage water is maintained at 1:1, keeping TDS below 1200 mg/L to support sustainable agricultural expansion (Gupta and Gupta 2015). Across Egypt, the potential for reusing drainage water for irrigation is estimated at 9 billion cubic meters per year, with an average electrical conductivity (EC) of 4.4 dS/m, indicating salinity challenges. Expanding agriculture in the Sinai region requires transporting irrigation water across the Suez Canal, a logistical challenge that the El-Salam Canal addresses. The canal extends 89 km west of the Suez Canal and 86 km east, crossing beneath the Suez Canal via a 1,300-m-long underground siphon, ensuring uninterrupted water flow (Aoubouazza 2003; Hafez 2005).

The quality of irrigation water is a critical factor in determining agricultural productivity and is vital in effective water management (Shaddad et al. 2025). High-quality water supports optimal crop yields, while poor-quality water can lead to soil salinization, impaired plant growth, and reduced productivity. These issues can degrade agricultural land without proper management, necessitating targeted reclamation and rehabilitation efforts (Ramadan et al. 2023). A sustainable irrigation system utilizing saline-sodic water presents opportunities for agricultural expansion while minimizing environmental risks. However, success requires a comprehensive approach, integrating efficient water use, crop selection, and adaptive management practices (Cuevas et al. 2019; Mezlini et al. 2024). The reclamation and rehabilitation of saline and sodic lands should prioritize enhancing water-use efficiency while ensuring long-term environmental sustainability (Elewa et al. 2024, 2023). In this context, land use planning remains a significant challenge, particularly in regions like the eastern Nile Delta, where environmental constraints and developmental needs often conflict. Addressing these issues requires strategic planning and sustainable management practices to optimize land and water resources (Elewa et al. 2024; Qadir and Oster 2004). This study aims to systematically monitor and assess the water quality of the El-Salam Canal system and its branches, which are vital sources of irrigation water for the surrounding region. A total of 26 water samples were collected monthly over one year from the El-Salam Canal and its feeding drains, at consistent points along their course, to provide a comprehensive evaluation of water quality. The primary objective is to identify temporal trends in water quality by analyzing key physicochemical parameters and to determine the water's suitability for agricultural use, thereby promoting sustainable irrigation practices for the region.

11.2 Scientific Background of the Study Area

11.2.1 El-Salam Canal Project Description

The Government of Egypt implemented the El-Salam Canal project to reuse drainage water, establish new communities along the canal, and redistribute the population to support sustainable development. The project aims to divert blended drainage and Nile water to newly reclaimed areas, irrigating approximately 643,560 acres of land in the north-eastern Delta and northern Sinai Peninsula. The government envisions reclaiming around 620,000 feddans of desert along the Mediterranean coast of Sinai by mixing agricultural drainage water with Nile water in a 1:1 ratio. Given Egypt's arid climate, characterized by minimal annual rainfall and high evapotranspiration rates, irrigation is essential for ensuring crop production (Alnaimy et al. 2022; Sassenrath et al. 2008).

While improving canal and field irrigation systems can contribute to water savings, future demands will likely necessitate the use of increasing amounts of lower-quality water (Hafez et al. 2008). To address this, the Ministry of Public Works and Water Resources has developed a strategy for reusing agricultural drainage water, alongside the 7.5 billion cubic meters (bcm) of Nile water currently discharged into the sea each year. The strategy includes expanding agricultural projects in Egypt's upstream regions and returning drainage water from Aswan to Cairo back into the Nile River, where it is reused in the Delta before being discharged into the Rosetta and Damietta branches (Wallace and Batchelor 1997).

Drainage water represents a readily available resource for irrigation, with the Ministry of Irrigation estimating Egypt's total annual drainage water discharge at 12 bcm. Currently, much of this water is disposed of in the Mediterranean Sea and the Northern Delta lakes. However, an intensive expansion program for drainage water reuse in agriculture requires proper precautions to mitigate salinity and alkalinity issues, preventing long-term soil degradation due to salt accumulation. One of Egypt's most promising projects for drainage water reuse is the El-Salam Canal, designed to collect and blend Eastern Delta drainage water with Nile water in a 1:1 ratio to irrigate new lands in northern Sinai. Given the scale of this initiative, monitoring the quality and quantity of drainage water in the Eastern Delta is of paramount importance to ensure the project's long-term sustainability (Mohamed 2013).

The El-Salam Canal project aims to transport irrigation water to the Sinai Peninsula to support horizontal agricultural expansion and reuse drainage water that was previously discharged into the sea. This is achieved by blending drainage water from the Bahr Hadous main drain—the largest drain in the Eastern Delta—and the lower Serw pumping station with freshwater from the Damietta branch of the Nile. The water is mixed in equal proportions, ensuring a balanced fresh-to-drainage water ratio suitable for irrigation. The project is expected to cultivate approximately 620,000 feddans of newly reclaimed land, with 220,000 feddans west of the Suez Canal and 400,000 feddans east of the Suez Canal in the Sinai Peninsula (Othman et al. 2012).

Beyond agriculture, the project seeks to reshape Egypt's population distribution, promote equitable income distribution, and create employment opportunities by settling smallholders and young graduates from overpopulated rural areas. Additionally, transferring Nile water into Sinai will help establish new communities along the canal while unlocking the region's vast economic potential. Sinai is home to valuable natural resources, including medicinal plants, wildlife, and rich mineral deposits, making the project a crucial step toward sustainable development and economic diversification in the region (Dalia et al. 2018).

11.2.2 Physiography and Water Resources of El-Salam Canal

The El-Salam Canal project is located in the northeastern Nile Delta, with a total length of 242 km. It sources its water from the right bank of the Damietta branch of the Nile at km 219 in Adlia City, just before Faraskour Dam, which was designed to regulate the Nile water level at 1.7 m, ensuring a supply of 1.6 m to the canal. The canal extends 89 km eastward from Damietta to the Suez Canal, passing through key hydraulic and drainage systems along its course (Hafez 2005).

The canal follows a southeastern route, passing near the Al-Attawe drainage and reaching the Al-Serw drain at km 17.850, where it collects approximately 2 million cubic meters of drainage water daily to be mixed with El-Salam Canal water. With a hydraulic slope of 5 cm/km, the canal's water level at km 22 aligns with both the natural topography and engineering requirements for efficient flow. The canal continues eastward, running behind pumping stations and parallel to the Al-Taweel Bahari drainage, with a 100-m buffer zone to accommodate filtration (Othman et al. 2012). At km 39.150, the canal intersects with Al-Taweel Bahari drainage, which directs wastewater flow before crossing Bahr Hadous drainage at km 51.200, where two pumping and mixing stations have been constructed. The canal is then diverted to cross Bahr El-Baqar drainage, passing through a culvert at km 19.650 and rejoining El-Salam Canal at km 77.325, continuing eastward until it reaches the Suez Canal at km 27.800. From there, water is transported through a 1,300-m-long underground tunnel beneath the Suez Canal to deliver Nile water to Sinai. In West Sinai, the canal is renamed the Sheikh Gabber Al-Sabah Canal, extending 155 km to irrigate 400,000 feddans of reclaimed agricultural land in Sahl El-Tina and along the coastal areas of Romanna, El-Arish, Ser, and Kawarer. This expansion is a key component of Egypt's efforts to develop agriculture and establish new communities in the Sinai Peninsula (Sassenrath et al. 2008).

11.3 Materials and Methods

11.3.1 Water Sampling

A total of 26 water samples were collected from different locations to assess the water quality in the El-Salam Canal system. These samples included Nile water (Location 1), four main drains—Faraskor Drain (Location 3), El-Atwi Drain (Location 4), El-Serw Drain (Location 7), and Hadous Drain (Location 16)—as well as multiple points along the El-Salam Canal (Locations 2, 5, 6, 8–15, and 17–26). The first three drains discharge their water into the El-Salam Canal, contributing to its overall composition. Water samples were collected monthly from the Damietta branch of the Nile, the drains, and the canal. Figure 11.1 illustrates the sampling locations along the drains and the canal's course.

All collected samples were filtered and subjected to chemical analysis to evaluate their composition and determine their suitability for irrigation.

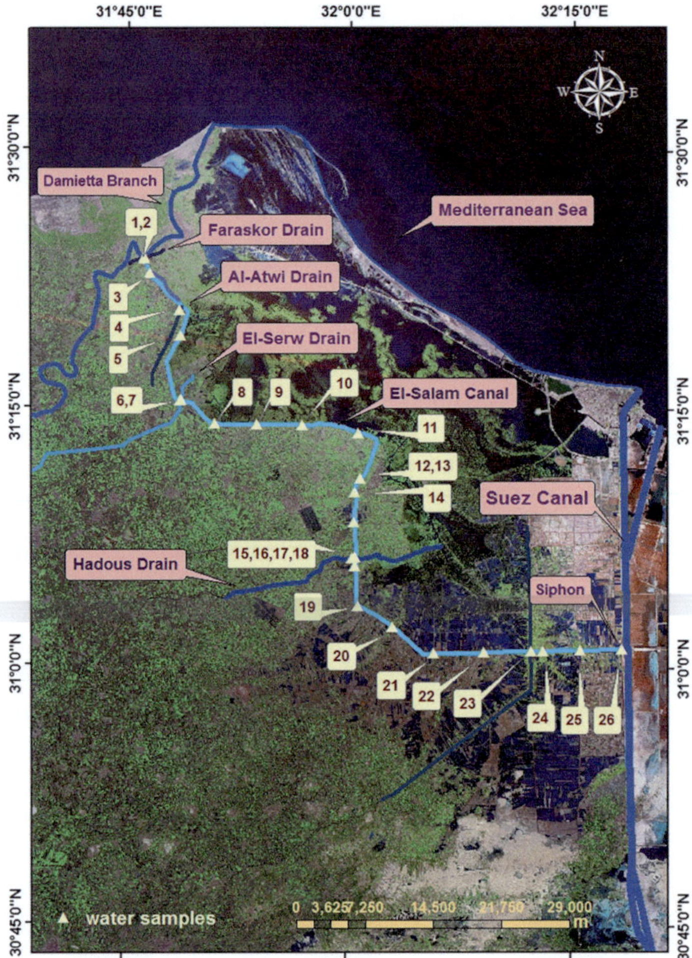

Fig. 11.1 Water sampling locations from El-Salam Canal and contributing drains: The 26 collected water samples include 18 from the El-Salam Canal and 8 from four main drains. The drain samples are distributed as follows: Sample No. 2 from Faraskour Drain, Sample No. 4 from Al-Atwi Drain, Samples Nos. 6 and 7 from El-Serw Drain, and Samples Nos. 15, 16, 17, and 18 from Hadous Drain (Alnaimy 2013)

11.3.2 Water Analysis

A total of 26 water samples were collected along the El-Salam Canal course and analyzed for various physicochemical parameters to assess water quality. Soil reaction (pH) was measured, while total soluble salts were determined using electrical conductivity (EC). Calcium and magnesium concentrations were analyzed using the Versenate method, with ammonium purpurate as an indicator for calcium and Eriochrome Black T for both calcium and magnesium. Sodium and potassium levels were determined photometrically using a Flame photometer, whereas carbonates and bicarbonates were identified through titration, using phenolphthalein as an indicator for carbonates and methyl orange for bicarbonates. Chloride content was measured using Mohr's method, while sulfate concentrations were calculated by subtracting the total soluble anions from the total soluble cations (Pal et al. 2009). Additionally, trace elements were analyzed using single dilution micro-methods with a Perkin-Elmer 290B Atomic Absorption Spectrophotometer (Perkin-Elmer 3300), Norwalk, CT. This comprehensive analysis provides essential insights into the chemical composition of the El-Salam Canal water, ensuring its suitability for irrigation and sustainable agricultural development (Page 1982).

11.3.3 Quality Indices

Using the above chemical analyses, the following quality indices were determined:

- Soluble sodium percentage (SSP) is calculated as:

$$\mathrm{SSP} = \frac{\mathrm{Na}}{\sum \mathrm{Cations}} \times 100 \tag{11.1}$$

  where ions are expressed as mmolc L^{-1}.
- The sodium adsorption ratio (SAR) is calculated as:

$$\mathrm{SAR} = \frac{\mathrm{Na}^{+}}{\sqrt{\mathrm{Ca}^{++} + \mathrm{Mg}^{++}/2}} \tag{11.2}$$

  where ions are expressed as mmolc L^{-1}.
- The adjusted Sodium Adsorption Ratio (adj. SAR) can be calculated from the procedure given by (Ayers and Westcot 1985):

$$\mathrm{Adj.SAR} = \mathrm{SAR}\left[1 + (8.4 - \mathrm{pH}\underline{\mathrm{C}})\right] \tag{11.3}$$

$$\mathrm{pH_c} = (\mathrm{pk_2} - \mathrm{PK_c}) + \mathrm{p}(\mathrm{Ca}^{2+} + \mathrm{Mg}^{2+}) + \mathrm{p}(\mathrm{A1K}) \tag{11.4}$$

– The equation for calculating adj.^{R} Na is as follows (Rhoades 1999):

$$\text{Adj.}^{R}\,\text{Na} = \frac{\text{Na}^{+}}{\sqrt{\text{Ca}_x{}^{2+} + \text{Mg}^{2+}/2}} \tag{11.5}$$

where Na and Ca in the irrigation water reported in mmolc L^{-1}, Ca_x is a modified calcium value in mmolc L^{-1}, Ca_x represents Ca in the applied irrigation water but modified due to salinity of the applied water (EC_w), its HCO_3/Ca ratio (HCO_3 and Ca in mmolc L^{-1}) and the estimated partial pressure of CO_2 in the surface few millimeters of soil ($PCO_2 = 0.0007$ atmospheres), and magnesium in the irrigation water in mmolc L^{-1}. To use the Ca_x, first determine the HCO_3 to Ca ratio (HCO_3/Ca) and EC_w from the water analysis, using HCO_3 and Ca in mmolc L^{-1} and the water salinity (EC_w) in deciSiemens per meter. Find the HCO_3/Ca ratio that falls nearest to the calculated HCO_3/Ca value for the subject water and read across to the EC_w column that most closely approximates the EC_w for the water being evaluated. The Ca_x value shown represents the mmolc L^{-1} of Ca that is expected to remain in solution in the soil water at equilibrium and is to be used in the equation. The adj.^{R} Na obtained is used in place of the SAR to evaluate the potential of the water to cause an infiltration problem if used for irrigation.

– The equation for calculating ESP is as follows (Seilsepour and Rashidi 2008):

$$\text{ESP} = \frac{100(-0.0126 + 0.001745\text{SAR})}{1 + (-0.0126 + 0.001745\text{SAR})} \tag{11.6}$$

– The permeability index is estimated as follows according to (Doneen 1964):

$$\text{PI} = \frac{\text{Na}^{+} + \sqrt{\text{HCO}_3{}^{-}}}{\text{Na}^{+} + \text{Ca}^{2+} + \text{Mg}^{2+}} \times 100 \tag{11.7}$$

where concentration in mmolc L^{-1}.

11.4 Results and Discussion

11.4.1 Spatial and Temporal Assessment of Water Quality Along El-Salam Canal

Three key factors were assessed to evaluate water suitability for irrigation: salinity and specific ion toxicities. According to (Tedeschi and Dell'Aquila 2005), the suitability of irrigation water is influenced by chemical reactions of dissolved salts, as well as soil properties, climatic conditions, and irrigation management practices. To accurately assess irrigation water quality, it is essential to identify critical characteristics that impact plant growth and determine their acceptable concentration levels to ensure optimal agricultural productivity (Organization 2002).

11.4.1.1 pH Values

The spatial variation of pH values in irrigation water samples collected from various locations along the El-Salam Canal over one year is presented in Fig. 11.2. In general, the pH values fall within the normal range, as classified by (Ayers and Westcot 1976). The lowest pH values were recorded at the end of the canal, west of the Suez Canal, near the siphon that transports water to Sinai at km 89.12 (Location 26), where the pH reached approximately 7.71. In contrast, the highest pH values were observed at km 49.87 (Location 14), with an average of 8.03. The lowest pH values were recorded during summer, averaging 7.4, whereas the highest values occurred in fall, with an average of 8.3. The increase in pH may be attributed to a slight rise in bicarbonate (HCO_3^-) ions and the higher proportion of drainage water mixed with freshwater during the winter closure period when Nile River water levels are at their lowest. Additionally, there is a notable increase in pH values after the mixing point with the Hadous drain, compared to the pH levels at the canal's inlet. This could be linked to pesticide use by farmers, as certain chemicals are known to increase water pH to aid in the removal of water hyacinths.

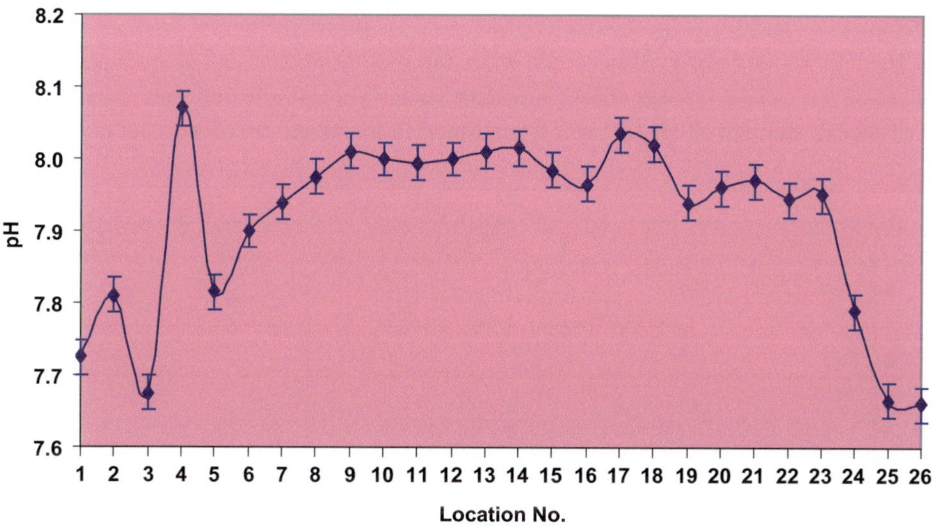

Fig. 11.2 Mean values of pH for water samples collected from different locations along the El-Salam Canal course (Alnaimy 2013)

11.4.1.2 Salinity Hazard

The spatial variation of electrical conductivity (EC) values in irrigation water samples collected along the El-Salam Canal over one year is presented in Fig. 11.3. The lowest EC value (0.46 dS/m) was recorded in fresh Nile water at El-Adliah village (Location 1, 0 km), while the highest salinity level (2.62 dS/m) was observed in water from Hadous Drain (Location 16, 53.67 km). A gradual increase in total dissolved solids (TDS) was noted from the canal's inlet at the Damietta Branch to its end at the Suez Canal siphon, indicating progressive mixing with drainage water. Seasonal variations in EC values showed lower salinity levels in summer compared to winter. This reduction may be attributed to the large volumes of irrigation water used for rice fields during summer, which dilutes salinity levels. The fluctuations in Total Dissolved Solids (TDS) following the mixing of water from the El-Serw and Hadous Drains may be influenced by several factors, including seasonal variations in water availability, climatic changes, and differences in the mixing ratios of Nile water and agricultural drainage water. The increase in TDS may also be due to high evaporation rates, wind effects in the desert region, and seepage of saline groundwater from surrounding salt-affected lands. The highest EC values were recorded in drainage water near El-Manzala Lake, with a salinity peak in January, likely due to reduced irrigation water flow during this period. These variations highlight the impact of seasonal irrigation patterns, particularly the dilution effect of summer irrigation for rice cultivation, on water salinity levels (Abdo et al. 2022; Alnaimy et al. 2020).

The high salinity observed in certain water samples may be attributed to several factors, including the canal's passage through recently reclaimed salt-affected soils, such as Sahil El-Hosaneia and San El-Hagar, which contribute to increased dissolved salts in the water

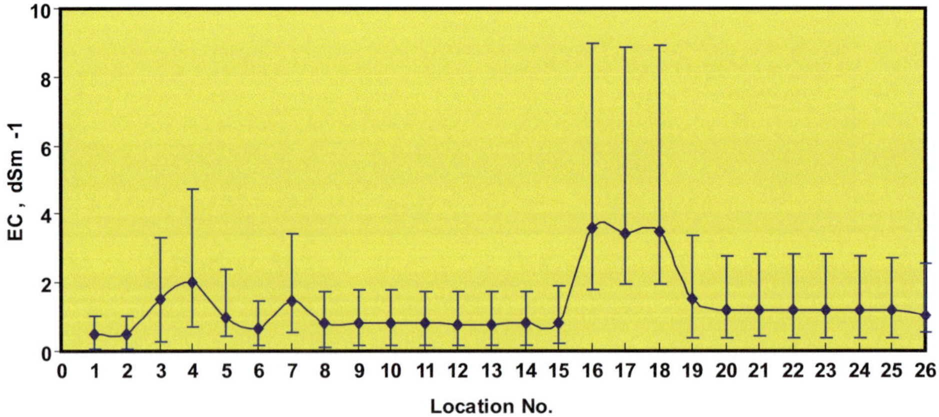

Fig. 11.3 Mean values of EC (dSm^{-1}) for water samples collected from different locations along the El-Salam Canal course (Alnaimy 2013)

(Alnaimy et al. 2018, 2023). Additionally, seasonal variations in water quantity, climatic fluctuations, and changes in the mixing ratio of Nile water with agricultural drainage water further influence salinity levels. The elevated EC values in some locations also suggest a higher proportion of drainage water in the final mix, particularly during the winter closure period, when Nile River freshwater levels are at their lowest. This trend is particularly evident after the mixing point with Bahr Hadous, where EC values show a notable increase compared to the canal's inlet. According to the water classification by (Ayers and Westcot 1976), the EC values of El-Salam Canal and its contributing drains—Faraskor Drain (1.51 dS/m), Al-Atwi Drain (1.87 dS/m), El-Serw Drain (1.25 dS/m), and Hadous Drain (2.62 dS/m)—fall within the range of 0.75–3.0 dS/m, indicating a growing salinity concern with prolonged use for irrigation. Based on the USDA (1954) classification, drainage water samples contributing to the El-Salam Canal are categorized as C3 (medium salinity) and C4 (high salinity), highlighting the potential risk of severe salinity buildup over time.

The spatial variation of cation concentrations along the El-Salam Canal is presented in Table 11.1. Sodium concentrations ranged from 1.76 mmolc/L in Nile water (Location 1) to 14.28 mmolc/L in Hadous drainage water (Location 16). Potassium concentrations averaged 0.21 mmolc/L in Nile water and 0.74 mmolc/L in Hadous drainage water (Location 16), as well as after mixing with Hadous drain (Location 18). Potassium availability in soil is influenced by fixation in clay colloids and leaching losses, particularly in clayey soils. Calcium concentrations varied from 1.94 mmolc/L in Nile water to 4.16 mmolc/L after mixing with Hadous drainage water (Location 18). The leaching of calcium depends on drainage water discharge, fertilization strategies, and gypsum application, which can increase calcium losses, especially in light-textured soils irrigated with high-quality water. These findings are consistent with (Jenkins and Jenkins 2014). Magnesium concentrations ranged from 2.18 mmolc/L at El-Adliah Bridge (Location 2) to 8.88 mmolc/L in Hadous drainage water (Location 16). Chloride concentrations varied between 1.72 mmolc/L in Nile water (Location 1) and 21.13 mmolc/L in Hadous drainage water (Location 16). According to (Ayers and Westcot 1985), chloride content in these water samples falls within the "no problem" or "increasing problem" classification, depending on concentration levels. Bicarbonate concentrations ranged from 3.08 mmolc/L in Nile water to 5.95 mmolc/L in Hadous drainage water, indicating potential precipitation of calcium (Ca^{2+}) in the soil, as influenced by pH conditions. Chlorides and bicarbonates were the dominant anions in water samples from the El-Salam Canal, while sulfate concentrations were relatively low. Overall, the water is suitable for irrigation but requires proper soil and water management to mitigate potential salinity and nutrient imbalances, ensuring sustainable agricultural use (Alnaimy et al. 2021).

Table 11.1 Mean values of dominant elements, some calculated indices, and classification for water samples collected from different locations on El-Salam Canal (Alnaimy 2013)

No	pH	EC, dSm^{-1}	Cations, mmolc L^{-1}				Anions, mmolc L^{-1}			Calculated indices								Water classification				%PI
			Na	K	Ca	Mg	Cl	HCO_3	SO_4	SAR	SSP	RSC	RSBC	ESP	pH_c	Adj SAR	$Adj.^R$ Na	SCAR	USDA	SAR/SCAR	ICAR	
1	7.86	0.46	1.76	0.21	1.94	2.26	1.72	3.08	1.37	1.29	30.22	−1.1	1.1	0.65	7.44	2.49	1.38	1.32	C_2S_1	0.98	C_1S_0	58.98
2	7.91	0.48	2.11	0.25	2.11	2.18	1.85	3.21	1.58	1.50	33.08	−1.1	1.1	0.97	7.42	2.96	1.64	1.51	C_2S_1	1.00	C_1S_0	61.01
3	7.83	1.51	7.72	0.41	3.64	4.38	11.26	4.30	0.58	4.01	48.44	−3.7	0.7	4.71	7.17	8.90	4.58	4.35	C_3S_1	0.95	C_2S_0	62.25
4	7.97	1.87	10.28	0.55	3.13	6.76	15.08	4.43	1.20	4.94	51.22	−5.5	1.3	6.08	7.13	11.08	5.39	6.15	C_3S_1	0.80	C_2S_0	61.43
5	7.91	0.81	4.17	0.30	2.47	2.49	5.05	3.46	0.92	2.81	42.66	−1.5	1.0	2.91	7.40	5.57	3.08	2.79	C_3S_1	1.01	C_1S_0	66.04
6	8.00	0.65	2.93	0.30	2.23	2.48	3.79	3.35	0.81	1.98	38.09	−1.4	1.1	1.67	7.40	3.93	2.16	2.01	C_3S_1	0.99	C_1S_0	62.27
7	7.97	1.25	6.70	0.45	2.51	4.21	8.53	4.20	1.14	3.92	48.05	−2.5	1.7	4.57	7.21	8.47	4.34	4.43	C_3S_1	0.90	C_1S_0	65.21
8	7.98	0.84	3.74	0.36	2.35	3.11	4.88	3.46	1.22	2.38	40.34	−2.0	1.1	2.27	7.36	4.80	2.58	2.50	C_3S_1	0.95	C_1S_0	60.88
9	8.02	0.81	3.48	0.28	2.28	3.10	4.58	3.58	0.98	2.23	38.99	−1.8	1.3	2.04	7.34	4.51	2.43	2.32	C_3S_1	0.94	C_1S_0	60.63
10	8.02	0.81	3.59	0.28	2.28	3.06	4.54	3.56	1.10	2.31	40.06	−1.8	1.3	2.17	7.37	4.63	2.51	2.41	C_3S_1	0.95	C_1S_0	61.38
11	8.00	0.81	3.67	0.32	2.27	3.06	4.57	3.63	1.12	2.36	40.34	−1.7	1.4	2.24	7.38	4.74	2.57	2.48	C_3S_1	0.95	C_1S_0	61.95
12	8.00	0.79	3.73	0.28	2.21	2.94	4.52	3.64	1.01	2.46	41.87	−1.5	1.4	2.39	7.36	4.94	2.67	2.57	C_3S_1	0.95	C_1S_0	63.51
13	8.02	0.78	3.93	0.35	2.28	2.98	4.67	3.68	1.18	2.56	41.98	−1.6	1.4	2.54	7.35	5.19	2.80	2.65	C_3S_1	0.96	C_1S_0	63.64
14	8.03	0.80	3.82	0.28	2.23	2.98	4.59	3.63	1.07	2.50	42.26	−1.6	1.4	2.45	7.36	5.05	2.72	2.61	C_3S_1	0.95	C_1S_0	63.47
15	7.94	0.84	4.00	0.37	2.14	3.33	4.73	3.66	1.45	2.53	41.61	−1.8	1.5	2.50	7.34	5.14	2.71	2.89	C_3S_1	0.89	C_1S_0	62.40
16	7.93	2.62	14.28	0.74	3.98	8.88	21.13	5.95	0.80	5.72	49.57	−6.9	2.0	7.25	6.88	14.61	6.30	7.30	C_4S_1	0.81	C_2S_0	61.60
17	7.97	2.55	13.60	0.73	4.13	8.54	20.13	5.83	1.04	5.46	49.07	−6.8	1.7	6.87	6.88	13.88	6.05	6.79	C_4S_1	0.83	C_2S_0	60.95
18	7.98	2.54	13.77	0.74	4.16	8.63	20.42	5.90	0.98	5.44	48.54	−6.9	1.7	6.83	6.88	13.93	6.04	6.78	C_4S_1	0.84	C_2S_0	60.98

(continued)

Table 11.1 (continued)

No	pH	EC, dSm^{-1}	Cations, mmolc L^{-1}				Anions, mmolc L^{-1}			Calculated indices								Water classification				%PI
			Na	K	Ca	Mg	Cl	HCO_3	SO_4	SAR	SSP	RSC	RSBC	ESP	pH_c	Adj SAR	$Adj.^{R}$ Na	SCAR	USDA	SAR/SCAR	ICAR	
19	7.94	1.41	6.95	0.43	2.86	5.44	9.98	4.59	1.12	3.77	45.90	−3.7	1.7	4.35	6.90	9.28	4.09	4.75	C_3S_1	0.83	C_2S_0	59.63
20	7.94	1.22	5.63	0.37	2.78	4.36	7.72	4.48	0.95	3.24	44.49	−2.7	1.7	3.56	7.12	7.20	3.55	3.64	C_3S_1	0.89	C_2S_0	60.66
21	7.94	1.21	5.51	0.38	2.78	4.35	7.71	4.45	0.86	3.18	43.66	−2.7	1.7	3.47	7.18	6.96	3.50	3.54	C_3S_1	0.89	C_2S_0	60.26
22	7.93	1.21	5.35	0.37	2.77	4.41	7.70	4.41	0.78	3.09	42.59	−2.8	1.6	3.33	7.19	6.69	3.40	3.45	C_3S_1	0.88	C_2S_0	59.48
23	7.91	1.20	5.33	0.39	2.70	4.35	7.46	4.38	0.93	3.06	43.02	−2.7	1.7	3.29	7.13	6.69	3.34	3.45	C_3S_1	0.88	C_2S_0	59.93
24	7.80	1.18	5.28	0.39	2.72	4.44	7.62	4.36	0.85	2.99	42.76	−2.8	1.6	3.18	7.13	6.54	3.26	3.40	C_3S_1	0.88	C_2S_0	59.22
25	7.76	1.19	5.18	0.37	2.78	4.50	7.59	4.38	0.85	2.90	41.93	−2.9	1.6	3.04	7.13	6.37	3.16	3.29	C_3S_1	0.88	C_2S_0	58.37
26	7.71	1.15	5.13	0.37	2.76	4.60	7.67	4.28	0.91	2.87	41.07	−3.1	1.5	3.01	7.17	6.35	3.13	3.29	C_3S_1	0.87	C_2S_0	57.62

11.4.1.3 Sodicity Hazard in El-Salam Canal

Soluble Sodium Percentage (SSP)

Sodium concentration in irrigation water plays a crucial role in soil structure stability, influencing clay dispersion, erosion, drainage, and permeability issues. The soluble sodium percentage (SSP) is a key indicator of sodium hazards, helping to assess water quality for agricultural use (Alnaimy et al. 2012). Excessive sodium levels can stunt plant growth and reduce soil permeability, particularly in alkali-affected soils. While SSP has historically been used to express sodium levels in irrigation water, the sodium adsorption ratio (SAR) is often preferred as it directly correlates with sodium retention in soil, making it a more reliable measure of alkali hazards. The spatial variation of SSP values in irrigation water samples collected over one year is presented in Table 11.1. The lowest SSP value (30.22%) was recorded in Nile freshwater at El-Adliah village (Location 1, 0 km), indicating a low alkali hazard. In contrast, the highest SSP value (51.22%) was observed in the Al-Atwi Drain (Location 4, 7.50 km), signifying a high alkali hazard. Seasonal variations in SSP show lower values in summer and higher values in winter, likely due to seasonal fluctuations in sodium concentration and water availability. Sodium (Na^+) levels in irrigation water are critical for evaluating water suitability, as high Na^+ concentrations negatively affect soil aeration, infiltration, and structure. Sodium causes clay and organic matter dispersion, leading to pore clogging, reduced water infiltration, and limited oxygen availability in the soil. These factors can severely impact crop growth and soil health, emphasizing the need for proper irrigation management to mitigate sodium-related risks (Bhandari and Joshi 2013).

Sodium Adsorption Ratio (SAR)

Irrigation water with high sodium (Na^+) content can displace exchangeable calcium (Ca^{2+}) and magnesium (Mg^{2+}) ions from soil clay minerals, leading to their replacement by sodium. This process causes soil dispersion, reduced permeability, and a decline in fertility, making the soil less suitable for cultivation (Abdel-Fattah et al. 2021). A high sodium adsorption ratio (SAR) further indicates the risk of sodium (alkali) accumulation, which can severely damage soil structure, reduce permeability, and lower crop productivity (Gupta and Gupta 2015). Several methods have been proposed to assess sodium hazard, with SAR being one of the most widely used indicators (McNeal 1981). SAR values are classified into four categories: S1 (<10, low sodium hazard), S2 (10–18, medium sodium hazard), S3 (18–26, high sodium hazard), and S4 (>26, very high sodium hazard). The spatial variation of SAR values in irrigation water samples collected over one year is presented in Fig. 11.4. The lowest SAR value (1.29 mmolc/L) was recorded in Nile freshwater at El-Adliah village (Location 1, 0 km), indicating a low alkali hazard. Conversely, the highest SAR value (5.72 mmolc/L) was observed in Hadous Drain (Location 16, 53.67 km), reflecting a higher alkali hazard. Seasonal variations also influence SAR levels, as fluctuations in drainage water contribution, irrigation practices, and climatic conditions impact sodium concentration in irrigation water. The increasing SAR values

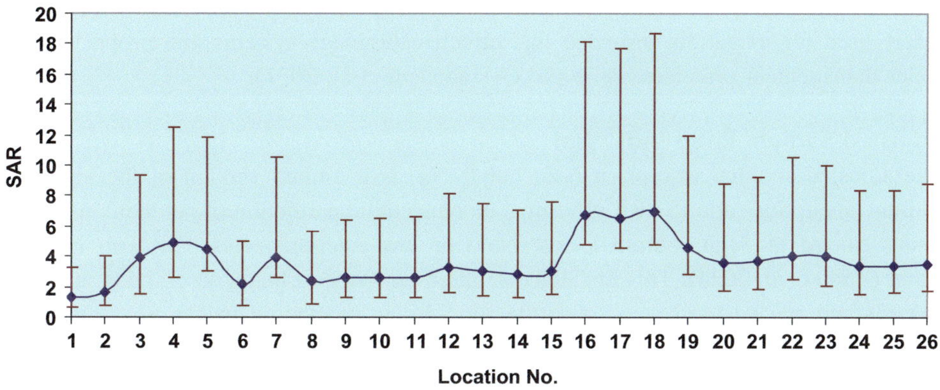

Fig. 11.4 Mean values of SAR (mmolc L^{-1}) for water samples collected from different locations along the El-Salam Canal course (Alnaimy 2013)

in certain locations highlight the risk of sodium accumulation, emphasizing the need for proper irrigation management, soil amendments, and drainage improvements to mitigate alkali hazards and maintain long-term soil productivity (Jalali et al. 2008).

Adjusted Sodium Adsorption Ratio (Adj. SAR)

The adjusted sodium adsorption ratio (adj. SAR) provides a more accurate assessment of sodium hazards in irrigation water by considering the effects of bicarbonate (HCO_3^-) and carbonate (CO_3^{2-}) ions on calcium (Ca^{2+}) and magnesium (Mg^{2+}) concentrations. These ions can precipitate calcium and magnesium, reducing their availability and increasing the sodium hazard. As a result, adj. SAR values are typically 10–15% higher than unadjusted SAR, making them a more reliable indicator of soil dispersion risks and permeability reduction. In irrigation water with low salinity but high SAR, soil permeability issues become more pronounced, making it essential to distinguish between SAR and adj. SAR when evaluating water quality (Bauder et al. 2008). The lowest adj. SAR value (2.49 mmolc/L) was recorded in Nile freshwater at El-Adliah village (Location 1, 0 km), indicating low alkali hazards, while the highest value (14.61 mmolc/L) was observed in Hadous Drain (Location 16, 53.67 km), indicating high alkali hazards (see Table 11.1). Seasonal fluctuations showed lower adj. SAR values in summer (2.1 mmolc/L) and higher values in winter (20.1 mmolc/L) due to changes in water availability, drainage contributions, and bicarbonate levels.

According to Gupta and Gupta (2019), most El-Salam Canal waterfalls within A1B1C1 (good water) and A2B1C2 (low sodium content) classifications, making it generally suitable for irrigation. However, some contributing drains, such as Hadous and El-Serw, exhibit higher adj. SAR (>10), requiring careful management when used on light-textured soils. Sodium hazards can be mitigated by increasing calcium levels through gypsum amendments or acidifying irrigation water to reduce bicarbonate levels (El Behairy et al.

2021). Overall, despite variations in salinity and SAR, the El-Salam Canal water can be safely used for irrigation, provided that effective drainage systems and proper soil and water management practices are in place (Dhaouadi et al. 2021).

Adjusted Sodium Ratio (Adj.^{R} Na)

The adjusted sodium adsorption ratio (adj.^{R} Na) is a refined version of the traditional sodium adsorption ratio (SAR), offering a more accurate prediction of potential infiltration issues caused by high sodium levels (Na^+) or low calcium levels (Ca^{2+}) in irrigation water (Zhou et al. 2024). This updated calculation provides a better assessment of sodium hazards and can be used as a substitute for SAR in evaluating irrigation water quality (Rhoades 1984; Suarez and Lebron 1993).

The lowest adj. R Na value (1.38 mmolc/L) was recorded in Nile freshwater at El-Adliah village (Location 1, 0 km), indicating a low alkali hazard. In contrast, the highest adj. R Na value (6.30 mmolc/L) was observed in the Hadous Drain (Location 16, 53.67 km), signifying a high alkali hazard (Table 11.1). Seasonal variations showed that the lowest adj. R Na values were recorded in summer, averaging 1.1 mmolc/L, while the highest value (8.5 mmolc/L) was also observed in summer, classifying the water as having low sodium content. These fluctuations highlight the impact of seasonal changes, water source variations, and drainage contributions on sodium levels in irrigation water.

Exchangeable Sodium Percentage (ESP)

With salinity becoming an increasing global concern, accurately measuring exchangeable cation concentrations in saline soils remains a challenge. Traditional laboratory tests for Exchangeable Sodium Percentage (ESP) are often time-consuming and labor-intensive, making them impractical for large-scale monitoring. To address this, researchers have explored alternative methods, including a linear regression model that predicts soil ESP from the Sodium Adsorption Ratio (SAR) of irrigation water, allowing for a more efficient and cost-effective assessment of soil sodicity (Seilsepour et al. 2009). The lowest ESP value (0.65%) was recorded in Nile freshwater at El-Adliah village (Location 1, 0 km), indicating a low alkali hazard. Conversely, the highest ESP value (7.25%) was observed in Hadous Drain (Location 16, 53.67 km), reflecting a higher alkali hazard (Table 11.1 and Fig. 11.5). Seasonal variations showed minimum ESP values in summer (0.3%) and maximum values in winter (10.4%), likely due to reduced water availability, increased evaporation, and greater sodium accumulation in winter months. These findings emphasize the importance of continuous monitoring and effective irrigation management to prevent long-term soil degradation in salt-affected areas.

11.4.1.4 Miscellaneous Elements Content

The concentrations of various trace elements in irrigation water samples collected from different locations along the El-Salam Canal were monitored spatially and temporally over one year. The obtained data, presented in (see Table 11.2), indicate that miscellaneous

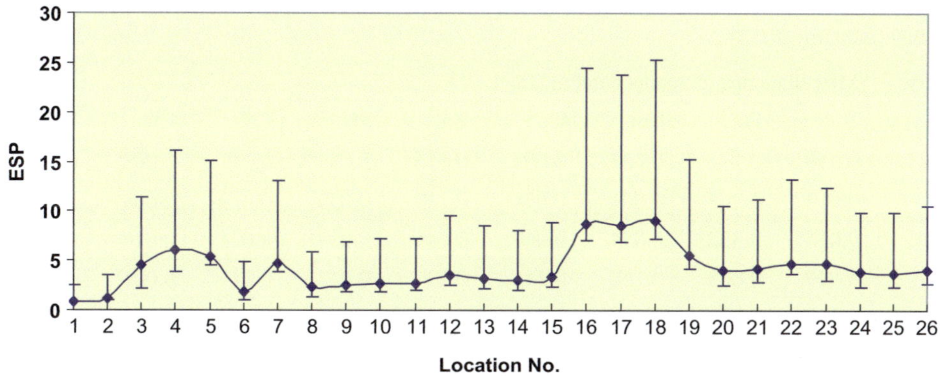

Fig. 11.5 Mean values of ESP (%) for water samples collected from different locations along the El-Salam Canal course (Alnaimy 2013)

element concentrations were generally low across all investigated samples. Boron is an essential micronutrient for plant growth, but its concentration in irrigation water must be carefully regulated, as excessive levels can be toxic to sensitive crops. While some plants require as little as 0.2 mg/L for optimal growth, concentrations above 1–2 mg/L can become harmful. Unlike sodium-related salinity issues, which often originate from the soil, boron toxicity is more commonly caused by irrigation water. To prevent crop damage, boron concentrations should generally remain below 2 mg/L (Gupta and Gupta 2019). According to (Ayers and Westcot 1985), irrigation water with boron concentrations below 0.75 mg/L is classified as safe, while levels between 0.75 and 2.0 mg/L indicate increasing risk, and values above 2 mg/L pose a severe hazard to plant health. The analysis revealed that boron levels ranged from 0.01 to 0.02 mg/L, classifying the water as B1 (normal water) with no risk of boron toxicity (see Table 11.2). The highest boron concentrations were recorded in winter, averaging 0.022 mg/L, yet still well below the recommended safety threshold for irrigation water (Gupta and Gupta 2019). These findings confirm that El-Salam Canal water is suitable for most crops and soil types, posing no significant boron-related risks, and making it a reliable water source for sustainable agriculture in the study area (Zare et al. 2014).

The copper content ranged from a minimum value of 0.003 mg/L, recorded at 27.43 and 53.56 km, to a maximum value of 0.282 mg/L at 54.0 km along the canal (Table 11.2). The elevated copper levels at this location may be attributed to the mixing of canal water with Hadous Drain water. Similarly, cobalt concentrations were lowest in Nile freshwater, averaging 0.004 mg/L, while the highest level (0.145 mg/L) was observed in Al-Atwi Drain water (Table 11.2). For nickel, the lowest concentration (0.741 mg/L) was detected in Nile freshwater, while the highest (1.769 mg/L) was recorded at 40.500 km along El-Salam Canal, likely due to mixing with Al-Serw Drain (Table 11.2). Cadmium levels were also minimal at 32.0 km (0.002 mg/L) but reached a maximum of 0.090 mg/L in Hadous

Table 11.2 Mean values of miscellaneous elements content for samples collected from El-Salam Canal (Alnaimy 2013)

Loc	Miscellaneous elements content (mg L^{-1})								
	B	Fe	Mn	Co	Zn	Pb	Cd	Cu	Ni
1	0.010	0.211	0.001	0.004	0.115	0.037	0.011	0.006	0.753
2	0.014	0.245	0.001	0.004	0.129	0.039	0.019	0.008	0.741
3	0.016	0.322	0.054	0.069	0.145	0.037	0.021	0.016	0.843
4	0.019	0.805	0.166	0.145	0.224	0.013	0.024	0.005	1.698
5	0.017	0.298	0.002	0.071	0.145	0.035	0.003	0.005	0.869
6	0.016	0.298	0.002	0.071	0.145	0.035	0.003	0.005	0.869
7	0.016	0.351	0.003	0.087	0.229	0.043	0.011	0.013	1.169
8	0.013	0.320	0.003	0.091	0.201	0.039	0.004	0.008	1.105
9	0.012	0.348	0.003	0.025	0.194	0.037	0.005	0.003	1.047
10	0.013	0.341	0.003	0.016	0.254	0.035	0.003	0.004	1.050
11	0.013	0.349	0.006	0.007	0.201	0.028	0.002	0.004	1.086
12	0.012	0.339	0.009	0.012	0.189	0.029	0.014	0.007	1.013
13	0.020	0.618	0.108	0.051	0.222	0.013	0.080	0.013	1.769
14	0.014	0.211	0.001	0.004	0.115	0.037	0.011	0.006	0.753
15	0.019	0.388	0.113	0.067	0.180	0.030	0.053	0.003	1.096
16	0.020	0.565	0.153	0.106	0.222	0.085	0.090	0.028	1.495
17	0.021	0.577	0.127	0.104	0.211	0.095	0.070	0.282	1.473
18	0.022	0.564	0.087	0.096	0.241	0.063	0.047	0.038	1.529
19	0.016	0.453	0.061	0.037	0.193	0.036	0.015	0.018	1.264
20	0.013	0.454	0.022	0.036	0.203	0.058	0.003	0.004	1.119
21	0.013	0.444	0.032	0.035	0.194	0.046	0.024	0.027	1.210
22	0.013	0.454	0.024	0.039	0.203	0.049	0.026	0.027	1.168
23	0.013	0.485	0.02	0.06	0.205	0.05	0.035	0.004	1.085
24	0.013	0.488	0.048	0.077	0.216	0.060	0.021	0.013	1.149
25	0.013	0.506	0.055	0.057	0.211	0.057	0.023	0.017	1.222
26	0.013	0.540	0.072	0.081	0.231	0.059	0.026	0.037	1.402

Drain water (Table 11.2). These findings highlight the variability of trace element concentrations in the El-Salam Canal system, largely influenced by drainage water contributions and mixing patterns.

The spatial and temporal distribution of trace elements in El-Salam Canal water indicates variability influenced by drainage water mixing. Lead concentrations ranged from a minimum of 0.013 mg/L at 46.0 km to a maximum of 0.095 mg/L in Hadous Drain water at 54.0 km. Zinc levels were lowest in Nile freshwater (0.115 mg/L) and peaked at 32.28 km (0.195 mg/L). Manganese concentrations ranged from 0.001 mg/L in Nile freshwater to 0.166 mg/L in Al-Atwi Drain water, while iron levels were lowest in Nile freshwater (0.211 mg/L) and highest in Al-Atwi Drain water (0.805 mg/L) (Table 11.2).

Seasonal variations showed that trace element concentrations peaked in January, coinciding with the winter closure period, when irrigation canals are temporarily shut down for maintenance. While miscellaneous element concentrations remained below the recommended maximum limits for irrigation water (Gupta and Gupta 2015). Prolonged use could lead to the accumulation of these elements in soil and plant tissues, potentially causing toxicity issues over time. The primary sources of contamination are industrial wastewater discharges into El-Serw and Hadous Drains, as well as agricultural runoff containing elevated trace element levels. These findings align with (Alnaimy et al. 2021), who reported higher concentrations of trace elements in agricultural drainage water compared to the River Nile, emphasizing the need for continuous monitoring to prevent long-term soil and crop contamination (Abdel-Azeem et al. 2015).

11.4.1.5 Permeability Index (PI)

The permeability index (PI) is a key parameter in assessing irrigation water quality, as it directly affects soil permeability and agricultural development. According to (Stanley and Warne 1998), irrigation water is classified into three categories (Class 1, Class 2, and Class 3) based on its permeability index and total salinity levels. To visualize these classifications, (Doneen 1964) developed a permeability index diagram, dividing water quality into three suitability classes for irrigation.

In this study, permeability index values ranged from 40 to 80%, with an average of 62% (Table 11.2). Most of the El-Salam Canal water samples fall within Class I and Class II, indicating that they are suitable for irrigation, particularly in soils with medium permeability. These findings confirm that the El-Salam Canal water is generally well-suited for agricultural use, provided that proper soil and water management practices are implemented (Alnaimy et al. 2021; Islam 2023).

11.5 Conclusion

This chapter highlights the transformative role of information technologies, such as precision agriculture, in modernizing agricultural practices. By integrating Geographic Information Systems (GIS), remote sensing, and geostatistical analysis, precision agriculture enables data-driven decision-making, optimizing crop production while addressing spatial and temporal variability in soil, water, and crop factors. This study focused on the El-Salam Canal region in the northeastern Nile Delta, employing advanced technologies to assess irrigation water quality, soil suitability, and crop production potential.

Key Findings from the Study Include

1. **Water Quality Assessment**
   - Spatial and temporal variations in pH, EC, SAR, and other parameters revealed significant fluctuations along the El-Salam Canal, influenced by seasonal changes, drainage water mixing, and agricultural practices.
   - While most water samples fell within acceptable irrigation ranges, elevated salinity, and sodium hazards were observed in areas receiving drainage water, particularly from Hadous and Al-Atwi drains.
2. **Trace Elements and Pollution Concerns**
   - Trace elements such as copper, cobalt, nickel, and iron were detected at low concentrations, posing minimal immediate risks.
   - However, long-term use of drainage-influenced water could lead to the accumulation of these elements, potentially harming soil health and crop productivity.

11.6 Recommendations

To address the challenges identified and leverage the opportunities offered by precision agriculture, the following recommendations are proposed:

1. **Enhancing Water Management**
   - Implement real-time monitoring systems using remote sensing and GIS to track water quality variations along the El-Salam Canal and its branches.
   - Optimize the mixing ratio of Nile water and drainage water to minimize salinity and sodium hazards, especially during the winter closure period.
   - Promote water-saving irrigation techniques, such as drip and sprinkler systems, to reduce dependence on drainage water and improve water-use efficiency.
2. **Pollution Control Measures**
   - Enforce stricter regulations on industrial wastewater discharge into agricultural drains to limit the influx of trace elements and pollutants.
   - Implement constructed wetlands or bioremediation systems to treat drainage water before its reuse in irrigation.
   - Encourage sustainable farming practices to prevent land resalinization and maintain soil health over time.
3. **Capacity Building and Awareness**
   - Train farmers and agricultural extension workers on the benefits and applications of precision agriculture technologies.
   - Raise awareness about the risks of long-term use of drainage-influenced water and the importance of sustainable soil and water management practices.

4. **Policy and Institutional Support**
   - Develop policies that incentivize the adoption of precision agriculture and sustainable irrigation practices.
   - Strengthen collaboration between research institutions, government agencies, and farmers to ensure the effective implementation of these strategies.

By implementing these recommendations, the Nile Delta region can enhance agricultural productivity, ensure long-term sustainability, and mitigate the environmental impacts of intensive farming. The integration of precision agriculture technologies provides a promising pathway to address water scarcity, soil degradation, and climate change, ultimately contributing to food security and rural development in the region.

References

Abdel-Azeem A, El-Morsy E, Nour El-Dein M, Rashad H (2015) Occurrence and diversity of mycobiota in heavy metal contaminated sediments of Mediterranean coastal lagoon El-Manzala, Egypt. Mycosphere 6:228–240

Abdel-Fattah MK, Mohamed ES, Wagdi EM, Shahin SA, Aldosari AA, Lasaponara R, Alnaimy MA (2021) Quantitative evaluation of soil quality using principal component analysis: the case study of El-Fayoum depression, Egypt. Sustainability 13:1824

Abdo AI, Deng Y, Sun D, Chen X, Alnaimy MA, El-Sobky E-SE, Wei H, Zhang J (2022) Maintaining higher grain production with less reactive nitrogen losses in China: a meta-analysis study. J Environ Manag 322:116018

Alnaimy M, Soliman K, Atia N, El-Naka E (2012) Spatial and temporal evaluation of El-Salam canal water resources for irrigation purposes. Zagazig J Agric Res 39

Alnaimy M (2013) Using new information technology for evaluating soil and water resources in some regions of El-Salam Canal area. MSc thesis, Zagazig University, Fac. of Agriculture, Zagazig University, 290 pp

Alnaimy M, Zelenakova M, Vranayova Z, Abu-Hashim M (2020) Effects of temporal variation in long-term cultivation on organic carbon sequestration in calcareous soils: Nile Delta, Egypt. Sustainability 12:4514

Alnaimy MA, Soliman K, El-Naka E (2018) Evaluation of land suitability for agricultural use in El-Nubaria region, Egypt. Zagazig J Agric Res 45:2031–2048

Alnaimy MA, Shahin SA, Vranayova Z, Zelenakova M, Abdel-Hamed EMW (2021) Long-term impact of wastewater irrigation on soil pollution and degradation: a case study from Egypt. Water 13:2245

Alnaimy MA, Shahin SA, Afifi AA, Ewees AA, Junakova N, Balintova M, Abd Elaziz M (2022) Spatio prediction of soil capability modeled with modified RVFL using Aptenodytes Forsteri Optimization and digital soil assessment technique. Sustainability 14:14996

Alnaimy MA, Elrys AS, Zelenakova M, Pietrucha-Urbanik K, Merwad A-RM (2023) The vital roles of parent material in driving soil substrates and heavy metals availability in arid alkaline regions: a case study from Egypt. Water 15:2481

Aoubouazza M (2003) Sustainable utilization of wastelands and saline groundwater for plant production iaea project (INT/5/144). In: Sustainable strategies for irrigation in salt-prone mediterranean region: a system approach, p 172

Ayers RS, Westcot DW (1976) Water quality for agriculture
Ayers RS, Westcot DW (1985) Water quality for agriculture, 29. Food and agriculture organization of the United Nations Rome
Bauder J, Bauder T, Waskom R, Scherer TF (2008) Assessing the suitability of water (quality) for irrigation-Salinity and Sodium. Western fertilizer handbook
Bhandari N, Joshi HK (2013) Quality of spring water used for irrigation in the Almora District of Uttarakhand, India. Chin J Geochem 32:130–136
Cipolla SS, Maglionico M, Masina M, Lamberti A, Daprà I (2019) Real-time monitoring of water quality in an agricultural area with salinity problems. Environ Eng Manag J 18:2229–2240
Cuevas J, Daliakopoulos IN, del Moral F, Hueso JJ, Tsanis IK (2019) A review of soil-improving cropping systems for soil salinization. Agronomy 9:295
Dalia S, Korany EA, El Saadi A (2018) Increasing the efficiency of El-Salam canal project using data driven water quality models. J Environ Sci 43:1-19
Dhaouadi L, Besser H, Karbout N, Al-Omran A, Wassar F, Wahba MS, Yaohu K, Hamed Y (2021) Irrigation water management for sustainable cultivation of date palm. Appl Water Sci 11:1–19
Doneen LD (1964) Notes on water quality in agriculture. Department of Water Science and Engineering, University of California, Davis
El-Saidi MT (2002) Use of saline drainage water for irrigation, reclamation and cultivation of new land areas in Sinai Peninsula, Prospects for Saline Agriculture. Springer, pp 57–67
El Behairy RA, El Baroudy A, Ibrahim MM, Shokr M (2021) Assessment and mapping of surface water quality index for irrigation purpose: Case study northwest of Nile Delta, Egypt. Menoufia J Soil Sci 6:163–182
Elewa HH, Nosair AM, Zelenakova M, Mikita V, Abdel Moneam NA, Ramadan EM (2023) Environmental sustainability of water resources in coastal aquifers, case study: El-Qaa Plain, South Sinai, Egypt. Water 15:1118
Elewa HH, Nosair AM, Ibrahim A, Zelenakova M, Pietrucha-Urbanik K, Habib HM, Moneam NAA, Ragab RM, Ramadan EM (2024) Use of remote sensing, spatial and geophysical modeling, and real recharging capabilities to identify suitable areas for groundwater exploitation in dry coastal areas. J Environ Manag 363:121243
Gupta I, Gupta S (2019) Crop production in salt affected soils. Scientific Publishers
Gupta P, Singh J, Verma S, Chandel AS, Bhatla R (2021) Impact of climate change and water quality degradation on food security and agriculture. In: Water conservation in the era of global climate change. Elsevier, pp 1–22
Gupta SK, Gupta I (2015) Management of saline & waste water in agriculture. Scientific Publishers
Hafez A (2005) Investigation of El-Salam Canal project in Northern Sinai, Egypt. In: The 9th international water technology conference, IWTC9, pp 953–970
Hafez A, Khedr M, El-Katib K, Alla HG, Elmanharawy S (2008) El-Salaam Canal project, Sinai II. Chemical water quality investigations. Desalination 227:274–285
Islam MS (2023) Irrigation water quality, hydrogeochemical evaluation and groundwater quality. Springer, pp 223–280
Jalali M, Merikhpour H, Kaledhonkar M, Van Der Zee S (2008) Effects of wastewater irrigation on soil sodicity and nutrient leaching in calcareous soils. Agric Water Manag 95:143–153
Jenkins TA, Jenkins V (2014) Use of Gypsum to reduce effluent and fertiliser nutrient losses to waterways. NZ J Agri Res 46:147–154
McNeal BL (1981) Evaluation and classification of water quality for irrigation. Salinity in irrigation and water resources. Marcel Dekker, NY, pp 21–46
Mezlini W, Amor RB, Beneduci A, Romdhane IB, Shammas MI, Almazroui M, Attia R (2024) Effects of irrigation water quality on soil physico-chemical proprieties: case study in North-West of Tunisia. Earth Syst Environ 1–21

Mohamed AI (2013) Irrigation water quality evaluation in El-Salam canal project. Int J Eng Appl Sci 3:2305–8269
Organization WH (2002) Guidelines for drinking-water quality. World Health Organization
Othman AA, Rabeh SA, Fayez M, Monib M, Hegazi NA (2012) El-Salam canal is a potential project reusing the Nile Delta drainage water for Sinai desert agriculture: microbial and chemical water quality. J Adv Res 3:99–108
Page AL (1982) Methods of soil analysis. Part 2. Chemical and microbiological properties
Pal D, Tarafdar J, Sahoo A (2009) Analysis of soil for soil survey and mapping. Soil Survey Manual
Qadir M, Oster J (2004) Crop and irrigation management strategies for saline-sodic soils and waters aimed at environmentally sustainable agriculture. Sci Total Environ 323:1–19
Ramadan EM, Badr AM, Abdelradi F, Negm A, Nosair AM (2023) Detection of groundwater quality changes in minia governorate, west nile river. Sustainability 15:4076
Rhoades J (1984) Use of saline water for irrigation. Calif Agric 38:42–43
Rhoades J (1999) Use of saline drainage water for irrigation. Agric Drain 38:615–657
Sassenrath GF, Heilman P, Luschei E, Bennett GL, Fitzgerald G, Klesius P, Tracy W, Williford JR, Zimba PV (2008) Technology, complexity and change in agricultural production systems. Renew Agric Food Syst 23:285–295
Seilsepour M, Rashidi M (2008) Modeling of soil cation exchange capacity based on soil colloidal matrix.
Seilsepour M, Rashidi M, Khabbaz BG (2009) Prediction of soil exchangeable sodium percentage based on soil sodium adsorption ratio. Am Eurasian J Agric Environ Sci 5:1–4
Shaddad S, Castrignanò A, Di Curzio D, Rusi S, Salem HSA, Nosair AM (2025) An integrated statistical, geostatistical and hydrogeological approach for assessing and modelling groundwater salinity and quality in Nile Delta aquifer. AgriEngineering 7:34
Stanley DJ, Warne AG (1998) Nile Delta in its destruction phase. J Coast Res 795–825
Suarez D, Lebron I (1993) Water quality criteria for irrigation with highly saline water. In: Towards the rational use of high salinity tolerant plants. Agriculture and forestry under marginal soil water conditions, vol 2, pp 389–397
Tedeschi A, Dell'Aquila R (2005) Effects of irrigation with saline waters, at different concentrations, on soil physical and chemical characteristics. Agric Water Manag 77:308–322
Wallace J, Batchelor C (1997) Managing water resources for crop production. Philos Trans R Soc Lond. Ser B Biol Sci 352:937–947
Zare M, Ordookhani K, Emadi A, Azarpanah A (2014) Relationship between soil exchangeable sodium percentage and soil sodium adsorption ratio in Marvdasht plain. Iran Int J Adv Biol Biom Res 2:2934–2939
Zhou Q, Bleam WF, Soldat DJ (2024) An alternative approach for estimating the sodium adsorption ratio of irrigation water. J Irrig Drain Eng 150:04024031

Part III

Hydrocarbons Resources, Challenges, and Management

Is the Nile Delta Prolific for Oil Exploration Prospects?

12

Bassem S. Nabawy

Abstract

The Nile Delta is considered among the most prolific hydrocarbon-bearing prospects in Africa and worldwide. It is primarily known for its huge gas reserves and production from both onshore and offshore fields. However, a question has been raised about the potential of the Nile Delta as an oil-prospective zone: Is it possible to produce oil? Due to these reservoirs' deep depths and location in the metagenetic gas zone, they are rarely recognized for oil/condensate production. Only two localized fields are known as oil/condensate producers: El Tamad and West Dikirnis Fields. The current review outlines the main litho-stratigraphy, petrophysical features, and structural settings. These two fields are studied in detail to delineate their structural setting and reservoir properties. Integration between seismic, well logging, and petrography was applied to study the main producing reservoirs in these fields: the Middle Miocene Sidi Salem Formation and the Upper Miocene Qawasim and Abu Madi formations. Sand lenses, tongues, and channels acting as stratigraphic traps within thick shale sequences primarily represent these reservoirs. However, structural closures, such as three-way dip closures, anticlines, and horsts, are also present. Despite the scarcity of oil traps in the Nile Delta, this review highlights their potential exploration on top of anticlines, horsts, and uplifted structures. This concept is also considered for the various exploration activities in the deltas, especially the fluvially dominated deltas.

B. S. Nabawy (✉)
Geophysical Sciences Department, National Research Centre, Cairo, Egypt
e-mail: bsnabawy@yahoo.co.uk

M. A. Khalifa and K. S. Gemail (eds.), *Geology of the Nile Deltas and Natural Resources*, Deltas of the World, https://doi.org/10.1007/978-3-031-96594-4_12

Keywords

Nile Delta • Oil-producing • El Tamad field • West Dikirnis field • Petrophysical properties • Structural setting

12.1 Introduction

Deltas generally represent the most complex and composite sand bodies (lenses, lobes, tongues, and channels). This complexity is due to different interplays of fluvial input, wave and tidal action, sediment dispersal, tectonic setting, and various other factors that control the deltaic sedimentation style. This complexity makes hydrocarbon exploration in the deltaic areas more difficult than in any other sedimentary environment, e.g., Nabawy and Shehata (2010), Khalifa et al. (2024). In this context, the Nile Delta of Egypt is considered among the most complex and challenging provinces for hydrocarbon exploration. Despite this difficulty, this province attracted attention three decades ago after discovering Abu Madi and Abu Qir Gas Fields in 1967 and 1969, respectively (Nabawy and Shehata 2010, 2015).

The Nile Delta is known to be one of the oldest hydrocarbon production areas worldwide. It has been among the most productive areas in Egypt. Since 1967 and until June 2019, 173 discoveries have been achieved in the Nile Delta with 387 drilled wells. Its Hydrocarbons production reached about 89.5 equivalent mmbbl in the fiscal year 2018/2019, representing about 16% of the total Egyptian hydrocarbons production according to the data published by the EGPC (Egyptian General Petroleum Corporation) and the EGAS (Egyptian Natural Gas Holding Company). Among this annual production, the crude oil and condensates were estimated to be 0.096 and 4.61 mmbbl/Y, respectively (Hussien et al. 2020; Fig. 12.1). Despite the important hydrocarbon discoveries achieved during the last few decades, the main hydrocarbon reserve of the Nile Delta still unrevealed and in its preliminary stages; it still poorly understood, Vandré et al. (2007) and Leila et al. (2016).

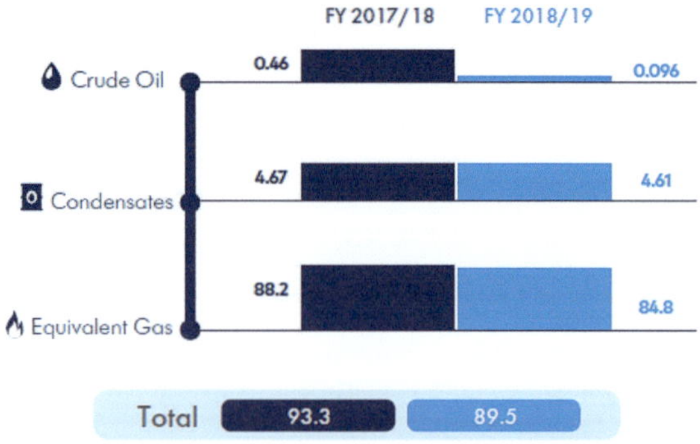

Fig. 12.1 The Nile Delta's Oil and condensate production in mmbbl/Y (Hussien et al. 2020)

In July 2014, the Italian Eni Company discovered the Nooros Field in the NE of Alexandria, in the NW of the Nile Delta, including onshore and shallow offshore activities. It was estimated to have some condensate reserves, besides about 530 bcf of natural gas. The Nidoco NW-1 exploratory well has been discovered in Nooros Field to produce hydrocarbon gas and condensates with a total net-pay of 100 m of hydrocarbon-bearing sandstones (50 m of the Pliocene Kafr-El-Sheik and 50 m of the Messinian Abu Madi formations). However, although the fact that the Nile Delta is a min hydrocarbon producer in Egypt and North Africa, the EGPC and the EGAS database states that the annual condensates and the oil production represents 5.2 and 0.11% of the Nile Delta's hydrocarbon production, respectively, during the FY 2018/2019 (Hussien et al. 2020). Therefore, a main question should be raised: can the Nile Delta be considered an oil and condensate producer? All discoveries reveal that the Nile Delta can produce a few mmbbl/years of condensates and oil. It seems to be located in the gas window (>120 °C, >4500 m) rather than the oil window (60–1 °C, 20 °C, 1500–4500 m). Therefore, if the oil were present in the past at relatively shallow depths, it is thermally cracked down into metagenetic natural gas and condensates with increasing depth due to moving down the stratigraphic traps from the oil window into the gas window. This movement is attributed to the regional hinge zone in the central parts of the delta, which threw the sequence down to the north by about 1500 m (Nabawy et al. 2023a).

However, some researchers referred that the pressure and temperature of some gas-producing fields in the Nile Delta suggest the situation within the oil window rather than the gas window, e.g., Zaghloul et al. (2001), Shaaban et al. (2006), Leila et al. (2016). No oil discoveries were achieved in the Nile Delta until 2005 when the onshore petroleum fields, El Tamad and West Dikirnis, were discovered in 2005 and 2007, respectively. These discoveries have changed the exploration strategy in the Nile Delta basin, as it may host some oil accumulation, especially in the horsts that are still situated in the oil window (1500–4500 m). A few examples of the oil and condensate-producing fields (i.e., El Tamad and West Dikirnis) in the Nile Delta will be given in the present chapter.

12.2 Geologic Setting

The Nile Delta of Egypt is characterized by a rolling regional surface sloping toward the Mediterranean, with the landforms controlled by some exogenic and endogenic factors. The endogenic morphological structures have subsequently been shaped by a variety of physical and chemical exogenic processes, including weathering and erosion. Faults have a prime implementation on their landscape, as folding has a limited implication on the topographic morphologies (Tarabees and Azam 2016). Following the geophysical studies, the thickness of the sedimentary sequence increases to the north, and it is estimated to reach up to 13,000 m along the Mediterranean shoreline (Abu El-Ata and Ibrahim 1988).

The depositional evolution and litho-stratigraphy of the Nile Delta have been studied by several authors, e.g., El Shazly (1977), Abu El Ella (1990), Rotaf et al. (1994), Nabawy and Shehata (2010, 2015), Nabawy et al. (2023a), Eysa et al. (2024), etc. (Fig. 12.2).

By the Early Oligocene, alternations between open marine muddy facies and fluvial coarse clastic facies are dominant, forming the Tineh Formation (Harms and Wary 1990; Guiraud and Bosworth 1990; Salem et al. 2005; Sakran et al. 2019; Nabawy et al. 2023a). During the Aquitanian–Burdigalian, a wide transgression covered the Nile Delta area and reached the Gulf of Suez area, creating a channel link between the Mediterranean and the Red Sea. During this time interval, a thick sequence of basinal marls and shales with subordinate coarse terrigenous sediments (Qantara Formation) was deposited during the closure of the communication between the Mediterranean Sea and the Gulf of Suez (Nabawy and Shehata 2010). Following these events, a large volume of sediment influx and a high subsidence rate caused the development of the thick terrigenous sequence of the Sidi Salem Formation during the Serravallian and Late Langhian. By the Late Serravallian, a strong erosive event, because of an additional set of positive movements, has dominated the area by the major unconformity of the Tortonian/Serravallian age between the Sidi Salem and the Qantara formations (Nabawy et al. 2018, 2023a).

The Tortonian sequence is similar to the Serravallian interval; it is presented by a shelf northward prograding regime with a turbiditic complex being deposited in its frontal area (Wakar Formation in El Temsah and Wakar areas, and Sidi Salem Formation in Baltim area). With the Early Messinian, an abrupt fall of the Mediterranean Sea level occurred, producing a strong erosion of all the outcropped formations with subareal incisions of some deep canyons. During this period, the marine water and the fluvial drainage regimes remained separated, so an evaporitic sedimentation developed. Elsewhere, the sequence of fluvial-deltaic sediments with lagoonal and/or restricted marine episodes of the Qawasim Formation was deposited. Following a new erosional period (a second major unconformity) in the Late Messinian, a fluvial estuarine sequence of sediments was deposited over the Abu Madi Formation. The Messinian interval in the studied area is composed mainly of sandstones interbedded with some shale streaks. It was followed by the deposition of evaporite Rosetta complexes that are considered a well-seismic marker (Nabawy and Shehata 2010).

By the beginning of the Pliocene, a widespread transgression led to the invasion of the entire area by the seawater that predominated during the deposition of the Early Pliocene Kafr El-Sheikh and El-Wastani formations (Nabawy and Shehata 2010). This transgression dominated until the deposition of the Late Pliocene El-Wastani Formation, which was distinguished by climatic changes and accompanied by a change from fluvio-marine deposition to fluviatile deposition (Ross and Uchupi 1977; Nabawy and Shehata 2010). They are characterized by the prevalence of shales intercalated with sand units, particularly in the Kafr El-Sheikh Formation, while the El-Wastani Formation is predominantly composed of sand. The Pleistocene Mit Ghamr and Bilqas formations are composed mainly of loose sands with minor claystone intercalations (Nabawy and Shehata 2010).

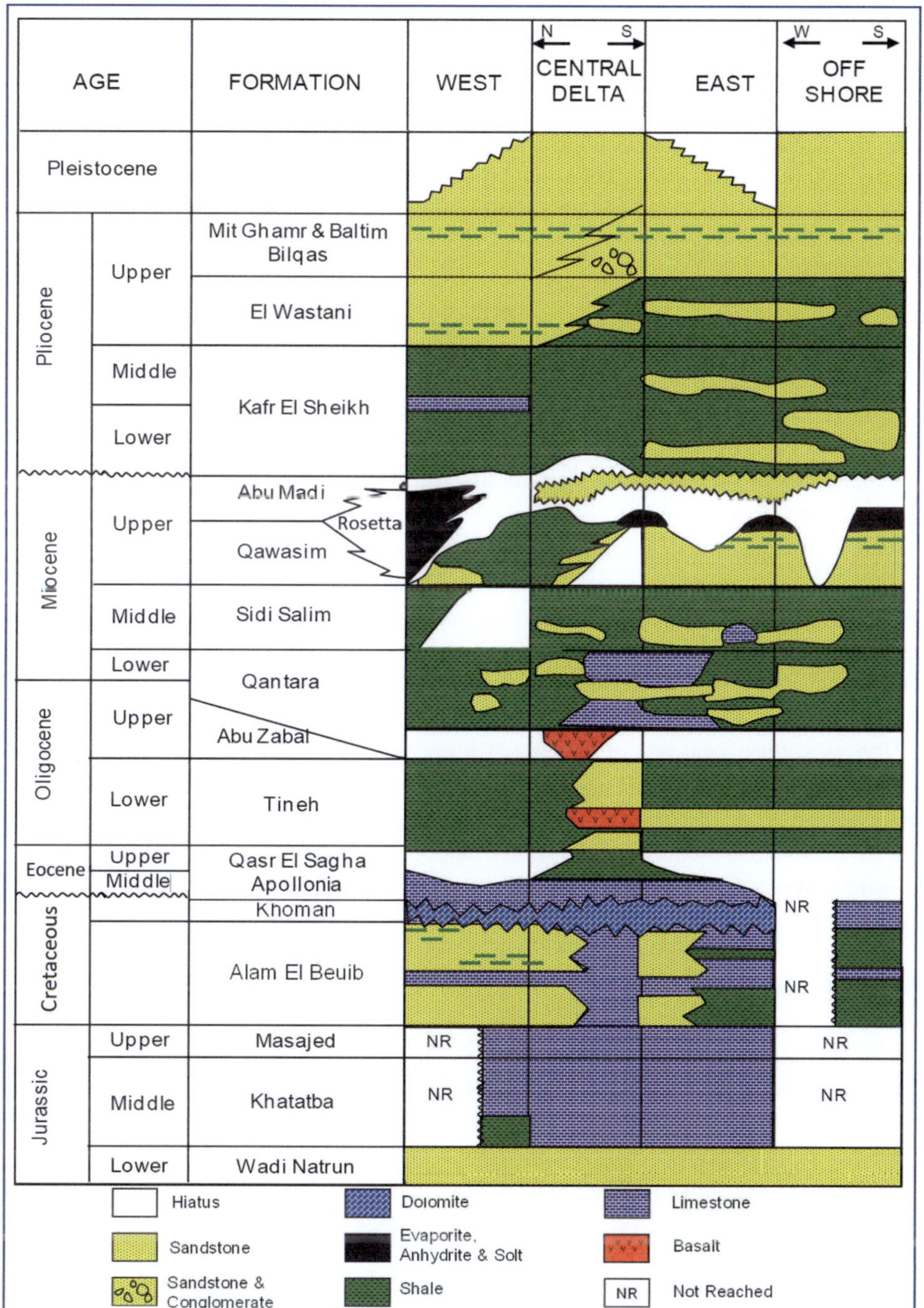

Fig. 12.2 Generalized litho-stratigraphic column of the Nile Delta of Egypt (EGPC 1994; Eysa et al. 2024)

12.3 Structural Configuration

The Nile Delta is situated on the northeastern deformed margin of the African Plate. It has been controlled by some regional structural features that configured its depositional patterns (Kamel et al. 1998; Sarhan and Hemdan 1994; Leila et al. 2016). The structural regime of the onshore Nile Delta was studied by many researchers (e.g., Abu el-Ata 1988; Abu el-Ata and Ibrahim 1988; Said 1990; Tarabees and Azam 2016). Based on this literature, it is stated that during the Pre-Miocene time, the Nile Delta was shaped by some regional geologic events that controlled the Northern province of Egypt (Abd El-Aal et al. 1994; Tarabees and Azam 2016). The Nile Delta is intensively fractured and faulted, and the North Nile Delta province is structurally isolated by a major fault zone (hinge line) from the middle province of the Nile Delta (Fig. 12.3a, b, Sestini 1989; Abd El-Aal et al. 1994; Barakat 2010; Eysa et al. 2024). This flexure zone (30–40 km wide) dates back to the Jurassic crustal breakup of the southern Neotethys. It separates the southern depositional Basin structurally from the northern one (Marzouk 1981). It represents a facies boundary zone separating the northern thick Tertiary marine facies from the southern Delta Cretaceous-Middle Eocene platform (Leila et al. 2016).

The Temsah and Rosetta fault trends, to the offshore northeastern and northwestern parts of the Nile Delta, respectively, subdivided its northern basin into western, central, and eastern offshore subbasins (Figs. 12.3 and 12.4). Therefore, it formulated the tectonic and litho-stratigraphic development of the Nile Delta (Leila et al. 2016; Nabawy et al. 2023a).

The subsurface sequence of the Nile Delta is dominated by multiple stacked sedimentary deposits, which are separated by two major unconformities, indicating the long-lived tectonic history of the Neotethyan continental margin in NE Africa. Most of the subsurface depositional sequences are gently dipping to the north with a set of listric faults trending toward the Mediterranean Sea (Fig. 12.4; Schlumberger 1984; Barakat 2010; Badawy et al. 2018; Fig. 12.5, Abd El-Aal et al. 1994; Sabry et al. 2019).

In general, structurally, the architecture of the Nile Delta is configured by three main faulting trends (Marzouk 1981):

(1) E-W faults (parallel to the hinge line trend),
(2) SE-NW faults (parallel to the Clysmic trend of the Gulf of Suez), and
(3) SW-NE faults (parallel to the Gulf of Aqaba trend).

Northern Egypt, in general, and the Nile Delta, in particular, have been configured by three tectonic phases (Abd El-Aal et al. 1994; Tarabees and Azam 2016):

(1) The Late Paleozoic-Early Mesozoic tectonic phase: It is represented by the deepest structures of the Nile Delta going province parallel to the pre-existing E-W faults,

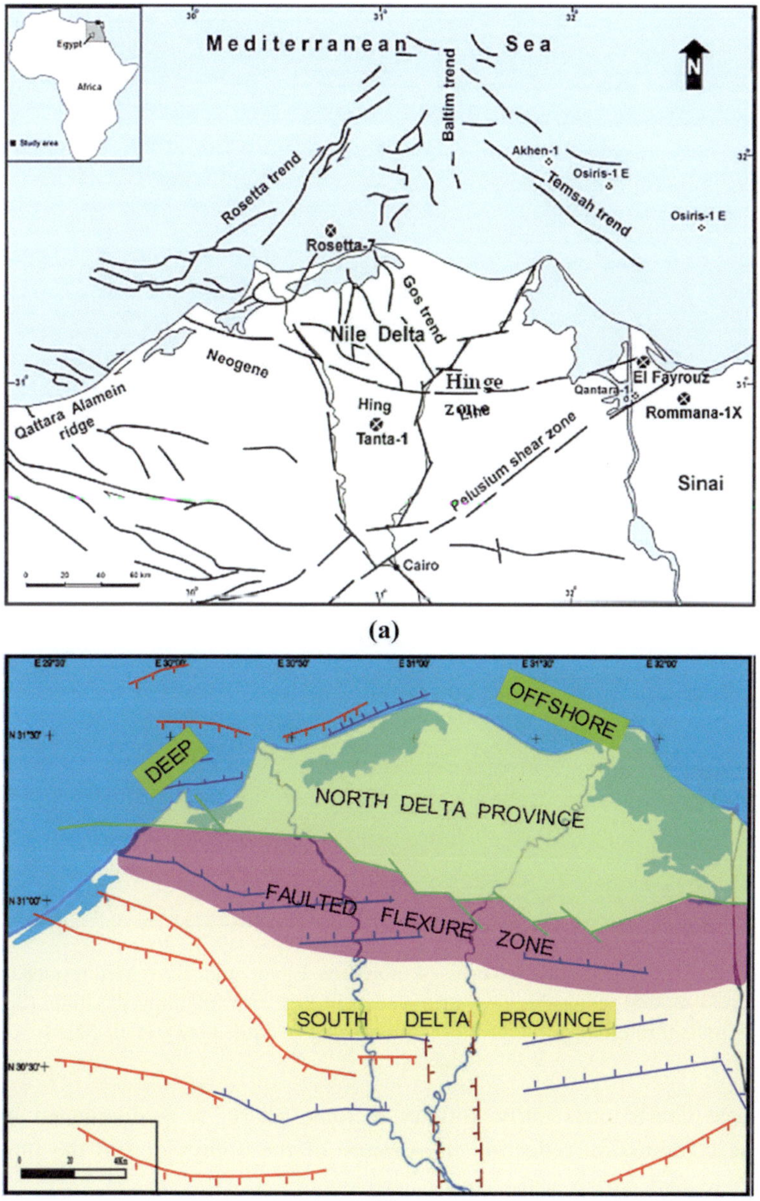

Fig. 12.3 **a** The Hinge line separating the northern Nile Delta from its southern part (Abd El-Aal et al. 1994; Eysa et al. 2024), and **b** the main subsurface tectonic sectors of the Nile Delta (Sestini 1989; Barakat 2010)

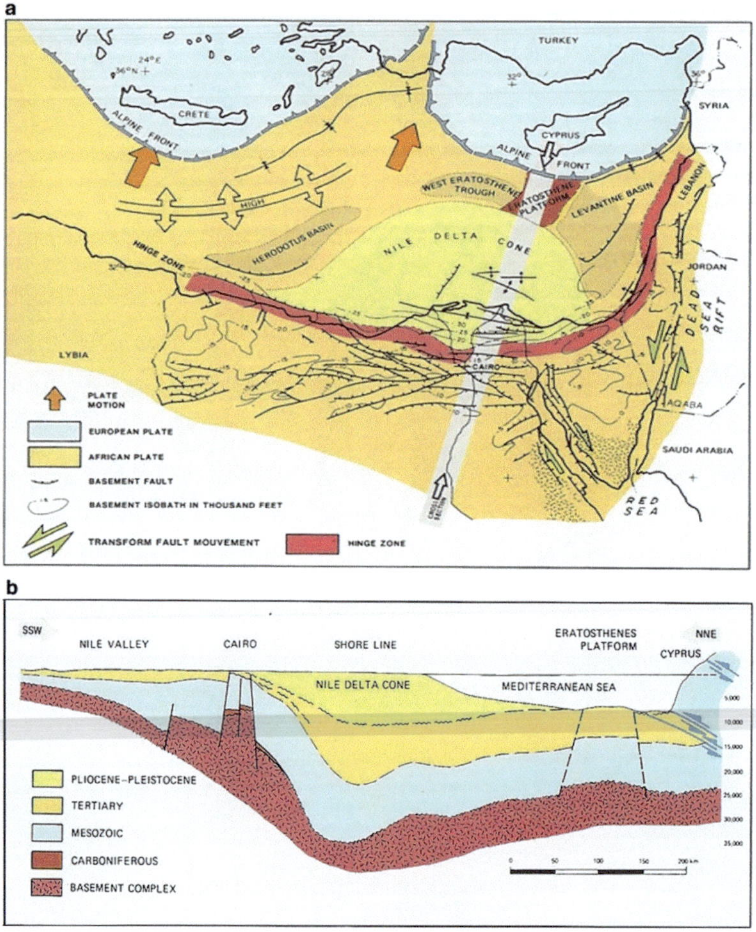

Fig. 12.4 **a** The main tectonic features of northern Egypt and East Mediterranean, and **b** a schematic cross-section along the grey strip shown in Fig. an explaining the main dipping of the various rock units through the Nile Delta (Schlumberger 1984; Badawy et al. 2018)

(2) The Late Cretaceous- Early Tertiary tectonic phase: It is dominated by SE-NW oblique compression reflecting the closure of the Tethys due to the movement of Eurasia toward Africa, and

(3) The Late Eocene to Recent tectonic phase: It can be subdivided into three fault trends (Abd El-Aal et al. 1994; Tarabees and Azam 2016):
- The structural N-S Baltim trend.
- The Late Eocene-Miocene faults are parallel to the Gulf of Suez Clysmic trend.
- The Miocene-Recent faults parallel to the Gulf of Aqaba Rift.

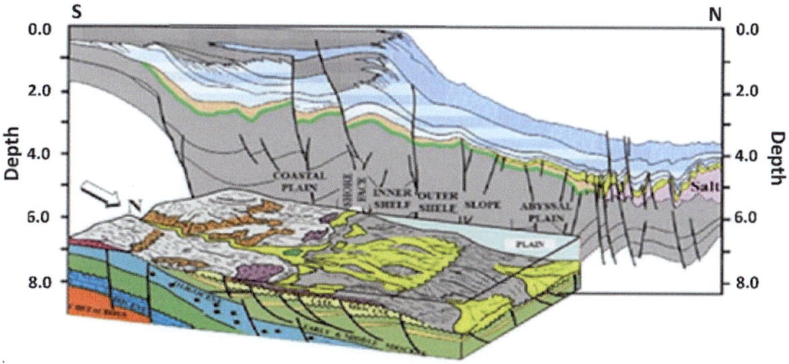

Fig. 12.5 The geologic section and stratigraphic model of the pre-Pliocene sedimentary sequence of the Nile Delta (Abdel Aal et al. 1994; Sabry et al. 2019)

12.4 Applied Techniques

For detecting the hydrocarbon prospectivity of El Tamad Field, the lithology is determined using the composite logs of five wells (El Tamad 1, 2, 3, 4, and 7), the total porosity of the Sidi Salem Formation was estimated from the neutron-density logs, using the Schlumberger's formula (1985), while the effective porosity was determined as effective porosity from the sonic logs (Nabawy and Shehata 2010). Shallow, deep, and micro-spherical resistivity logs were used for the resistivity determination and estimating the water saturation (Sw) in the uninvaded zone using Archie's equation (1942), while the average permeability values were calculated in terms of the valid effective porosity values and Sw using Timur's equation (1968) (Nabawy and Shehata 2010; Baouche and Nabawy 2021).

Prediction of the hydrocarbon-water contact depends on the Repeat Formation test (RFT) data, which allows repeated pressure measurements to be made during a single descent in the borehole (Dake 1978; Chapman 1983; Badry et al. 1993; Lee et al. 1998; Nabawy and Shehata 2010; Radwan and Nabawy 2022; Elmahdy et al. 2023; etc.).

To reveal the facies types, mineral composition, internal pore and grain fabrics, clay types and distribution, and the prevailing diagenetic features a representative set of petrographic thin sections (21 samples) impregnated with blue dye and stained with Alizarin Red-S, images of scanning electron microscopy (SEM of 18 samples), and X-ray diffraction (XRD of 10 samples) data were available for the Qawasim Formation in El Tamad-2 well. In addition, some core data, including the horizontal and vertical permeabilities and the helium porosity, were available for some plug samples (Leila et al. 2016).

For the Qawasim sequence in West Dikirnis Field, a complete set of conventional well logs (gamma-ray, density, PE, neutron, shallow and deep resistivity, and sonic logs) and core data (porosity and permeability) as well as petrographical studies were available for only four wells (WD-1, 2, 3 and 7) (Nabawy et al. 2023a). They considered the shale parameters as the neutron porosity = 21.5%, the true electric resistivity = 3.1 ohmm, the bulk density = 2.54 g/cm^3, and the transit time = 79.5 μs/ft, while Archie's parameters are assumed as a = 1.0, m = 2.1, n = 2.3, and Rw = 0.021 ohmm at 215° F formation temperature. They applied 5% as a porosity cutoff value, 40% for shale volume, and 65% for water saturation (Nabawy et al. 2023a). In addition, a total of cleaned and dried 435 core samples derived from the Qawasim reservoir were available for the petrophysical study. Helium porosity, density, nitrogen vertical and horizontal permeabilities, as well as water and oil saturations, were measured for these samples. The measuring techniques and workflow were published in detail by many researchers (Amaefule et al. 1993; Kassab et al. 2015; Abuhagaza et al. 2021; Fea et al. 2022; Fallah-Bagtash et al. 2022; Cherana et al. 2022). For revealing the sedimentological facies and the various pore types of the Qawasim reservoir, a set of thin sections was prepared and dyed blue by impregnation with Araldite for the petrographical studies (Nabawy et al. 2023a).

12.5 Case Studies

12.5.1 El Tamad Field

El Tamad Field is a local limited Field with a few drilled wells producing oil from the Sidi Salem and Qawasim formations, so little literature has been published on it (e.g., Nabawy and Shehata 2010; Leila et al. 2016). It is located in the El Mansura concession to the east of the offshore Nile Delta (Fig. 12.6).

12.5.1.1 Results

I. **Structural Setting**

El Tamad area to the south of the El Mansoura concession in northern Egypt has been studied for a long time using a 2-D seismic survey. A fast-track, post-stack migration/ conventional processing was applied, and a special AVO processing was applied as well. Due to the complexity of the structural pattern and situation of the deep target Sidi Salem Formation and the overlying rocks in El Tamad Oil Field, it was necessary to conduct an additional 3-D seismic survey, which has contributed significantly to more accurate structural and stratigraphic interpretations. Carrying out the 3-D seismic interpretation enabled minimizing the possibility of dry holes. The obtained 3-D seismic data were very heterogeneous due to the complexity of the structural setting in the main stratigraphic trap in El Tamad Field, but at least the obtained data enabled, for the first time, the detection and interpretation of the top Sidi Salem unconformity.

From the 2-D seismic studies, the El Tamad structure is a three-way dip closure in the down dip of an E-W major normal fault trend (growth fault), which bounds the structure from the south in the El Tamad area (Nabawy and Shehata 2010). This fault probably reactivated during the Sidi Salem sandstone deposition. The reservoir sequence in the

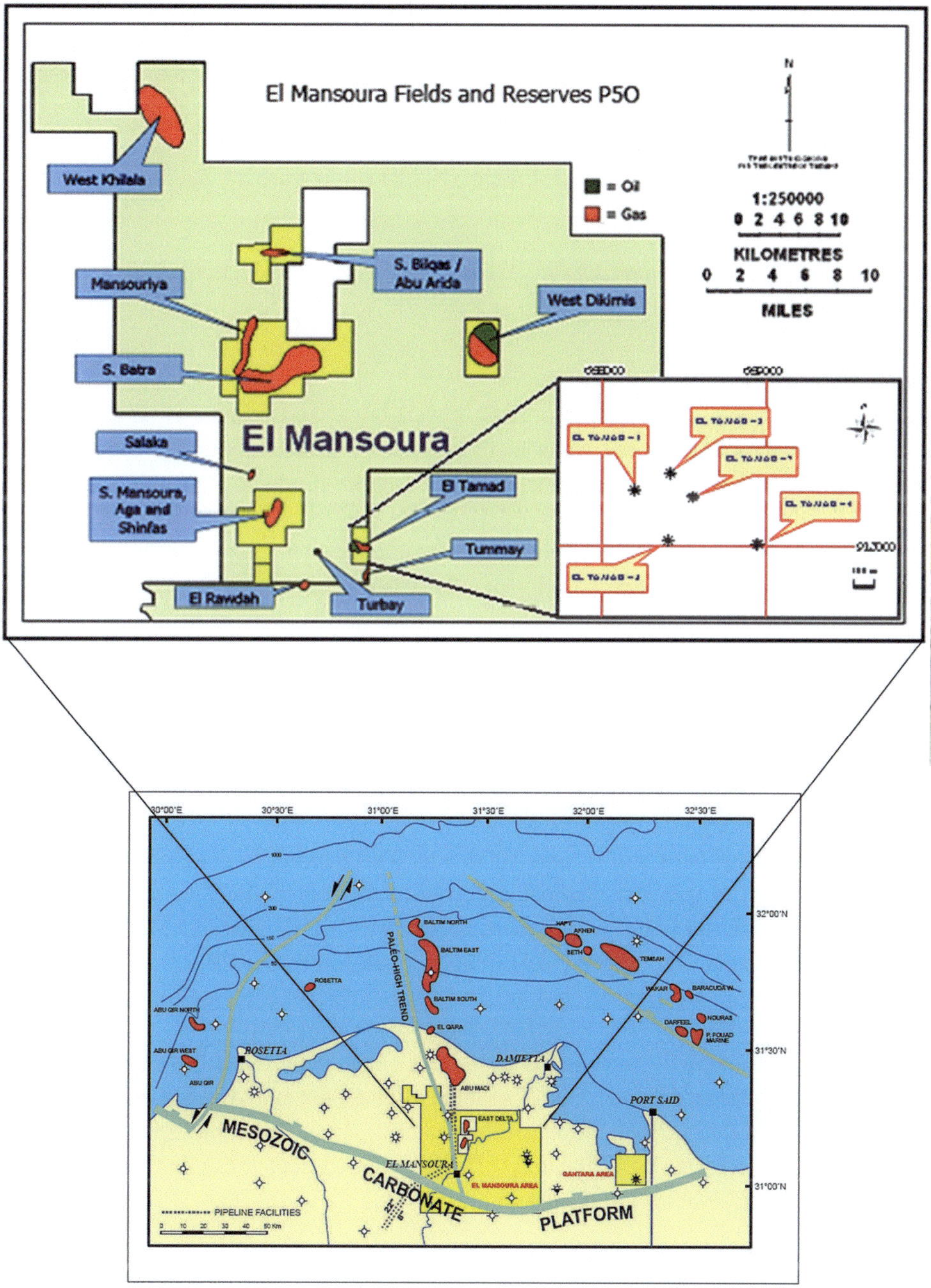

Fig. 12.6 El Tamad Oil field in El Mansoura concession to the east of the onshore Nile Delta, the northern Nile Delta from its southern part (Melrose 2005, 2006; Nabawy and Shehata 2010)

studied area is a culmination bounded in the southern part by that fault, forming three-way trapping. The entrapment has been configured by a rollover anticline on the down dip throw. The anticline has zones represented in the tilted fault block due to some minor faults affecting Sidi Salem anticline, which has a gentle slope towards the NE and NW directions, where three wells were drilled in the up-dip of this anticline near the core channel; namely El Tamad 1, 3, and 4 wells (Nabawy and Shehata 2010).

II. **Hydrocarbon Potentiality**

The petroleum system of El Tamad structure was studied by Nabawy and Shehata (2010) as follows.

Source rocks

According to El Nady (2007), the organic geochemical and biomarker analyses of the Miocene source rocks of some wells in the onshore Nile Delta suggested that the Sidi Salem Formation has fair to good mature source rocks producing oil and gas originating mainly from marine organic sources. However, the geochemistry of the condensates revealed that the Sidi Salem condensates were derived from more contributions of terrestrial organic matter. The probability of source rock for the El Tamad reservoir may extend to the Oligocene (Tineh Formation) and the pre-Oligocene sequences (Nabawy and Shehata 2010).

Reservoir rocks

The main reservoir in El Tamad Oil Field is the deep target Sidi Salem Formation, which is predominantly a shale sequence; it produces light oil of API = 46°, gas, and condensates from sand lobes and channels of effective porosity varying from 17 to 26% and permeability ranging between 15 and 1000 mD. The maximum net-pay thickness (26–33 ft) encountered in the core channel occurred to the east of El Tamad 3 and 4 wells, whereas the lowest thickness was present to the west of El Tamad 1, 2, and 7 wells, penetrating the channel tail (Table 12.1) (Nabawy and Shehata 2010).

Seal rocks

The El Tamad structure is surrounded by a set of sealed faults acting as the main trap of the structure. Moreover, the post-Tortonian Sidi Salem and Qawasim shales act as stratigraphic traps due to their impermeable nature. Moreover, some opinions extended upward to add the thick Kafr El-Sheikh shales as good seal rocks for the Sidi Salem sandstone reservoir (Nabawy and Shehata 2010).

III. **Lithofacies Types**

The core sequence of the El Tamad Oil Field well is represented mainly by the lower parts of the Kafr El-Sheikh and the upper parts of the Qawasim formations. On the core scale, the Kafr El-Sheikh Formation consists mainly of dark grey mudstones with minor

Table 12.1 The average petrophysical parameters of the Sidi Salem sandstone reservoir in the El Tamad Oil field (Nabawy and Shehata 2010)

Well	Reservoir zone (TVDSS, feet)		Net-pay (feet)	Øe (%)	Rt (ohm m)	Sw (%)	k (mD)
	Top	Base					
El Tamad 1	6400	6420	12	26	10	25	366.1
El Tamad 2	6400	6422	20	20	7	39	47.42
El Tamad 3	6394	6427	33	25	30	15	855.6
El Tamad 4	6400	6426	26	26	30	15	1016.8
El Tamad 7	6413	6422	8	17	7	46	16.67

* Effective porosity (Øe) was estimated from sonic logs, and Rw = 0.043 ohmm

interbeds of fossiliferous limestone and seems not promising (Leila et al. 2016). Downward, the core sequence of the Messinian Qawasim reservoir is divided into three types of subarkosic arenites: massive to laminated sandstone, fossiliferous dolomitic sandstone, and fossiliferous calcareous sandstone. In addition, some samples of algal dolostone and fossiliferous limestone are less frequently noticed (Leila et al. 2016).

Petrographically, the Qawasim reservoir is mostly composed of massive to laminated sandstone (subarkosic arenite), which consists of fine to very fine quartz grains with significant amounts of altered feldspars that are partly deteriorating the intergranular pore connectivity. A few amounts of framboidal pyrite and iron oxide patches are also noticed (Fig. 12.7a). The intergranular porosity of these facies is good to very good due to the poor cementation (Fig. 12.7b, c).

The fossiliferous dolomitic sandstone consists of quartz and feldspar grains (5–20% plagioclase and potash feldspars) with some glauconite pellets and rock fragments. Fossils are represented by shell fragments, echinoderm spines, and red algae cemented by some pseudo-dolostone patches (Fig. 12.7d, e). It seems that the precursor cement has been replaced by dolomite during the dolomitization process. The interstitial pore spaces of this facies are blocked by dolomite rhombs, reducing its pore connectivity (Fig. 12.7d, e), i.e., its permeability and visual porosity values.

Finally, the fossiliferous calcareous sandstone is composed of coarse bioclasts (foraminifera, benthic, red algae, and bivalve), with many fine to very fine quartz grains. It is described as algal packstone, which is tightly cemented by pseudo- to microsparite. These facies seem to be deposited in the highly agitated depositional environment (shelf lagoon) due to the abundance of terrigenous quartz grains and the roundness of the fossil fragments (Leila et al. 2016). Some algal dolostone and fossiliferous limestone samples are also described by the same authors (Fig. 12.7f; Leila et al. 2016).

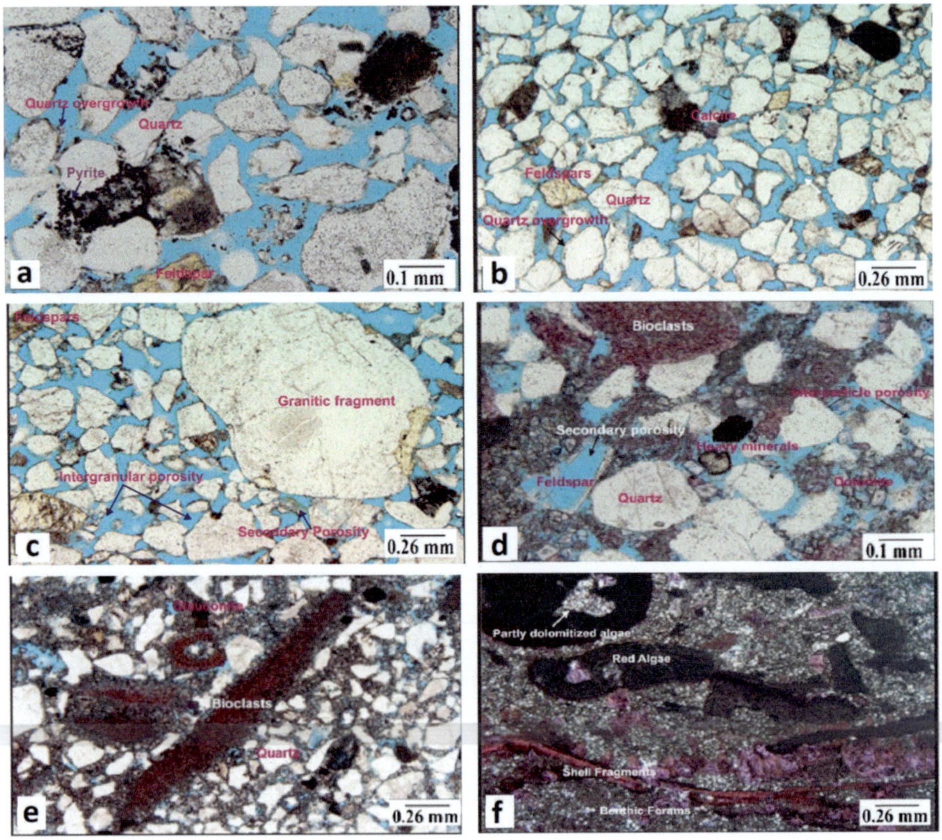

Fig. 12.7 Photomicrographs showing the fabric of the massive to laminated sandstone facies: **a** Framboidal pyrite partially filled the intergranular pore spaces of sandstone facies, and **b**, **c** Poorly cemented subarkose arenite having good to very good intergranular porosity. The fabric of the bioclastic dolomitic sandstone facies: **d**, **e** A few dolomite rhombs scattered within the pore spaces, and **f** Algal dolostone facies showing partially dolomitized algae, and dolomite rhombs filling the pore spaces (Leila et al. 2016)

IV. **Petrophysical Properties**

On the plug scale, considering the core analysis data derived from El Tamad-2 well, the helium porosity (Ø) is higher for the massive to laminated sandstone and the dolomitic sandstone ($Ø_{av.} \approx 30.30\%$) than the other facies, with the lowest values assigned to the bioclastic limestone ($Ø_{av.} \approx 16.3\%$). Also, the average horizontal permeability values are assigned for the laminated and massive sandstones (kav. = 1770.4 mD and 1223.4 mD, respectively), while lower permeability values are assigned for the dolomitic sandstones (kav. = 355.88 mD) due to the partial plugging of the interstitial pores, and much less for the bioclastic limestone (kav. = 1.70 mD). The horizontal and the vertical permeability values are primarily contributed by the helium porosity ($R^2 = 0.85–0.87$), indicating the verod pore connectivity (Leila et al. 2016). The authors stated that the laminated and massive sandstone facies have the best reservoir prospectivity.

On the wellbore scale, the Sidi Salem Formation was studied by Nabawy and Shehata (2010) using a set of conventional well log data (gamma-ray, spontaneous potential, neutron, density, sonic, PEF, and shallow and deep resistivities) derived from wells (El Tamad 1, 2, 3, 4, and 7). Locations of these drilled wells were chosen, according to the available seismic data, to be drilled in the core and tail of the sand channels of the Sidi Salem deep target (Nabawy and Shehata 2010).

In a complementary routine work, the hydrocarbon/water contacts in the wells drilled through the core and the tail of the deep target sandy trap have been achieved using the RFT tool, which helped detect the position of the net-pay zone in the target reservoir. There was an agreement between the hydrocarbon/water contact detected from the log data and that detected from the RFT data plots. It is achieved that the net-pay zone was detected further benefiting from the petrophysical well-log data, where the maximum average porosity values (Ø) and the minimum water saturation values (Sw) were recorded in El Tamad-1, -3 & -4 wells (Fig. 12.8), near the core channel. Also, the minimum average porosity values and the maximum water saturation values were estimated for El Tamad-2 & -7 wells, near the tail channel. It is also achieved from the present studies that, the target structure is slightly tilted to the NE, where El Tamad-2 and -7 wells (the tail channel) were drilled, so the gas is completely absent in these wells, whereas the water is absent in El Tamad-1 well, that has been drilled through the up-dip structure (Fig. 12.8). It is stated that oil is found in economic reserve amounts in El Tamad with a good gas cap. The petrophysical analyses of the obtained well logs of the various boreholes indicate the presence of a good thickness of pay zone in this field (Fig. 12.9). The average porosity values range from 17 to 26%. The maximum average values recorded in El Tamad 1, 3, and 4 wells are 26, 25, and 26% (Table 12.1), respectively, near the core channel, whereas the minimum average values recorded in El Tamad 2 and 7 wells are 20 and 17%, respectively. The average Sw values range from 15 to 46% (Table 12.1). The average Sw_{min} is recorded as 15% in El Tamad-4 & -3 wells (near the core channel), while the average Sw_{max} (46%) is recorded in El Tamad 7 well, respectively, to the north down dip, near the tail channel (Nabawy and Shehata 2010).

There is a good agreement between the hydrocarbon/water contacts detected from the log data and those detected from the RFT data plots. Depths of the free oil and free water levels in the various boreholes were determined using the slopes of the pressure-depth best-fit lines intersections obtained from the RFT logs (Fig. 12.9). The oil gradient fluctuates between 0.25 and 0.28 psi/ft for the different boreholes, the gas gradient is in the range of 0.062–0.079 psi/ft, while the water gradient varies between 0.42 psi/ft and 0.46 psi/ft. The RFT and litho-saturation plots indicate the absence of a gas cap in El Tamad 2 and 7 wells, only an oil reservoir with no gas cap, where they have been drilled in the down dip of the studied anticline structure; whereas a gas cap is detected in the

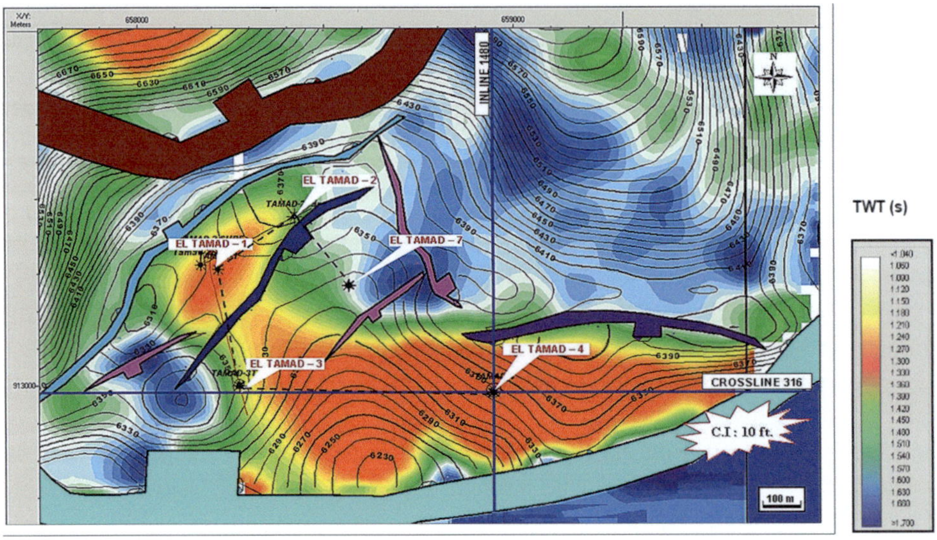

Fig. 12.8 Structure depth contour map on top of the Sidi Salem pay zone in El Tamad oil field (Nabawy and Shehata 2010)

other wells that were drilled in the up-dip flank (El Tamad 3 well, Fig. 12.10, Nabawy and Shehata 2010).

The iso-parametric maps of the petrophysical parameters of El Tamad Oil Field indicate the best porosity values near the core channel (25–26% at El Tamad 1, 3, and 4 wells), whereas the minimum values are recorded to the northeast down dip, near the tail channel (17–20% at El Tamad 2 and 7 wells), where the lithofacies is tight. The lowest Sw values are recorded in El Tamad 3 and 4 wells (15%, Table 12.1), while the highest values are in wells in El Tamad 1, 2, and 7 wells (25–46%, Table 12.1), to the north down dip, near the tail channel. Permeability values are ranked as very good reservoir quality for El Tamad 1, 3, and 4 wells, while it is fair to good for El Tamad 7 and 2 wells (Table 12.1).

For El Tamad Field, permeability can be estimated by considering the effective porosity and Sw using the following model (Nabawy and Shehata 2010).

$$\mathrm{Log\,k} = 9.2\,\mathrm{Log}\,\varnothing\mathrm{e} - 10.2 \quad (\mathrm{r} = 0.963)$$

$$\mathrm{Log\,k} = -3.4\,\mathrm{Log\,Sw} + 7.1 \quad (\mathrm{r} = -0.976)$$

In general, the petrophysical values detected for El Tamad wells through the core channel are prospective, where the permeability values increase with increasing porosity and decreasing water saturation. Furthermore, the electric resistivity values increase toward the up-dip structure/the core channel, indicating the presence of a good accumulation

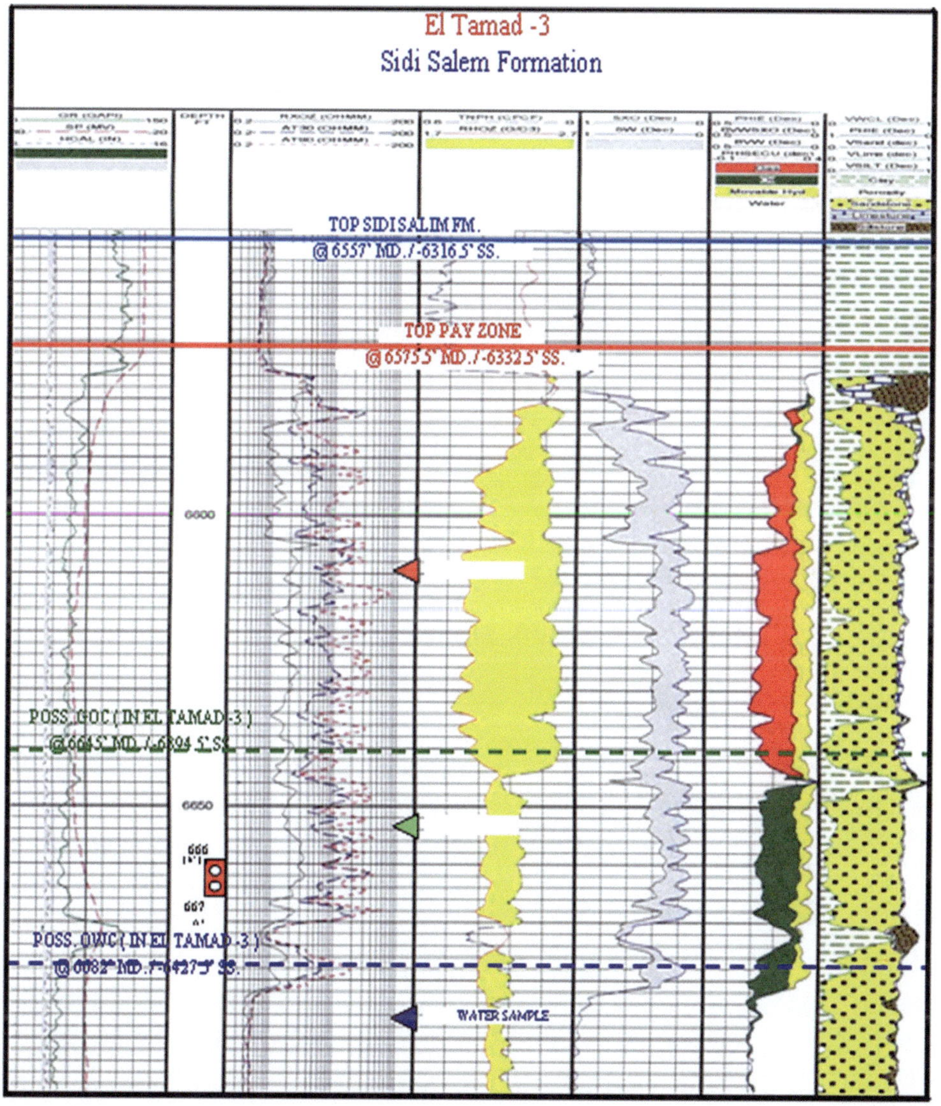

Fig. 12.9 A litho-saturation log of well El Tamad 3, Sidi Salem formation, El Mansoura Concession (Nabawy and Shehata 2010)

of oil to the west. Therefore, from these complementary and compatible studies, it was expected to have the maximum net-pay zone thickness to the south in the center of the deep target Sidi Salem channel and decrease to the northeast toward the channeling tail.

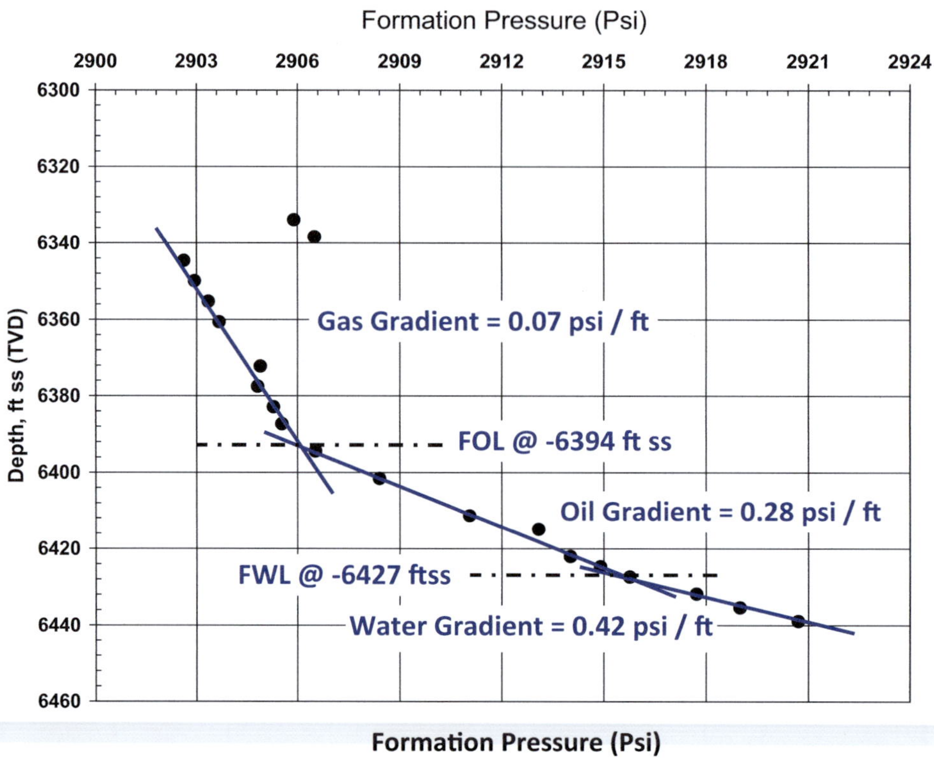

Fig. 12.10 The RFT pressure analysis of El Tamad-3 well indicates the free oil level (FOL) and the free water level (FWL) (Nabawy and Shehata 2010)

12.5.1.2 Discussion

I. ***Implementation of the Primary Factors on the Reservoir Quality***

The Late Cenomanian Eonile acted as a deep canyon in which the Abu Madi and Qawasim formations were accumulated due to the transgression and regression of the Mediterranean (Said 1990; Leila et al. 2016). The siliciclastic composition of the Sidi Salem and the Qawasim sandstone sequence, with the dominance of quartz, feldspars, and igneous rock fragments, declares the derivation of these components from the basement complex of the Red Sea province (El-Sisi et al. 1996; Leila et al. 2016). The microfacies association, composition, and sedimentary structures indicate a transition from a shallow depositional marine setting to an open marine setting with some fluvial alternations. The alternation between the laminated and the massive sandstone facies refers to the predominance of alternated low and high-energy turbidity currents, resulting in the accumulation of high-density and low-density turbidites. Following Leila et al. (2016), this statement is consistent with the achievement of Rizzini et al. (1978), who concluded that typical high-energy deposits are represented by a thick massive sandstone sequence at the base of the Qawasim reservoir. Moreover, the gradual change between the laminated and the massive siliciclastics, the fossiliferous dolomitic and calcareous sandstones, and the fossiliferous limestones ensures the intensive change in the depositional configuration of the Qawasim reservoir. This heterogeneity is supported by the dominance of the bivalve red algae and the rhodolith fragments that are scattered in a dolomitic matrix of the fossiliferous dolomitic sandstone facies. These alternations between the clastic and non-clastic facies (the calcareous and the dolomitic sandstones, and the laminated and massive sandstones) resulted in a high heterogeneity of the entire sequence and gave a chance for a highly promising sequence (Leila et al. 2016).

II. ***Implementation of the Diagenetic Factors on the Reservoir Quality***

Diagenetic factors are the main post-depositional processes controlling the reservoir quality. They may act positively by supporting the primary petrophysical properties (dissolution and fracturing) or negatively by deteriorating the petrophysical properties (compaction, cementation, authigenic minerals, and neomorphism), while dolomitization may be considered a deteriorating or enhancing factor based on its impact grade (e.g., Nabawy and El-Aal 2019a, b; Nabawy et al. 2020a, 2022b, 2023b; Mebrouki et al. 2024; Kassem et al. 2024).

The fossiliferous limestone facies are primarily controlled by compaction and cementation as deteriorating features, while the leaching out and dolomitization act as improving reservoir quality factors. However, this facies seems to be buried at shallow depths with a slight overload pressure impact (Leila et al. 2016). On the other side, the siliciclastic facies of the Qawasim reservoir have been slightly controlled by a slight overload pressure as indicated by the predominance of long and point grain-to-grain contacts (Fig. 12.7a–d). The reservoir quality of the promising Qawasim sandstones is also mostly affected by the calcareous, dolomitic, and siliceous cementation, e.g., the fossiliferous dolomitic sandstones and fossiliferous calcareous sandstone facies. The fossiliferous dolomitic sandstones are highly cemented by dolomite, deteriorating and occluding the volume and connectivity of the interstitial pore spaces (Fig. 12.7e, f). The siliceous cement is primarily represented by the quartz overgrowth, which is developed due to a slight overload pressure, i.e., its implementation is slight. The presence of authigenic clay content and pyrite is another main deteriorating-reservoir quality factor affecting the Qawasim sandstone facies. The clay content is dominated by kaolinite and illite in shaley zones and is mostly attributed to the feldspar alteration in these zones. The scattered framboidal pyrite crystals are noticed through the pore channels of the laminated sandstones to slightly block the interstitial pores (Fig. 12.7a). They are likely accumulated in the pores due to the invasion by marine water in the phreatic zone causing the reduction of SO_4^{2-} into sulfides forming the iron sulfides FeS (Berner 1981; Leila et al. 2016). The authigenic kaolinite and illite are represented by booklets and hair-like fibers partially lining and blocking the pores to mimic the pore connectivity and fluid flow, causing an effective reduction in the permeability (Leila et al. 2016).

On the other side, the laminated and the massive sandstone facies are poorly cemented by dolomite and calcite and still have many primary interstitial pores (Fig. 12.7a–d). They are highly affected by the dissolution and leaching out of the unstable rock components, especially the feldspars and calcareous fossil fragments, and the cement as a main enhancing-reservoir quality factor improving the pore volume and connectivity of these facies due to the invasion by a low Ca-bearing solution (Mathisen 1984; Nabawy 2013; Leila et al. 2016; Nabawy and Géraud 2016). Dissolving and leaching out the feldspars, leaving relics behind, helped in evolving and supporting the secondary porosity values represented by intergranular and vuggy pore spaces (Nabawy 2013; Nabawy and Wassif 2017; Safa et al. 2021).

12.5.2 West Dikirnis Field

Similar to El Tamad Field, the West Dikirnis Field is located in El Mansoura onshore concession, which is built up of thick Tertiary deposits structurally formulated by horst structures trapping the hydrocarbon content in the oil window (EGPC 1994; Abuamarah et al. 2019; Nabawy et al. 2023a; Eysa et al. 2024).

12.5.2.1 Results

I. ***Structural Setting***

The structure-dip contour map on top of the Qawasim reservoir shows that this field is dissected by a local hinge zone trending E-W, throwing down the Pliocene-Miocene reservoirs into the north (Fig. 12.11). Some additional N-S faults throw down the sequence to the east & the west (Harms and Wary 1990; Nabawy et al. 2023a), giving rise to the WNW-ESE four-way closure of this field (Abd El-Aal et al. 1994; Nabawy et al. 2023a). These high stands are dominant in the field (Fig. 12.12; Leila and Moscariello 2018; Nabawy et al. 2023a); this active syn-depositional tectonic event controlled and formulated the presence of the Messinian sandstone reservoirs (Nabawy et al. 2023a). The deeply-seated E-W faults form suitable trapping zones for entrapping the hydrocarbons, while the Jurassic, Cretaceous, and Early Tertiary sedimentary sequences represent potential source rocks for charging the Messinian reservoirs by their hydrocarbon reserves (Leila and Moscariello 2018; Nabawy et al. 2023a) (Fig. 12.13).

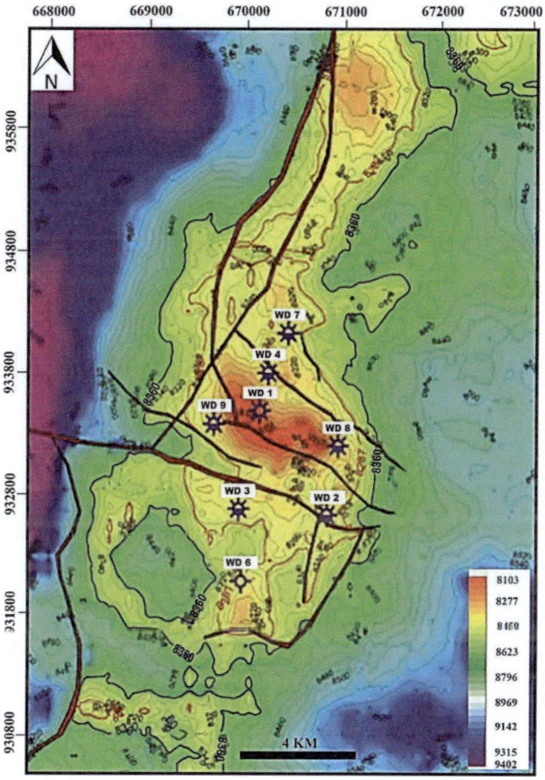

Fig. 12.11 Structure contour map on the top surface of the Qawasim formation in West Dikirnis field (C.I. = 86.57 ft) (Nabawy et al. 2023a)

II. ***Litho-stratigraphic Setting***

The revealed stratigraphic sequence starts with the Tineh Formation, which consists of clastic fluvial-dominated facies alternated with open marine argillaceous facies (Salem et al. 2005; Sakran et al. 2019; Nabawy et al. 2023a). During the Miocene time (represented by the Qawasim, Sidi Salem, Qantara, and the Abu Madi formations) and the Pliocene sediments (represented by El-Wastani and the Kafr El-Sheikh formations) (Fig. 12.2), two regional unconformity surfaces are detected due to the critical changes in the depositional rate and the uplifting of the onshore Delta basin. The first unconformity is found between the Sidi Salem and Qantara formations, while another, related to the main Messinian salinity crisis, separates the Messinian Abu Madi Formation from the Pliocene Kafr El-Sheikh Formation (Fig. 12.2). During the deposition of the Kafr El-Sheikh Formation in the Early Pliocene, the Nile Delta experienced a marine transgression that continued until the deposition of the El-Wastani Formation in the Late Pliocene, marking a shift from a fluvio-marine to a fluviatile-dominated depositional system (Fig. 12.2) (Ross and Uchupi 1977; Nabawy et al. 2023a).

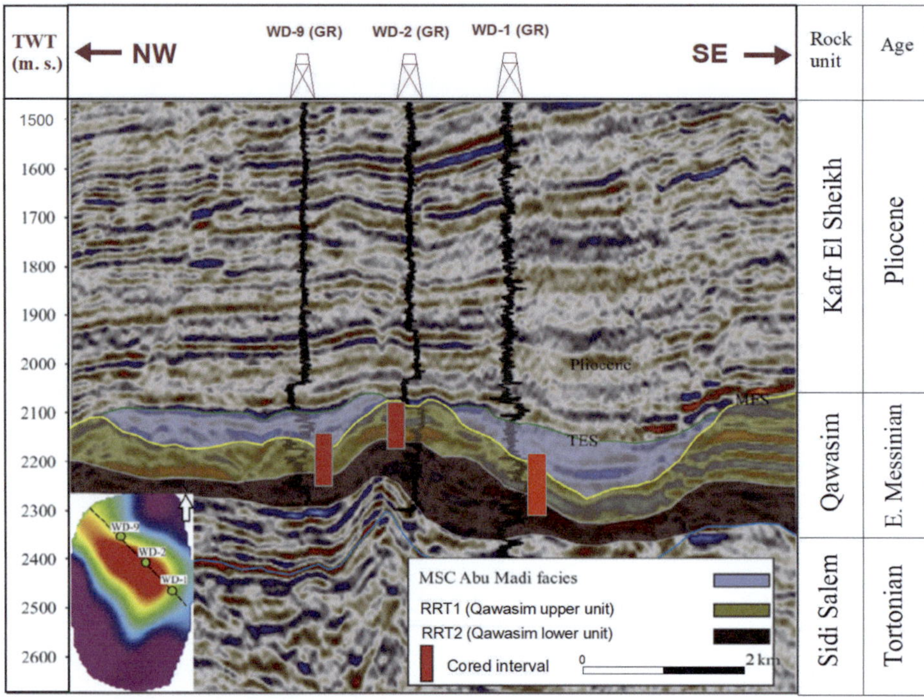

Fig. 12.12 2D NW–SE seismic section along the West Dikirnis field showing the main pre-Messinian stratigraphic zones including the Qawasim formation (Leila and Moscariello 2018; Nabawy et al. 2023a)

12.5.2.2 Discussion

I. ***Reservoir Zonation and Rock Typing***

Nabawy et al. (2023a) stated that in the West Dikirnis Oil Field, the Qawasim sandstone reservoir has a wide range of grain density (2.60–2.72 g/cm^3), indicating a high heterogeneity. Also, a wide range of storage and flow capacities are detected, indicating the need for classifying this reservoir into 5 RRTs (reservoir rock types): RRT5 (sandy mudstone/siltstone), RRT4 (glauconitic quartz wacke), RRT3 (subarkose arenite and glauconitic quartz wacke), RRT2 (sublithic arenite and subarkose arenite), and RRT1 (sublithic arenite). They attributed the relatively low density of the RRT1 to the dominance of some clay content, while the increase in the density of the other rock types is due to the dominance of some iron oxides. Considering the depth and the grain size change, the Qawasim clastic reservoir has a fluvial-deltaic nature. It is a coarsening upward sequence that changes from the RRT1 and RRT2 coarse sandstone at the top to the RRT5 laminated siltstones/mudstones. Thereby, the reservoir quality decreases downward from very good reservoir quality at the top of the sequence ($1600 \leq k_H \leq 10{,}011$ mD, $3.04 \leq RQI \leq 5.66$ µm) to

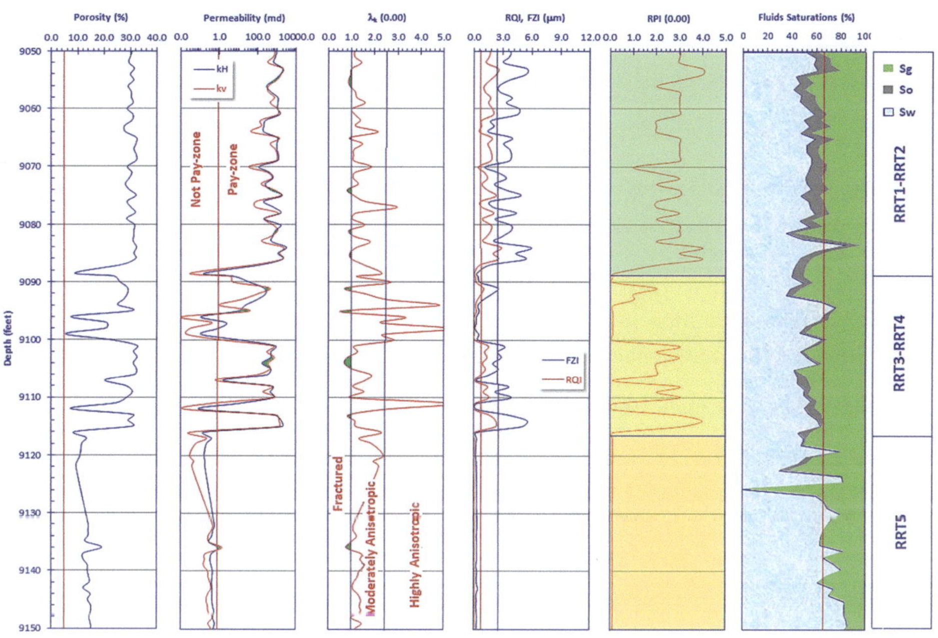

Fig. 12.13 Plotting the reservoir quality and the petrophysical data of the WD-7 well vertically including the porosity (track-1), vertical and horizontal permeability values (k_V and k_H, track 4), permeability anisotropy (λ_k, track-3), reservoir potentiality index (RPI), flow zone indicator (FZI), reservoir quality index (RQI), as well as gas, oil and water saturations (Sg, So, Sw, respectively) as a function of depth (Nabawy et al. 2023a)

a relatively tight/poor reservoir quality downward ($0.085 \leq k_H \leq 40.8$ mD, $0.04 \leq$ RQI ≤ 0.38 μm) (Nabawy et al. 2023a).

They also stated that plotting the helium porosity of the cored Qawasim vertically in WD-7 well shows that it fluctuates between 8 and 30% values downward in a spiky feature while consistently upward around 30%. The spiky behavior is due to the dominance of microfractures as revealed from the permeability anisotropy values (Fig. 12.12, track 3, $\lambda_k < 1.0$, Fig. 12.1) and shale streaks ($\lambda_k > 2.5$). Increasing the k_H to k_V causes λ_k to be > 1.0; it is primarily attributed to the dominance of the primary depositional fabrics and some shale streaks. On the y, decreasing the permeability anisotropy less than one means that the k_V is higher than k_H due to the presence of vertical micro fractures (Nabawy 2015, 2018; Nabawy et al. 2015, 2023a; Elgendy et al. 2020; El Sawy et al. 2020; Abuhagaza et al. 2021). Therefore, discriminating the sequence into hydraulic flow units (HFUs) and RRTs is critically needed to build a mathematical static model of high certainty (Nabawy et al. 2023a).

Nabawy et al. (2023a) added that the lower part of the Qawasim reservoir has an argillaceous composition; its permeability is 1.0 mD, i.e., it is not promising. On the contrary, the permeability of the topmost parts is composed of pebbly and coarse sandstone having excellent permeability values (av. $k_H = 6592$ mD), i.e., it is highly promising. The authors stated that the variation in the petrophysical properties through the Qawasim sequence caused a spiky fluctuation in the reservoir quality, varying between tight and excellent reservoir quality, particularly in the middle parts of the cored interval (RQI > 5.0 μm, and FZI > 15.0 μm). They also attributed this reservoir quality variation to the changes in the effective pore radius between the nano and mega pore radii. The relatively high water saturation in some zones (25.2–86.7%), especially the lower sequence, is contributed to by many shale streaks (Nabawy et al. 2023a).

II. ***Implementation of Facies Types and Porosity on the Reservoir Quality and Permeability***

Nabawy et al. (2023a, c) stated that, although the reservoir rock typing is primarily attributed to the percentage of the permeable pore spaces, the presence of authigenic minerals and the non-systematic spatial distribution of pore spaces disturb both the storage and flow capacities between the vertical and horizontal directions. Therefore, they plotted k_H and k_V of the Qawasim reservoir versus porosity and stated that the RRT1 (sublithic arenite) and the RRT2 (subarkose arenite) are characterized by the best storage and flow capacities (k > 1000 mD) (Fig. 12.14a, b). This plot states a high variance between k_H and k_V of the RRT3 & RRT4 samples (Fig. 12.14c, d). The authors introduced a set of models of high reliability to estimate k_H and k_V in terms of the porosity of the Qawasim reservoir.

The water and gas breakthrough during production because gas and water are less viscous than oil, i.e., their mobility is much higher than oil. Therefore, the presence of horizontal permeability barriers is highly required to prevent this breakthrough, i.e., the presence of horizontal shale streaks and depositional fabrics is highly appreciated ($\lambda_k > 1.0$), indicating the presence of some pressure streaks (Abd El-Aal et al. 2021; Nabawy 2021; Nabawy et al. 2023a; Radwan et al. 2023; Rahmouni et al. 2023).

Also, it is stated that λ_k is the highest for the RRT5 sandy mudstone/siltstone samples (max $\lambda_k = 4.85$) and also the glauconitic quartz wake of the RRT4 ($1.69 \leq \lambda_k \leq 4.75$), while it is the lowest for the massive sublithic arenite of the RRT1 samples ($0.87 \leq \lambda_k \leq 1.57$) (Nabawy et al. 2023a). The high λ_k of the RRT5 and RRT4 plugs is contributed by some flow/pressure barrier streaks as proposed by the classification of permeability anisotropy (Nabawy 2018).

III. ***The Dominant Reservoir Quality Attributes***

Depositionally, the reservoir quality of Qawasim and the predominance of the permeability barriers in its lower unit are contributed by the dominant argillaceous streaks and the depositional setting of the RRT3-RRT5 (glauconitic quartz wacke, subarkose arenite, and sandy mudstone/siltstone) in pro- and distal delta, while the high reservoir quality of the

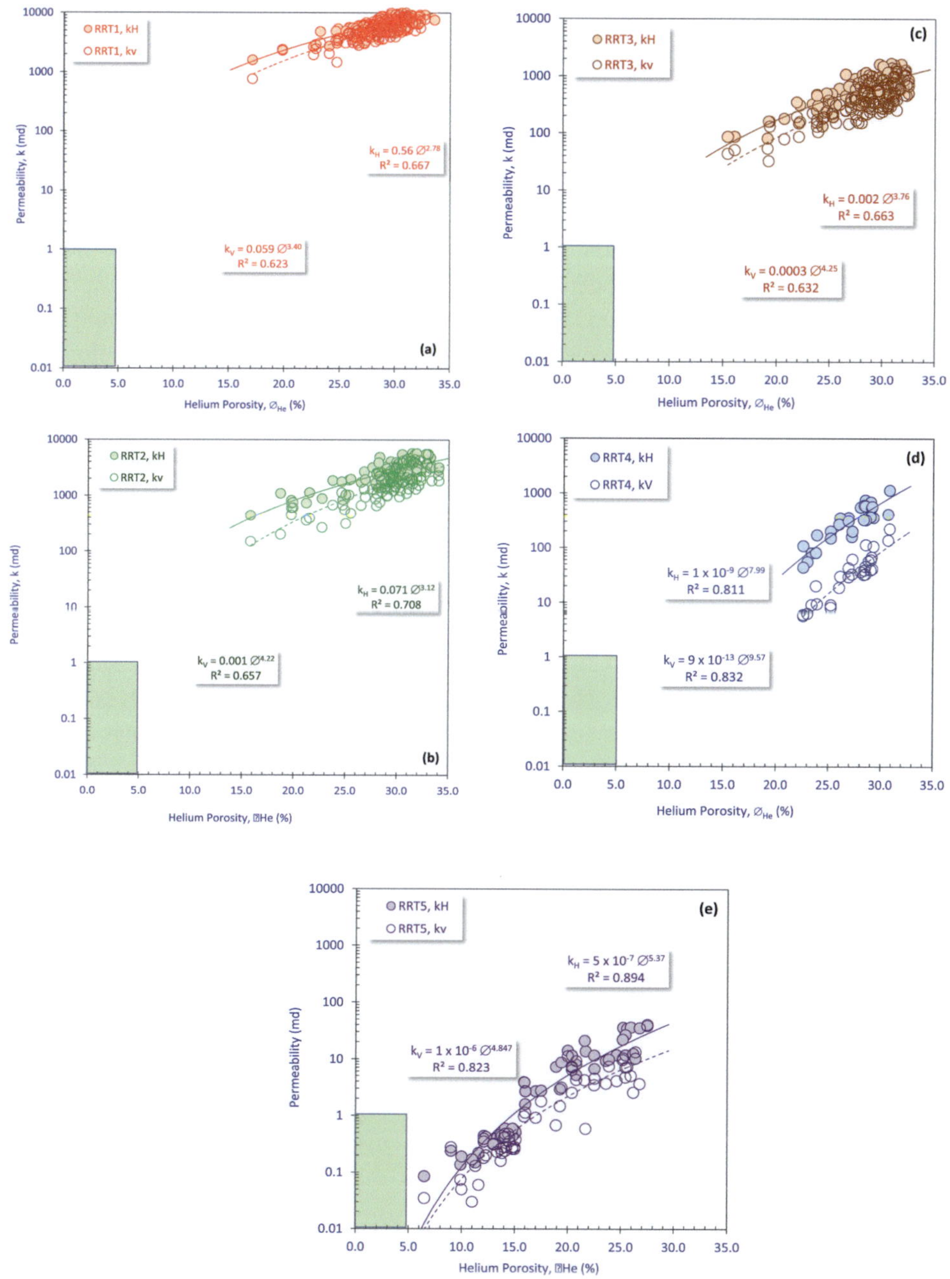

Fig. 12.14 Plotting k_V & k_H versus $Ø_{He}$ of the various rock types of the Qawasim reservoir in West Dikirnis field: **a** RRT1, **b** RRT2, **c** RRT3, **d** RRT4, and **e** RRT5 samples. The green shaded areas refer to the non-promising values that are less than the porosity and permeability cutoff values that were applied by the authors, $Ø_{He} < 5\%$ and $k < 1.0$ mD (Nabawy et al. 2023a)

RRT1 and RRT2 (sublithic arenite and subarkose arenite) is attributed to its depositional setting in proximal delta-plain and front (Nabawy et al. 2023a).

Petrophysically, the reservoir quality (RQ) is primarily contributed by the permeability, pore sizes, and porosity in descending order. The authors introduced some reliable models for estimating the reservoir quality index (RQI) in terms of permeability ($R^2 \geq 0.953$). Due to this high reliability, the authors introduce a general model to estimate the RQI using k as follows.

$RQI = 0.08\ k^{0.47}$ ($R^2 = 0.997$) (Nabawy et al. 2023a).

They added that the dependence of RQI on porosity is much less reliable than the permeability because of the variation in the interstitial pore types, the different composition of the included rock fragments (especially in the sublithic arenites), the distribution of the authigenic minerals, and the pore size distribution.

In addition, the variation in the mineral composition and the predominant diagenetic features from rock type to another are other factors that improved the quality of the RRT1-2 samples that mostly form the upper unit the Qawasim reservoir and the tightness and compactness of the RRT3-5 that form the lower parts (not promising) of the reservoir (Nabawy et al. 2023a).

IV. ***Implementation of Ø on the BVW***

Following many authors (e.g., Dally et al. 2023; Nabawy et al. 2023a, d; Mohamed et al. 2023) stated that increasing the pore volume and pore throat size are the main reason for decreasing the pore connectivity and increasing the irreducible water saturation, i.e., decreasing the bulk volume of water (BVW) and the water breakthrough during the hydrocarbons production (Abdeen et al. 2021; Nabawy et al. 2020b, 2022a, b; Haque et al. 2022; Azab et al. 2024). Therefore, studying the implication of Ø on Sw is highly recommended for revealing the reservoir quality (Teama and Nabawy 2016; Ashoor et al. 2023; Aboelhassan et al. 2024). For the Qawasim reservoir, Nabawy et al. (2023a) plotted the Sw versus Ø (Fig. 12.15). BVW is a critical parameter referring to the ability of producing water during the hydrocarbons retrieving, especially when BVW is higher than 0.05; it is the multiplication result of the Sw and Ø values (Buckles 1965; Holmes et al. 2009). Plotting the Sw and Ø indicates their scattering along a wide range of Sw and Ø because of the variation in the pore types and pore size, and clay content. Also, it revealed that the BVW of most samples is higher than 0.75, i.e., the hydrocarbon production is accompanied by high water content (Nabawy et al. 2023a).

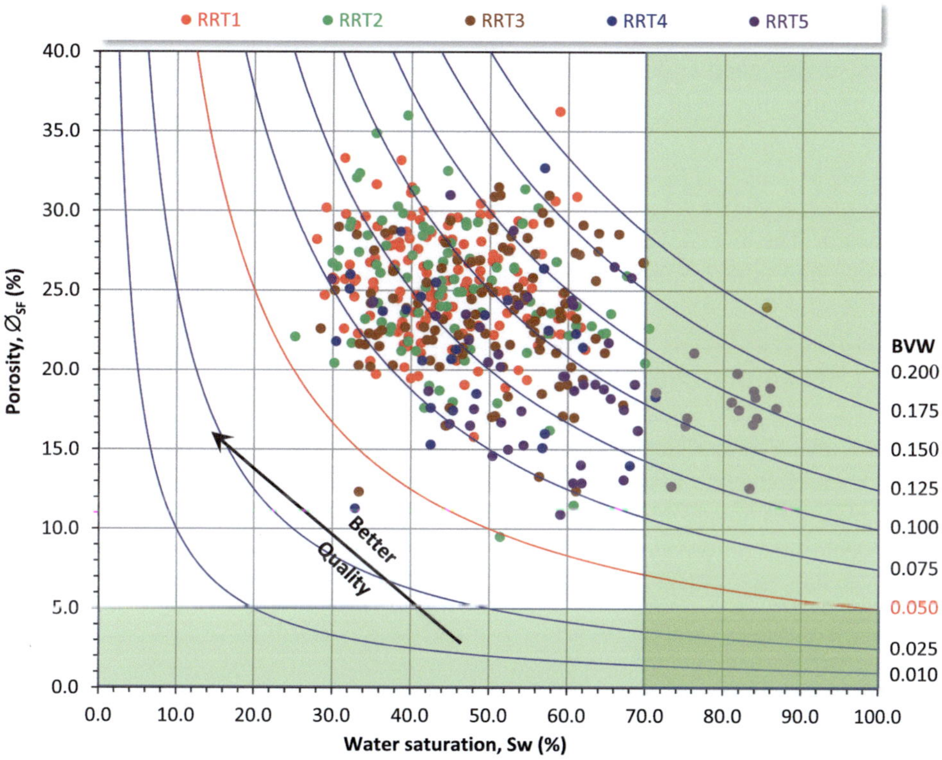

Fig. 12.15 Presenting the summation fluids porosity as a function of Sw Shaded green area refers to non-prospective samples ($Ø_{cutoff} = 5.0\%$, $Sw_{cutoff} = 70\%$) (Nabawy et al. 2023a). The red BVW hyperbola line indicates the BVW cutoff value (0.050) that is referred by (Buckles 1965; Holmes et al. 2009; Nabawy et al. 2023a) to produce oil without water at BVW values less than it

12.6 Conclusions

Based on the current review and previous literature, the deep-seated structural setting of the Nile Delta has resulted in the definition of only a few fields as oil/condensate producers. The 3D seismic survey is an advanced technique that allows for more accurate detection of volatile oil and condensate in horsts and anticlines.

The only two discovered oil/condensate producing fields in the Nile Delta province, El Tamad and West Dikirnis fields, are located in the El Mansoura concession in the eastern portion of the onshore Nile Delta Basin as a result of the formation of some structural horsts and three-way dip closures. These fields are primarily producing oil/condensate from the Abu Madi, Qawasim, and Sidi Salem formations.

The studied reservoirs are primarily dominated by structurally controlled stratigraphic traps, including sand lobes, lenses, channels, and tongues included in thick shale sequences. These siliciclastic reservoirs are characterized by excellent to very good porosity (>20%) and excellent permeability (>1000 mD), i.e., they are characterized by excellent to very good reservoir quality.

Petrographically, these siliciclastic reservoirs consist of subarkose arenite, sublithic arenites, sandy mudstone/siltstone, and glauconitic quartz wacke microfacies.

Porosity is mainly attributed to the predominance of the intergranular pore spaces that are governed by the dissolution and leaching out as a reservoir quality improving diagenetic feature. The low clay content, mostly kaolinite, may be a result of fines leaching out during the fluid flow, i.e., it has a minor reducing implication on the reservoir quality.

12.7 Recommendations

Based on this review, which responds to the raised question, can we find any promising areas for oil production in the Nile Delta of Egypt? It is highly recommended to follow up and match the horsts, which may be found in some limited areas in this Delta and can be considered for oil production. This could be achieved by more detailed and advanced seismic studies and utilizing the seismic attributes.

References

Abu El Ella R (1990) The neogene-quaternary section in the Nile Delta, Egypt: geology and hydrocarbon potential. J Petrol Geol 13:329–340

Abd El-Aal A, Price RJ, Vaitl DJ, Sharallow JA (1994) Tectonic evolution of the Nile Delta, its impact on sedimentation and hydrocarbon potential. In: 12th EGPC, petroleum exploration and production conference Cairo, pp 19–34

Abd El-Aal AK, Zakhera M, Khalifa MA, Talha Qadri SM, Nabawy BS (2021) Carbonate strength classification based on depositional textures and fossil content of the Lower Eocene Drunka Formation, Assiut Area, central Egypt. J Petrol Sci Eng 207:109061. https://doi.org/10.1016/j.petrol.2021.109061

Abdeen MM, Ramadan FS, Nabawy BS, El Saadawy O (2021) Subsurface structural setting and hydrocarbon potentiality of the Komombo and Nuqra Basins, South Egypt: a seismic and petrophysical integrated study. Nat Resour Res 30(5):3575–3603. https://doi.org/10.1007/s11053-021-09898-2

Aboelhassan N, Tarabees E, El Bastawesy M, Mogren S, Nabawy BS, Ibrahim E, Qadri SMT (2024) Integrated seismic, petrophysical, and geochemical studies for evaluating the petroleum system of the Upper Bahariya-Abu Roash G sequence in the Karama Field, Abu Gharadig Basin, North Western Desert, Egypt. J Afr Earth Sci 219:105413. https://doi.org/10.1016/j.jafrearsci.2024.105413

Abu El-Ata ASA (1988) The relation between the local tectonics of Egypt and the plate tectonics of the surrounding regions, using geophysical and geological data. In: EGS proceeding of the 6th annual meeting, pp 92–112

Abu El-Ata ASA, Ibrahim MA (1988) Structural-stratigraphic inferences of the Nile Delta of Egypt, based on regional and detailed gravity stripping. Bull ICESS 15:119–144

Abuamarah BA, Nabawy BS, Shchata AM, Kassem OMK, Ghrefat H (2019) Integrated geological and petrophysical characterization of Oligocene deep marine unconventional poor to tight sandstone gas reservoir. Mar Pet Geol 109:868–885. https://doi.org/10.1016/j.marpetgeo.2019.06.037

Abuhagaza AA, El Sawy MZ, Nabawy BS (2021) Integrated petrophysical and petrographical studies for characterization of reservoirs: a case study of Muglad Basin, North Sudan. Environ Earth Sci 80:171. https://doi.org/10.1007/s12665-021-09489-7

Amaefule JO, Altunbay M, Tiab D, Kersey DG, Keelan DK (1993) Enhanced reservoir description: using core and log data to identify hydraulic (flow) units and predict permeability in uncored intervals/wells. In: SPE 26436. Presented at the annual technical conference and exhibition, Houston, TX, pp 3–6. https://doi.org/10.2118/26436-MS

Archie GE (1942) The electrical resistivity log as an aid in determining some reservoir characteristics. J Petrol Technol 5:54–62. SPE-942054-G. https://doi.org/10.2118/942054-G

Ashoor R, Nabawy BS, Kamel A (2023) Formation evaluation and reservoir quality assessment of the Albian clastic sequence in the SWM Oilfield, Southwest the Mesopotamian Basin, Iraq. J Petrol Explor Prod Technol 13(2):553–576. https://doi.org/10.1007/s13202-022-01563-5

Azab AA, Nabawy BS, Mogren S, Saqr Kh, Ibrahim E, Qadri SMT, Barakat MK (2024) Structural assessment and petrophysical evaluation of the pre-Cenomanian Nubian sandstone in the October Oil Field, central Gulf of Suez, Egypt. J Afr Earth Sci 218:105351. https://doi.org/10.1016/j.jafrearsci.2024.105351

Badawy A, Hegazi M, Gaber H, Korrat I (2018) Crustal structure of northern Egypt from joint inversion of receiver functions and surface wave dispersion velocities. J Seismol 22:697–719. https://doi.org/10.1007/s10950-018-9729-x

Badry R, Head E, Morris Ch, Traboulay I (1993) New wireline formation tester: techniques and applications. In: Transactions of society of professional well log analysts, 34th annual logging symposium. Calgary, Alberta, Canada, pp ZZ1–ZZ15

Barakat MKhA (2010) Modern geophysical techniques for constructing a 3D geological model on the Nile Delta, Egypt. PhD thesis, Planen Bauen Umwelt der Technischen Universität Berlin, 158 p

Baouche R, Nabawy BS (2021) Permeability prediction in argillaceous sandstone reservoirs using fuzzy logic analysis: a case study of Triassic sequences, Southern Hassi R'Mel gas field, Algeria. J Afr Earth Sci 173:104049. https://doi.org/10.1016/j.jafrearsci.2020.104049

Berner R (1981) A new geochemical classification of sedimentary environments. J Sed Petrol 51:359–365. https://doi.org/10.1306/212F7C7F-2B24-11D7-8648000102C1865D

Buckles RS (1965) Correlating and averaging connate water saturation data. J Can Pet Technol 9(1):42–52. https://doi.org/10.2118/65-01-07

Chapman RE (1983) Petroleum geology, 2nd edn. Elsevier Publication Company, Amsterdam

Cherana A, Aliouane L, Doghmane MZ, Ouadfeul S-A, Nabawy BS (2022) Lithofacies discrimination of the Ordovician unconventional gas-bearing tight sandstone reservoirs using a subtractive fuzzy clustering algorithm applied on the well log data: Illizi Basin, the Algerian Sahara. J Afr Earth Sc 196:104732. https://doi.org/10.1016/j.jafrearsci.2022.104732

Dake LP (1978) Fundamentals of reservoir engineering. Elsevier Publication Company, Amsterdam

Dally NHE, Youssef MS, Abdel Aal MH, Metwalli FI, Nabawy BS (2023) Delineating the main structural outlines and the petrophysical properties of the Albian-upper Cretaceous Reservoirs using seismic and well log data, Shushan Basin, northwestern Desert, Egypt. J Petrol Explor Prod Technol 13(4):1009–1030. https://doi.org/10.1007/s13202-022-01603-0

EGPC (Egyptian General Petroleum Corporation) (1994) Nile Delta and North Sinai fields, discoveries and hydrocarbon potentials (a comprehensive overview). Egyptian General Petroleum Corporation, Cairo, Egypt, 387 p

El Nady MM (2007) Organic geochemistry of source rocks, condensates, and thermal geochemical modeling of Miocene sequence of some wells, onshore Nile Delta, Egypt. Pet Sci Technol 25:791–818. https://doi.org/10.1080/10916460600803629

El Sawy MZ, Abuhagaza AA, Nabawy BS, Lashin A (2020) Rock typing and hydraulic flow units as a successful tool for reservoir characterization of Bentiu-Abu Gabra sequence, Muglad basin, southwest Sudan. J Afr Earth Sc 171:103961. https://doi.org/10.1016/j.jafrearsci.2020.103961

El-Sisi Z, Sharaf L, Yehia H, Dawood A, Hassouba A (1996) Source of clastic supply to the Sidi Salem and Abu Madi reservoirs of the Nile Delta: a province study. In: Proceedings of the 13th EGPC exploration and production conference, Cairo, Egypt, vol 1, pp 291–296

El Shazly CM (1977) The geology of the Egyptian region in the ocean basins and margins. J East Mediterr. Plenum Press. New York, London, vol 4A. https://doi.org/10.1007/978-1-4684-3036-3_10

Elgendy NTH, Abuamarah BA, Nabawy BS, Ghrefat H, Kassem OMK (2020) Pore fabric anisotropy of the Cambrian-Ordovician Nubia sandstone in the onshore Gulf of Suez, Egypt: a surface outcrop analog. Nat Resour Res 29(2):1307–1328. https://doi.org/10.1007/s11053-019-09520-6

Elmahdy M, Radwan AA, Nabawy BS, Abdelmaksoud A, Nastavkin AV (2023) Integrated geophysical, petrophysical and petrographical characterization of the carbonate and clastic reservoirs of the Waihapa Field, Taranaki Basin, New Zealand. Mar Pet Geol 151:106173. https://doi.org/10.1016/j.marpetgeo.2023.106173

Eysa EA, Nabawy BS, Ghoneimi A, Saleh AH (2024) Petrophysical rock typing based on the digenetic effect of the different microfacies types of Abu Madi clastic reservoir in Faraskur Gas Field, onshore Nile Delta, Egypt. J Petrol Explor Prod Technol 14(2):381–406. https://doi.org/10.1007/s13202-023-01727-x

Fallah-Bagtash R, Adabi MH, Nabawy BS, Omidpour A, Sadeghi A (2022) Integrated petrophysical and microfacies analyses for a reservoir quality assessment of the Asmari Dolostone sequence in the Khesht Field, SW Iran. J Asian Earth Sci 223:104989. https://doi.org/10.1016/j.jseaes.2021.104989

Fea I, Abioui M, Nabawy BS, Jain S, Bruno DZ, Kassem AA, Benssaou M (2022) Reservoir quality discrimination of the Albian-Cenomanian reservoir sequences in the Ivorian basin: a lithological and petrophysical study. Geomech Geophys Geo-Energy Geo-Resour 8(1):1. https://doi.org/10.1007/s40948-021-00297-8

Guiraud R, Bosworth W (1990) Phanerozoic geodynamic evaluation of northeastern Africa and the northwestern Arabian platform. Tectonophysics 315:73–104. https://doi.org/10.1016/S0040-1951(99)00293-0

Haque AE, Qadri SMT, Bhuiyan MAH, Navid M, Nabawy BS, Hakimi MH, Abd-El-Aal AK (2022) Integrated wireline log and seismic attribute analysis for the reservoir evaluation: a case study of the Mount Messenger Formation in Kaimiro Field, Taranaki Basin, New Zealand. J Natl Gas Sci Eng 99:104452. https://doi.org/10.1016/j.jngse.2022.104452

Harms J, Wary J (1990) Nile Delta. In: Said R (ed) Geology of Egypt. A. Balkema, pp 329–343

Holmes M, Holmes D, Holmes A (2009) Relationship between porosity and water saturation: methodology to distinguish mobile from capillary bound water. In: AAPG annual convention, Denver, Colorado June 7–10, Article 110108

Hussien A, Gamal R, Madi T (2020) A guide to the Nile Delta's oil and gas resources. Egypt Oil & Gas Newspaper. February 2020, Issue 158, pp 32–35

Kamel H, Eita T, Sarhan M (1998) Nile delta hydrocarbon potentiality, Egypt. In: Proceedings of the 14th EGPC exploration and production conference, Cairo, Egypt (1), 485–503

Kassab MA, Teama MA, Cheadle BA, El-Din ES, Mohamed IF, Mesbah MA (2015) Reservoir characterization of the Lower Abu Madi formation using core analysis data: El-Wastani gas field, Egypt. J Afr Earth Sci 110:116–130. https://doi.org/10.1016/j.jafrearsci.2015.06.008

Kassem AA, Nabawy BS, Mogren S, Abdel-Fattah MI, Radwan AA, Ibrahim E (2024) Implications of the depositional and diagenetic attributes on the reservoir properties of the siliciclastic mangahewa formation, Taranaki Basin, New Zealand: integrated petrographical and petrophysical studies. Mar Pet Geol 164:106823. https://doi.org/10.1016/j.marpetgeo.2024.106823

Khalifa MA, Nabawy BS, Abu-Hashish MF, Al-Shareif AW, Hassan NM (2024) Identification of the low resistivity-low contrast (LRLC) gas-bearing pay zones in Shaly sand reservoirs using acoustic data: a case study from the Messinian Abu Madi formation, onshore Nile Delta, Egypt. J Petrol Explor Prod Technol 14:2973–2993. https://doi.org/10.1007/s13202-024-01864-x

Lee J, Kasap E, Shaker A (1998) Pressure test analysis of gas bearing formations. In: Transactions of society of professional well log analysts, 34th annual logging symposium. Houston, Texas, paper EE, 9p

Leila M, Moscariello A (2018) Depositional and petrophysical controls on the volumes of hydrocarbons trapped in the Messinian reservoirs, onshore Nile Delta, Egypt. Petroleum 4:250–267. https://doi.org/10.1016/j.petlm.2018.04.003

Leila M, Kora M, Ahmed M, Ghanem A (2016) Sedimentology and reservoir characterization of the Upper Miocene Qawasim Formation, El Tamad oil Field onshore Nile Delta, Egypt. Arab J Geosci 9(17):1–13. https://doi.org/10.1007/s12517-015-2088-9

Mathisen M (1984) Diagenesis of Plio-Pleistocene non-marine sandstones, Cagayan Basin, Philipines: early development of secondary porosity in volcanic sandstones. In: Clastic diagenesis, AAPG Memoir 37, pp 177–193. https://doi.org/10.1306/M37435C11

Marzouk MA (1981) Subsurface geological and geophysical studies in the Northern Area of the Nile Delta. MSc thesis, Fac. Science Alex. University

Mebrouki N, Nabawy B, Hacini M, Abdel-Fattah MI (2024) Deciphering the implication of microfacies types and diagenesis on the reservoir quality of the Cambrian sequence in Hassi Messaoud field, Algeria. Mar Petrol Geol 162:106650. https://doi.org/10.1016/j.marpetgeo.2023.106650

Melrose Resources Plc. (2005) Interim results, September 2005, published report, http://miranda.hemscott.com/ir/mrs/pdf/ar2005.pdf, 32 p

Melrose Resources Plc. (2006) Interim results, November 2006, published report, http://www.oilbarrel.com/conference/nov06/melrose.pdf, 29 p

Mohamed AS, Omran AA, Mohamed MT, Nabawy BS (2023) Petrophysical analysis and hydrocarbon potential of the Matulla formation in the Muzhil Field, central part of the Gulf of Suez, Egypt. Min Miner Depos 17(2):121–139. https://doi.org/10.33271/mining17.02.121

Nabawy BS (2013) Impacts of dolomitization on the petrophysical properties of El-Halal Formation, North Sinai, Egypt. Arab J Geosci 6(2):359–373. https://doi.org/10.1007/s12517-011-0369-5

Nabawy BS (2015) Impacts of the pore- and petro-fabrics on porosity exponent and lithology factor of Archie's equation for carbonate rocks. J Afr Earth Sc 108:101–114. https://doi.org/10.1016/j.jafrearsci.2015.04.014

Nabawy BS (2018) Impacts of fossil anisotropy on the electric and permeability anisotropy of highly fossiliferous limestone: a case study. Mar Geophys Res 39(4):537–550. https://doi.org/10.1016/j.jafrearsci.2015.04.014

Nabawy BS (2021) An improved stratigraphic modified lorenz (ISML) plot as a tool for describing efficiency of the hydraulic flow units (HFUs) in clastic and non-clastic reservoir sequences. Geomech Geophys Geo-Energy Geo-Resour 7(3):67. https://doi.org/10.1007/s40948-021-00264-3

Nabawy BS, Shehata AM (2010) Reservoir characterization of the Miocene, Deep Target, Sidi Salem Sandstones, El Tamad Oil Field, Nile Delta, Egypt. J Appl Geophys 9(1):239–253

Nabawy BS, Shehata AM (2015) Integrated petrophysical and geological characterization for the Sidi Salem-Wakar sandstones, off-shore Nile Delta, Egypt. J Afr Earth Sci 110:160–175. https://doi.org/10.1016/j.jafrearsci.2015.06.017

Nabawy BS, Géraud Y (2016) Impacts of pore- and petro-fabrics, mineral composition and diagenetic history on the bulk thermal conductivity of sandstones. J Afr Earth Sc 115:48–62. https://doi.org/10.1016/j.jafrearsci.2015.11.025

Nabawy BS, Wassif NA (2017) Effect of the mineralogical composition on the petrophysical behavior of the amygdaloidal and vesicular basalt of Wadi Wizr, Eastern Desert, Egypt. J Afr Earth Sci 134:613–625. https://doi.org/10.1016/j.jafrearsci.2017.07.009

Nabawy BS, El Aal AA (2019a) Effects of diagenesis on the geomechanical and petrophysical aspects of the Jurassic Bir El-Maghara and Masajid carbonates in Gebel El-Maghara, North Sinai, Egypt. Bull Eng Geol Environ 78(7):5409–5429. https://doi.org/10.1007/s10064-019-01471-9

Nabawy BS, El Aal AA (2019b) Impacts of the petrophysical and diagenetic aspects on the geomechanical properties of the dolomitic sequence of Gebel El-Halal, Sinai, Egypt. Bull Eng Geol Environ 78(4):2627–2640. https://doi.org/10.1007/s10064-018-1264-z

Nabawy BS, Sediek KN, Nafee SA (2015) Pore fabric assignment using electrical conductivity of some Albian-Cenomanian sequences in north Eastern Desert, Egypt. Arab J Geosci 8(8):5601–5615. https://doi.org/10.1007/s12517-014-1631-4

Nabawy BS, Basal AMK, Sarhan MA, Safa MG (2018) Reservoir zonation, rock typing and compartmentalization of the Tortonian-Serravallian sequence, Temsah gas field, offshore Nile Delta, Egypt. Mar Petrol Geol 92:609–631. https://doi.org/10.1016/j.marpetgeo.2018.03.030

Nabawy BS, Mansour AS, Rashed MA, Afify WSM (2020a) Implementation of sedimentary facies and diagenesis on the reservoir quality of the Aquitanian-Burdigalian Rudeis Formation in the Gulf of Suez, Egypt: a comparative surface and subsurface study. Geol J 55(6):4543–4563. https://doi.org/10.1002/gj.3683

Nabawy BS, Khalil HM, Fathy MS, Ali F (2020b) Impacts of microfacies type on reservoir quality and pore fabric anisotropy of the Nubia sandstone in the central Eastern Desert, Egypt. Geol J 55(6):4507–4524. https://doi.org/10.1002/gj.3690

Nabawy BS, Abudeif AM, Masoud MM, Radwan AE (2022a) An integrated workflow for petrophysical characterization, microfacies analysis, and diagenetic attributes of the lower Jurassic type section in northeastern Africa margin: implications for subsurface gas prospection. Mar Petrol Geol 140:105678. https://doi.org/10.1016/j.marpetgeo.2022.105678

Nabawy BS, Lashin AA, Barakat MK (2022b) Implementation of lithofacies and microfacies types on reservoir quality and heterogeneity of the Late Cretaceous Upper Bahariya Member in the Shurouk field, Shoushan Basin, North Western Desert, Egypt. J Asian Earth Sci 224:105014. https://doi.org/10.1016/j.jseaes.2021.105014

Nabawy BS, Abd El Aziz EA, Ramadan M, Shehata AA (2023a) Implication of the micro- and lithofacies types on the quality of a gas-bearing deltaic reservoir in the Nile Delta, Egypt. Sci Rep 13(1):8873. https://doi.org/10.1038/s41598-023-35660-0

Nabawy BS, El-Bialy M, Hamimi Z, Khamis HA, Abdel Wahed SA, Osman RA, Afify AM (2023b) Implication of the diagenetic evolution, litho- and microfacies types on the storage capacity of the carbonate rocks in West Esh El Mallaha area, SW onshore Gulf of Suez, Egypt. J Afr Earth Sci 204:104971. https://doi.org/10.1016/j.jafrearsci.2023.104971

Nabawy BS, Mostafa A, Radwan AA, Kotb AG, Leila M (2023c) Seismic reservoir characterization of the syn-rift lower Miocene Rudeis formation in the July oilfield, Gulf of Suez basin, Egypt: implication for reservoir quality assessment. Geoenergy Sci Eng 226:211797. https://doi.org/10.1016/j.geoen.2023.211797

Nabawy BS, Mohamed AS, Omran AA, Mohamed MT (2023d) Applying the improved stratigraphic modified Lorenz technique for dividing the highly heterogeneous clastic reservoirs into hydraulic flow units. Sci Rep 13(1):20465. https://doi.org/10.1038/s41598-023-47709-1

Radwan AA, Nabawy BS (2022) Hydrocarbon prospectivity of the miocene-pliocene clastic reservoirs, Northern Taranaki basin, New Zealand: integration of petrographic and geophysical studies. J Petrol Explor Prod Technol 12(7):1945–1962. https://doi.org/10.1007/s13202-021-01451-4

Radwan AA, Nabawy BS, Kassem AA, Elmahdy M (2023) An integrated workflow for seismic interpretation, petrophysical and petrographical characterization for the clastic Mangahewa reservoir in Pohokura gas field, Taranaki Basin, New Zealand. Geoenergy Sci Eng 229:212117. https://doi.org/10.1016/j.geoen.2023.212117

Rahmouni A, Boulanouar A, El Rhaffari Y, Hraita M, Zaroual A, Géraud Y, Sebbani J, Rezzouk A, Nabawy BS (2023) Impacts of anisotropy coefficient and porosity on the thermal conductivity and P-wave velocity of calcarenites used as building materials of historical monuments in Morocco. J Rock Mech Geotech Eng 15(7):1687–1699. https://doi.org/10.1016/j.jrmge.2023.02.008

Rizzini A, Vezzani F, Coccocetta V, Milad G (1978) Stratigraphy and sedimentation of neogene-quaternary section in the Nile Delta area, (A.R.E). Mar Geol 27:327–348

Ross D, Uchupi E (1977) Structure and sedimentary history of northeastern Mediterranean Sea-Nile cone area. AAPG 61:872–902. https://doi.org/10.1306/C1EA4397-16C9-11D7-8645000102C1865D

Rotaf A, Palmieri G, Quaglioroli F (1994). Nile Delta Basin: geological and structural setting. IEOC, Internal Report, Cairo, Egypt

Sabry S, Helal A, Farag KS (2019) Imaging the Pliocene channel system using 3D seismic attributes at the west El Qantara region, Nile Delta Basin, Northern Egypt. Egypt J Geol 63:53–62

Safa MG, Nabawy BS, Basal AMK, Omran MA, Lashın A (2021) Implementation of a petrographical and petrophysical workflow protocol for studying the impact of heterogeneity on the rock typing and reservoir quality of Reefal limestone: a case study on the Nullipore carbonates in the Gulf of Suez. Acta Geol Sin (English Edition) 95(5):1746–1762. https://doi.org/10.1111/1755-6724.14700

Said R (1990) The geological evolution of the River Nile. Springer, New York, Heidelberg, Berline, p 151. https://doi.org/10.1007/978-1-4612-5841-4

Sakran Sh, Shehata AA, Osman O, El Sherbiny M (2019) Superposed tectonic regimes in West Beni Suef basin, Nile Valley, Egypt: implications to source rock maturation and hydrocarbon entrapment. J Afr Earth Sc 154:1–19. https://doi.org/10.1016/j.jafrearsci.2019.03.010

Sarhan M, Hemdan K (1994) North Nile delta structural setting trapping and mechanism. In: Egypt Proceedings of the 12th EGPC exploration and production conference, Cairo, Egypt, vol 1, pp 1-18

Schlumberger (1984). Well Evaluation Conference, Egypt. Schlumberger Middle East Surface.

Schlumberger (1985) Well evaluation conference. Schlumberger well services, USA, 122p

Salem A, Ketzer JM, Mora S, Rizk RR, Al Aasm IS (2005) Diagenesis and reservoir quality evaluation on incised – valley sandstone: evidence from the Abu Madi gas reservoir (Upper Miocene), The Nile Delta Basin. Egypt J Sediment Res 75:572–584. https://doi.org/10.2110/jsr.2005.047

Sestini G (1989) Nile Delta: depositional environments and geological history. In: Pickering K, Whateley T (eds) Deltas: sites and traps for fossil fuel. - Geol. Soc. London Spec. Publ. 41:99–128, Blackwell Scientific Publications, London.

Shaaban F, Lutz R, Littke R, Bueker C, Odisho K (2006) Source rock evaluation and basin modeling in the NE Egypt (NE Nile Delta and Northern Sinai). J Pet Geol 29(2):103–124. https://doi.org/10.1111/j.1747-5457.2006.00103.x

Teama MA, Nabawy BS (2016) Petrophysical evaluation of the hydrocarbon potential of the Lower Cretaceous Kharita clastics, North Qarun oil field, Western Desert, Egypt. J Afr Earth Sci 121:62-71. https://doi.org/10.1016/j.jafrearsci.2016.05.022

Tarabees EA, Azam SS (2016) Structural effect on hydrocarbon trapping, North Western Nile Delta, Egypt. Int J Geosci 7:353–366. https://doi.org/10.4236/ijg.2016.73028

Timur A (1968) An investigation of permeability, porosity, & residual water saturation relationships for sandstone reservoirs. The Log Analyst IX(4). SPWLA-1968-vIXn4a2

Vandré C, Cramer B, Gerling P, Winsemann J (2007) Natural gas formation in the western Nile Delta (Eastern Mediterranean): thermogenic versus microbial mechanisms. Org Geochem 38:523–539. https://doi.org/10.1016/j.orggeochem.2006.12.006

Zaghloul Z, Elgamal M, Shaaban F, Yousef A (2001) Plates interactions and petroleum potentials in the Nile Delta. In: Proceedings of the 1st conference on the deltas, modern and ancient, pp 41–53

Role of Machine Learning in the Exploration and Development of Gas Fields, Offshore Nile Delta Region, Egypt

13

Ashraf Ghoneimi, Muhammad Nabih, Mohamed Abd Elaziz, and Ahmed Bakry

Abstract

Improving petroleum exploration and development is important worldwide for reducing drilling risks and understanding reservoir characteristics. This study tackles these challenges by presenting new methods that use machine learning algorithms to analyze complex seismic and well-log data in Egypt's West Nile Delta Deep Marine (WNDDM) concession. The main goal of this study is to enhance reservoir rock characterization by integrating information from seismic attributes and well logs. To achieve this, a novel approach combining fluid substitution and machine learning techniques is used to assess how pore fluids affect reservoir elastic properties. Initially, gas-bearing zones were excluded to minimize the negative impact of gas on Poisson's ratio predictions across three different machine learning models. These gas zones were

A. Ghoneimi (✉) · M. Nabih · A. Bakry
Faculty of Science, Department of Geology, Zagazig University, Zagazig, Egypt
e-mail: aeghoneimy@science.zu.edu.eg

M. Nabih
e-mail: mnabih@science.zu.edu.eg

A. Bakry
e-mail: ambakry@science.zu.edu.eg

M. A. Elaziz
Artificial Intelligence Research Center (AIRC), College of Engineering and Information Technology, Ajman University, Ajman, United Arab Emirates

Faculty of Computer Science and Engineering, Galala University, Suez, Egypt

Department of Electrical and Computer Engineering, Lebanese American University, Byblos, Lebanon

M. A. Khalifa and K. S. Gemail (eds.), *Geology of the Nile Deltas and Natural Resources*, Deltas of the World, https://doi.org/10.1007/978-3-031-96594-4_13

replaced with oil and water equivalents, resulting in a more accurate prediction process that demonstrated improved results, validated by higher correlation coefficients. This thorough method allowed for a detailed examination of how gas, oil, and water affect Poisson's ratio estimation. The study employed the Random Vector Functional Link (RVFL) model, fine-tuned using the Cheetah Optimizer (CO) and Wild Geese Algorithm (WGA), to investigate how pore fluid saturation affects reservoir elastic properties. The findings emphasized the significant impact of gas-bearing zones on the accuracy of Poisson's ratio predictions, showing notable improvement when gas was replaced with water rather than oil. Regression analysis validated these results, revealing better correlation coefficients for the CO-RVFL and WGA-RVFL models after excluding gas-bearing well-log data and further enhancing when gas zones were substituted with oil or water equivalents. Furthermore, the study utilized seismic data and well logs to predict shale volume and effective porosity through the Random Forest (RF) algorithm. The results demonstrated the efficiency of this machine learning approach in estimating vital petrophysical parameters, underscoring the potential of artificial neural networks in advancing reservoir characterization methods.

Keywords

Machine learning (ML) • Nile Delta • Seismic attributes • Petrophysical characterization • Random Forest (RF) algorithm

13.1 Introduction

Machine learning (ML) techniques have demonstrated exceptional utility across various domains of geoscience. For example, Dorrington and Link (2004) used a genetic algorithm combined with an artificial neural network (ANN) to predict reservoirs in the Stratton field using seismic data. Chaki et al. (2018) utilized an ANN and a neuro-fuzzy system for reservoir evaluation. Additionally, Priezzhev et al. (2019) integrated well logs and seismic data using several ML models, including random forest, linear regression, and ANN, to identify petrophysical parameters. Farsi et al. (2021) applied a three-hybrid machine learning (ML) optimization approach utilizing a multi-layer perceptron (MLP) neural network to predict pore pressure. Feng (2021) demonstrated the effectiveness of random forest as a ML technique to address uncertainties in well-log classification. Zahmatkesh et al. (2021) successfully identified seismic facies in the Iranian Mansuri field using an artificial neural network (ANN). Additionally, various ML techniques have been implemented for well logs, as documented by Abd Elaziz et al. (2022), Ali et al. (2021), Garia et al. (2021), Iturrarán-Viveros et al. (2021), Nabih et al. (2022), and Yasin et al. (2021). Several studies have been conducted in the Western Delta Deep Marine (WDDM) region to identify gas channels and predict elastic properties using ANN and amplitude variation with offset (AVO) attributes (Ismail et al. 2020, 2022; Mohamed et al., 2015).

Recently, machine learning methods have gained significant traction in addressing real-world applications across diverse fields, such as geoscience, medicine, and industry (Priezzhev et al. 2019; Dramsch 2020; Farouk et al. 2021; Ismail et al. 2022; Abd Elaziz et al. 2023). The importance of these techniques lies in their ability to minimize human intervention and their ease of implementation.

Furthermore, machine learning techniques can be utilized to predict petrophysical and rock physics parameters, including Poisson's ratio, shale volume, and effective porosity, from seismic traces and well-log data. The innovation of this study lies in employing an unconventional approach, specifically machine learning, to estimate petrophysical parameters such as Poisson's ratio directly from seismic data, even in the absence of well-log data like shear wave velocity. Consequently, this methodology can be adapted and applied to other regions with similar geological conditions.

This study is applied on the Scarab field located in the concession of the Western Delta Deep Marine (WDDM) (Fig. 13.1). This area covers 6150 km2 (Samuel et al. 2003). At depths of 1600–1900 m, the field consists of a sequence of deep marine slope channels as revealed from the four wells in the study area. The main hydrocarbon-bearing zone is the Late Pliocene El-Wastani Formation. The main reservoirs are in two channels (Ch-1, Ch-2) in El-Wastani Formation (Abd El-Gawad et al. 2019; Ghoneimi et al. 2021; Nabih et al. 2022; Bakry et al. 2023).

In this study, machine learning and rock physics analysis are employed using raw well-log data from four wells in the Scarab gas field within the WDDM concession, Egypt. The petrophysical parameters are shale content (Vsh), effective porosity (PHIE), water saturation (Sw), and hydrocarbon saturation (Sh). These parameters are estimated from well-log analysis and serve as both input data and comparable factors. Furthermore, to investigate the potential of ML algorithms in predicting rock physics parameters, seismic

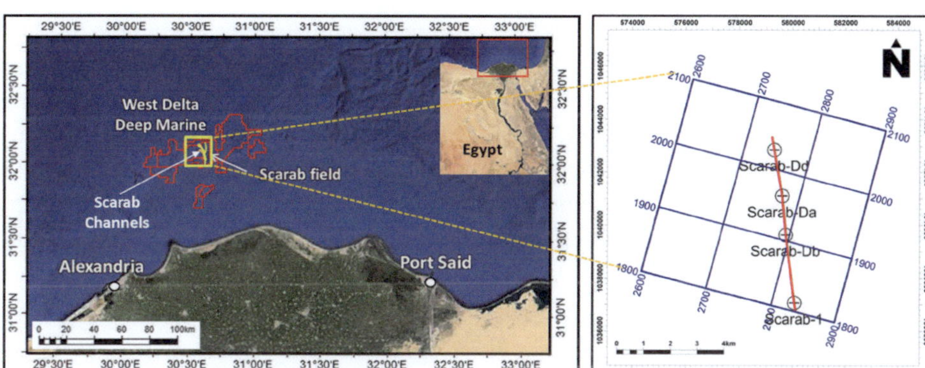

Fig. 13.1 WDDM concession location map, containing the Scarab field (left) and the well data and seismic survey (right). (Universal Transverse Mercator (UTM) WGS-84 zone 36 N)

traces extracted from a dedicated seismic survey at the well locations were employed as input data.

13.2 Methodology

13.2.1 Poisson's Ratio

Poisson's ratio is a crucial mechanical parameter used to predict geo-mechanical behavior during well drilling and subsequent recovery activities. The magnitude of this ratio can significantly influence strength parameters, impacting sand production, hydraulic fracturing, and well instability (Kumar 1976; Zhang and Bentley 2005). Additionally, changes in reservoir volume due to injection or production, and the resulting surface uplift or subsidence, can be substantial (Stancliffe and Kooij 2001). Accurate predictions of Poisson's ratio are essential inputs for numerical modeling to forecast its effects on wells and surface installations. Consequently, making precise predictions is vital for the successful execution of oilfield operations.

Poisson's ratio, a significant geo-mechanical property, is defined as the ratio of transverse strain to axial strain. This property exhibits a broad range of both positive and negative values and is influenced by various factors. The elastic properties of rocks are frequently inferred from Poisson's ratio (Belyadi et al. 2019). The average Poisson's ratio of sedimentary rocks is approximately 0.25. Determining Poisson's ratio through in-situ field geophysical methods, such as seismic surveys, necessitates the measurement of both P-wave and S-wave velocities. However, recording shear waves using standard seismic techniques presents challenges due to the need for specialized wave sources and receivers. This study attempts to predict Poisson's ratio from seismic data combined with well-log data, utilizing machine learning techniques.

In this study, fluid substitution and ML techniques are performed using the available borehole logs in Scarab-Db well in the reservoir of Scarab field. The well logs analyzed in these wells were the GR, LLD, RHOB, APLC, and DTCO and DTSM logs. The petrophysical parameters (shale content (VSH), effective porosity (PHIE), water saturations (SW), and net pay (PayFlag)), calculated using well-log interpretation (Abd Elaziz et al. 2022**;** Bakry 2021 and 2024) are used as input in the ML and fluid substitution estimations. The seismic amplitude readings were derived from the seismic traces extracted from the SEGY seismic cube (Fig. 13.1) using the open-source software SeiSee (https://seisee.software.informer.com/).

13.2.2 Machine Learning

A designed approach was adopted for model development, with three wells (Scarab-1, Scarab-Da, and Scarab-Dd) selected as training wells to establish robust predictive relationships. The fourth well (Scarab-Db) is subsequently designated as a test well to evaluate the model's generalizability and predictive accuracy.

Figures 13.2 and 13.3 present the detailed methodology adopted to achieve the objective of this study. The framework of the developed method is given in Fig. 13.2, where the data are splitted into two parts, namely training and testing sets, which represent 70 and 30% of the data, respectively, (Bakry 2021, 2024; Abd Elaziz et al. 2023).

In the proposed CO–RVFL method, the main target is to determine the best parameters and structure of the RVFL by considering it as an optimization problem and using the operators of the CO to handle this problem. This enhanced version of the RVFL to improve the performance of prediction the fluid substitution. The framework of the developed method is given in Fig. 13.3, where the first step is to split the data into two parts, namely training and testing sets, which represent 70% and 30% of the data, respectively. The next process is to construct the set of solutions X that represents the structure of the RVFL model.

13.2.2.1 Training Data Step

1. Three wells (Scarab-1, Scarab-Da, and Scarab-Dd) were employed as training data. To facilitate model development and evaluation, the dataset was partitioned into training and testing subsets, comprising 70 and 30% of the total data, respectively. The input features consisted of GR, seismic trace, DTCO, and Poisson's ratio, with the

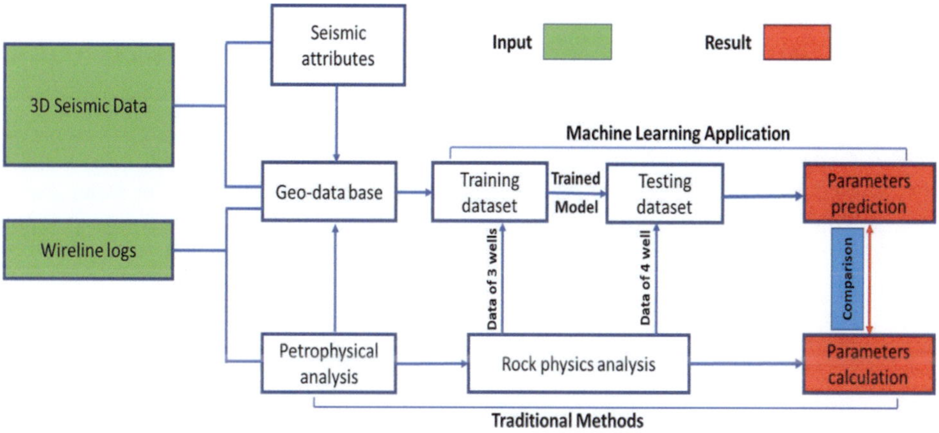

Fig. 13.2 Proposed workflow of the machine learning procedure

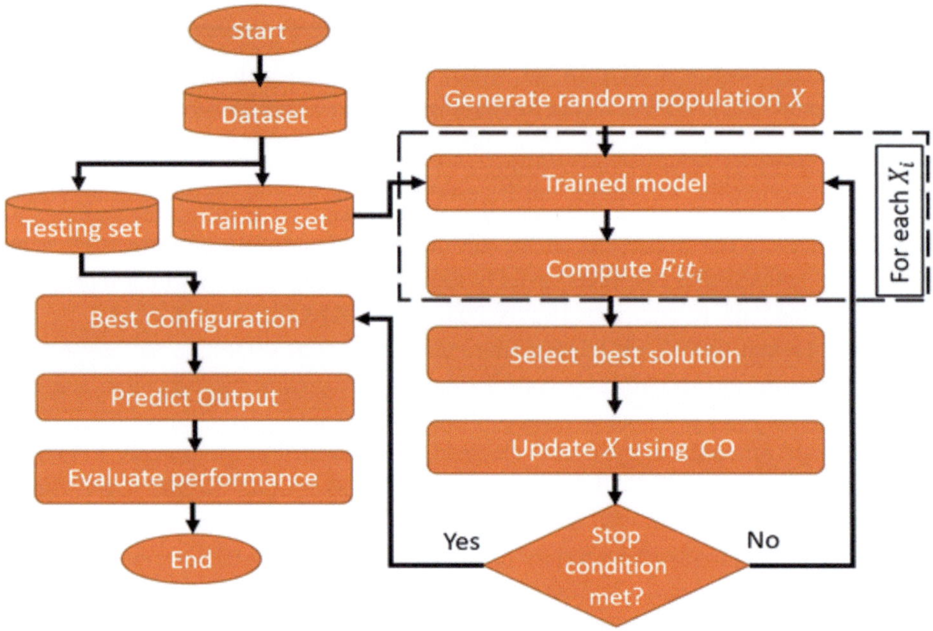

Fig. 13.3 Proposed CO–RVFL for predicting rock physics parameters of gas-bearing reservoirs

latter serving as the target variable. A machine learning model was trained on 70% of the dataset (GR, seismic trace, DTCO, and Poisson's ratio) to predict Poisson's ratio. Subsequently, the trained model was applied to the remaining 30% of the dataset (GR, seismic trace, DTCO) for testing purposes. The results obtained from this step demonstrated the exceptional performance of the proposed models.

2. After the fluid substitution step, we applied the same model and used the same input data GR and seismic trace but the DTCO and Poisson's ratio were changed to DTCO-sub and PoisRatio-sub.

13.2.2.2 Testing Data Step

The derived models are applied on the fourth well (Scarab-Db) and the input curves are GR, Seismic trace, and DTCO. The acoustic impedance (AI), velocity ratio (Vp/Vs), Poisson's ratio, bulk modulus, Lambda-Rho, and Mu-Rho are calculated and utilized for rock physics analysis.

13.3 Results

Poisson's Ratio Prediction Using the Wild Geese Algorithm (WGA)

In this section, we predict Poisson's ratio by excluding the majority of log data from gas zones within the well-log dataset. This prediction is accomplished using a modified version of the Random Vector Functional Link (RVFL) model, incorporating the operators of the WGA.

To assess the correlation between Poisson's ratio values derived from log data and those predicted by the proposed RVFL-WGA, RVFL-AEO, RVFL-MRFO, and RVFL-SCA models, a regression analysis was conducted using MATLAB (2020b). When applied to the fourth test well (Scarab-Db), the RVFL-WGA model exhibited a strong correlation between the two sets of Poisson's ratio values, with a correlation coefficient (r) of 0.8 (Fig. 13.4a). In contrast, the correlation coefficients for the RVFL-AEO, RVFL-MRFO, and RVFL-SCA models were notably lower, at 0.67, 0.61, and 0.65, respectively.

To further validate the robustness of these models, they were applied to a different channel (channel-2) in various wells. The RVFL-WGA model continued to demonstrate a strong correlation with a coefficient of $r = 0.76$, while the other models yielded significantly lower correlation coefficients ($r = 0.05$, $r = 0.15$, and $r = 0.65$ for RVFL-AEO, RVFL MRFO, and RVFL-SCA, respectively) (Fig. 13.4b). The lower correlation coefficient for channel-2 compared to channel-1 may be attributed to geological variations between the two channels.

These results were obtained using the following configurations, as determined by the respective optimization algorithms: WGA: $X_i = [100, 1, 2]$, AEO: $X_i = [170, 1, 2]$, MRFO: $X_i = [130, 2, 1]$, and SCA: [200, 1, 1].

The correlation coefficients (r) calculated for the test well (Scarab-Db) are summarized in Table 13.1. A comparative analysis of the table and histograms unequivocally indicates the superior performance of the proposed RVFL-WGA model over the other three models (RVFL-AEO, RVFL-MRFO, and RVFL-SCA) in terms of correlation coefficient values.

Furthermore, Poisson's ratio, estimated through machine learning techniques using the WGA, was employed to calculate shear wave velocity. A strong correlation ($r = 0.86$) was observed between the shear wave velocity derived from log data and the proposed shear wave velocity (Fig. 13.5). To conduct a comprehensive rock physics analysis, acoustic impedance (AI), velocity ratio (Vp/Vs), Poisson's ratio, bulk modulus, Lambda-Rho, and Mu-Rho were computed from both the original log data and the log data derived from machine learning (Poisson's ratio-WGA and Vs-WGA). These derived properties were subsequently utilized to establish correlations among them.

The Scarab-Db well, previously employed as a test well in ML techniques, intersects channel-1. This alignment enables a direct comparison between the results derived from well-log data and ML outputs. To facilitate rock physics analysis, various parameters, including P-impedance, Poisson's ratio, $\lambda\rho$ (Lambda-Rho), $\mu\rho$ (Mu-Rho), and Vp/

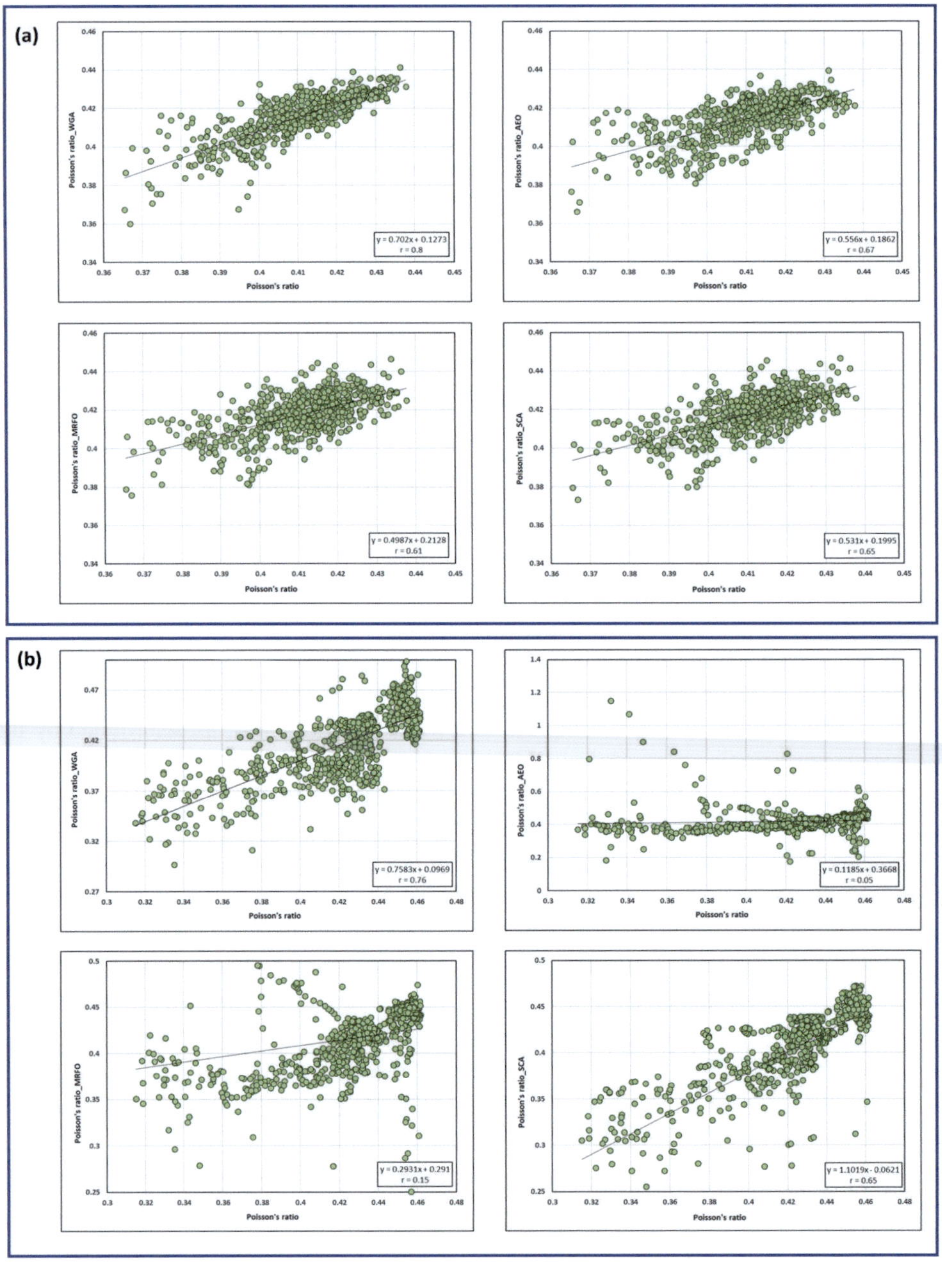

Fig. 13.4 Relationship between Poisson's ratio obtained from conventional well-log analysis and Poisson's ratio predicted by machine learning techniques for the test wells

Table 13.1 Correlation coefficient (r) values for the test well Scarab-Db

	RVFL-WGA	RVFL-AEO	RVFL-MRFO	RVFL-SCA
Test well	0.8	0.67	0.61	0.65

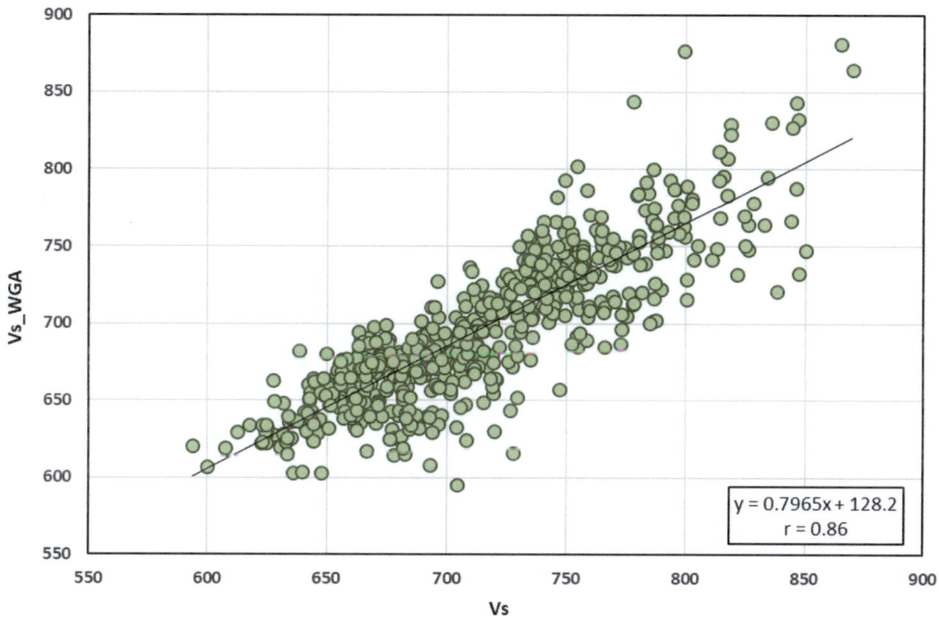

Fig. 13.5 Relationship between conventional shear wave velocity from well-log analysis and predicted shear wave velocity from machine learning techniques

Vs ratio, were estimated from both well log and ML data for channel-1. These parameters were subsequently cross-plotted to visualize relationships and trends. The cross-plots, color-coded by shale volume and hydrocarbon saturation, were overlaid with rock physics templates to assess their ability to differentiate between lithologies and fluid content, specifically distinguishing shale and gas-bearing sand intervals.

The plots of P-impedance (X-axis) versus Poisson's ratio (Y-axis) and P-impedance (X-axis) versus Poisson's ratio-WGA (Y-axis), both utilizing color-coding to denote shale content and hydrocarbon saturation (Sh) (Figs. 13.6 and 13.7), reveal distinct patterns. Specifically, P-impedance alone demonstrates less differentiation between shale points and gas-sand points. In contrast, Poisson's ratio offers a clearer distinction between these points. Across all plots, shale points exhibit higher Poisson's ratio values compared to gas-sand points. Gas zones are characterized by lower values of both impedance and Poisson's ratio.

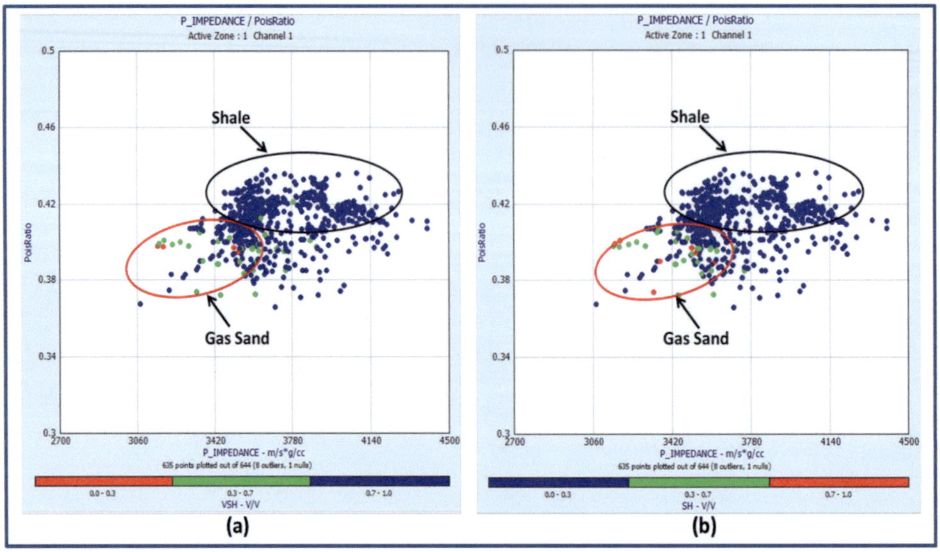

Fig. 13.6 Cross plots of P-impedance versus Poisson's ratio for the Db well, color-coded by shale volume and hydrocarbon saturation

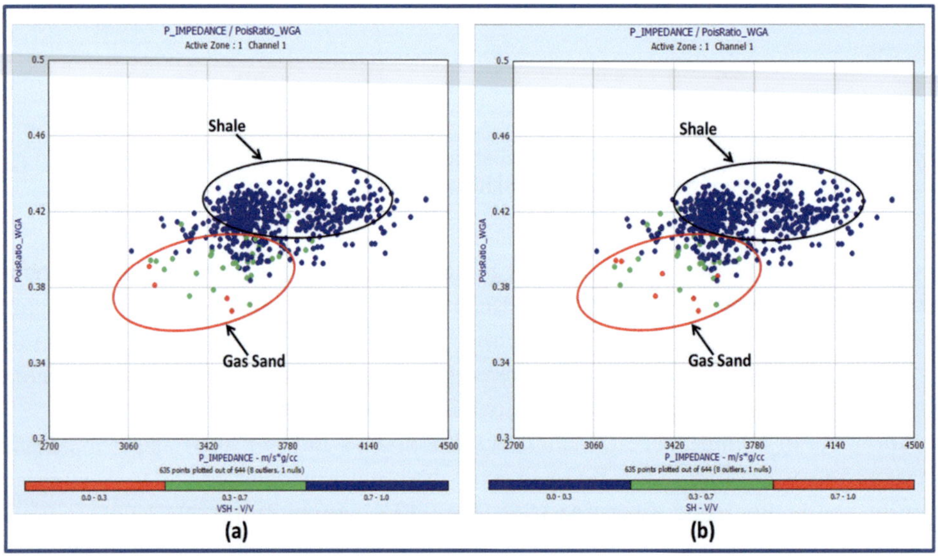

Fig. 13.7 Cross plots of P-impedance versus Poisson's ratio-WGA for the Db well, color-coded by shale volume and hydrocarbon saturation

The cross-plots presented in Figs. 13.8 and 13.9 display the relationships between Lambda-Rho (X-axis) and Mu-Rho (Y-axis), as well as Lambda-Rho-WGA (X-axis) and Mu-Rho-WGA (Y-axis). These plots employ color-coding to indicate Vsh and Sh. Notably, Lambda-Rho effectively differentiates between shale points and gas-sand points, while Mu-Rho does not provide a clear distinction. In all plots, shale points consistently show higher Lambda-Rho values compared to gas-sand points. In contrast, gas-sand points exhibit slightly elevated Mu-Rho values relative to shale points.

The cross-plots shown in Figs. 13.10 and 13.11 illustrate the relationship between P-impedance (X-axis) and the Vp/Vs ratio (Y-axis), as well as P-impedance (X-axis) and the Vp/Vs ratio-WGA (Y-axis). These plots are color-coded to represent Vsh and Sh. P-impedance and the Vp/Vs ratio effectively distinguish between shale and gas-sand points. As shale content increases, acoustic impedance (AI) and the velocity ratio (Vp/Vs) also rise.

Figures 13.12, 13.13, and 13.14 highlight a strong agreement between the rock physics parameters estimated from well-log data and those predicted using the ML method for the Scarab-Db test well. These Figs demonstrate that the values for Poisson's ratio, Lambda-Rho, bulk modulus, and velocity ratio (Vp/Vs) exhibit a decrease in gas-bearing zones, while shear wave velocity shows a slight increase due to its dependence on the rock framework. The close correspondence between the original and predicted logs confirms the efficacy and reliability of the ML tool in estimating shear wave velocity in wells where this parameter is unavailable. Moreover, a comparison of these results with those from cross-plot analysis reveals a strong correlation and agreement between rock physics parameters derived from well-log data and those obtained through ML.

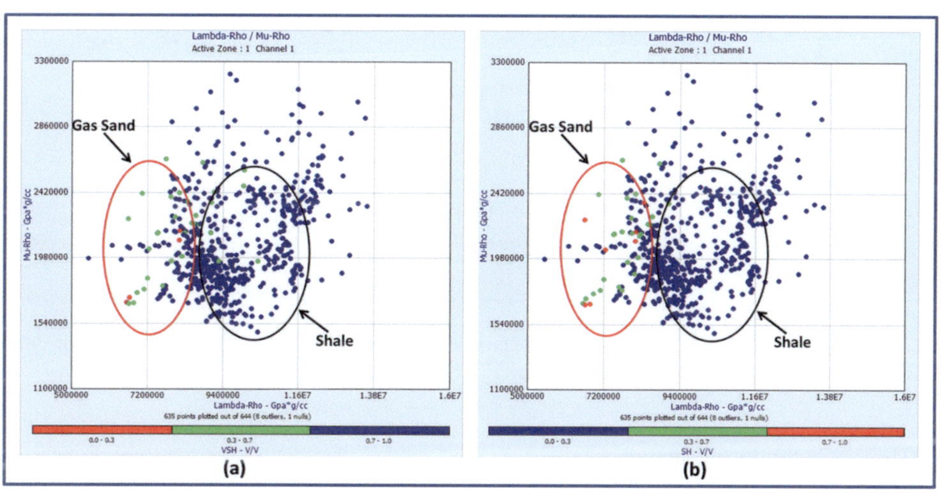

Fig. 13.8 Cross plots of Lambda-Rho versus Mu-Rho for the Db well, color-coded by shale volume and hydrocarbon saturation

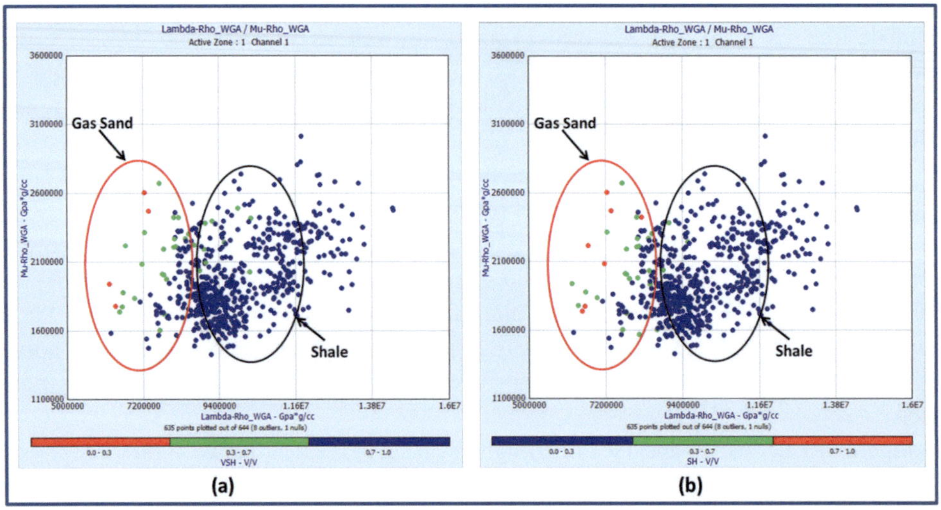

Fig. 13.9 Cross plots of Lambda-Rho-WGA versus Mu-Rho-proposed for the Db well, color-coded by shale volume and hydrocarbon saturation

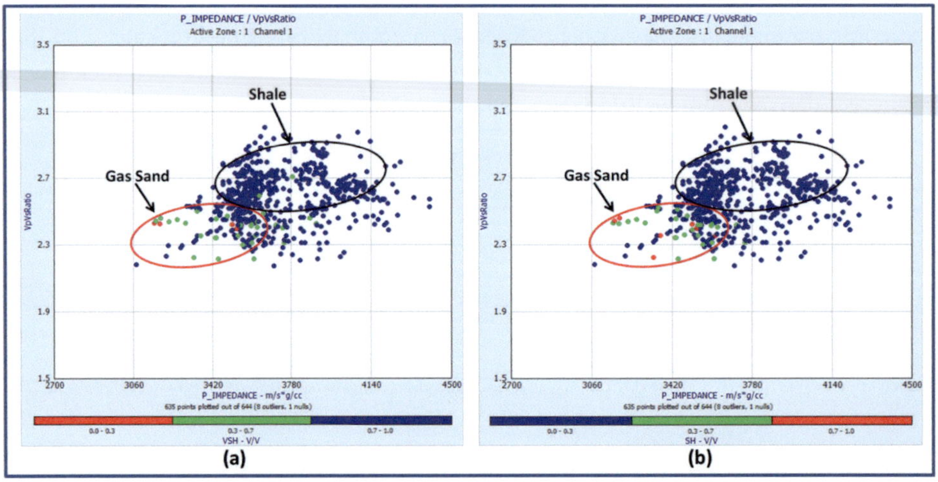

Fig. 13.10 Cross plots of P-impedance versus velocity ratio (Vp/Vs) for the Db well, color-coded by shale volume and hydrocarbon saturation

The predictive performance of the proposed model was evaluated using a set of standard metrics: the coefficient of determination (R^2), root mean squared error (RMSE), and mean absolute error (MAE). An optimal model is characterized by a high R^2 value and

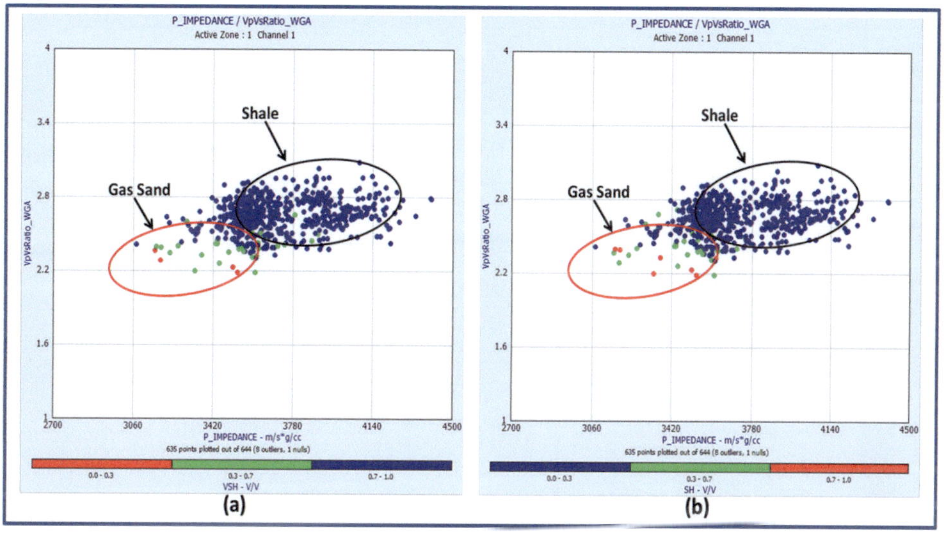

Fig. 13.11 Cross plots of P-impedance versus velocity ratio (Vp/Vs-WGA) for the Db well, color-coded by shale volume and hydrocarbon saturation

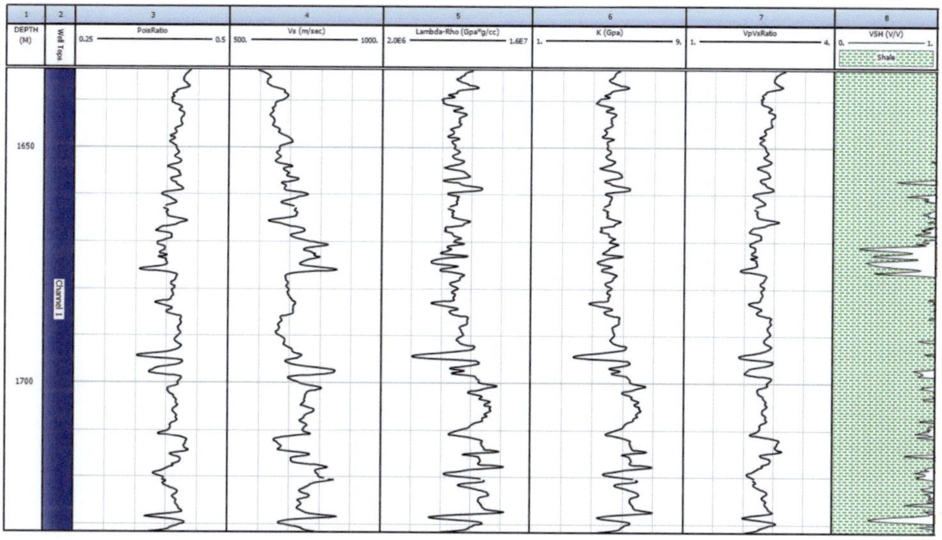

Fig. 13.12 Input data for Scarab-Db well: from left—Poisson's ratio, Shear wave velocity (Vs), Lambda-Rho, Bulk modulus (K), Vp/Vs ratio, and Shale volume (Vsh)

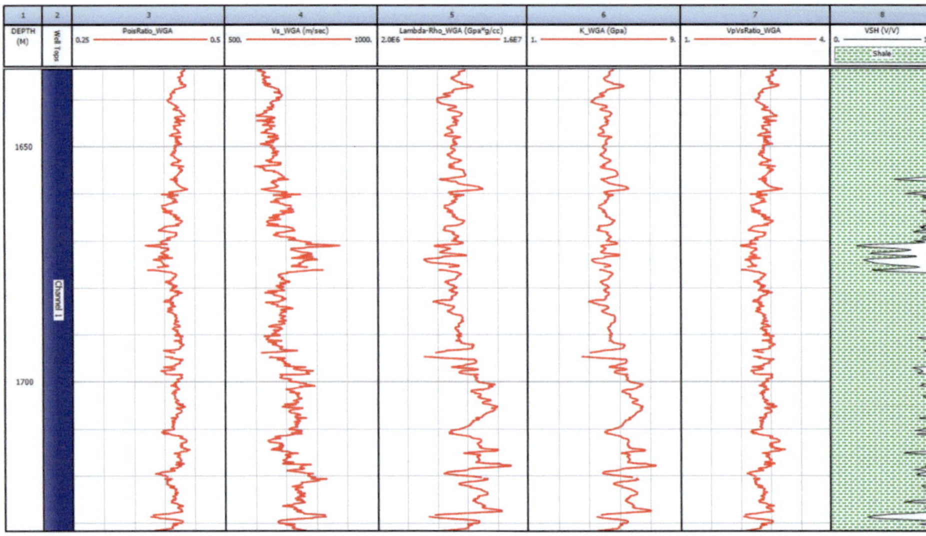

Fig. 13.13 Input data for Scarab-Db well: from left—Poisson's ratio (WGA), Shear wave velocity (Vs, WGA), Lambda-Rho (WGA), Bulk modulus (K, WGA), Vp/Vs ratio (WGA), and Shale volume (Vsh)

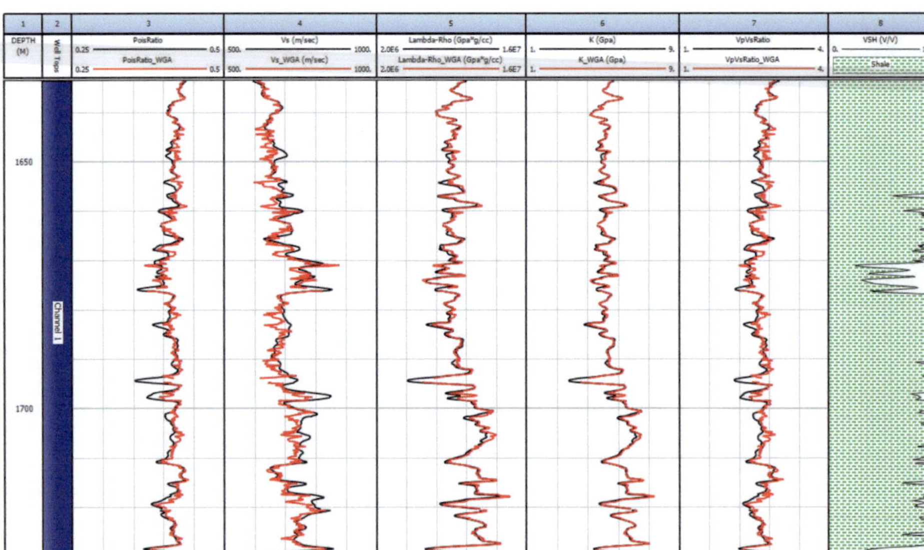

Fig. 13.14 Correlation among various parameters in the Scarab-Db well: from left to right—Poisson's ratio, Poisson's ratio (WGA), Shear wave velocity (Vs), Shear wave velocity (Vs, WGA), Lambda-Rho, Lambda-Rho (WGA), Bulk modulus (K), Bulk modulus (K, WGA), Vp/Vs ratio, Vp/Vs ratio (WGA), and Shale content (Vsh)

Table 13.2 Statistical evaluation of the proposed models

	RVFL-WGA	RVFL-AEO	RVFL-MRFO	RVFL-SCA
R^2	0.9995	0.9993	0.9990	0.9991
RMSE	0.0095	0.0109	0.0130	0.0126
MAE	0.0068	0.0082	0.0101	0.0098

low RMSE and MAE values. Table 13.2 presents a comparative analysis of the performance of the developed RVFL-WGA model in predicting Poisson's ratio, alongside other competitive algorithms such as RVFL-AEO, RVFL-MRFO, and RVFL-SCA. The results demonstrate the superior performance of the RVFL-WGA model, achieving the highest R^2 and the lowest RMSE and MAE values. Based on these findings, it can be concluded that the RVFL-WGA model is a reliable and effective tool for predicting Poisson's ratio.

Based on the previous discussion, it is evident that the integration of the WGA with the RVFL model demonstrates considerable potential in determining its configuration. However, the proposed method requires further improvement, as the exploration capabilities of WGA surpass its exploitation capabilities. Additionally, the time complexity of the developed method presents a limitation. This limitation can be addressed by implementing parallel processing techniques.

Poisson's Ratio Prediction Using Cheetah Optimizer (CO)

In this section, we investigate the influence of gas zones on the efficacy of ML models. A modified version of the RVFL model, incorporating the operators of the CO, is employed to address this challenge. To further explore the impact of fluid saturation on rock properties, a fluid substitution method is implemented to analyze various saturation scenarios and predict Poisson's ratio parameter under different fluid types.

In the context of gas-bearing reservoirs, the regression analysis method was employed to assess the reliability of the relationships between log-calculated Poisson's ratio ($\boldsymbol{\sigma}_{\mathbf{G}}$) and predicted Poisson's ratio ($\boldsymbol{\sigma}_{\mathbf{G1}}, \boldsymbol{\sigma}_{\mathbf{G2}}, \boldsymbol{\sigma}_{\mathbf{G3}}$) obtained from the SCA-RVFL, WOA-RVFL, and CO-RVFL models, respectively, (Table 13.3), when applied to the fourth test well (Scarab-Db). The correlation analysis between σ_G and σ_{G1}, σ_{G2}, and σ_{G3} revealed low r values of 0.11, 0.22, and 0.3, respectively, (Fig. 13.15a, b, c). These results indicate a weak correspondence between the log-calculated Poisson's ratio in gas zones and the Poisson's ratio predicted using ML techniques for the Scarab-Db test well (Fig. 13.15d).

Given the initial low correlation coefficient values, the log data from gas zones were excluded from the well-log dataset, and the models were reapplied. After the exclusion of gas zone data, the correlation coefficient between the log-calculated Poisson's ratio ($\boldsymbol{\sigma}_{\mathbf{Gr}}$) and the predicted values ($\boldsymbol{\sigma}_{\mathbf{Gr1}}$ and $\boldsymbol{\sigma}_{\mathbf{Gr2}}$) increased to r = 0.24 and r = 0.33, respectively, (Fig. 13.16a and b). However, the correlation between $\boldsymbol{\sigma}_{\mathrm{Gr}}$ and the predicted value $\boldsymbol{\sigma}_{\mathbf{Gr3}}$ exhibited a significantly higher correlation coefficient of r = 0.83 (Fig. 13.16c). Figure 13.14d visually demonstrates the strong agreement between the

Table 13.3 Definitions and annotations used for Poisson's ratio

PRG	σ_G	Poisson's ratio derived from original well-log data
PRGSCA	σ_{G1}	Poisson's ratio predicted from original well-log data utilizing the SCA-RVFL model
PRGWOA	σ_{G2}	Poisson's ratio predicted from original well-log data utilizing the WOA-RVFL model
PRGCO	σ_{G3}	Poisson's ratio predicted from original well-log data utilizing the CO-RVFL model
PRGR	σ_{Gr}	Poisson's ratio derived from well-log data after excluding gas zones
PRGRSCA	σ_{Gr1}	Poisson's ratio predicted after excluding the well-log data of gas zones utilizing the SCA-RVFL model
PRGRWOA	σ_{Gr2}	Poisson's ratio predicted after excluding the well-log data of gas zones utilizing the WOA-RVFL model
PRGRCO	σ_{Gr3}	Poisson's ratio predicted after excluding the well-log data of gas zones utilizing the CO-RVFL model
PRO	σ_O	Poisson's ratio derived after fluid substitution with oil
PROSCA	σ_{O1}	Poisson's ratio predicted after fluid substitution with oil utilizing the SCA-RVFL model
PROWOA	σ_{O2}	Poisson's ratio predicted after fluid substitution with oil utilizing the WOA-RVFL model
PROCO	σ_{O3}	Poisson's ratio predicted after fluid substitution with oil utilizing the CO-RVFL model
PRW	σ_W	Poisson's ratio derived after fluid substitution with water
PRWSCA	σ_{W1}	Poisson's ratio predicted after fluid substitution with water utilizing the SCA-RVFL model
PRWWOA	σ_{W2}	Poisson's ratio predicted after fluid substitution with water utilizing the WOA-RVFL model
PRWCO	σ_{W3}	Poisson's ratio predicted after fluid substitution with water utilizing the CO-RVFL model

log-calculated Poisson's ratio ($\boldsymbol{\sigma}_{\mathbf{Gr}}$) and the Poisson's ratio predicted using the machine learning technique in the proposed CO model ($\boldsymbol{\sigma}_{\mathbf{Gr3}}$). In contrast, the predictions from the sine cosine algorithm (SCA) and whale optimization algorithm (WOA) ($\boldsymbol{\sigma}_{\mathbf{Gr1}}$ and $\boldsymbol{\sigma}_{\mathbf{Gr2}}$) remain relatively weak.

As evidenced by the preceding results, the accuracy of machine learning predictions is compromised in the presence of gas zones. To address this, a fluid substitution technique was employed to examine the specific fluid-induced responses before and after fluid type modifications. Figure 13.17 illustrates a comparison of Poisson's ratio logs under three distinct fluid saturation scenarios: fully gas-saturated, fully oil-saturated, and fully water-saturated conditions.

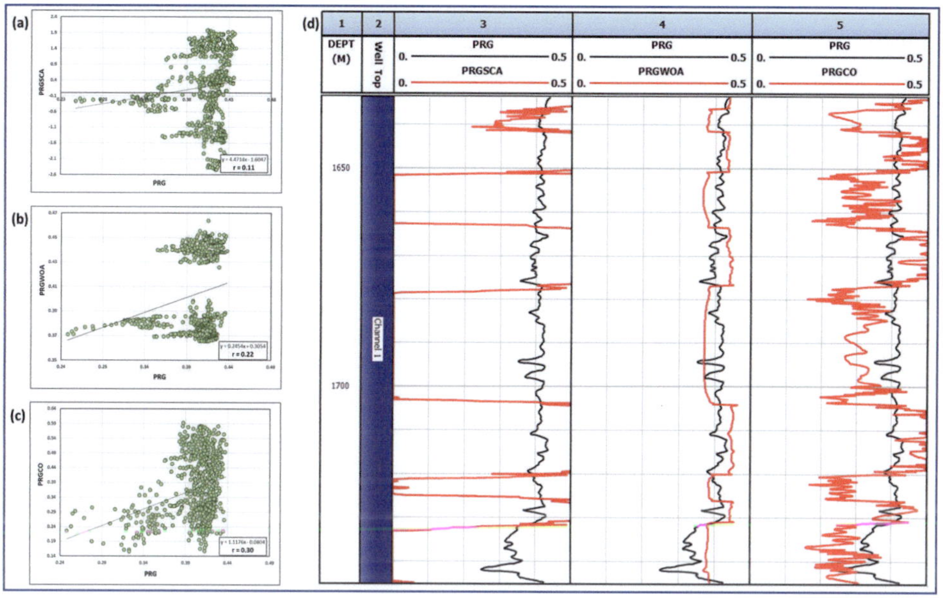

Fig. 13.15 Regression analysis and resulting logs of calculated and predicted Poisson's ratio in the presence of gas. **a**, **b**, **c** relationships and correlation coefficients. **d** correlation between output Poisson's ratio

Following the substitution of reservoir gas with oil and water, machine learning algorithms were applied to these two scenarios. The results indicated an improvement in model performance for oil and water cases compared to the original gas-saturated scenario.

The relationship between the log-calculated Poisson's ratio after gas substitution with oil ($\boldsymbol{\sigma_O}$) and the predicted $\boldsymbol{\sigma_{O3}}$ exhibited an excellent correlation coefficient of r = 0.99 (Fig. 13.18c). Conversely, the correlation coefficients between $\boldsymbol{\sigma_O}$ and the predicted $\boldsymbol{\sigma_{Gr1}}$ and $\boldsymbol{\sigma_{Gr2}}$ (Table 13.3) decreased to r = 0.63 and r = 0.65, respectively, (Fig. 13.18a and b). Figure 13.18d further emphasizes the strong alignment between the log-calculated Poisson's ratio in oil-saturated zones and the machine learning-predicted Poisson's ratio within the CO model.

Similarly, in the case of gas replacement with water, the relationship between $\boldsymbol{\sigma_W}$ and the predicted $\boldsymbol{\sigma_{W3}}$ displayed an excellent correlation coefficient of r = 0.99 (Fig. 13.19c). In contrast, the correlation coefficients between $\boldsymbol{\sigma_W}$ and the predicted $\boldsymbol{\sigma_{W1}}$ and $\boldsymbol{\sigma_{W2}}$ increased to r = 0.66 and r = 0.67, respectively, (Fig. 13.19a and b). Figure 13.19d highlights the significant alignment between the log-calculated Poisson's ratio and the machine learning-predicted Poisson's ratio within the CO model. Notably, the results for the water-saturated reservoir demonstrated a slight improvement over the oil-saturated scenario.

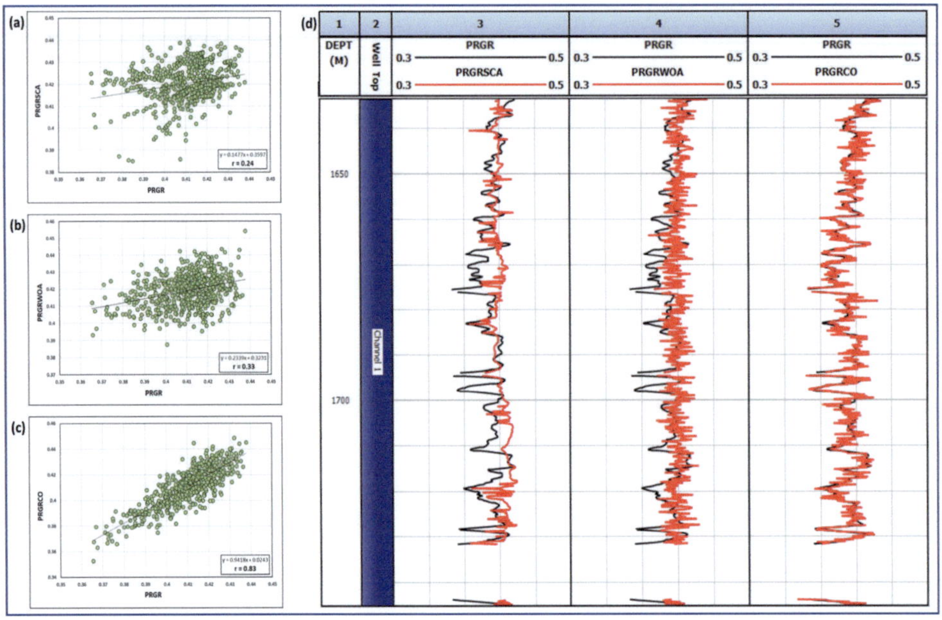

Fig. 13.16 Regression analysis and the resulting logs of calculated and predicted Poisson's ratio in the case of gas removal. **a**, **b**, **c** Relationships and correlation coefficients. **d** Correlation between output Poisson's ratio

Table 13.4 presents the correlation coefficients calculated for the Scarab-Db test well under four different scenarios. The results indicate an enhancement in correlation coefficients upon the exclusion of gas zone log data. This trend continues with further improvements observed when oil and water log data are substituted for the original gas data. Moreover, a comparative analysis of the correlation coefficients (r) reveals the superiority of the proposed CO model over the SCA and WOA models, as demonstrated in both the table and histograms.

Shale Volume

In recent years, machine learning techniques have seen increased utilization for predicting shale volume from seismic traces. The shale content is a critical petrophysical parameter for evaluating hydrocarbon reservoirs. This study introduces an innovative approach for calculating shale content through the application of machine learning to seismic traces, particularly in the context of insufficient well-log data.

Shale volume is a crucial quantitative parameter in log data analysis. Shale serves as a source rock and significantly influences the quality of the reservoir rock; an increase in reservoir quality is associated with a decrease in shale volume. Shale volume can be estimated using various logs, including self-potential, gamma-ray (GR), density, neutron,

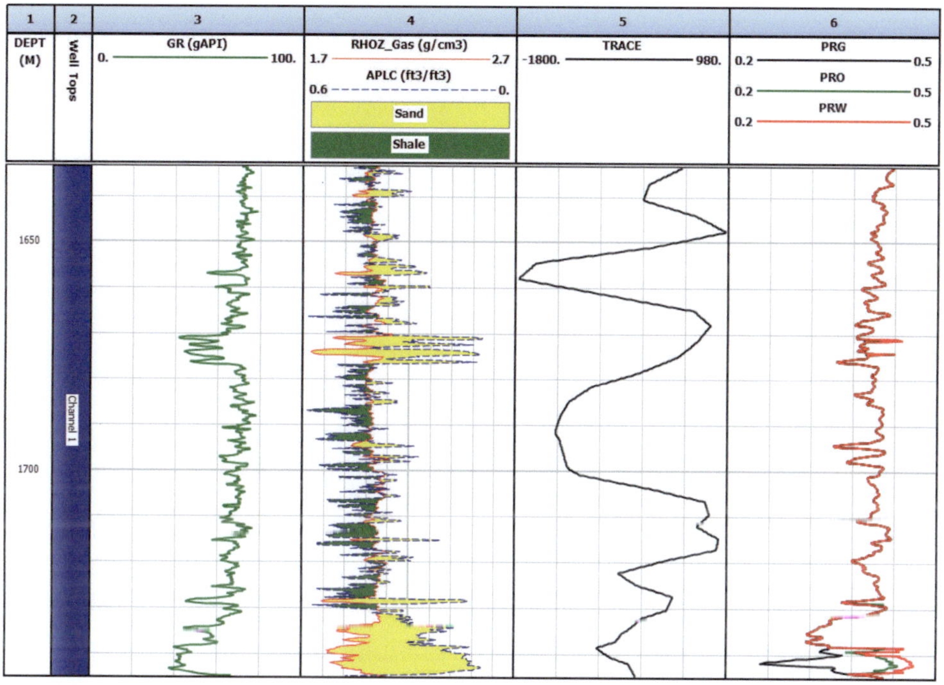

Fig. 13.17 Input data for the Scarab-Db well, including gamma-ray (GR), neutron-density, and seismic trace (from left). Additionally, fluid substitution is illustrated by Poisson's ratio under different fluid conditions: gas zones (black), oil (green), and water (red)

resistivity, and velocity deviation (Kamel and Mabrouk 2003; Nabih 2021). Researchers have endeavored to identify and validate different methods for calculating shale volume to determine the minimum possible values (Crain 1986; Nabih and Al-Alfy 2018).

In this study, the shale volume was predicted using ML techniques, specifically the Random Forest (RF) algorithm. Two approaches were employed: one utilizing raw well-log data from four wells in the WDDM area, and the other incorporating seismic traces and attributes extracted from a seismic survey at the well locations. Three wells (Scarab-1, Scarab-Da, and Scarab-Dd) were utilized as training data, while the fourth well (Scarab-Db) served as a test well to evaluate the model's predictive performance.

Shale Volume from Log Data and Seismic Amplitude

For the training purposes of the ML random forest algorithm, three wells of data (Scarab-1, Scarab-Da, and Scarab-Dd) are used. The input data are resistivity (LLD), compressional sonic (DTCO), density (RHOZ), seismic trace and shale volume is the main target. We applied the model to the data based on the following parameters of the RF algorithm (Estimator: 100, Max depth: 50). The model is applied to 2273 samples,

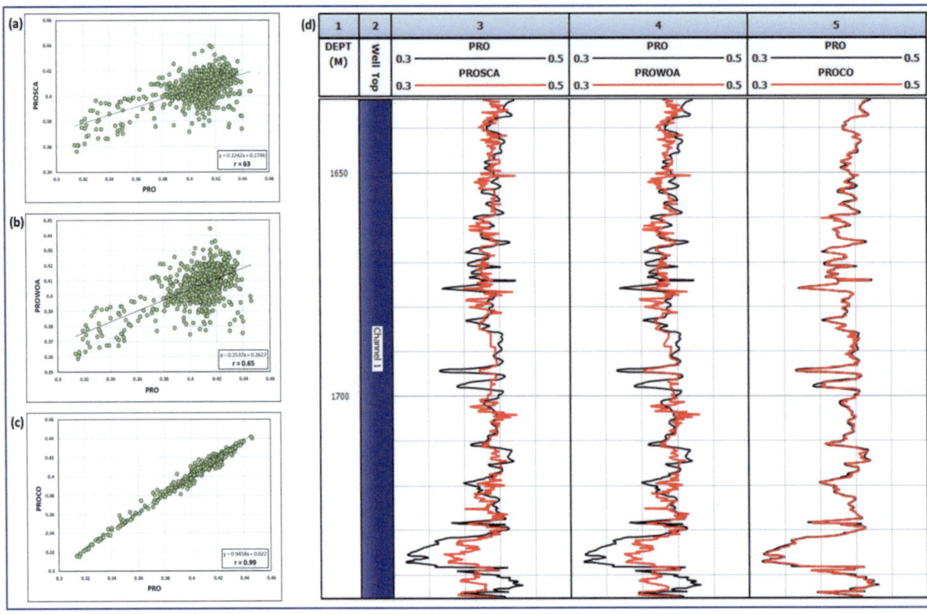

Fig. 13.18 Regression analysis and the resulting logs of calculated and predicted Poisson's ratio in the scenario of fluid substitution with oil. **a**, **b**, **c** relationships and correlation coefficient. **d** correlation between output Poisson's ratio logs

trained on 1820 samples (80%), and validated on 453 samples (20%). The results of the model on the validating data are excellent, where the correlation coefficient with the validation data is 0.96, and then the model is applied to the test well (Scarab-Db). The results of the model in this step are excellent.

A regression analysis was conducted to assess the effectiveness of the relationships between the shale volume calculated from GR and the shale volumes predicted using the RF algorithm. The model was implemented using OpendTect Pro (6.6) software for academic research purposes. Figure 13.20 illustrates the relationship between VSH (derived from GR) and VSH (predicted by RF) for the training wells. The correlation coefficients for wells Scarab-1, Scarab-Da, and Scarab-Dd were found to be 0.98, 0.99, and 0.98, respectively.

When the RF is used on the fourth test well (Scarab-Db), there is a very high correlation between shale volume (from log data) and shale volume (Predicted from RF) is r = 0.91 (Fig. 13.21). Figure 13.22 reveals good matches between VSH estimated from well-log data, and VSH predicted from the machine learning RF method of the Scarab-Db test well. This good match between the original and predicted logs indicates that the machine learning tool is an effective and reliable method for predicting shale volume in other wells.

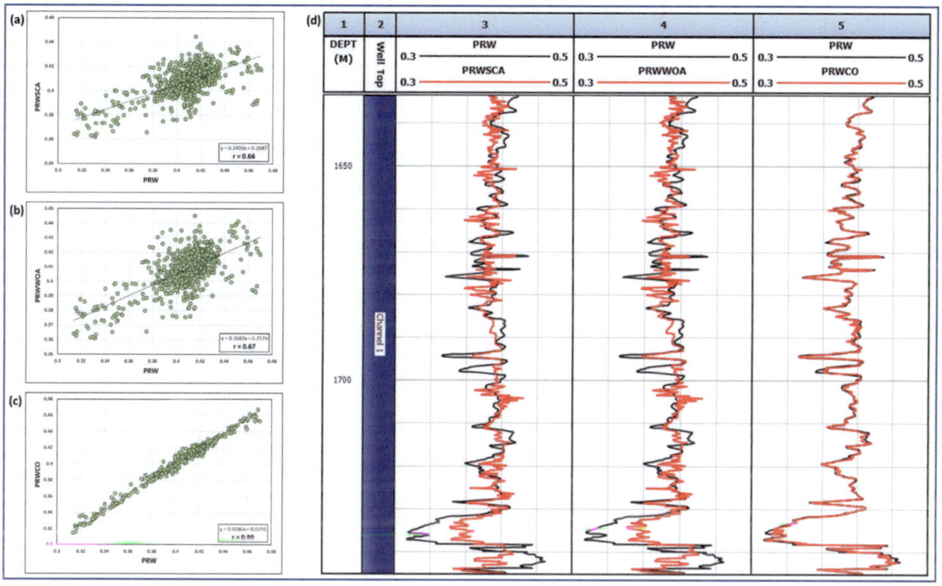

Fig. 13.19 Regression analysis and the resulting logs of calculated and predicted Poisson's ratio in the scenario of fluid substitution with water. **a**, **b**, **c** Relationships and correlation coefficients. **d** Correlation between output Poisson's ratio logs

Table 13.4 Values of correlation coefficients (r) for the Scarab-Db test well

Predicted		PRSCA (model 1)	PRWOA (model 2)	PRCO (model 3)
Calculated				
PRG	σ_G	0.11	0.22	0.3
PRGR	σ_{Gr}	0.24	0.33	0.83
PRO	σ_O	0.63	0.65	0.99
PRW	σ_W	0.66	0.67	0.99

Shale Volume from Log Data and Seismic Attributes

In this case, we used three wells as train data (Scarab-1, Scarab-Da and Scarab-Dd); the input data are spectral decomposition (F-10, F-45), Relative acoustic impedance (RAI), root mean square amplitude (RMS), RMS/RAI, resistivity (LLD) and shale volume is the main target. We applied the model to the data based on the following parameters of the RF algorithm (Estimator: 100, Max depth: 50). The model is applied to 2273 samples, training on 1820 samples (80%), and validating on 453 samples (20%). The model's results on the validating data are excellent, where the correlation coefficient with the validation data

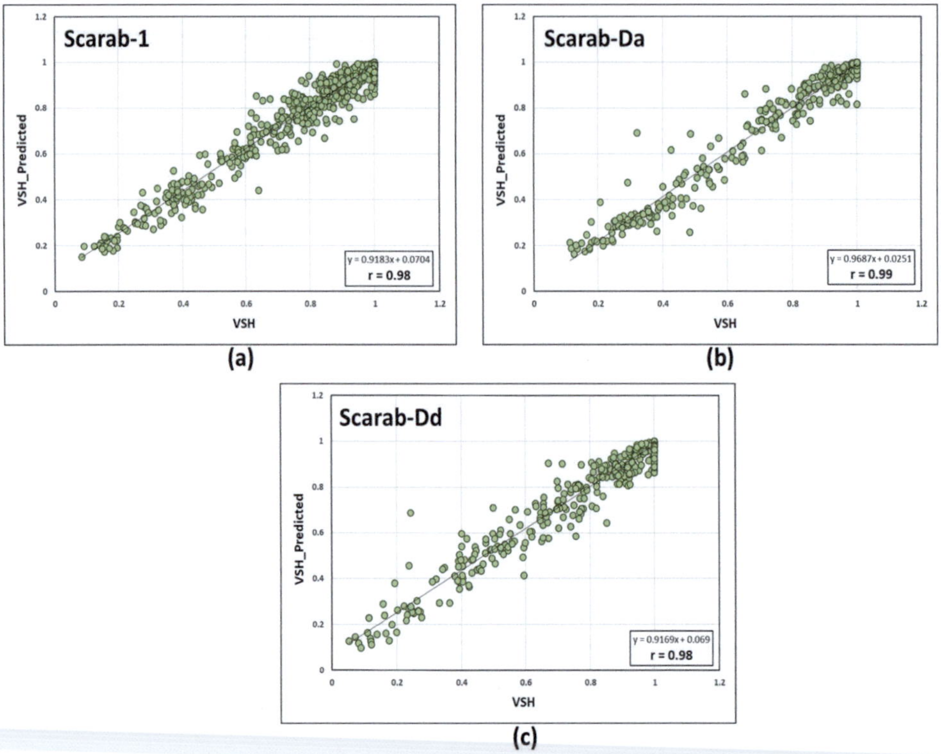

Fig. 13.20 Correlation between the conventional VSH derived by well-log analysis and the VSH predicted by machine learning technique for the training wells using the original seismic amplitude data with well logs

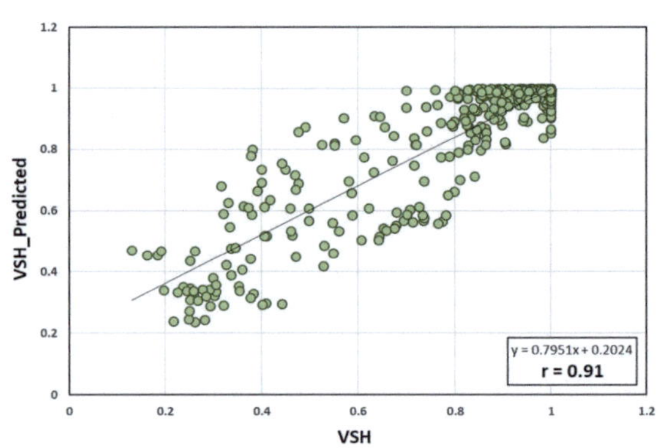

Fig. 13.21 The relationship between conventional VSH from well log analysis and the derivative predicted VSH from machine learning technique for test well Scarab-Db using the original seismic amplitude data with well logs

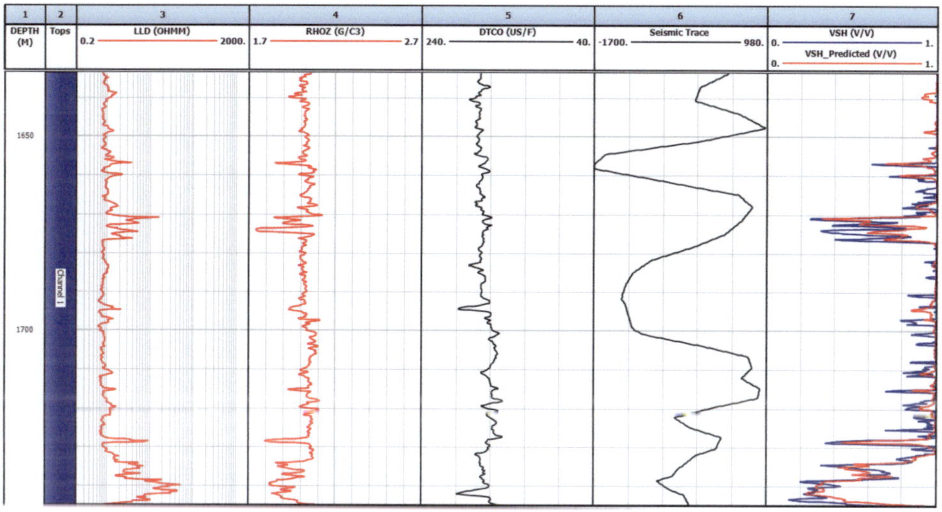

Fig. 13.22 Scarab-Db well input data (from left, LLD, RHOZ, DTCO, seismic trace) and correlation between VSH from GR (blue color) and VSH predicted form RF (red color)

is 0.97, and then the model is applied to the test well (Scarab-Db). The model's results in this step are very good.

A regression analysis was conducted to assess the effectiveness of the relationships between the shale volume calculated from GR and the shale volumes predicted using the RF algorithm. The model was implemented using OpendTect Pro (6.6) software for academic research purposes. Figure 13.23 illustrates the relationship between VSH (derived from GR) and VSH (predicted by RF) for the training wells. The correlation coefficients for wells Scarab-1, Scarab-Da, and Scarab-Dd were found to be 0.98, 0.99, and 0.97, respectively.

When the RF was applied to the fourth test well (Scarab-Db), a significant correlation was observed between the shale volume estimated from log data and Poisson's ratio predicted by the RF model, with a correlation coefficient of r = 0.88 (Fig. 13.24). This high correlation underscores the effectiveness of the RF algorithm in accurately predicting Poisson's ratio from shale volume data, demonstrating its potential utility in geophysical applications.

Figure 13.25 demonstrates a strong correlation between the shale volume (VSH) estimated from well-log data and the VSH predicted using the machine learning methodology applied to the Scarab-Db test well. The close alignment between the original and predicted logs underscores the effectiveness and reliability of the machine learning tool for predicting shale volume in other wells. This validation highlights the potential of machine learning techniques to provide accurate and consistent predictions in geophysical applications.

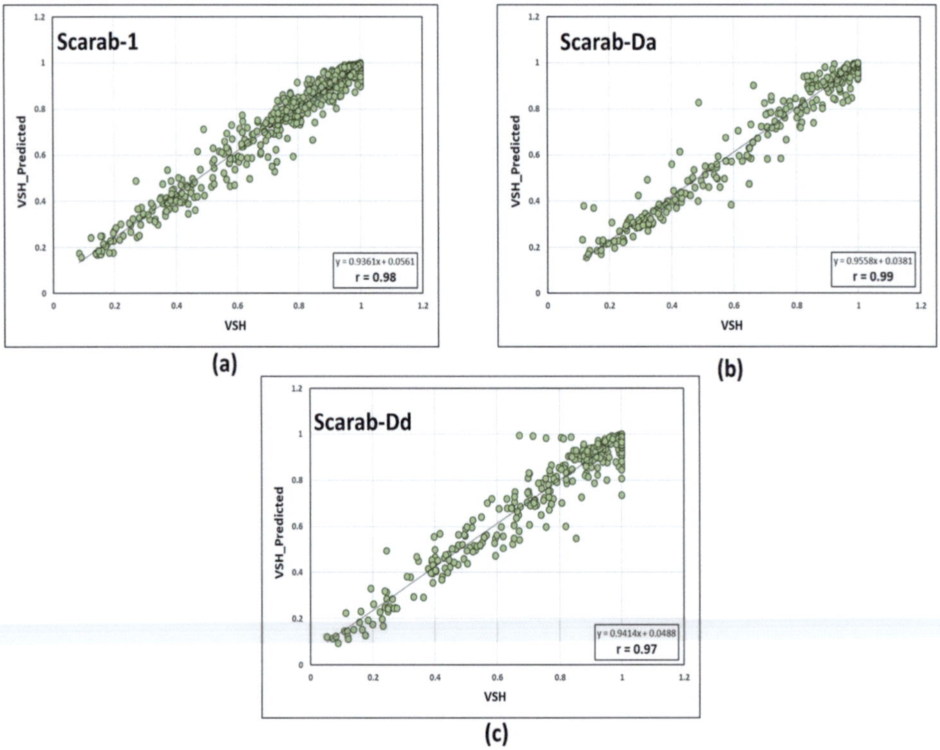

Fig. 13.23 Correlation between the conventional VSH derived by well-log analysis and the VSH predicted by machine learning technique for the training wells using the derived seismic attributes with well logs

Effective Porosity

Porosity is a critical parameter for evaluating reservoir characteristics. The porosity derived from a single porosity log remains reliable if the lithology is constant. However, when lithology varies, formation porosity can be more accurately determined by combining the readings from two porosity logs. The most effective combination of tools for this purpose is the density and neutron logs, which yield reliable results in various contexts. In this study, effective porosity is determined using the density-neutron log combination (Asquith et al. 2004; Bateman 2012).

Three wells (Scarab-1, Scarab-Da, and Scarab-Dd) were employed as training data for the Random Forest (RF) algorithm. The input features included LLD, DTCO, GR, and seismic trace, with effective porosity serving as the primary target variable. The RF algorithm was conFig.d with 100 estimators and a maximum depth of 50. The dataset, comprising 2273 samples, was divided into training and validation sets, with 80% (1820 samples) allocated for training and 20% (453 samples) reserved for validation. The model

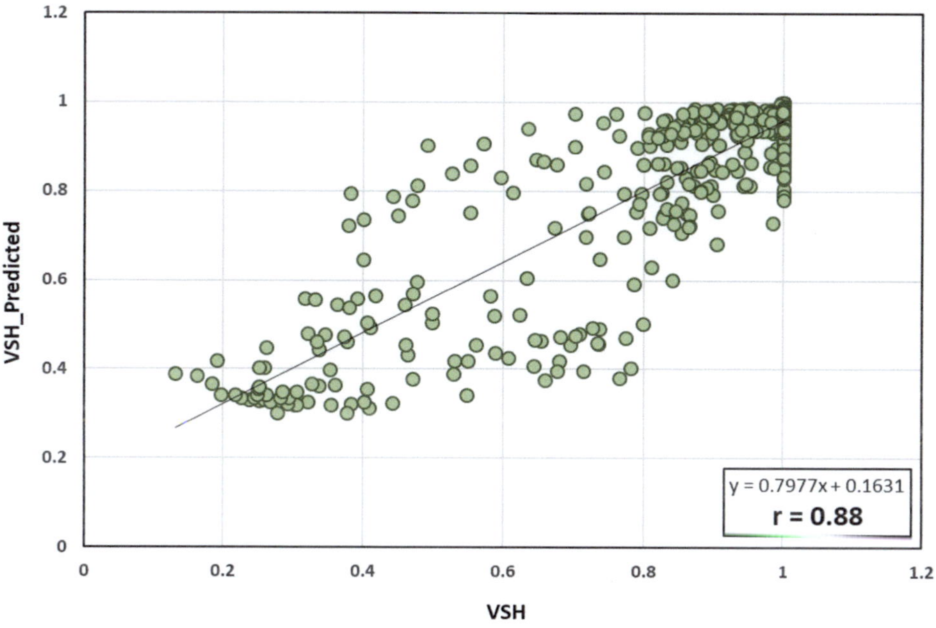

Fig. 13.24 The relationship between conventional VSH from well log analysis and the derivative predicted VSH from machine learning technique for test well Scarab-Db using the derived seismic attributes with well logs

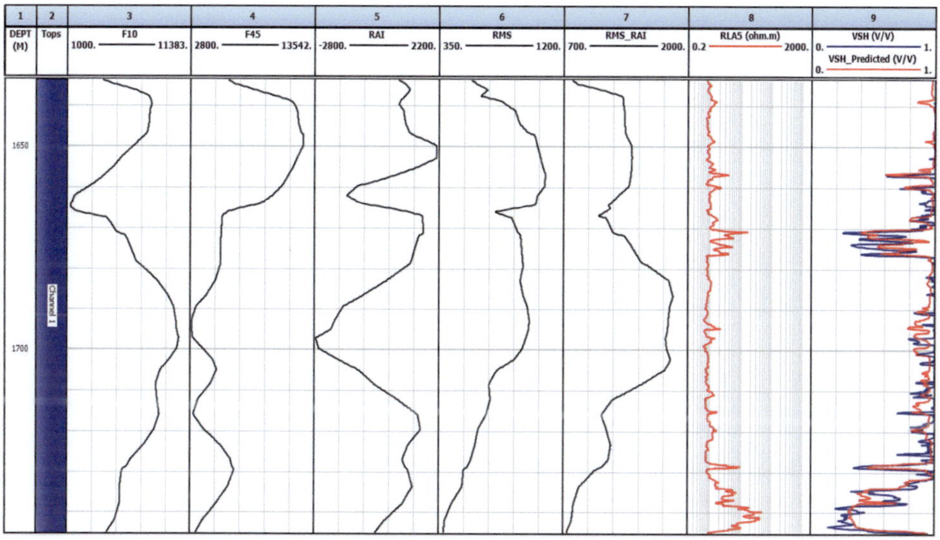

Fig. 13.25 Db well input data (from left, LLD, RHOZ, DTCO, seismic trace) and correlation between VSH from GR (blue color) and VSH predicted form RF (red color)

exhibited a strong correlation coefficient of r = 0.9791 with the validation data. Subsequently, the trained model was applied to the test well (Scarab-Db), yielding promising results.

To evaluate the efficacy of the RF algorithm in predicting effective porosity, a regression analysis was performed. The RF model was implemented within the OpendTect Pro 6.6 software framework, utilizing a comprehensive training dataset. The results, visualized in Fig. 13.26, reveal exceptionally high correlation coefficients (exceeding 0.98) between the actual effective porosity values, as determined from density-neutron logs, and the predicted values generated by the RF model for the training wells. These findings underscore the robustness and reliability of the RF model for predicting effective porosity in this specific geological setting.

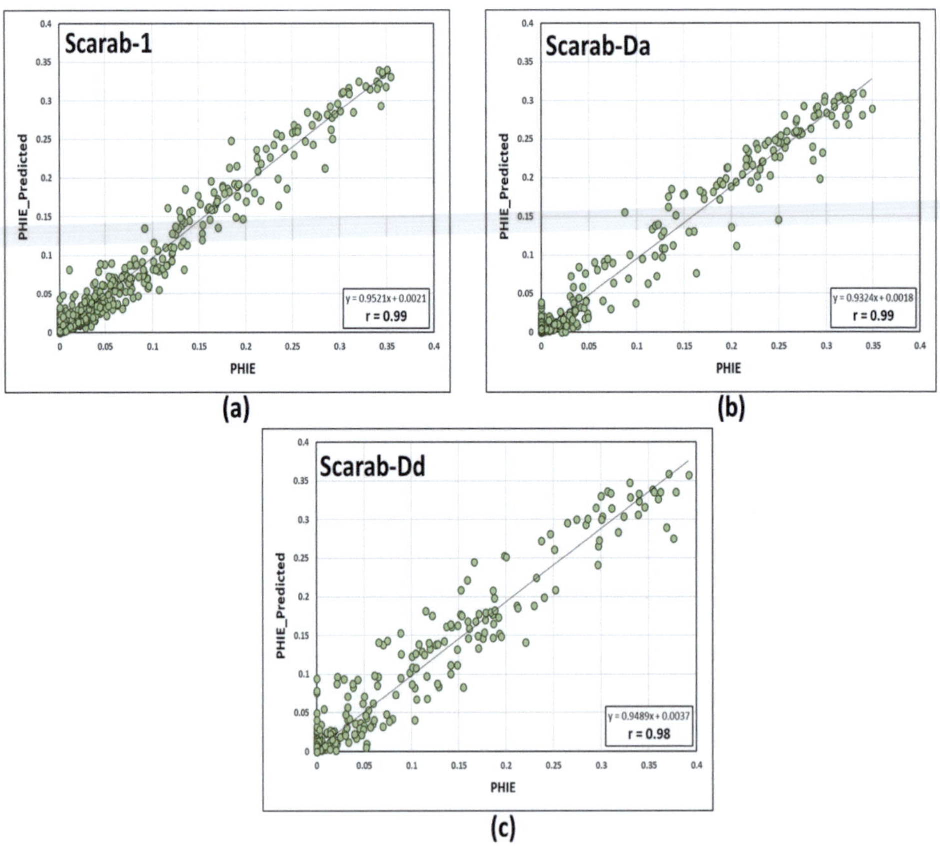

Fig. 13.26 Correlation between the conventional PHIE calculated by well-log analysis and the derived PHIE obtained by machine learning technique for the training wells

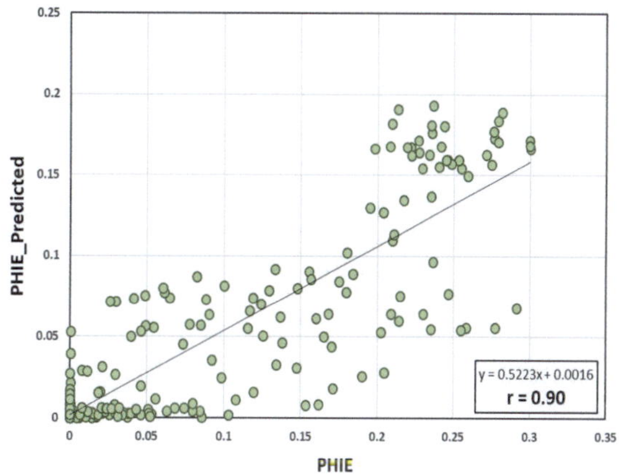

Fig. 13.27 Relationship between conventional PHIE from well-log analysis and the derivative predicted PHIE from machine learning technique for test well Scarab-Db

When the RF was applied to the fourth test well (Scarab-Db), an excellent correlation was observed between the effective porosity estimated from log data and the effective porosity predicted by the RF model, with a correlation coefficient of $r = 0.91$ (Fig. 13.27). Furthermore, Fig. 13.28 illustrates a strong agreement between the effective porosity (PHIE) calculated from well-log data and the PHIE predicted using the machine learning method for the Scarab-Db test well. This high correlation between the logs indicates that the RF model is highly efficient and reliable for predicting effective porosity.

13.4 Summary and Conclusions

13.4.1 Summary

Machine learning (ML) methods are playing an increasingly important role in reducing the risk of drilling operations in petroleum exploration. This is because ML can provide and predict precise petrophysical information for the reservoirs more than conventional interpretation methods of well log and seismic data. Conventional interpretation methods are often time-consuming and labor-intensive and can be inaccurate. ML methods, on the other hand, can quickly and accurately analyze large amounts of data to identify potential drilling targets and predict production rates, and this reduces the risk of drilling dry wells. ML methods are also being used to optimize the production of existing reservoirs, which can further reduce costs and increase profits.

This research presents a novel workflow that synergizes well-log and multi-attribute seismic data with ML tools to predict key petrophysical and rock physics parameters, including Poisson's ratio, shale volume, and effective porosity, within Egypt's WDDM basin. This integrated approach aims to enhance reservoir characterization and facilitate informed decision-making in hydrocarbon exploration and production.

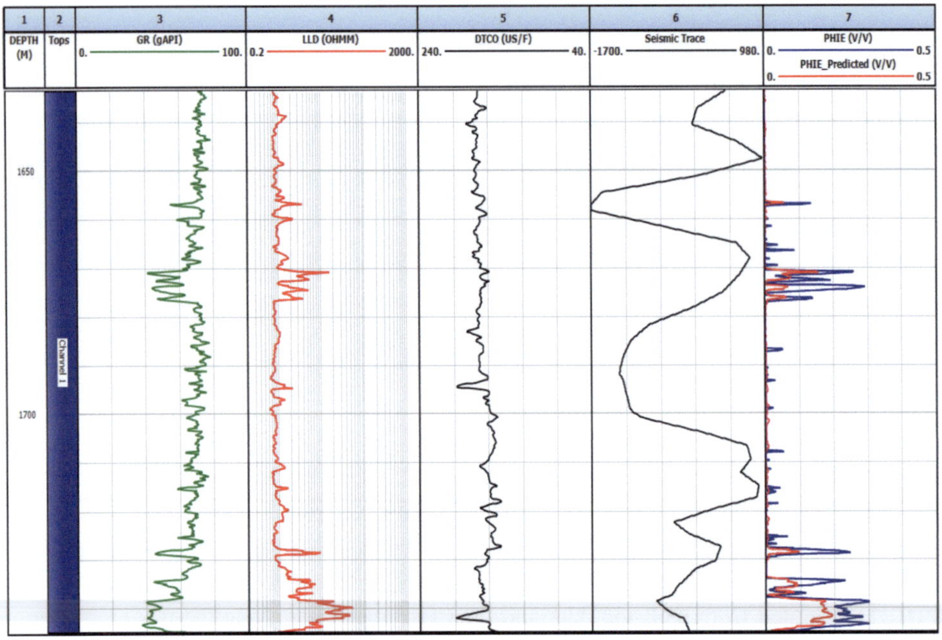

Fig. 13.28 Db well input data (from left, LLD, DTCO, GR, seismic trace) and correlation between PHIE from density-neutron (blue color) and PHIE predicted form RF (red color)

13.4.2 Conclusions

From the findings, the following conclusions and recommendations are obtained:

- **Strong correlation**: A strong correlation (r = 0.8) was observed between the Poisson's ratio values derived from well-log data and those predicted by the RVFL-WGA model.
- **Accurate shear wave velocity estimation**: The Poisson's ratio values predicted by the RVFL-WGA model were effectively utilized to estimate shear wave velocity, resulting in a high correlation coefficient of 0.86.
- **Robust model validation**: A robust agreement was found between the petrophysical and rock physics parameters calculated from well-log data and those predicted using the ML approach.
- **Impact of fluid substitution**: Gassmann's fluid substitution technique significantly enhanced the prediction of Poisson's ratio behavior. A notable increase in Poisson's ratio was observed for fully saturated oil and water conditions.
- **Synergistic approach**: The integration of fluid substitution and ML tools yielded improved accuracy and reliability in the results.

- **Accurate shale volume estimation**: The correlation coefficients between shale volume (VSH) derived from gamma-ray (GR) logs and that predicted by the RF model were excellent, with r values of 0.91 for well-log data and 0.88 for seismic attributes.
- **Precise porosity prediction**: The correlation coefficient between effective porosity (PHIE) calculated from neutron-density (N-D) logs and that predicted by the RF model was excellent, with an r value of 0.9.
- **Machine learning efficacy**: The study demonstrated the potential of ML methods to accurately model petrophysical and rock physics parameters.

13.4.3 Recommendations

- **Future applications**: To further validate the accuracy of the ML models in predicting gas channels and chimneys, it is recommended to apply them to a larger dataset from diverse geological settings.
- **Broader applicability**: The algorithms employed in this study have the potential to be applied to a wider range of fields and petroleum basins, offering valuable insights into reservoir characterization.
- **Future research directions**: Encourage interdisciplinary research that combines geology, petrophysics, and machine learning to develop more comprehensive models for hydrocarbon exploration. This collaboration could lead to innovative solutions and methodologies in the industry.

References

Abd Elaziz M, Ghoneimi A, Elsheikh AH, Abualigah L, Bakry A, Nabih M (2022) Predicting shale volume from seismic traces using modified random vector functional link based on transient search optimization model: a case study from Netherlands North Sea. Nat Resour Res 31(3):1775–1791. https://doi.org/10.1007/s11053-022-10049-4

Abd Elaziz M, Ghoneimi A, Nabih M, Bakry A, Al-Betar MA (2023) Contribution of fluid substitution and cheetah optimizer algorithm in predicting rock-physics parameters of gas-bearing reservoirs in the Eastern Mediterranean sea, Egypt. Nat Resour Res:1–19. https://doi.org/10.1007/s11053-023-10219-y

Abd El-Gawad E, Abdelwahhab M, Bekiet M, Nooh AZ, Abd El-Aziz NM, Fouda AEH (2019) Reservoir quality determination through petrophysical analysis of El Wastani formation in scarab field, offshore Nile Delta, Egypt. Al-Azhar Bull Sci 30(1):1–12. https://doi.org/10.21608/absb.2019.13007.1010

Ali M, Jiang R, Ma H, Pan H, Abbas K, Ashraf U, Ullah J (2021) Machine learning—A novel approach of well-log similarity based on synchronization measures to predict shear sonic logs. J Pet Sci Eng 203. https://doi.org/10.1016/j.petrol.2021.108602

Asquith G, Krygowski D, Henderson S, Hurley N (2004) Basic well-log analysis. Am Assoc Pet Geol. https://doi.org/10.1306/Mth16823

Bakry A, Ghoneimi A, Nabih M, Abd Elaziz M (2023) Petrophysical and rock physics analysis using well log data for gas detection in scarab gas field, West Delta Deep Marine Concession, Egypt. Bull Fac Sci, Zagazig Univ (2):91–112. https://doi.org/10.21608/bfszu.2023.177130.1210

Bakry A (2021) Optimization of rock physics recognition using integrated seismic and well log data. MSc Thesis. Faculty of Science, Zagazig University, Zagazig, Egypt

Bakry A (2024) Optimization of reservoir characterization from seismic and well log data using machine learning in the WDDM area, Egypt. Ph.D. Thesis. Faculty of Science, Zagazig University, Zagazig, Egypt

Bateman RM (2012) Openhole log analysis and formation evaluation. Second Edition Society of Petroleum Engineers (SPE). https://doi.org/10.2118/9781613991565

Belyadi H, Fathi E, Belyadi F (2019) Hydraulic fracturing in unconventional reservoirs: Theories, operations, and economic analysis. Gulf Professional Publishing

Chaki S, Routray A, Mohanty WK (2018) Well-log and seismic data integration for reservoir characterization: a signal processing and machine-learning perspective. IEEE Signal Process Mag 35:72–81. https://doi.org/10.1109/MSP.2017.2776602

Crain ER (1986) Log analysis handbook. PennWell Books, Tulsa, OK

Dorrington KP, Link CA (2004) Genetic algorithm/neural network approach to seismic attribute selection for well-log prediction. Geophysics 69. https://doi.org/10.1190/1.1649389

Dramsch JS (2020) 70 years of machine learning in geoscience in review, pp 1–55. https://doi.org/10.1016/bs.agph.2020.08.002

Farouk S, Sen S, Ganguli SS, Abuseda H, Debnath A (2021) Petrophysical assessment and permeability modeling utilizing core data and machine learning approaches—A study from the Badr El Din-1 field, Egypt. Mar Pet Geol 133:105265. https://doi.org/10.1016/j.marpetgeo.2021.105265

Farsi M, Mohamadian N, Ghorbani H, Wood DA, Davoodi S, Moghadasi J, Ahmadi Alvar M (2021) Predicting formation pore-pressure from well-log data with hybrid machine-learning optimization algorithms. Nat Resour Res 30. https://doi.org/10.1007/s11053-021-09852-2

Feng R (2021) Improving uncertainty analysis in well-log classification by machine learning with a scaling algorithm. J Pet Sci Eng 196. https://doi.org/10.1016/j.petrol.2020.107995

Garia S, Pal AK, Ravi K, Nair AM (2021) Prediction of petrophysical properties from seismic inversion and neural network: A case study. In: EGU general assembly conference abstracts, pp EGU21–11824. https://doi.org/10.5194/egusphere-egu21-11824

Ghoneimi A, Farag AE, Bakry A, Nabih M (2021) A new deeper channel system predicted using seismic attributes in Scarab gas field, west delta deep marine concession, Egypt. J Afr Earth Sci 177:1–18. https://doi.org/10.1016/j.jafrearsci.2021.104155

Ismail A, Ewida HF, Al-Ibiary MG, Zollo A (2020) Application of AVO attributes for gas channels identification, west offshore Nile Delta, Egypt. Pet Res 5:112–123. https://doi.org/10.1016/j.ptlrs.2020.01.003

Ismail A, Ewida HF, Nazeri S, Al-Ibiary MG, Zollo A (2022) Gas channels and chimneys prediction using artificial neural networks and multi-seismic attributes, offshore West Nile Delta, Egypt. J Pet Sci Eng 208:109349. https://doi.org/10.1016/j.petrol.2021.109349

Iturrarán-Viveros U, Muñoz-García AM, Castillo-Reyes O, Shukla K (2021).Machine learning as a seismic prior velocity model building method for full-waveform inversion: a case study from Colombia. Pure Appl Geophys 178. https://doi.org/10.1007/s00024-021-02655-9

Kamel MH, Mabrouk WM (2003) Estimation of shale volume using a combination of the three porosity logs. J Petrol Sci Eng 40(3–4):145–157. https://doi.org/10.1016/S0920-4105(03)00120-7

Kumar J (1976) The effect of Poisson's ratio on rock properties. In: All days. SPE. https://doi.org/10.2118/6094-MS

Mohamed IA, El-Mowafy HZ, Fathy M (2015) Prediction of elastic properties using seismic prestack inversion and neural network analysis. Interpretation 3:T57–T68. https://doi.org/10.1190/INT-2014-0139.1

Nabih M (2021) Reliability of velocity-deviation logs for shale content evaluation in clastic reservoirs: a case study, Egypt. Arab J Geosci 14(6):507. https://doi.org/10.1007/s12517-021-06886-x

Nabih M, Al-Alfy IM (2018) New approach for releasing uranium radiation impact on shale content evaluation in shaly sand formations: A case study, Egypt. Appl Radiat Isot 141:33–43. https://doi.org/10.1016/j.apradiso.2018.08.002

Nabih M, Ghoneimi A, Bakry A, Chelloug SA, Al-Betar MA, Abd Elaziz M (2022) Rock physics analysis from predicted Poisson's ratio using RVFL based on Wild Geese Algorithm in Scarab gas field in WDDM concession, Egypt. Mar Pet Geol:105949. https://doi.org/10.1016/j.marpetgeo.2022.105949

Priezzhev II, Veeken PCH, Egorov SV, Strecker U (2019) Direct prediction of petrophysical and petroelastic reservoir properties from seismic and well-log data using nonlinear machine learning algorithms. Lead Edge 38:949–958. https://doi.org/10.1190/tle38120949.1

Samuel A, Kneller B, Raslan S, Sharp A, Parsons C (2003) Prolific deep-marine slope channels of the Nile Delta, Eqypt. AAPG Bull 87(4):541–560

Stancliffe RPW, Kooij MW (2001) The use of satellite-based radar interferometry to monitor production activity at the Cold Lake heavy oil field, in Alberta, Canada. Am Assoc Pet Geol Bull 85

Yasin Q, Sohail GM, Khalid P, Baklouti S, Du Q (2021) Application of machine learning tool to predict the porosity of clastic depositional system, Indus Basin, Pakistan. J Pet Sci Eng 197. https://doi.org/10.1016/j.petrol.2020.107975

Zahmatkesh I, Kadkhodaie A, Soleimani B, Azarpour M (2021) Integration of well-log-derived facies and 3D seismic attributes for seismic facies mapping: A case study from Mansuri oil field, SW Iran. J Pet Sci Eng 202. https://doi.org/10.1016/j.petrol.2021.108563

Zhang JJ, Bentley LR (2005) Factors determining Poisson's ratio. CREWES Res Rep 17:1–15

Geomechanics Workflow Challenges and Hydrocarbon Potential in the Nile Delta

14

Ali E. Farag and Rafik A. Hammad

Abstract

In recent years, the Nile Delta has emerged as a promising frontier for oil and gas exploration, drawing significant attention from industry experts. This chapter emphasizes the importance of understanding the subsurface mechanical behavior of the region's geological formations, which directly influences drilling operations, wellbore stability, and reservoir productivity. We present a comprehensive geomechanical workflow tailored for oil and gas exploration in the Nile Delta, focusing on the findings related to rock mechanics. Through an exploration of key concepts, principles, and applications of geomechanics, this chapter includes case studies that illustrate the unique challenges and solutions encountered in drilling operations within the Nile Delta's petroleum system. These insights underscore the vital role of rock mechanics in optimizing drilling strategies and enhancing the overall success of exploration efforts in this geologically complex area.

Keywords

Nile Delta • Rock mechanics • Petroleum system • Reservoir productivity • Petroleum exploration

A. E. Farag (✉)
The British University in Egypt, El-Sherouk City, Egypt
e-mail: Ali.hammad@bue.edu.eg

R. A. Hammad
Maadi, Cairo Governorate, Egypt
e-mail: Rafik.hammad@bakerhughes.com

M. A. Khalifa and K. S. Gemail (eds.), *Geology of the Nile Deltas and Natural Resources*, Deltas of the World, https://doi.org/10.1007/978-3-031-96594-4_14

14.1 Geomechanical Advantages in the Nile Delta

The characterization of sedimentary deposits and rock formations in the Nile Delta serves as a foundational element for geomechanical studies. A comprehensive understanding of stress distribution, rock strength, and the mechanical properties of these formations is essential for assessing drilling risks and potential wellbore instability. Additionally, evaluating the presence of natural faults and fractures is crucial for identifying possible pathways for hydrocarbon migration.

In this context, geomechanics emerges as a vital tool for oil and gas exploration in the Nile Delta region. Its implications extend beyond wellbore stability and drilling efficiency to encompass reservoir characterization, environmental safety, and production optimization. By thoroughly investigating the geomechanical features of the area, exploration and production companies can make informed decisions, mitigate drilling risks, and unlock the hydrocarbon potential of this promising region.

14.2 Geomechanics: A Base for Successful Operations

Geomechanics is the relationship between the Rock (in terms of stress), Fluid (in terms of pressure), Residual tectonic (in terms of strain), and interpreted as rock failures (in terms of shear and tensile). In the oilfield's context, geomechanics is an interpretation of all data from geophysics, geology, drilling, petrophysics, and production operations to avoid rock failure during drilling or manage rock failure in hydrofracturing operations. The oil and gas business relies heavily on geomechanics, a foundational discipline that has an impact on many facets of exploration, drilling, production, and reservoir management. It entails the investigation of the mechanical behavior, stress response, and fluid interaction of subsurface rocks. Throughout the lifecycle of oil and gas projects, understanding geomechanics is essential for making the right decisions and optimizing operations. Some significant geomechanics values in the oil and gas industry include (1) Exploration and characterization of the reservoir. Geomechanical studies shed light on the structural integrity of reservoir rocks and establish the geological characteristics crucial to hydrocarbon reserves. By examining the mechanical behavior of these formations, geoscientists and reservoir engineers can forecast the appearance of faults, fractures, and other geological features that affect reservoir connectivity and productivity. (2) Drilling Operations Geomechanics significantly helps drilling operations. To maintain wellbore stability and avoid problems while drilling, such as wellbore collapse, it is essential to understand the distribution of stresses in the subsurface. To identify the best drilling fluids, build the best well trajectories, and minimize the hazards of drilling through challenging formations. (3) Risk Assessment and Mitigation. Geomechanics is essential for risk assessment and mitigation. Operators can detect and mitigate drilling and production risks to ensure more secure and reliable operations by understanding the mechanical behavior of rocks and anticipated subsurface stress changes. (4) Fracking. Additionally referred to as hydraulic fracturing, this technique is utilized to boost production in unconventional reservoirs like shale gas and tight oil deposits. Geomechanical investigations are required to comprehend how the rock responds to fracturing fluids and to guarantee that fractures propagate appropriately to enhance hydrocarbon recovery. (5) Reservoir management and enhanced oil recovery. Geomechanics is still useful for managing reservoirs and production. Changes in subsurface stress brought on by fluid extraction may have an impact on reservoir behavior and well performance. By concentrating on Geomechanical factors, operators can more successfully use enhanced oil recovery techniques and optimize production schedules. (6) Subsurface Storage Geomechanics is crucial for assessing the ability of subsurface rock formations to properly store and retain fluids for long periods in the context of carbon capture and storage (CCS) or natural gas storage in depleted reservoirs. (7) Seismic Events and Seismicity That Are Caused. The possibility of induced seismicity, which can be caused by human actions like hydraulic fracturing or changes in reservoir pressure, is determined in part by geomechanics. Managing and lowering seismic risks can be aided by understanding changes in subsurface stress.

14.3 Geomechanics Workflow

There are various processes involved in creating a geomechanics workflow. The best approach to do it is as follows, with sequential steps.

(1) The project's objectives and scope of work should be defined first.
(2) Gathering Relevant Data and compiling all the information needed for the analysis. This could comprise reservoir characteristics, boundary conditions, seismic data, well logs, and geological information. Table 14.1 shows the Geomechanics workflow and the source of information to build a geomechanics model. Data gathering and QC is an important step for a lot of benefits to help define the scope of work, Identification of missing data required for pore pressure, classification of drilling events, and diagnosis of the root cause of drilling problems.
(3) Choosing the Correct Software and Tools that will work best for the necessary Geomechanical analysis.
(4) Preprocess gathered Data, which would entail activities like formatting, data integration, and data cleaning. Utilize software and tools to preprocess the data and arrange it in an appropriate format.
(5) Building up a Geomechanics Model, Fig. 14.1 shows the 1D geomechanics model based on the goals and information at hand. This model needs to be a precise representation of the reservoir being investigated. Constructing a 3D model using software depends on data consistency while considering factors like the characteristics of the rock, the distribution of the stress, and the boundary conditions.

Table 14.1 Geomechanics workflow and source of information modified after Akbar et al. 2003)

Step	Property profiled	Source log	Other sources
1	Stratigraphy	GR, density, resistivity, Vp	Cuttings, caving, sequence stratigraphy
2	Overburden stress	Bulk density	Cuttings
3	Pore pressure	Vp, Check shot survey, Resistivity	Interval Velocity from Seismic, FIT, DDR
4	Stress direction	Oriented multiarm calipers, borehole image, oriented velocity anisotropy	Structure Maps, 3D seismic data
5	Minimum horizontal stress σ_h	Vp, Vs, wireline stress tool	PP, LOT, XLOT, microfrac, step-rate injection tests, DDR, modeling, database
6	Maximum horizontal stress σ_H	Borehole images	PP, $\sigma_{h,}$ rock strength database, Wellbore stress model
7	Elastic parameters (E, G, v)	Vp, Vs, bulk density	Database, cavings, laboratory core tests
8	Rock strength parameters (UCS, ϕ)	Vp, Vs, bulk density, stratigraphy	Database, cavings, laboratory core tests
9	Failure mechanisms	Borehole image, orientated multi-arm calipers	DDR, cavings

(6) The appropriate mechanical and physical attributes, such as Young's modulus, Poisson's ratio, cohesion, internal friction angle, permeability, and fluid characteristics, should be chosen based on the field experience and static model from core sample analysis when assigning model parameters to different layers of the Geomechanical model.

(7) Integration of geological interpretation in the workflow to predict consequences like rock deformation, stress variations, pore pressure variations, and likely failures like fault sliding or sand production may be important.

(8) Conducting the required Geomechanical analysis by running numerical simulations or analytical computations. For example, stability analysis, mechanical reaction to fluid injection, fracture propagation, sand production prediction, or stress analysis.

(9) Analysis and interpretation of the results of the geomechanics model output. Compare the findings to the stated objectives to determine whether they are satisfactory or require further investigation.

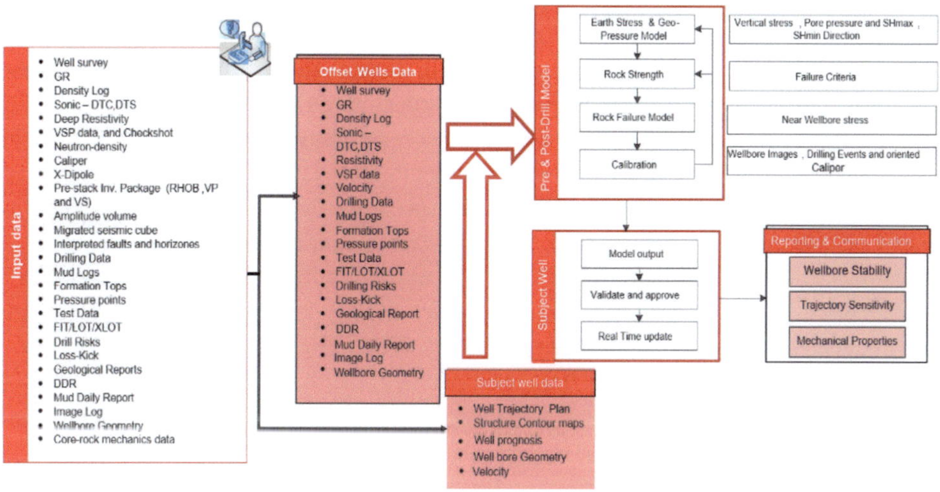

Fig. 14.1 1D Geomechanics earth model workflow shows the input borehole logs data and the rock mechanics modeling

(10) Calibration and Optimization based on how the results are interpreted. It may need to improve the Geomechanical model, change the data inputs, or alter the analysis parameters until the desired results are achieved.

(11) Presentation and Documentation The geomechanics model should be documented and presented, including all data inputs, modeling stages, simulation settings, analysis results, and interpretations. Stakeholders should be briefed and easily understand the results and recommendations.

(12) The accuracy of the Geomechanics model and the study's findings are Calibrated and updated by comparing them to real-time data or field observations like well tests (such as mini-fracs or injection tests). Calibration and sensitivity assessments are crucial to improve model accuracy if gaps are found. Keeping the geomechanics model up to date and up to date as new data or information becomes available. Geomechanics engineers and geoscientists should be involved to get the best results.

14.4 Construction and Data Challenges for a 1D Geomechanical Earth Model

14.4.1 Overburden Pressure

For the start of constructing a 1D Geomechanical earth model, the density log RHOB (bulk density) is necessary for calculating overburden pressure using the equation:

$$\text{Overburden pressure (P in KPa)} = \text{RHOB} * \text{g} * \text{Depth} \tag{14.1}$$

where RHOB in g/cc denotes the bulk density, g denotes the gravity acceleration (about 9.8 m/s^2), and Depth is the formation of true vertical depth in meters.

It is essential to emphasize that the accuracy of overburden pressure calculation using RHOB is dependent on several parameters, including log data quality, rock qualities, and local conditions. Additionally, Geomechanics specialists and related industry literature should be consulted for site-specific calibration factors and considerations.

Gardner et al. (1974), Bellotti et al. (1979), and Lindseth (1979) provided methods to create a density curve using sound log data. When density log information is unavailable, it is frequently determined from P-wave velocity (Vp) using an empirical relationship.

The density-from-sonic transit time methodology has the advantage of being able to be utilized before drilling by translating seismic interval velocities (V, in ft/sec) to sonic transit times (1/V *106, in usec/ft).

Gardner's Equation is expressed as

$$\rho b = a{*}vp^{b} \tag{14.2}$$

where *rb* is Bulk density in g/cc,*a* equals 0.23, b equals 0.25, and *vp* is interval velocity (ft/sec)

Lindseth' equation is represented as

$$\rho b = \frac{(vp - 3460)}{(0.308vp)} \tag{14.3}$$

where: vp is interval velocity (ft/sec) and ρb is Bulk density (g/cc)

Bellotti (1979) proposed two equations, one for consolidated formation and the other for an unconsolidated formation

$$\rho b = 3.8 - \frac{\Delta t}{89} \tag{14.4}$$

$$\rho b = 2.75 - 2.11\frac{(\Delta t - 47)}{(\Delta t + 200)} \tag{14.5}$$

where: Δt sonic transit times (usec/ft) and ρb is Bulk density (g/cc)

Terry Miller obtained the shallow density correlation based on the measured porosity and density data. (Zhang et al. 2008), Miller's near-mudline density correlation is expressed as follows.

$$\rho = \rho_{matrix}(1 - \emptyset) + \rho_w\emptyset \tag{14.6}$$

with

$$\emptyset = \emptyset_a + \emptyset_b e^{\left(-k(\mathrm{TVDBML})^{\frac{1}{n}}\right)} \tag{14.7}$$

where

ρ = bulk density (g/cc).

ρmatrix = matrix density (default 2.65 g/cc).

ρw = density of pore water (default 1.03 g/cc).

φa = sediment porosity at great depth (fraction).

φb = sediment porosity fitting parameter equal to mudline porosity minus φa (fraction).

K = porosity decline parameter (fitting parameter).

N = curvature parameter (fitting parameter).

Several empirical relations have been used to describe the overburden gradient such as the Marty Traugott of Amoco's technique (1997), which uses the expression where ρ_{avg} has units of ppg and D is true vertical depth, W is water depth, and A is the air Gap (vertical distance between the Kelly pushing and the sea surface), have units of ft to estimate the average sediment density between the sea floor and the depth of interest. Only sand/shale sequences without unloading or uplift are acceptable for this straightforward compaction relationship. An empirical exponent, the value is 0.6.

$$\rho_{avg} = 16.3 + \left[\left(\frac{D - W - A}{3125}\right)\right]^{0.6} \tag{14.8}$$

The overburden gradient (OBG) can be expressed as follows:

$$OBG = 8.5 * W + \rho_{avg}\left(\frac{D - W - A}{D}\right) \tag{14.9}$$

Assumed sea water density is 8.5 lbs./gal.

14.4.2 Pore Pressure Concepts

The pressure that fluids place on sedimentary or other rock formations is referred to as formation pressure, sometimes known as pore pressure. Based on how it relates to hydrostatic pressure, formation pressure can be divided into three different classes. These groups consist of Pressures below the hydrostatic pressure threshold that are classified as subnormal pressure. Normal pressure is the pressure level that is normally anticipated at a certain depth and is determined by the density of the pore fluid. Overpressure, often referred to as abnormal pressure, is a term used to describe pressures that are higher than the hydrostatic pressure and are frequently limited by overburden pressure. The figure labeled (2) illustrates the concept of Terzaghi's effective stress principle, which expresses that the Overburden is equal to the sum of the effective vertical stress and the pore pressure (Fig. 14.2).

$$\left(\sigma \mathrm{v} = \sigma' \mathrm{v} + \mathrm{Pp}\right). \tag{14.10}$$

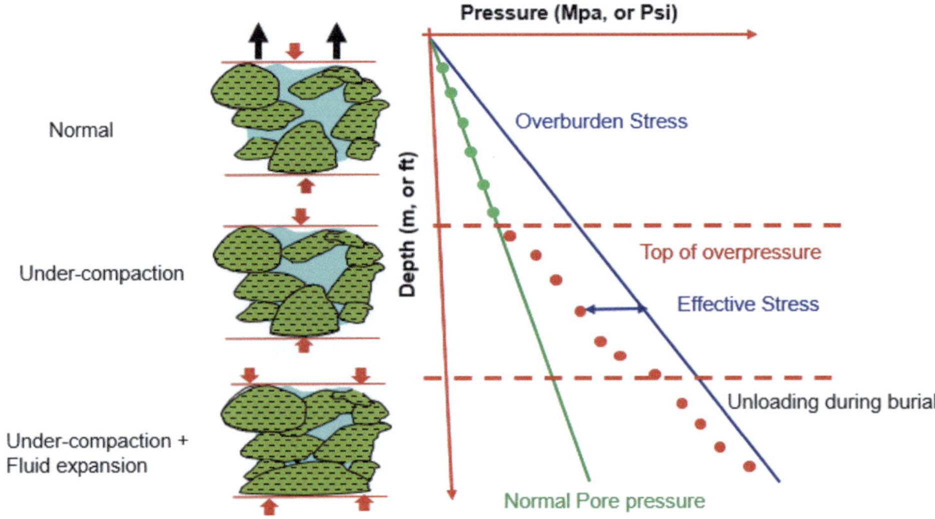

Fig. 14.2 Terzaghi's effective stress principle, where the overburden equals effective vertical stress plus pore pressure

Pore pressure plays a pivotal role in various geological and engineering contexts, making it a fundamental concept in the fields of geophysics, geology, and petroleum engineering. Understanding and accurately assessing pore pressure is of paramount importance for several reasons.

Primarily, pore pressure directly impacts the safety and efficiency of drilling operations, particularly in the oil and gas industry. Accurate pore pressure predictions are crucial in preventing drilling hazards such as blowouts, wellbore instability, and fluid influxes, which can have catastrophic consequences. By knowing the pore pressure, drilling engineers can design wellbore strategies that minimize risks and optimize drilling practices.

Pore pressure is also instrumental in reservoir characterization and hydrocarbon exploration. Accurate assessments of pore pressure help geoscientists and reservoir engineers delineate the boundaries of hydrocarbon reservoirs, estimate the number of recoverable resources, and design effective reservoir management strategies. Furthermore, it aids in understanding subsurface fluid flow and migration patterns, which are essential for successful reservoir production.

Beyond the energy industry, pore pressure has applications in geotechnical engineering, civil engineering, and environmental science. It influences soil stability, foundation design, and the management of underground water resources, especially in the Nile Delta. Accurate pore pressure data is indispensable for making informed decisions in construction, infrastructure development, and groundwater management.

In conclusion, pore pressure is a critical parameter with wide-ranging implications across Drilling and production disciplines. Its accurate assessment and understanding are essential for ensuring the optimized casing program, optimizing the mud program, reducing well control incidents, improving drilling efficiency, maintaining high-quality wellbores, optimizing hole sizes for production, maintaining casing integrity, managing sand production, optimizing reservoir management, and addressing geotechnical and environmental challenges. As technology and research continue to advance, our ability to harness the importance of pore pressure will only grow, contributing to safer, more efficient, and more sustainable practices in the fields where it is applied.

14.4.2.1 Overpressure Types

Pore Pressure Type

The mechanism of overpressure Type 1, sometimes referred to as "under-compaction pressure," occurs when the pore pressure within an underground rock formation is greater than the expected hydrostatic pressure gradient corresponding to its depth. The main cause of this specific type of overpressure, which is regularly observed in sedimentary basins, is the rapid compaction of sedimentary rocks over long geological timescales. When shale beds are thick, the shale may find it difficult to release enough water, which could cause its porosity to drop and its pore pressure to increase to values that are higher than those predicted for that depth. The natural permeability is typically minimal, and the rocks have a significant degree of compaction. In general, porosity and this kind of overpressure are closely related. This type is common in the Nile Delta of Egypt.

Pore Pressure Type

Pore Pressure Type 2 is a geological phenomenon that greatly deviates from the hydrostatic pressure gradient expected to be linked with the depth of a subsurface rock formation. It is sometimes referred to as "repressurization." Type 2 overpressure is characterized by faster and more dynamic processes as opposed to Type 1, which is caused by the gradual compaction of sedimentary strata. It can happen because of several geological occurrences, including tectonic movements and the uplift of shale, fluid migration, a supply of fluid to recharge, and the absence of a fluid escape route. These elements may trap fluids within the rock matrix and raise pore pressure above typical values for a given depth. There is a degree of uncertainty stemming from the complexity and lack of correlation between type 2 overpressure and porosity.

14.4.2.2 Origins of Pore Pressure

Geology and petroleum engineering use the term "abnormal pressure" to describe subsurface pressure conditions that differ significantly from what is considered normal or hydrostatic pressure. Complex geological, tectonic, and sedimentary processes are responsible for the development of anomalous pressure. Since, aberrant pressure circumstances can affect drilling operations, well stability, and reservoir behavior, understanding these

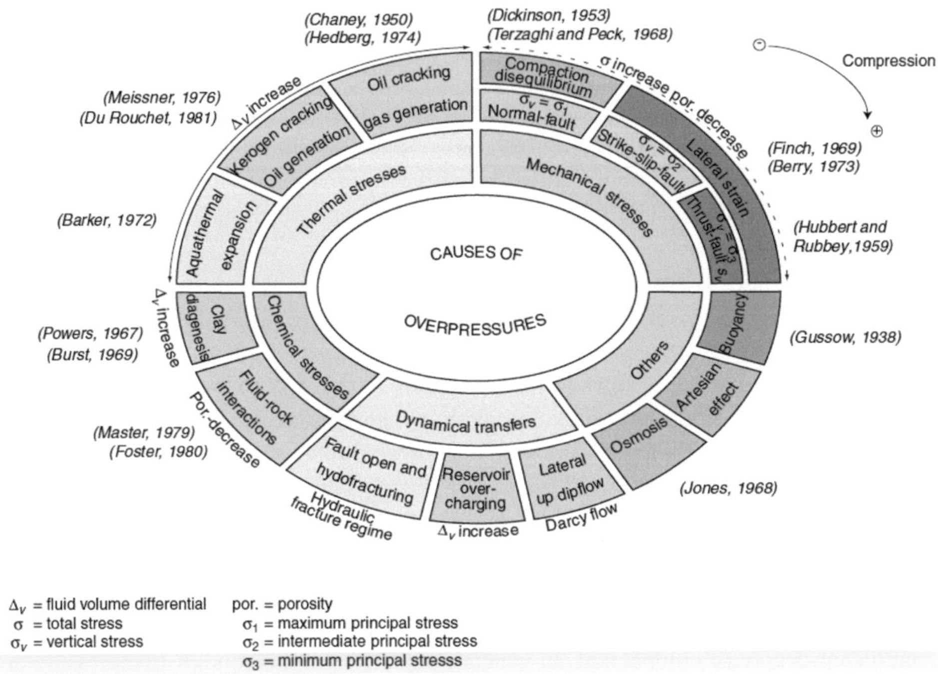

Fig. 14.3 Main overpressure causal mechanisms (Grauls 1997)

origins is essential for a variety of sectors, including oil and gas exploration. Porosity loss and fluid expansion are the primary causes of anomalous pressure, as shown in Fig. 14.3.

Dickinson (1953) and Terzaghi and Peck (1968) discussed the Compaction Disequilibrium as one of the main reasons for anomalous pressure. Due to the weight of the stuff above them, sediments become compacted as they build up throughout geological time. The pressure inside the sedimentary rock increases because of the compaction, which lowers the pore space. However, in deltaic environments such as the Nile Delta or during rapid deposition, where sedimentation rates are unusually high, compaction does not have enough time to adjust to the new sediment load, resulting in abnormal overpressures. In the Middle Miocene age, in a deep marine environment (outer shelf), the Sidi Salem formation was formed because of the significant flexure that affected pre-Miocene formations extending from east to west over the mid-Delta area Rizzini et al., 1978; and Said 1962). The formation consists of a massive body of shale (about 700 m thick), with only a few sand layers at the top. The fluid was trapped in the shale because it is what makes up most of the material, and because that area was subjected to successive overburden loads (+3400 m), the pore pressure increased above the hydrostatic value. Like the Sidi Salem Formation, the bottom portion of Kafr El Sheikh was formed by sedimentation and consists of thick shale with a small amount of sand. Therefore, the overburden effect caused the formation's pore pressure to rise above its hydrostatic value. The sand/shale ratio influences the amount of overpressure through Kafr El Sheikh. Kafr El Sheikh's upper portion was deposited in a shallow marine environment. As a result of shale compaction, the percentage of sand increases, certain fluids disperse through the interbedded sand layers, and the pressure anomalies are less than those in the lower part of the layer that achieves normal hydrostatic pressure at the top. Hussein and Taha (1993).

Aberrant pressure conditions can be caused by tectonic processes such as plate collisions or the development of fault zones, according to Finch (1969), Berry (1973), Hubber and Rubbey (1959), and others. Rocks can be fractured and deformed by these processes, creating pathways for fluid flow and trapping. These flaws or folds could trap fluids, leading to pressure buildup and abnormal pressures. Apart from certain tectonic shifts occurring in the late Messenian era, which led to tectonic pressures, faulting, and the intrusion of Sidi Salem shale into the Qwasem and Abu Madi formations, the rapid subsidence of the basin during the deposition of underlying deltaic sediments from the Qwasem formation resulted in the formation of differential compaction gravitational faults Said (1962). The Nile Delta exhibits a typical gravitational growth fault that inclines the basin inward, indicating that sediment deposition happens more quickly than subsidence, leading to the gradual development of an anti-regional dip over time and with increasing depth.

Oil cracking and gas generation were discussed by Chaney (1950) and Hedberg (1974), and kerogen cracking and oil generation were mentioned by Meissner (1976) and Du Rouchet (1981). The maturation of organic materials in sedimentary basins, combined with other fluids, results in the production of oil and gas. The production and transit of these hydrocarbons may contribute significantly to the creation of abnormal pressures. As hydrocarbons pass through porous rocks and become trapped in low-permeability seals or cap rocks, pressure in the reservoir below may increase.

A significant component to the creation of mechanical stresses and the source of pore pressure is thought to be the existence of salt diapirs. Because salt deposits have a lower density than surrounding sedimentary layers, they can occasionally rise through certain geological settings. Salt domes or diapirs can deform nearby rocks and produce specific areas of aberrant pressure as they push their way higher. In regions where salt tectonics is active, these pressure anomalies are particularly concerning. Common geological features like faults can divide reservoirs into distinct compartments. If an anomaly acts as an effective seal, it might confine fluids inside a compartment and cause excessive pressure to rise. Conversely, fluids may move if the fault loses its sealing integrity, lowering the tension in the impacted area. Additionally, changes in the Earth's crust brought on by subsidence (sinking) or uplift (elevation) can affect the subsurface pressure environment. Due to the delay in sediment compaction, areas that are rapidly subsiding may experience unusual overpressures. As sediments have had more time to equilibrate, areas that are experiencing uplift may, in contrast, have lower-than-normal pressures.

In the Nile Delta, overpressure is frequently attributed to diagenesis, which is the post-depositional transformation of one rock or mineral into another. The most important theory of normal fluid pressure is called clay diagenesis. A commonly held belief is that excessive pressure results from diagenetic processes occurring in specific clay types, either directly or indirectly. Illite forms because of the transformation of the montmorillonite structure. By adding potassium and causing an irreversible loss of water, the temperature is raised to between 220 and 230 0F. According to Freed and Peacor (1989), normal pressure halted when the transition from smectite to illiter started and rose throughout the zone of increasing illiter.

Powers (1967) proposed that the interlayer water, assumed to be denser than the pore water, would cause a rise in pressure as it attempted to expand during the clay transformation process. As clay releases interlayer water, a reduced volume results, while overall porosity remains unchanged. This occurs because interlayer water shifts from an interlayer position to an intergranular pore position, maintaining a consistent total fluid and solid volume. Several reactions have been proposed to explain the transition from montmorillonite to illite.

Hussein and Taha (1993) conducted an analysis of clay minerals in the Kafr El Sheikh, Abu Madi, Qwasem, and Sidi Salem formations. Their research revealed that over 65% of the shale in the Nile Delta consisted of smectite (montmorillonite), while the rest consisted of kaolinite and illite.

14.4.2.3 Pore Pressure Estimation

Various methods have been used worldwide to estimate pore pressure (i.e., Bowers 1995; 2001; Zhang 2011; Dutta et al. 2021; Muhsan et al. 2024). Direct methods for pore pressure establish a direct correlation between the deviation of a pore pressure indicator from its typical trend line and the pore pressure gradient at that specific depth as described by Hottman and Johnson (1965), Pennebaker and McClure (1968) Effective Stress Methods, propose that the consolidation of geological materials is primarily governed by the contrast between the overall confining pressure and the pressure exerted by the pore fluid. Vertical effective stress methods were examined by Bellotti and Giacca (1978), Hart and Flemings (1995), Bryant (1989), and Alixant & Desbrandes (1991), along with horizontal approaches by Eaton (1972a, b), Weakley (1989), and Rasmus and Stephens (1991).

The effective stress and pore pressure together make up the total stress on a rock mass, according to Eaton's effective stress, which is based on Terzaghi's Principle. While the pore pressure impacts the material's strength and deformation properties, the effective stress is the stress that determines how the material behaves mechanically. Eaton offered numerous equations to estimate the effective stress following the field observations. In the case of Interval Transit Time ($\sigma' v = \sigma \mathrm{n}\ [\Delta \mathrm{Tn}/\Delta \mathrm{T}]^3$) Sonic Velocity ($\sigma' v = \sigma \mathrm{n}\ [V/Vn]^3$) Resistivity ($\sigma' v = \sigma \mathrm{n}\ [R/Rn]1.2$) Conductivity ($\sigma' v = \sigma \mathrm{n}\ [Cn/C]^{1.2}$), and the Drilling exponent corrected ($\sigma' v = \sigma \mathrm{n}\ [\mathrm{Dxc/Dxcn}]^{1.2}$) where ($\sigma' v$) is the effective stress, σn denotes the normal effective stress, and the exponents 3 and 1.2 are Eaton exponents.

Eaton's resistivity approach, which is frequently used to forecast pore pressure, proved to be a reliable tool for pressure formation assessment. However, it comes with several limitations and potential sources of mistakes, just like any scientific technique. The sensitivity to changes in the electrical characteristics of the rock is one of Eaton's resistivity method's notable drawbacks. This includes elements like porosity and saturation (salinity changes in formation fluids have an impact), mineral composition (Shale is conductive, and lithology affects resistivity), and temperature of the formations under study. Resistivity measurements can be inaccurate due to inhomogeneities or changes in these qualities, which can then affect how the subsurface pressure regime is estimated.

Eaton's Transit time method is frequently used in pore pressure estimation. However, they come with certain limitations when applied to predict pore pressure. Eaton's method assumes that the geological formations being investigated have consistent properties, including acoustic velocity, throughout the measurement area. Environmental effects on wireline and measuring while drilling (LWD) measurements should be considered, and larger hole sizes affect sonic reading and compressional/shear ratio. and shear wave from LWD can be suspected. These variations can cause errors in the transit time measurements and, as a result, in the estimates of pore pressure. Therefore, while Eaton's transit time methods can provide valuable insights into pressure estimation, it's essential to be mindful of these limitations and consider them when applying the technique for pore pressure modeling.

Fertl (1976) regarded wireline sonic logs as the most reliable data source for estimating pore pressure because these logs are relatively insensitive to variations in hole size, formation temperature, and pore water salinity.

During the 1960s, one of the predominant approaches for identifying abnormal pressure was monitoring the drilling penetration rate. Nonetheless, the accuracy of this measurement was influenced by several variables. In 1966, Jordan and Shirley introduced the concept of the "d" exponent as an alternative method to estimate pore pressure.

$$d = log_{10}\frac{\frac{ROP}{60 \times \mathrm{RPM}}}{\frac{12\times \mathrm{WOB}}{10^6\times \mathrm{Diam}}} \tag{14.11}$$

where

d drillability.

ROP Rate of Penetration in ft/hr.

WOB Weight on Bit in Klbs.

Diam Bit Diameter in inches.

Subsequently, Rehmad and McClendon (1971) made improvements to this approach by accounting for the impact of mud weight, introducing the corrected "d" exponent, referred to as "dc." Practical experience has demonstrated that this correction is efficient; however, it may not be feasible when the mud weight significantly exceeds the pore pressure.

$$d_c = d \times \frac{\text{Normal Pore Pressure}}{\text{ECD}} \tag{14.12}$$

where ECD stands for effective circulating density, and dc contributes to the "d" exponent is for mud weight fluctuations.

During the mid-1970s, AGIP and Geoservices collaborated to create the sigma log, a tool designed to address the limitations of the "d" exponent in determining total rock strength. Bellotti et a1. (1976)

$$\sqrt{\sigma_t} = \frac{\text{WOB}^{0.5}\ .\ \text{RPM}^{0.25}}{Diam\ .\ \text{ROP}^{0.25}} \tag{14.13}$$

where

$\sqrt{\sigma t}$ Raw Sigma or total rock strength (dimensionless).

ROP Rate of Penetration in mt/hr.

WOB Weight on Bit in Mtons.

Diam Bit Diameter in inches.

14.5 Nile Delta Pressure Profile

As shown in Figs. 14.4 and 14.5, the Kafr El Sheikh formation is where the overpressurized zones start. The Kafr El Sheikh shale has a slight overpressure that rises gradually from the top and grows with depth and sand percentage reduction, peaking at the bottom. It was found that throughout the Kafr El Sheikh formation close to the Nile Delta, the pore pressure profile changed laterally with sand concentration. Although it reaches its highest position in the northwest (Port Fouad Marine), this value declines to around 1.25 to 1.30 g/cc. across Abu Madi and El Qara, and to 1.10 to 1.15 g/cc. In Nidoco and East Delta wells (Taha 1993).

The pressure gradient is constant throughout the Abu Madi and Qwasem formations before it abruptly drops to the standard gradient at the Abu Madi formation. The maximum value varies according to the formation burial depth and the in-place tectonic stresses; it is sharply increased at the top of the Sidi Salem formation. Rizzini et al. (1978) indicated environmental cycle is generally consistent with the pore pressure profile of the Nile Delta.

14.5.1 Fracture Pressures

In the literature, an extensive array of many distinct methods for calculating fracture gradients was identified. Bowers (1999) identified commonalities, and these methods were categorized into four groups, depending on their fundamental approach.

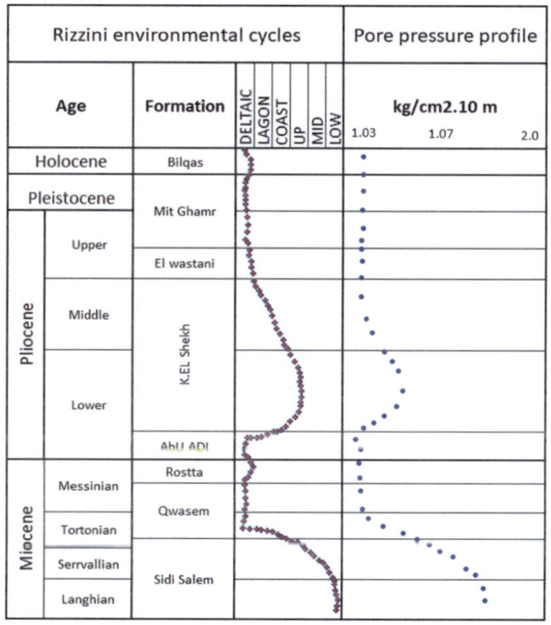

Fig. 14.4 Demonstration of the alignment of the pore pressure profile in the Nile Delta with the environmental cycles modified after Rizzin et al. (1978)

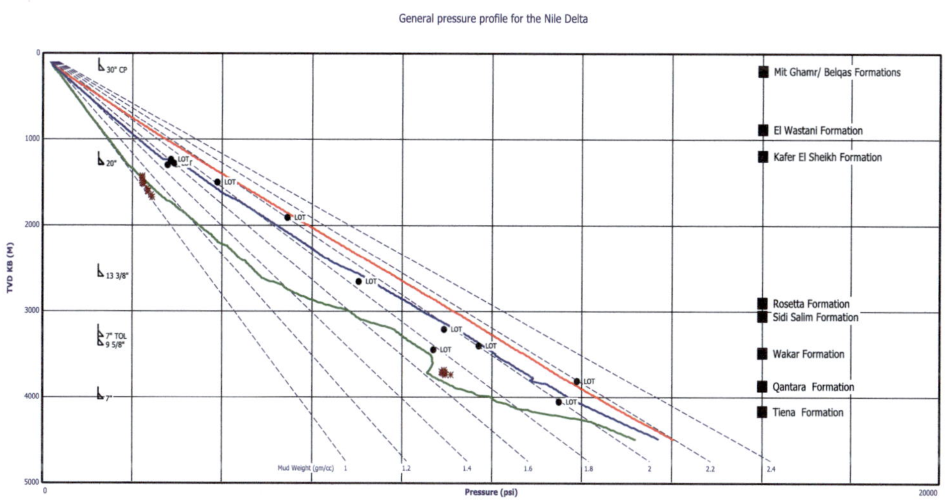

Fig. 14.5 General pore pressure profile of the Nile Delta (Data source from Mazzoni et al, 1997)

- Minimum Stress Methods (long cracks)
- Hoop Stress Methods (short cracks)
- Fracture Mechanics Methods (all length cracks)
- Direct Methods.

Fracture pressure is the pressure required for fracture propagation away from the wellbore, potentially resulting in lost circulation, and is closely associated with the minimum principal stress. It can sometimes be conflated with the pressure required to initiate a fracture at the wellbore, which may be either higher or lower in magnitude. The lower fracture limit, known as the fracture propagation pressure, acts as a guide for a safe upper limit for the circulating pressure while drilling. The overburden pressure frequently corresponds to the upper fracture limit. Pore fluid pressure (P), effective vertical (σv), and horizontal (σh) stresses all impact the fracture pressures. The PP/FG model's fracture pressure factor is calibrated to effective vertical stress rather than depth and is determined for each well rather than from a regional profile. This allows a more precise and direct measurement of fracture pressures. The effective stress idea is extended utilizing Drcker and Prager's (1952) failure criteria in the PP/FG technique of fracture pressure evaluation. These are lithology-dependent relationships between effective vertical (σv) and horizontal (σh) stresses. The lowest pressure necessary for existing vertical fractures to propagate is known as the fracture propagation pressure (Fpv), and it is equal in magnitude to the product of the pore pressure (P) and the effective horizontal matrix stress (σh). According to Hubbert and Willis (1957), there is no fluid invasion, and the predominant matrix stress is vertical, as it is in relaxed tectonic zones.

$$F_{PV} = P + \sigma_h \tag{14.14}$$

by the findings of Haimson and Fairhurst (1967), the pressure required to initiate a vertical fracture in a borehole with a smooth surface is denoted as:

$$F_{iv} = \sigma_h + \sigma_\theta + P \cong 2\sigma_h + P \tag{14.15}$$

where 2σh is the borehole tangential stress (σθ) concentration factor at the borehole wall. However, the fracture initiation pressure (Fiv) is usually equal to the overburden stress in tectonically relaxed basins, Pilkington (1978).

$$\sigma_v < 2\sigma_h or \sigma_h + \sigma_\theta,\ then F_{ih} = S = \sigma_v + P \tag{14.16}$$

When σv is less than σh and the tensile strength of the bedding plane are low, one would typically propagate a horizontal fracture (i.e., open a bedding plane by lifting the overburden) close to the borehole wall that will rotate vertically after leaving the near-borehole stress concentration area.

LOT results typically fall between the two fracture pressure limits, depending on the occurrence of vertical fractures close to the borehole. Since sediments lack vertical fissures

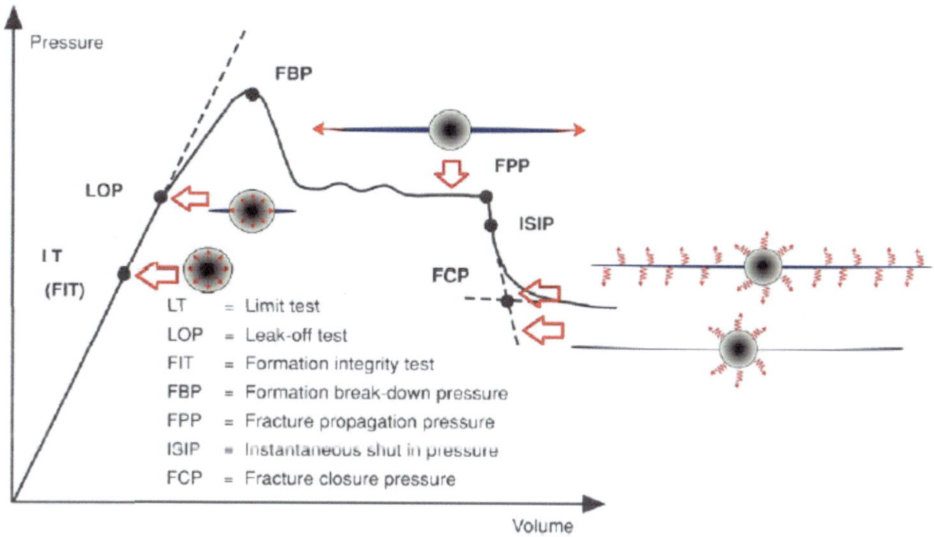

Fig. 14.6 Extended Leak-off Test (LOT), (Modified after Gaarenstroom et al. 1993)

and occasionally bedding planes and may deform plastically, shallow LOTs frequently sit near or above the overburden stress. Due to the existence of pre-existing sub-vertical fractures, LOTs (Direct Methods, Fig. 14.6) tend to be located along the fracture propagation gradient at depth.

The fracture pressure is assumed equal to the minimum horizontal stress. The ratio of minimum horizontal effective stress to vertical effective stress (k) is defined as follows:

$$\boldsymbol{K} = \frac{FG - PP}{OBG - PP} \tag{14.17}$$

where: FG is the fracture gradient is the predicted pressure gradient, BG is the overburden gradient, and K is the matrix stress coefficient. Except for the overburden constant methods, the methods used to predict the fracture gradient rely on the calculation of the matrix stress coefficient.

Many authors discussed fracture gradient prediction models, utilized in the oilfield industry. Hubbert and Willis' method (1957) assumes the rock is at failure, where $K = (1 - \sin \text{FANG})/(1 + \sin \text{FANG})$ and FANG donates to the rock friction angle.

$$P_{f(min)} = \left[\frac{S - P_p}{3}\right] + P_p, \ \text{or}\, P_{f(max)} = \left[\frac{S - P_p}{2}\right] + P_p \tag{14.18}$$

where: P_f Is Fracture pressure, S is the overburden pressure (Psi), P_p is the Pore pressure (Psi). The widely used Eaton's fracture gradient prediction model, where Eaton (1972a, b), gave the fracture gradient prediction model as follows:

$$P_f = \frac{3v}{2(1-v)} - (\sigma v - P_p) + P_p \tag{14.19}$$

where: σv is the vertical tension (vertical overburden stress) in psi, and v is the Poisson ratio is given by Eaton as:

$$\boldsymbol{\nu} = \frac{\frac{1}{2}\left(\frac{Vp}{Vs}\right)^2 - 1}{\left(\frac{Vp}{Vs}\right)^2 - 1} \text{ or } \boldsymbol{\nu} = \frac{\frac{1}{2}\left(\frac{DTS}{DTC}\right)^2 - 1}{\left(\frac{DTS}{DTC}\right)^2 - 1} \tag{14.20}$$

14.5.2 Rock Mechanical Properties

Accurate assessment of in-situ stress and Geomechanical evaluations, which encompass wellbore stability analysis, sand production prediction and management, hydraulic fracturing design, fault stability and reactivation analysis, and other Geomechanical applications, heavily rely on rock mechanical properties. These properties are pivotal for developing a Geomechanical model based on the linear Mohr–Coulomb failure criterion. Khaksar et al. (2009) presented crucial rock mechanical parameters necessary for this purpose, specifically the Unconfined Compressive Strength (UCS or Co), in Table 14.2, 14.3 and 14.4). Friction angle (θ), or Coefficient of internal friction (μ, where μ = tanθ) Table 14.5. Additionally, Thick Wall or hollow Cylinder strength (TWC) may be necessary for sanding evaluation and calibration (Table 14.6), collectively referred to as rock strength parameters. Elastic moduli are also crucial rock mechanical properties, with Poisson's ratio (v) and Young's modulus (E) being the most needed constants. These parameters serve as the foundation for deriving other elastic moduli, like shear and bulk moduli. It's worth noting that while rock elastic moduli can be computed from well logs, accurate quantitative data on rock strength parameters can only be obtained through specific-depth laboratory tests conducted on core samples. Furthermore, laboratory measurements of elastic moduli on core samples under in-situ stress conditions are essential for calibrating dynamic elastic moduli derived from logs to static values obtained from core samples.

Stress analysis of the Nile Delta is critical in the exploration and development of oil and gas Fields. The Geomechanics behavior of rocks and the loads acting on them is critical for field development overall. A full review of the geomechanical characteristics of the Nile Delta is important; in addition to stress magnitude, stress orientation is mandatory. An in-situ stress regime is critical for assessing wellbore stability and fracture propagation. Borehole breakout analysis, hydraulic fracturing tests, and overburden stress estimation are all part of this process. The orientation and amplitude of maximum horizontal stresses must be determined to design drilling trajectories and optimize hydraulic fracturing operations, in-situ stress measurements play a crucial role in gaining valuable insights into stress conditions at both broad and localized scales., these estimations and

Table 14.2 UCS models for sandstone

Model and reference	Equations	Remarks
Dt-McNally (McNally 1987)	$C_0 = 185213e^{-0.067\,Dt}$	Low to medium porosity sandstones, 65 < Dt < 100 μs/ft and UCS > 3000 psi, Permo-Triassic age SE Australia
Dt-Mod McNally (Modified McNally)	$C_0 = 838825e^{-0.057\,Dt}$	A modified McNally equation for unconsolidated and high porosity sandstones with UCS < 3000 psi
Dt-HRDS (Khaksar et al. 2008)	$C_0 = 40487e^{-0.0268\,Dt}$	Tertiary sandstones, offshore gas field, South Asia
Dt-FORMEL (Raaen et al. 1996)	$C_0 = 145 \times (140 - 2.1Dt + 0.0083Dt^2)$	90 < Dt < 140 μs/ft
ϕ- FORMEL (Raaen et al. 1996)	$C_0 = 145 \times (43 - 140\,\varphi + 63\varphi^2)$	$0.2 < \varphi < 0.35$
Dt Cubed-Sand (Chang et al. 2006)	$C_0 = 2.05 \times 10^9 Dt^{-3}$	Gulf of Mexico, weak and unconsolidated rocks
Dt-Freyburg (Freyburg 1972)	$C_0 = 1.55 \times 10^6 /Dt - 4567.5$	Consolidated Thuringia sandstones, Germany
ϕ-Sarda (Sarda et al. 1993)	$C_0 = 16172e^{-11.6\phi}$	Germany-sous-Couombs reservoir, with $\varphi \approx 0.35$
ϕ-Vernik (Vernik et al. 1993)	$C_0 = 36{,}830 \times (1 - 2.7\phi)^2$	Reasonable for consolidated sandstones with $\varphi < 0.30$
ϕ V_{clay}-Vernik	$C_0 = 145 \times (254 - 204 \times V_{clay}) \times (1 - 2.7\phi)^2$	Modified Vernik equation with V_{clay} for shaly sandstones with $\varphi < 0.30$
ϕ-Literature1 (Chang et al. 2006)	$C_0 = 40165e^{-10\phi}$	UCS between 300 and 52,000 psi and $\varphi < 0.33$
M-Bongkot (McPhee et al. 2000)	$C_0 = 0.001182M - 1436$	Bongkot Field, Gulf of Thailand, for UCS < 5000 psi
M-Hemlock (Moos et al. 1999)	$C_0 = 1.745 \times 10^{-3}\,M - 3045$	Cook Inlet, Alaska, unconsolidated fine to coarse-grained low-strength sandstones, 10,000 ft depth
M-GOM (Chang et al. 2006)	$C_0 = 561.15e^{7.862 \times 0.0000001\,M}$	Gulf of Mexico

(continued)

Table 14.2 (continued)

Model and reference	Equations	Remarks
M-Browse (Chang et al. 2006)	$C_0 = 6104.5e^{1.31 \times 0.0000001\ M}$	Consolidated sandstone with $0.05 < \varphi < 0.12$ and UCS > 12,000 psi, Browse Basin, Australia
E-Plumb (Bradford et al. 1998)	$C_0 = 330.7 + 0.0041E_{sta}$	Worldwide for 725 < UCS < 29,000 psi
E-Everest (Bradford et al. 1998)	$C_0 = 330.7 + 1.177 \times 10^{-14}\ E^{2.7}{}_{dyn}$	Another form of the E-Plumb equation with dynamic Young's modulus
E-Literature1 (Chang et al. 2006)	$C_0 = 6700e^{-0.0186 \times 0.0000001\ E}$	Based on static Young's modulus
E_{stat}-C&D (Coates and Denoo 1981)	$C_0 = 4.54 \times 10^{-3} \times E_{stat}$	Linear relation between C_0 and E_{stat}
BRUCE (Bruce 1990)	$C_0 = A \times 0.026 \times 10^{-6} E_{dyn}\ E_b\ (10.0045 + 0.0035V_{clay})$	Applicable to UCS > 4350 psi with $A = 2\ X\ \text{Cos}\Theta/\ (1\text{-sin}\Theta)$
W&P (Weingarten and Perkins 1995)	$C_0 = 145 \times 10^{-12}(114 + 97\ V_{clay})K_b E_{dyn}$	Unconsolidated sandstones, gas fields in the USA
MECHPRO1 (Fjaer et al. 1992)	$C_0 = 8.7 \times 10^{-12} KE_{dyn}\ (1 + 0.78\ Vclay)$	Sandstones with UCS > 4350 psi
MECHPRO2 (Fjaer et al. 1992)	$C_0 = 2.27 \times 10^{-10} M^2 \times [(1 + v)/(1 - v)]^2 (1 - 2v\ (1 + 0.78\ V_{clay})$	Sandstones with UCS > 4350 psi
ϕ-Travis peak	$C_0 = 4697\phi^{-0.466}$	Tight sandstone with $0.01 < \varphi < 0.18$
M-Travis peak	$C_0 = 3648e3.65 \times 0.0000001M$	Tight sandstone with $0.01 < \varphi < 0.18$
E-Travis peak	$C_0 = 3668e4.145 \times 0.0000001M$	Tight sandstone with $0.01 < \varphi < 0.18$

measurements contribute to a better understanding of how crustal stresses correspond to observed stress regimes, as documented in various geodetic and geological studies (e.g., Zoback et al. 1987; Magee and Zoback 1993; Townend and Zoback 2006).

Wellbore image data has been proven to be the most valuable. Wellbore breakouts, on the other hand, are enlargements that take place in vertical wells toward the direction of the far-field minimum principal stress due to compressive shear failure of the wellbore wall (Fig. 14.7). As a result, when breakouts are discovered, information on the orientations of the minimum horizontal stress can be obtained from their orientations. Breakouts develop out of deviated wells in a direction that depends on stress directions, borehole trajectory, and stress magnitudes.

Table 14.3 UCS models for shales

Model and reference	Equation	Remarks
Dt-Horsrud (Horsrud 2001)	$C_0 = 111.65 \times (304.8/Dt)^{2.93}$	High porosity North Sea Tertiary shales
Dt-GOM (Chang et al. 2006)	$C_0 = 62.35 \times (304.8/Dt)^{3.22}$	Pliocene and younger shales
Dt-Global (Chang et al. 2006)	$C_0 = 195.75 \times (304.8/Dt)^{2.6}$	Globally applicable
Dt Cubed-Shale (Chang et al. 2006)	$C_0 = 72.5 \times (304.8/Dt - 1)^3$	Gulf of Mexico
Dt-Lal (Lal 1999)	$C_0 = 1450 \times (304.8/Dt\text{-}1)$	High porosity Tertiary shales
E-Horsrud (Horsrud 2001)	$C_0 = 0.0232\ E^{0.91}$	High porosity North Sea Tertiary shales
E-Literature1 (Chang et al. 2006)	$C_0 = 0.221\ E^{0.712}$	Strong and compacted shales
ϕ\phi-L&D (Lashkaripour and Dusseault 1993)	$C_0 = 145.1\phi^{1.143}$	Compacted shales ($\phi < 0.10$)
ϕ\phi-Horsrud (Horsrud 2001)	$C_0 = 124.7\phi^{0.912}$	High porosity North Sea Tertiary shales
ϕ\phi-Literature1 (Chang et al. 2006)	$C_0 = 41.7\phi^{1.762}$	Shales with $\phi > 0.27$
Rhob-shale	$C_0 = 0.0123e^{4.89\rho b}$	Developed from published data for density < 2.4 g/cc

Table 14.4 UCS models for carbonates

Model and reference	Equation	Remarks
Dt-M&S (Militzer and Stoll 1973)	$C_0 = 7682/Dt^{1.82}$	Limestones
Dt-G&R (Golubev and Rabinovich 1976)	$C_0 = 10^{(2.44+109.14/Dt)}$	Limestones
ϕ\phiϕ-Rzhevski (Chang et al. 2006)	$C_0 = 40{,}020(1-3\phi)^2$	Similar to the Vernik formula with different constants
ϕ\phiϕ-Limestone1 (Chang et al. 2006)	$C_0 = 19{,}705.5e^{-4.8\phi}$	Strong limestones with low porosity (0.06 on average)
ϕ\phiϕ-Limestone2 (Chang et al. 2006)	$C_0 = 20851e^{-6.95\phi}$	UCS > 4900 psi in a field in the Middle East
E-Limestone (Chang et al. 2006)	$C_0 = 4.66E^{0.51}$	Moderately to very strong limestones (UCS > 2000 psi)
E-Dolomite (Chang et al. 2006)	$C_0 = 64E^{0.34}$	Dolomite with 8700 < UCS < 14,500 psi

Table 14.5 Friction angle models

Model and reference	Equation	Remarks
FANG-Dt (Lal 1999)	$\theta = \sin^{-1}[(304{,}878/Dt - 1000)/(304{,}878/Dt + 1000)]$	Shales
FANG-M (McPhee et al. 2000)	$\theta = 1.0691 \times 10^{-6}M + 28.51$	Sandstone, Bongkot Field, Gulf of Thailand
FANG-V_{clay}–1 (Plumb 1994)	$\theta = 26.5 - 37.4(1 - \phi - V_{clay}) + 62.1(1 - \phi - V_{clay})^2$	Both sandstones and shales
FANG-V_{clay}–2	$\theta = 20.5 + 15(1 - V_{clay})$	Sandstones
FANG-ϕ1 (Weingarten and Perkins 1995)	$\theta = 57.8 - 105\phi$	Sandstones
FANG-ϕ2 (Perkins and Weingarten 1988)	$\theta = 58 - 135\phi$	Weak sandstones
FANG-rb	$\tan\theta = 0.1{r_b}^{2.85}$	Sandstones

Table 14.6 TWC models

Model and reference	Equation	Remarks
TWC-UCS	$TWC = 80.8765 \times {C_0}^{0.58}$	Global for sandstones
TWC-M (Khaksar et al. 2008)	$TWC = 10^{-8}M^{1.77}$	Tertiary sandstones, a gas field in South Asia
TWC-ϕ	$TWC = 20.62\phi^{-3.54}$	Weak sandstones

In general, and as shown in Fig. 14.8b, they form over long sections of the well bore in very consistent orientations defined by the far-field stresses. Breakouts occur only where the circumferential compressive hoop stress at the wellbore wall exceeds the compressive strength of the rock. Thus, breakout occurrence and the width of the failed zone can be used to constrain the in-situ stress magnitudes and effective rock strength. While it is not possible to determine the stress.

Induced fracture drilling, also known as borehole-induced fractures, refers to the creation of fractures in the rock surrounding a borehole during drilling operations. These fractures are formed by the interaction of the drilling technique with subsurface stress conditions. Tensile is useful while drilling because it offers crucial information on the maximum horizontal stress. Geoscientists and drilling engineers can enhance their understanding of the in-situ stress field by analyzing the orientation of maximum horizontal stress and the properties of induced fractures. This evaluation is vital for assessing wellbore stability, predicting fault behavior, and optimizing drilling operations.. Furthermore, the presence of produced fractures is essential for evaluating reservoir properties, such as permeability, and for assessing the potential for hydraulic fracturing in the oil and gas industry..

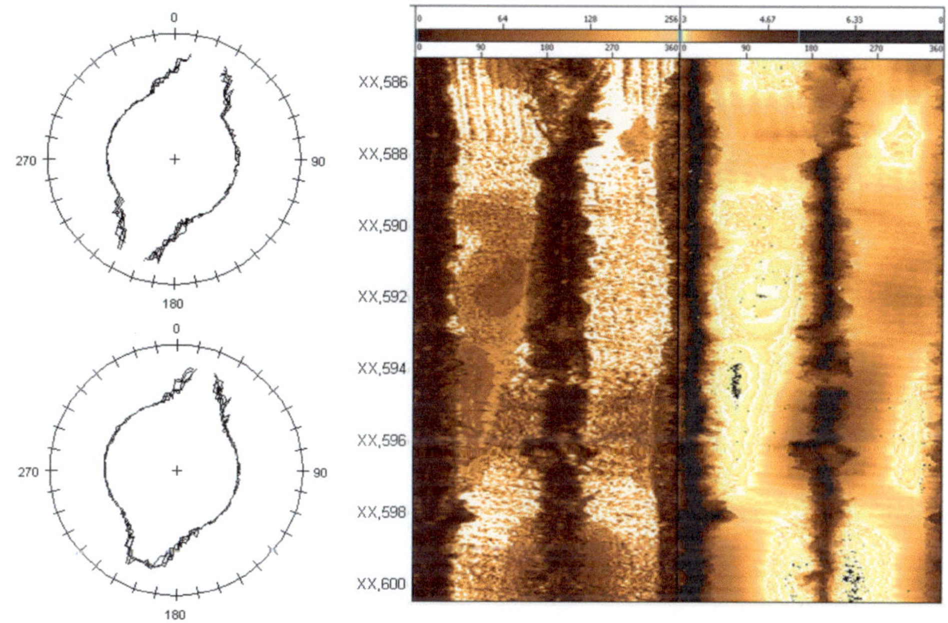

Fig. 14.7 Wellbore breakouts visible in an acoustic image (Barton and Moos 2008)

Tensile fractures induced by drilling occur on the borehole wall when the circumferential stress is negative (indicating extension) and surpasses the tensile strength of the rock. These small-scale features are confined to the borehole wall due to localized stress concentration and typically do not extend significantly into the surrounding formation unless the mud weight applied exceeds the least principal stress. In the case of a vertical well, where the vertical stress is one of the principal stresses, these fractures align parallel to the borehole's axis, following the far-field orientation of the maximum horizontal stress (SHmax), as observed in studies by Moos and Zoback and Peska and Zoback (1995). Figure 14.4a illustrates such a drilling-induced tensile wall fracture aligned with the axis of a vertical well.

The current orientations of SHmax stress in the Nile Delta (Fig. 14.9) were determined using drilling-induced fractures and borehole breakout analysis, as reported by Tingay et al. (2012). Below or to the south of the Messinian evaporites (represented by blue symbols), the western part of the Nile Delta exhibits a SHmax orientation roughly aligned with NNE-SSW, the central region displays an orientation close to ESE-WNW, and the eastern sector demonstrates an orientation approximately along NNE-SSW (Fig. 14.9). These findings generally correspond to the observations made by Bosworth in 2006 (indicated by red crosses). While somewhat scattered, SHmax orientations predominantly trend in the NE-SW direction within the sequences underlying the Messinian evaporites (indicated

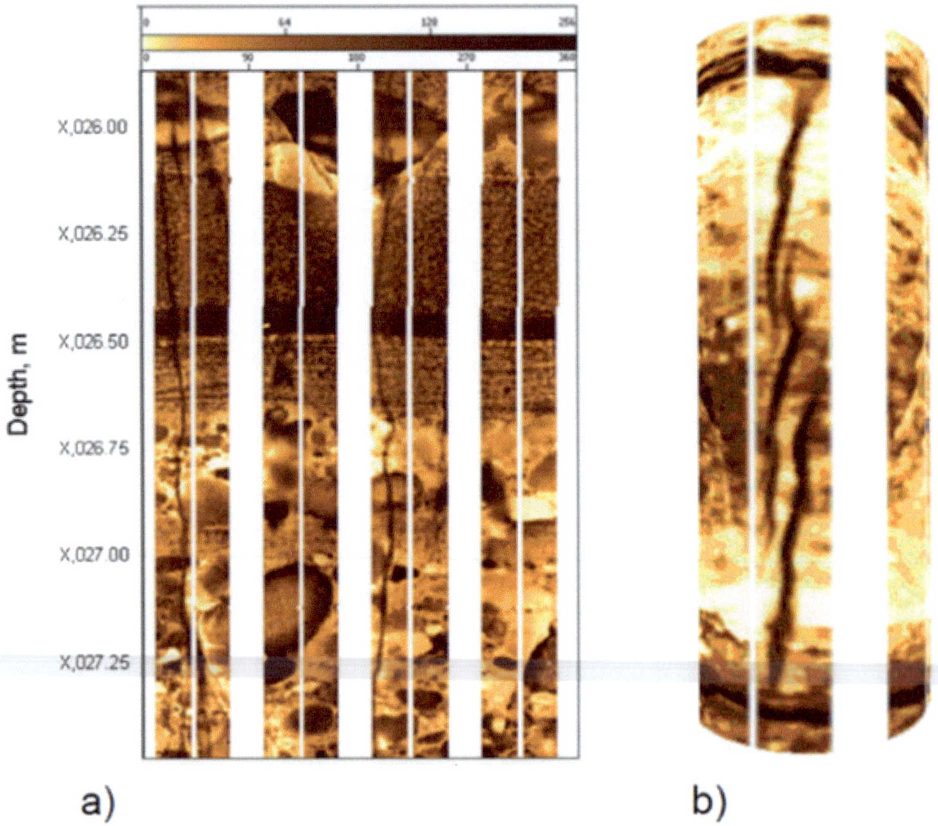

Fig. 14.8 Examples of drilling-induced tensile wall failure developed in a vertical (**a**) and deviated (**b**) wellbore (Barton and Moos 2008)

by yellow symbols). Consequently, the average SHmax orientation in sequences above the evaporites is approximately parallel to that observed below the evaporite layers.

Unconfined Compressive Strength (UCS) Tests, commonly conducted under zero confining pressure, provide a direct measure of rock strength. However, these tests often underestimate the true strength of the rock due to the formation of micro-cracks during coring and sample preparation, leading to premature failure under uniaxial loading. This limitation makes it challenging to determine the internal friction angle accurately.

Young's modulus, bulk modulus, shear modulus, and Poisson's ratio are some of the mechanical properties that are used to characterize stress–strain relationships. Shear modulus (G) and Young's modulus (E) are constants that connect shear stresses to strains and normal stress to strains, respectively. Poisson's ratio is lateral expansion to longitudinal contraction under uniaxial stress, whereas the bulk modulus depends on both Young's modulus and Poisson's ratio. Out of these four moduli, only two are independent, and

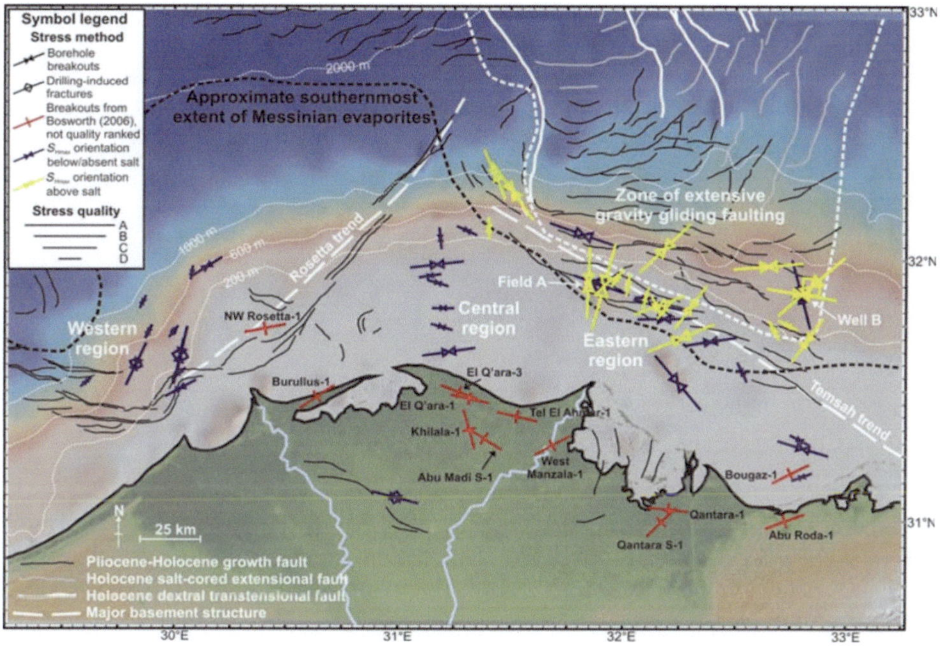

Fig. 14.9 Nile Delta present-day stress map (Tingay et al. 2012)

the other two can be derived from them. Through measurements of stress-induced strain, the laboratory analysis of rock samples produces mechanical characteristics known as static elastic constants. On the other hand, dynamic elastic constants are determined by determining the velocities of elastic waves inside the rock, a procedure made easier by acoustic logging and waveform analysis. Compressional and shear velocities can be continuously measured using these approaches. They make it possible to determine the rock's in-situ mechanical characteristics when combined with measurements of bulk density. In Table 7, Schlumberger (1991) outlines the correlation between elastic moduli and bulk density. Elastic moduli are stated in terms of elastic wave velocities (or transit durations).

7 Dynamic Elastic Properties (Modified after Schlumberger 1991), where: ρ_b is the Bulk density of the formation (g/cm^3), DTCO is the Compressional slowness of the bulk formation (μs/ft), and DTSM is the Shear slowness of the bulk formation (μs/ft).

v	Poisson's ratio	$\frac{0.5\left(\frac{DTSM}{DTCO}\right)^2-1}{\left(\frac{DTSM}{DTCO}\right)^2-1}$	Lateral strain Longitudinal strain
G	Shear modulus	$\left(\frac{\rho b}{DTSM^2}\right)*a$	Applied stress Shear strain

(continued)

(continued)

ν	Poisson's ratio	$\frac{0.5\left(\frac{DTSM}{DTCO}\right)^2-1}{\left(\frac{DTSM}{DTCO}\right)^2-1}$	Lateral strain Longitudinal strain
E	Young's modulus	$2G(1+\upsilon)$	Applied uniaxial strain Normal strain
K	Bulk modulus	$\rho b\left(\frac{1}{DTCO^2}-\frac{4}{3DTSM^2}\right)*a$	Hydrostatic pressure Volumetric strain
α	Biot elastic constant	$1-\frac{K(Skeltob)}{K(Soild)}$	Pore pressure Proportionality

14.6 Conclusions

In conclusion, the application of geomechanics in oil and gas exploration within the Nile Delta region emerges as a pivotal tool for enhancing reservoir characterization, ensuring environmental safety, optimizing production, and improving drilling efficiency. By delving into the intricacies of rock mechanics, exploration and production entities can make informed decisions, mitigate drilling risks, and unlock the hydrocarbon potential of this promising locale. Notably, the prevalence of overpressure in the region, often linked to diagenetic processes such as clay transformation, underscores the dynamic geological conditions influencing drilling operations.

The Kafr El Sheikh formation stands out as a focal point for over-pressurized zones, with pressures gradually escalating with depth and decreasing sand content, peaking toward the base. Pore pressure profiles exhibit lateral variations, reaching maximum values in specific areas like Port Fouad Marine before tapering off across other regions. The pressure gradients in formations like Abu Madi and Qwasem demonstrate consistent behavior before transitioning to standard gradients, influenced by burial depths and tectonic stresses.

Moreover, insights from studies like those by Taha (1993) and Rizzini et al. (1978) align with the environmental cycles of the Nile Delta, shedding light on the interconnected nature of geological processes and pressure dynamics in this complex geological setting. This chapter underscores the indispensable role of geomechanics in navigating the challenges and harnessing the opportunities presented by the unique geological landscape of the Nile Delta, ultimately contributing to the success and sustainability of oil and gas exploration endeavors in the region.

14.7 Recommendations

Based on the geomechanical insights and findings presented in the review regarding oil and gas exploration in the Nile Delta region, the following recommendations can be considered:

1. Sustained geomechanical studies should be conducted to deepen the understanding of the stress distribution, rock strength, and mechanical properties of sedimentary deposits and rock formations in the Nile Delta. This ongoing analysis is crucial for mitigating drilling risks and ensuring wellbore stability.
2. Given the presence of natural faults and fractures in the region, it is essential to incorporate comprehensive assessments of these geological features into exploration and drilling plans. Understanding these structures can help identify potential pathways for hydrocarbon migration and optimize drilling strategies.
3. Considering the prevalence of overpressure zones in formations like Kafr El Sheikh, exploration and production companies should implement effective strategies for managing overpressure conditions. This includes monitoring pore pressure profiles, adapting drilling techniques accordingly, and ensuring safety measures are in place.
4. Recognizing the correlation between environmental cycles and pore pressure profiles in the Nile Delta, environmental factors should be carefully integrated into geomechanical assessments. This holistic approach can enhance environmental safety protocols and contribute to sustainable exploration practices.
5. Encouraging collaboration and data sharing among industry experts, researchers, and stakeholders can foster a more comprehensive understanding of the geomechanical aspects specific to the Nile Delta. This collaborative approach can lead to innovative solutions, improved decision-making, and enhanced exploration outcomes.
6. Investing in training programs and skill development initiatives focused on geomechanics can empower professionals working in the oil and gas sector in the Nile Delta. Enhanced expertise in geomechanical principles and applications can drive efficiency, safety, and success in exploration and production activities.

References

Akbar AHA et al (2003) Watching rocks change—mechanical earth modelling. Oilfield Review, Summer

Alixant JL, Desbrandes R (1991) Explicit pore-pressure evaluation: concept and application. SPE Drilling Engineering

Barker JW, Wood TD (1997) Estimating shallow below mudline deepwater Gulf of mexico fracture gradients. Houston AADE Annual Technical Forum

Barton C, Moos D (2008) Geomechanical wellbore imaging, key to managing the asset life cycle. In: AAPG memoir on diameter and borehole image log technology, pp 1–50
Bellotti P et al (1976) Instantaneous log indicates porosity and pore pressure. World Oil
Bellotti P, Di Lorenzo V, Giacca D (1979) Overburden gradient from sonic log Trans. SPWLA, London
Bellotti P, Giacca D (1978) Seismic data can detect overpressures in deep drilling. Oil and Gas J
Berry FAF (1973) High fluid potential in California Coast ranges and their tectonic significance. Am Assoc Pet Geol Bull 57:1219–1249
Bowers GL (1995) Pore pressure estimation from velocity data; accounting for overpressure mechanisms besides undercompaction. In: SPE drilling and completions, June
Bowers GL (1999) State of the art in pore pressure estimation. DEA-119 Report No 1
Bowers GL (2001) Determining an appropriate pore pressure estimation strategy. In: Offshore technology conference, p 13042
Bryant TM (1989) A dual pore pressure detection technique, SPE 18714
Chaney PE (1950) Abnormal pressure and lost circulation. World Oil 130:122–126
Chang C, Zoback MD, Khaksar A (2006) Empirical relations between rock strength and physical properties in sedimentary rocks. J Pet Sci Eng 51(3–4):223–237
Coates GR, Denoo S (1981) The producibility answer product. Schlumberger Tech Rev 29:55–63
Daines SR (1980) The prediction of fracture pressures for wildcat wells—SPE 9254
Dickinson G (1953) Geological aspects of reservoir pressure in Gulf Coast Louisiana. Am Assoc Pet Geol Bull 37:410–432
Drcker DC, Prager W (1952) Soil mechanics and plastic analysis for limit design. Q Appl Math 10:157–165
Du Rouchet J (1981) Stress fields, a key to oil migration. Bull Am Assoc Pet Geol 65:74–85
Dutta NC, Bachrach R, Mukerji T(2021) Methods for pore pressure detection: Well logging and drilling parameters. In: Quantitative analysis of geopressure for geoscientists and engineers. Cambridge University Press, pp 310–337
Eaton BA (1969) Fracture gradient prediction and its application in oil field operations. J Pet Tech 21:1353–1360
Eaton BA, Eaton TL (1997) Fracture gradient prediction for the new generation. World Oil
Eaton BA (1972a) The effect of overburden stress on geopressure prediction from well logs. J Pet Technol
Eaton BA (1972b) Graphical method predicts geopressures worldwide. World Oil 182(6):51–56
Ehsan M, Manzoor U, Chen R, Hussain M, Abdelrahman K, Radwan AE, Ullah J, Iftikhar MK, Arshad F (2024) Pore pressure prediction based on conventional well logs and seismic data using an advanced machine learning approach. J Rock Mech Geotech Eng. ISSN 1674–7755.https://doi.org/10.1016/j.jrmge.2024.09.049
Fertle H (1976) Abnormal formation pressure(1978). Elsevier Scientific Publishing Co., Amsterdam—New York
Finch WC (1969) Abnormal Pressure in Antelope Field, North Dakota. JPT 213:170–179
Fjaer E, Holt RM, Horsrud P, Raaen AM (1992) Petroleum related rock mechanics. Elsevier, Amsterdam
Freed RL, Peacor DR (1989) Geopressured shale and sealing effect of smectite to illite transition. The American Association of Petroleum Geologists Bulletin, vol 73, no. 10, pp 1223–1232
Gaarenstroom L,Tromp RAJ, deJong MC, Brandenburg AM (1993) Overpressures in the Central North Sea: implications for trap integrity and drilling safety. In: Petroleum Geology of Northwest Europe: Proceedings of the 4th conference, vol 4, pp 1305–1313
Gardner GLF, Gardner LW, Gregory AR (1974) Formation velocity and density—the diagnostic basics for stratigraphic traps. Geophysics 39:770–780

Golubev AA, Rabinovich GY (1976) Resultaty primeneia appartury akusticeskogo karotasa dlja predeleina proconstych svoistv gornych porod na mestorosdeniaach tverdych isjopaemych. Prikl. Geofiz. Moskva 73:109–116

Grauls D (1997) Minimum principal stress as control of overpressures in sedimentary Basins. In: Hendry JP (ed) Geofluids II'97—extended abstracts, pp 219–222

Hart BS, Flemings PB, Deshpande A (1995) Porosity and pressure: Role of compaction disequilibrium in the development of geopressures in a Gulf Coast Pleistocene Basin. Geology 23:1995

Hedberg HD (1974) Relation of methane generation to undercompacted shales, shale diapirs, and mud volcanoes. Am Assoc Pet Geol Bull 58:668–673

Horsrud P (2001) Estimating mechanical properties of shale from empirical correlations. SPE Drill Complet 16:68–73

Hottman CE, Johnson RK (1965) Estimation of formation pressures from log-derived shale properties. J Pet Technol

Hubbert MK, Rubey WW (1959) Mechanics of fluid filled porous solids and its application to overthrust faulting. 1, role of fluid pressure in mechanics of overthrust faulting. Geol Soc Am Bull 70:115–166

Hussein NK, Taha ALAE (1993) Hole stability in Nile Delta Area: Abu Madi, Nidoco, and El-Qara Wells. In: Presented at the SPE Middle East oil technical conference & exhibition held in Bahrain, 3–6 April 1993, SPE 25544

Khaksar A, Rahman K, Ghani J, Mangor H (2008) Integrated geomechanical study for hole stability, sanding potential and completion selection: a case study from South East Asia. SPE 115915

Khaksar A, Taylor PG, Fang Z, Kayes T, Salazar A, Rahman K, Bowers GL (2009) Rock strength from core and logs: Where we stand and ways to go: SPE 121972

Lal M (1999) Shale stability: Drilling fluid interaction and shale strength. In: Asia pacific oil and gas conference and exhibition. https://doi.org/10.2118/54356-MS

Lashkaripour GR, Dusseault MB (1993) A statistical study on shale properties; relationship among principal shale properties. In: Proc. conference on probabilistic methods in geotechnical engineering. Canberra, Australia, pp 195–200

Lindseth RO (1979) Synthetic sonic logs—a process for stratigraphic interpretation. Geophysics 44(1):3–26. Meissner FF (1976) Abnormal electric resistivity and fluid pressure in bakken formation, Williston Basin, and its relation to petroleum generation, migration, and accumulation. AAPG Bulletin 60(8):1403–1404

Magee ME, Zoback MD (1993) Evidence for a weak interplate thrust fault along the northern Japan subduction zone and implications for the mechanics of thrust faulting and fluid expulsion. Geology 21:809–812

Matthews WR, Kelly J (1967) How to predict formation pressure and fracture gradient. Oil & Gas J 65:92–106.

Mazzoni R, Wahdan T, Bassem A, Ward CD (1997) Real-Time Pore and Fracture Pressure Prediction With FEWD in the Nile Delta. Paper Presented at the SPE/IADC Drilling Conference, Amsterdam, Netherlands, March. https://doi.org/10.2118/37669-MS

McPhee CA, Lemanczyk ZR, Helderle P, Thatchaichawalit D, Gongsakdi N (2000) Sand management in Bongkot field, Gulf of Thailand: an integrated approach. In: SPE Asia pacific oil and gas conference and exhibition. Brisbane, Australia, SPE-64467-MS. https://doi.org/10.2118/64467-MS

Miller TW, Luk CH, Olgaard DL (2002) The interrelationships between overpressure mechanisms and in-situ stress. Press Regimes Sediment Basins Their Predict: AAPG Mem 76:13–20

Moos D, Zoback MD, Bailey L (1999) Feasibility study of the stability of openhole multilaterals, Cook Inlet, Alaska, 1999. SPE Mid-Continent Operations Symposium held in Oklahoma City, Oklahoma, 28–31 March 1999, SPE 52186

Mouchet J-P, Mitchell A (1989) Abnormal pressures while drilling—manuels techniques 2 Elf Aquitaine

Pennebaker ES (1968) An engineering interpretation of seismic data. SPE 2165

Perkins TK, Weingarten JS (1988) Stability and failure of spherical cavities in unconsoli-dated sand and weakly consolidated rock. Paper no. 18244, 63rd Annual Technical Conference, Houston, TX, October 2–5

Plumb RA (1994) Influence of composition and texture on the failure properties of clastic rocks. In: Rock mechanics in petroleum engineering. https://doi.org/10.2118/28022-MS

Powers MC (1967) Fluid-release mechanisms in compacting marine mudrocks and their importance in oil exploration. Am Assoc Pet Geol Bull 51(7):1240–1254

Rasmus JC, Gray Stephens DMR (1991) Real-time pore-pressure evaluation from MWD/LWD measurements and drilling-derived formation strength. SPE Drilling Engineering

Rehmad B, McClendon R (1971) Measurement of formation pressure from drilling data. In: SPE of AIME Annual fall meeting. New Orleans, LA, USA (SPE 4001). https://doi.org/10.2118/4001-MS

Rizzini A et al (1978) Stratigraphy and sediment of a Neogene-Quaternary section in the Nile Delta. Mar Geol 27(1978):327–348

Said R (1992) The geology of Egypt. Elsevier, New York 1962

Sayers (2006) An Introduction to velocity-based pore pressure estimation. Lead Edge, 25(12):1496–1500

Schlumberger(1991): Log interpretation principles/applications; 3rd printing, May 1991. Schlumberger Educational Services, p 223

Simmons EL, Rau WE (1988) Predicting deepwater fracture pressures: A proposal (SPE 180250). Pore pressure and fracture gradients—SPE Reprint Series No.49 (1999 edition)

Terzaghi K, Peck (1968) Soil mechanics in engineering practice. John Wiley and Sons, New York, N.Y.

Tingay M, Bentham P, De Feyter A, Kellner A (2012) Evidence for non-andersonian faulting above evaporites in the Nile delta. In: Healy D, Butler RWH, Shipton ZK, Sibson RH (eds) Faulting, fracturing and igneous intrusion in the earth's crust. Geol Soc 367:155–170

Townend J, Zoback MD (2006) Stress, strain, and mountain building in central Japan. J Geophys Res: Solid Earth 111(3):1–11. https://doi.org/10.1029/2005JB003759

Traugott M (1997) Pore pressure and fracture pressure determinations in deepwater—deepwater technology supplement to world oil

Weakley RR (1989) Use of surface seismic data to predict formation pore pressure (Sand shale depositional environments, SPE 18713

Zhang J (2011) Pore pressure prediction from well logs: methods, modifications, and new approaches. Earth Sci Rev 108(50):63

Zhang J, Standifird W, Lenamond C (2008) Casing ultradeep, ultralong salt sections in deep water: a case study for failure diagnosis and risk mitigation in record-depth well. Paper SPE 114273

Zoback MD, Peska P (1995) In-situ stress and rock strength in the GBRN/DOE Pathfinder well, south Eugene Island. Gulf Mex: J Pet Technol 47(7):582–585

Zoback MD, Zoback ML, Eaton JP, Mount VS, Suppe J (1987) New evidence on the state of stress of the San Andreas Fault. Science 238:1105–1111. https://doi.org/10.1126/science.238.4830.1105

MIX
Papier aus verantwortungsvollen Quellen
Paper from responsible sources
FSC® C105338
FSC
www.fsc.org

If you have any concerns about our products,
you can contact us on
ProductSafety@springernature.com

In case Publisher is established outside the EU,
the EU authorized representative is:
Springer Nature Customer Service Center GmbH
Europaplatz 3, 69115 Heidelberg, Germany

Printed by Libri Plureos GmbH
in Hamburg, Germany